Taphonomy

Releasing the Data Locked in the Fossil Record

TOPICS IN GEOBIOLOGY

Series Editors: F. G. Stehli, DOSECC, Inc., Gainesville, Florida
D. S. Jones, University of Florida, Gainesville, Florida

Taphonomy

Releasing the Data Locked in the Fossil Record

Edited by
Peter A. Allison
University of Reading
Reading, United Kingdom

and
Derek E. G. Briggs
University of Bristol
Bristol, United Kingdom

Plenum Press • New York and London

Library of Congress Cataloging-in-Publication Data

Taphonomy : releasing the data locked in the fossil record / edited by
Peter A. Allison and Derek E.G. Briggs.
p. cm. -- (Topics in geobiology ; v. 9)
Includes bibliographical references and index.
ISBN 0-306-43876-3
1. Taphonomy. 2. Fossilization. I. Allison, Peter A.
II. Briggs, D. E. G. III. Series.
QE721.2.F6T37 1991
560--dc20 91-29242
CIP

ISBN 0-306-43876-3

A Division of Plenum Publishing Corporation
233 Spring Street, New York, N.Y. 10013

Printed in the United States of America

Contributors

Peter A. Allison Postgraduate Research Institute for Sedimentology, The University, Whiteknights, Reading RG6 2AB, United Kingdom

Anna K. Behrensmeyer Department of Paleobiology, National Museum of Natural History, MRC NHB 121, Smithsonian Institution, Washington, D.C. 20560

Daniel W. J. Bosence Department of Geology, Royal Holloway and Bedford New College, University of London, Egham, Surrey TW20 OEX, United Kingdom

Carlton E. Brett Department of Geological Sciences, University of Rochester, Rochester, New York 14627

Derek E. G. Briggs Department of Geology, University of Bristol, Bristol BS8 1RJ, United Kingdom

Donald E. Canfield NASA–Ames Research Center, MS 239–4 Moffett Field, California 94035; *present address:* Georgia Institute of Technology, School of Earth and Atmospheric Science, Atlanta, Georgia 30332-0340

Greg A. Carson Geochem Group Ltd., Chester CH4 8RD, United Kingdom

Matthew J. Collins Department of Geology, University of Bristol, Bristol BS8 1RJ, United Kingdom

Geoffrey Eglinton Organic Geochemistry Unit, School of Chemistry, University of Bristol, Bristol BS8 1TS, United Kingdom

Susan M. Kidwell Department of Geophysical Sciences, University of Chicago, Chicago, Illinois 60637

Graham A. Logan Organic Geochemistry Unit, School of Chemistry, and Department of Geology, University of Bristol, Bristol BS8 1RJ, United Kingdom

Jacques Lucas Centre de Géochimie de la Surface, Centre National de la Recherche Scientifique, Institut des Sciences de la Terre, Université Louis Pasteur, 67084 Strasbourg Cedex, France

Lilian E. Prévôt Centre de Géochimie de la Surface, Centre National de la Recherche Scientifique, Institut des Sciences de la Terre, Université Louis Pasteur, 67084 Strasbourg Cedex, France

Robert Raiswell Department of Earth Sciences, University of Leeds, Leeds LS2 9JT, United Kingdom

Stephen E. Speyer Department of Geosciences, University of Arizona, Tucson, Arizona 85721; *present address:* Department of Geology, Arizona State University, Tempe, Arizona 85287

Robert A. Spicer Department of Earth Sciences, Oxford University, Oxford OX1 3PR, United Kingdom

Preface

The term *taphonomy* was coined in 1941 by the Russian paleontologist Efremov to embrace the variety of biological, physical, and chemical processes which determine the extent and style of fossil preservation. That is not to say, however, that taphonomic research began at that point. Far from it in fact—researchers have mused upon the mode of formation of fossils almost since the first one was discovered. Nevertheless, it was not until the first half of this century that German paleontologists such as Richter, Weigelt, Hecht, and later Schäfer began to consider preservational problems in earnest, and implemented actualistic studies. Schäfer's classic treatise *Aktuo-paläontologie nach Studien in der Nordsee* was published in 1962 but did not become very widely known outside Germany until the English translation appeared in 1972 (as *Ecology and Palaeoecology of Marine Environments*). Most taphonomically oriented researchers are aware of the work of Hecht and Weigelt, but few have read the original German texts. Weigelt's (1929) *magnum opus, Rezente Wirbeltier-leichen und ihre paläobiologische Bedeutung,* however, has recently been published in English (1989: *Recent Vertebrate Carcasses and Their Paleobiological Implications*).

The language barrier has been responsible for a considerable time lag in the dissemination of the taphonomic data in these classic German publications. The time lag results not only from different languages; it can also be exacerbated by the language of science. The evolution of classification schemes and the proliferation of scientific nomenclature can alienate those unfamiliar with a particular area of study and inhibit interdisciplinary exchange. Taphonomic research is especially multidisciplinary, involving aspects of the biological, earth, and chemical sciences. Its literature is enormously wide ranging. Publications in oceanography dealing with the rate of carrion utilization by scavengers; botanical literature on the rate of degradation of plant detritus on the forest floor; papers in environmental chemistry on microbial utilization of natural biochemicals; learned volumes for lawyers on forensic medicine; archeological reports on early hominid scavenging; all this is taphonomic data by another name. This breadth of literature is already being used by taphonomists—as evidenced by the diversity of citation sources in the chapters of this book. The continued expansion of this interchange of ideas will be a fertile source of taphonomic insights.

When considering fossilization, it is convenient to group organisms according to their preservational response to taphonomic processes. The major categories in such a scheme are: soft-bodied organisms, plants, shelled organisms, and vertebrates. The first six chapters of this book follow such a *taphonomic taxonomy*, detailing the way that different biological, physical, and chemical processes affect these major groupings.

Chapters 1 and 2 deal particularly with the fate of nonmineralized tissues (both in organisms with some mineralized skeletal elements and in those which are wholly soft-bodied). Chapter 1 reviews the taphonomy of the organic biomolecules which make up soft-tissues, and provides a review of this aspect of the burgeoning field of biomolecular paleontology. The preservation potential of molecules varies with type (nucleic acid, protein, polysaccharide, lipid, etc.), depositional environment, and diagenetic history. Molecular fossils have been used for some time to determine the origin of organic matter in sediments and as a source of environmental data (biomarkers). Recently, it has been shown that molecular fossils may also yield, albeit rarely, data of potential taxonomic significance, even DNA sequences.

Chapter 2 considers the factors which result in decay inhibition and the processes which lead to the preservation of nonmineralized tissues, reviewing the environmental conditions which promote the fossilization of soft-parts. Until recently it was generally accepted that soft-tissues are rarely preserved, and then only in anoxic sediments. The role of anoxia in soft-part preservation is now more fully understood, and it is known that soft-tissues are more commonly fossilized than was previously thought. This chapter reviews the distribution of exceptionally preserved biotas (*Konservat-Lagerstätten*) in the fossil record. They cluster in certain environments at particular intervals of geologic time (forming "taphonomic windows"). Soft-bodied organisms account for 60–70% of the diversity of modern marine communities. Hence a better understanding of the controls on their preservation promises to expand the basis for investigations of evolutionary patterns in the fossil record.

Chapter 3 reviews the processes involved in the preservation of plants and the taphonomic characteristics of specific depositional environments. Apart from the intrinsic importance of plant fossils to evolutionary and paleoecological studies, they can also be very sensitive indicators of past climate. A knowledge of plant taphonomy is critical to allowing the definition of the confidence limits of interpretations based on paleobotanical data.

Chapters 4 and 5 deal with shelly fossils, and Chapter 6 with vertebrates. Chapter 4 discusses the ways in which benthic communities become modified as they are incorporated into the fossil record. A range of processes, hydraulic, biogenic, and geochemical, reduce the fidelity of the death assemblage to the original living fauna. The controls on time-averaging and the criteria which can be used to estimate the degree to which it has occurred are reviewed. Chapter 5 focuses on the origins and significance of shell concentrations (coquinas). Such concentrations not only are a source of paleontologic data, but are increasingly being used as a key to sedimentary dynamics. The analysis of time-averaging has potential as one of the few methods of estimating the amount of time represented in a particular sedimentary unit. The distribution of coquinas can be related to idealized transgressive and regressive sequences. These chapters provide extensive tables of Recent and fossil examples of shelly accumulations.

Chapter 6 provides a complementary approach to the taphonomy of vertebrates, concentrating on the terrestrial realm. In addition, it advocates a multivariate approach to the study of vertebrate assemblages. Data on the assemblage (numbers of specimens, individuals, species, relative abundance, size, etc.), the distribution of specimens, and the degree of bone modification (breakage, abrasion, etc.) can be plotted to provide a graphic comparison between assemblages.

Chapters 7–10 review the major diagenetic phases involved in fossilization (pyrite, phosphate, carbonate, silica)—from the conditions favoring the preservation of original mineralogy to the environments which promote diagenetic precipitation and alteration. The role of microbes in authigenic mineralization is emphasized in each case, and particularly in phosphates (Chapter 8), where *in vivo* precipitation by bacteria has been documented. Chapter 7 provides a detailed model for the processes involved in pyritization and Chapter 9 for the preservation of original carbonate mineralogy and for the growth of carbonate concretions. Chapter 10 reviews the variety of depositional circumstances that can lead to preservation in silica.

Finally, Chapter 11 reviews recurrent associations of preservational style and particular sedimentary facies (the taphofacies concept). The dual roles of background and episodic processes are emphasized. Background processes occur over long periods of time, in contrast to episodic events, which are geologically instantaneous. The interplay of the two allows the definition of taphofacies which are characteristic of different depositional environments in epeiric seas.

Taphonomy is no longer perceived simply as the study of post-mortem information loss; as the chapters in this book demonstrate, taphonomic research goes far beyond the recognition of bias in the fossil record. When the race to move from descriptive paleontology to the search for evolutionary patterns in the fossil record really took off in the mid-1970s it left our understanding of the fossilization process, the nature of the data base, far behind. Taphonomy is now catching up, but in addition to prompting a reassessment of patterns already recognized, it is allowing others to be identified through a much more rigorously understood and adjusted data base.

The classic German literature remains relevant, because of the insights the observational data afford. But modern taphonomy increasingly involves experimental and analytical approaches. Future advances will rely even more on the interdisciplinary style of research now widely adopted. The investigation of decay and diagenesis (both organic and inorganic) will require approaches more microbiological and chemical than paleontological. Unraveling the history of fossil accumulations will be led by sedimentological considerations. The results of these multifaceted approaches to taphonomic research will be fundamental to making the analysis of paleontological data a much more rigorous and exact endeavor. We will continue to rely on taphonomy to release the data locked in the fossil record.

Peter A. Allison
Derek E. G. Briggs

Contents

Chapter 4 • Taphonomy and Time-Averaging of Marine Shelly Faunas

Susan M. Kidwell and Daniel W. J. Bosence

Chapter 5 • The Stratigraphy of Shell Concentrations

Susan M. Kidwell

Chapter 6 • Terrestrial Vertebrate Accumulations

Anna K. Behrensmeyer

Chapter 1

Preservation of Organic Biomolecules

GRAHAM A. LOGAN, MATTHEW J. COLLINS, and GEOFFREY EGLINTON

1. Introduction

Living organisms are composed of nucleic acids, proteins, carbohydrates, lipids, and a variety of other organic molecules containing C, H, N, O, S, and

GRAHAM A. LOGAN • Organic Geochemistry Unit, School of Chemistry, and Department of Geology, University of Bristol, Bristol BS8 1RJ, United Kingdom. MATTHEW J. COLLINS • Department of Geology, University of Bristol, Bristol BS8 1RJ, United Kingdom. GEOFFREY EGLINTON • Organic Geochemistry Unit, School of Chemistry, University of Bristol, Bristol BS8 1TS, United Kingdom.

Taphonomy: Releasing the Data Locked in the Fossil Record, Volume 9 of *Topics in Geobiology*, edited by Peter A. Allison and Derek E. G. Briggs, Plenum Press, New York, 1991.

P. The fossil record contains large quantities of organic carbon derived from these biologically synthesized organic molecules. The new field of molecular paleontology applies the high-sensitivity, high-resolution analytical techniques from the fields of organic geochemistry, biochemistry, and molecular biology to access this store of information (Eglinton, 1983; Runnegar, 1986; Curry, 1987; Runnegar and Schopf, 1988; Farrimond and Eglinton, 1990). Molecular paleontology, although still in its infancy, has demonstrated the potential to reveal information on deposition, decay, and diagenesis within the fossil record.

The first stage in the development of the field was the application by the oil industry of organic chemical techniques to determine the origin of oils (Calvin, 1969). Successes with this approach prompted investigation of less diagenetically altered biomolecules using both organic geochemical and biochemical techniques.

2. Organic Matter in the Geosphere

The potential for preservation of biomolecules in the fossil record is generally low and varies with the class of molecule, the depositional environment, and the diagenetic history. Nevertheless, a surprising proportion of molecules is preserved; recognizably biosynthetic molecules occur in sediments as old as late Precambrian (Jackson *et al.*, 1986; Summons *et al.*, 1988). Important sources of biomolecules include (in decreasing order of abundance) kerogen, coal, organic-rich muds and shales, and even (albeit in small quantities) fossil shells and bones.

Although coal, produced by the burial of plant matter under reducing conditions, is perhaps the best-known form of insoluble organic matter in the fossil record, it is not the most abundant. In excess of 80% of all organic matter in immature sediments is kerogen. The term kerogen refers to all poorly characterized, highly heterogeneous, amorphous, insoluble polymeric material found in the geological record. Its composition is dependent on the source matter and the thermal history of the sediment, but its formation remains a matter of debate. One hypothesis is that kerogen derives from the condensation products of amino acids and sugars (e.g., Larter and Douglas, 1980; Engel *et al.*, 1985). An alternative theory is that the most resistant biopolymers from plant cuticle, and algal and bacterial cell walls, are selectively enriched relative to more labile (but initially more abundant) biomolecules (Hatcher *et al.*, 1983; Goth *et al.*, 1988; Tegelaar *et al.*, 1989). The complex heteropolymeric nature of kerogen forms an ideal environment for lipid preservation, and its chemical degradation has led to the recovery of almost unaltered biolipids (Mycke and Michaelis, 1986; Mycke *et al.*, 1987) (Table I).

Soluble organic matter comprises a much smaller proportion of the total organic matter in the sediment. Geolipids are by far the most abundant

TABLE I. Selective Degradation Techniques Used to Determine the Components of Messel Oil Shale and Isolated Kerogen

Treatment	Sample	Compound released	Reference
BBr_3 followed by $LiAlH_4$	Messel oil shale	Long-chain diols of higher plant origin, isoprenoids of methanogenic bacterial cell walls	Chappe *et al.* (1979)
Rh/H_2 on charcoal	Messel kerogen	Lignin-derived components, from vascular plant tissue	Mycke and Michaelis (1986)
Catalytic hydrogenolysis	Messel kerogen	Bacteriohopanetetrol	Mycke *et al.* (1987)

component of the soluble fraction and are estimated to account for up to 5% of total preserved organics; the most common forms are alkanes, fatty acids, alcohols, terpenoids, and steroids (Brassell *et al.*, 1983). Geolipids represent the preserved decay products of glycolipids, phospholipids, sphyngolipids, and waxes, which can often be related back to their precursor molecule. They form a particularly convenient group to study because they can be extracted using organic solvents, and identified using standard analytical organic chemistry techniques (Section 4.1).

Carbohydrate, protein, and nucleic acids represent a vanishingly small proportion of the total organic matter in the fossil record, and are typically restricted to environments (such as shells and bones) protected from microbial decay. However, their recovery is of particular interest since they represent valuable stores of (evolutionary) information.

3. Organic Matter Contributing to the Fossil Record

The preservation potential of the four major classes of organic molecules differ considerably (Table II). This has consequences both for recovery and for the extent to which diagenetic modifications can be interpreted usefully.

3.1. Nucleic Acids

Ribonucleic acid (RNA) and deoxyribonucleic acid (DNA) are complex macromolecules formed from sequences of purine and pyrimidine bases on a backbone of sugar–phosphate units. Typically, RNA is concerned with translating DNA, the cell's "recipe book," into the structure of proteins.

Some of the bonds that hold nucleic acids together are very reactive, and consequently the macromolecule is rapidly hydrolyzed when the cell dies. However, using a very powerful amplification technique (Section 4.3),

TABLE II. Potential for Preservation of Biopolymers in Depositional Environments[a]

Biomolecule	Source	Preservation potential[b]	Conditions believed to enchance preservation
Nucleic acid			
DNA	All organisms	–	In bone/bound to clays
RNA	All organisms	–	
Protein			
Albumin	Animals	–	In frozen tissue, bone
Collagen	Animals	–/+	As intracrystalline inclusions
Osteocalcin	Animals; bones and teeth	+	
Keratin	Animals	–/+	In peat
Skeletal glycoproteins	Animals; shells, bones, teeth	+	As intracrystalline inclusions
Amino acids	All organisms	–	In kerogens?
Carbohydrate			
Starch	Vascular plants; some algae; bacteria	–	
Glycogen	Animals	–	
Cellulose	Vascular plants; some fungi	–/+	
Chitin	Arthropods; fungi; algae	+	
Simple sugars	All organisms	–	In kerogens?
Lipid			
Glycolipids	Plants; algae; eubacteria	+ +	Bound to kerogen
Lipopolysaccharides	Gram-negative eubacteria	+ +	Bound to kerogen
Resistant biopolymer			
Resins	Vascular plants	+/+ +	
Lignins	Vascular plants	+ + + +	
Sporopollenin	Vascular plants	+ + +	
Cutan	Vascular plants	+ + + +	
Suberan	Vascular plants	+ + + +	

[a]Modified from Tegelaar (1990).
[b]Estimate of the preservation potential of verious biopolymers, ranging from – (very labile) to + + + + (not believed to be degraded in any depositional environment).

original DNA has been detected and sequenced, most spectacularly from Miocene leaves of the Clarkia deposit, Idaho (Golenberg *et al.*, 1990).

3.2. Proteins

Proteins are formed from long chains of amino acid residues linked to each other through peptide bonds. These bonds are formed by a condensation reaction between the free carboxyl group of one amino acid and the free amino group of the next. The 20 amino acids which make up almost all proteins are distinguished by different side chains; the interaction of these side chains with each other and with water causes the protein chains to fold into unique conformations which can be further stabilized via cross-linking. The three-dimensional structure of a protein, coupled with the distribution of atoms and charges on its surface, determines the characteristics and properties of the molecule.

Tight packing, hydrogen bonding, and cross-linking (principally via disulfide bonds) are common features of fibrous structural proteins (e.g., collagen and keratin). The features which confer structural rigidity also enhance post-mortem stability relative to more labile proteins, such as enzymes and serum proteins. Although little consideration has been given to the diagenetic stability of proteins, it is usually assumed that structural proteins survive longer in the geological record, and there is some evidence to support this view (Table II). However, given that biodegradation and chemical hydrolysis rapidly destroy even structural proteins, interest has centered on proteins enclosed within, and stabilized by, biominerals (see Section 6.3).

3.3. Carbohydrates

The simplest carbohydrates are monosaccharides, which can be linked by condensation reactions between the C-1 carbonyl group of one unit and (typically) the C-4 or C-6 hydroxyl groups of the next, to form long (often branched) polysaccharide chains. Polysaccharides are principally used either for storage or structure, and these alternate functions are reflected in the differing characteristics of starch and cellulose. Starch and cellulose are biologically by far the most important polysaccharides, and while both are composed entirely of glucose units, one (starch) is used for storage and the other (cellulose) for structure.

Starch is composed of long helical chains of up to 4000 glucose units with almost no branching. Its use as a storage form of glucose derives from the fact that it is water soluble and the glucoside linkages are amenable to hydrolysis. Cellulose differs from starch in the configuration of the glucoside (β rather than α) linkage; this minor difference alters the secondary and tertiary structure of the molecule. Cellulose forms into long straight chains rather than helices, and these lie in bundles hydrogen bonded to each other through the numerous neighboring hydroxyl groups. The bundles are twisted together to form ropelike structures which themselves are grouped together to form fibers. Cellulose is the most important carbohydrate in woody tissue as well as the main structural component of green plants.

The more chemically resistant (structural) polysaccharides should survive longest in the fossil record (see Section 5.1). Yet given the widespread biological importance of structural polysaccharides (cellulose is probably the most abundant macromolecule in the biosphere), it is generally assumed that these are rapidly recycled by microorganisms. Few studies have attempted to isolate carbohydrates from pre-Cenozic rocks, but intact saccharides have been isolated from Mesozoic sediments (e.g., Michaelis *et al.*, 1986; Moers, 1989), while saccharide residues have been claimed from Cambrian and Precambrian rocks (Swain *et al.*, 1970).

3.4. Lignin and Other Highly Resistant Biopolymers

Lignin, although it is not regarded as one of the major classes of organic molecules, is an important structural component of vascular plants. In wood, ''ropes'' of cellulose are embedded in lignin to give a composite structure similar to that of reinforced concrete. Lignin is believed to be one of the most resistant biopolymers; vertical profiles through peat show a selective concentration of lignin residue relative to cellulose with depth (Boon *et al.*, 1986).

Due to its resistance to chemical degradation, the details of the structure of lignin have yet to be defined, but the major component is thought to be the polymerization products of three phenolic alcohols. Special degradation techniques, such as CuO oxidation and pyrolysis (Section 4.1), have been used to examine the structure of lignin, and have revealed compositional differences between the lignin of gymnosperms and that of angiosperms (Logan and Thomas, 1985). Pyrolysis of lignin derivatives in the fossil record has also led to the discovery of other highly resistant plant biopolymers (Table II) (Nip *et al.*, 1986). These biopolymers, which are thought to be composed of predominantly polyalkyl chains (e.g., cutan and suberan), are considered to be selectively concentrated in kerogen (Goth *et al.*, 1988; Tegelaar *et al.*, 1989).

3.5. Lipids

Lipids are synthesized by all cells and used in a variety of different functions, most notably membrane structure and energy storage. The breakdown products of biolipids are of particular interest, as many of them can survive microbial action to be buried in the sediment and preserved as geolipids. Geolipids are relatively easy to extract and analyze. Some are specific to one particular group of organisms [e.g., dinosterol to dinoflagellates (Robinson *et al.*, 1984)], while total lipid extracts of both Recent and ancient samples often give a general idea of the relative abundance of higher plant, algal, and bacterial inputs.

Unlike polysaccharides or proteins, which are degraded to water-soluble components, lipids tend to be resistant to chemical attack. They are insoluble in water and can be incorporated into kerogens (Larter and Douglas, 1980), thereby increasing their preservation potential. The conversion of biolipids to geolipids involves information loss, through structural alteration (e.g., hydrogenation of double bonds, aromatization of rings, loss of functional groups). The original class of lipid can often be identified even after these diagenetic changes, which can themselves yield useful data on the history of the sediments following burial (Brassell, 1985).

4. Analytical Techniques and Procedures

The quantities of original organic molecules preserved in any sediment or fossil are usually small and consequently contamination can prove a serious problem. Weathering of rock samples and permeation of fluids through microcracks and fissures can introduce water-soluble organic matter. Bacteria, fungi, algae, and lichen growing on samples leave behind their own (modern) signature, which can sometimes be identified on the basis of the concentrations of differing stereo isomers of the same molecule.

Contamination may be introduced during collection, from finger grease, skin, and plastic materials; therefore extreme care should be taken. Samples should be handled as little as possible. They should be wrapped in aluminum foil (which has been prewashed in distilled solvents) and stored in Teflon bags or in clean glass jars; plastic bags may introduce plasticizers (i.e., organic compounds used in the formulation of plastics). In the laboratory, samples should first be washed in organic solvents to remove external contamination such as fingerprints and grease. Only high-grade solvents should be used, and glassware must be cleaned of all organic matter. The number of analytical steps should be kept to a minimum since contamination can be introduced at any stage.

The analytical techniques selected will depend on the material to be examined and the compounds to be studied (Eglinton, 1969). It is important that the matrix is analyzed in the same way as the fossil, to identify contamination; molecules introduced by ground water, for example, should be present in both fossil and sediment.

4.1. Organic Chemistry

Unbound lipids can be extracted using suitable organic solvents, the efficiency being improved by either ultrasonic vibration or continual extraction in hot solvent. However, the bulk of sedimentary organic matter is bound up as heterogeneous macromolecules, some of whose bonds must be broken to release component structures. The gentlest form of chemical attack is acid or base hydrolysis, which cleaves amide and ester-linked acids or alcohols, but hydrolysis is not suitable for kerogens. A variety of alternative techniques have been used successfully to degrade kerogens selectively, releasing characteristic molecules (Table I) which can then be analyzed using standard organic chemical techniques (see below).

One alternative approach used in the analysis of insoluble organic matter is pyrolysis (Meuzelaar *et al.*, 1982). Extremely rapid heating thermally cleaves bonds, and the pyrolysis products are either passed directly into a mass spectrometer (PY-MS) or first separated by gas chromatography (GC) before MS (PY-GC-MS).

Separations of complex mixtures can be performed using a variety of different techniques, by far the most common of which is chromatography. Chromatography takes advantage of the different distribution of components between two phases, one stationary, the other mobile. Separation of a mixture occurs as the components partition differently between the two phases and therefore migrate at different rates. The three types commonly used in the separation of organic molecules are thin-layer (TLC), gas (GC), and liquid (LC) chromatography, and within each a wide variety of different mobile and stationary phases can be used to obtain the desired separation (e.g., Willet, 1987).

Gas chromatographs are often directly linked to mass spectrometers (GC-MS), which enables complex mixtures to be separated into individual components and then identified (Fig. 1). Mass spectrometry (MS) is one of the most powerful techniques in the identification of molecules. Molecules are ionized and fragmented in a variety of ways, and the resulting mass spectra of the ions used to identify compounds or help to determine the structure of an unknown (Eglinton *et al.*, 1985; Philp and Gilbert, 1986). A major development has been the use of higher-resolution mass spectrometers to determine the carbon isotopic composition ($\delta^{13}C$) of individual molecules. Biomolecules become progressively enriched in light carbon at

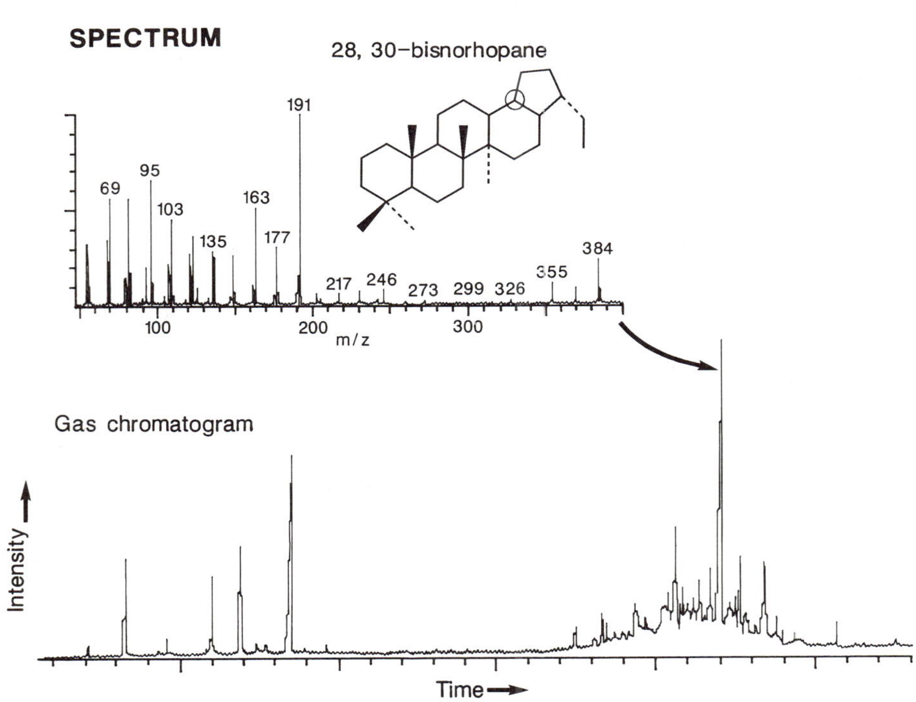

FIGURE 1. The use of GC-MS in the identification of individual compounds from a complex mixture. The mixture is first separated by gas chromatography, and each peak is then passed directly into the mass spectrometer. The mass spectrum is used to identify the compound present in the mixture on the basis of its fragmentation pattern.

each step in the food chain. Therefore $\delta^{13}C$ can be used to identify specific sources of molecules and reconstruct carbon flows within an environment (e.g., Hayes *et al.*, 1987; DeNiro, 1987).

A variety of nondestructive spectroscopic techniques can be used to obtain structural information. Of these, the most important are infrared spectroscopy (IR) and nuclear magnetic resonance (NMR). Both techniques can reveal detail about the structure of molecules and the types of bond present. NMR is a nondestructive technique and has been used to examine the gross structure of coal and insoluble kerogen (Barwise *et al.*, 1984; Davidson, 1986).

4.2. Biochemical Techniques

Most biochemical techniques utilize properties of a whole macromolecule for separation and/or identification (e.g., net charge, molecular weight, degree of hydrophobicity). These techniques are therefore not suitable for the heterogeneous mixtures of variously altered molecules isolated from the fossil record. However, modifications to extraction procedures, such as the selective isolation of intracrystalline molecules from biominerals (DeNiro and Weiner, 1988a; Collins *et al.*, 1991a), and the use of highly specific enzymic cleavage [e.g., collagenase (Armstrong *et al.*, 1983; DeNiro and Weiner, 1988b)] will increase the power and reproducibility of the few techniques which have proved valuable.

Amino acid analysis was the earliest of these biochemical approaches. The ratios of amino acid concentrations can be determined following hydrolysis of the peptide bonds in concentrated acid or alkali. The type of information obtained from amino acid analysis is limited, however, to the concentrations of the different types of acid relative to one another; the sequence of the amino acid residues in the protein (which gives it its unique characteristics) is lost. The technique can be likened to comparing the chapters of this book on the basis of the number of times the different letters of the alphabet occur in each. Many authors now caution against the analyses of amino acids from total extracts (e.g., Robbins and Brew, 1990), because of differences in the rate of diagenesis and the interference of diagenetic products in the analysis (Hoering, 1980). The technique can still prove useful, however, if the protein of interest has a very characteristic amino acid "signature" [e.g., collagen (Armstrong *et al.*, 1983; Gurley *et al.*, 1991)] or contains an unusual residue (e.g., osteocalcin; see Section 6.3).

Electrophoresis and chromatography generally have not proved successful in the separation of fossil extracts, due to the diagenetic modification of macromolecules (e.g., Tuross *et al.*, 1980). Robbins and Brew (1990), however, were able to use two-dimensional electrophoresis, and successfully detected two distinct protein fractions by radiolabeling proteins extracted from 300,000-year-old foraminiferal tests.

An alternative approach which has met with some success is immunology. This exploits the high-specificity recognition of antibodies to small domains (''epitopes'') on larger macromolecules. Since epitopes are small regions of much larger molecules, they can be recognized by antibodies even when the original macromolecule has undergone some degradation (Grisley and Boyle, 1988). The approach, pioneered by groups in Holland and the United States, has demonstrated that original organic material may retain antigenic properties over considerable periods of time (e.g., de Jong *et al.*, 1974; Prager *et al.*, 1980; Lowenstein, 1981; Shoshani *et al.*, 1985; Rowley *et al.*, 1986; Ulrich *et al.*, 1987; Muyzer *et al.*, 1988; Baird and Rowley, 1990; Cattaneo *et al.*, 1990; Collins *et al.*, 1991a,b). Labeled antibodies have been used to find the specific location of epitopes within a fossil matrix (Muyzer and Westbroek, 1989) and to establish relationships among extinct species (e.g., Lowenstein *et al.*, 1981; Lowenstein and Ryder, 1985; Lowenstein, 1988). The use of antibodies may prove particularly applicable to the study of decay, since different epitopes have been shown to degrade at different rates (Muyzer *et al.*, 1984, 1986).

4.3. Molecular Biological Techniques

DNA and RNA have a very low preservation potential, and all early attempts to recover nucleic acids from fossil samples met with failure. The situation has been radically altered, however, with the advent of the polymerase chain reaction (PCR), which can be used to amplify specific nucleic acid sequences (Saiki *et al.*, 1988; Desforges, 1990). This technique theoretically permits the amplification of just a single copy of the target sequence, and can even ''patch together'' the sequence from a series of fragments with overlapping regions (Pääbo *et al.*, 1990). Contamination is a major problem because of the sensitivity of PCR, and great care must be taken to avoid the inadvertent amplification of Recent DNA.

The technique is already being used on a wide variety of samples, ranging from museum pelts to archeological remains, and even Miocene leaves, and in a wide variety of studies, including analysis of migration, speciation, and phylogeny (e.g., Pääbo, 1989; Golenberg *et al.*, 1990; Thomas *et al.*, 1989, 1990).

5. The Fate of Organic Matter

An understanding of molecular taphonomy is vital to an interpretation of molecular data obtained from fossils. This can be considered under two headings: (1) the relative susceptibility of different classes of organic matter to molecular degradation; and (2) the temporal and spatial zonation of decay.

5.1. Molecular Degradation

Rates of decay of biomolecules differ considerably. Metabolites are rapidly destroyed, at one extreme, and yet it is believed that some highly resistant biopolymers suffer no post-mortem modification at all (Tegelaar, 1990).

Hydrolysis is the first step in the degradation of nucleic acids, proteins, carbohydrates, and lipids. The building blocks of the macromolecules liberated by hydrolysis tend to be very reactive and are rapidly metabolized or react with other molecules to form large complex heteropolymers. Within the different classes of organic molecules, those used for storage are usually broken down more rapidly than structural molecules. Klok *et al.* (1984) observed that cyanobacterial storage carbohydrates were rapidly degraded following deposition, while structural polysaccharides were much more resistant, particularly in anaerobic environments. In bone, collagen appears to be slightly more stable than albumin (Shoshani *et al.*, 1985). The chemical instability of nucleic acids and the failure of the cell to offer them protection from hydrolysis lead to the belief that they have the lowest preservation potential of the major classes of organic molecules. Studies have now shown that the extracellular stability of nucleic acids is greater than predicted, due in part to the stabilization of these macromolecules on mineral surfaces (e.g., Lorenz *et al.*, 1981; Romanowski *et al.*, 1991). The advent of very high sensitivity amplification technology (Section 4.3) has enabled a number of fossil samples ranging from bones to seeds and leaves to be successfully probed for fragments of original genes.

In comparative studies, proteins and carbohydrates are degraded at a much faster rate than either lipids or lignin (Leo and Barghoorn, 1970; Meyers *et al.*, 1980; Boon *et al.*, 1982; Benner *et al.*, 1984; Hedges *et al.*, 1985; Fukushima *et al.*, 1987; Kendricks and Edwards, 1988). However, the constituent monomers of both proteins and carbohydrates appear to be surprisingly stable (e.g., Curry, 1988; Moers, 1989). The stability of these monomers in the lithosphere may be the result of condensation reactions (Larter and Douglas, 1980; Engel *et al.*, 1985). An alternative mechanism may be the absorption of these polar organic molecules by inorganic substrates. Amino acids have been reported to be absorbed by calcite, quartz, and clays (Carter, 1978; Hedges and Hare, 1987).

Lignin has one of the highest preservation potentials of any plant molecule. Lignin derivatives have been obtained from Carboniferous fossil plant compressions by breaking down the polymeric structure of the macromolecule (Logan and Thomas, 1987). Sigleo (1978) used PY-GC-MS to identify lignin derivatives in 200-million year (My)-old silicified wood from a petrified forest in Arizona.

Cyclic lipids, such as steroids and terpenoids, undergo structural changes during decay and diagenesis. These compounds are particularly important as biomarkers, because the precursor biolipid can often be identified on the basis of the structure of the more stable geolipid (MacKenzie *et al.*, 1982; MacKenzie, 1984; Johns, 1986).

Sterols and stanols are natural products found in cell membranes, and they are also commonly recorded in surface sediments. They are known to alter to geolipids called steranes by various pathways (Fig. 2) (Gagosian *et al.*, 1979; de Leeuw and Baas, 1986; Wolf *et al.*, 1986). Double bonds between carbon atoms in sterols are rapidly hydrogenated in the microbially active top layers of sediment (Gaskell and Eglinton, 1975; Nishimura and Koyama, 1977). The pathway by which sterols are converted to stanols differs between aerobic and anaerobic environments. In an aerobic lake the stanol is produced directly by hydrogenation of the sterol double bond. In an anoxic lake, on the other hand, the sterols are converted, via a steroidal ketone stage, to the stanol (Gagosian and Heinzer, 1979; Mermoud *et al.*, 1984).

Unbound lipids, freed from their amide, ester, or ether linkages, are relatively easily altered by microbes and by chemical reactions requiring

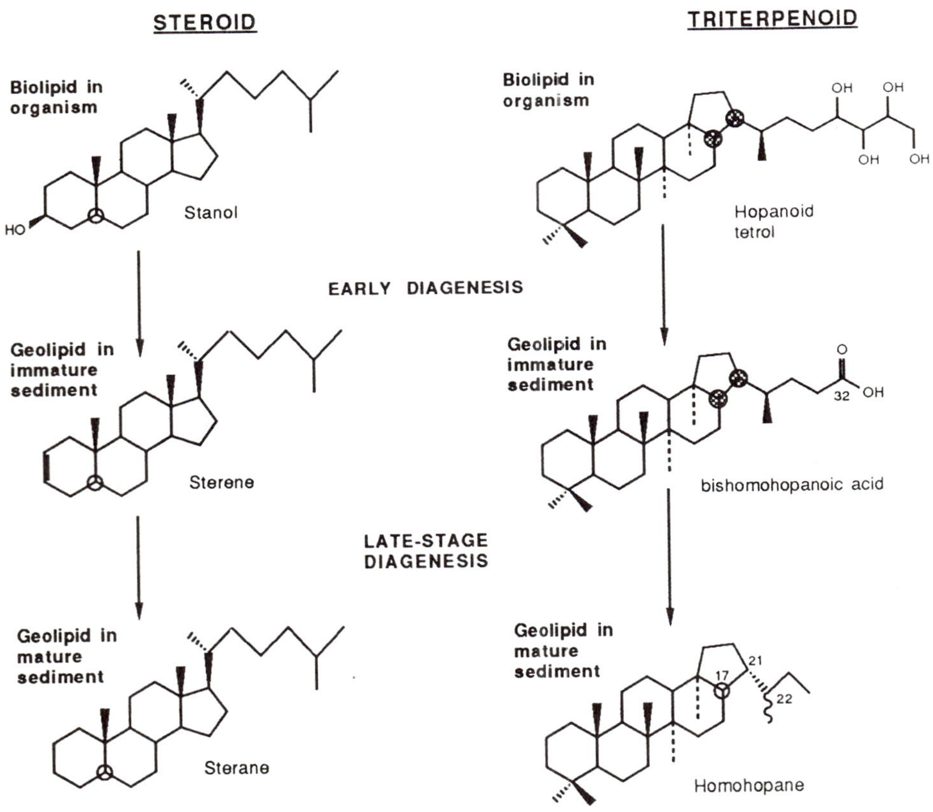

FIGURE 2. The study of source organisms and sediments at various diagenetic stages is leading to an understanding of the diagenetic alteration of biolipids to geolipids. The diagram illustrates postulated diagenetic pathways commencing with a stanol and a hopanoid tetrol. Although the final geolipids have lost a certain amount of information, they can still be related to the precursor biolipids on the basis of their carbon skeletons and, in particular, their stereochemical features.

aqueous conditions (Nishimura, 1977; de Leeuw *et al.*, 1976). The binding of lipids into kerogens may play an important role in lipid preservation. Heating of these kerogens during diagenesis can then release the bound lipids to form the geolipids found in crude oils. It has been demonstrated that pyrolysis of microbially decayed plant material can generate hydrocarbons (Fukushima *et al.*, 1989).

Concentrations of sterols were observed to decrease with depth in cores of Recent sediment (spanning a time range of 130 years), as a result of progressive concentration of sterols into the bound fraction, and also of their degradation to nonsterol products (Lee *et al.*, 1977). Unbound sterols are preferentially degraded compared to unbound stanols (Nishimura and Koyama, 1977).

Sterols are not the only lipid class to show a bias in preservation. Shorter-chain lipids ($<C_{20}$), which tend to be more soluble in water, are preferentially lost as a consequence of microbial activity (Fukushima *et al.*, 1987), while lipids with double bonds are more susceptible to microbial degradation than their saturated counterparts (Matsuda and Koyama, 1977; Kawamura *et al.*, 1987).

Unbound lipids in freshwater lakes show a decrease in preservation potential in the following order: *n*-alkanes, alkan-2-ones, sterols, *n*-alkanoic acids, *n*-alkanols, *n*-alkenoic acids (Cranwell, 1981). Preservation reflects the source as well as the type of lipid. Cutin-derived lipids are less susceptible to degradation than cellular and wax lipids from decaying leaf samples in a mangrove swamp (Wannigama *et al.*, 1981). Furthermore, the presence of large quantities of organic matter may itself promote preservation. Harvey *et al.* (1986) demonstrated that the anaerobic decay of lipids is slowed by the presence of humic acids.

Even after decay and diagenesis, interpretation of data can be further complicated by weathering processes at the surface. Surface oxidation and soil formation can cause further chemical changes in the compounds preserved in weathered rocks (Clayton and Swetland, 1978).

5.2. The Temporal and Spatial Zonation of Decay

Rates of decay of biomolecules differ considerably even within the same organism, depending upon the attendant microbial community and the environment of deposition.

Decay processes have been extensively studied in plant material (see this volume, Chapter 3), which form an important organic input into many lakes and marine shelf sequences. Decomposition of plant material begins with the process of senescence, which is regulated by the plant. Leaching of water-soluble waxes occurs in the first few days after senescence in moist or aqueous environments (Nykvist, 1961; Bridson, 1985). The decomposition of leaf litter is a complex interactive process involving heterotrophic communities of scavenging macrofauna, fungi, and bacteria. Colonization by microorganisms is rapid and initiates microbial degradation. As decay progresses, successive, different microorganisms utilize different chemicals derived from the decaying matter (Kjoller and Struwe, 1980; Kjoller *et al.*, 1985). Mason (1976) showed that initially bacteria and fungi were almost equally responsible for decomposition of *Phragmites* leaves. After 35 days, however, the fungal population declined more rapidly than the bacteria. Brock (1984) recorded that decay of a floating leafed macrophyte, *Nymphoides peltata*, was dominated by bacteria for the first 19 days, followed by fungi over the next 12 days.

The pattern of decay is also dictated by a variety of environmental conditions, including pH, oxygen tension, and salinity, which can completely alter the nature of the decomposer community.

Fungi, which play an important role in aerobic decay, are rare in anaerobic environments (Emerson and Natvig, 1981), and once oxygen is depleted, most decomposition is bacterially mediated. Anaerobic decay is dominated by two types of bacteria, sulfate reducers and methanogens. In marine sediments the most important group are the sulfate reducers, which use sulfate ions in the pore water as their terminal electron acceptor (Jorgensen, 1982). Once the sulfate ions have been depleted, methanogenic bacteria become the dominant group of decay organisms (Capone and Kiene, 1988). Because of the high concentration of sulfate ions in marine pore waters, methanogenesis only occurs below a broad zone of sulfate reduction (Reeburgh, 1980; Ward and Winfrey, 1985).

In freshwater environments a similar stratification occurs, but because of the lower concentration of salts (and hence electron acceptors), it is much more compressed. Consequently, methanogenic bacteria become the dominant decomposers of organic matter much nearer the oxic/anoxic boundary (Saunders, 1976), and therefore play a more significant role.

Organic matter deposited in high-stress environments is an important source of molecular fossils. Only specialized microorganisms are adapted to such environments and a full complement of degradation pathways may not be present. In eutrophic environments (e.g., ox-bow lakes) the high diversity of microorganisms ensures that macromolecules are rapidly leached into solution or mineralized to CO_2. Conversely, in acidic environments, such as moorland pools, only a reduced diversity of decomposers can survive the low-pH conditions, which enhance the preservation of macromolecules such as cellulose (Brock *et al.*, 1985). The dense brine pools which form by the dissolution of evaporites in sea floor sediment are another extreme depositional environment. Low bacterial diversity may account for the accumulation of organic matter in these deposits and their anomalous isotopic composition (Klinkhammer and Lambert, 1989).

6. The Molecular Fossil Record

Although the field of molecular paleontology is still very new, investigations have covered a wide range of different organisms and stratigraphic levels. Studies of microorganisms have concentrated on the identification of biomarkers, or the structural modifications undergone by specific molecules during diagenesis. Fossil plant material has also been analyzed for both taxonomically diagnostic chemicals and diagenetic modifications. Taxonomic assignment is generally not a problem with the bones, teeth, and shells which commonly form the basis of animal studies, and questions have centered on evolutionary relationships and dietary analysis.

6.1. Microorganisms

Heterotrophic microorganisms consume the products of organisms higher up the microbial food chain, modifying the molecular signatures of allochthonous organic inputs. They also constitute an important source of environmental information.

The presence of sulfate-reducing bacteria in Recent sediments can be identified by the distribution of certain saturated and unsaturated C_{14}–C_{18} fatty acids (Bobbie and White, 1980; Parkes and Taylor, 1985; Parkes and Calder, 1985). The value of these lipids in geological samples is limited due to the reduction of double bonds and the preferential degradation of these compounds. Another common group of bacterial lipids are the hydroxy fatty acids, which have an increased preservation potential because the presence of hydroxyl functional groups allows binding to insoluble organic matter (Kawamura and Ishiwatari, 1982). The amide-linked β-hydroxy fatty acids are of particular importance because they are believed to be specific to gram-negative bacteria and have been recovered in Pleistocene sediments (Goossens *et al.*, 1986).

Methanogenic and thermophilic bacteria produce characteristic acyclic isoprenoidal lipids based on branched units of five carbon atoms (Holzer *et al.*, 1979; Brassell *et al.*, 1981; Risatti *et al.*, 1984). The range of hopanoid hydrocarbons in crude oil is thought to be due in part to the degradation products of bacteriohopanetetrol produced in the cell wall of bacteria (Ourisson *et al.*, 1979, 1982). This particular lipid has been identified in chemically degraded kerogen (Table I).

6.2. Plants

Fossil plant material has often been analyzed in a search for chemicals diagnostic of particular taxonomic groups (Thomas, 1986). Decay and diagenesis lead to changes in the chemistry of plant tissue which can be studied by examining material of known age and environment.

Studies of Recent material show that fatty acids are lost relatively rapidly over the first few hundred years of decay (Priestley *et al.*, 1981). A 10,000-year-old spruce tree buried in a lake revealed a decrease in fatty acid concentration of 90%, whereas decrease in concentrations of sterols and hydrocarbons was much less marked. In the same study, concentrations of cellulose were shown to have decreased relative to lignin, which was the least altered component in the sample (Meyers *et al.*, 1980). Needles of the spruce showed a different pattern of alteration to the rest of the tree; the fatty acids and sterols were equally degraded, but the hydrocarbons were much less affected. Thus, chemotaxonomy requires a knowledge of the taphonomic response of different component parts of the same plant.

Unbound *n*-alkanes, hydroxy fatty acids, steranes, and triterpenoids have been reported from Miocene fossil leaves (Niklas and Giannasi, 1977, 1978). Unsaturated fatty acids have also been reported from 30- to 50-My-old angiosperm fruits and seeds (Niklas *et al.*, 1982). Specific compounds have been found in green angiosperm leaf impressions from Eocene brown coals from eastern Germany, including chlorophyll derivatives (Dilcher *et al.*, 1970). A Triassic fossil horsetail *Equisetum* (Knoche and Ourisson, 1967) was found to have the same odd-over-even ratio of C_{23}–C_{29} alkanes observed in living examples.

Flavanoids are found as pigments of flowers in most angiosperm families and as pigments in many higher plants (Hahlbrock, 1981). Members of this group of compounds have been isolated from Miocene leaf tissue (Giannasi and Niklas, 1977, 1981; Niklas and Giannasi, 1978; Rieseberg and Soltis, 1987).

Swain *et al.* (1967) described monosaccharides from Paleozoic plant fossils, but were uncertain whether these carbohydrates represented plant or bacterial decay products. Fossil leaves have been found to contain carbohydrates (Wilson *et al.*, 1987), but in a much reduced concentration compared to that in living plant material (Hedges *et al.*, 1985; Bates and Hatcher, 1989; Hatcher *et al.*, 1989). Cellulose and saccharides are much diminished compared to hydrocarbons over a period of several years (Fukushima *et al.*, 1987), probably due to fungal and bacterial activity, and progressive loss during diagenesis.

Fossil amber, such as that of the Baltic Eocene, yields not only exceptionally preserved fossils, but also molecules that are normally extensively degraded during diagenesis. An ether-soluble fraction obtained from Baltic amber and analyzed by GC-MS contained many monoterpenoid, diterpenoid, and sesquiterpenoid compounds. The ether-insoluble fraction was a copolymer similar to that formed by modern conifer resins (Mills *et al.*, 1984). A diagenetically produced hydrocarbon, cedrane, is preserved in some Tertiary fossil resins, indicating that the amber underwent a period of diagenesis under reducing conditions (Grantham and Douglas, 1980). Diterpenoids which can yield aromatized hydrocarbons on diagenesis are dominant in higher plant resins (Simoneit *et al.*, 1986).

6.3. Animals

DNA provides the most complete phylogenetic information of any molecule, but unless exceptional circumstances prevail, the nucleotide sequences only survive for a short period of time after the death of the organism. The potential for sequencing is therefore restricted to young or exceptionally well-preserved samples. DNA has been sequenced from dried muscle tissue of the quagga, an extinct member of the horse family, but even after just a century only 1% of the DNA was preserved (Higuchi *et al.*, 1984). Fragments of original DNA have been detected in human tissues from 3000-year-old Egyptian mummies, and from 7000-year-old brain tissue preserved in a swamp in Florida (Pääbo, 1985; Doran *et al.*, 1986).

Proteins are thought to be more stable than nucleic acids, particularly when protected within bones and shells. Westbroek *et al.* (1979) used immunological techniques to compare the reactivity of macromolecules isolated from belemnite guards with those from Recent cephalopod shells, and demonstrated recognizable epitopes in the former. Using electrophoretic techniques, Robbins and Brew (1990) detected two distinct fossil proteins preserved in the tests of 300,000-year-old Foraminifera. Claims have been made for the detection of dried blood on stone implements dating back 100,000 years (Loy, 1983); albumins have been detected immunologically from some of these samples (Lowenstein, 1988).

Proteins isolated from bones and teeth, most notably collagen, have been used in a variety of different studies, including taxonomic affinities (Lowenstein, 1981), diagenesis of proteins (Armstrong *et al.*, 1983), and interpretation of paleodiet (DeNiro 1987; Tuross *et al.*, 1988). Although ultrastructural detail characteristic of collagen has even been reported from Ordovician graptolite periderm (Towe and Urbanek, 1972), collagen epitopes only survive for a few million years (Lowenstein, 1981). Interest has therefore also focused on osteocalcin, an extremely acidic protein involved in calcium regulation, which binds strongly to hydroxyapatite of teeth and bone. It is believed that small size and intimate association with the mineral phase will partially protect osteocalcin from diagenetic modification (Huq *et al.*, 1985). In addition, osteocalcin possesses an unusual (and unstable) amino acid (γ-carboxy glutamic acid, GLA) which can be used as a biomarker in fossils. Immunological techniques have been used to detect osteocalcin in the teeth of Miocene mammals (Ulrich *et al.*, 1987). However, although osteocalcin appears to be the most stable protein found in bone, its taxonomic applications are very limited due to its conservative structure.

7. Conclusions

The study of organic matter in the fossil record is beginning to reveal valuable information not attainable from other sources. The interpretation of these new data requires an understanding of molecular taphonomy—the processes of decay in various environments and the biases in preservation that they cause.

Knowledge of the molecular fossil record is still very limited. Advances in analytical instruments now allow very low concentrations of organic molecules to be studied, and future refinements are sure to further increase sensitivity. The cost of instrumentation should fall as more advanced equipment replaces present systems, and this will widen availability of the techniques, while increased automation will also reduce the need for specialist training.

Experiments to simulate decay in different environments will result in a greater understanding of preservational biases. However, the processes of decay and the development of decomposer communities are very complex. Experiments cannot hope to duplicate these complex interactive processes in more than general terms, so studies must be coupled with examination of both Recent and paleoenvironments. The information gained from these studies will be invaluable in the interpretation of preserved organic matter. It will also help in understanding paleoenvironments and in the development of models to predict the sedimentary conditions under which exceptional preservation is likely to occur.

References

Armstrong, W. G., Halstead, L. B., Reed, F. B., and Wood, L., 1983, Fossil proteins in invertebrate calcified tissues, *Phil. Trans. R. Soc. Lond. B* **301**:301–343.

Baird, R. F., and Rowley, M. J., 1990, Preservation of avian collagen in Australian Quaternary cave deposits, *Palaeontology* **33**:447–451.

Barwise, A. J. G., Mann, A. L., Eglinton, G., Gowar, A. P., Wardroper, A. M. K., and Gutteridge, C. S., 1984, Kerogen characterization by ^{13}C spectroscopy and pyrolysis-mass spectrometry, *Org. Geochem.* **6**:343–349.

Bates, A. L., and Hatcher, P. G., 1989, Solid-state ^{13}C NMR studies of a large fossil gymnosperm from the Yallourn open cut, Latrobe valley, Australia, *Org. Geochem.* **14**:609–617.

Benner, R., Maccubbin, A. E., and Hodson, R. E., 1984, Anaerobic biodegradation of the lignin and polysaccharide componemts of liginocellulose and synthetic lignin by sediment microflora, *Appl. Environ. Microbiol.* **47**:998–1004.

Bobbie, R. J., and White, D. C., 1980, Characterization of benthic microbial community structure by high resolution gas chromatography of fatty acid methyl esters, *Appl. Environ. Microbiol.* **39**:1212–1222.

Boon, J. J., Wetzel, R. G., and Godshalk, G. L., 1982, Pyrolysis mass spectrometry of some *Scirpus* species and their decomposition products, *Limnol. Oceanogr.* **27**:839–848.

Boon, J. J., Dupont, L., and de Leeuw, J. W., 1986, Characterization of a peat bog profile by Curie point pyrolysis-mass spectrometry combined with multivariant analysis and by pyrolysis-gas chromatography-mass spectrometery, in: *Peat and Water* (C. H. Fuchsman, ed.), Elsevier, Amsterdam, pp. 215–239.

Brassell, S. C., 1985, Molecular changes in sediment lipids as indicators of systematic early diagenesis, *Phil. Trans. R. Soc. Lond. A* **315**:57–75.

Brassell, S. C., Wardroper, A. M. K., Thomson, I. D., Maxwell, J. R., and Eglinton, G., 1981, Specific acyclic isoprenoids as biological markers of methanogenic bacteria in marine sediments, *Nature* **290**:693–696.

Brassel, S. C., Eglinton, G., and Maxwell, J. R., 1983, The geochemistry of terpenoids and steroids, *Biochem. Soc. Trans.* **11**:575–586.

Bridson, J. N., 1985, Forest litter lipids decomposition, *Soil Biol. Biochem.* **17**:285–290.

Brock, T. C. M., 1984, Aspects of the decomposition of *Nymphoides peltata* (Gmel.) O. Kuntze (Menyantaceae), *Aquat. Bot.* **19**:131–156.

Brock, T. C. M., Boon, J. J., and Paffen, B. G. P., 1985, The effects of the season and of water chemistry on the decomposition of *Nymphaea alba* L.; Weight loss and pyrolysis mass spectrometry of the particulate matter, *Aquat. Bot.* **22**:197–229.

Calvin, M., 1969, *Chemical Evolution*, Oxford University Press, Oxford.

Capone, D. G., and Kiene, R. P., 1988, Comparison of microbial dynamics in marine and freshwater sediments: Contrasts in anaerobic carbon catabolism, *Limnol. Oceanogr.* **33**:725–749.

Carter, P. W., 1978, Adsorption of amino acid-containing organic matter by calcite and quartz, *Geochim. Cosmochim. Acta* **42**:1239–1242.

Cattaneo, C., Gelsthorpe, K., Phillips, P., and Sokol R. J., 1990, Blood in ancient human bone, *Nature* **347**:339.

Chappe, B., Michaelis, W., and Albrecht, P., 1979, Molecular fossils of Archaebacteria as selective degradation products of kerogen, in: *Advances in Organic Geochemistry* (A. G. Douglas and J. R. Maxwell, eds.), Pergamon Press, Oxford, pp. 265–274.

Clayton, J. L., and Swetland, P. J., 1978, Subaerial weathering of sedimentary organic matter, *Geochim. Cosmochim. Acta* **24**:305–312.

Collins, M. J., Muyzer, G., Curry, G. B., Sandberg, P. A., and Westbroek, P., 1991a, Macromolecules in brachiopod shells: characterization and diagenesis, *Lethaia* in press.

Collins, M. J., Muyzer, G., Westbroek, P., Curry, G. B., Sandberg, P. A., Xu, S. J., Quinn, R., and MacKinnon, D., 1991b, Preservation of fossil biopolymeric structures: conclusive immunological evidence, *Geochem. Chosmachem. Acta.* in press.

Cranwell, P. A., 1981, Diagenesis of free and bound lipids in terrestrial detritus deposited in a lacustrine sediment, *Org. Geochem.* **3**:79–89.

Curry, G. B., 1987, Molecular palaeontology: New life for old molecules, *Trends Ecol. Evol.* **2**:161–165.

Curry, G. B., 1988, Amino acids and proteins from fossils, in: *Paleontological Society Short Courses in Paleontology, Number 1* (B. Runnegar and J. W. Schopf, eds.), Paleontological Society, pp. 20–33.

Davidson, R. M., 1986, Nuclear Magnetic Resonance Studies of Coal, Report No. ICTIS/TR32, IEA Coal Research, London, United Kingdom.

de Jong, E. W., Westbroek, P., Westbroek, J. F., and Bruning, J. W., 1974, Preservation of antigenic properties of macromolecules over 70 myr old, *Nature* **252**:63–64.

de Leeuw, J. W., and Baas, M., 1986, Early-stage diagenesis of steroids, in: *Biological Markers in the Sedimentary Record* (R. B. Johns, ed.), Elsevier, Amsterdam, pp. 101–124.

de Leeuw, J. W., Rijpstra, W. I. C., Boon, J. J., de Lange, F., and Schenck, P. A., 1976, The relationship between lipids from *Fontinalis antipyetica*, its detritus and the underlying sediment: The fate of wax esters and sterol esters, in: *Interactions Between Sediment and Freshwater* (H. L. Golterman, ed.) Amsterdam, September 6–10, 1976), Dr. W. Junk B. V. Publishers, The Hague.

DeNiro, M. J., 1987, Stable isotopes and archaeology, *Am. Sci.* **75**:182–191.

DeNiro, M. J., and Weiner, S. J., 1988a, Organic matter within crystalline aggregates of hydroxyapatite: A new substrate for stable isotopic and possibly other biogeochemical techniques, *Geochim. Cosmochim. Acta* **52**:2415–2423.

DeNiro, M. J., and Weiner, S. J., 1988b, Use of collagenase to purify collagen from prehistoric bones for stable isotope analysis, *Geochim. Cosmochim. Acta* **52**:2425–2431.

Desforges, J. F., 1990, The polymerase chain reaction—A new method of using molecular genetics for medical diagnosis, *N. Engl. J. Med.* **322**:178–183.

Dilcher, D. L., Paulick, R. J., and Mitchell, J., 1970, Chlorophyll derivatives in Middle Eocene sediments, *Science* **168**:1447–1449.

Doran, G. H., Dickel, D. N., Ballinger, W. E., Agee, O. F., Laipis, P. J., and Hauswirth, W. W., 1986, Anatomical, cellular and molecular analysis of 8000 year old human brain tissue from the Windover archaeological site, *Nature* **323**:803–806.

Eglinton, G., 1969, The organic chemist's approach, in: *Organic Geochemistry* (G. Eglinton and M. T. J. Murphy, eds.), Longman, London, pp. 20–73.

Eglinton, G., 1983, Organic molecules as chemical fossils—The molecular fossil record, in: *Cosmochemistry and the Origin of Life* (C. Ponnamperuma, ed.), Reidel, Dordrecht, pp. 323–359.

Eglinton, G., Evershed, R. P., Gill, J. P., and Hein, C. S., 1985, Analytical methods in organic geochemistry, *Anal. Proc.* **22**:263–266.

Emerson, R., and Natvig, D. O., 1981, Adaption of fungi to stagnant waters, in: *The Fungal Community* (D. T. Wicklow and G. C. Carroll, eds.), Marcel Dekker, New York, pp. 109–128.

Engel, M. H., Rafalska-Bloch, J., Schiefelbein, C. F., Zumberge, J. E., and Serban, A., 1985, Simulated diagenesis and catagenesis of marine kerogen precursors: Melanoidins as model systems for light hydrocarbon generation, *Org. Geochem.* **10**:1073–1079.

Farrimond, P., and Eglinton, G., 1990, The record of organic matter and the nature of source rocks, in: *Palaeobiology—A Synthesis* (D. E. G. Briggs and P. R. Crowther, eds.), Blackwell, Oxford, pp. 217–222.

Fukushima, K., Yamamoto, S., Uzaki, M., Morinaga, S., and Ishiwatari, R., 1987, Characterization of the microbial degradation products of a submerged plant with particular reference to the production of the kerogen-like material, *Chem. Geol,* **64**:169–179.

Fukushima, K., Moringa, S., Uzaki, M., and Ochinai, M., 1989, Hydrocarbons generated by pyrolysis of insoluble kerogen-like materials isolated from microbially degraded plant residues, *Chem. Geol.* **76**:131–141.

Gagosian, R. B., and Heinzer, F., 1979, Stenol and stanols in the oxic and anoxic waters of the Black Sea, *Geochim. Cosmochim. Acta* **43**:461–486.

Gagosian, R. B., Smith, S. O., Lee, C., Farrington, J. W., and Frew, N. M., 1979, Steroidal transformations in Recent marine sediments, in: *Advances in Organic Geochemistry* (A. G. Douglas and J. R. Maxwell, eds.), Pergamon Press, Oxford, pp. 407–419.

Gaskell, S. T., and Eglinton, G., 1975, Rapid hydrogenation of sterols in a contemporary lacustrine sediment, *Nature* **245**:209–211.

Giannasi, D. E., and Niklas, K. J., 1977, Flavanoids and other chemical constituents of Miocene *Celtis* and *Ulmus* (Succor Creek Flora), *Science* **197**:765–767.

Giannasi, D. E., and Niklas, K. J., 1981, Comparative paleobiochemistry of some fossil and extant Fagaceae, *Am. J. Bot.* **68**:762–770.

Golenberg, E. M., Giannassi, D. E., Clegg, M. T., Smiley, C. J., Durbin, M., Henderson, D., and Zurawski, G., 1990, Chloroplast DNA sequences from a Miocene *Magnolia* species, *Nature* **344**:656–658.

Goossens, H., Irene, W., Rijpstra, R., Duren, R. R., de Leeuw, J. W., and Schenck, P. A., 1986, Bacterial contribution to sedimentary organic matter; A comparative study of lipid moieties in bacteria and Recent sediments, *Org. Geochem.* **10**:683–696.

Goth, K., de Leeuw, J. W., Putlmann, W., and Tegelaar, E. W., 1988, Origin of Messel Oil Shale kerogen, *Nature* **336**:759–761.

Grantham, P. J., and Douglas, A. G., 1980, The nature and origin of sesquiterpenoids in some Tertiary fossil resins, *Geochim. Cosmochim. Acta* **44**:1801–1810.

Grisley, M. S., and Boyle, P. R., 1988, Recognition of food in *Octopus* digestive tract, *J. Exp. Mar. Biol. Ecol* **118**:7–32.

Gurley, L. R., Valdez, J. G., Spall, W. D., Smith, B. F., and Gillett, D. D., 1991, Proteins in the fossil bone of the dinosaur, *Seismosaurus, J. Protein Chem.* **10**: 75–90.

Hahlbrock, K., 1981, Flavanoids, in: *The Biochemistry of Plants,* Volume 7, *Secondary Plant Products* (P. K. Stumpf and E. E. Conn, eds.), Academic Press, New York, pp. 425–455.

Harvey, H. R., Fallon, R. D., and Patton, J. S., 1986, The effect of organic matter and oxygen on the degradation of bacterial membrane lipids in marine sediments, *Geochim. Cosmochim. Acta* **50**:795–804.

Hatcher, P. G., Spiker, E. C., Szeverenyi, N. M., and Maciel, G. E., 1983, Selective preservation and origin of petroleum-forming aquatic kerogen, *Nature* **305**:498–501.

Hatcher, P. G., Lerch, H. E., Bates, A. L., and Verheyen, T. V., 1989, Solid-state ^{13}C nuclear magnetic resonance studies of coalified gymnosperm xylem tissue from Australian brown coals, *Org. Geochem.* **14**:355–364.

Hayes, J. M., Takiguiku, R., Ocampo, R., Callot, H. J., and Albrecht, P., 1987, Isotopic compositions and probable origins of organic molecules in the Eocene Messel Shale, *Nature* **329**:48–51.

Hedges, J. T., and Hare, P. E., 1987, Amino acid adsorption by clay minerals in distilled water, *Geochim. Cosmochim. Acta* **51**:255–259.

Hedges, J. T., Cowie, G. L., Ertel, J. R., Barbour, R. J., and Hatcher, P. G., 1985, Degradation of carbohydrates and lignins in buried woods, *Geochim. Cosmochim. Acta* **49**:701–711.

Higuchi, R., Bowan, B., Freuberger, M., Ryder, O. A., and Wilson, A. C., 1984, DNA sequences from the quagga, an extinct member of the horse family, *Nature* **312**:282–284.

Hoering, T. C., 1980, The organic constituents of fossil mollusc shells, in: *Biogeochemistry of Amino Acids* (P. E. Hare, T. C. Hoering, and K. King, eds.), Wiley, New York, pp. 193–201.

Holzer, G., Oro, J., and Tornabene, T. G., 1979, Gas chromatographic-mass spectrometric analysis of neutral lipids from methanogenic and thermophilic bacteria, *J. Chromatogr.* **186**:795–809.

Huq, L. N., Rambaud, S. M., Teh, L. C., Davies, A. D., McCulloch, B., Trotter, M. M., and Chapman, G. E., 1985, Immunochemical detection and characterization of osteocalcin from *Moa* bone, *Biochem. Biophys. Res. Commun.* **129**:714–720.

Jackson, M. J., Powell, T. G., Summons, R. E., and Sweet, I. P., 1986, Hydrocarbon shows and petroleum source rocks in sediments as old as 1.7 ^{a} 10^{9} years, *Nature* **322**:727–729.

Johns, R. B. (ed.), 1986, *Biological Markers in the Sedimentary Record*, Elsevier, Amsterdam.

Jorgensen, B. B., 1982, Mineralization of organic matter in the sea bed—The role of sulphate reduction, *Nature* **296**:643–645.

Kawamura, K., and Ishiwatari, R., 1982, Tightly bound β-hydroxy acids in a Recent sediment, *Nature* **297**:144–145.

Kawamura, K., Ishiwatari, R., and Ogura, K., 1987, Early diagenesis of organic matter in the water column and sediments: Microbial degradation and resynthesis of lipids in Lake Haruna, *Org. Geochem.* **11**:251–264.

Kendricks, P., and Edwards, D., 1988, The anatomy of Lower Devonian *Gosslingia breconensis* Heard based on pyritized axes, with some comments on the permineralization process, *Bot. J. Linn. Soc.* **97**:95–123.

Kjoller, A., and Struwe, S., 1980, Microfungi of decomposing Red Alder leaves and their substrate utilization, *Soil Biol. Biochem.* **12**:425–431.

Kjoller, A., Struwe, S., and Vesterg, K., 1985, Bacterial dynamics during decomposition of Alder litter, *Soil Biol. Biochem.* **17**:463–468.

Klinkhammer, G. P., and Lambert, C. E., 1989, Preservation of organics during salinity excursions, *Nature* **339**:271–274.

Klok, J., Cox, H. C., Bass, M., de Leeuw, J. W., and Schenck, P. A., 1984, Carbohydrates in Recent marine sediments—II. Occurrence and fate of carbohydrates in a Recent stromatolitic deposit: Solar Lake, Sinai, *Org. Geochem.* **7**:101–109.

Knoche, H., and Ourisson, G., 1967, Organic compounds in fossil plants (*Equisetum;* Horsetails), *Ang. Chem. Int. Ed.* **6**:1085.

Larter, S. R., and Douglas, A. G., 1980, Melanoidins—Kerogen precursors and geochemical lipid sinks: A study using pyrolysis gas chromatography (PGC), *Geochem. Cosmochim. Acta* **44**:2087–2095.

Lee, C., Gagosian, R. B., and Farrington, J. W., 1977, Sterol diagenesis in Recent sediments from Buzzards Bay, Massachusetts. *Geochim. Cosmochim. Acta* **41**:985–992.

Leo, R. F., and Barghoorn, E. S., 1970, Phenolic aldehydes: Generation from fossil woods and carbonaceous sediments by oxidative degradation, *Science* **168**:582–584.

Logan, K. J., and Thomas, B. A., 1985, Distributions of lignin derivatives in plants, *New Phytol.* **99**:571–585.

Logan, K. J., and Thomas, B. A., 1987, The distribution of lignin derivatives in fossil plants, *New Phytol.* **105**:157–173.

Lorenz, M. G., Aardema, B. W., and Krumbein, W. E., 1981, Interaction of marine sediments with DNA and DNA availability to nucleases, *Mar. Biol.* **644**:225–230.

Lowenstein, J. M., 1988, Immunological methods for determining phylogenetic relationships, in: *Paleontological Society Short Courses in Paleontology, Number 1* (B. Runnegar and J. W. Schopf, eds.), Paleontological Society, pp. 12–19.

Lowenstein, J. M., 1981, Immunological reactions from fossil material, *Phil. Trans. R. Soc. Lond. B* **292**:143–149.

Lowenstein, J. M., and Ryder, D. A., 1985, Immunological systematics of the quagga (Equidae), *Experienta* **41**:1192–1193.

Lowenstein, J. M., Sarich, V. M., and Richardson, B. J., 1981, Albumin systematics of the extinct mammoth and Tasmanian wolf, *Nature* **291**:409–411.

Loy, T. H., 1983, Prehistoric blood residues: detection on tool surfaces and identification of species of origin, *Science* **220**:1269–1270.

MacKenzie, A. S., 1984, Applications of biological markers in petroleum geochemistry, *Adv. Petroleum Geochem.* **1**:115–213.

MacKenzie, A. S., Brassell, S. G., Eglinton, G., and Maxwell, J. R., 1982, Chemical fossils: The geological fate of steroids, *Science* **217**:491–504.

Mason, C. F., 1976, Relative importance of fungi and bacteria in the decomposition of *Phragmites* leaves, *Hydrobiology* **51**:65–69.

Matsuda, H., and Koyama, T., 1977, Early diagenesis of fatty acids in lacustrine sediments—2. A statistical approach to changes in fatty acids composition from Recent sediments and some source material, *Geochim. Cosmochim. Acta* **41**:1825–1834.

Mermoud, F., Wunsche, L., Clerc, O., Gulacar, F. O., and Buchs, A., 1984, Steroidal ketones in the early diagenetic transformation of Δ^5 sterols in different types of sediment, *Org. Geochem.* **6**:25–29.

Meuzelaar, H. L. C., Haverkamp, J., and Hileman, F. D., 1982, Pyrolysis mass spectrometry of Recent and fossil biomaterials, in: *Techniques and Instrumentation in Analytical Chemistry*, Volume 3, Elsevier, Amsterdam, pp. 1–293.

Meyers, P. A., Leenheer, M. J., Erstfeld, K. M., and Bourbonniere, R. A., 1980, Changes in spruce composition following burial in lake sediments for 10,000 years, *Nature* **287**:534–535.

Michaelis, W., Myke, B., and Richnow, H. H., 1986, Organic chemical indicators for reconstructions of Angola Basin sedimentation processes, *Mitt. Geol. Paläontol. Inst. Univ. Hamburg* **60**:90–113.

Mills, J. S., White, R., and Gough, L., 1984, The chemical composition of Baltic Amber, *Chem. Geol.* **47**:15–39.

Moers, M. E. C., 1989, Occurrence and fate of carbohydrates in Recent and ancient sediments from different environments of deposition, Ph. D. thesis, Delft Technical University.

Muyzer, G., and Westbroek, P., 1989, An immunohistochemical technique for the localization of preserved biopolymeric remains in fossils, *Geochim. Cosmochim. Acta* **53**:1699–1702.

Muyzer, G., Westbroek, P., de Vrind, J. P. M., Tanke, J., Vrijheid, T., de Jong, E. W., Bruning, J. W., and Wehmiller, J. F., 1984, Immunology and organic geochemistry, *Org. Geochem.* **6**:847–855.

Muyzer, G., de Koster, S., van Zijl, Y., Boon, J. J., and Westbroek, P., 1986, Immunological studies in microbial mats from Solar Lake (Sinai). A contribution to the organic geochemistry of sediments, *Org. Geochem.* **10**:697–704.

Muyzer, G., Westbroek, P., and Wehmiller, J. F., 1988, Phylogenetic implication and diagenetic stability of macromolecules from Pleistocene and Recent shells of *Mercenaria mercenaria* (Mollusca, Bivalvia), *Hist. Biol.* **1**:135–144.

Mycke, B., and Michaelis, W., 1986, Molecular fossils from chemical degradation of macromolecular organic matter, *Org. Geochem.* **10**:847–858.

Mycke, B., Narjes, F., and Michaelis, W., 1987, Bacteriohopanetetrol from chemical degradation of an oil shale kerogen, *Nature* **326**:179–181.

Niklas, K. J., and Giannasi, D. E., 1977, Flavanoid and other chemical constituents of fossil Miocene *Zelokova* (Ulmaceae), *Science* **196**:877–878.

Niklas, K. J., and Giannasi, D. E., 1978, Angiosperm paleobiochemistry of the Succor Creek flora (Miocene), Oregon, USA, *Am. J. Bot.* **65**:943–952.

Niklas, K. J., Tiffney, B. H., and Leopold, A. C., 1982, Preservation of unsaturated fatty acids in Palaeogene angiosperm fruits and seeds, *Nature* **296**:63–64.

Nip, M., Tegelaar, E. W., Brinkuis, H., de Leeuw, J. W., Schneck, P. A., and Holloway, P. J.,

1986, Analyses of modern and fossil plant cuticles by Curie point PY-GC and Curie point PY-GC-MS: Recognition of a new, highly aliphatic and resistant biopolymer, *Org. Geochem.* **10:**769–778.

Nishimura, M., 1977, The geochemical significance in early sedimentation of geolipids obtained by saponification of lacustrine sediments, *Geochim. Cosmochim. Acta* **41:**1817–1823.

Nishimura, M., and Koyama, T., 1977, The occurrence of stanols in various living organisms and the behaviour of sterols in contemporary sediments, *Geochim. Cosmochim. Acta* **41:**379–385.

Nykvist, N., 1961, Leaching and decomposition of litter—3. Experiments on leaf litter of *Betula verrucosa, Oikos* **12:**249–263.

Ourisson, G., Albrecht, P., and Rohmer, M., 1979, The hopanoids—Palaeochemistry and biochemistry of a group of natural products, *Pure Appl. Chem.* **51:**709–729.

Ourisson, G., Albrecht, P., and Rohmer, M., 1982, Predictive microbial biochemistry—From molecular fossils to procaryotic membranes, *Trends Biochem. Sci.* **7:**236–239.

Pååbo, S., 1985, Molecular cloning of ancient Egyptian mummy DNA, *Nature* **314:**644–645.

Pååbo, S., Irwin, D. M., and Wilson, A. C., 1990, DNA damage promotes jumping between templates during enzymic amplification, *J. Biol. Chem.* **265:**4718–4721.

Pååbo, S., Higuchi, R. G., and Wilson, A. C., 1989, Ancient DNA and the polymerase chain reaction, *J. Biol. Chem.* **264:**9709–9712.

Pååbo, S., Gifford, J. A., and Wilson, A. C., 1988, Mitochondrial DNA sequences from a 7,000 year old brain, *Nucleic Acid Res.* **16:**9775–9787.

Parkes, R. J., and Calder, A. G., 1985, The cellular fatty acids of three strains of *Desulfobulbus,* a propionate-utilising sulphate bacterium. *Microbiol. Ecol.* **31:**361–363.

Parkes, R. J., and Taylor, J., 1985, Characterization of microbial populations in polluted marine sediments, *J. Appl. Bacteriol. Symp. Suppl.* **1985:**155s–173s.

Philp, R. P., and Gilbert, T. D., 1986, The detection and identification of biological markers by computerized-gas chromatography-mass spectrometery, in: *Biological Markers in the Sedimentary Record* (R. B. Johns, ed.), Elsevier, Amsterdam, pp. 227–248.

Prager, E. M., Wilson, A. C., Lowenstein, J. M., and Sarich, V. M., 1980, Mammoth albumin, *Science* **209:**287–289.

Priestley, D. A., Galinat, W. C., and Leopold, A. C., 1981, Preservation of polyunsaturated fatty acid in ancient Anasazi maize seed, *Nature* **292:**146–148.

Reeburgh, W. S., 1980, Anaerobic methane oxidation: Rate depth distributions in Skan Bay sediments, *Earth Planet. Sci. Lett.* **47:**345–352.

Rieseberg, L. H., and Soltis, D. E., 1987, Flavanoids of fossil Miocene *Platanus* and its extant relatives, *Biochem. Syst. Ecol.* **15:**109–112.

Risatti, J. B., Rowland, S. J., Yon, D. A., and Maxwell, J. R., 1984, Stereochemical studies of acyclic isoprenoids—XII. Lipids of methanogenic bacteria and possible contributions to sediments, *Org. Geochem.* **6:**93–104.

Robbins, L. L., and Brew, K., 1990, Proteins from the organic matrix of core-top and fossil planktonic Foraminifera, *Geochim. Chosmochim. Acta* **54:**2285–2292.

Robinson, N., Eglinton, G., Brassell, S. C., and Cranwell, P. A., 1984, Dinoflagellate origin for sedimentary 4α-methyl steroids and 5α(H) stanols, *Nature* **308:**439–441.

Romanowski, G., Lorenz, M. G., and Wackernagel, W., 1991, Adsorption of plasmid DNA to mineral surfaces and protection against DNase I, *Appl. Environ. Microbiol.* **57:**1057–1061.

Rowley, M. J., Rich, P. V., Rich, T. H., and Mackay, I. R., 1986, Immunoreactive collagen in avian and mammalian fossils, *Naturwissenschaften* **43:**620–623.

Runnegar, B., 1986, Molecular palaeontology, *Palaeontology* **29:**1–24.

Runnegar, B., and Schopf, J. W. (eds.), 1988, Molecular evolution and the fossil record, in: *Paleontological Society Short Courses in Paleontology, Number 1,* Paleontological Society.

Saiki, R. K., Gelfand, D. H., Stoffel, S., Scarf, S. J., Higuchi, R., Horn, G. T., Mullis, K. B., and Erlich, H. A., 1988, Primer-directed enzymatic amplification of DNA with a thermostable DNA polymerase, *Science* **239:**487–491.

Saunders, G. W., 1976, Decomposition in freshwater, in: *The Role of Terrestrial and Aquatic Organisms in Decomposition Processes* (J. M. Anderson and A. Macfadgen, eds.), Blackwell, London, pp. 341–374.

Shoshani, J., Lowenstein, J. M., Walz, D. A., and Goodman, M., 1985, Proboscidean origins of mastodont and woolly mammoth demonstrated immunologically, *Paleobiol.* **11**:429–437.

Sigleo, A. C., 1978, Organic geochemistry of silicified wood, Petrified Forest National Park, Arizona, *Geochim. Cosmochim. Acta* **42**:1397–1405.

Simoneit, B. R. T., Grimalt, J. O., Wang, T. G., Cox, R. E., Hatcher, P. G., and Nissenbaum, A., 1986, Cyclic terpenoids of contemporary resinous plant detritus and of fossil woods, ambers and coals, *Org. Geochem.* **10**:877–889.

Summons, R. E., Brassell, S. C., Eglinton, G., Evans, E., Horodyski, R. J., Robinson, N., and Ward, D. M., 1988, Distinctive hydrocarbon biomarkers from fossiliferous sediment of the Late Proterozoic Walcott Member, Chur Group, Grand Canyon, Arizona, *Geochim. Cosmochim. Acta* **52**:2625–2637.

Swain, F. M., Bratt, J. M., and Kirkwood, S., 1967, Carbohydrate components of some Paleozoic plant fossils, *J. Palaeontol.* **41**:1549–1554.

Swain, F. M., Bratt, J. M., and Kirkwood, S., 1970, Carbohydrates from Precambrian and Cambrian rocks and fossils, *Geol. Soc. Am. Bull.* **81**:499–504.

Tegelaar, E. W., 1990, Duripartically preserved constituents of kerogen: A key to the relationship between biomass and fossil fuels, Ph. D. thesis, Delft Technical University.

Tegelaar, E. W., de Leeuw, J. W., Derenne, S., and Largeau, C., 1989, A reappraisal of kerogen formation, *Geochim. Cosmochim. Acta* **53**:3103–3106.

Thomas, B. A., 1986, The biochemical analysis of fossil plants and its use in taxonomy and systematics, in: *Systematic and Taxonomic Approaches in Palaeobotany* (R. A. Spicer and B. A. Thomas, eds.), Oxford University Press, Oxford, pp. 39–51.

Thomas, R. H., Schaffner, W., Wilson, A. C., and Pååbo, S., 1989, DNA phylogeny of the extinct marsupial wolf, *Nature* **340**:465–467.

Thomas, W. K., Pååbo, S., Villablanca, F. X., and Wilson A. C., 1990, Spatial and temporal continuity of kangaroo rat populations shown by sequencing mitochondrial DNA from museum specimens, *J. Mol. Evol.* **31**:101–112.

Towe, K. M., and Urbanek, A., 1972, Collagen-like structures in Ordovician graptolite periderm, *Nature* **237**:443–445.

Tuross, N., Eyre, D. R., Holtrop, M. E., Gilmicher, M. J., and Hare, P. E., 1980, Collagen in fossil bone, in: *Biogeochemistry of Amino Acids* (P. E. Hare, T. C. Hoering, and K. King, eds.), Wiley, New York, pp. 53–63.

Tuross, N., Fogel, M. Y., and Hare P. E., 1988, Variability in the preservation of the isotopic composition of collagen from fossil bone, *Geochim. Cosmochim. Acta* **52**:929–935.

Ulrich, M. M. W., Perizonius, W. R. K., Spoor, C. F., Sandberg, P., and Vermeer, C., 1987, Extraction of osteocalcin from fossil bones and teeth, *Biochem. Biophys. Res. Commun.* **149**:712–719.

Wannigama, G. P., Volkman, J. K., Gillan, F. T., Nichols, P. D., and Johns, R. B., 1981, A comparison of lipid components of the fresh and dead leaves and pneumatophores of the mangrove *Avicennia marina*. *Phytochemistry* **20**:659–666.

Ward, D. M., and Winfrey, M. R., 1985, Interactions between methanogenic and sulfate-reducing bacteria in sediments, *Adv. Aquat. Microbiol.* **3**:141–179.

Westbroek, P., van der Meide, P. H., van der Wey-Kloppers, J. S., van der Sluis, R. J., de Leeuw, J. W., and de Jong, E. W., 1979, Fossil macromolecules from cephalopod shells: Characterization, immunological response and diagenesis, *Paleobiology* **5**:151–167.

Willet, J. E., 1987, *Gas Chromatography, analytical chemistry by Open Learning*, Wiley, London.

Wilson, M. A., Verheyen, T. V., Vassallo, A. M., Hill, R. S., and Perry, G. J., 1987, Selective loss of carbohydrates from plant remains during coalification, *Org. Geochem.* **11**:265–271.

Wolf, G. A., Lamb, N. A., and Maxwell, J. R., 1986, The origin and fate of 4-methyl steroid hydrocarbons. I. Diagenesis of 4-methyl steranes, *Geochim. Cosmochim. Acta* **50**:335–342.

Chapter 2

Taphonomy of Nonmineralized Tissues

PETER A. ALLISON and DEREK E. G. BRIGGS

PETER A. ALLISON • Postgraduate Research Institute for Sedimentology, The University, Whiteknights, Reading, RG6 2AB, United Kingdom. DEREK E. G. BRIGGS • Department of Geology, University of Bristol, Bristol BS8 1RJ, United Kingdom.

Taphonomy: Releasing the Data Locked in the Fossil Record, Volume 9 of *Topics in Geobiology*, edited by Peter A. Allison and Derek E. G. Briggs, Plenum Press, New York, 1991.

1. Introduction

Soft-tissues are a rich source of nutrients whether consumed by predators, scavengers, or microbes. For this reason, nonmineralized organic tissues are comparatively rare in the fossil record. The rapidity of consumption/ degradation does not merely impact upon the preservation potential of wholly soft-bodied forms—although this is the most obvious manifestation of bias—but also affects preservation of hard-parts. For example, skeletal elements may remain articulated during transport where soft-parts have not decayed (Allison, 1986). Conversely, following decay, disarticulation occurs with minimal transport (Kidwell and Baumiller, 1990). Information loss through decay limits our knowledge not only of fossil communities, but also of the affinities of certain fossil groups.

1.1. Community Preservation

Attempts to estimate variation in the preservation potential of soft tissue and shelly material in the fossil record have usually involved actualistic observations based on Recent biotas (e.g., Jones, 1969; Lawrence, 1968; Dorjes, 1972; Driscoll and Swanson, 1973; Macdonald, 1976; Schopf, 1978). If the possession of mineralized hard-parts were a prerequisite for fossilization, 66–79% of species in marine communities would not normally be preserved (Table I).

It would seem reasonable to expect the soft-parts of infaunal animals to be preserved more readily than those of other organisms because they are already buried. However, this is not always the case (Schäfer, 1962, 1972). As death approaches, ability to absorb oxygen from pore-waters is reduced.

TABLE I. Estimates of Preservation Potential of Modern Nearshore Invertebrates Based on Percentage of Species and Individuals Which Include Hard-Parts[a]

	Number of species (% of group that is preservable where data are available)		
	Southern California 0—22 km offshore, 0—180 m depth (Jones, 1969)	Georgia shelf (Dorjes, 1972)	Living epifauna colonizing bivalves, Buzzards Bay, Massachusetts (Driscoll and Swanson, 1973)
Polychaetes	523	76 (56)	57 (4)
Crustaceans	419	67 (15)	4 (25)
Molluscs	408	90 (3.5)	8 (12)
Echinoderms	64	—	1 (0)
Others	59	—	36 (80)
Total	1473	268	106
Likely % of preservable fauna	<33%	21%	39%

[a]From Allison and Briggs (1991).

Thus, some organisms leave their burrows to seek the higher oxygen content of the overlying water, dying on the surface, where they may be destroyed by predation, the activity of scavengers, or transport. However, environmental stress, such as anoxia, elicits stasis rather than flight in some organisms and this may lead to the formation of mass mortality horizons.

1.2. Interaction of Decay and Transport: Bloat and Float

Live and freshly killed organisms are remarkably resistant to damage, even in turbulent transport regimes (Allison, 1986). Decomposition, however, produces several gaseous phases which substantially alter the hydrodynamic properties of organic carcasses. In experiments (Allison, 1986) decaying shrimps were rapidly buoyed by decay gases to the surface of an experimental vessel where they disarticulated to produce a "carpet" of skeletal elements upon the vessel floor. Similarly, Schäfer (1962, 1972) recorded that the decomposing carcasses of marine mammals in the North Sea which floated became disarticulated during buoyancy-induced transport. These observations have been used to explain preservational features of ichthyosaurs in the Lower Oxford Clay of the East Midlands of England (Martill, 1986).

The propensity of an organism to float during decay is controlled by tissue strength, rate of gas production, and hydrostatic pressure. There comes a point when tissues become so weakened by decay that expanding gas production breaks the carcass apart. The gas is then released and the carcass sinks. The site where the gas builds up may be significant; it is easier for gas to escape from coelomic cavities than from the interstices of flesh.

Rate of gas production is linked to rate of decay, but also varies with decay pathway (Table II). Mass/surface area constraints prevent diffusion of oxygen into large carcasses and promote methanogenesis and fermentation (Allison, 1988a). Determining the size of carcass at which this effect becomes important requires further research, but it is likely that most large fish, reptiles, and cetaceans decay fermentatively. Most animals are negatively buoyant and sink upon death, only refloating after the generation of sufficient gas. As hydrostatic pressure increases with depth, the volume of decay gas and its ability to create buoyancy are reduced. The relative values for this process are unknown, but with future work it may be possible to use preservation of articulated carcasses as a paleobathymetric indicator (Allison *et al.*, 1991).

1.3. Modification of Taxonomically Significant Features

As organic tissues are weakened by scavenging and decomposition they become less able to hold the more resistant parts of a carcass together and disarticulation may occur. If this process is accompanied by transport, the

TABLE II. Idealized Stratification of Bacterial Reduction Zones[a,b]

Aerobic Decay
$(CH_2O)_{106}(NH_3)_{16}H_3PO_4 + 106O_2 \rightarrow 106\ CO_2 + 16NH_3 + H_3PO_4 + 106H_2O$
Anaerobic Decay
Manganese reduction
$(CH_2O)_{106}(NH_3)_{16}H_3PO_4 + 212MnO_2 + 332CO_2 + 120H_2O$ $\rightarrow 438HCO_3^- + 16NH_4^+ + HPO_4^{2-} + 212Mn_4^{2+}$
Nitrate reduction
$(CH_2O)_{106}(NH_3)_{16}H_3PO_4 + 84.8NO_3^-$ $\rightarrow 7.2CO_2 + 98.8HCO_3^- + 16NH_4^+ + 42.4N_2 + HPO_4^{2-} + 49H_2O$
Iron reduction
$(CH_2O)_{106}(NH_3)_{16}H_3PO_4 + 424Fe(OH)_3 + 756CO_2$ $\rightarrow 862HCO_3^- + 16NH_4^+ + HPO_4^{2-} + 424Fe^2 + 304H_2O$
Sulfate reduction
$(CH_2O)_{106}(NH_3)_{16}H_3PO_4 + 53SO_4^{2-}$ $\rightarrow 39CO_2 + 67HCO_3^- + 16NH_4^+ + HPO_4^{2-} + 53HS^- + 39H_2O$
Methanogenesis (carbonate reduction)
$(CH_2O)_{106}(NH_3)_{16}H_3PO_4 + 14H_2O \rightarrow 39CO_2 + 14HCO_3^- + 53CH_4 + 16NH_4^+ + HPO_4^{2-}$
Fermentation[c]
$12\ (CH_2O)_{106}(NH_3)_{16}H_3PO_4$ $\rightarrow 106CH_3CH_2COOH + 106CH_3COOH + 212CH_3CH_2OH + 318CO_2$

[a]Reactions at the top of the column yield more free energy than those lower in the column.
[b]Carbon composition is from Redfield (1958).
[c]Dominant beyond the zone of carbonate reduction, but occurs throughout the sediment profile.

different components of a carcass may be separated. Due to contrasts in morphology, some organisms are more likely to disarticulate than others. Nowhere is this effect more widespread and pronounced than in plants (Spicer, 1981; and this volume, Chapter 3). Plants may produce a host of disarticulated elements, which include leaves, pollen, seeds, and fruit, as a normal part of the life cycle. On death the stem of the plant may be separated from its roots, and during transport the outer skin or bark may be stripped from the stem. As a result of disarticulation, fossil plants are often difficult to reconstruct and form-taxa are frequently used.

This bias also affects animals, particularly those which are predominantly soft-bodied. A good example is provided by *Anomalocaris canadensis* from the Middle Cambrian Burgess Shale of British Columbia (Whittington and Briggs, 1985). For many years this animal was known only from disarticulated elements. The body was named *Laggania cambria* (Walcott, 1911) and identified as a holothurian, the mouth was classified as a medusoid, *Peytoia nathorsti* (Walcott, 1911), and the appendages were thought to be the body of a crustacean (Whiteaves, 1892). However, the discovery of complete specimens (Whittington and Briggs, 1985) showed that these "different" taxa were actually all parts of the one animal. Disarticulated remains of *A. canadensis* represent those carcasses which had decayed before and during transport. Complete specimens, on the other hand, probably originated from carcasses which were transported and deposited prior to substantial decay (Briggs, 1979).

More frequently encountered examples of decay and scattering include spicule-bearing soft-bodied forms such as ascidians, sponges, and holothurians, and toothed creatures such as polychaetes and conodonts. The organisms are known principally from hard-part morphology and in many cases the range of tooth and spicule morphology which occurs within one taxon is unknown; form-taxa are often used for identification.

Decay may also modify ornament and sculpture. Arthropods are particularly susceptible because their cuticle comprises several layers. Bacterial decay of successive layers can produce preservational artifacts similar to natural sculpture. For example, removal of crustacean epicuticle exposes a nodular sculpture of cuticular polygons upon the upper surface of the exocuticle (Allison, 1990a) [see Briggs *et al.* (1991, Fig. 2d) for a possible Carboniferous example].

2. Degradation of Nonmineralized Tissue

2.1. Death and Scavenging

Different modes of death—senility, disease, predation, and accident—have different effects on the preservation potential of an organism.

Few animals die of senility; most individuals are weakened by the process of aging and become more susceptible to accidents, predation, and disease. As a cause of death, senility rarely exerts a preservational bias, although there are potential exceptions. For example, some organisms aggregate to breed and then die *en masse*, producing natural mass mortality horizons (e.g., salmon and some cephalopods). The sudden deposition of large numbers of carcasses may locally reduce oxygen levels and could raise the anoxic/oxic boundary above the sediment/water interface, thereby inhibiting scavenging and bioturbation. [Bodies in the middle of mass graves are better preserved than those at the periphery (Mant 1984, 1987).] However, breeding sites are usually well-aerated environments capable of fulfilling the oxygen demands of developing larvae.

Disease may increase rate of decay and thereby reduce preservation potential. Septicemia, peritonitis, and jaundice, for example, accelerate decomposition in human cadavers (Stone and Johnson, 1987). Conversely, disease can in some circumstances inhibit decay. For instance, so-called "dry gangrene" leads to tissue desiccation and subsequent mummification with the exclusion of putrefaction (Anderson, 1987).

During predation, soft-tissues are removed and hard-parts may be disarticulated and fragmented. The effect of this bias is controlled principally by the size relationship between predator and prey. If the predator is smaller than the prey, the skeleton may be preserved even though the soft-parts are removed. Mortal wounding brought about by attempted predation can accelerate decay and thus reduce preservation potential. For example, severe puncture wounds may introduce microbes into tissues and initiate decay prior to death (Anderson, 1987; Rubin and Farben, 1988).

Some forms of accidental death significantly enhance the preservation potential of an organism. Preservational traps (*sensu* Seilacher, 1970) provide some of the best known exceptional preservations: insects in amber, mammals in tarsands, mammoths in permafrost, and vertebrates in peat deposits. A more common example in aquatic environments is death by asphyxiation or poisoning brought about by the turnover of a stratified water body with subsequent deposition in anaerobic conditions.

Scavenging by macroorganisms is a common event in almost every environment. As in the case of predation, the degree of preservational bias imparted by scavenging is dependent upon the size of the scavenger and the skeletal morphology and size of the organism being scavenged. Carcasses in terrestrial systems are more likely to disarticulate as a result of scavenging by large vertebrates than by invertebrates, although the latter are extremely efficient in removing soft-tissues. In terrestrial systems removal of soft-tissues by saprophagous insects and microbes is between one and two orders of magnitude more rapid than by microbes alone (Payne, 1965). Many terrestrial scavengers may be excluded by burial if it precedes insect settlement, or by submergence in water—although emergent surfaces of floating carcasses may be colonized (K. G. V. Smith, 1986). However, aquatic systems contain their own scavenging biota, which may include mammals, fish, crustaceans, echinoderms, molluscs, and annelids. In deep-water habitats, carrion may supply up to 10% of the carbon budget to benthic macroorganisms (C. R. Smith, 1985). The response to carcass falls can be rapid. Experiments on the seabed of the Santa Catalina Basin off the coast of California demonstrated that fish could be stripped of soft-parts and disarticulated in 2–3 days (C. R. Smith, 1985). The principal scavenger in this cold (4°C), dysaerobic (O_2 content 0.4 ml/liter), high-pressure (110 atm) environment is an *ophiuroid!* Normal densities of ophiuroids in the area were $17/m^2$, although this was increased around the fish carcasses to over $700/m^2$ by the time the fish had been scavenged.

Scavenging can only be prevented if the carcass is deposited in an inhospitable environment characterized, for example, by catastrophic burial or anoxia.

2.2. Death and the Role of Microbes in Decomposition

Upon death the temperature of a cadaver equilibrates with the environment, vital enzymatic processes fail (while autolytic enzymes such as phosphatases become active), and cellular excretion ceases. One of the most immediate manifestations is *rigor mortis,* which is brought about by the reduction of ATP in muscle, resulting in heightened lactic acid content. In human cadavers the resulting stiffness may appear after 5–12 hours and disappears with putrefaction (Mason, 1983). Putrefaction is the result of an anaerobic microbial fermentation reaction which is initiated upon death, although it is manifest only after a few days. The activity of maggots may raise the temperature (Payne, 1965; Mant, 1984). After 1 week, tissues become swollen and blistered, with disfigurement occurring in 3 weeks and partial liquefaction in about 1 month. Different parts of the body decompose at different rates: intestines and stomach, liver, spleen, and brain decompose early, lungs, heart, kidneys, bladder, prostrate, testis, uterus, and ovaries later (Mant, 1984; Gordon *et al.*, 1988). Rate is strongly dependent upon environmental factors and can be substantially slowed if the cadaver is immersed in water, since body heat is lost more quickly (Simpson and Knight, 1985). Mummification will also retard the decay process. This can occur naturally in dry environments and is particularly prevalent in newborn animals, since they lack a natural gut microbiota (Mason, 1983; Simpson and Knight, 1985; K. G. V. Smith, 1986); stillborn animals also lack lung contamination (Janssen, 1984). Preservation by mummification is not permanent, since decay can be reactivated with moisture. Preservation is therefore controlled by climate and is ephemeral.

The rate at which microbes respire and multiply, breaking down organic carbon, is generally considered to be controlled by: (1) the nature of the organic carbon; (2) temperature; (3) oxidant supply; (4) depositional setting; and (5) rate of burial.

2.2.1. Nature of Organic Carbon

Organic carbon in natural systems is composed principally of the elements C, O, H, N, and P, which occur in variable proportions and chemical configurations to give a variety of different molecular types. Some molecules, termed volatiles, are rapidly consumed during microbial respiration, while others, refractories, are degraded more slowly. For example, certain molecular configurations such as carbon–carbon bonds or structural branching may impart decay resistance. Polymers which include a single bond type (e.g., proteins) and those composed of only one type of monomer (e.g., chitin and cellulose) are readily degraded. Complex molecules, however, which include both a variety of building blocks and bond types, tend to be more stable, since a consortium of microbial enzymes is required for degradation (Alexander, 1973, p. 639).

The mechanical strength and inherent resistance to decay of some refractory compounds account for their occurrence as protective exoskeletal coverings in nonmineralized taxa. Organisms thus equipped are more likely to be preserved as fossils. Preservation of the tissue formed of such compounds (Table III) does not imply or require preservation of unaltered molecules. For instance, ultrastructural detail of fossil arthropod cuticle

Table III. The Occurrence of Recalcitrant Carbon Molecules in Living Organisms and Examples of Molecular and Tissue Preservation in the Geological Record

Tissue/molecule	Occurrence	Class of compound	Geological example of tissue	Geological example of molecule
Chitin	Arthopod cuticle, fungal hyphae	Polysaccharide carbohydrate	Trilobite cuticle (Tiegler and Towe, 1973); ceratiocaridid Crustacea (Rolfe, 1962)	Cambrian *Hyolithellus*? (Carlisle, 1964)
Cellulose	Plants	Polysaccharide carbohydrate	Devonian land plants from Rhynie Chert (Kidston and Lang, 1917)	*Derivatives* in Carboniferous *Calamites* (Swain *et al.*, 1969)
Lignin	Plants	Polyaromatic	Devonian land plant *Gosslingia* (Kenrick and Edwards, 1988)	*Derivatives* in Jurassic conifer (Sigleo, 1978)
Melanin	Pigment in fungi and cephalopod ink	Polyaromatic	Jurassic ink sacs of cephalopods (Allison 1988b)	Triassic ink sacs of cephalopods (Mathur, 1977)
Sporopollenin	Walls of plant spores	Oxidative polymer of carotenoid or carotenoid esters	Ordovician *Tasmanites* spores (Brooks, 1971)	Ordovician *Tasmanites* spores (Brooks, 1971)
Collagen	Skin and matrix of bone; graptolite periderm	Protein polymer	Cellular tissue in Ordovidian graptoblast (Crowther *et al.*, 1987)	*Possible* denatured *remnant* in Cretaceous dinosaur bone (Armstrong *et al.*, 1983, p. 313)

(presumed chitinous) may be preserved even though the chitin molecule is no longer present (Rolfe, 1962; Tiegler and Towe, 1973). Similarly, microscopic structure of three-dimensional graptolite periderm (presumed collagenous) may be preserved even though collagen itself is not (Crowther and Rickards, 1977; Armstrong *et al.*, 1984). A combination of degradation and chemical transformation (including cross-linking) of the original molecule may lead to the preservation of the tissue as condensed insoluble organic matter. Certain plants, however, contain biomacromolecules which are even more resistant intrinsically to degradation [algaenan in algae, cutan and suberan in vascular plants (Tegelaar *et al.*, 1989)] and give these cuticles a very high preservation potential.

The preservation potential of particular tissues may be increased *in vivo* by organismic adaptations. For instance, alteration of insect chitin by quinones such as sclerotin can significantly retard decay (Alexander, 1973). The sealing of a tissue, such as collagen, within a relatively inert mass, such as bone phosphate, may also inhibit decay. The preservation potential of recalcitrant tissues, however, may vary with depositional environment [lignin, for example, is more decay resistant under anaerobic than aerobic conditions (Benner *et al.*, 1984)].

2.2.2. Temperature

All chemical reactions are temperature sensitive; microbial degradation of carbon via enzyme-catalyzed reactions is no exception. Rate of carbon degradation follows an Arrhenius curve (Stanier *et al.*, 1987). Rate increases with temperature to an optimum above which enzymatic reactions become inactivated, and cell proteins and structures such as membranes are altered. Different microorganisms have different growth temperature optima and on this basis they can be classed as either thermophiles (above 55°C), mesophiles (20–45°C), or psychrophiles (grow well at 0°C). In general, a twofold increase in decay rate can be expected for a temperature increase of 10°C (Swift *et al.*, 1979), although the response of natural microbial communities to temperature changes may vary with season. However, other controlling factors may vary with increased temperature, resulting in a three- or fourfold increase in rate (Bunnel *et al.*, 1975). These observations relate to homogeneous accumulations of organic carbon (leaf litter, etc.), but the same principle also applies to heterogeneous accumulations such as carcasses (Payne and King, 1972; Cahenzli and Staffeldt, 1976; Mason, 1983; G. R. Smith and Elder, 1985; Kidwell and Baumiller, 1990).

The effect of temperature on decay rate led Kidwell and Baumiller (1990) to propose the existence of a latitudinal/water depth gradient in preservation of multielement articulated skeletons based upon known ocean temperatures. Such a model requires further testing, since it is possible for saprophagous bacteria to adapt to climatic temperature regimes (e.g., Nedwell and Floodgate, 1971). As rate increases, so does nutrient demand.

If this exceeds supply, then increase in rate may be constrained [see Boddy, (1983) for a study of wood in a terrestrial system].

2.2.3. Oxidant Supply

The most fully understood degradation pathway for organic carbon is that of aerobic respiration. During this process organic carbon is combined with oxygen and broken down to yield carbon dioxide and water. If the rate of carbon deposition exceeds that of oxygen supply, however, anoxia results. Under anaerobic conditions microbes utilize a series of alternative electron acceptors (Table II). In an idealized sediment profile these reactions are stratified (Berner, 1981) with decreasing free-energy yields from the top of the sediment pile down. However, very few sediment profiles are ''ideal'' and many sediment pore waters are deficient in some of these oxidants. The most obvious departure from the ideal concerns sulfates, which are abundant in marine pore waters but occur in very low concentrations in most fresh waters. Conversely, nitrates are present in higher concentrations in fresh water than they are in marine environments. Thus, the thickness of the reduction zones varies according to pore-water ion concentrations, and this in turn reflects depositional environment. In addition, microbial activity within microenvironments often blurs the idealized stratification.

The association of anaerobic conditions with accumulations of sapropelic sediments is generally accepted as proof that anoxia inhibits decay and is an instrumental factor in the preservation of organic carbon in sediment (e.g., Curtis, 1980). This is supported by rate measurements such as those of Jahnke *et al.* (1982), which have shown that aerobic degradation in deep-sea sediments is faster than anaerobic decay. However, laboratory experiments using fresh (Foree and McCarty, 1970) and aged organic matter (Westrich and Berner, 1984) suggest that rate of aerobic decay is broadly similar to that of anaerobic decay. This view is supported by Henrichs and Reeburgh (1987), who believe that reduced rate of decay in anaerobic deep-sea sediments is merely a function of molecular recalcitrance. The aerobic breakdown of carbon settling through the water column is more rapid because this carbon is more volatile. Thus, the reduced rate of carbon utilization reflects not the degradative pathway, but molecular composition. This view is contested by Emmerson and Hedges (1988), who believe that anaerobic decay *can be* slower than aerobic decay if the carbon source is *refractory*. This hypothesis is supported by reports on the decomposition of lignin, which degrades much more slowly under anaerobic conditions than in an aerobic system (Benner *et al.*, 1984). Cannfield (1989) showed that only a small reduction in rate between aerobic and anaerobic conditions is necessary to account for the accumulations of organic matter in anoxic and euxinic sediments.

Research into decomposition in sediment has focused upon the fate of disseminated organic carbon [see reviews by Alexander (1973) and Emerson and Hedges (1988)]. Very few studies have attempted to assess rates of degradation of heterogeneous accumulations of organic matter such as macroorganisms. Early studies suggested that anoxia did indeed reduce the rate of decay (e.g., Hecht, 1933; Zangerl and Richardson, 1963; Zangerl, 1971). However, more recent experiments on the decay of crustaceans in jars filled with anoxic mud (Plotnick, 1986; Allison, 1988a) showed that anaerobic decomposition was rapid, thereby demonstrating the ineffectiveness of anoxia as a precursor to the fossilization of soft-parts. This is supported by mass balance calculations (Allison, 1988a) which demonstrate the high oxygen requirement for aerobic decomposition (1 g dry weight of soft-tissue requires 671 cm^3 of oxygen for complete respiration), while anaerobic sulfate reduction will decompose twice as much organic matter, mole for mole, as an equivalent amount of aerobic respiration. Jørgenson (1977) calculated that even in oxygen-saturated seawater a particle greater than 2 mm in diameter would be anaerobic at the center. Indeed, given size/surface area constraints on oxygen diffusion, it is obvious that any large carcass must decompose essentially anaerobically.

2.2.4. Depositional Setting

Some of the physical and chemical properties outlined above may vary according to sediment lithology. Temperature, for instance, may be lower beneath sediment than on the surface in terrestrial and intertidal settings (K. G. V. Smith, 1986). Permeability variations may affect the transport of pore-water nutrients to degradative bacteria; lower permeability in mudrocks impedes oxygen supply to microbes, thereby promoting anoxia. Supply of nutrients is also affected by bioturbation, since this can increase the effective surface area of a sediment, thereby promoting heightened diffusion of nutrients (Aller, 1980). Porosity may vary by a factor of two or more in ''seemingly uniform'' sediment and this may promote a similar heterogeneity in rate of decay (Allison, 1990b).

The lithological characteristics reviewed so far may retard decay, but are unlikely to promote long-term preservation. Significant inhibition of decay, however, may occur in sediments with toxic pore waters. This is the case in peat bogs, where tannic acids and phenolic derivatives are generated by decomposing plant matter. These compounds are toxic to microbes and thereby reduce decay, but may also chemically alter proteins such as collagen, imparting increased molecular recalcitrance (Connolly *et al.*, 1986). The peat bogs of northern Europe have yielded a number of human cadavers, some over 5000 years old, which preserve skin, muscle, and internal organs. However, the hard-parts of some of these carcasses have dissolved due to the acidity of the enclosing pore waters (Glob, 1969; Brothwell, 1987).

Destruction of tissues by decay may be impeded by the hydrogenization of oily unsaturated acids such as oleic acid into higher fatty acids such as stearic acid. The resultant fat, termed adipocere, is hard and opaque and may adhere to bone long after the decay of skin (Mant, 1957; Den Dooren de

Jong, 1961; Hueck, 1971). Pigmented muscle groups have been preserved for as long as 5 years when surrounded by adipocere, and although cellular structure is lost after this period, recognizable human soft-part images may be preserved in adipocere for up to 40 years (De Fourcroy, 1790; Den Dooren de Jong, 1961). The formation of adipocere is brought about by anaerobic microbes [e.g., *Clostridium* (Mant, 1957)] in the presence of moisture, but it only affects fatty tissues (Den Dooren de Jong, 1961). It can occur in less than a month in warm moist climates, but several months are usually required in temperate climates for carcasses immersed in water or buried in moist soil (Simpson and Knight, 1985). Adipocere formation may also increase the chance of a skeleton remaining articulated. Simpson and Knight (1985, p. 16) figured a human cadaver which had lain in a lake for over 7 years and was still largely articulated with recognizable soft-parts. The head, however, had been lost; presumably its weight had caused it to detach during flotation (as happens in floating cetaceans: Schäfer, 1972). The acids which give rise to adipocere may in some cases react with pore-water calcium to form natural salts (Berner, 1968; Allison, 1988a) which may also retard the destruction of soft-parts. Naturally occurring concretionary masses of these calcium salts have been found enclosing fish and replacing muscle (Sondheimer *et al.*, 1966; Berner, 1968). Equivocal radiometric evidence suggests that recognizable tissues may be preserved in this fashion for over 1000 years (Sondheimer *et al.*, 1966, p. 221).

Hydrogen sulfide is toxic, killing humans at concentrations of 650–2000 ppm in air. At higher concentrations it is also toxic to many sediment bacteria. Such concentrations are rarely attained even in modern anoxic basins such as the Black Sea. This does not preclude the possibility that higher concentrations of dissolved sulfide may have existed in the geological past. Such a system could promote long-term decay inhibition, *but may inhibit mineralization*. For example, pyritization in euxinic basins is iron limited and is therefore restricted to iron-bearing substrates (Berner and Raiswell, 1983). The iron present in soft-parts is insufficient to support pyrite formation (Raiswell and Canfield, this volume, Chapter 7).

2.2.5. Rate of Burial

Catastrophic burial has long been considered a means of preserving organic soft-parts (e.g., Conway Morris, 1986; Seilacher, 1970) since it distances a carcass from scavengers and bioturbators and tends to induce anoxia. The Middle Cambrian Burgess Shale of British Columbia, for example, has yielded a diversity of exceptionally preserved soft-bodied organisms enclosed within rapidly deposited turbidites (Piper, 1972) or tempestites (Ludvigsen, 1989).

There is a marked reduction in bacterial numbers with depth (ZoBell and Anderson, 1936; Parkes *et al.*, 1990). This is a reflection of the supply of oxidant and the availability of the organic carbon present. Introduction of additional sulfate or fresh organic substrate will result in an increase in bacterial biomass (Parkes *et al.*, 1990). Thus, rapid burial *per se* does not inhibit decay by placing a carcass in a zone of lower bacterial population (Conway Morris, 1986, p. 431). Bacterial degradation at normal rates of burial rapidly destroys volatile carbon and then progresses more slowly with less reactive forms—thus, sulfate reduction is slowed and bacterial populations are reduced [the "G model" (Berner, 1980, p. 155)]. Introduction of more reactive carbon to this sediment increases rate of degradation to that higher in the sediment pile [as demonstrated by Westrich and Berner (1984) by injecting more reactive carbon into a sediment core at depth].

3. Preservational Mechanisms

In original form most nonmineralized tissues are readily degraded by microbes. Many examples of soft-tissue preservation are associated with early diagenetic mineralization (Allison, 1988d) which may occur rapidly under anoxic conditions (Allison, 1988a), obviating the necessity for long-term decay inhibition (Fig. 1).

Decay inhibition in peat bogs and permafrost is ephemeral on a geological time scale. Although preservational enhancement of fatty tissues by adipocere formation occurs in several settings and only requires the presence of water, it confers no long-term stability unless it functions as a precursor to diagenetic mineralization. The decay resistance of some biomacromolecules (e.g., lignin, cutan, suberan, algaenan), however, imparts considerable preservation potential to certain plant cuticles (Tegelaar *et al.*, 1989). Preservation potential varies, however, with depositional conditions. Even lignin may be degraded under anaerobic conditions, albeit at a reduced rate (Hedges *et al.*, 1985).

A range of nonmineralized tissues are commonly fossilized in the absence of early diagenetic mineral precipitation. Plant organs such as leaves and wood are most common, although arthropod cuticle (Bartram *et al.*, 1987; Shear *et al.*, 1984) and graptolite periderm (Armstrong *et al.*, 1984; Crowther *et al.*, 1987) may also be preserved in this way. However, although the tissues may even retain ultrastructural detail, the molecules have undergone biodegradation and chemical transformation to resistant macromolecular structures. The absence of fungal and microbial decay structures in such recalcitrant tissues suggests that alteration has been an abiotic reaction mediated by pore-water chemistry. Where more volatile soft-tissues are preserved as carbon films, cellular detail is rarely evident. The carbon may simply be a chemical residue of the mass of carbon in the organism and in the degradative bacteria, incorporating several tissue types. Alternatively, it may represent only the carbonaceous residue of degradative bacteria. At least one example of this type of preservation actually includes carbonaceous fossil bacteria (Martill, 1987a). The microbial residues pseudomorphing organic soft-parts may be more recalcitrant than the organism itself.

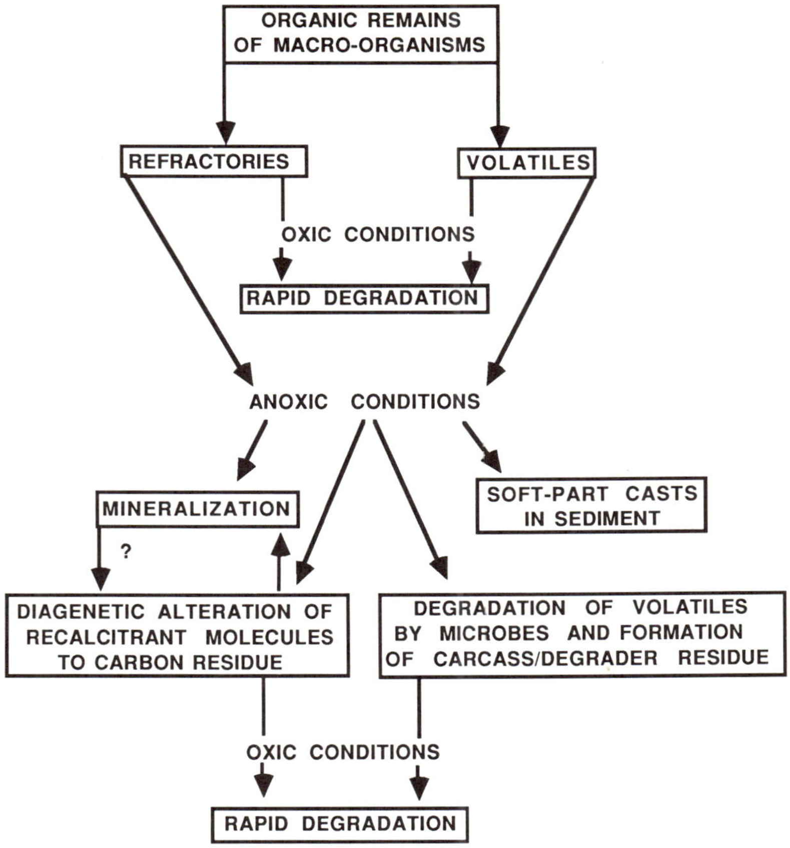

FIGURE 1. Flow diagram summarizing the pathways leading to the preservation of nonmineralized tissues.

More volatile tissues may be preserved as outlines: e.g., Silurian protochordates (Ritchie, 1985), Triassic medusoids (Gall, 1990), Jurassic ichthyosaurs (Martill, 1987b), and Pleistocene mice (Allison *et al.*, in preparation). These are invariably compressions and have obviously undergone considerable bacterial degradation, since little original carbon is retained.

In all cases where nonmineralized tissues are preserved as residues, decay must have been halted. Insufficient data exist to define the conditions necessary for this type of preservation, although the absence of oxygen must be a prerequisite. Most tissue residues are recovered from rocks deposited in oxygen-deficient or euxinic systems and the residues themselves may sometimes be further degraded by microbes upon exposure to oxygen [e.g., hydrocarbons are broken down by heterotrophs in aqueous culture (Alexander, 1973)].

The conditions required for the formation of carbonized fossils originating from either refractory or volatile tissues remain largely enigmatic. Their elucidation requires further investigation of the chemistry of tissue degradation and the factors controlling decay inhibition, and of the sedimentology and geochemistry of the host sediment.

4. Evidence of Decay

Decay is ubiquitous (except in naturally sterile systems such as peat bogs) and arrested only where the breakdown of recalcitrant biomacromolecules would yield insignificant energy, or where they have become stable through chemical condensation and transformation. Information loss may be prevented by early diagenetic mineralization (Allison, 1988a). On the rare occasions when the organisms responsible for degradation are preserved, they can provide data on the decay history of the carcass. The most informative data on decay pathways, however, are obtained from the molecular, elemental, and isotopic chemistry of the fossils and their associated diagenetic minerals.

4.1. Degree of Preservation

The nature of preservation provides an indication of degree of decay prior to fossilization. There are five broad preservational styles: carbonized volatile soft-tissues (Fig. 2C), diagenetically mineralized three-dimensional tissue (Fig. 2A), tissue outlines (Fig. 2B), carbonized refractory tissues (Fig. 2D), and originally mineralized tissues (shells, bones) (Allison, 1988d, Fig. 6).

If decay is inhibited and mineralization is rapid, then three-dimensional soft-tissue preservation may occur. The most spectacular examples are associated with phosphatization (Allison, 1988b; Martill, 1988; Müller, 1985), which often preserves fine details of even volatile tissues such as muscle fibers. Pyritization of soft-tissues is rare; the best examples occur in Beecher's Trilobite Bed in the Ordovician Frankfort Shale, New York State (Cisne, 1973) and in the Devonian Hunsrück Slate of Germany (Stürmer and Bergström, 1973). Although fine details of the exoskeleton are preserved, internal soft-parts are rare, appearing as ghostly images.

If mineralization is delayed and decay continues, internal details of soft-parts will be destroyed, but outlines may remain, as well as more refractory tissues. Examples of this type of preservation include the Precambrian Ediacaran biotas such as those in the Pound Quartzite of the Flinders Range of Australia (Wade, 1968), the Carboniferous Francis Creek Shale of Illinois (Nitecki, 1979), and the Jurassic Solnhofen Limestone of Germany (Barthel, 1978; Barthel *et al.*, 1990). Further advancement of decay leads to complete destruction of volatile tissues, so that only more complex macromolecules such as chitin, cellulose, and finally lignin are preserved (Fig. 1,

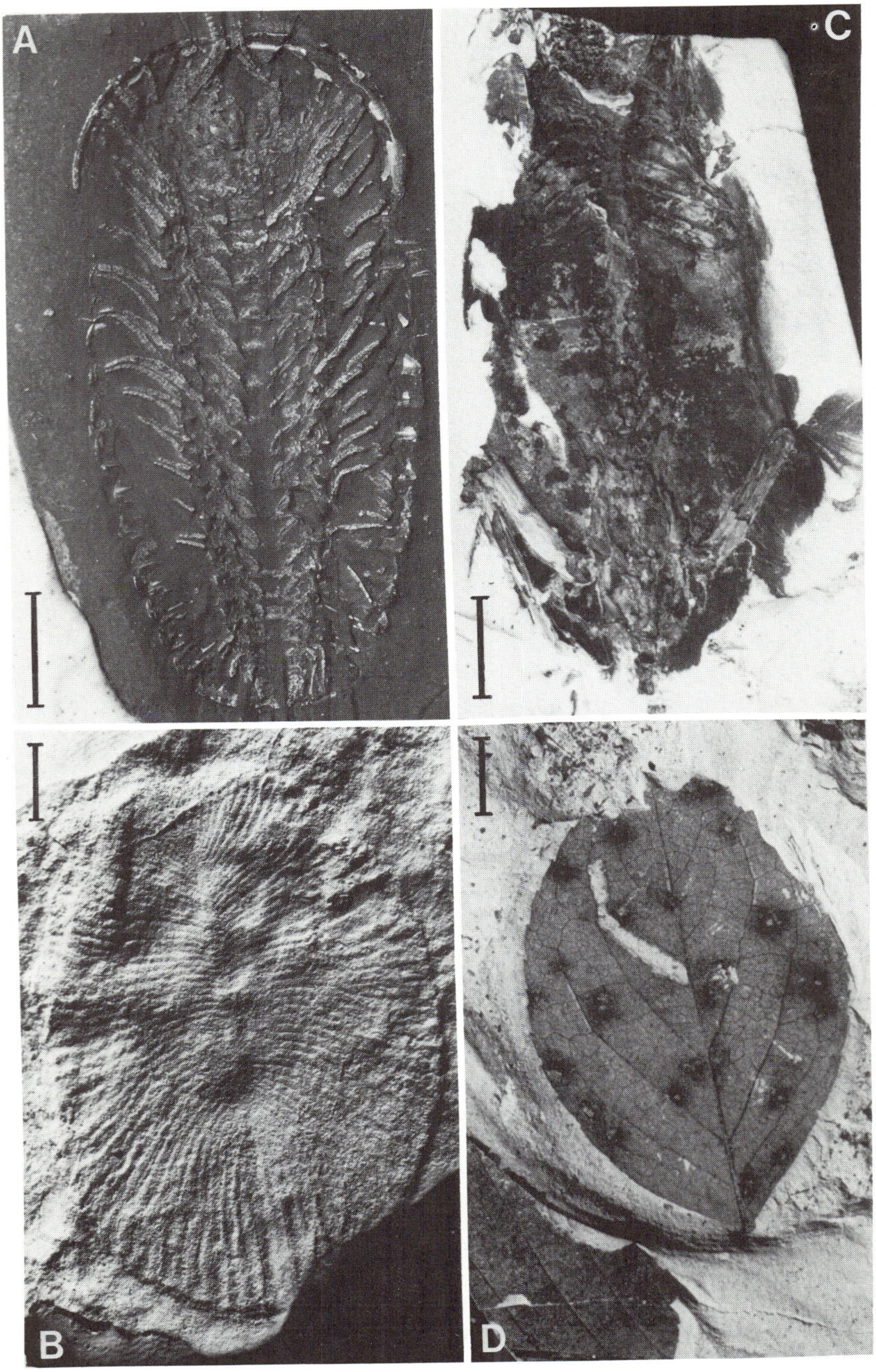
A
C
B
D

Table II). However, where decay continues unabated, organic carbon is completely destroyed and only hard-parts such as shell and bone remain.

4.2. Preservation of Microbes

The most direct evidence of decay is the preservation of microbial organisms. Examples include sideritized bacterial rods pseudomorphing the outline of a frog from the Eocene Messel Oil Shale of Darmstadt in Germany (Wuttke, 1983); bacterial rods and filaments of uncertain affinity associated with the soft-part outlines of ichthyosaurs from the Oxford Clay of central England (Martill, 1987b); possible bacterial rods from the Cretaceous Santana Formation of Brazil (Martill, 1988); and fungal hyphae associated with decay structures in fossil leaves (Dilcher, 1965). Spectacular as this form of preservation is, it gives very little information on the community of decomposers or the preservational history of the fossil. Microbial morphologies are strongly controlled by environment; thus, particular forms give little data on decay history. Upon death microbes undergo autolysis and may be destroyed within hours or days of mortality; their preservation therefore usually implies very rapid mineralization (Allison, 1988c). Rate of autolysis depends both on bacterial type and prevailing conditions. Opportunistic bacteria, which are fast growing, may be degraded rapidly after death. Autochthonous slow-growing bacteria, which are active *in situ*, may survive for long periods [as evidenced by deep sediment samples (Parkes *et al.*, 1990)].

There is also a risk of contamination by Recent bacteria [which may resemble fossilized forms following dehydration and exposure to SEM vacuum (Allison, 1990a)]. Bacteria can self-mineralize (Ennever *et al.*, 1981); hence, it may even be possible for Recent forms to become mineralized on fossils.

4.3. Chemical Markers

Where authigenic mineralization occurs in association with soft-bodied fossils, the elemental and isotopic chemistry of the minerals may provide a way of elucidating the decomposition pathway. The respiratory reactions

←

FIGURE 2. Examples of the preservation of nonmineralized tissues. (A) Pyritized trilobite *Triarthrus eatoni* from the Late Ordovician Beecher's Trilobite Bed, New York State (scale bar 0.5 cm). (B) *Dickinsonia* from the Late Precambrian (Vendian) Ediacara Hills of South Australia preserved as a tissue imprint/outline (scale bar 1 cm). (C) Mouse from the Pleistocene Shiobara biota, Honshu, Japan. The carbon residues preserving the soft-part outline include silica microspheres (scale bar 2 cm). (D) Leaf from the Pleistocene Shiobara biota, Honshu, Japan, preserving traces of original epiphyllous fungi as dark areas (scale bar 1 cm) (Allison *et al.*, in preparation). The leaf carbon is absent, although fungal filaments are preserved, presumably due to their high melanin content (see Table II).

used by microbes involved in decomposition yield a number of reactive phases which may be incorporated into mineral precipitates to provide inorganic chemical signatures characteristic of individual decay pathways (Curtis, 1980; Berner, 1981; Coleman, 1985). Reduction of insoluble manganese oxide by bacteria, for example, produces soluble manganese ions and bicarbonate. These may react to form rhodochrosite, or combine with calcite minerals to form manganous calcite. Iron reduction produces bicarbonate and iron ions, and sulfate reduction produces bicarbonate, carbon dioxide, and hydrogen sulfide; these may form siderite, ferroan calcite, and/or pyrite. Methanogenesis produces additional bicarbonate which can be utilized in further precipitation of carbonate minerals.

Stable isotopes of elements with low atomic number may also be indicators of decomposition pathway and depositional environment (Maynard, 1982; Coleman, 1985). The relative proportion of certain isotopes (e.g., $^{12}C/^{13}C$; $^{34}S/^{32}S$) in modern environments is assumed to have remained roughly the same throughout geological time. Fractionation of these isotopes occurs when they are precipitated. If they are incorporated into mineral phases, a permanent record of the degree of fractionation, which reflects the depositional environment and decay pathway, is retained (e.g., Coleman and Raiswell, 1981).

5. The Distribution of *Konservat-Lagerstätten*

promote mineralization

A number of factors favor soft-tissue preservation (Section 2.2). Anoxia is insignificant in inhibiting decay, but it may promote mineralization. Rapid burial excludes scavengers, and fine sediment, in particular, may cast the outline of soft-tissues. Very low temperatures (below 4°C) may halt bacterial degradation. Naturally abiotic systems such as peat and permafrost halt decay, but are ephemeral. Additional factors such as sediment chemistry and the degradative pathway affect the relative timing of the onset of anoxicity and redox potential. The processes of decay and authigenic mineralization interact to produce different fossilization styles which may be characteristic of different depositional environments.

Deposits characterized by exceptional preservation are termed *Konservat-Lagerstätten*, or conservation deposits, following the terminology of Seilacher (1970; see also Seilacher *et al.*, 1985). They can be categorized in different ways, most obviously on the basis of environmental setting (Seilacher *et al.*, 1985) and diagenesis (Allison, 1988d). Here we review the distribution of *Konservat-Lagerstätten* in space and time, and build on these "classifications" to highlight a number of "taphonomic windows." *Konservat-Lagerstätten* range from those in which the soft-parts have decayed but which preserve complete articulated skeletons, to those in which evidence of the soft-tissues survives. It is the latter which are most significant in the context of decay inhibition and which are considered here.

The nonmineralized tissues of organisms are preserved in a range of depositional environments through the Phanerozoic (Table IV), but particular preservational environments are dominant at certain times. The most important of these taphonomic windows (Fig. 3) occur in the following environments: (1) deep-water slope/basin; (2) restricted offshore marine; (3) deltaic; (4) lacustrine; (5) terrestrial (particularly in amber).

Precambrian biotas are not considered in this review. Ediacaran fossils may represent a different type of organismic construction to that of later life forms (Seilacher, 1989), or they may be primitive early metazoans. In either event the large number of occurrences preserving traces of nonmineralized tissue (Glaessner, 1984; Conway Morris, 1990) in relatively coarse clastics suggests that the system may have been unique taphonomically. In contrast, earlier Proterozoic fossil occurrences are mainly microbiotas preserved in chert (Knoll, 1985).

5.1. Deeper-Water Environments

Preservation in deeper-water environments predominates in the Lower Paleozoic. In the Lower and Middle Cambrian Burgess Shale-type faunas, and in Beecher's Trilobite Bed (Upper Ordovican) and the Hunsrückschiefer (Lower Devonian), rapid burial is an important preservational factor. It is unlikely that anoxicity extended above the sediment–water interface prior to the onset of decay.

Significant soft-bodied preservation in deeper-water deposits is unknown through most of the Upper Paleozoic, Mesozoic, and Tertiary. Preservation can only occur where bioturbators are excluded by anoxicity, or where the thickness of event beds exceeds the depth to which infaunal organisms burrow. The depth of bioturbation and degree of biological disturbance of sediment increased through the Phanerozoic (Thayer, 1983; Sepkoski and Sheehan, 1983; Bottjer and Ausich, 1986; Bottjer *et al.*, 1988) [but see Sheehan and Schiefelbein (1984) and Miller and Byers (1984) for exceptions]. So did the diversity of deep-water bioturbators (Seilacher, 1977). The diversification of deep burrowers into deeper-water habitats [e.g., *Ophiomorpha* and *Zoophycos* (Bottjer *et al.*, 1988)] may account for the paucity of *Konservat-Lagerstätten* in this environment after the Devonian.

The postulated relationship between sedimentation rate, depth of burrowing, and soft-tissue preservation in deeper-water environments is supported by the Miocene Morozaki Formation of the Chita Peninsula of Honshu, Japan (1c, Fig. 3). Here bioturbation is inhibited by very high rates of sedimentation associated with rapid continental uplift on a tectonically active plate margin. The deep-water biota is preserved in fine muds and turbidites. Deposition of the muds was probably noncatastrophic, although articulated hard-parts of echinoderms predominate, perhaps indicating a fluctuating oxycline [the exaerobic environment of Savrda and Bottjer (1987)]. The turbidites yield worms (possibly echiurans), soft-part outlines of fish, and flowers.

TABLE IV. Major Phanerozoic *Konservat-Lagerstätten*[a]

Shiobara	Pleistocene	Honshu, Japan	Lacustrine	Onoe (1989)
Willershausen	Late Pliocene	Western Germany	Lacustrine	Westphal (1967), Meischner and Paul (1982)
Barstow	Mid–Late Miocene	California	Lacustrine	Palmer (1957), Park (1990)
Oeningen	Miocene	Western Germany	Lacustrine	Jörg (1966), Westphal (1978)
Kabutoiwa	Miocene	Honshu, Japan	Lacustrine	Koshimizu (1984)
Chita	Mid Miocene	Honshu, Japan	Slope-basin	Hachiya *et al.* (1988), Oji (1990)
Clarkia	Mid Miocene	Idaho	Lacustrine	Smiley *et al.* (1975), Smiley (1985)
Rott	Late Oligocene, Early Miocene	Western Germany	Lacustrine	Koenigswald (1989)
Florissant	Early Oligocene	Colorado	Lacustrine	MacGinitie (1953), McLeroy and Anderson (1966), Wilson (1978)
Baltic amber	Late Eocene–Early Oligocene	Baltic, northwestern Europe	Terrestrial	Dietrich (1975), Larsson (1978), Schlüter (1990), Langenheim (1990)
Geiseltal	Mid Eocene	Eastern Germany	Lacustrine	Krumbiegel *et al.* (1983), Voigt (1988)
Messel	Mid Eocene	Western Germany	Lacustrine	Franzen and Michaelis (1988), Schaal and Ziegler (1988), Franzen (1990)
Horsefly	Mid Eocene	British Columbia, Canada	Lacustrine	Wilson (1977a–c)
Monte Bolca	Early Eocene	Northern Italy	Subtidal lagoon	Sorbini (1968), Massari and Sorbini (1975)
Green River	Early–Mid Eocene	Wyoming, Utah, Colorado	Lacustrine	Grande (1984, 1989)
Hvar	Late Cretaceous, Cenomanian-Turonian	Yugoslavia	Intertidal	Langer (1961), Hemleben and Freels (1977a)
Hakel	Late Cretaceous, Cenomanian	Lebanon	Offshore (restricted)	Roger (1946), Hückel (1970)
Saskatchewan amber	Late Cretaceous	Manitoba, Canada	Terrestrial	McAlpine and Martin (1969)
Santana, Romualdo	Early Cretaceous, Aptian	Northeastern Brazil	Offshore	Mabesoone and Tinico (1973), Martill (1988), Maisey (1988, 1991)

TABLE IV. *(Continued)*

Santana, Crato	Early Cretaceous, Aptian	Northeastern Brazil	Lacustrine	Grimaldi (1990)
Montsech	Early Cretaceous, Berriasian–Valanginian	Northeastern Spain	Peritidal	Schairer and Janicke (1970), Barale *et al.* (1984), Whalley and Jarzembowski (1985), Selden (1990)
Solnhofen	Late Jurassic, Tithonian	West Germany	Offshore (restricted)	Barthel (1970, 1978), Barthel *et al.* (1990), Seilacher *et al.* (1985), Viohl (1985)
Cerin	Late Jurassic, Kimmeridgian	Jura, France	Offshore (restricted)	Bernier (1985), Gall *et al.* (1985)
Cordillera de Domeyko	Late Jurassic, Oxfordian	Northern Chile	Offshore	Schultze (1989)
Voulte-sur-Rhône	Middle Jurassic, Callovian	Southeastern France	Offshore	Dietl and Mundlos (1972), Fischer and Riou (1982a,b)
Christian Malford	Middle Jurassic, Callovian	England	Offshore	Allison (1988b)
Holzmaden	Early Jurassic, Toarcian	Western Germany	Offshore (restricted)	Hauff and Hauff (1981), Kauffman (1979, 1981), Seilacher (1982), Riegraf (1984), Wild (1990)
Osteno	Early Jurassic, Sinemurian	Northern Italy	Offshore	Pinna (1985)
Culpeper	Late Triassic, Rhaetian	Virginia	Lacustrine	Gore (1986, 1988a,b)
Montral-Alcover	Mid Triassic, Ladinian	Northeastern Spain	Offshore (restricted)	Hemleben and Freels (1977b), Via Boada *et al.* (1977), Beltan *et al.* (1989)
Grès à Voltzia	Mid Triassic, Anisian	Vosges, north-eastern France	Fluvial to coastal plain delta	Gall (1971, 1985), Gall and Grauvogel-Stamm (1984)
Odernheim	Early Permian	Saar-Nahe Basin, western Germany	Lacustrine	Boy and Hartkopf, (1983), Boy (1987), Willems and Wuttke (1987)
Hamilton	Pennsylvanian, Gzelian, Noginskian	Kansas	Subtital	Mapes and Mapes (1988)
Blanzy-Montceau	Pennsylvanian, Kasimovian-Gzelian	Massif Central, France	Deltaic–lacustrine	Rolfe *et al.* (1982), Heyler (1986), Heyler and Poplin (1988)

(continued)

TABLE IV. *(Continued)*

Mazon Creek	Pennsylvanian, Moscovian, Myachkovskian	Illinois	Deltaic	Nitecki (1979), Baird *et al.* (1985a,b)
Mecca/Logan	Pennsylvanian, Moscovian, Podolskian-Myachkovskian	Indiana	Deltaic/coastal plain	Zangerl and Richardson (1963)
Bear Gulch	Mississippian, Serpukhovian, Arnsbergian	Montana	Offshore (restricted)	Williams (1983), Horner (1985), Horner and Lund (1985)
Bearsden	Mississippian Serpukhovian, Pendleian	Southern Scotland	Subtidal	Wood (1982), Clark (1989)
East Kirkton	Mississippian, Visean, Brigantian	Southern Scotland	Lacustrine	Rolfe (1988), Rolfe *et al.* (1990)
Granton	Mississippian, Visean, Holkerian-Asbian	Southern Scotland	Deltaic	Briggs and Clarkson (1983), Cater (1987), Briggs *et al.* (1991)
Glencartholm	Mississippian, Visean, Holkerian	Southern Scotland	Deltaic, shallow marine	Schram (1983), Cater *et al.* (1989)
Foulden	Mississippian, Tournaisian, Courceyan	Southeast Scotland	Lacustrine	Wood and Rolfe (1985), Clarkson (1985)
Cleveland Shale	Late Devonian, Famennian	Ohio	Offshore	Hannibal (1989)
Gilboa	Mid Devonian, Givetian	New York	Deltaic, fluvial	Shear *et al.* (1984, 1987), Banks *et al.* (1985)
Achanarras	Mid Devonian, Eifelian	Northwest Scotland	Lacustrine	Trewin (1986), Hamilton and Trewin (1988)
Hunsrück	Early Devonian, Emsian	Western Germany	Open shelf–basin	Stürmer *et al.* (1980), Bartels and Brassel (1990), Bergström (1990)
Alken	Early Devonian, Ermsian	Western Germany	Subtidal	Størmer (1970, 1975)
Rhynie	Early Devonian, Pragian	Eastern Scotland	Lacustrine	Rice and Trewin (1988), Trewin (1989)
Saaremaa	Silurian, Wenlock Gleedonian	Estonia	Offshore (restricted)	Kaljo (1970, 1972), Selden (1981)
Waukesha	Silurian, Llandovery, Telychian	Wisconsin	?Peritidal	Mikulic *et al.* (1985a,b)

TABLE IV. *(Continued)*

Lesmahagow	Silurian, Llandovery, Telychian	Southern Scotland	Offshore (restricted)	Rolfe (1973), Ritchie (1968, 1985)
Beecher's Bed	Late Ordovician, Caradoc, ?Pusgillian	New York State	Slope-basin	Cisne (1973), Briggs *et al.* (1990)
Orsten	Late Cambrian, Maentwrogian-Dolgellian	Southern Sweden	Offshore	Müller (1985)
Marjum Formation	Mid Cambrian, Menevian	Utah	Slope-basin	Rogers (1984), Briggs and Robison (1984), Conway Morris and Robison (1986)
Wheeler Formation	Mid Cambrian, Menevian	Utah	Slope-basin	Rogers (1984), Briggs and Robison, (1984), Conway Morris and Robison (1986)
Burgess Shale	Mid Cambrian, Solvan	British Columbia, Canada	Slope-basin	Whittington (1985), Conway Morris (1986, 1990)
Spence Shale	Mid Cambrian, Solvan	Utah	Slope-basin	Rogers (1984), Briggs and Robison (1984), Conway Morris and Robison (1986)
Emu Bay Shale	Early Cambrian, Lenian	Kangaroo Island, South Australia	Subtidal	Conway Morris (1985), Conway Morris and Jenkins (1985)
Kinzers Formation	Early Cambrian, Lenian	Pennsylvania	Slope-basin	Conway Morris (1985)
Buen Formation	Early Cambrian, Atdabanian	Peary Land, Greenland	Outer shelf	Conway Morris *et al.* (1987), Conway Morris and Peel (1990)
Chengjiang	Early Cambrian, Atdabanian	Yunnan Province, China	?Slope-basin	Zhang and Hou (1985), Hou and Sun (1988), Chen *et al.* (1989), Hou (1989)

[a]This list of 61 examples is not exhaustive. Two important categories of soft-part preservation, at opposite ends of the geological time scale, are omitted: (1) mammoths, bison, and other Pleistocene mammals in the permafrost, and (2) Precambrian Ediacaran biotas. Many of the *Konservat-Lagerstätten* listed represent the most completely preserved members of taphonomic gradients which include many lesser examples (see text). Geological ages are standardized in terms of the Harland *et al.* (1990) geological time scale. In some cases assignments to stage are tentative, either because of difficulties in correlating between continents or due to uncertainties of dating. The references have been selected to help the reader gain access to the literature on individual *Konservat-Lagerstätten;* they do not necessarily represent original sources of taphonomic data.

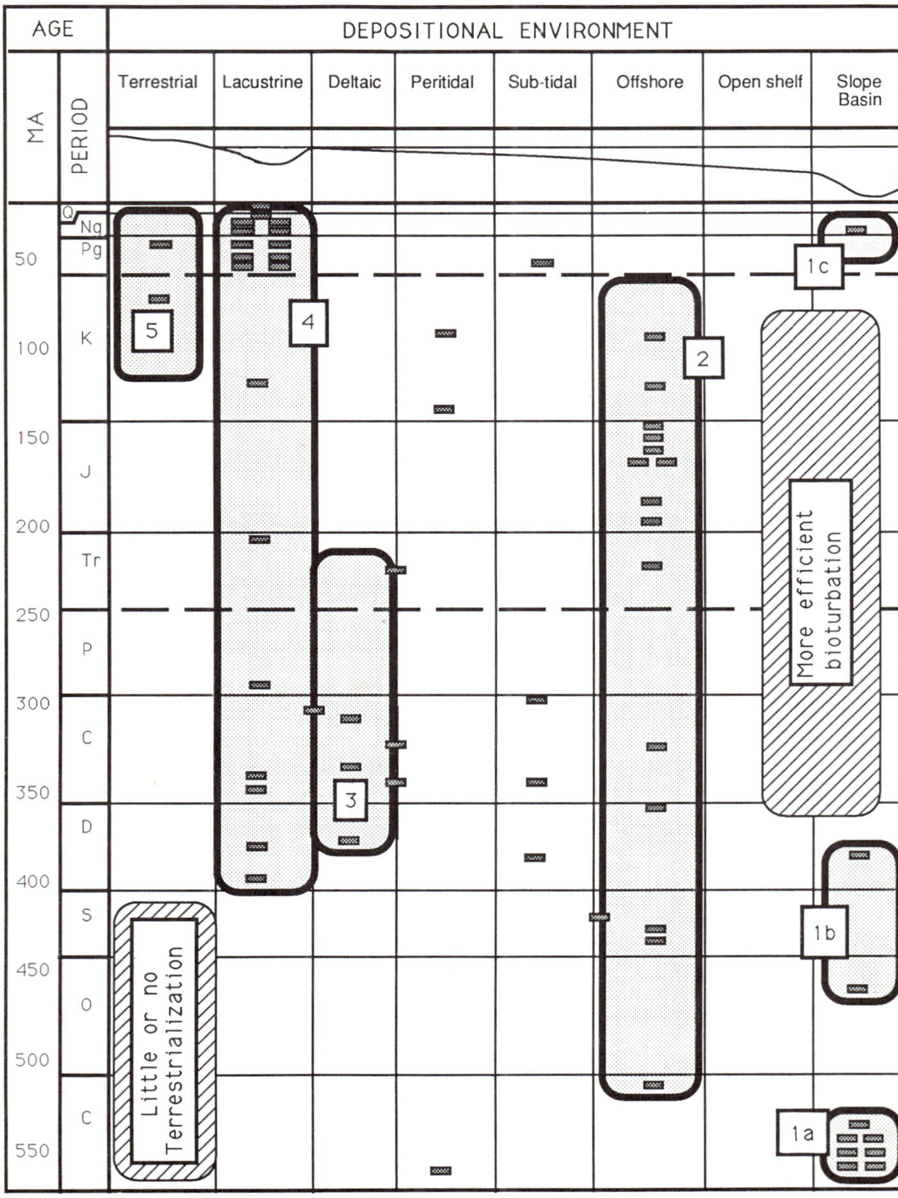

FIGURE 3. The distribution of major *Konservat-Lagerstätten* (Table IV) (represented by stippled bars) in the main categories of sedimentary environment through the Phanerozoic. In some cases the interpretation of environment is generalized; the sedimentological context of a number of exceptionally preserved biotas has not been fully investigated. The separation of "basin" from "open shelf" by a "slope" may not be appropriate in some geological contexts; the Hunsrückschiefer, for example, may have been deposited in a basin on the open shelf. Where symbols traverse the line separating two environmental categories this indicates that the *Konservat-Lagerstätte* is transitional between the two. The numbered fields represent taphonomic "windows," their position constrained in space and time by a range of physical and biotic factors (see text for discussion).

5.1.1. Cambrian

Many of the more important deeper-water Lower–Middle Cambrian *Konservat-Lagerstätten*, including the Burgess Shale, occur on the ocean-ward side of the median carbonate belt which flanked the Laurentian craton during much of the Cambrian [see Conway Morris (1985) for a review]. They are therefore designated as slope-basin (1a, Fig. 3). These faunas show a striking taxonomic overlap with Walcott's original Burgess Shale fauna quarried from the Phyllopod Bed on Fossil Ridge near Field, British Columbia, and they have been designated Burgess Shale-type faunas (Conway Morris, 1989a,b). The evidence is inadequate to determine whether the Burgess Shale taxa were gradually displaced or suffered a mass extinction at the end of the Middle Cambrian (Conway Morris, 1989a). It is likely, however, that their relatively abrupt disappearance reflects taphonomic factors rather than a major extinction. Many of the major Burgess Shale soft-bodied taxa which *must* have have had representatives in subsequent periods (e.g., nonmineralized arthropods, polychaetes, priapulids) are poorly represented in younger rocks. The combination of circumstances which promoted their preservation must have disappeared. The increase in infaunal tiering which developed in the Ordovician (Bottjer and Ausich, 1986) may have been a contributory factor.

Burgess Shale-type biotas are known from at least 26 Lower to Middle Cambrian horizons (Conway Morris, 1989a) in addition to the eight listed in Table IV. They range in age from Lower Cambrian *Fallotaspis* Zone (northeastern Poland) to close to the top of the Middle Cambrian (Nevada) (Conway Morris, 1989a, Fig. 1, and Appendix); rare later occurrences of a small number of taxa are known. The major Burgess Shale-type assemblages in North America (Kinzers Formation, Spence Shale, Burgess Shale, Wheeler Formation, Marjum Formation) were deposited near the outer edge of the carbonate belt which separated the inner detrital belt from the deeper-water outer detrital belt beyond (Palmer, 1960; Robison, 1960). This threefold division around the craton does not apply throughout the Cambrian in all parts of North America, however, and a number of localities are facies equivalents of the inner detrital belt [e.g., Eager Formation, Latham Shale (Briggs and Mount, 1982)], and therefore more appropriately considered open-shelf faunas (as these do not qualify as major *Konservat-Lagerstätten*, however, they are not included in Table IV and therefore do not appear on Fig. 3).

Neither the environmental setting nor the diagenetic history of the Burgess Shale-type occurrences are well known in detail (Conway Morris, 1985). Even the Walcott Quarry itself is currently the subject of controversy. The widely accepted model of the preservation of the Burgess Shale near the foot of the steep-faced Cathedral Escarpment, some 170 m high, where the animals were buried by turbidity currents (Whittington, 1985; Conway Morris, 1986), has recently been challenged. Ludvigsen (1989) argued that the biostratigraphic basis for the escarpment is suspect, the boundary between zonal trilobites in the off-reef shales and on top of the reef reflecting facies control rather than a time line. On this basis Ludvigsen suggested that the Burgess Shale was deposited on the distal part of a ramp, the animals buried essentially *in situ* by tempestites. There are equivocal aspects of both models, including whether the sediment was deposited as turbidites or tempestites (Fritz, 1990; Aitken and McIlreath, 1990; Ludvigsen, 1990). The controversy clearly illustrates that the escarpment model cannot be adopted uncritically as a paradigm for the preservation of Burgess Shale-type faunas [at least one of which, the Emu Bay Shale, was deposited in a near-shore environment (Conway Morris and Jenkins, 1985; Conway Morris, 1985)].

Not only has the environmental interpretation of the Burgess Shale been challenged (Ludvigsen, 1989, 1990), so have previous interpretations of the significance of diagenesis. Delicate processing of a range of Burgess Shale fossils yielded organic carbon (Butterfield, 1990). The reflectivity of these carbon films led Butterfield (1990) to suggest that the Burgess Shale fossils are preserved as kerogenized organic remains, rather than by diagenetic replacement by aluminosilicates (Conway Morris, 1986). Butterfield (1990) suggested that the intimate association of clay particles with the Burgess Shale organisms, which resulted from turbulent transport, may have contributed to decay inhibition. But before diagenetic replacement can be discounted completely in favor of organic preservation, more Burgess Shale fossils need to be processed. A systematic study of the mineralogy/petrography of other Burgess Shale-type occurrences is also necessary. The processes leading to the preservation of fossils in the Phyllopod Bed may not be typical of Burgess Shale-type *Konservat-Lagerstätten*. Glaessner (1979), for example, reported that the soft-tissues of annelids in the shallow-water Lower Cambrian Emu Bay Shale are preserved in iron-stained calcite, indicating that the diagenetic history of at least this Burgess Shale-type fauna is fundamentally different to that of Walcott's original discovery.

5.1.2. Ordovician–Devonian

The two most significant post-Cambrian *Konservat-Lagerstätten* from basinal environments (1b, Fig. 3), the Late Ordovician Beecher's Trilobite Bed of New York State, and the Early Devonian Hunsrückschiefer of western Germany, are preserved in pyrite. Pyrite is a common component of fine-grained organic-rich marine sediments, where it forms through the agency of sulfate-reducing bacteria (Berner, 1970, 1984). It occurs in association with fossils through the geological record, but it is very rarely involved in the extensive preservation of volatiles, or even of refractory organic tissue. Where it occurs in the Burgess Shale, for example, it is largely restricted to mineralized tissues (Conway Morris, 1986).

Beecher's Trilobite Bed is a 4-cm-thick silt-grade microturbidite. The trilobites are concentrated near the base of the bed in a layer a few millimeters thick (Cisne, 1973). They are almost exclusively carcasses and therefore provided a concentration of volatile tissue as a substrate for sulfate-reducing bacteria. An investigation of sulfur isotopes (Briggs *et al.*, 1990) shows that pyritization persisted for longer in the trilobite layer than in the adjacent sediment. Thus rapid burial of localized high concentrations of organic matter (e.g., carcasses) within a sequence otherwise very low in volatiles appears to be a requirement for the preservation of soft-tissues (Allison, 1988d).

The sedimentology and diagenetic history of the Hunsrückschiefer, which was also deposited from distal density currents or turbidites, is broadly similar to that of Beecher's Trilobite Bed. The biota, however, is much more diverse (Mittmeyer, 1980; Bartels and Brassel, 1990) and includes rare organisms with tissues more volatile than arthropod cuticle [e.g., worms (Bergström, 1990) and ctenophores (Stanley and Stürmer, 1983, 1987)]. The fossils are scattered, individual carcasses providing the requisite concentration of organic material. The depositional environment of the Hunsrückschiefer, like Beecher's Trilobite Bed, is plotted as slope/basin (1b, Fig. 3) although the basin may have been on the open shelf (Bergström, 1990).

The rarity of pyritized soft-tissues indicates that a very precise suite of circumstances is required to allow volatiles to be preserved in this way. The large number of occurrences of pyritized graptolites from Ordovician and Silurian sequences represent a similar mode of preservation, but traces of tissues more volatile than the periderm are almost unknown [Rickards and Stait (1984) reported traces of pyritized graptolite zooids].

5.2. Offshore Environments

Offshore marine *Konservat-Lagerstätten* (2, Fig. 3) are characterized by laminated sediments deposited under conditions of restricted circulation. In some cases these conditions correspond to an oceanic anoxic event (Jenkyns, 1980). The Toarcian event, for example, coincides with the exceptional preservation of the Posidonia Shales of Holzmaden. In most cases, however, conditions were more localized, corresponding to the development of local barriers, such as reefs. Such local geographic factors were equally important, as evidenced by the lack of *Konservat-Lagerstätten* in association with the Late Cretaceous oceanic anoxic events. The preservation of tiny phosphatized arthropods in calcareous nodules in some offshore settings (typified by the Upper Cambrian Orsten of Sweden) occurs sporadically throughout the Phanerozoic.

5.2.1. "Restricted Circulation"

A range of *Konservat-Lagerstätten* occur in offshore marine environments due to restricted circulation and the resulting stratification of the water body [see review by Seilacher *et al.* (1985)]. The best known are the Holzmaden Posidonia Shale and the Solnhofen Lithographic Limestone, both in the Jurassic of western Germany. The Holzmaden black shale corresponds to the widespread Toarcian anoxic event. Anoxic bottom conditions were dominant, although abundant benthic organisms became established when major storms occasionally mixed the stratified water. Nektonic and epibenthic forms are characteristic.

The Solnhofen Limestone *Konservat-Lagerstätte* is also the product of a stratified water body, probably a halocline (Keupp, 1977), with associated oxygen deficiency in the bottom water. The area of deposition is much less extensive than in the case of Holzmaden; it was confined to depressions between sponge reefs. Fossils are rare, apart from the pelagic crinoid *Saccocoma;* benthic elements were washed in from shallower water. The original organic content was lower than in the bituminous Holzmaden Shale (and may have been further reduced by weathering).

Although the Holzmaden and Solnhofen *Konservat-Lagerstätten* differ in detail, in both cases preservation relies on a stratified water body and on episodic influxes of sediment which lead to rapid burial (Seilacher *et al.*, 1985). Microbial films [prokaryotic scum (Seilacher *et al.*, 1985)] may also have played a role in the case of Solnhofen (see also Gall, 1990). A number of similar *Konservat-Lagerstätten* occur, particularly in the Mesozoic. Examples of exceptional preservation in the plattenkalk lithology characteristic of Solnhofen range from the Mississippian (Bear Gulch) to Early Eocene (Monte Bolca) and include a number of significant Triassic (Montral-Alcover), Jurassic (Cerin), and Cretaceous (Montsech, Hakel) examples. A similar environment may be represented by the Silurian eurypterid beds at Saaremaa (Wenlock), and in North America (Pridoli; e.g., the Bertie "waterlime" of New York). Other deposits, like the Early Jurassic Osteno example from Italy, are intermediate in character between Solnhofen and Holzmaden (Seilacher *et al.*, 1985). The style of preservation in Holzmaden and Solnhofen clearly differs from that in *Konservat-Lagerstätten* where concretions form (e.g., Santana-Romualdo, Cordillera de Domeyko; see Section 5.5). This contrast reflects differences in diagenesis rather than in major environmental parameters.

5.2.2. "Orsten"

The Upper Cambrian Orsten of southern Sweden are bituminous nodular limestone lenses which occur within the highly condensed organic-rich shales of the Alum Shale Formation. The nodules yield a fauna of tiny three-dimensional phosphatized arthropods (e.g., Müller and Walossek, 1986, 1987, 1988). The majority of the exceptionally preserved specimens come from a relatively small number of nodules (D. Walossek, personal communication). The source of phosphate in the Orsten is uncertain. The high organic content of the sequence suggests an organismic source, although Müller (1985) suggested the possibility of a derivation from adjacent landmasses. Phosphatization of nonmineralized tissues within the

Orsten is confined to arthropod cuticle, and none of the fossils (either entire or fragments of individuals) exceeds 2 mm in length.

Preservation of microarthropods in calcium phosphate within calcareous nodules (Müller, 1985) is not confined, however, to the Upper Cambrian of Sweden. Examples are known from the Lower Cambrian of Shropshire, England (Hinz, 1987), the Upper Cambrian of Poland (Walossek and Szaniawski, 1991), the Lower Ordovician of Öland, Sweden (Andres, 1989), the Upper Devonian of the Carnic Alps (Müller, 1982), the Lower Triassic of Spitsbergen (Weitschat, 1983a,b), and the Lower Cretaceous (Santana Formation) of Brazil (Cressey and Boxshall, 1989; Wilby and Martill, 1991). The phosphatization of such delicate structures in three dimensions requires very rapid, almost *in vivo,* mineralization, and indicates that diagenesis occurred prior to significant decay. While benthic conditions led to phosphatization in the Cambrian Orsten, the source of phosphate in some of the younger examples of this kind of preservation was apparently an associated vertebrate carcass. Thus, a mode of life as scavenger or parasite (Cressey and Boxshall, 1989), or occurrence as prey in the gut of a fish (Wilby and Martill, 1991), enhances preservation potential.

Apatite also may be involved in soft-tissue preservation in larger organisms, but here different circumstances account for the phosphatization. Some form the nucleus of concretions: e.g., concavicarids from the Upper Devonian of Australia (Briggs and Rolfe, 1983), some fish from the Santana Formation (Martill, 1988). Others occur in fine-grained sediments: e.g., shrimps from the Carboniferous of Scotland (Briggs and Clarkson, 1983, 1985), concavicarids from the Jurassic of Osteno, Italy (Pinna, 1985), squid from Osteno and from the Oxford Clay of England (Donovan, 1983; Allison, 1988b). Three-dimensional soft-tissue in carbonate concretions may be composed of minerals other than apatite. The muscles of fish are preserved in calcite in nodules from the Upper Devonian Cleveland Shale of Ohio (Criss *et al.*, 1988), the Upper Jurassic Cordillera de Domeyko of Chile (Schultze, 1989), and commonly in the Lower Cretaceous Santana Formation of Brazil (Maisey, 1988, 1991).

5.3. Deltaic Environments

A number of major Carboniferous *Konservat-Lagerstätten* occur in environments which are categorized as *broadly* deltaic (Fig. 3, 3). Low-lying land areas in tropical latitudes were characterized by rapid sedimentation (e.g., crevasse splays) in paralic basins. High levels of organic carbon accumulation promoted local anoxia (e.g., in interdistributary bays). These *Konservat-Lagerstätten* yield a range of taxa from fully terrestrial to nearshore marine. The best known and most diverse is the Pennsylvanian Mazon Creek biota of Illinois (Nitecki, 1979; Baird *et al.*, 1985a). The fossils are preserved as a result of rapid burial and the formation of early diagenetic siderite concretions (Richardson and Johnson, 1971; Woodland and Stenstrom, 1979). The most productive concretion horizon occurs in silty to sandy mudstone in the basal few meters of the Francis Creek Shale, where it overlies the Colchester Coal. The release of bicarbonate ions during decay of the peat may have been a controlling factor. Two separate assemblages are recognized, a nonmarine Braidwood Fauna, and a more marine-influenced Essex Fauna.

Mazon Creek fossils have been collected from outcrops of the Francis Creek Shale in over 270 different localities within the Mazon Creek area (Baird *et al.*, 1985a). At least six other areas have yielded fossil-bearing sideritic concretions in the Francis Creek Shale—both elsewhere in Illinois, and farther afield in Missouri and Oklahoma (Baird *et al.*, 1985b, 1986). Soft-bodied fossils are also preserved in siderite concretions from the Late Pennsylvania of Blanzy-Montceau, France (Heyler and Poplin, 1988), and in the English Coal Measures around Rochdale (Woodward, 1907, 1908, 1911) and Lancaster (Bolton, 1905).

Not all *Konservat-Lagerstätten* in Carboniferous deltaic environments are preserved in siderite nodules. Most of the Mississippian shrimp faunas of northern Britain are preserved in original organic material and/or with associated phosphate (Cater *et al.*, 1989). Deltas are a particularly important site for exceptional preservation in the Carboniferous, due to the prevalence of broad coastal plain environments. A comparable example occurs in the Triassic, however, where it is represented by the Grès à Voltzia of northeastern France (Briggs and Gall, 1990), although the fossils are preserved in laminated silts rather than siderite nodules.

The taphonomic window represented by these Late Palaeozoic, Early Mesozoic deltaic environments is not controlled by a specific type of diagenetic mineralization, but reflects the environmental conditions which prevailed. A broad coastal delta plain provided a setting in which restricted water bodies such as interdistributary bays and lagoons could form. Salinity was variable (due to mixing of normal marine and fresh water) in a tidally influenced estuarine-deltaic environment such as that of Mazon Creek (Baird *et al.*, 1986), and may have been instrumental in inhibiting decay [a tidal influence also may have prevailed in other North American Carboniferous *Konservat-Lagerstätten* (H. R. Feldman, personal communcation)]. Sedimentation was episodic. These deltaic *Konservat-Lagerstätten* form a faunal continuum, together with exceptionally preserved biotas in other contemporaneous environments (Schram, 1979; Briggs and Gall, 1990), which can be traced from Lower Carboniferous to Triassic.

5.4. Lacustrine Environments

Lacustrine environments are an important site of exceptional preservation. Deeper lakes readily become stratified (meromictic), with cold anoxic bottom conditions which inhibit decay and promote the preservation of refractory organic compounds. Soft-tissue outlines may also be replicated by diagenetic minerals (e.g., siderite, vivianite). The concentration of lacustrine *Konservat-Lagerstätten* in Tertiary rocks [Figs. 3, 4; Table IV; and in previous compilations (Conway Morris, 1981, 1986)] reflects the fact that they are less likely to have been eroded than in older sequences. In addition, more ancient lacustrine sediments may be difficult to identify.

The most extensive lacustrine *Konservat-Lagerstätte* is the Green River Formation, which was deposited during the Late Paleocene to Late Eocene over a large area of Wyoming, Colorado, and Utah. The best preserved of the Green River biotas (Early to Middle Eocene in age) include a high diversity of freshwater and terrestrial organisms (Grande, 1984). Environments varied from ephemeral playa and shallow lacustrine settings subject to evaporation (Bucheim and Surdam, 1981) to deeper, more permanent conditions (Desborough, 1978). The only generalization than can be made about the sedimentological context of the exceptional preservation (Grande, 1989) is that carbonates with organic-rich laminae are characteristic. While the Green River Formation yields a diversity of vertebrates as well as plants and insects, fish, many of which display soft-tissue outlines, are dominant.

The Green River Formation is part of a continuum of Tertiary lake deposits in western North America which range from Paleocene to Oligocene in age, and extend from Colorado, Utah, and Nevada, through Wyoming, Montana, Idaho, and Washington, to British Columbia and Alberta. The proportions of taxa in different formations at different localities varies [see Grande (1984, 1989) for vertebrate localities in the Green River Formation; Wilson (1978) for Paleogene insect localities (28 in number); Smiley *et al.* (1975) for Miocene localities]. Volcanic activity was prevalent in western North America in Tertiary times (associated with the Laramide Orogeny), and the occurrence of tuffs in some of the fossil-bearing strata (McLeroy and Anderson, 1966; Wilson, 1977a; Smiley and Rember, 1985) may have promoted preservation. It has proved possible to sequence DNA from leaves from the Mid-Miocene Clarkia localities in northern Idaho (Golenberg *et al.*, 1990). Some of the older *Konservat-Lagerstätten* in this area provide obvious targets for future investigations of molecular preservation.

The Tertiary lake biotas of western North America are well known, but by no means unique. Lake basins associated with other orogenic belts also yield exceptionally preserved fossils (e.g., the Miocene deposits at Oeningen in southern Germany). Lakes may also form in fault-bounded basins; examples of *Konservat-Lagerstätten* in this context include the Middle Eocene examples at Messel and Geiseltal.

Older examples of lacustrine *Konservat-Lagerstätten* are more common than is usually recognized, attracting less attention because of their lower diversity. Even though insects require exceptional conditions for preservation (Carpenter and Burnham, 1985), biotas yielding only insects and plants are omitted from consideration due to the large numbers involved. Insects reached a peak of familial diversity in the Permian, a level which was only regained (and greatly exceeded) in the Tertiary (e.g., Sepkoski and Hulver,

1985). Inclusion of the Permian insect deposits [with important faunas from the Soviet Union, Czechoslovakia, the United States, and Australia (Carpenter and Burnham, 1985)] would significantly increase the number of *Konservat-Lagerstätten* in this period.

Some of the more diverse, or better-documented, older lake sequences are included in Table IV to emphasize the continuum of exceptional preservation in this environment. The Triassic Culpeper Basin of Virginia includes numerous lake sequences deposited in an elongate half graben (Gore, 1988a). Nine lacustrine facies have been described representing deposition in offshore, nearshore, and mudflat environments. Two of these yield fossilized soft-tissues: phosphatized fish occur in organic-rich laminated black shales; notostracan crustaceans, insects, fish, and plants in olive-gray to grayish-red laminated mudshale and clay-shale facies (Gore, 1986, 1988a).

Fish, shrimps, other arthropods, and plants are preserved in just over 1 m of the Mississippian Foulden Fish Bed of northeastern England, probably representing a brief interval of stratification followed by a rapid sediment influx (Wood and Rolfe, 1985; Anderton, 1985). The slightly younger East Kirkton biota of southern Scotland includes a range of terrestrial organisms; the introduction of hot spring waters into the freshwater lake was instrumental in promoting preservation (Rolfe *et al.*, 1990).

Exceptionally preserved lake biotas in the Silurian and Devonian, which precede the radiation of the insects and also lack diverse macrofloras, appear unremarkable compared to later examples (this is reflected in the position of the lower boundary of taphonomic window 4 in Fig. 3). Achanarras, from the Middle Old Red Sandstone of Scotland, is included here as one of the most completely documented examples of a Devonian fish assemblage (Trewin, 1986; Hamilton and Trewin, 1988), which includes soft-tissue outlines. There are many others, however, even in the Old Red Sandstone of Scotland (Mykura, 1983; Trewin, 1986).

5.5. Terrestrial Environments

The results of strictly terrestrial (as opposed to aquatic) agents of preservation are ephemeral (hence, exceptionally preserved terrestrial organisms normally occur in aquatic settings). Freezing of Pleistocene mammals in permafrost, or drying/mummification in caves, merely halts decay, which can be reactivated by thawing and wetting (Guthrie, 1990). Preservation in such naturally abiotic systems is therefore confined to the Quaternary and Neogene. Conservation traps (*sensu* Seilacher *et al.*, 1985), such as amber and asphalt, encase the organism in a medium which can protect it from decay on a time scale of millions as opposed to thousands of years, but removal of the organism from the enclosing substance can initiate decay. (This applies in varying degrees to any organically preserved fossil.)

The most significant terrestrial mode of preservation of nonmineralized tissue is trapping in plant resins. The occurrence of amber in the fossil record reflects the distribution and number of resin-producing plants and the preservation potential and hydrodynamic properties of the amber itself. Amber may be preserved *in situ* in coal deposits, or transported and concentrated in a range of aquatic environments (Dietrich, 1975). The resins which form amber harden as a result of cross-linking and isomerization (Mills *et al.*, 1984) thus becoming stable and highly resistant to subsequent alteration, particularly in a reducing environment.

Amber is renowned for the preservation of plants, insects, and other arthropods, and rare vertebrates. Preservation in some ambers has yielded details of cell organelles (Poinar and Hess, 1982) and even traces of DNA (Pååbo, *et al.*, 1989). The best known examples are from the Paleogene of the Baltic (Larsson, 1978) and the Dominican Republic, and the Upper Cretaceous of Manitoba, Canada (McAlpine and Martin, 1969). Occurrences range from Carboniferous to Pleistocene and occur in hundreds of localities worldwide [Langenheim (1990) lists 30 major deposits worldwide; Dietrich (1975) maps occurrences in Europe; McAlpine and Martin (1969) list over 50 localities in Canada and Alaska]. Late Paleozoic plants did not exude large quantities of resin, and this mode of preservation is rare prior to the Cretaceous when amber became much more abundant, the major source being Araucariaceae [as determined by chemical analysis (Langenheim and Beck, 1965)]. Later ambers were derived from a variety of botanical sources (Langenheim, 1990). Production of resins is most profuse in tropical and subtropical latitudes and this is reflected in the distribution of amber. Thus although this taphonomic window was initiated in the Carboniferous it was not until the Cretaceous (Fig. 3, 5), with the rise of Araucaciaceae, that this mode of preservation extensively sampled truly terrestrial biotas (mainly insects and plants) in tropical and subtropical latitudes.

6. Conclusion

A range of factors inhibit decay and promote the preservation of soft-tissues. The amount of detail preserved in a fossil is determined by how far information loss (through decay, scavenging, and mechanical processes) proceeds. A polychaete, for example, decays in a sequence which leads to loss of muscle, gut, exocuticle, setae, and jaws in order—depending on their resistance to decay. Such differences in preservation potential allow a series of taphonomic thresholds to be defined, using the occurrence of different tissue types, or organisms, in *Konservat-Lagerstätten* (Allison and Briggs, 1991). The taphonomic threshold reached varies among *Konservat-Lagerstätten*. The conditions required for soft-tissue preservation are often very precise, and therefore only very locally developed. Hence changes in conditions can lead to differences in the degree of preservation within individual *Konservat-Lagerstätten*.

The examples of *Konservat-Lagerstätten* listed in Table IV and discussed in this chapter are only a representative selection. Many of them represent the tip of an iceberg, the most completely preserved or best known examples of a large number of similar examples. More systematic analyses (in preparation) will require the application of the concept of taphonomic thresholds to specific environmental categories. Further investigation of the taphonomy and paleobiology of *Konservat-Lagerstätten* will allow a more comprehensive genetic classification of these important fossil occurrences to be developed. The identification and refinement of taphonomic windows will allow the potential of *Konservat-Lagerstätten* as a data base for studying patterns in evolution to be realized.

ACKNOWLEDGMENTS. We are grateful to Drs. A. J. Kear and R. J. Parkes for discussion and critical comments. Prof. H. B. Whittington and Dr. R. Goldring kindly provided the photographs in Figs. 2A and B, respectively. D.E.G.B.'s research on the taphonomy of exceptionally preserved fossil biotas is funded by NERC GR3/7235. (This is University of Reading PRIS contribution number 129.)

References

Aitken, J. D., and McIlreath, I. A., 1990, Comment on "The Burgess Shale: Not in the shadow of the Cathedral Escarpment", *Geoscience* **17**:111–115.

Alexander, M., 1973, Nonbiodegradable and other recalcitrant molecules, *Biotechnol. Bioeng.* **15**:611–647.

Aller, R. C., 1980, Diagenetic processes near the sediment–water interface of Long Island Sound. 1. Decomposition and nutrient element geochemistry, *Adv. Geophys.* **22**:237–350.

Allison, P. A., 1986, Soft-bodied animals in the fossil record; the role of decay in fragmentation during transport, *Geology* **14**:979–981.

Allison, P. A., 1988a, The decay and mineralization of proteinaceous macrofossils, *Paleobiology* **14**:139–154.

Allison, P. A., 1988b, Soft-bodied squids from the Jurassic Oxford Clay, *Lethaia* **21**:403–410.

Allison, P. A., 1988c, Taphonomy and diagenesis of the Eocene London Clay biota, *Palaeontology* **31**:1079–1100.

Allison, P. A., 1988d, *Konservat-Lagerstätten:* Cause and classification, *Paleobiology* **14**:331–344.

Allison, P. A., 1990a, The origin of encrusting microbes on Eocene fossils, *Mem. Fac. Sci. Kochi Univ. Ser. E Geol.* **11**:23–34.

Allison, P. A., 1990b, Variation in rate of decay and disarticulation of Echinodermata: Implications for taphonomic experiments, *Palaios* **5**:432–440.

Allison, P. A., and Briggs, D. E. G., 1991, The taphonomy of soft-bodied animals, in: *Fossilization: the Processes of Taphonomy* (S. K. Donovan, ed.), Belhaven Press, pp. 120–140.

Allison, P. A., Smith, C. R., Demming J., Kukert, H., and Bennet, B., 1991, Deep water taphonomy of vertebrate carcasses: a whale skeleton in the Santa Catalina Basin, *Paleobiology* **17**:78–89.

Allison, P. A., Maeda, H., and Hamada, T., in preparation, Taphonomy of an exceptionally preserved Pleistocene lake biota.

Anderson, J. R. (ed.), 1987, *Muir's Textbook of Pathology*, Edward Arnold, London.
Anderton, R., 1985, Sedimentology of the Dinantian of Foulden, Berwickshire, Scotland, *Trans. R. Soc. Ed. Earth Sci.* **76**:7–12.
Andres, D., 1989, Phosphatisierte Fossilien aus dem unteren Ordoviz von Südschweden, *Berl. Geowiss. Abh. A* **106**:9–19.
Armstrong, W. G., Halstead, L. B., Reed, F. B., and Wood, L., 1983, Fossil proteins in vertebrate calcified tissues, *Phil. Trans. R. Soc. Lond. B* **301**:301–343.
Armstrong, W. G., Noel, P., and Urbanek, A., 1984, Collagen in the pterobranch coenecium and the problem of graptolite affinities, *Lethaia* **17**:145–152.
Baird, G. C., Shabica, C. W., Anderson, J. L., and Richardson, E. S., Jr., 1985a, Biota of a Pennsylvanian muddy coast: Habitats within the Mazonian Delta Complex, northern Illinois, *J. Paleontol.* **59**:253–281.
Baird, G. C., Sroka, S. D., Shabica, C. W., and Beard, T. L., 1985b, Mazon Creek-type fossil assemblages in the U.S. midcontinent Pennsylvanian: Their recurrent character and palaeoenvironmental significance, *Phil. Trans. R. Soc. Lond. B* **311**:87–98.
Baird, G. C., Sroka, S. D., Shabica, C. W., and Kuecher, G. J., 1986, Taphonomy of middle Pennsylvanian Mazon Creek area fossil localities, northeast Illinois: Significance of exceptional fossil preservation in syngenetic concretions, *Palaios* **1**:271–285.
Banks, H. P., Grierson, J. D., and Bonamo, P. M., 1985, The flora of the Catskill clastic wedge, *Geol. Soc. Am. Spec. Pap.* **201**:125–141.
Barale, G., Blanc-Louvel, C., Buffetaut, E., Courtinat, B., Peybernes, B., Via Boada, L., and Wenz, S., 1984, Les gisements de calcaires lithographiques du Crétacé inférieur du Montsech (province de Lérida, Espagne). Considérations paléoécologiques. *Geobios Mém. Spécial* **8**:275–283.
Bartels, C., and Brassel, G., 1990, *Fossilien im Hunsrückschiefer—Dokumente des Meereslebens im Devon*, Museum Idar-Oberstein.
Barthel, K. W., 1970, On the deposition of the Solnhofen lithographic limestone (Lower Tithonian, Bavaria, Germany), *Neues Jahrb. Geol. Paläontol. Abh.* **135**:1–18.
Barthel, K. W., 1978, *Solnhofen. Ein Blick in die Erdgeschichte*, Ott-Verlag, Thun.
Barthel, K. W., Swinburne, N. H. M., and Conway Morris, S., 1990, *Solnhofen—A Study in Mesozoic Palaeontology*, Cambridge University Press, Cambridge.
Bartram, K. M., Jeram, A. J., and Selden, P. A., 1987, Arthropod cuticles in coal, *J. Geol. Soc. Lond.* **144**:513–517.
Beltan, L., Calzada, S., Via Boada, L., and Villalta, J. F., 1989, On Ladinian biotas from northeastern Spain, in: *28th International Geological Congress, Washington, 1989, Abstracts*, Volume 1, pp. 121–122.
Benner, R. A., Macubbin, A. D., and Hodson, R. E., 1984, Anaerobic biodegradation of the lignin and polysaccharide components of lignocellulose and synthetic lignin by sediment microflora, *Appl. Environ. Microbiol.* **47**:998–1004.
Bergström, J., 1990, Hunsrück Slate, in *Palaeobiology: A Synthesis* (D. E. G. Briggs and P. R. Crowther, eds.), Blackwell, Oxford, pp. 277–279.
Berner, R. A., 1968, Calcium carbonate concretions formed by the decomposition of organic matter, *Science* **159**:195–197.
Berner, R. A., 1970, Sedimentary pyrite formation, *Am. J. Sci.* **268**:1–23.
Berner, R. A., 1980, *Early Diagenesis, a Theoretical Approach*, Princeton University Press, Princeton, New Jersey.
Berner, R. A., 1981, Authigenic mineral formation resulting from organic matter decomposition in modern sediments, *Fortschr. Mineral.* **59**:117–135.
Berner, R. A., 1984, Sedimentary pyrite, an update, *Geochim. Cosmochim. Acta* **48**:605–615.
Berner, R. A., and Raiswell, R., 1983, Burial of organic carbon and pyrite sulfur in sediments over Phanerozoic time: A new theory, *Geochim. Cosmochim. Acta* **47**:855–862.
Bernier, P. (ed.), 1985, *Cerin, une lagune tropicale au temps des dinosaures*, Centre Nationale de la Recherche Scientifique, Musée de Lyon, Lyon, France.
Boddy, L., 1983, Carbon dioxide release from decomposing wood: Effect of water content and temperature, *Soil Biol. Biochem.* **15**:501–510.

Bolton, H., 1905, Notes on the geological horizon and palaeontology of "Soapstone Bed" in the Lower Coal Measures, near Cole, Lancashire, *Geol. Mag.* **5:**433–444.
Bottjer, D. J., and Ausich, W. I., 1986, Phanerozoic development of tiering in soft substrata suspension-feeding communities, *Paleobiology* **12:**400–420.
Bottjer, D. J., Droser, M. L., and Jablonski, D., 1988, Palaeoenvironmental trends in the history of trace fossils, *Nature* **333:**252–255.
Boy, J. A., 1987, Die Tetrapoden-Lokalitäten des saarpfälzischen Rotliegenden (?Ober-Karbon–Unter-Perm; SW-Deutschland) und die Biostratigraphie der Rotliegend-Tetrapoden, *Mainz. Geowiss. Mitt.* **16:**31–65.
Boy, J. A., and Hartkopf, C., 1983, *Paläontologie des saarpfälzischen Rotliegenden* (Paläontologische Gesellschaft, 53. Jahresversammlung in Mainz, Exkursion C, 85 S., 17 Abb.), Frankfurt am Main, Germany.
Briggs, D. E. G., 1979, *Anomalocaris*, the largest known Cambrian arthropod, *Palaeontology* **22:**631–664.
Briggs, D. E. G., and Clarkson, E. N. K., 1983, The Lower Carboniferous Granton "shrimp-bed", Edinburgh, *Spec. Pap. Palaeontol.* **30:**161–177.
Briggs, D. E. G., and Clarkson, E. N. K., 1985, The Lower Carboniferous shrimp *Tealliocaris* from Gullane, East Lothian, Scotland, *Trans. R. Soc. Ed. Earth Sci.* **76:**173–201.
Briggs, D. E. G., and Gall, J.-C., 1990, The continuum in soft-bodied biotas from transitional environments: A quantitative comparison of Triassic and Carboniferous Konservat-Lagerstätten, *Paleobiology* **16:**204–218.
Briggs, D. E. G., and Mount, J. D., 1982, The occurrence of the giant arthropod *Anomalocaris* in the Lower Cambrian of southern California, and the overall distribution of the genus, *J. Paleontol.* **56:**1112–1118.
Briggs, D. E. G., and Robison, R. A., 1984, Exceptionally preserved nontrilobite arthropods and *Anomalocaris* from the Middle Cambrian of Utah, *Univ. Kansas Paleontol. Contrib.* **111:**1–23.
Briggs, D. E. G., and Rolfe, W. D. I., 1983, New Concavicarida (new Order: ?Crustacea) from the Upper Devonian of Gogo, Western Australia, and the palaeoecology and affinities of the group, *Spec. Pap. Palaeontol.* **30:**249–276.
Briggs, D. E. G., Bottrell, S. H., and Raiswell, R., 1990, Pyritization of soft-bodied fossils; isotope study of Beecher's Trilobite Bed, Late Ordovician, New York State, *Geol. Soc. Am. Abstr. Progr.* **22**(7):A81–82.
Briggs, D. E. G., Clark, N. D. L., and Clarkson, E. N. K., 1991, The Granton "shrimp-bed", Edinburgh—A Lower Carboniferous Konservät-Lagerstätte, *Trans. R. Soc. Ed. Earth Sci.* **82:**65–85.
Brooks, J., 1971, Some chemical and geochemical studies on sporopollenin, in: *Sporopollenin* (J. Brooks, P. R. Grant, M. Muir, P. van Gijzel, and G. Shaw, eds.), Academic Press, London, pp. 351–407.
Brothwell, D., 1987, *The Bog Man and the Archaeology of People*, Harvard University Press, Cambridge, Massachusetts.
Bucheim, P. H., and Surdam, R. R., 1981, Paleoenvironments and fossil fishes of the Laney Member, Green River Formation, Wyoming, in: *Communities of the Past* (J. Gray, A. J. Boucot, and W. B. N. Berry, eds.), pp. 415–452.
Bunnel, F. L., Maclean, S. F., Jr., and Brown, J., 1975, Barrow, Alaska, U.S.A., *Ecol. Bull.* **20:**73–124.
Butterfield, N. J., 1990, Organic preservation of non-mineralizing organisms and the taphonomy of the Burgess Shale, *Paleobiology* **16:**272–286.
Cahenzli, M., and Staffeldt, E. E., 1976, Microbial decomposition of animal muscle buried in desert soil, in: *Proceedings of the Third International Biodegradation Symposium*, Sessions IV, XIV (J. M. Sharpley and A. M. Kaplan, eds.), Applied Science Publishers, London, pp. 523–533.
Canfield, D. E., 1989, Sulfate reduction and oxic respiration in marine sediments: Implications for organic carbon preservation in euxinic environments, *Deep-Sea Res.* **36:**121–138.
Carlisle, D. B., 1964, Chitin in a Cambrian fossil *Hyclithellus*, *Biochem. J.* **90:**1–2.

Carpenter, F. M., and Burnham, L., 1985, The geological record of insects, *Annu. Rev. Earth Planet. Sci.* **13**:297–314.

Cater, J. M. L., 1987, Sedimentology of part of the Lower Oil-Shale Group (Dinantian) sequence at Granton, Edinburgh, including the Granton "shrimp-bed", *Trans. R. Soc. Ed. Earth Sci.* **78**:29–40.

Cater, J. M. L., Briggs, D. E. G., and Clarkson, E. N. K., 1989, Shrimp-bearing sedimentary successions in the Lower Carboniferous (Dinantian) Cementstone and Oil Shale Groups of northern Britain, *Trans. R. Soc. Ed. Earth Sci.* **80**:5–15.

Chen, J.-Y., Hou, X.-G., and Lu, H.-Z., 1989, Early Cambrian netted scale-bearing worm-like sea animal, *Acta Palaeontol. Sin.* **28**(1):1–16.

Cisne, J. L., 1973, Beecher's Trilobite Bed revisited: Ecology of an Ordovician deepwater fauna, Postilla **160**:25 pp.

Clark, N. D. L., 1989, A study of a Namurian crustacean-bearing shale from the western Midland Valley of Scotland, Ph. D. thesis, University of Glasgow.

Clarkson, E. N. K., 1985, Palaeoecology of the Dinantian of Foulden, Berwickshire, Scotland, *Trans. R. Soc. Ed. Earth Sci.* **76**:97–100.

Coleman, M. L., 1985, Geochemistry of diagenetic non-silicate minerals: Kinetic considerations, *Phil. Trans. R. Soc. Lond.* **315A**:39–54.

Coleman, M. L., and Raiswell, R., 1981, Carbon, oxygen and sulphur isotope variations in concretions from the Upper Lias of N. E. England, *Geochim. Cosmochim. Acta* **45**:329–340.

Connolly, R. C., Evershed, R. P., Embery, G., Stanbury, J. B., Green, D., Beahan, P., and Shortall, J. B., 1986, The chemical composition of some body tissue, in: *Lindow Man: The Body in the Bog* (I. M. Stead, J. B. Bourke, and D. Brothwell, eds.), British Museum, London, pp. 72–77.

Conway Morris, S., 1981, The nature and interpretation of fossils, in: *The Cambridge Encyclopaedia of Earth Sciences* (D. G. Smith, ed.), Cambridge University Press, Cambridge, pp. 363–370.

Conway Morris, S., 1985, Cambrian Lagerstätten; their distribution and significance, *Phil. Trans. R. Soc. Lond. B* **311**:49–65.

Conway Morris, S., 1986, The community structure of the Middle Cambrian phyllopod bed (Burgess Shale), *Palaeontology* **29**:423–458.

Conway Morris, S., 1989a, The persistence of Burgess Shale-type faunas: Implications for the evolution of deeper-water faunas, *Trans. R. Soc. Ed. Earth Sci.* **80**:271–283.

Conway Morris, S., 1989b, Burgess Shale faunas and the Cambrian explosion, *Science* **246**:339–346.

Conway Morris, S., 1990, Late Precambrian and Cambrian soft-bodied faunas, *Annu. Rev. Earth Planet. Sci.* **18**:101–122.

Conway Morris, S., and Jenkins, R. J. F., 1985, Healed injuries in Early Cambrian trilobites from South Australia, *Alcheringa* **9**:167–177.

Conway Morris, S., and Peel, J. S., 1990, Articulated halkieriids from the Lower Cambrian of north Greenland, *Nature* **345**:802–805.

Conway Morris, S., and Robison, R. A., 1986, Middle Cambrian priapulids and other soft-bodied fossils from Utah and Spain, *Univ. Kansas Paleontol. Contrib.* **117**:22 pp.

Conway Morris, S., Peel, J. S., Higgins, A. K., Soper, N. J., and Davis, N. C., 1987, A Burgess Shale-like fauna from the Lower Cambrian of North Greenland, *Nature* **326**:181–183.

Cressey, R., and Boxshall, G., 1989, *Kabatarina pattersoni*, a fossil parasitic copepod (Dichelesthiidae) from a Lower Cretaceous fish, *Micropaleontology* **35**:150–167.

Criss, R. E., Cooke, G. A., and Day, S. D., 1988, An organic origin for the carbonate concretions of the Ohio Shale, *U. S. Geol. Surv. Bull.* **1836**:21 pp.

Crowther, P. R., and Rickards, R. B., 1977, Cortical bandages and the graptolite zooid, *Geol. Palaeontol.* **11**:9–46.

Crowther, P. R., Rickards, R. B., and Urbanek, A., 1987, Possible cellular tissue in an Ordovician graptoblast, *Geol. Mag.* **124**:67–72.

Curtis, C. D., 1980, Diagenetic alteration in black shales, *J. Geol. Soc. Lond.* **137**:189–194.

De Fourcroy, A. F., 1790, Mémoire sur les differens étâts des cadavres trouvés dans les fiuilles du cimetiere des Innocens en 1786 et 1787, *Ann. Chemie* **5**:165–185.

Den Dooren de Jong, L. E., 1961, On the formation of adipocere from fats: Contribution to the microbiology of systems containing two liquid phases, *Antonie van Leeuwenhoek* **27**:337–361.

Desborough, G. A., 1978, A biogenic-chemical stratified lake model for the origin of oil shale of the Green River Formation: An alternative to the playa-lake model, *Geol. Soc. Am. Bull.* **89**:961–971.

Dietl, G. von, and Mundlos, R., 1972, Ökologie und Biostratinomie von *Ophiopinna elegans* (Ophiuroidea) aus dem Untercallovium von la Voulte (Südfrankreich), *Neues Jahrb. Geol. Paläontol. Monatsh.* **8**:449–464.

Dietrich, H.-G., 1975, Zur entstehung und erhaltung von Bernstein-Lagerstätten—1: Allgemeine aspekte, *Neues Jahrb. Geol. Paläontol. Abh.* **149**:39–72.

Dilcher, D. L., 1965, Epiphyllous fungi from Eocene deposits in Western Tennessee, U.S.A., *Palaeontographica B* **116**:1–54.

Donovan, D. T., 1983, *Mastigophora* Owen 1856: A little known genus of Jurassic coleoids, *Neues Jahrb. Geol. Paläontol. Abh.* **165**:484–495.

Dorjes, J., 1972, Distribution and zonation of macrobenthic animals, *Senckenb. Marit.* **4**:183–216.

Driscoll, E. G., and Swanson, R. A., 1973, Diversity and structure of epifaunal communities on mollusc valves, Buzzard Bay, Massachusetts, *Palaeogeogr. Palaeoclimatol. Palaeoecol.* **14**:229–247.

Emmerson, S., and Hedges, J. I., 1988, Processes controlling the organic carbon content of open ocean sediments, *Paleoceanography* **3**:621–634.

Ennever, J., Streckfuss, J. L., and Goldschmidt, M. C., 1981, Calcifiability comparison among selected micro-organisms, *J. Dent. Res.* **60**:1793–1796.

Fischer, J.-C., and Riou, B., 1982a, Les teuthoîdes (Cephalopoda, Dibranchiata) du Callovien inférieur de la Voulte-sur-Rhône (Ardèche, France), *Ann. Paléontol. (Vertebr.-Invertebr.)* **68**:295–325.

Fischer, J.-C., and Riou, B., 1982b, Le plus octopode connu (Cephalopoda, Dibranchiata): *Proteroctopus ribeti* nov. gen., nov. sp., du Callovien de l'Ardèche (France), *C. R. Acad. Sci. Paris II* **295**:277–280.

Foree, E. G., and McCarty, P. D., 1970, Anaerobic decomposition of algae, *Environ. Sci. Technol.* **4**:842–849.

Franzen, J. L., 1990, Grube Messel, in *Palaeobiology: A Synthesis* (D. E. G. Briggs and P. R. Crowther, eds.), Blackwell, Oxford, pp. 289–294.

Franzen, J. L., and Michaelis; W. (eds.), 1988, Der eozäne Messelsee—Eocene Lake Messel, *Cours Forschungsinst. Senckb.* **107**:1–452.

Fritz, W. H., 1990, In defense of the escarpment near the Burgess Shale fossil locality, *Geosci. Can.* **17**:106–110.

Gall, J.-C., 1971, Faunes et paysages du Grès à Voltzia du nord des Vosges. Essai paleoecologique sur le Buntsandstein Supérieur, *Mém. Serv. Carte Géol. Alsace Lorraine* **34**:1–318.

Gall, J.-C., 1985, Fluvial depositional environment evolving into deltaic setting with marine influences in the Buntsandstein of northern Vosges (France), in: *Aspects of Fluvial Sedimentation in the Lower Triassic Buntsandstein* (D. Mader, ed.), Springer-Verlag; Berlin, pp. 449–477.

Gall, J.-C., 1990, Les voiles microbiens: Leur contribution à la fossilisation des organismes au corps mou, *Lethaia* **23**:21–28.

Gall, J.-C., and Grauvogel-Stamm, L., 1984, Genèse des gisements fossilifères du Grès à Voltzia (Anisien) du nord des Vosges (France), *Géobios Mém. Special* **8**:293–297.

Gall, J.-C., Bernier, P., Gaillard, C., Barale, G., Bourseau, J.-P., Buffetaut, E., and Wenz, S., 1985, Influence du développement d'un voile algaire sur la sédimentation et la taphonomie des calcaires lithographiques. Exemple du gisement de Cerin (Kimmeridgien supérieur, Jura méridional francais), *C. R. Acad. Sci. Paris II* **301**:547–552.

Glaessner, M. F., 1979, Lower Cambrian Crustacea and annelid worms from Kangaroo Island, South Australia, *Alcheringa* **3**:21–31.

Glaessner, M. F., 1984, *The Dawn of Animal Life. A Biohistorical Study*, Cambridge University Press, Cambridge.

Glob, P. V. 1969, *The Bog People*, Faber, London.

Golenberg, E. M., Giannasi, D. E., Clegg, M. T., Smiley, C. J., Durbin, M., Henderson, D., and Zurawski, G., 1990, Chloroplast DNA sequence from a Miocene *Magnolia* species, *Nature* **344**:656–658.

Gordon, I., Shapiro, H. A., and Berson, S. D., 1988, *Forensic Medicine: A Guide to Principles*, 3rd ed., Churchill Livingston, Edinburgh.

Gore, P. J. W., 1986, Triassic notostracans in the Newark Supergroup, Culpeper Basin, northern Virginia, *J. Paleontol.* **60**:1086–1096.

Gore, P. J. W., 1988a, Lacustrine sequences in an early Mesozoic rift basin: Culpeper Basin, Virginia, U.S.A., in: Lacustrine Petroleum Source Rocks (A. J. Fleet, K. Kelts, and M. R, Talbot, eds.), *Geol. Soc. Lond. Spec. Publ.* **40**:247–278.

Gore, P. J. W., 1988b, Paleoecology and sedimentology of a Late Triassic Lake, Culpeper Basin, Virginia, U.S.A., *Palaeogeogr. Palaeoclimatol. Palaeoecol.* **62**:593–608.

Grande, L., 1984, Paleontology of the Green River Formation, with a Review of the Fish Fauna, Geological Survey of Wyoming Bulletin No. 63, 2nd edn.

Grande, L., 1989, The Eocene Green River lake system, Fossil Lake, and the history of the North American fish fauna, in: *Mesozoic/Cenozoic Vertebrate Paleontology: Classic Localities, Contemporary Approaches* (J. Flynn, ed.), 28th International Geological Congress guidebook T322, American Geophysical Union, Washington, D.C., pp. 18–28.

Grimaldi, D. A. (ed.), 1990, Insects from the Santana Formation, Lower Cretaceous, of Brazil, *Bull. Am. Mus. Nat. Hist.* **195**:1–191.

Guthrie, R. D., 1990, *Frozen Fauna of the Mammoth Steppe. The Story of Blue Babe*, Chicago University Press, Chicago.

Hamilton, R. F. M., and Trewin, N. H., 1988, Environmental controls on fish faunas of the Middle Devonian Orcadian Basin, in: *Devonian of the World, Proceedings of the Second International Symposium on the Devonian System, Calgary, Canada* (N. J. McMillan, A. F. Embry, and D. J. Glass, eds.), Canadian Society of Petroleum Geologists, pp. 589–600.

Hannibal, J. T., 1989, The Cleveland Shale and associated strata exposed in the vicinity of Cleveland, Ohio, in: *Seventh Mid-Continent Paleobotany Colloquium, Fieldtrip Guidebook*, pp. 1–12.

Harland, W. B., Armstrong, R. L., Cox, A. V., Craig, L. E., Smith, A. G., and Smith, D. G., 1990, *A Geologic Time Scale 1989*, Cambridge University Press, Cambridge.

Hauff, B., and Hauff, R. B., 1981, *Das Holzmadenbuch*, Hauff and Hauff, Holzmaden.

Hachiya, K., Yamaoka, M., and Mizuno, Y., 1988, Deep sea fauna from the Middle Miocene Morozaki Group in the Chita Peninsula, Aichi Prefecture, Central Japan. *J. Growth* **27**:119–139.

Hecht, F., 1933, Der Verbleib der Organischen Substanz der Tiere bei Meerischer Einbettung, *Senckenbergiana* **15**:165–249.

Hedges, J. I., Cowie, G. L., Ertel, J. W., Barbour, R. J., and Hatcher, P. G., 1985, Degradation of carbohydrates and lignins in buried woods, *Geochim. Cosmochim. Acta* **49**:701–711.

Hemleben, C., and Freels, D., 1977a, Algen-laminierte und gradierte Plattenkalke in der Overkreide Dalmatiens (Jugoslawien), *Neues Jahrb. Geol. Paläontol. Abh.* **154**:61–93.

Hemleben, C., and Freels, D., 1977b, Fossilführende dolomitisierte Plattenkalke aus dem "Muschelkalk superior" bei Montral (Prov. Tarragona, Spanien), *Neues Jahrb. Paläontol. Abh.* **154**:186–212.

Henrichs, S. M., and Reeburgh, W. S., 1987, Anaerobic mineralization of marine sediment organic matter: Rates and the role of anaerobic processes in the oceanic carbon economy, *Geomicrobiol. J.* **5**:191–237.

Heyler, D., 1986, Table ronde internationale du C.N.R.S. sur le gisement Stéphanien de Montceau-les-Mines. Conclusions, *Bull. Trimest. Soc. Hist. Nat. Amis Mus. Autun* **117**:161–173.

Heyler, D., and Poplin, C., 1988, The fossils of Montceau-les-Mines, *Sci. Am.* **256**:104–110.

Hinz, I., 1987, The Lower Cambrian microfauna of Comley and Rushton, Shropshire, England, *Palaeontogr. Abt. A* **198**:41–100.

Horner, J. R., 1985, Stratigraphic position of the Bear Gulch Limestone (Lower Carboniferous) of central Montana, in: *Compte Rendu Neuvième Congrès International de Stratigraphie et de Géologie du Carbonifère*, Volume 5, Southern Illinois University Press, Carbondale, pp. 427–436.

Horner, J. R., and Lund, R., 1985, Biotic distribution and diversity in the Bear Gulch Limestone of central Montana, in: *Compte Rendu Neuvième Congrès International de Stratigraphie et de Géologie du Carbonifère*, Volume 5, Southern Illinois University Press, Carbondale, pp. 437–442.

Hou, X.-G., 1989, Features and significance of Chengjiang fauna: Exceptionally preserved soft-bodied fauna from Lowest Cambrian in Yunnan, China, in: *28th International Geological Congress Abstracts*, Volume 2, Washington, D.C., p. 75.

Hou, X.-G., and Sun, W.-G., 1988, Discovery of Chengjiang fauna at Meishucun, Jinning, Yunnan, *Acta Palaeontol. Sin.* **27:**1–12.

Hückel, U., 1970, Die Fischschiefer von Haqel und Hjoula in der Oberkreide des Libanon, *Neues Jahrb. Geol. Paläontol. Abh.* **135:**113–149.

Hueck, H. J., 1971, Biodeterioration and environmental pollution, *Int. Biodetn. Bull.* **7:**81–85.

Jahnke, R. S., Emerson, S., and Murray, J., 1982, A model of oxygen reduction, denitrification, and organic matter mineralization in marine sediments, *Limnol. Oceanogr.* **27:**610–623.

Janssen, W., 1984, *Forensic Histopathology*, Springer-Verlag, Berlin.

Jenkyns, H. C., 1980, Cretaceous anoxic events: From continents to oceans, *J. Geol. Soc. Lond.* **137:**171–188.

Jones, G. F., 1969, The benthic macrofauna of the mainland shelf of Southern California, *Allan Hancock Monogr. Mar. Biol.* **4:**1–219.

Jörg, E., 1966, Die Öhninger Steinbrüche, in: *Dorf und Stift Öhningen* (H. Berner, ed.), Ortsangabe, pp. 29–38.

Jørgenson, B. B., 1977, Bacterial sulfate reduction within reduced microniches of oxidized marine sediments, *Mar. Biol.* **41:**7–17.

Kaljo, D., 1970, *Silur Estonii*, Institut Geologii Akademii Nauk Estonickoi CCP, Tallin.

Kaljo, D., 1972, Facies control of the faunal distribution in the Silurian of the Baltic region, in: *24th International Geological Congress*, Volume 7, pp. 544–548, Montreal, Canada.

Kauffman, E. G., 1979, Benthic environments and paleoecology of the Posidonienschiefer (Toarcian), *Neues Jahrb. Geol. Paläontol. Abh.* **157:**18–36.

Kauffman, E. G., 1981, Ecological reappraisal of the German Posidonienschiefer (Toarcian) and the stagnant basin model, in: *Communities of the Past* (J. Gray, A. J. Boucot, and W. B. N. Berry, eds.), Hutchinson Ross, Stroudsburg, Pennsylvania, pp. 311–381.

Kenrick, P., and Edwards, D. 1988, The anatomy of Lower Devonian *Gosslingia breconensis* Heard based on pyritized axes, with some comments on the permineralization process, *Bot. J. Linn. Soc.* **97:**95–123.

Keupp, H., 1977, Ultrafazies und genese der Solnhofer Plattenkalke (Oberer Malm, südliche Frankenalb), *Abh. Naturhist. Ges. Nürnberg* **37:**1–128.

Kidston, R., and Lang, W. H., 1917, An Old Red Sandstone plant showing structure from the Rhynie Chert Bed, Aberdeenshire, *Trans. R. Soc. Ed.* **51:**761–784.

Kidwell, S. M., and Baumiller, T., 1990, Experimental disintegration of regular echinoids: Roles of temperature, oxygen and decay thresholds, *Paleobiology* **16:**247–271.

Knoll, A. H., 1985, Exceptional preservation of photosynthetic organisms in silicified carbonates and silicified peats, *Phil. Trans. R. Soc. Lond. B* **311:**111–122.

Koenigswald, W. von (ed.), 1989, *Fossillagerstätte Rott bei Hennef am Siebengebirge*, Rheinlandia Verlag, Siegburg.

Koshimizu, T., 1984, Neogene avian, amphibian and arthropod fossils collected from the Kabutoiwa Plant Bed in Nagano and Gunma Prefectures of Japan, *Geosci. Mag. Nippon Chigaku Kenkyu-Kai* **35:**73–87.

Krumbiegel, G., Ruffle, L., and Haubold, H., 1983, *Das eozäne Geiseltal, ein mitteleuropaisches Braunkohlenvorkommen und seine Pflanzen- und Tierwelt*, Die Neue Brehm-Bucherei, A. Ziemsen Verlag, Wittenberg, Lutherstadt.

Langenheim, J., 1990, Plant resins, *Am. Sci.* **78:**16–24.

Langenheim, J., and Beck, C. W., 1965, Infrared spectra as a means of determining botanical sources of amber, *Science* **149:**52–55.

Langer, W., 1961, Über das Alter der Fischschiefer von Hvar-Lesina (Dalmatien), *Neues Jahrb. Geol. Paläontol. Monatsh.* **1961:**329–331.

Larsson, S. G., 1978, *Baltic Amber—A Paleobiological Study*, Scandanavian Science Press, Klampenborg, Denmark.

Lawrence, D. R., 1968, Taphonomy and information loss in fossil communities, *Bull. Geol. Soc. Am.* **79:**1315–1330.

Ludvigsen, R., 1989, The Burgess Shale: Not in the shadow of the Cathedral Escarpment, *Geosci. Can.* **16:**51–59.

Ludvigsen, R., 1990, Reply to comments by Fritz *and* Aitken and McIlreath, *Geosci. Can.* **17:**116–118.

Mabesoone, J. M., and Tinoco, I. M., 1973, Palaeoecology of the Aptian Santana Formation (Northeastern Brazil), *Paleogeogr. Palaeoclimatol. Palaeoecol.* **14:**97–118.

Macdonald, K. B., 1976, Paleocommunities: Towards some confidence limits, in: *Structure and Classification of Paleocommunities* (R. W. Scott and R. R. West, eds.), Dowden, Hutchinson and Ross, Stroudsburg, Pennsylvania, pp. 87–106.

MacGinitie, H. D., 1953, Fossil plants of the Florissant beds, Colorado, *Carnegie Inst. Wash. Publ. Contrib. Paleontol.* **599:**1–198.

Maisey, J. G., 1988, Diagenetic features and the origin of calcareous concretions from the Santana Formation (N.E. Brazil), *J. Vertebr. Paleontol.* **8:**21A.

Maisey, J. G., ed., 1991, *Santana Fossils: An Illustrated Atlas*, Tropical Fish Hobbyist, Neptune City, N.J.

Mant, A. K., 1957, Adipocere—A review, *J. Forensic Med.* **1:**18–35.

Mant, A. K., 1984, Post-mortem changes, in: *Taylor's Principles and Practice of Medical Jurisprudence*, 13th ed. (A. K. Mant, ed.), Churchill Livingston, Edinburgh, pp. 128–155.

Mant, A. K., 1987, Knowledge acquired from post-war exhumations, in: *Death, Decay and Reconstruction—Approaches to Archaeology and Forensic Science* (A. Boddington, A. N. Garland, and R. C. Janaway, eds.), Manchester University Press, pp. 65–78.

Mapes, G., and Mapes, R. H. (eds.), 1988, Regional Geology and Paleontology of the Upper Paleozoic Hamilton Quarry Area in Southeastern Kansas, Kansas Geological Survey Guidebook Series, No. 6.

Martill, D. M., 1986, The stratigraphic distribution and preservation of fossil vertebrates in the Oxford Clay of England, *Mercian Geol.* **10:**161–188.

Martill, D. M., 1987a, Prokaryote mats replacing soft-tissues in Mesozoic marine reptiles, *Mod. Geol.* **11:**265–269.

Martill, D. M., 1987b, A taphonomic and diagenetic case study of a partially articulated ichthyosaur, *Palaeontology* **30:**543–555.

Martill, D. M., 1988, Preservation of fish in the Cretaceous Santana Formation of Brazil, *Palaeontology* **31:**1–18.

Mason, J. K., 1983, *Forensic Medicine for Lawyers*, Butterworths, London.

Massari, F., and Sorbini, L., 1975, Aspects sedimentologiques des couches à poissons de L'Eocene de Bolca (Verone-Nord Italie), in: *9th Congres International de Sedimentologie*, pp. 55–61, Nice, France.

Mathur, A. C., 1977, Über Ammonien der Kössener Schichten und der Nachweis der Tintenbeutel-Substanz Melanin, Doctoral dissertation, University of Munich.

Maynard, J. B., 1982, Extension of Berner's "New geochemical classification of sedimentary environments" to ancient sediments, *J. Sediment. Petrol.* **52:**1325–1331.

McAlpine, J. F., and Martin, J. E. H., 1969, Canadian amber, a palaeontological treasure-chest, *Can. Entomol.* **101**(8):819–839.

McLeroy, C. A., and Anderson, R. Y., 1966, Laminations in the Oligocene Florissant Lake deposits, Colorado, *Geol. Soc. Am. Bull.* **77:**605–618.

Meischner, D., and Paul, J., 1982, Die pliozäne Fossilfundstätte Naturdenkmal Tongrube Willershausen, *Cours Forschungsinst. Senckenb.* **56:**147–152.

Mikulic, D. G., Briggs, D. E. G., and Kluessendorf, J., 1985a, A Silurian soft-bodied biota, *Science* **228:**715–717.

Mikulic, D. G., Briggs, D. E. G., and Kluessendorf, J., 1985b, A new exceptionally preserved biota from the Lower Silurian of Wisconsin, U.S.A., *Phil. Trans. R. Soc. Lond. B* **311**:75–85.

Miller, M. F., and Byers, C. W., 1984, Abundant and diverse early Paleozoic infauna indicated by the stratigraphic record, *Geology* **12**:40–43.

Mills, J. S., White, R., and Gough, L. J., 1984, The chemical composition of Baltic amber, *Chem. Geol.* **47**:15–39.

Mittmeyer, H.-G., 1980, Vorläufige gesamtliste der Hunsrückschiefer-fossilien, *Kleine Senckenb. Reihe* **11**:34–39.

Müller, K. J., 1982, Weichteile von Fossilien aus dem Erdaltertum, *Naturwissenschaften* **69**:249–254.

Müller, K. J., 1985, Exceptional preservation in calcareous nodules, *Phil. Trans. R. Soc. Lond. B* **311**:67–73.

Müller, K. J., and Walossek, D., 1986, *Martinssonia elongata* gen. et sp. n., a crustacean-like euarthropod from the Upper Cambrian "Orsten" of Sweden, *Zool. Scripta* **15**:73–92.

Müller, K. J., and Walossek, D., 1987, Morphology, ontogeny, and life-habit of *Agnostus pisiformis* (Linnaeus, 1757) from the Upper Cambrian of Sweden, *Fossils Strata* **19**:1–124.

Müller, K. J., and Walossek, D., 1988, External morphology and larval development of the Upper Cambrian maxillopod *Bredocaris admirabilis, Fossils Strata* **23**:1–70.

Mykura, W., 1983, Old Red Sandstone, in: *Geology of Scotland* (G. Y. Craig, ed.), Scottish Academic Press, Edinburgh, pp. 205–251.

Nedwell, D. B., and Floodgate, G. D., 1971, The seasonal selection by temperature of heterotrophic bacteria in an intertidal sediment, *Mar. Biol.* **1**:306–310.

Nitecki, M. H. (ed.), 1979, *Mazon Creek Fossils,* Academic Press, New York.

Oji, T., 1990, Miocene Isocrinidae (stalked crinoids) from Japan and their biogeographic implication, *Trans. Proc. Palaeontol. Soc. Japan N.S.* **157**:412–429.

Onoe, T., 1989, Palaeoenvironmental analysis based on the Pleistocene Shiobara flora in the Shiobara volcanic basin, central Japan, *Geol. Surv. Japan Rep.* **269**:1–207.

Pååbo, S., Higuchi, R. G., and Wilson, A. C., 1989, Ancient DNA and the polymerase chain reaction, *J. Biol. Chem.* **264**:9709–9712.

Palmer, A. R., 1957, Miocene arthropods from the Mojave Desert, California, *U. S. Geol. Surv. Spec. Pap.* **294-G**:237–280.

Palmer, A. R., 1960, Some aspects of the early Upper Cambrian stratigraphy of White Pine County, Nevada and vicinity, in: *Geology of East Central Nevada. Intermountain Association Petroleum Geologists, 11th Annual Field Conference, 1960 Guide Book,* pp. 53–58.

Park, L. E., 1990, Stratigraphic context and paleoenvironmental significance of arthropod-bearing concretions of the Barstow Formation, southern California, *Geol. Soc. Am. Abstr. Progr.* **22**(7):A364.

Parkes, R. J., Cragg, B. A., Fry, J. C., Herbert, R. A., and Wimpenny, J. W. T., 1990, Bacterial biomass and activity in deep sediment layers from the Peru margin, *Phil. Trans. R. Soc. Lond. A* **331**:139–153.

Payne, J. A., 1965, A summer carrion study of the baby pig *Sus scrofa* Linnaeus, *Ecology* **46**:592–602.

Payne, J. A., and King, E. W., 1972, Insect succession and decomposition of pig carcasses in water, *J. Georgia Entomol. Soc.* **7**:153–162.

Pinna, G., 1985, Exceptional preservation in the Jurassic of Osteno, *Phil. Trans. R. Soc. Lond. B* **311**:171–180.

Piper, D. J. W., 1972, Sediments of the Middle Cambrian Burgess Shale, *Lethaia* **5**:169–175.

Plotnick, R. E., 1986, Taphonomy of a modern shrimp: Implications for the arthropod fossil record, *Palaios* **1**:286–293.

Poinar, G. O., and Hess, R., 1982, Ultrastructure of 40-million-year-old insect tissue, *Science* **215**:1241–1242.

Redfield, A. C., 1958, The biological control of chemical factors in the environment, *Am. Sci.* **48**:206–226.

Rice, C. M., and Trewin, N. H., 1988, A Lower Devonian gold-bearing hot spring system, Rhynie, Scotland, *Trans. Inst. Min. Metall. B: Appl. Earth Sci.* **97**:141–144.

Richardson, E. S., Jr., and Johnson, R. G., 1971, Mazon Creek faunas, in: *Proceedings 1st North American Paleontological Convention*, Volume I, pp. 1222–1235.

Rickards, R. B., and Stait, B. A., 1984, *Psigraptus*, its classification, evolution and zooid, *Alcheringa* **8**:101–111.

Riegraf, W., 1984, *Der Posidonienschiefer. Biostratigraphie, Fauna und Fazies des südwestdeutschen Untertoarciums*, Enke, Stuttgart.

Ritchie, A., 1968, New evidence on *Jamoytius kerwoodi* White, an important ostracoderm from the Silurian of Lanarkshire, Scotland, *Palaeontology* **11**:21–39.

Ritchie, A., 1985, *Ainiktozoon loganense* Scourfield, a protochordate? from the Silurian of Scotland, *Alcheringa* **9**:117–142.

Robison, R. A., 1960, Lower and Middle Cambrian stratigraphy of the eastern Great Basin, in: *Geology of East Central Nevada. Intermountain Association of Petroleum Geologists, 11th Annual Field Conference, 1960 Guide Book*, pp. 43–52.

Roger, J., 1946, Les invertébrés des couches à poissons du Crétacé supérieur du Liban, *Mém. Soc. Géol. Fr. N. S. 23* **51**:1–92.

Rogers, J. C., 1984, Depositional environments of paleoecology of two quarry sites in the Middle Cambrian Marjum and Wheeler formations, House Range, Utah, *Brigham Young Univ. Geol. Stud.* **31**:97–115.

Rolfe, W. D. I., 1962, The cuticle of some Middle Silurian ceratiocaridid Crustacea from Scotland, *Palaeontology* **5**:30–51.

Rolfe, W. D. I., 1973, Silurian arthropod and fish assemblages from Lesmahagow, Lanarkshire, in: *Excursion Guide to the Geology of the Glasgow District* (B. J. Bluck, ed.), Geological Society of Glasgow, pp. 119–126.

Rolfe, W. D. I., 1988, Early life on land—The East Kirkton discoveries, *Earth Sci. Conserv.* **25**:22–28.

Rolfe, W. D. I., Schram, F. R., Pacaud, G., Sotty, D., and Secretan, S., 1982, A remarkable Stephanian biota from Montceau-les-Mines, France, *J. Paleontol.* **56**:426–428.

Rolfe, W. D. I., Durant, G. P., Fallick, A. E., Hall, A. J., Large, D. J., Scott, A. C., Smithson, T. R., and Walkden, G. M., 1990, An early terrestrial biota preserved by Visean vulcanicity in Scotland, *Geol. Soc. Am. Spec. Pap.* **244**:13–24.

Rubin, E., and Farben, J. W., 1988, *Pathology*, Lippincott, London.

Savrda, C. E., and Bottjer, D. J., 1987, The exaerobic zone, a new oxygen-deficient marine biofacies, *Nature* **327**:54–56.

Schaal, S., and Ziegler, W. (eds.), 1988, *Messel—Ein schaufenster in die Geschichte der Erde und des Lebens*, Verlag Waldemar Kramer, Frankfurt am Main.

Schäfer, W., 1962, *Aktuo-paläontologie nach Studien in der Nordsee*, Kramer, Frankfurt am Main.

Schäfer, W., 1972, *Ecology and Paleoecology of Marine Environments*, University of Chicago Press, Chicago.

Schairer, G., and Janicke, V., 1970, Sedimentologisch-paläontologische Untersuchungen an den Plattenkalken der Sierra de Montsech (Prov. Lérida, NE-Spanien), *Neues Jahrb. Geol. Paläontol. Abh.* **135**:171–189.

Schlüter, T., 1990, Baltic amber, in: *Palaeobiology: A Synthesis* (D. E. G. Briggs and P. R. Crowther, eds.), Blackwell, Oxford, pp. 294–297.

Schopf, T. J. M., 1978, Fossilization potential of an intertidal fauna: Friday Harbor, Washington, *Paleobiology* **4**:261–270.

Schram, F. R., 1979, The Mazon Creek biotas in the context of a Carboniferous faunal continuum, in: *Mazon Creek Fossils* (M. H. Nitecki, ed.), Academic Press, New York, pp. 159–190.

Schram, F. R., 1983, Lower Carboniferous biota of Glencartholm, Eskdale, Dumfriesshire, *Scot. J. Geol.* **19**:1–15.

Schultze, H.-P., 1989, Three-dimensional preservation in Jurassic fishes of Chile, *Rev. Geol. Chile* **16**(2):183–215.

Seilacher, A., 1970, Begriff and bedeutung der Fossil-Lagerstätten, *Neues Jahrb. Geol. Paläontol. Abh.* **1970:**34–39.

Seilacher, A., 1977, Evolution of trace fossil communities, in: *Patterns of Evolution* (A. H. Hallam, ed.), Elsevier, Amsterdam, pp. 359–376.

Seilacher, A., 1982, Ammonite shells as habitats in the Posidonia Shales of Holzmaden—Floats or benthic islands?, *Neues Jahrb. Geol. Paläontol. Monatsh.* **1982:**98–114.

Seilacher, A., 1989, Vendozoa: Organismic contruction in the Proterozoic biosphere, *Lethaia* **22:**229–239.

Seilacher, A., Reif, W.-E., and Westphal, F., 1985, Sedimentological, ecological and temporal patterns of fossil Lagerstätten, *Phil. Trans. R. Soc. Lond. B* **311:**5–23.

Selden, P. A., 1981, Functional morphology of the prosoma of *Baltoeurypterus tetragonophthalmus* (Fischer) (Chelicerata: Eurypterida), *Trans. R. Soc. Ed. Earth Sci.* **72:**9–48.

Selden, P. A., 1990, Lower Cretaceous spiders from the Sierra de Montsech, north-east Spain, *Palaeontology* **33:**257–285.

Sepkoski, J. J., Jr., and Hulver, M. L., 1985, An atlas of Phanerozoic clade diversity diagrams, in: *Phanerozoic Diversity Patterns, Profiles in Macroevolution* (J. W. Valentine, ed.), Princeton University Press, Princeton, New Jersey, pp. 11–39.

Sepkoski, J. J., Jr., and Sheehan, P. M., 1983, Diversification, faunal change, and community replacement during the Ordovician radiations, in: *Biotic Interactions in Recent and Fossil Benthic Communities* (M. J. S. Tevesz and P. L. McCall, eds.), Plenum Press, New York, pp. 673–717.

Shear, W. A., Bonamo, P. M., Grierson, J. D., Rolfe, W. D. I., Smith, E. L., and Norton, R. A., 1984, Early land animals in North America: Evidence from Devonian age arthropods from Gilboa, New York, *Science* **224:**492–494.

Shear, W. A., Selden, P. A., Rolfe, W. D. I., Bonamo, P. M., and Grierson, J. D., 1987, New terrestrial arachnids from the Devonian of Gilboa, New York (Arachnida, Trigonotarbida), *Am. Mus. Novit.* **2901:**1–74.

Sheehan, P. M., and Schiefelbein, D. R. J., 1984, The trace fossil *Thalassinoides* from the Upper Ordovician of the eastern Great Basin: Deep burrowing in the early Paleozoic, *J. Paleontol.* **58:**440–447.

Sigleo, A. C., 1978, Organic geochemistry of silicified wood, Petrified Forest, National Park, Arizona, *Geochim. Cosmochim. Acta* **42:**1397–1405.

Simpson, K., and Knight, B., 1985, *Forensic Medicine*, Edward Arnold, London.

Smiley, C. J. (ed.), 1985, *Late Cenozoic History of the Pacific Northwest*, Pacific Division, AAAS, San Francisco.

Smiley, C. J., and Rember, W. C., 1985, Physical setting of the Miocene Clarkia Fossil Beds, northern Idaho, in: *Late Cenozoic history of the Pacific Northwest* (C. J. Smiley, ed.), Pacific Division, AAAS, San Francisco, pp. 11–31.

Smiley, C. J., Gray, J., and Huggins, L. M., 1975, Preservation of Miocene fossils in unoxidized lake deposits, Clarkia, Idaho, *J. Paleontol.* **49:**833–844.

Smith, C. R., 1985, Food for the deep sea: Utilization, dispersal, and flux of nekton falls at the Santa Catalina Basin floor, *Deep Sea Res.* **32:**417–442.

Smith, G. R., and Elder, R. L., 1985, Environmental interpretation of burial and preservation of Clarkia fishes, in: *Late Cenozoic History of the Pacific Northwest* (C. J. Smiley, ed.), AAAS, San Francisco, pp. 85–93.

Smith, K. G. V., 1986, *A Manual of Forensic Entomology*, British Museum (Natural History), London.

Sondheimer, E., Dence, W. A., Mattick, L. R., and Silverman, S. R., 1966, Composition of combustible concretions of the alewife, *Alosa pseudoharengus*, *Science* **152:**221–223.

Sorbini, L., 1968, Contributo alla sedimentologia della "Pesciara" di Bolca, *Mem. Mus. Civ. Stor. Nat. Verona* **15:**213–221.

Spicer, R. A., 1981, The sorting and deposition of allochthonous plant material in a modern environment at Silwood Lake, Silwood Park, Berkshire, England, *U. S. Geol. Surv. Prof. Pap.* **1143:**1–77.

Stanier, R. Y., Ingraham, J. L., Wheelis, M. L., and Painter, P. R., 1987, *General Microbiology*, Macmillan, London.

Stanley, G. D., Jr., and Stürmer, W., 1983, The first fossil ctenophore from the Lower Devonian of West Germany, *Nature* **303**:518–520.

Stanley, G. D., Jr., and Stürmer, W., 1987, A new fossil ctenophore discovered by X-rays, *Nature* **327**:61–63.

Stone, E., and Johnson, H., 1987, *Forensic Medicine,* Waterlow, London.

Størmer, L., 1970, Arthropods from the Lower Devonian (Lower Emsian) of Alken an der Mosel, Germany. Part 1: Arachnida, *Senckenb. Lethaea* **51**:335–369.

Størmer, L., 1975, Arthropods from the Lower Devonian (Lower Emsian) of Alken an der Mosel, Germany. Part 5: Myriapoda and additional forms, with general remarks on fauna and problems regarding invasion of land by arthropods, *Senckenb. Lethaea* **57**:87–183.

Stürmer, W., and Bergström, J., 1973, New discoveries on trilobites by X-rays, *Paläontol. Z.* **47**:104–141.

Stürmer, W., Schaarschmidt, F., and Mittmeyer, H.-G., 1980, Versteinertes Leben im Rontgenlicht, *Kleine Senckenb. Reihe* **11**:1–80.

Swain, F. M., Bratt, J. M., Kirkwood, S., and Tobback, P., 1969, Carbohydrate components of Paleozoic plants, in: *Advances in Organic Geochemistry, 1968,* (P. A. Schenck and I. Havenaar, eds.), Plenum Press, London, pp. 167–180.

Swift, M. J., Heal, O. W., and Anderson, J. M., 1979, *Decomposition in Terrestrial Ecosystems,* Blackwell, Oxford.

Tegelaar, E. W., de Leeuw, J. W., Derenne, S., and Largeau, C., 1989, A reappraisal of kerogen formation, *Geochim. Cosmochim. Acta* **53**:3103–3106.

Thayer, C. W., 1983, Sediment-mediated biological disturbance and the evolution of marine benthos, in: *Biotic Interactions in Recent and Fossil Benthic Communities* (M. J. S. Tevesz and P. L. McCall, eds.), Plenum Press, New York, pp. 479–625.

Tiegler, D. J., and Towe, K. M., 1973, Microstructure and composition of the trilobite exoskeleton, *Fossils Strata* **4**:137–151.

Trewin, N. H., 1986, Palaeoecology and sedimentology of the Achanarras fish bed of the Middle Old Red Sandstone, Scotland, *Trans. R. Soc. Ed. Earth Sci.* **77**:21–46.

Trewin, N. H., 1989, The Rhynie hot-spring deposit, *Earth Sci. Conserv.* **26**:10–12.

Via Boada, L., Villalta, J. F., and Esteban Cerdá, M., 1977, Paleontologia y paleoecologia de los yacimientos fosiliferos del muschelkalk superior entre Alcover y Montral (Montanas de prades, Provincia de Tarragona), *Cuad. Geol. Ibér.* **4**:247–256.

Viohl, G., 1985, Geology of the Solnhofen Lithographic limestone and the habitat of *Archaeopteryx,* in: *The Beginnings of Birds. Proceedings of the International Archaeopteryx Conference* (M. K. Hecht, J. H. Ostrom, G. Viohl, and P. Wellnhofer, eds.), Eichstätt, pp. 31–44.

Voigt, E., 1988, Preservation of soft tissues in the Eocene Lignite of the Geiseltal near Halle/S, *Cours. Forschungsinst. Senckenb.* **107**:325–343.

Wade, M., 1968, Preservation of soft-bodied animals in Pre-Cambrian sandstones at Ediacara, South Australia, *Lethaia* **1**:238–267.

Walcott, C. D., 1911, Cambrian geology and paleontology. II. Middle Cambrian holothurians and medusae, *Smithson. Misc. Collect.* **57**:41–68.

Walossek, D., and Szaniawski, H., 1991, *Cambrocaris balticus,* n. gen., n. sp., a possible stem-lineage crustacean from the Upper Cambrian of Poland, *Lethaia,* in press.

Weitschat, W., 1983a, Ostracoden (O. Myodocopida) mit Weichkörper-Erhaltung aus der Unter-Trias von Spitzbergen, *Paläontol. Z.* **57**:309–323.

Weitschat, W., 1983b, On *Triadocypris spitzbergensis* Weitschat, in: *Stereo-Atlas of Ostracod Shells,* Volume 10, British Micropalaeontological Society, Llandudno, Wales, pp. 127–138.

Westphal, F., 1967, Erster Nachweis des Riesensalamanders (Andrias, Urodela, Amphibia) im Europäischen Jungpliozän, *Neues Jahrb. Geol. Paläontol. Monatsh.* **2**:67–73.

Westphal, F., 1978, *Tylototriton* (Amphibia, Urodela) aus dem Obermiozän von Öhningen, *Neues Jahrb. Geol. Paläontol. Monatsh.* **1978**:491–501.

Westrich, J. T., and Berner, R. A., 1984, The role of bacterial sulfate reduction: The G model tested, *Limnol. Oceanogr.* **29**:236–249.

Whalley, P. E. S., and Jarzembowski, E. A., 1985, Fossil insects from the Lithographic

Limestone of Montsech (late Jurassic–early Cretaceous), Lérida Province, Spain, *Bull. Br. Mus. Nat. Hist. (Geol.)* **38**:381–412.

Whiteaves, J. F., 1892, Description of a new genus and species of phyllocarid crustacean from the Middle Cambrian of Mount Stephen, British Columbia, *Can. Rec. Sci.* **5**:205–208.

Whittington, H. B., 1985, *The Burgess Shale*, Yale University Press, New Haven, Connecticut.

Whittington, H. B., and Briggs, D. E. G., 1985, The largest Cambrian animal, *Anomalocaris*, Burgess Shale, British Columbia, *Phil. Trans. R. Soc. Lond.* **309B**:569–609.

Wilby, P., and Martill, D. M., 1991, Exceptional preservation in the guts of fossil fish, in: *Historical Biology*, in press.

Wild, R., 1990, Holzmaden, in: *Palaeobiology: A Synthesis* (D. E. G. Briggs and P. R. Crowther, eds.), Blackwell, Oxford, pp. 282–285.

Willems, H., and Wuttke, M., 1987, Lithogenese lakustriner Dolomite und mikrobiell induzierte "Weichteil-Erhaltung" bei Tetrapoden des Unter-Rotliegenden (Perm, Saar-Nahe-Becken, SW-Deutschland), *Neues Jahrb. Geol. Paläontol. Abh.* **174** (2):213–238.

Williams, L. A., 1983, Deposition of the Bear Gulch Limestone: A Carboniferous Plattenkalk from central Montana, *Sedimentology* **30**:843–860.

Wilson, M. V. H., 1977a, Paleoecology of Eocene lacustrine varves at Horsefly, British Columbia, *Can. J. Earth. Sci.* **14**:953–962.

Wilson, M. V. H., 1977b, New records of insect families from the freshwater Middle Eocene of British Columbia, *Can. J. Earth. Sci.* **14**:1139–1155.

Wilson, M. V. H., 1977c, Middle Eocene freshwater fishes from British Columbia, *R. Ont. Mus. Life Sci. Contrib.* **113**:1–61.

Wilson, M. V. H., 1978, Paleogene insect faunas of Western North America, *Quaest. Entomol.* **14**:13–34.

Wood, S. P., 1982, New basal Namurian (Upper Carboniferous) fishes and crustaceans found near Glasgow, *Nature* **297**:574–577.

Wood, S. P., and Rolfe, W. D. I., 1985, Introduction to the palaeontology of the Dinantian of Foulden, Berwickshire, Scotland, *Trans. R. Soc. Ed. Earth Sci.* **76**:1–6.

Woodland, B. G., and Stenstrom, R. C., 1979, The occurrence and origin of siderite concretions in the Francis Creek Shale (Pennsylvanian) of northeastern Illinois, in *Mazon Creek Fossils* (M. H. Nitecki, ed.), Academic Press, New York, pp. 69–103.

Woodward, H., 1907, Further notes on the Arthropoda of the British Coal Measures, *Geol. Mag.* **2**:539–549.

Woodward, H., 1908, Some Coal Measures crustaceans with modern representatives, *Geol. Mag.* **5**:385–396.

Woodward, H., 1911, On a large form of *Anthrapalaemon* from the clay-ironstone nodules of the Middle Coal Measures, Sparth Bottoms, Rochdale, *Geol. Mag.* **8**:361–366.

Wuttke, M., 1983, Weichteil-Erhaltung durch lithifizierte Microorganismen bei mitteleozänen Vertebraten aus den Ölschiefern der Grube Messel bei Darmstadt, *Senckenb. Lethaea* **64**:509–527.

Zangerl, R., 1971, On the geological significance of perfectly preserved fossils, in: *Proceedings 1st North American Paleontological Convention*, Volume I, pp. 1207–1222.

Zangerl, R., and Richardson, E. S., Jr., 1963, The paleoecological history of two Pennsylvanian black shales, *Fieldiana Geol. Mem.* **4**:352 pp.

Zhang, W.-T., and Hou, X.-G., 1985, Preliminary notes on the occurrence of the unusual trilobite *Naraoia* in Asia, *Acta Palaeontol. Sin.* **24**(6):591–595.

ZoBell, C. E., and Anderson, D. Q., 1936, Vertical distribution of bacteria in marine sediments, *Bull. Am. Assoc. Petroleum Geol.* **20**:258–269.

Chapter 3

Plant Taphonomic Processes

ROBERT A. SPICER

ROBERT A. SPICER • Department of Earth Sciences, Oxford University, Oxford OX1 3PR, United Kingdom.

Taphonomy: Releasing the Data Locked in the Fossil Record, Volume 9 of *Topics in Geobiology*, edited by Peter A. Allison and Derek E. G. Briggs, Plenum Press, New York, 1991.

1. Introduction: The Quality of the Plant Fossil Record

Plants are the primary producers upon which all animal life on Earth depends. They were responsible for oxygenating Earth's early reducing atmosphere, thereby creating an environment in which animal life as we know it could evolve, and since then have been recycling the atmosphere and providing both food and ecospace for animals. Because of their essential place in the physical and biological evolution of this planet, an accurate understanding of the evolution of plant lineages and communities, and their changing interactions with their environment, is critical to understanding a wide range of Earth systems. In the course of their existence land plants have developed many adaptations that reflect their environment. Today the abundant plant fossil record and the environmental signal that it yields is proving of major importance in paleoclimate reconstruction and studies of global change.

Photosynthetic land plants are spatially fixed after germination. Because they have to process gases such as carbon dioxide and oxygen, intercept light energy, and utilize water in an efficient a manner as possible in order to compete successfully with their neighbors, they have developed an intimate relationship with their immediate physical environment and the climate to which they are exposed. In the course of evolution, plants, which are relatively simple organisms, have developed a suite of morphological characteristics that reflect their contemporaneous climate and its fluctuations (Spicer, 1989). Nevertheless, this primary climatic signal, a signal strongly displayed in living climax vegetation (e.g., Wolfe, 1979) is likely to become distorted by transport, burial, and diagenetic processes inherent in fossilization. Thus, a sound taphonomic understanding is critical to the correct interpretation of paleoclimatic signals revealed by plant assemblages.

Communities are consistent aggregations of individuals of a variety of species that are recognizably different from other such aggregations. Each community has a restricted spatial range and although the transitions between communities may be sharp, they are more likely to be gradational. The transitional zones are termed ecotones. In any regional depositional basin a large number of different communities may occur.

In an ecological sense communities are highly plastic and typically the climax state (in which there is a steady-state equilibrium of composition with respect to the physical environment) is reached by way of a series of intermediate seres. For a variety of reasons, particularly in paleoclimate studies, distinguishing temporary seres from the climax state is essential. Most depositional sites represent situations in which sediment is being moved, and thus communities local to those sites are likely to be disturbed to some degree or another, often resetting the seres and delaying or preventing the establishment of the climax state. The identification of climax communities therefore often necessitates sampling many sedimentary facies across a regional basin (Spicer and Parrish, 1986). Plant communities have also changed with time and current species associations cannot be relied upon as a guide to past community composition. Thus, the plant taphonomist has to deal with both spatial and temporal ecotones over a wide range of scales. Since the mid-1970s significant advances have enhanced our ability to interpret the fossil record in terms of the structure and composition of ancient communities in spatial and temporal relationship to each other and the physical environment.

A plant fossil assemblage is an accumulation of plant parts, derived from one or several individuals, that is entombed within a volume of sediment laid down under essentially the same conditions (Spicer, 1989). Superficially this definition is identical to that which might be applied to faunal accumulations. However, important differences exist between plants and animals in the way that they are represented in the fossil record. Plants are composed of, and produce throughout their lifetime, an indeterminate number of organs each of which is shed into the environment individually, and each of which has greatly differing potential for transport, deposition, and preservation. For example, spores and pollen grains are extremely durable, easily dispersed, and produced in huge numbers, whereas floral organs are usually delicate, ephemeral structures, produced in much smaller quantities. Usually only one trunk is produced by a tree, but leaves are produced in large numbers irrespective of whether the tree is deciduous or evergreen.

The different types of plant organ each have different preservation and reworking potentials. The durability of spores and pollen grains means that they may be deposited, exhumed, and redeposited several times, in some cases over a period of millions of years, each time being mixed with contemporaneous palynological debris and/or that derived from eroding rocks of differing ages. Time-averaging in palynology is potentially a serious problem. On the other hand, leaves, although robust enough to be relatively easily preserved, have almost no reworking potential and therefore can be extremely good stratigraphic and climatological indicators.

To make matters worse, different plant organs evolve at different rates and a morphological change occurring in, for example, a leaf may not be reflected in a simultaneous change in flower or pollen morphology. This process, termed heterobathmy (mosaic evolution), takes place because different organs experience different selection pressures. For example, factors affecting leaf evolution are likely to be related to the physical environment, while those affecting the reproductive system may have more to do with evolution in biotic pollinators. Thus, rates of evolutionary change in isolated organs cannot be extrapolated to the whole organism, lineage, or community in any meaningful way, and uniformitarian approaches to community reconstructions based on biological variables are rendered utterly invalid for most of terrestrial plant history.

It should be obvious from the foregoing that whole plants are rarely found in the fossil record and that reconstructing ancient plant forms is often a laborious process requiring great skill. Because the plant fossil record is one of isolated organs, paleobotanists have evolved systematic and taxonomic approaches capable of dealing with such data [see Spicer and Thomas (1986) for a number of papers dealing with this topic]. Although some ancient plants are known more or less in their entirety [e.g., *Archeopteris* (Beck, 1962), *Cordaixylon* (Rothwell and Warner, 1984), *Pseudofrenelopsis* (Alvin, 1983)], in practice very few whole plants are, or can ever be, reconstructed, and yet it is whole plants that comprise communities.

Because it is an accumulation of debris shed through time, a plant fossil assemblage represents a distorted measure of the capacity of the source vegetation to produce litter. Inasmuch as plant litter production is related to biomass and productivity, it is influenced by factors such as soil conditions, climate, exposure, community composition, and seasonality (Bray and Goreham, 1964; Bonnevie-Svendsen and Gjems, 1957; Krassilov, 1969; Spicer, 1981). Litter production by any given defined unit or stand of vegetation varies with time in relation to the seral development of the stand, climatic conditions, and disturbance. Although locally catastrophic events such as floods or volcanic eruptions may bury plant debris produced predominantly within a single year, less catastrophic deposition results in the gradual accumulation of plant debris over periods of time sometimes lasting as long as several millions of years. It is therefore extremely difficult to equate assemblage organic abundance to measures of living biomass.

The complexity of the plant fossil record and the limited lateral continuity of continental (terrestrial) sedimentary environments means that the treatment of plant taphonomy is justifiably separate from studies concerned with the formation of animal fossil assemblages. It is not yet possible to make many broad generalizations about plant assemblage formation because relatively few nonmarine depositional environments have been studied in great detail. Tabulation of what one might perceive to be general taphonomic patterns would at this stage only serve to trivialize and oversimplify assemblage structure and obscure the opportunities for detailed analytical studies. Similarly, although facies/depositional environment models are useful for recognizing consistently produced patterns of assemblage formation in the field, these patterns are the result of *processes*. By understanding these processes, assemblage patterns produced under any circumstances, not just those of the typical "landmark" models, can be readily interpreted. In previous reviews, such as that of Spicer (1989), patterns of plant deposition in various types of depositional environment have been described. Here, however, processes will be emphasized.

2. Wind Dispersal

In all but catastrophically produced autochthonous assemblages arising from sudden inundation of a standing plant community, aerial transport either by falling or blowing through the air is the first stage in the journey from the biosphere to the geosphere for all above-ground plant organs. Most studies have concentrated on the dispersal of naturally abscised organs such as fruits or seeds (hereafter known as diaspores) and leaves. Of all plant organs, leaves carry the strongest climate signal because they have evolved morphological characteristics that enable them to photosynthesize as efficiently as possible in the environments to which they are exposed (Spicer, 1989). This inherent climate signal and their abundance in the fossil record make leaves particularly important in paleobotany and consequently they have received most taphonomic attention.

2.1. Factors Affecting Fall Velocity

Of the characteristics intrinsic to the leaves, weight per unit area appears to be the most important in determining fall velocity (Ferguson, 1985; Spicer, 1989). In general, fall velocity determines the length of time a leaf will remain in the air column and the strength of updrafts necessary to keep it aloft. Hence, light leaves have a greater chance of being blown further from any given height than do leaves with a higher weight per unit area.

Leaf shape is only significant in that leaves with a long, narrow form rotate about the primary length axis (Spicer, 1981) and extend their fall time.

Although leaf size will in part be related to weight per unit area (Ferguson, 1985), size alone has little influence on leaf dispersal unless the leaves are particularly large and in a forested environment. Large leaves will tend to be caught by trunk and branch space obstacles more frequently than small leaves.

2.2. Dispersal and Sorting

Leaves growing at or near the top of a tree crown have a tendency to be rather small and thick and thus have a high weight per unit area. These so-called "sun" leaves were found to be more widely dispersed than the larger, thinner "shade" leaves of the subcanopy (Spicer, 1981). At first sight this observation is contrary to what is expected from the air-fall experiments, but the explanation of this anomaly is that the sun leaves are born higher on the tree and are exposed to higher wind energies than the shade leaves. This differential sorting can be a significant problem in open lake situations, where assemblages in deeper water, further from the shore, might well be enriched by small wind-transported sun leaves. This bias toward small, thick leaves could be erroneously interpreted as representing a cooler, dryer climate than that actually experienced by the source plants.

The dispersal of leaves into a lake from the shoreline follows a negative exponential model of the form

$$Z_x = Z_r \exp[-k(r - x)]$$

where x is the distance from the lake center, Z_x is the deposition occurring at distance x, r is the distance from the lake center to the shoreline, Z_r is the deposition at the shoreline, and $k = -(r - x)^{-1} \ln(Z_x Z_r^{-1})$ (Rau, 1976).

This ideal relationship may be distorted by local topography, but it has been shown to be reliable in natural situations (Rau, 1976; Spicer, 1981).

Dispersal from a point source within a forest is rather restricted. In conifers 90% of the needles fall within a 20° cone projected from the tip of the tree (Dunwiddie, 1987), and in angiosperms dispersion is similarly limited (Ferguson, 1985). These observations indicate that there is enormous scope for reconstructing the three-dimensional structure of ancient forests where they are represented by autochthonous assemblages. Mapping the distribution of different leaf forms over exposed bedding surfaces representing autochthonous deposition would enable the structure of ancient vegetation to be reconstructed. This could be achieved by mapping the concentrations of various leaf taxa; the center of each contagion would approximate to the position of the source tree (even if the actual trunk remains were not preserved), while the diameter of the contagion would indicate the height of the source. Even cursory sampling could yield sufficient information to reconstruct the number of stories in a multistratal forest and their main constituents. Strongly heterogeneous litter samples have been noted from back levee sites in modern paratropical forest (Burnham, 1989). Such sites have a high preservation potential due to burial by crevasse splays (Gastaldo *et al.*, 1987).

3. Decay

3.1. Decay Processes

The progress of decay varies enormously depending on the nature of the plant organ, the species of plant, and the type of environment to which it is exposed. Understanding the decay process, however, is at the heart of taphonomy. In spite of the plethora of variables involved, some aspects of the decay process are beginning to be understood well enough to be used reliably in paleoecological studies.

Decay is essentially due to the activity of bacteria and fungi, both of which may predispose a plant organ to attack by invertebrates and other large-particle feeders (Petersen and Cummins, 1974). Because decay is a biological process, both moisture and relative warmth have a controlling influence over the rate and type of decay that takes place. Provided a plant organ remains dry, decay only proceeds at a very slow rate, if at all, although slow oxidation and/or autolysis may well render tissues brittle. Once the organ is wet, however, microbial infestation is rapid. For this reason most studies of decay have concentrated on the processes that take place in wet soils or aquatic environments.

Routinely shed organs provide a convenient method of disposal of metabolic by-products that would otherwise be toxic to the plant producing them. Thus, leaves often contain substances such as tannins that, by being phytotoxic, delay or inhibit the activity of fungi and/or bacteria. However, within a few hours after entry into water, leaves and other plant debris begin to lose soluble sugars, tannins, and elements such as potassium, and within days may reach equilibrium with the water chemistry (Nykvist, 1962; Spicer, 1975). Those leaves not containing condensed or insoluble tannins are therefore rendered highly susceptible to microbial attack. The rate of decay depends on a variety of factors, particularly the temperature, pH, and nutrient status of the water body (Ferguson, 1985; Hanlon, 1982; Newbold *et al.*, 1983; Mathews and Kawaczewski, 1969; Kaushik and Hynes, 1971). In general, leaves containing relatively high levels of biologically important minerals otherwise deficient in lake waters will be selectively degraded (Ferguson, 1985). Thus, in oligotrophic lakes species degradation is likely to be highly selective, whereas in eutrophic systems there is a greater probability that all species will be represented to some extent in any potential fossil deposit.

It has long been axiomatic that thick, leathery leaves have a higher probability (all other factors being equal) of entering the fossil record than do thin leaves (Chaney and Sanborn, 1933). However, when Ferguson (1971) cut disks of fresh leaves from 11 species with different leaf thicknesses and tumbled them in a rotating drum there was little difference in wear among the species. It therefore seems unlikely that leaf thickness alone, in the absence of microbial preconditioning, plays a major role in determining preservation potential.

Leaf cuticles are far more resistant to decay than cellulose or lignin (White, 1933), but cuticles alone do not present an impermeable barrier to microbiological colonization of leaf tissues, because decay often begins within the tissues while the cuticle is intact. Presumably, microbial entry is through stomatal apertures or petiole ends. Thick cuticles tend to have an inherent strength and often remain intact long after both cellulosic and lignified tissues have been destroyed (Spicer, 1981) and where cuticle degradation does take place it begins with the surface originally in contact with the epidermal cell walls (De Vries *et al.*, 1967).

Aerobic decomposition inevitably leads first to the loss of proteins, sugar, and fats comprising the living cytoplasm of a plant cell, followed by cellulose, lignin, then cuticle, and finally the walls of spores and pollen composed of sporopollenin. In complex tissues such as wood, which is composed both of cellulose and lignin, the preferential loss of one component (cellulose) leads to enrichment by another (lignin) (Rolfe and Brett, 1969). The exclusion of air or a reduction of temperature to near or below freezing slows decay processes significantly and thus enhances the preservation potential of the tissues.

3.2. Mechanical Degradation

Within days of immersion in water, microbial action considerably weakens the tissues and renders them susceptible not only to small- and large-particle invertebrate feeders (Petersen and Cummins, 1974), but also to mechanical degradation.

Tumbling experiments with leaves (Spicer, 1981; Ferguson, 1985) have demonstrated the robustness of fresh material (Fig. 1), but even a few days of decay leads to substantial mechanical fragmentation if the water is flowing at all. The occurrence of whole leaves rolled and poorly bedded within medium to fine fluvial sandstone is a common feature of some sediments and has been used to argue that the leaf assemblages represented recently abscised (fresh) leaves from locally growing deciduous plants (Spicer and Parrish, 1986; Spicer, 1987). Conversely, angular tears and breaks in leaf tissue point to mechanical damage during transport, while skeletonization or rounded holes or rounded margin damage suggests biological degradation which may be independent of transport (Spicer, 1981).

Microbial action is not necessarily deleterious to the potential fossil record. In some instances bacteria may precipitate mineral films on (or in) plant matter even before deposition and burial (Spicer, 1977). This mineral film has two effects: (1) it protects the plant from abrasion or invertebrate

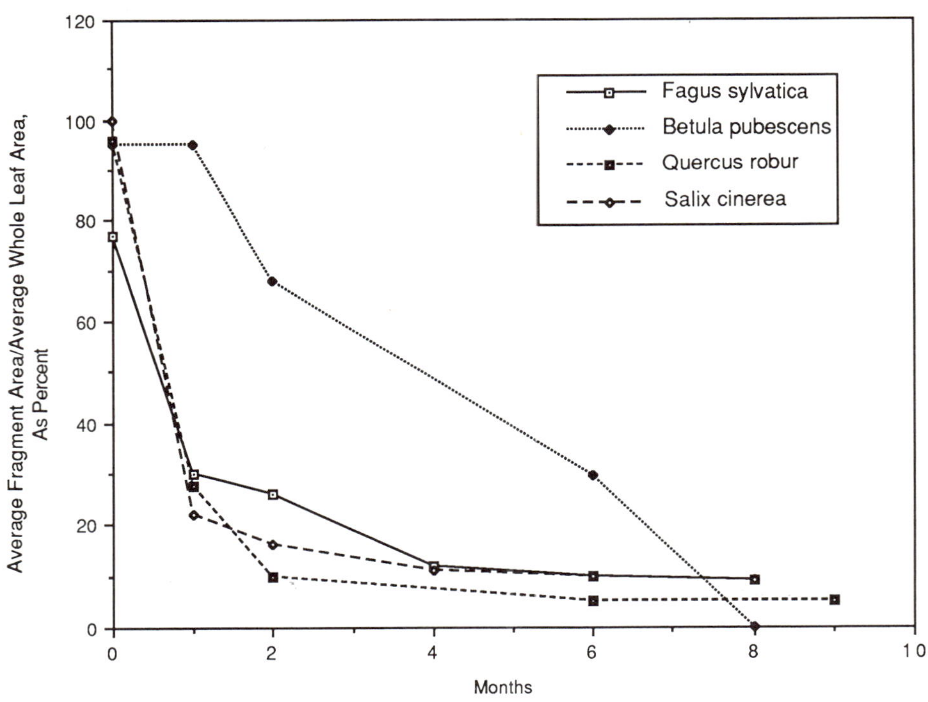

FIGURE 1. Experimentally induced leaf fragmentation after exposure to lake bed conditions at Silwood Lake, Berkshire, England. Partly decayed leaves were tumbled in a rotating drum. See Spicer (1981) for additional experimental details. Most leaves lose their mechanical strength within a short time of entering a natural aquatic environment, but some leaves (e.g., *Betula pubescens*), by virtue of their small size, are less easily mechanically degraded.

attack, and (2) it forms a fine-grained template which may preserve details of the plant surface that would otherwise not be preserved due to coarse sediment grain size.

3.3. Correcting for Selective Decay

Some correction for selective degradation may be made from detailed observations on the sequence of leaf deposition in relation to changes in sediment type. Cores through a modern fluviolacustrine delta showed that leaves of taxa growing local to the depositional site were rare in lake bottom and deltaic sediments except where a high-energy flood event had flushed relatively fresh leaves and sediment into the system (Spicer, 1981). The coarser inorganic sediment fraction settled out as a sand layer, but this was overlain by leaves rich in local, normally readily degraded, taxa, which in turn were overlain by the finer fractions of the inwashed sediment pulse (Fig. 2). Here rapid burial preserved vulnerable taxa. Clearly, if detailed sampling of this kind were to be adopted in fossil situations, not only might otherwise seemingly rare taxa be recovered, but a more complete picture of the vegetation would emerge. It may be that taxa frequently introduced into aquatic systems (i.e., those locally abundant to a depositional site) are always preferentially degraded due to sympathetic adaptations among detritivores to the most reliable (consistently available) food supply. If similar preferential degradation of local taxa took place in the past, then in many slowly accumulated plant fossil assemblages the most locally abundant component of the ancient source vegetation will be rare.

4. Water Transport

4.1. Entry into the System

Relatively fresh plant material may enter an aquatic environment by a number of routes. Storm damage may lead to traumatic removal of individual leaves or whole branches and, coupled with flooding, may lead to whole trees entering fluvial or lacustrine systems. However, less violent processes such as river bank undercutting can also introduce whole fresh plants into the aquatic system (Fig. 3).

Floating times for many deciduous leaves are less than a few days (Fig. 4), so most degradation takes place after they have sunk. In the case of evergreen leaves, however, significant microbial colonization and tissue degradation can take place while they remain afloat. Wave action, suspension, and bedload transport therefore might be expected to destroy a higher proportion of evergreen leaves compared to those of deciduous species.

Not all material enters an aquatic environment fresh. Undoubtedly in a

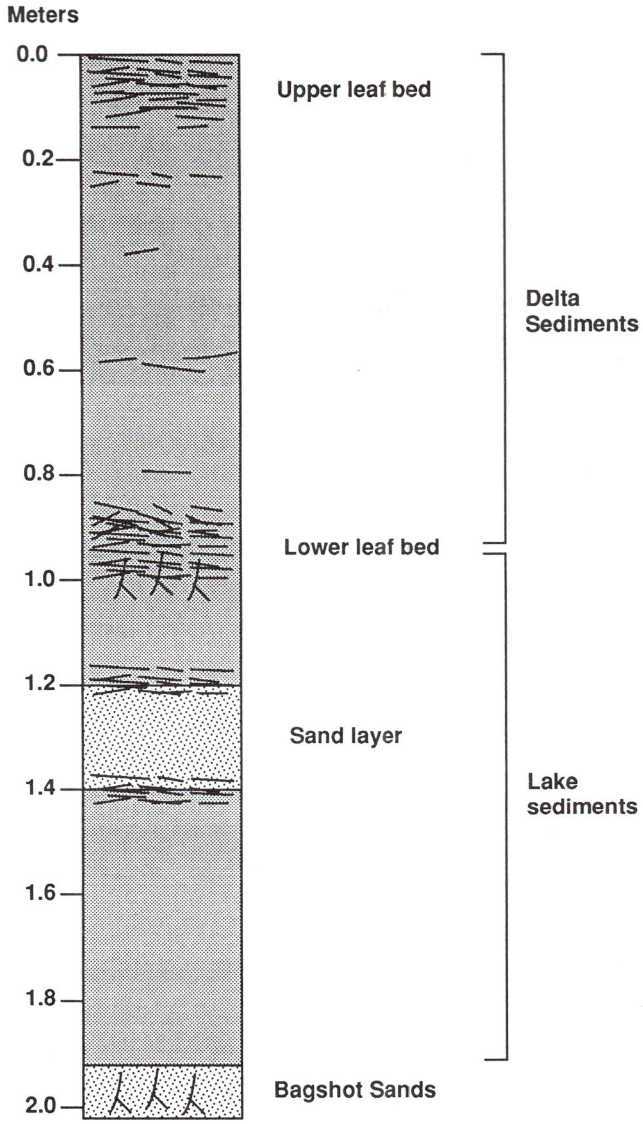

FIGURE 2. Composite core through a small temperate fluviolacustrine delta at Silwood Park, Berkshire, England. Leaves of the most abundant lakeside plant, *Alnus glutinosa*, are rarely preserved except where rapid influxes of sediment buried them quickly. This occurred both below and above the sand layer between 1.2 and 1.4 m depth. In fossil beds close examination of millimeter-scale bedding and changes in grain size can provide data on otherwise apparently rare taxa. [From Spicer (1981).]

natural system much debris enters by stream erosion of forest floor accumulations. Such plant litter will be in various stages of decay and saturation, so its subsequent behavior is difficult to predict, except that its potential for long-distance transport must be diminished by virtue of its reduced buoyancy and susceptibility to degradation.

FIGURE 3. Entry of fresh plant material into rivers by bank collapse during normal undercutting at the outside of meander loops. In this instance *Salix* thickets bordering the Colville River, Alaska.

4.2. Floating

Freshly abscised leaves exhibit a range of floating times (Fig. 4). Laboratory experiments (Spicer, 1981; Ferguson, 1985) demonstrate few general trends except that evergreen species with thick cuticular or wax coatings may float for many weeks, whereas deciduous chartaceous leaves often float for only several days. The entry of water into the tissues appears to be enhanced by tears or wounds (Ferguson, 1985). This suggests that there is great potential for differential long-distance transport, particularly among evergreen taxa, but these results are difficult to apply in any meaningful way to fossil assemblages because of additional variables.

The range of floating times exhibited by diaspores exceeds that of leaves; many diaspores remain afloat longer than leaves of the same species. However, floating times do not appear to be related to diaspore size or the habit of the parent plant (Collinson, 1983a). Thus, retrodictive use of relative floating times is difficult for the paleontologist, but these observations help explain the frequent mutual exclusion of leaves and diaspores in fossil assemblages, particularly in open-water depositional environments.

Large wood fragments, in particular logs, often have floating times exceeding several years (Spicer, 1989; Coffin, 1983). No comprehensive survey of floating times with respect to species exists, partly because in natural situations floating times depend on log size, the presence of root systems (which enhance water uptake), and the degree of damage the log has sustained prior to entry into the water. For this reason tree ring studies based

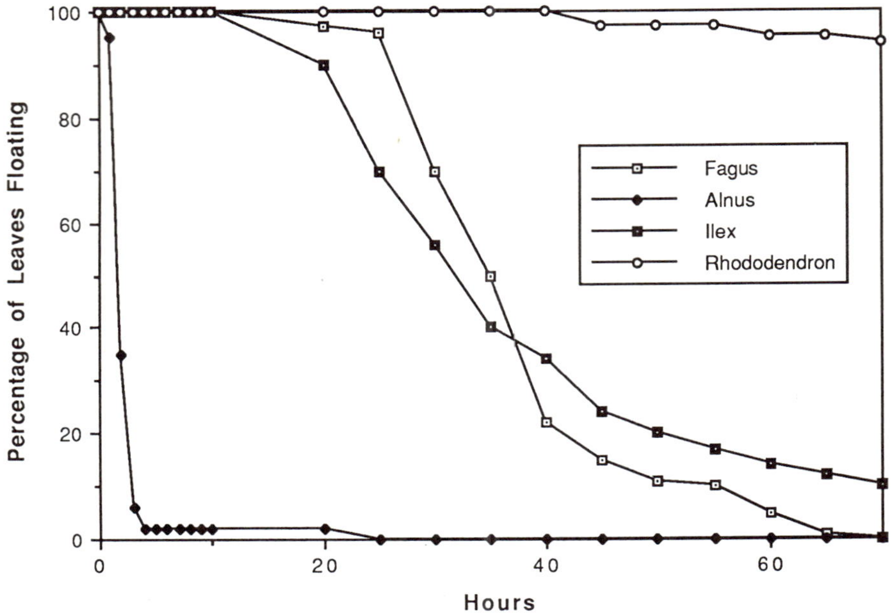

FIGURE 4. Floating times of leaves under laboratory conditions with agitation of surface water. Considerable difference in floating times is evident between species. [From Spicer (1981).]

on isolated logs in marine sediments can give highly spurious allochthonous climate signals (Spicer, 1989). Isolated logs in nonmarine settings, however, must have been growing within the drainage basin upstream of the point of deposition—a more or less definable area determined by the size of the channel complexes involved.

4.2.1. Floating and Entrapment

In order to test floating-leaf transport in a natural system under ideal conditions, 1000 air-dried *Fagus sylvatica* leaves were introduced into a stream system in the posteruption landscape at Mount Saint Helens, Washington (Spicer and Greer, 1986). The stream traversed the barren landscape to the north of the volcano and was devoid of fluvial or riparian vegetation. *F. sylvatica* leaves were chosen because there was no risk of them being confused with local taxa, and their natural light brown/orange color made them visible against the gray ash.

Entrapment of unsaturated leaves was mostly by stranding on the stream edges. Over time the median point of the population gradually moved downstream, indicating their entrapment was metastable, but even after 48 hours only 0.2% of the unsaturated leaves had traveled 1.2 km. Thus, even under conditions of relatively high-energy stream flow devoid of in-stream vegetation, downstream transport was minimal before microbial colonization began to weaken the leaf tissues. If the stream had been wider, it is possible that more leaves would have traveled the 1.2 km because the ratio of stream edge length to water surface area would have been less. This remains to be tested. In some circumstances when leaf shedding is synchronous abundant rafts of floating leaves can form, which probably enhances transport distances (Berry, 1906).

The buoyancy of unsaturated or partially saturated plant parts clearly plays an important role in forming plant assemblages in estuarine or marine environments. Large quantities of fruits and seeds were observed off the mouth of the River Ambernoe, New Guinea last century (Moseley, 1892) and it has been suggested that such a phenomenon may account for the marine fruit and seed assemblages of the Eocene London Clay (Chandler, 1964). Floating debris can traverse open water in a lake, perhaps going from shore to shore several times before sinking, but in anything other than very large lake systems this will be of relatively minor consequence unless fine-scale vegetation reconstructions are required.

4.3. Submergence and Transport in the Water Column

By treating inorganic (usually quartz) grains as spheres, sedimentologists have long been able to describe mathematically their fall in still fluids. Many fluid dynamics equations require that the particle under consideration be ascribed an effective geometric dimension, which for a sphere is the diameter. In this case the effective dimension remains the same irrespective of the orientation of the particle to the fluid flow. For a two-dimensional object such as a leaf an effective dimension is difficult to determine because orientation of the leaf tends to change as it falls. Instability such as lateral oscillations during fall are produced by the formation of a turbulent wake, or series of vortices, behind the object.

Predicting when this wake will form is possible by calculating a dimensionless number, the Reynolds number, which is proportional to the ratio between inertial and viscous forces.

The Reynolds number is given by

$$\mathrm{Re} = UL_{\rho}/\mu$$

where U is the fall velocity, L is the effective dimension, $_{\rho}$ is the density of the particle, and μ is the dynamic viscosity of the fluid. At low Reynolds numbers fluid flow around the particle is laminar, resulting in stable particle descent. Such conditions exist when the particle is extremely light or the fluid very viscous. At higher Reynolds numbers, however, a turbulent wake is generated that imparts instability to the falling object.

Difficulties in assigning a single length dimension to a particle with a constantly changing orientation during fall has resulted in the use of various measures that combine dimensions such as length and width (e.g., Allen,

1984; Clift *et al.*, 1978). None, however, are entirely satisfactory, because simplifications that may yield more or less accurate predictions of behavior under one set of conditions fail to perform well when the conditions change. Furthermore, the progressive absorption of water that takes place after the plant part has ceased to float results in steadily increasing fall velocities. The difficulties in predicting fall behavior by use of theory have forced a more empirical approach (e.g., Ferguson, 1985; Spicer, 1981).

In spite of the variable shape, density, two-dimensional aspect, and flexibility of many plant parts, experimental determinations of fall velocity show that statistical populations of leaves harvested from trees belonging to single species display a remarkably coherent range of fall velocities when fully saturated (Spicer and Greer, 1986).

Thus, in spite of the range of leaf shapes and sizes exhibited by even a single tree, the hydraulic properties of such leaves are surprisingly predictable. By use of a 2-m^3 settling tank, it has also been shown that no statistically significant difference exists in the fall velocity *between* populations of leaves of *different* broad-leaved taxa, including *Ginkgo biloba* and fern pinnules. Thus, most leaves with a broad lamina also behave in a predictable and similar manner. However, significant differences are detectable between the leaves of broad-leaved taxa and leaves with a more linear and three-dimensional shape such as conifer needles. Furthermore, leaves of different conifer species have hydrodynamic properties that are significantly different from each other (Spicer and Greer, 1986). It follows from these observations that differential sorting during stream transport is likely to be a common phenomenon and will be particularly strongly displayed in the gradient flow fields associated with fluviolacustrine delta development. At its simplest level, conifer and broad-leaved taxa are likely to be most strongly represented in different parts of the delta due to differential hydraulic sorting of their leaves. Although this inevitably complicates the reconstruction of pattern in the source vegetation, the coherence of the behavior offers considerable potential for the development of computer models for simulating assemblage composition under a variety of depositional circumstances.

4.4. Entrainment and Deposition

Experimental flume studies have demonstrated the importance of bedforms in the trapping and burial of plant debris (Spicer and Greer, 1986). Only those plant elements which have a size comparable to the scale of the bedform will be trapped and buried. Thus, in the lee of 1-cm climbing ripples conifer needles are easily trapped and buried, whereas larger elements such as angiosperm leaves are swept through the system. Under these circumstances the resulting fossil assemblage would yield a highly misleading sample of the source vegetation. However, an assemblage rich in plant parts with a wide range of sizes and lacking evidence of bedforms could be regarded as not suffering this kind of selection. Considerably more experimental work is required in this context, particularly with fine-grained sediments.

4.4.1. Flow Orientation

Flow orientation is common in nature, and MacDonald and Jefferson (1985) have observed that (a) shape eccentricity is the dominant factor controlling orientation parallel to flow—that is, long, narrow shapes are more easily oriented, (b) wood fragments that are more or less equidimensional, and therefore with a high order of symmetry, can be oriented transverse to currents without lodgement in bedforms, (c) strong velocities result in strong orientation, and (d) large specimens can be deposited over a wider range of current velocities than smaller ones and thus adopt a wide range of orientations.

5. Burial

It is generally accepted that rapid burial enhances the preservation of plant matter and this has been demonstrated in modern sediments in relation to changing grain size (Spicer, 1981) (Section 3.3).

In a series of flume experiments, Rex (1985) studied burial processes with respect to hollow stems. Using fine sand, Rex determined that for 1.5-cm-diameter stems aligned horizontal and parallel to flow the structure of the fill was dependent on the length of the stem, with short stems (5 cm) filling almost completely, but long (35 cm) stems filling only at their ends (Fig. 5).

Stem diameter clearly is important in the infilling process, as larger stems (2.5 cm) are more completely filled and it seems likely that the degree of infilling is related to the maximum flow velocity possible in the stem. In a hollow tube, flow resistance is proportional to both the tube length and the fourth power of its radius and it is the resistance to flow that diminishes flow velocity and therefore the sediment-carrying capacity of the water. Once obstruction of the ends of the tubes begins, resistance increases at these points, leading to the formation of sediment wedges at the tube ends. In horizontal tubes oriented normal to flow sediment, wedges build up at the open ends and quickly seal the tube from further infilling. In Rex's experiments, however, only short stems in conjunction with low flow velocities could be studied, as high flow velocities tended to reorient the stems parallel to flow.

Although Rex's experiments mark a significant advance in our understanding of the processes of void infilling, they cannot be said to be comprehensive in that it is likely that the phenomena she observed were restricted to relatively coarse grain sizes. It is likely that clays and silts might exhibit significantly different infilling processes in that lower flow velocities are necessary to transport the individual grains, the smaller grain size would permit the penetration of smaller cavities, and the sediment couldmore easily flow under compaction into voids not filled at the time of deposition. Similar phenomena are seen in animals (Briggs and Williams, 1981).

To model stems with internal obstructions such as nodal diaphragms, Rex (1985) experimented with tubes sealed at one end. The effect of sealing

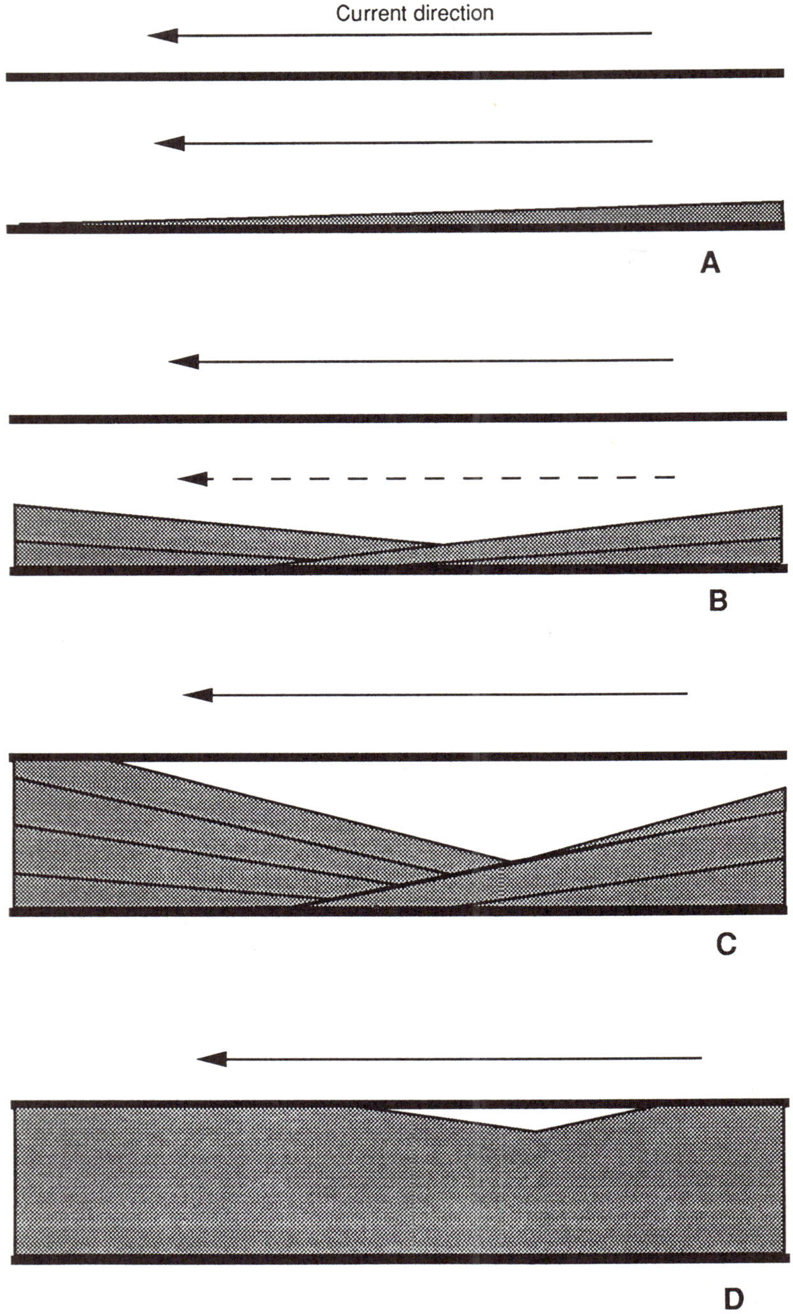

FIGURE 5. Experimental infilling of hollow stems. (A) Infilling progresses as wedges of sediment and occurs from both open ends (B) even when the stem is aligned parallel to flow. Infilling from the downstream end is caused by flow partitioning and backflow. (C) As the downstream end becomes blocked, sediment may still enter from the upstream end. (D) Long stems (in relation to stem diameter) will almost never completely fill, due to blockage of the ends by sediment. [After Rex (1985).]

the upstream or downstream ends was the same; a sediment wedge formed at the open end and sealed the stem from further infilling. Rapid filling of the stem ends leaving the central part hollow also occurred in stems buried by bedform movement. Similar patterns of infill are seen in fossil material (Rex, 1985).

Rex's experiments demonstrate that with relatively coarse grain sizes infilling of hollow stems is rarely complete and as a consequence substantial compactional distortion is inevitable as the hollow regions of the stem collapse. Distortions introduced by this collapse may have implications for reconstruction of original morphology, understanding the biology of the plant, and taxonomic assignment. Further experiments are needed to understand the processes of fine sediment infiltration into voids and injection phenomena under compaction.

6. Diagenesis

Schopf (1975) categorized the most common modes of plant preservation as (a) compressions/impressions, (b) permineralizations and petrifactions, (c) duripartic or hard-part preservation, and (d) casts and molds. These modes are of course not mutually exclusive and often a particular specimen will exhibit intergradations between two or more of these categories.

The preservation of plant components in an essentially unaltered state [duripartic preservation of Schopf (1975)] is very common and takes the form of palynomorphs and cuticles. The walls of pollen grains and spores are composed of sporopollenin, which is remarkably resistant to degradation, but will eventually decay in aerobic conditions (Havinga, 1984). Although not as resistant as sporopollenin, cutin, a major component of cuticles, also has a long residence time in anaerobic environments. This resistance to decay means that palynomorphs have a very high reworking potential, therefore smearing temporal and spatial resolution. Compression/impression fossils are, however, almost immune to this phenomenon due to their weak mechanical strength, but they pose other taphonomic problems.

6.1. Impressions, Compaction, and Compressions

Most plant megafossils are preserved as compressions/impressions where the original plant material has been compressed into a coaly film. Depending on the degree of compaction, cellular detail may be retained, although external details, often at the cellular level, are preserved as impressions in the surrounding matrix. Where the rock has been weathered the organic material may have been oxidized away, leaving only the impression. High-quality impressions can form within days after entry into a suitable depositional environment and *prior to burial*. Perhaps the most common mineral coating is iron oxide/hydroxide, which forms a red-brown film by bacterial action wherever dissolved iron is readily available in stream water. This coating not only provides a fine template for impressions of the plant surface, but also may limit degradation by invertebrate feeders (Spicer, 1977).

Similar impressions can occur in fine-grained sediment, but may form prior to burial or even abscission if leaves are coated by wind-blown sediment. Burnham and Spicer (1986) reported the impressions of *Theobroma cacao* (Cocoa) in fine ash in otherwise relatively coarse volcaniclastics resulting from eruptions of El Chichón 2 years prior to their investigations. All organic matter had decayed, but the high-resolution impressions, although unconsolidated, remained (Fig. 6).

After burial, plant matter inevitably undergoes some degree of compaction or flattening as additional sediment is deposited above it. The only

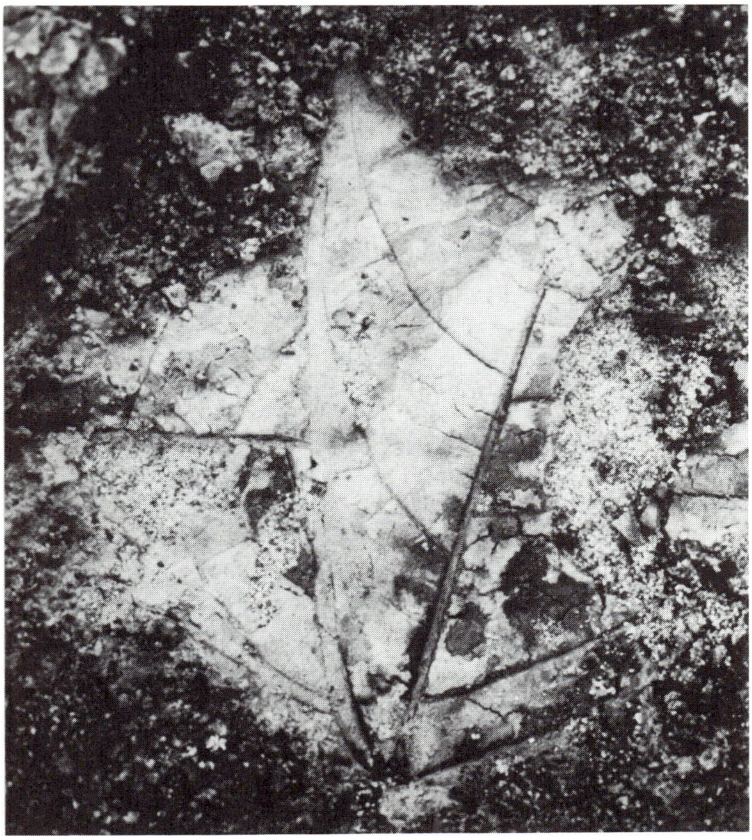

FIGURE 6. Leaves of *Theobroma cacao* preserved as impressions in fine-grained air-fall ash 2 years after the 1982 eruption of El Chichón, southern Mexico. The organic matter has all but rotted away in that time, but the template for potential fossil impressions was formed by very fine ash sticking to the wet surface of the leaf prior to abscission. It is embedded in coarser-grained crystal tuff, which shows through the impression where the finer material was accidentally removed during excavation.

exception to this is where mineralization occurs almost immediately and before significant sediment loading has taken place. Unlike the situation in animals, where there is often an extreme difference in the compressibility of skeletal and soft tissues, and soft tissues may undergo extreme distortion (Briggs and Williams, 1981), all terrestrial plant cells have either a cellulosic or lignified wall, so that there is a degree of similarity and resistance to crushing independent of tissue type. However, even with this narrower range of tissue differences there have been few attempts to consider the way that compression might distort plant material. Walton (1936) published a theoretical account of the processes involved that subsequent experimental work (Harris, 1974; Rex, 1983; Rex and Chaloner, 1983) has shown to be largely correct. In general, compression leads to distortion in the vertical plane, but minimal lateral expansion ($<<10\%$). As compression of a hollow cylinder proceeds, compaction of the vertical sides of the cylinder give rise to lateral "wings" of coaly material known as compression borders (Walton, 1936; Rex and Chaloner, 1983) (Fig. 7), and vertically oriented spines or ridges tend to retain their thickness, while horizontally oriented features become flattened (Fig. 8). However, the situation is not as straightforward as it might appear. By examining the stems of Carboniferous

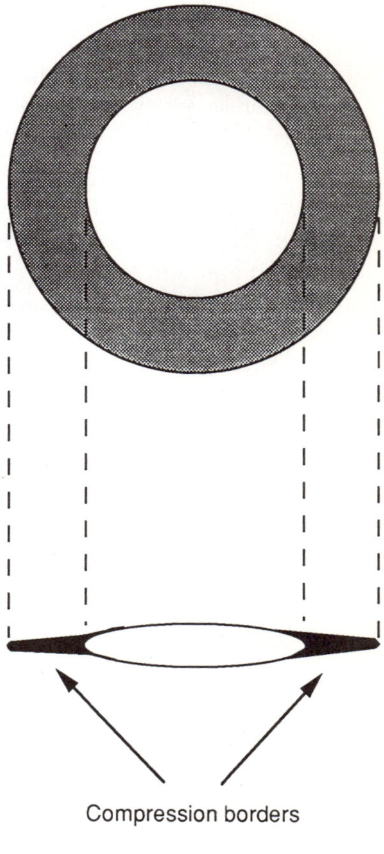

FIGURE 7. Experimental compression of cylinders produces minimal lateral expansion, but compression in the vertical plane to form "compression borders." [After Rex and Chaloner (1983).]

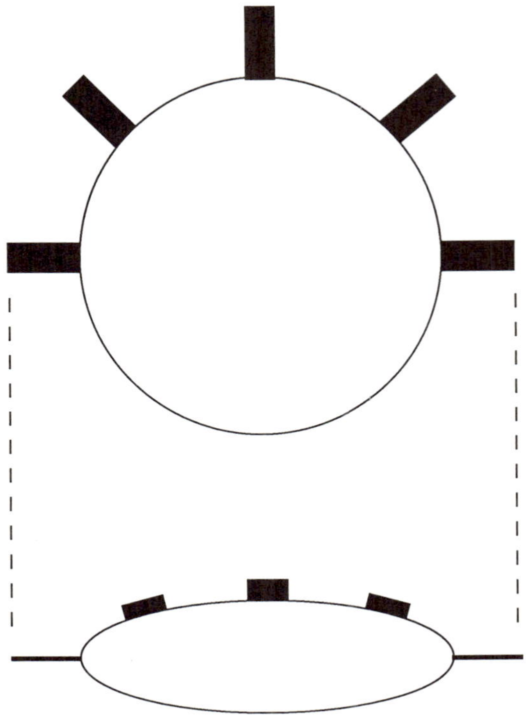

FIGURE 8. Experimentally produced compressions of plants with spines produces pronounced flattening in the vertical plane and thinning in the horizontal. Such distortions can lead to taxonomically significant artifacts. [After Rex and Chaloner (1983).]

arborescent lycopods which have regularly spaced leaf cushions, Thomas (1986) and Leary and Thomas (1989) have shown convincingly that large, hollow stems can expand significantly in the horizontal plane, and even approach the theoretical maximum of a 57% enlargement of their original diameter (Fig. 9). This is achieved because the trunks may collapse, possibly under their own weight, or after partial loading by sediment, before significant burial and under these circumstances lateral expansion may not be inhibited by surrounding sediment.

Compression of leaves and reproductive structures is more complicated, but experimental results have proved particularly useful in interpreting these structures in compressed form (Rex, 1983, 1986). Of critical importance in this context is the final plane of fracture that exposes the fossil in relation to the compressed three-dimensional form (Fig. 10). Knowledge of compaction artifacts has enabled taxonomic revisions of Carboniferous lycopod leaves to be made (Rex, 1983) and is clearly critical to the correct interpretation of many compression/impression plant fossils.

At the microscopic scale compaction leads to destruction of cellular detail as water is expelled from the plant material and opposing cell walls are forced into contact. If compaction is extreme, the organic matter may become amorphous as chemical alteration of the organic matter takes place.

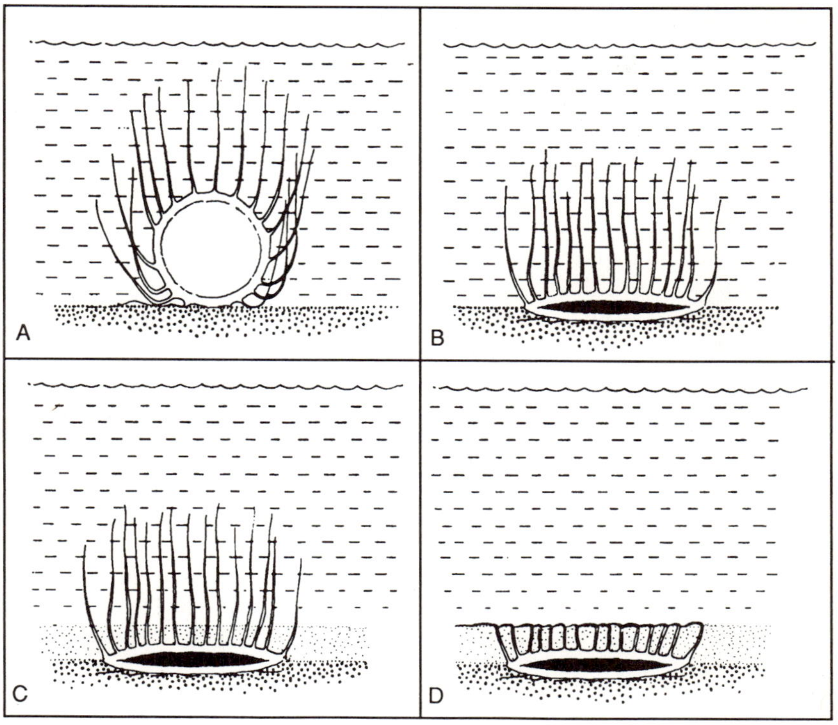

FIGURE 9. Model for the preburial collapse of hollow stems, in this case a Carboniferous *Lepidodendron* stem, that leads to lateral expansion of the stem diameter and partial preservation of external details. (A) Stem resting submerged on the sediment, (B) partly infilled and decay-weakened stem collapses under its own weight, expanding laterally, (C) leaf bases surrounded by sediment limiting decay, (D) exposed parts of leaves partly decay before continued burial. [After Leary and Thomas (1989).]

Increasing temperature and pressure lead to an increase in the percentage of carbon in the plant remains as hydrogen and oxygen are driven off in the form of water, carbon dioxide, and gases such as methane. As the percentage of carbon increases, so does the rank of the organic matter and its resistance to decay. This process is known as coalification and is summarized in Table I.

6.2. The Preservation of Tissues

Tissue preservation, while it is not as common as duripartic or compression/impressions, does occur remarkably often in the plant fossil record and takes several forms. Apart from "mummification," which under exceptional circumstances preserves original organic matter with minimal alteration (e.g., Christophel and Lys, 1986), permineralization or petrifaction yields the highest quality biological information. Early mineralization prevents

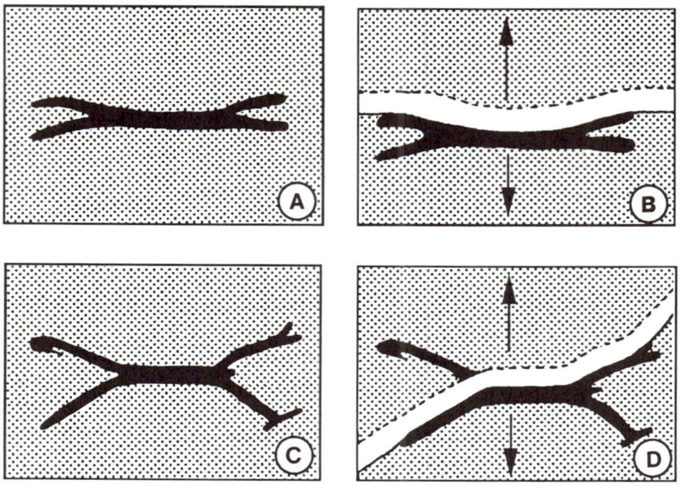

FIGURE 10. Diagram showing how compressional artifacts, coupled with differences in the line of rock splitting during collecting, can influence the interpretation of morphology and therefore taxonomic assignment. (A, C) Compressed leaves as seen in cross section in the rock matrix; (B, D) the line of splitting of the two different leaf forms may yield apparently similar morphologies. [After Rex (1983).]

tissue compaction and thus the three-dimensional organization of the plant is preserved. Some of the most ephemeral structures, such as delicate female reproductive organs (archegonia) or cell contents, can be permineralized (Brack-Haynes, 1978; Taylor and Millay, 1977) and permineralizations provide much of the evidence of early life forms on Earth (e.g., Tyler and Barghoorn, 1954; Barghoorn and Tyler, 1965).

6.2.1. Silicification

The preservation of three-dimensional tissue structure involves the permeation of the tissues by mineralizing fluids to eventually yield either a permineralization, in which the cell voids are filled with mineral, leaving the cells walls with a high organic component, or a petrifaction, in which the cell walls also are mineralized. The form of mineralization is largely dependent on the depositional environment. In volcanic terrains, hot, silica-laden fluids may inundate a plant community and preserve the community almost intact. Such a scenario has been traditionally envisaged for the Devonian Rhynie Chert flora (Kidston and Lang, 1917–1921). The precise chemistry of such a process is not well understood, but it is clear that under certain circumstances preservation can be extremely rapid (Knoll, 1985; Carson, this volume, Chapter 10), while in others the onset of silicification may take tens of thousands of years even in high-silica environments (Karowe and Jefferson, 1987). The silicification of wood in otherwise chert-poor sediments is common, an observation that led Leo and Barghoorn (1976) to propose a model for silicification that depends upon the presence of organic material. They suggested that in partly decayed organic

TABLE I. Effects of Increasing Coalification Rank on Plant Tissues[a]

Rank stages[b]	Percent reflectance of vitrinite	Important misroscopic characteristics	Percent C in vitrinite	Volatile matter[a]	Percent water *in situ*
Peat		Large pores	50		
		Details of initial plant material still recognizable			
		Free cellulose			
Brown coal		No free cellulose	60		~75
Soft brown coal		Plant structures still recognizable			
		(cell cavities frequently empty)			
		Marked gelification and compaction takes place			
Hard brown coal	~0.3			~53	~35
Dull brown coal		Plant structures still partly recognizable	70	~49	~25
Bright brown coal					
		(cell cavities filled with collinite)			
Hard coal	~0.5	Exinite becomes markedly lighter in color	80	~45	~8–10
Bituminous hard coal					
		("coalification jump")			
Anthracite	~2.5	Exinite no longer distinguishable from vitrinite in reflected light		30	
				10	
Graphite	11.0	Reflectance anisotropy	90		

[a]Coalification rank is a product of burial depth (pressure and temperature) and time. Based on ICCP (1963).
[b]Rank increases as one moves downward in the table.
[c]Percent dry ash free basis in vitrinite.

matter hydroxyl and other functional groups bond with mono- or polysilicic acid in solution, forming a template for subsequent silica precipitation. Thus, partial decay appears necessary for silica preservation, but the degree of decay may be so small as to be negligible from a morphological viewpoint. Because decay almost at the molecular level is all that is required for initiation of silicification, the preservation of logs often appears to begin at the external surfaces where penetration of the silica solution begins. The inner parts of the logs are therefore progressively partly sealed from further penetration, and decay may proceed to destruction of the tissues, the resultant cavities filling with crystalline quartz.

6.2.2. Pyritization

Mineralization by pyrite is also common in the plant fossil record. It is particularly common in woody material preserved in marine sediments [for example, the woods and seeds of the Eocene London Clay flora (Collinson, 1983b; Wilkinson, 1983; Allison, 1988), but is not restricted to such environments. Most pyrite forms as a result of the activity of sulfate-reducing bacteria (Berner, 1970, 1984); thus, decay (necessary to liberate hydrogen sulfide) is a prerequisite of preservation. Allison (1988, text Fig. 6a) figures a specimen of wood from the London Clay in which the easily degraded thin-walled spring or earlywood is preserved in pyrite, but the thick-walled and more decay-resistant latewood is preserved as a coalified residue. The anoxia required may be very local and even at the scale of surface depressions on logs (Allison, 1988).

6.2.3. Carbonate Mineralization

Carbonate mineralization is also a common form of plant, particularly wood, preservation. It can occur in both marine and nonmarine settings and may occur very early in diagenesis. As with the previously discussed minerals, it may be that partial decay actually enhances calcium carbonate precipitation (Raiswell, 1971, 1976). Thus, calcite nodules would tend to nucleate around organic material. The extent to which this occurs in nonmarine environments will depend on the availability of calcium ions.

The formation of calcium/magnesium carbonate concretions in coals, commonly known as "coal balls," has been the subject of much controversy and speculation [see Scott and Rex (1985) for a recent review]. The source of the carbonate is generally assumed to be seawater, but the paradox of carbonate precipitation in an acidic peat has required some explanation. Models have included marine transgressions of coal-forming mires (Stopes and Watson, 1908), but not all coal-bearing units are overlain by marine units. Breaching of beach barriers by storm action has been suggested (Mamay and Yochelson, 1962), but this would introduce marine organisms into the mire and disturbance of the swamp community by changes in pH and salinity—phenomena that are only evident in a small number of cases. Feliciano (1924), Evans and Amos (1961), and Phillips and DiMichele (1981) have all suggested that nonmarine, carbonate-rich groundwaters might be the source of carbonate, but this does not explain the marine debris found at the core of some coal balls.

Spicer (1989) suggested an alternative model that goes some way to explaining most coal ball formation, and one which also accommodates storm breaching and the occasional introduction of marine organisms into the mire. In this model the subsidence of fluviomarine deltas upon which most Carboniferous coal mires seem to have developed may have led to flooding of the pore spaces by marine waters. The penetration of the marine water wedge and the thickness of the freshwater cap above would depend on the rate of subsidence, proximity to the delta edges, and precipitation. These factors would determine the depth at which mineralization would take place and hence the degree of compaction of organic material exhibited in the coal ball. The marine water penetration would effect a change in buried peat chemistry without necessarily altering the vegetation ecology above. This model allows for and predicts that rapid subsidence would lead to storm breaching in some instances, but not necessarily produce overlying marine sediments.

6.2.4. Charcoalification

Tissues, particularly if they are woody, may also be preserved by charcoalification by wildfires or burial in pyroclastic flows from volcanoes. The occurrence of fossil charcoal in the rock record is extensive (Cope and Chaloner, 1980), but some authors (e.g., Schopf, 1975) have not considered all charcoal-like material, collectively termed fusain, to be the unique product of pyrolysis. However, the properties of fusain (White, 1933; Raistrick and Marshall, 1939; Stutzer and Noé, 1940; Cope, 1980; and numerous references in Scott, 1989) are identical to that of charcoal produced by burning, and advocates of a cold diagenetic origin for fusain production have been unable to suggest a suitable or even likely process. Fusain consists of 77–94% carbon and 2–3% hydrogen (Scott, 1989) and the high carbon content renders it biologically inert. However, both fusain and present-day charcoal retain three-dimensional tissues. Typically there is a shrinkage of between 20 and 50% depending on cell wall type and temperature (E. A. McGinnes *et al.*, 1971, 1976), and burning above 300°C causes the middle lamella between cells to disappear: a phenomenon that may be useful as a paleopyrometer (Prior and Alvin, 1983). Even delicate structures such as subcellular details of wood tracheids remain exquisitely preserved (Scott, 1989), and leaves (Alvin, 1974; Harris, 1981) or even flowers (Friis and Skarby, 1981) with internal structure are not an uncommon feature of some sediments. Although brittle and easily crushed, charcoal is light, floats easily, and can be transported over long distances, by river action or in turbidity flows or both (Spicer, 1989).

7. Plant Assemblages: Taphonomic vs. Ecological Factors

The composition and structure of any fossil assemblage are the result of a combination of taphonomic factors (transport, degradation, diagenesis) and ecological pattern (community composition and structure). No longer is the paleobotanist confined to the construction of species lists and analogy for paleoecological studies, for taphonomy (albeit still in its infancy) provides key insights for isolating and quantifying the relative importance of those factors. However, much of the above discussion has involved small-scale or laboratory observations and it is often difficult, or even misleading, to extrapolate too readily to larger-scale natural systems. Based on laboratory experiments one might assume that the complexities of differential transport would lead to assemblages unrelated to the composition or structure of the source vegetation, but this assumption needs to be evaluated in real systems. The only way to understand the integration of transport variables at larger scales is to make field observations.

7.1. Fluvial Transport and the Question of Transport Distances

Older paleobotanical literature abounds with speculation concerning the distances over which plant remains might be transported prior to deposition (Black, 1929) and some of the earliest taphonomic observations related to this problem (e.g., Berry, 1906).

Spicer and Wolfe (1987) examined plant debris transport in six drainage systems flowing into Trinity Lake, northern California. Here, although the composition of debris accumulations in the fluviolacustrine deltas reflected well the differing vegetational types growing in the six drainage systems, only one specimen in over 1 million potential megafossil specimens studied could be identified as having been transported from vegetation 1000 m above the lake. The maximum distance this specimen (an abraded cone scale of *Pinus balfouriana*) could have traveled was only 1.5 km. Part of the explanation for the rarity of high-altitude material is the dilution effect by locally derived debris.

The above example is concerned with steep gradient transport over substrates that are coarse in relation to plant particle size. The advantages of high flow rates are offset by a high proportion of in-stream obstacles and a high channel perimeter-to-volume ratio which increases trapping and stranding potential. In studies in the Mobile Delta, Alabama, Gastaldo *et al.* (1985) noted that crevasse splay deposits yielded a high proportion (34 out of 37 taxa identifiable to species) of material that was not assignable to local community members. Most of the debris represented ''pine savanna–bay forest upland'' communities of the upper delta plain. From this it would appear that the larger stream size of delta floodplains favors a higher proportion of long-distance transport than the small, high-energy systems of high-relief areas. The probability of finding truly high-altitude vegetation (e.g., >1000 m) represented in fluviomarine deposits is, however, so small as to be virtually nonexistent.

7.2. Lacustrine Deposition

Lake geometry plays a very important role in determining the extent to which lake assemblages represent accurately the composition and structure of source vegetation. Large shallow lakes tend to be well oxygenated and thermal overturn in seasonal environments tends to destroy lake bottom organics. However, lakes that are deep in relation to their area and do not overturn tend to have organic-rich bottom sediments because oxygen diffusion rates are proportional to the area of the water/air interface (Hutchinson, 1957). In addition, oxygen-producing rooted aquatic macrophytes are limited to shallow water. Equally, lakes with a high perimeter-to-surface-area ratio will tend to be low in oxygen because the area of water/air interface is limited, and all parts of the lake are a short distance from the lake margin communities and thus will receive a high organic input. Where the lake diameter is significantly greater than the height of the surrounding vegetation differential wind transport in favor of sun leaves will take place, leading to skewed assemblage composition.

7.2.1. Fluviolacustrine Deltas

7.2.1a. Low-Energy Systems. In most large lake systems the only place in which plant material has any significant preservation potential is where burial rates are enhanced in the vicinity of inflowing streams. The dynamics of these fluviolacustrine deltas offers significant potential for reconstructing not only the positions of source taxa relative to the delta, but also seral development and hence vegetation dynamics.

A model describing vertical two-dimensional flow fields over fluviolacustrine delta profiles (Jopling, 1963, 1965a,b) was applied to the deposition of leaf material by Spicer (1981) and Spicer and Wolfe (1987). The model recognizes three flow fields: (a) residual stream flow, (b) a zone of mixing, and (c) a zone of backflow (Fig. 11). The trajectory of a sedimentary particle, whether it be a sand grain or a leaf, can be predicted from the stream depth in relation to lake depth, the vertical position of the particle in the stream water column in residual stream flow, its settling velocity, and the stream flow velocity.

In spite of the complex behavior of many, essentially two-dimensional, plant parts at various states of saturation, Jopling's model can be used to describe the formation of plant debris concentrations in fluviolacustrine deltas. Spicer (1981) has shown that, under low-energy conditions where

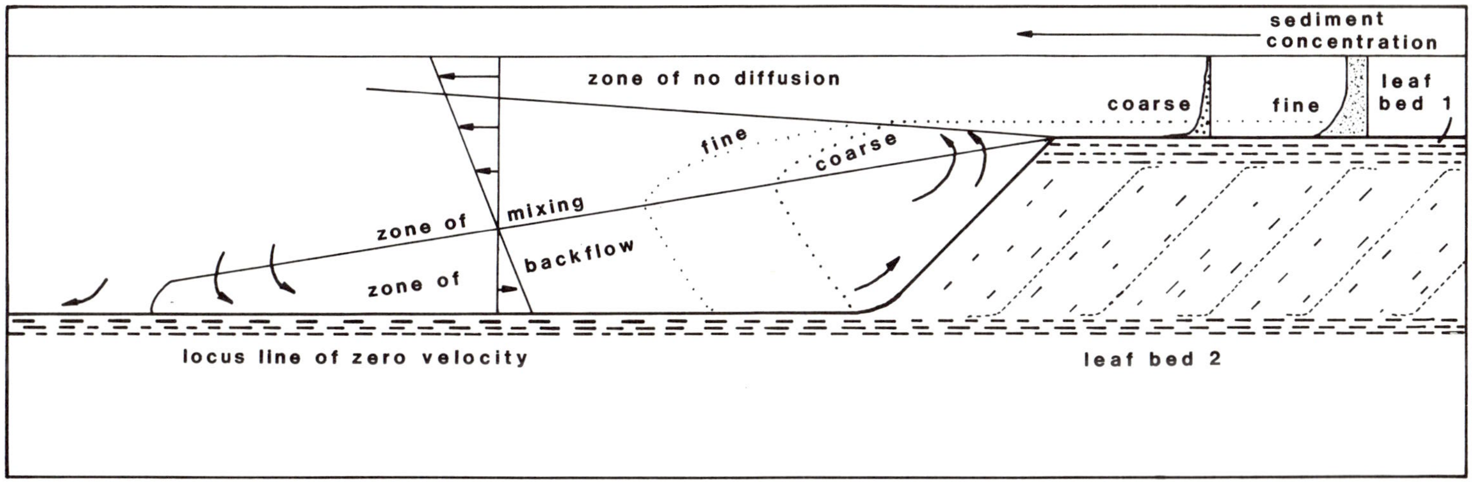

FIGURE 11. Diagram of the vertical flow fields generated over a fluviolacustrine delta front. The zone of no diffusion is made up of residual stream flow, the size of the zone of mixing depends on the height of the foreset slope and stream velocity, and the zone of backflow sweeps material into the toesets. [After Jopling (1963).]

the settling velocity of very fine to colloidal inorganic particles is less than that of the plant debris, two discrete plant assemblages or horizons develop that reflect spatial and temporal pattern in the source vegetation (Fig. 12). A lower, mostly lake bed horizon is formed by windblow and floating transport into the lake. The initial composition of this horizon reflects to a large extent that of the lake margin vegetation, but differential degradation can result in significant species biases. However, this effect is lessened when, due to delta advance, sedimentation rates of both organic and inorganic particles increase in the toeset region. In core the uppermost parts of this lower (lake bed) horizon are likely to be better preserved and contain more stream-transported taxa than the underlying debris because of the higher sedimentation rates due mostly to the deltaward sweeping of the backflow field. In the foreset region inorganic sedimentation rates are high relative to that of the plant material, although isolated plant fragments occur frequently. At the top of the foreset, however, sedimentation rates of stream-transported plant debris are high because, as the energy of the stream flow lessens, the plant debris is the first to settle out. This produces an upper plant horizon rich in taxa growing extralocal to the lake upstream and, because of mechanical degradation, many plant parts will be fragmented. Thus, there is a vertical differentiation of taxa between the two horizons which reflects spatial, not temporal, pattern in the source communities.

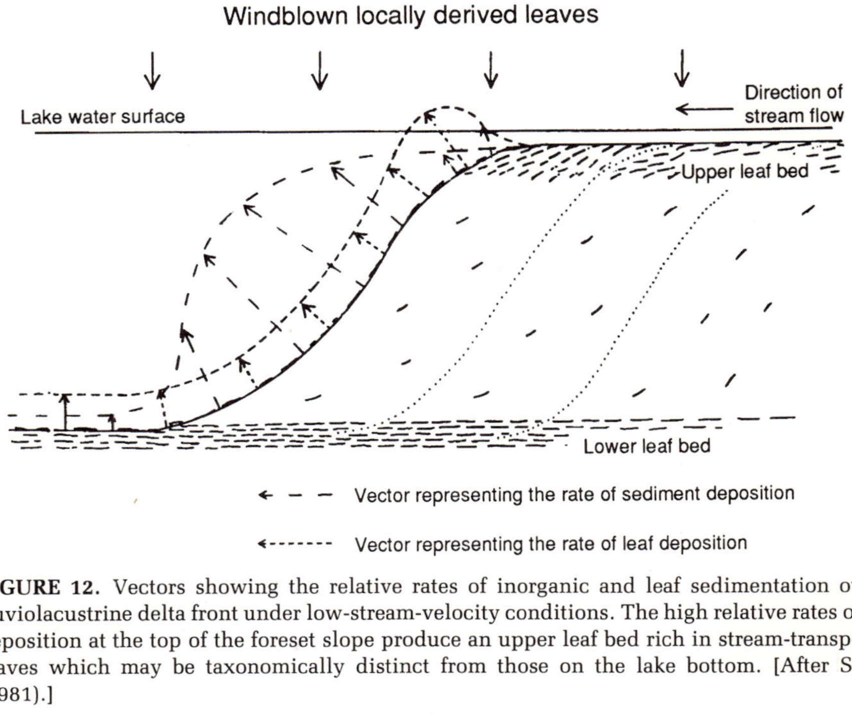

FIGURE 12. Vectors showing the relative rates of inorganic and leaf sedimentation over a fluviolacustrine delta front under low-stream-velocity conditions. The high relative rates of leaf deposition at the top of the foreset slope produce an upper leaf bed rich in stream-transported leaves which may be taxonomically distinct from those on the lake bottom. [After Spicer (1981).]

The species (in the paleobotanical sense of isolated organs) composition of the two horizons is not constant, however. As the delta advances, its surface will become colonized by a succession of communities that will filter out stream-transported (extralocal) debris (McQueen, 1969) and contribute their own litter to the system. Thus, as lake infilling progresses, the composition of the two horizons will converge. A transect along the upper horizon, starting in the oldest part of the delta, parallel to the direction of infilling, will reveal a change in species composition reflecting the temporal change in the delta flora as the colonizing communities develop.

7.2.1b. High-Energy Systems. In high-energy environments this binary horizon pattern breaks down (Spicer and Wolfe, 1987) because here the stream is capable of transporting larger inorganic clasts with higher fall velocities. The fall velocities of the bulk of the plant material will remain within the usual limits because the size of the clasts is determined by the abscission units of the plants and not erosive processes. According to the Jopling model, under these high-energy conditions the plant debris should be in suspension as it passes over the top of the foresets and much of it will be swept well clear of the foreset slope in residual stream flow, and reside longer in the zone of mixing. Some may be swept back into the toeset area (Fig. 12). The foresets, by definition, receive most of the inorganic sediment, but very high deposition rates may lead to slumping, tangential contacts with the bottomsets, and rapid burial of plant debris in the toeset position. Plant debris with high fall velocities (fully saturated diaspores and woody debris) also tends to accumulate in the foresets, while more delicate material with low fall velocities is swept lakeward (Fig. 13).

Thus, high-energy systems lead to a breakdown in the binary assemblage pattern with a loss of attendant spatial information as regards the ralative growing positions of the source taxa, and pronounced hydraulic sorting. However, taken overall, they can yield a statistically more valid sample of the vegetation within the drainage basin than is preserved in the low-energy system, which preserves more locally derived, yet partially degraded material. This is because high-energy flow is often associated with storm conditions which introduce fresh traumatically removed plant parts, erosion of forest floor litter, and rapid transport and burial. Not only are a wide range of plant parts simultaneously introduced into a depositional setting, but selective biological breakdown is minimized by rapid burial (Spicer, 1980).

Although the foregoing is presented in terms of fluviolacustrine deltas, the processes involved are those common to most "hydraulic jump" situations and may be found in fluvial channels or even marine settings. Under these circumstances, however, interpretation of changes in species composition could not be made in terms of spatial or temporal patterns in the source vegetation, but they could be used to design sampling strategies that would, for example, represent maximum species diversity.

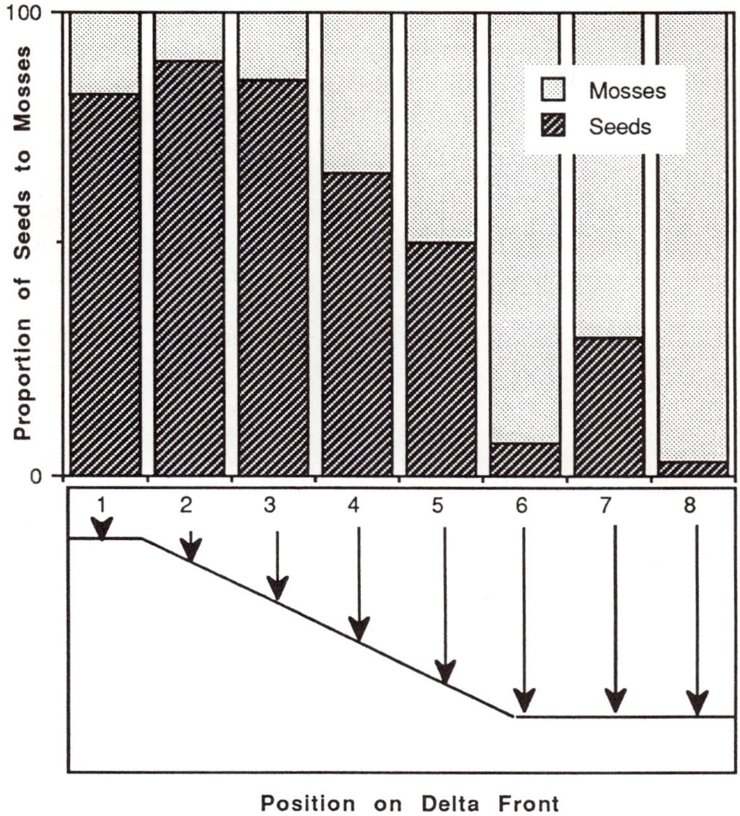

FIGURE 13. Under high stream velocities plant parts with a high saturated settling velocity (such as seeds) will accumulate on the delta slope, whereas particles with low settling velocities (such as moss fragments) are carried into the lake. [After Spicer (1980).]

7.3. Vulcanism and the Plant Fossil Record

The richness of the plant fossil record is due in large part to vulcanism. Communities may be preserved essentially intact by the influx of mineralizing fluids of volcanic origin, for example, the classic Devonian Rhynie Chert (Kidston and Lang, 1917–1921) or the Early Carboniferous Oxroad Bay floras (Alvin, 1965), or inundation by volcaniclastics, as in the fossil forests of Yellowstone National Park (Fisk, 1976) or the Antarctic Peninsula (Jefferson, 1982; Creber and Chaloner, 1984). Still more assemblages are allochthonous and owe their existence to the rapid influx of fine ash. Examples here include innumerable Tertiary leaf assemblages of the Pacific Northwest of the United States (e.g., Axelrod, 1964, 1966, 1985; Wolfe, 1964, 1969; Wolfe and Wehr, 1987; MacGinitie, 1933, 1941, 1953).

Detailed plant taphonomic studies in modern volcanic terrains are currently limited to those based on the 1980 eruptions of Mount Saint Helens (Coffin, 1983; Fritz, 1986; Fritz and Harrison, 1985; Karowe and Jefferson, 1987; Spicer, 1989), the 1982 eruptions of El Chichón (southern Mexico) (Burnham and Spicer, 1986; Spicer, 1989) and the 1985 eruptions of Nevado del Ruis (Colombia) (Fritz, 1986), although observations relevant to plant fossilization are recorded earlier (e.g., Dorf, 1945). All these volcanoes are of the steep-sided explosive type that generate air-fall tephra, pyroclastic flows, and lahars, but seldom lava. Consequently, vegetation proximal to the active vents suffers catastrophic devastation by explosive destruction of the cone and/or surges produced by eruptive column collapse.

7.3.1. Ash-Fall Effects: Evidence from Mount Saint Helens and El Chichón

Before May 18 1980 Mount Saint Helens (MSH) was a steep-sided, symmetrical volcano rising to a height of 2950 m, while the surrounding foothills-type topography rarely exceeded 1250 m. Principal drainages from MSH were the Toutle River to the west and the Muddy River to the east. The lower slopes of the mountain and the surrounding countryside supported a mature mixed conifer forest with angiosperm trees mostly confined to lake and river margin sites. Catastrophic failure of the northern flank of MSH resulted in a lateral explosive eruption in which the collapsing mountainside was propelled forward initially at 200 m/sec and in 13 min had fanned out over an area of 600 km^2 (Lipman and Mullineaux, 1981, and papers therein).

Proximal vegetation was inundated first by relatively cool cone slope material. The high energy of this impact stripped leaves and branches from trees, and trunks were shattered and/or snapped from the rootstock. At MSH within 8–12 km of the vent even the stumps and soil were removed down to resistant country rock (Spicer, 1989). Nevertheless, relatively low temperatures resulted in little charring except at the edges of the blast fan. Subsequent hot pyroclastic flows baked the underlying cone debris, charcoalifying some of the buried plant matter, thus increasing its preservation potential. Unless fusainized, the long-term preservation potential of plant material is low, as the porosity of the blast and debris flow material is great and therefore permits oxygen penetration into the sediments. However, large organic masses such as tree stumps buried by previous eruptions have survived long enough for mineralization to be initiated (Karowe and Jefferson, 1987).

Approximately 8 km from the vent the energy of the blast fan had decreased sufficiently for topography to play an important role in influencing its effect on the vegetation. On hillsides facing away from the vent some trees remained standing to a height of several meters but were shattered. Beyond 12 km only branches and bark on the sides of the trees facing the blast were removed from standing trees. At the edge of the blast fan (approximately 30 km from the vent) convective forces lifted the blast cloud above tree height, but significant tree kill occurred from temperatures hot enough to melt epicuticular waxes (Winner and Casadevall, 1981).

Ash fall often has surprisingly little effect on living vegetation provided that the ash is cool when it comes into contact with the plants and the thickness of ash deposition is small compared to the size of the plants (Spicer, 1989). Unless ash inundation is heavy and repeated, air-fall ash is unlikely to substantially alter vegetation structure or composition. In some situations ash fall may actually enhance plant growth by providing nutrients and temporarily reducing predator populations.

In the case of explosive eruptions like that of Mount Saint Helens hot lateral blasts can carry leafy debris high into the atmosphere by convection (Waitt, 1981). Thus air-fall ash may contain some allochthonous charred plant matter, but this is likely to represent only a small proportion of the ash volume.

Ash fall also plays a major role in plant preservation. Low-stature groundcover plants tend to be buried rapidly and even delicate plants such as *Selaginella* can be preserved even in tropical environments (Burnham and Spicer, 1986). Ash coatings often stick to leaf surfaces either by melting epicuticular waxes or by adhering to surface moisture (volcanic eruptions often generate local precipitation, so ash fall is usually accompanied by rain). Canopy leaves are particularly vulnerable in this respect. The ash coating reduces the photosynthetic efficiency of the leaf to a point where food production is exceeded by respiration and the leaf becomes predisposed to abscission. Subsequent ash falls may dislodge such leaves, which rapidly become incorporated in the ash deposits. Sequential ash falls affect different structural components of the vegetation at different times, with canopy leaves receiving most ash fall and thus falling first, followed by subcanopy leaves when those of the canopy have been lost. This gives rise to a series of leaf layers each with characteristic species compositions. By careful analysis of the pattern of deposition the structural organization of the vegetation may be deduced (Burnham and Spicer, 1986) (Fig. 14).

Ash layers are frequently encountered in alpine lakes of the Pacific Northwest of the United States (e.g., Dunwiddie, 1986) and undoubtedly these play a role in preserving lake-bottom organics by restricting oxygen and limiting the action of detritivores. Such inundations of ash may be responsible for preservation of the mire community in Cretaceous coals of North America. Normally compression destroys all plant megafossil evidence in coals and the only identifiable remains are those associated with water-transported clastics. However, such plant remains could have been washed into the mire and do not necessarily represent the mire community itself. Air-fall ash layers, on the other hand, are rarely the source of exotic plant debris and by infiltrating between leaves prevent compressional homogenization. Thus, ash layers in coal provide important windows into mire community composition.

7.3.2. Lacustrine Deposition in Volcaniclastic Terrains

The rapid emplacement of large quantities of old and new volcaniclastic material into the drainages surrounding a volcano may seal preexisting

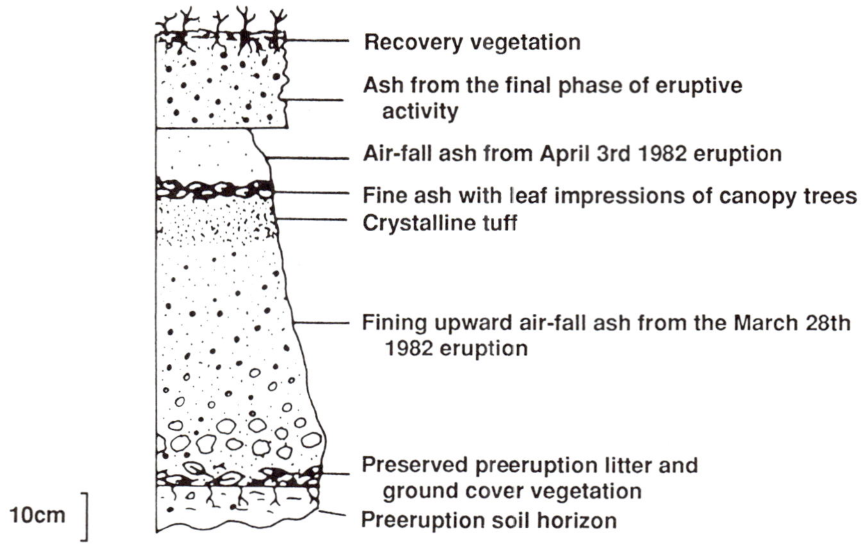

FIGURE 14. Vertical section through air-fall ashes near El Chichón, southern Mexico, after the 1982 eruptions. Initial air-fall ash resulted in the burial of ground cover plants such as *Selaginalla*, while subsequent falls defoliated canopy trees, thus preserving a record of the vertical structure of the vegetation. [After Burnham and Spicer (1986).]

proximal deposits while simultaneously creating new depositional basins. In the first few months after the MSH eruption proximal drainage patterns became reestablished and in so doing reworked much of the eruption-buried plant debris into channel lags and fluviolacustrine deltas. Lateral lakes, some 4 km long, formed in tributary valleys leading to the drainages choked by the debris flow (Fig. 15). Many of these newly created lakes drain naturally as they overtop the debris dams and downcut through the unconsolidated volcaniclastics (Spicer, 1989), but sometimes they remain intact long enough to record the development of posteruption recovery vegetation (the pre-1980 Spirit Lake was one such system). Should the lakes survive and accumulate organic debris from preeruption and recovery vegetation, subsequent eruptions may seal the deposits and enhance preservation. Such a process appears to have occurred at Florissant, Colorado, where a Tertiary intermontain basin has preserved extremely high-quality insect and plant assemblages (MacGinitie, 1953).

Within these lakes fluviolacustrine deltas may occur in a similar fashion to that described in Section 7.2.1. The important difference in posteruption systems is, however, the rapid influx of unconsolidated ash/blast sediment following an eruption. The deltaic units therefore tend to be poor in plant debris due to dilution by volcaniclastics (Spicer, 1989) and floating, and therefore different hydraulic behavior of fusainized material will complicate interpretations. Nevertheless, as vegetation recovery proceeds, the capacity of deltaic systems to preserve spatial and temporal pattern in source commu-

FIGURE 15. Coldwater North lateral lake formed after the May 18 1980 eruption of Mount Saint Helens, Washington. The lake is 4 km long and 40 m deep and did not exist prior to 1980. It was formed by a debris flow traveling from right to left in the foreground damming the stream valley now drowned by the lake. Subsequent reworking of volcaniclastics and buried plant debris is forming a delta shown in the center of the photograph.

nities offers the potential of recording vegetation dynamics in considerable detail.

7.3.3. Vegetation Dynamics

Posteruption vegetation recovery has traditionally been a favorite study topic of botanists and a rich literature exists documenting the arrival of taxa to otherwise isolated areas (e.g., Eggler, 1948, 1959, 1963; Gadow, 1930; Hendrix, 1981; Adams and Adams, 1981; Manko, 1975, and references therein; Halpern *et al.*, 1990). However, few studies have been conducted from the point of view of the paleobotanical record.

The extent to which any given volcanic event will record vegetation dynamics depends in large part on the prevailing climate and the timing of eruptions with respect to any seasonal cycle. In a comparative study of Mount Saint Helens and El Chichón, Spicer (1989) noted that there were considerable differences not only in the rate of recovery, but also in the differences between the successional and climax vegetation that could be related to climatic differences between the two sites. Mount Saint Helens is situated at latitude 46°N, where the natural climax vegetation is a mixed coniferous forest and the effective growing season limited to 3 months or so.

El Chichón is at 17°N and is surrounded by a highly diverse paratropical rainforest rich in angiosperms actively growing throughout the year.

In both cases the primary colonizer was not an angiosperm. At El Chichón the young sporophytes of the fern *Pityrogramma calamelenos* were seen to occupy even well-drained pumice slopes of the remnant cone itself 2 years after the eruption (Spicer *et al.*, 1985). This was interpreted to be due to the high spore productivity and dispersal capacity of fertile fronds arising from regenerating ash-buried rhizomes on the volcanic slopes, coupled with constant high humidity that allowed completion of the life cycle at the gametophyte stage. In seasonally dryer environments around MSH sexual reproduction of pteridophytes was limited, but *Equisetum* (horsetail) was exceptionally successful at colonizing valley bottoms and slopes within the blast zone by vegetative reproduction from blast-fragmented rhizomes.

Longer-term recovery also appears to be significantly different. At MSH herbaceous and bushy angiosperms had become well established 4 years after the main eruption. Taxa commonly found in sites disturbed by fire or logging, such as *Epilobium*, *Sambucus*, and *Rubus*, together with a few grasses covered most hill slopes and were typically 1 m or so in height, while naturally regenerating conifers were barely 3 cm tall. This underscores the advantages of the rapid angiosperm life cycle compared to that of conifers. Another important factor in angiosperm recovery at MSH was the timing of the eruption in relation to the season. Evergreen conifers were killed by heat and traumatic damage resulting from the high blast (wind) resistance offered by the presence of leaves. The angiosperms, on the other hand, were still in bud; the leaves were protected and the bare branches suffered little blast damage. If the eruption had occurred in summer, the angiosperms would undoubtedly have suffered more.

It is clear from these preliminary studies at MSH that during the early stages of succession the nature of the vegetation may be changed fundamentally and one might expect a radically different paleoclimatic signal from any resulting fossil assemblage. The restricted growing season and lower mean annual temperature must be significant factors in reducing diversity in the climax communities and thereby constraining the pool of taxa from which recovery can take place.

At El Chichón the higher diversity of the preeruption vegetation ensured a rich pool of taxa available for recovery and within 2 years of the eruption woody angiosperms 4 m high had colonized devastated areas. Although convectively rising surges at the edge of the devastated zone at El Chichón had killed canopy trees, climbers survived. These climbers were actively invading bare ash surfaces within 2 years of the eruption and would have considerably biased any fossil leaf population, and paleoclimate signal, preserved in those areas. Species selection due to the timing of the eruption was not a factor at El Chichón; but this is not necessarily a situation common to all past low-latitude eruptions, because throughout much of the Phanerozoic equatorial vegetation appeared to have experienced seasonality in the form of dry/wet variations (Spicer and Chapman, 1990).

From these two brief studies it is obvious that the effects of eruptions on vegetation will depend largely on the timing of the eruption with respect to the growth season, the composition of the preeruption vegetation, and the climate under which that vegetation has developed. This interplay will be reflected in the resulting fossil record. Fortunately for taphonomists, there is ample evidence that the pattern and processes of assemblage formation appear to have been highly similar between the Mount Saint Helens and El Chichón eruptions even though the direction, magnitude, and mechanics of the eruptions were quite different. By understanding these processes it should be possible to isolate ecological from taphonomic variables and learn a great deal about not only the pattern of ancient vegetation but the dynamics of vegetation composed of extinct plant forms.

8. Conclusions

I have attempted to concentrate on processes rather than patterns of assemblage formation. The fragmentary nature of plant remains, the spatial and temporal discontinuity of nonmarine deposits, and the variability of the physical and chemical aspects of diagenesis provide a complex interplay of processes that are vital to understand if plant fossil assemblages are to be correctly interpreted. To merely present patterns to be looked for without understanding the processes that produce them would lead to very restricted and probably inaccurate interpretations and be little better than the purely intuitive interpretations that existed prior to the development of plant taphonomy.

The requirement that potential fossil material be deposited in water and isolated from oxygen not only limits the proportion of the plant biomass alive at any one time that can be preserved, but it also implies the paradox that some degree of decay is necessary in some or all of the potential assemblage before preservation can begin. This may be a simple case of selective degradation by detritivores or microbial agents (some taxa are preferentially decayed, thus enhancing the preservation potential of others by increasing their residence time in the sediments), a case of altering pH and Eh conditions by decay processes so that further decay is slowed, or it may be the more subtle case of generating molecular templates for the initiation of mineralization.

It is clear that different plant organs have differing preservation potentials, depending on their chemical and tissue composition, the composition of their physical, chemical, and biological environments at the time of deposition, the numbers in which they are produced by a unit of vegetation, and the proximity of that vegetational unit to a site of deposition. In general, macroscopic plant parts have a very limited potential for significant transport away from the point of growth if they are to be found in any reasonable concentration in a fossil deposit. Thus, the spatial resolution available for reconstructing ancient vegetation is high. As regards reworking, with the exception of wood and some woody seeds, plant megafossils have a very limited reworking (time averaging) potential and thus also yield data of a high temporal resolution.

Taphonomic studies have shown that while there may be differential preservation in many instances, these biases may be understood and sampling strategies devised to recover rarely preserved components. The potential exists for reconstructing past vegetation in intricate spatial and temporal detail, notwithstanding the complexities of the nonmarine sedimentary record. Indeed, it is these complexities that often provide the mosaic of depositional environments that capture a wide variety of taphonomic windows on the past.

Most taphonomic research has been conducted in temperate environments. More work is needed in both polar and tropical conditions to reflect the range of possible taphonomic influences that have existed in the past. In particular, considerably more research is needed on decay and mineralization processes. Similarly, the sedimentology of palynodebris (including cuticles) is not well understood and yet such work is vital to making full use of the abundant palynological record. At the molecular level we are only just beginning to understand some of the more common and simpler processes, although the next few years should see major advances in this area.

References

Adams, A. B., and Adams, V. D., 1981, in: *Mount Saint Helens: One Year Later* (S. A. C. Keller, ed.), Eastern Washington University Press, Cheney, Washington, pp. 93–103.

Allen, J. R. L., 1984, Experiments on the settling, overturning and entrainment of bivalve shells and related models, *Sedimentology* **31**:227–250.

Allison, P. A., 1988, Taphonomy of the Eocene London Clay biota, *Palaeontology* **31**:1079–1100.

Alvin, K. L., 1965, A new fertile lycopod from the lower Carboniferous of Scotland, *Palaeontology* **8**:281–293.

Alvin, K. L., 1974, Leaf anatomy of Weichselia based on fusainised material, *Palaeontology* **17**:587–598.

Alvin, K. L., 1983, Reconstruction of a Lower Cretaceous conifer, *Bot. J. Linn. Soc.* **86**:169–176.

Axelrod, D. I., 1964, The Miocene Trapper Creek Flora of southern Idaho, *Univ. Calif. Publ. Geol. Sci.* **51**:1–119.

Axelrod, D. I., 1966, The Eocene Copper Basin flora of northeastern Nevada, *Univ. Calif. Publ. Geol. Sci.* **59**:1–125.

Axelrod, D. I., 1985, Miocene Floras from the Middlegate Basin West Central Nevada, *Univ. Calif. Publ. Geol. Sci.* **129**:1–279.

Barghoorn, E. S., and Tyler, S. A., 1965, Microorganisms from the Gunflint Chert, *Science* **147**:536–577.

Beck, C. B., 1962, Reconstruction of *Archeopteris* and further consideration of its taxonomic position, *Am. J. Bot.* **49**:373–382.

Berner, R. A., 1970, Sedimentary pyrite formation, *Am. J. Sci.* **268**:1–23.

Berner, R. A., 1984, Sedimentary pyrite, an update, *Geochim. Cosmochim. Acta* **48**:605–615.

Berry, E. W., 1906, Leaf rafts and fossil leaves, *Torreya* **6**:246–249.

Black, M., 1929, Drifted plant-beds of the Upper Estuarine Series of Yorkshire, *Q. J. Geol. Soc.* **85**:389–439.

Bonnevie-Svendsen, C., and Gjems, O., 1957, Amount and chemical composition of litter from larch, beech, Norway spruce and Scots pine stands and its effect on the soil, *Medd. Norske Skogsfors Ves.* **48**:111–174.

Brack-Haynes, S. D., 1978, On the megagametophytes of two lepidodendracean cones, *Bot. Gaz.* **139**:140–146.

Bray, J. R., and Goreham, E., 1964, Litter production in forests of the world, *Adv. Ecol. Res.* **2**:101–158.

Briggs, D. E. G., and Williams, S. H., 1981, The restoration of flattened fossils, *Lethaia* **14**:157–164.

Burnham, R. J., 1989, Relationships between standing vegetation and leaf litter in a paratropical forest: Implications for paleobotany, *Rev. Palaeobot. Palynol.* **58**:5–32.

Burnham, R. J., and Spicer, R. A., 1986, Forest litter preserved by volcanic activity at El Chichón, Mexico: A potentially accurate record of the preeruption vegetation, *Palaios* **1**:158–161.

Chandler, M. E. J., 1964, *The Lower Tertiary Floras of Southern England IV. A Summary and Survey of Findings in the Light of Recent Botanical Observations*, British Museum (Natural History), London.

Chaney, R. W., and Sanborn, E. I., 1933, The Goshen Flora of west central Oregon, *Carnegie Inst. Wash. Publ.* **439**:1–103.

Christophel, D. C., and Lys, S., 1986, Mummified leaves of two new species of Myrtaceae from the Eocene of Victoria, Australia, *Aust. J. Bot.* **34**:649–662.

Clift, R., Grace, J. R., and Weber, M. E., 1978, *Bubbles, Drops and Particles*, Academic Press, New York.

Coffin, H. G., 1983, Erect floating stumps in Spirit Lake Washington, *Geology* **11**:298–299.

Collinson, M. E., 1983a, Accumulations of fruits and seeds in three small sedimentary environments in southern England and their palaeoecological implications, *Ann. Bot.* **52**:583–592.

Collinson, M. E., 1983b, *Fossil Plants of the London Clay* (Palaeontological Association Filed Guide to Fossils No. 1), Palaeontological Association, London.

Cope, M. J., 1980, Physical and chemical properties of coalified and charcoalified phytoclasts from some British Mesozoic sediments: An organic geochemical approach to palaeobotany, in: *Advances in Organic Geochemistry* (A. C. Douglas and J. R. Maxwell, eds.), Pergamon Press, Oxford, pp. 663–677.

Cope, M. J., and Chaloner, W. G., 1980, Fossil charcoal as evidence of past atmospheric composition, *Nature* **283**:647–649.

Creber, G. T., and Chaloner, W. G., 1984, in: *Fossils and Climate* (P. J. Brenchley, ed.), Wiley, New York, pp. 49–74.

De Vries, H., Bredemeijer, G., and Heinen, W., 1967, Decay of cutin and cuticular compounds by soil microorganisms in their natural environment, *Acta Bot. Neerl.* **16**:102–110.

Dorf, E., 1945, Observations on the preservation of plants in the Paricutin area, *Trans. Am. Geophys. Union* **26**:257–260.

Dunwiddie, P. W., 1986, A 6000-year record of forest history on Mount Rainier, Washington, *Ecology* **67**:58–68.

Dunwiddie, P. W., 1987, Macrofossil and pollen representation of coniferous trees in modern sediments from Washington, *Ecology* **68**:1–11.

Eggler, W. A., 1948, Plant communities in the vicinity of the volcana El Paricutin in Mexico after 2½ years of eruption, *Ecology* **29**:415–436.

Eggler, W. A., 1959, Manner of invasion of volcanic deposits by plants, with further evidence from Paricutin and Jorullo, *Ecol. Monogr.* **29**:267–284.

Eggler, W. A., 1963, Plant life of Paracutin volcano, eight years after activity ceased, *Am. Midl. Nat.* **69**:38–68.

Evans, W. D., and Amos, D. H., 1961, An example of the origin of coal balls, *Proc. Geol. Assoc.* **72**:445–454.

Feliciano, J. M., 1924, The relation of concretions to coal seams, *J. Geol.* **32**:230–239.

Ferguson, D. K., 1971, The Miocene flora of Kreuzau, Western Germany, I. The leaf remains, *Verh. K. Ned. Akad. Wet. Afd. Natuurkd. Reeks 2* **60:**1–297.

Ferguson, D. K., 1985, The origin of leaf-assemblages—New light on an old problem, *Rev. Palaeobot. Palynol.* **46:**117–188.

Fisk, L. H., 1976, The Gallatin "Petrified Forest": A review, *Montana Bur. Mines Geol. Spec. Publ.* **73:**53–72.

Friis, E. M., and Skarby, A., 1981, Structurally preserved angiosperm flowers from the Upper Cretaceous of southern Sweden, *Nature* **291:**485–486.

Fritz, W. J., 1986, Plant taphonomy in areas of explosive volcanism, in: Land Plants, *Univ. Tenn. Dep. Geol. Sci. Stud. Geol.* **15:**1–9.

Fritz, W. J., and Harrison, S., 1985, Transported trees from the 1982 Mount St. Helens sediment flows: Their use as paleocurrent indicators, *Sediment. Geol.* **42:**49–64.

Gadow, H., 1930, *The History of the Volcano of Jorullo and the Reclamation of the Devastated District by Animals and Plants*, Cambridge University Press, Cambridge.

Gastaldo, R. A., Carroll, S. M., and Douglass, D. P., 1985, A model for the incorporation of plant detritus within clastic accumulating interdistributary bays, *Geol. Soc. Am. Abstr. Prog.* **17:**588.

Gastaldo, R. A., Douglass, D. P., and McCarroll, S. M., 1987, Origin, characteristics, and provenance of plant macrodetritus in a Holocene crevasse splay, Mobile Delta, Alabama, *Palaios* **2:**229–240.

Halpern, C. B., Frenzen, P. M., Means, J. E., and Franklin, J. F., 1990, Plant succession in areas of scorched and blown-down forest after the 1980 eruption of Mount St. Helens, Washington, *J. Vegeta. Sci.* **1:**181–194.

Hanlon, R. D. G., 1982, The breakdown and decomposition of allochtonous and autochthonous plant litter in an oligotrophic lake (Llyn Frongoch), *Hydrobiologia* **88:**281–288.

Harris, T. M., 1974, *Williamsoniella ligneri:* Its pollen and the compression of spherical pollen grains, *Palaeontology* **17:**125–149.

Harris, T. M., 1981, Burnt ferns from the English Wealden, *Proc. Geol. Assoc. Lond.* **92:**47–58.

Havinga, A. J., 1984, Pollen decay and pollen preservation, *Pollen Spores* **26:**541–558.

Hendrix, L. B., 1981, Post eruption succession on Isla Fernandia, Galapagos, *Madroño* **28:**242–254.

Hutchinson, D. E., 1957, *A Treatise on Limnology*, 1, Wiley, New York.

ICCP, 1963, *International Committee for Coal Petrology Handbook*, 2nd ed., CRNS, Paris.

Jefferson, T. H., 1982, Fossil forests from the Lower Cretaceous of Alexander Island, Antarctica, *Palaeontology* **25:**681–708.

Jopling, A. V., 1963, Hydraulic studies on the origin of bedding, *Sedimentology* **2:**115–121.

Jopling, A. V., 1965a, Hydraulic factors controlling the shape of laminae in laboratory deltas, *J. Sediment. Petrol.* **35:**777–791.

Jopling, A. V., 1965b, Laboratory study of the distribution of grain sizes in crossbedded deposits, *Soc. Econ. Paleontol. Mineral. Spec. Publ.* **12:**53–65.

Karowe, A. L., and Jefferson, T. H., 1987, Burial of trees by eruptions of Mount St Helens, Washington: Implications for the interpretation of fossil forests, *Geol. Mag.* **124:**191–204.

Kaushik, N. K., and Hynes, H. B. N., 1971, The fate of dead leaves that fall into streams, *Arch. Hydrobiol.* **68:**465–515.

Kidston, R., and Lang, W. H., 1917–1921, On Old Red Sandstone plants showing structure, from the Rhynie Chert Bed, Aberdeenshire, Parts I–V, *Trans. R. Soc. Ed.* **51:**761–784; **52:**603–627, 643–680, 831–854, 855–902.

Knoll, A. H., 1985, Exceptional preservation of photosynthetic organisms in silicified carbonates and silicified peats, *Phil. Trans. R. Soc. Lond. B.* **311:**111–122.

Krassilov, V. A., 1969, Types of palaeofloristic succession and their causes, *Paleontol. Zh.* **3:**296–308.

Leary, R. L., and Thomas, B. A., 1989, *Lepidodendron aculeatum* with attached foliage: Evidence of stem morphology and fossilization processes, *Am. J. Bot.* **76:**283–288.

Leo, R. F., and Barghoorn, E. S., 1976, Silicification of wood, *Bot. Mus. Leaflts Harv. Univ.* **25:**1–46.

Lipman, P. W., and Mullineaux, D. R. (eds.), 1981, The 1980 Eruptions of Mount St. Helens, *U. S. Geol. Surv. Prof. Pap.* **1250:**1–844.
MacDonald, D. I. M., and Jefferson, T. H., 1985, Orientation studies of waterlogged wood: A paleocurrent indicator?, *J. Sediment. Petrol.* **55:**235–239.
MacGinitie, H. D., 1933, The Trout Creek Flora of southeastern Oregon, *Carnegie Inst. Wash. Publ.* **416:**21–68.
MacGinitie, H. D., 1941, A middle Eocene flora from the central Sierra Nevada, *Carnegie Inst. Wash. Publ.* **534:**1–178.
MacGinitie, H. D., 1953, Fossil plants of the Florissant beds, Clorado, *Carnegie Inst. Wash. Publ.* **599:**1–198.
Mamay, S. H., and Yochelson, E. L., 1962, Occurrence and significance of marine animal remains in American coal balls, *U.S. Geol. Serv. Prof. Pap.* **354I:**193–224.
Manko, Yu. L., 1975, in: *Thirteenth Pacific Science Congress, Vancouver, British Columbia*, pp. 203–204.
Mathews, C. P., and Kawalczewski, A., 1969, Disappearance of leaf litter and its contribution to production in the River Thames, *J. Ecol.* **57:**543–552.
McGinnes, E. A., Harlow, C. A., and Beall, F. C., 1976, Use of scanning electron microscopy and image processing in wood charcoal studies, in: *Scanning Electron Microscopy 1976*, Part VII, IIT Research Institute, Chicago, Illinois, pp. 543–548.
McGinnes, R. A., Kandeel, S. A., and Szopa, P. S., 1971, Some structural changes observed in the transformation of wood into charcoal, *Wood Fibre* **3:**77–83.
McQueen, D. R., 1969, Macroscopic plants remains in Recent lake sediments, *Tuatara* **17:**13–19.
Moseley, H. N., 1892, *Notes by a Naturalist on the "Challenger"*, 2nd ed., London.
Newbold, J. D., Elwood, J. W., Schulze, M. S., Stark, R. W., and Barmeier, J. C., 1983, Continuous ammonium enrichment of a woodland stream: Uptake kinetics, leaf decomposition, and nitrification, *Freshwater Biol.* **13:**193–204.
Nykvist, N., 1962, Leaching and decomposition of litter, *Oikos* **13:**232–248.
Petersen, R. C., and Cummins, K. W., 1974, Leaf processing in a woodland stream, *Freshwater Biol.* **4:**343–368.
Phillips, T. L., and DiMichele, W. A., 1981, Paleoecology of Middle Pennsylvanian age in southern Illinois-Herrin Coal Member at Sahara Mine no. 6, in: *Paleobotany, Paleoecology and Evolution* (K. J. Niklas, ed.), Praeger, New York, pp. 231–284.
Prior, J., and Alvin, K. L., 1983, Structural changes on charring woods of *Dichrostachys* and *Salix* from Southern Africa, *IAWA Bull. N. S.* **4:**197–206.
Rau, G. H., 1976, Dispersal of terrestrial plant litter into a subalpine lake, *Oikos* **27:**153–160.
Raistrick, A., and Marshall, C. E., 1939, *The Nature and Origin of Coal Seams*, English University Press, London.
Raiswell, R., 1971, The growth of Cambrian and Liassic concretions, *Sedimentology* **17:**147–171.
Raiswell, R., 1976, The microbiological formation of carbonate concretions in the Upper Lias of N.E. England, *Chem. Geol.* **18:**227–244.
Rex, G., 1983, The compression state of preservation of Carboniferous lepidodendrid leaves, *Rev. Palaeobot. Palynol.* **39:**65–85.
Rex, G., 1985, A laboratory flume investigation of the formation of fossil stem fills, *Sedimentology* **32:**245–255.
Rex, G., 1986, in: Systematic and Taxonomic Approaches in Palaeobotany (R. A. Spicer and B. A. Thomas, eds.), *Syst. Assoc. Spec. Vol.* **34:**17–38.
Rex, G. M., and Chaloner, W. G., 1983, The experimental formation of plant compression fossils, *Palaeontology* **26:**231–252.
Rolfe, W. D. I., and Brett, D. W., 1969, Fossilisation processes, in: *Organic Geochemistry* (G. Eglington and M. J. T. Murphy, eds.), Germany, pp. 213–244.
Rothwell, G. W., and Warner, S., 1984, *Cordaixylon dumusum* n. sp. (Cordaitales). 1. Vegetative structures, *Botanic. Gaz.* **145:**275–291.
Schopf, J. M., 1975, Modes of fossil preservation, *Rev. Palaeobot. Palynol.* **20:**27–53.

Scott, A. C., 1989, Observations on the nature and origin of fusain, *Int. J. Coal Geol.* **12**:443–475.

Scott, A. C., and Rex, G., 1985, The formation and significance of Carboniferous coal balls, *Phil. Trans. R. Soc. Lond. B* **311**:123–137.

Spicer, R. A., 1975, The sorting of plant remains in a recent depositional environment, Ph.D. thesis, University of London.

Spicer, R. A., 1977, The pre-depositional formation of some leaf impressions, *Palaeontology* **20**:907–912.

Spicer, R. A., 1980, The importance of depositional sorting to the biostratigraphy of plant megafossils, in: *Biostratigraphy of Fossil Plants* (D. L. Dilcher and T. N. Taylor, eds.), Dowden, Hutchinson and Ross, Stroudsburg, Pennsylvania, pp. 171–183.

Spicer, R. A., 1981, The sorting and deposition of allochthonous plant material in a modern environment at Silwood Lake, Silwood Park, Barkshire, England, *U. S. Geol. Surv. Prof. Pap.* **1143**:1–77.

Spicer, R. A., 1987, The significance of the Cretaceous flora of Northern Alaska for the reconstruction of the climate of the Cretaceous, *Geol. Jahrb. Reihe A* **96**:265–291.

Spicer, R. A., 1989, The formation and interpretation of plant fossil assemblages, *Adv. Bot. Res.* **16**:95–191.

Spicer, R. A., and Chapman, J. L., 1990, Climate change and the evolution of high-latitude terrestrial vegetation and floras, *Trends Ecol. Evol.* **5**:279–284.

Spicer, R. A., and Greer, A. C., 1986, Plant taphonomy in fluvial and lacustrine systems, in: Land Plants, *Univ. Tenn. Dep. Geol. Sci. Stud. Geol.* **15**:10–26.

Spicer, R. A., and Parrish, J. T., 1986, Paleobotanical evidence for cool North Polar climates in the mid-Cretaceous (Albian-Cenomanian), *Geology* **14**:703–706.

Spicer, R. A., and Thomas, B. A. (eds.), 1986, Systematic and Taxonomic Approaches in Palaeobotany, *Syst. Assoc. Spec. Vol.* **31**:1–321.

Spicer, R. A., and Wolfe, J. A., 1987, Taphonomy of Holocene deposits in Trinity (Clair Engle) Lake, Northern California, *Paleobiology* **13**:227–245.

Spicer, R. A., Burnham, R. J., Grant, P., and Glicken, H., 1985, *Pityrogramma calomelanos*, the primary, post eruption colonizer of Volcán Chichonal, Chiapas, Mexico, *Am. Fern J.* **75**:1–5.

Stopes, M. C., and Watson, D. M. S., 1908, On the present distribution and origin of the calcareous concretions in coal seams known as 'coal balls', *Phil. Trans. R. Soc. Lond. B* **200**:167–218.

Stutzer, O., and Noé, A. C., 1940, *Geology of Coal*, University of Chicago Press, Chicago, Illinois.

Taylor, T. N., and Millay, M. A., 1977, Structurally preserved fossil cell contents, *Trans. Am. Microsc. Soc.* **96**:390–393.

Thomas, B. A., 1986, The formation of large diameter plant fossil moulds and the Walton theory of compaction, *Geologic. J.* **21**:381–385.

Tyler, S. A., and Barghoorn, E. S., 1954, Occurrence of structurally preserved plants in Precambrian rocks from the Canadian Shield, *Science* **119**:606–608.

Waitt, R. B., Jr, 1981, Devastating pyroclastic density flow and attendant air fall of May 18—Stratigraphy and sedimentology of deposits, in: The 1980 Eruptions of Mount St. Helens (P. W. Lipman and D. R. Mullineaux, eds.), *U. S. Geol. Surv. Prof. Pap.* **1250**:439–458.

Walton, J., 1936, On the factors which influence the external forms of fossil plants: With descriptions of the foliage of some species of the Palaeozoic Equisitalaean genus *Annularia* Sternberg, *Phil. Trans. R. Soc.* **226**:219–237.

White, D., 1933, Role of water conditions in the formation and differentiation of common (banded) coals, *Econ. Geol.* **28**:556–570.

Wilkinson, H. P., 1983, Starch grain casts and moulds in Eocene (Tertiary) fossil mangrove hypocotyls, *Ann. Bot.* **52**:39–45.

Winner, W. E., and Casadevall, T. J., 1981, Fir leaves as thermometers during the May 18 eruption, in: The 1980 Eruptions of Mount St. Helens (P. W. Lipman and D. R. Mullineaux, eds.), *U. S. Geol. Surv. Prof. Pap.* **1250**:315–320.

Wolfe, J. A., 1964, Miocene floras from Fingerrock Wash, southwestern Nevada, *U. S. Geol. Surv. Prof. Pap.* **454-N:**N1–N36.

Wolfe, J. A., 1969, Neogene floristic and vegetational history of the Pacific Northwest, *Mondroño* **20:**83–110.

Wolfe, J. A., 1979, Temperature parameters of humid to mesic forests of eastern Asia and relation to forests of other regions of the northern hemisphere and Australasia, *U. S. Geol. Surv. Prof. Pap.* **1106:**1–37.

Wolfe, J. A., and Wehr, W., 1987, Middle Eocene dicotyledonous plants from Republic, northeastern Washington, *U. S. Geol. Surv. Bull.* **1597:**1–25.

Chapter 4

Taphonomy and Time-Averaging of Marine Shelly Faunas

SUSAN M. KIDWELL and DANIEL W. J. BOSENCE

SUSAN M. KIDWELL • Department of Geophysical Sciences, University of Chicago, Chicago, Illinois 60637. DANIEL W. J. BOSENCE • Department of Geology, Royal Holloway and Bedford New College, University of London, Egham, Surrey TW20 OEX, United Kingdom.

Taphonomy: Releasing the Data Locked in the Fossil Record, Volume 9 of *Topics in Geobiology*, edited by Peter A. Allison and Derek E. G. Briggs, Plenum Press, New York, 1991.

1. Introduction

Shelly macroinvertebrate fossils are common, locally dominant components of sedimentary rocks and their differential preservation concerns many aspects of paleontological and geological analysis. Evaluation of post-mortem effects is essential, for example, to determine the fidelity of the fossil record as an archive of biological information, whether that information is used for paleoecological, evolutionary, or biostratigraphical purposes. In addition, the post-mortem modification of shelly fossils can provide valuable insights into paleohydraulic and diagenetic conditions, including short-term cycles of deposition, erosion, and omission on the seafloor.

The abundance and composition of shelly remains are governed by the complex interaction of many factors. These can be reduced in their simplest form to (a) the source and rate of shell supply, i.e., the raw material, (b) the inherent and highly specific susceptibility of those shells to post-mortem destruction or modification, (c) the taphonomic conditions that shells are exposed to in their environment of accumulation, and (d) the time scale of accumulation (Fig. 1). The longer the period of accumulation, the more likely that the taxonomic and age-class composition of the shell assemblage will be modified by differential preservation, that shells from successive generations will be mixed (time-averaged), and that the source, rate of supply, and local environment of accumulation will change. Some of these processes will tend to inflate the diversity of the death assemblage relative to community composition at a single moment in time (for example, time-averaging), whereas others will tend to diminish diversity and variation among assemblages (for example, the removal of fragile and mineralogically unstable forms). This ''taphonomic equation'' (Fig. 1) also presents many opportunities for feedback among factors, most notably because of the potential of accumulating shells to influence pore-water chemistry, seafloor mass properties, and benthic colonization.

In this review we attempt to assess environmental and temporal patterns in post-mortem modification, emphasizing actualistic evidence. These data are unavoidably biased in several ways: (a) toward mollusks, which dominate many modern shelly faunas; (b) toward physical processes of shell modification, owing to their experimental tractability relative to biological processes; and (c) toward the most accessible shallow-water environments. Conclusions on patterns and especially rates of shell accumulation thus should be extrapolated to other groups and settings with caution. Such extrapolations become increasingly problematic as one moves from Cenozoic to Mesozoic and Paleozoic biotas, owing to sweeping evolutionary change among hard-part producers and destroyers. Nevertheless, Recent systems, because they are amenable to controlled experimentation, can provide a valuable baseline for determining principles and the kinds of pathways that can underlie taphonomic patterns.

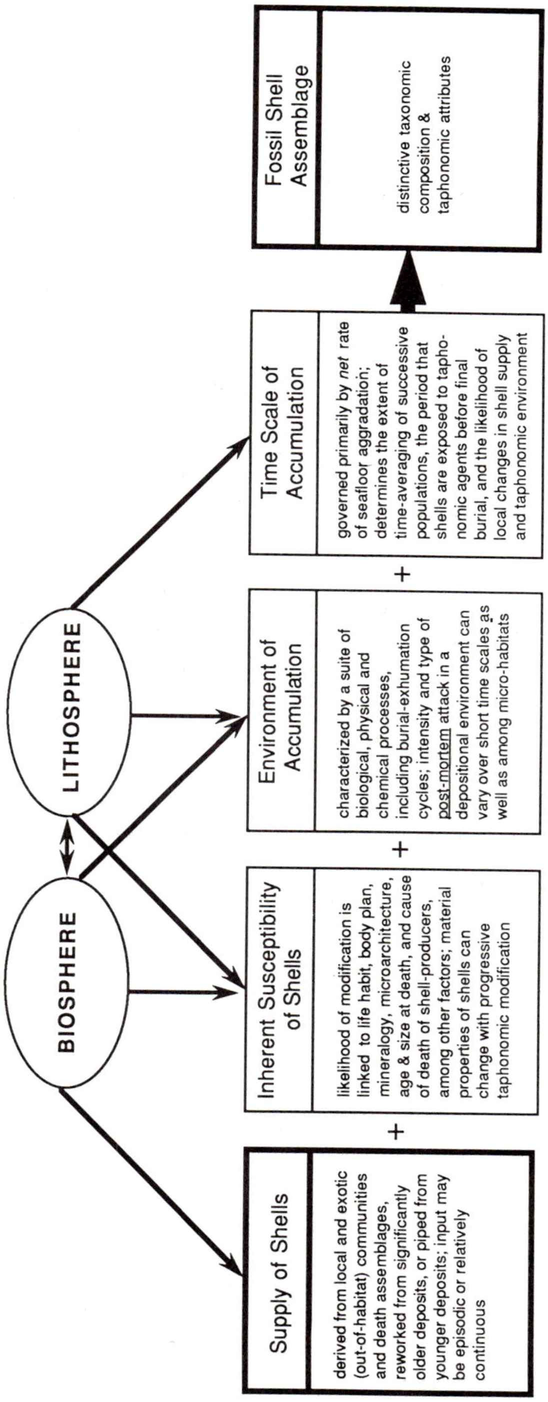

FIGURE 1. Major factors affecting the formation of shell assemblages.

1.1. Terminology

Post-mortem modification is a virtually unavoidable fact of fossilization: few faunas enter the record without some alteration, however minor, even those in spectacular *Lagerstätten* (Whittington and Conway Morris, 1985). Consequently, rather than reject imperfect records, paleontologists need reliable means to discriminate among degrees of bias and thus identify taphonomically comparable assemblages. Recognition of such ''isotaphonomic'' assemblages (*sensu* Behrensmeyer, 1988) would facilitate paleobiological interpretation by holding at least some biasing factors constant.

This is the rationale behind most classification schemes for fossil assemblages, whether they recognize (a) descriptive categories based on shell diagenesis (Koch and Sohl, 1983) and biostratinomic features (Johnson, 1960; Brandt, 1989) or (b) inferential categories based on likely source of shells, scale of time-averaging, or complexity of post-mortem history (Table I).

Ever since Walther's (1893–1894) pioneering work on macroinvertebrate biofacies, many terms have been introduced, and variously redefined, to categorize live and dead shelly faunas (Table I). To avoid confusion, we will use the term *community* only to denote recurrent groups of living organisms, Recent or ancient. *Assemblage* will only be used for accumulations of nonliving material, that is, death assemblages. Assemblages can be of any age (e.g., modern, subfossil, or fossil) and can be composed of any mixture of transported and local shells; the unmodified term is also noncommittal with respect to ecological interactions among the former organisms. For assemblages thought to reflect ecological relationships among the preservable (i.e., shelled) portion of former communities, we prefer the paleoecological term *association* [*sensu* Fürsich (1977); following Kauffman and Scott (1976)]. This term is less ambiguous than ''life assemblage,'' and presumes less about the biological significance of cooccurrences than terms such as ''paleocommunity'' and ''fossil biocoenosis'' (Table I).

2. Spatial and Temporal Variation in Recent Benthic Communities

Among many definitions of communities, that of Mills (1969) is most relevant for practical treatment of benthic faunas: ''a community is a group of organisms occurring in a particular environment, presumably interacting with each other and with the environment and separable by ecological survey from other groups.'' The distribution of such communities commonly can be related to areas of differing substratum (Jones, 1950; Hartnoll, 1983) or to environmental gradients in salinity (Cadée, 1968), temperature,

Järvinen *et al.* (1986)	Kidwell *et al.* (1986)	Fürsich (1990)	Terms used in this chapter
Community		Life assemblage	Community
Taphocoenosis (after Quenstedt, 1930)		Fossil assemblage	Death or fossil assemblage
Thanatocoenosis		Death assemblage = thanatocoenosis	Indigenous assemblage
Paleocommunity	Autochthonous assemblage		Census assemblage
Paleocommunity = recurrent assemblage	Parautochthonous assemblage	Taphocoenosis	Within-habitat time-averaged assemblage
Symmigy	Mixed (parautoch-allochthonous) assemblage	Taphocoenosis	Mixed (indigenous-exotic) assemblage
Symmigy		Taphocoenosis	Multi-habitat time-averaged or environmentally consensed assemblage
Symmigy			Biostratigraphically condensed assemblage
	Allochthonous assemblage		Exotic or allochthonous assemblage
	Remanié		Remanié assemblage
	Leaked assemblage		Piped assemblage

	R.W. Scott (1970)	Hallam (1972)	Fürsich (1975, 1977, 1978)	Yanin (1983)
		Life assemblage		Paleocoenosis
ge			Fossil assemblage	Taphocoenosis, thanatocoenosis, or *orictocoenosis*
				Taphocoenosis or thanatocoenosis
ity	In-place assemblage	Census (assemblage)		Taphocoenosis or thanatocoenosis
or	Disturbed neighborhood assemblage	Death assemblage	Fossil association (recurrent assemblage); time-averaged community relic	Taphocoenosis or thanatocoenosis
	Mixed assemblage			Taphocoenosis or thanatocoenosis
			Environmentally or faunally condensed assemblage	
			Faunally condensed assemblage	
il	Transported assemblage			

Orlov (1959)	Brouwer (1959)	Johnson (1960)	Craig and Hallam (1963)	Fagerstrom (1964)
Biocenosis	Biocoensosis	Life assemblage	Living community	
Orictocenosis	Thanatocoenosis	Death assemblage	Death or fossil assemblage (thanatocoenosis)	Fossil assembla
	Autochthonous thanatocoenosis			
Thanatocenosis	Autochthonous thanatocoenosis	Model I	Life assemblage (fossil community)	Fossil census (perfect) or fossil commu
	Autochthonous thanatocoenosis	Model II	Indigenous assemblage	Residual censu residual fossi assemblage
	Thanatocoenosis of mixed character	Model II and III		Mixed fossil assemblage
		Model III		Mixed fossil assemblage
		Model III		Mixed fossil assemblage
	Allochthonous thanatocoenosis	Model III	Exotic assemblage	Transported fos assemblage
			Remanié	

Table I. Synonymy of Classifications for Shelly Fossil Assemblages[a]

Classification	Wasmund (1926)	Boucot (1953)	Craig (1953)
I. Group of cooccurring live organisms	Biocoenosis		Community
II. Group of cooccurring dead organisms	Thanatocoenosis	Fossil death assemblage	Fossil assemblage
. . . that are thought to represent: A. Production by the local fauna			
. . . and are thought to provide 1. a nearly perfect record of live shelled taxa with respect to species relative abundances, size–frequency distributions, and life positions		Fossil life assemblage	Fossil community
or 2. a somewhat modified record owing to post-mortem disturbance of shells (disarticulation, reorientation, loss of fragile taxa and size classes) and/or time-averaging of generations			Fossil assemblage
. . . or to have been modified by 3. the addition of shells from other habitats			Mixed fossil assemblage
or by 4. the addition of shells from later, ecologially distinct communities that occupy the site; community change may have been driven by physical environmental changes or by live/dead interactions			Mixed fossil assemblage
or by 5. admixture with shells of significantly greater age			
B. Shells that have been transported out of their life habitat			Assemblage of foreign fossils
C. Shells reworked from significantly older deposits			
D. Shells introduced from significantly younger deposits			

[a]Italics indicates term used only for fossil examples.

light, physical disturbance (Hiscock, 1983; Probert, 1984), and subaerial exposure (Price *et al.*, 1980).

Communities are recognized by visual or statistical grouping of samples on the basis of similarity in taxonomic composition and/or relative abundance of species. Unfortunately, no standard quantitative approach has yet been adopted by marine biologists and actuopaleontologists [see Gauch (1982) for review], even though multivariate methods are used to deal with the large data sets obtained from multiple sampling. These often contain tens to hundreds of taxon occurrences among tens to hundreds of sample sites.

Notwithstanding these technical problems, two models of benthic community structure have arisen, largely by analogy with terrestrial plant communities [Gray (1981) and Erwin (1983), based on Whittaker (1975)]. One model is that species abundance peaks more or less coincide across a region (Fig. 2A). In benthic systems, such a view derives from large-scale (1–10 km) studies based on widely spaced samples; community boundaries and habitat boundaries appear to be closely linked [e.g., early studies of the open shelf (Petersen, 1914; Jones, 1950)]. The alternative model is that species distributions vary independently and peak in abundance at different sites across a region (Fig. 2B). This view reflects finer-scale (10–100 m), transect-style studies [e.g., benthic zonations documented for sandy and rocky intertidal environments (Mills, 1969; Hughes and Thomas, 1971)]. This latter view has gained considerable acceptance with regard to both littoral and sublittoral settings with steeper environmental gradients (Stephenson *et al.*, 1971; Gray, 1981; Erwin, 1983).

Both models may be valid for a single environment, depending upon the scale of observation. For example: given a broad shelf with one region of rippled clean sand adjacent to another with shell gravel, it is likely (e.g., J. B. Wilson, 1986) that the two substrates would have faunas that could be distinguished by a grab or box-core kilometer-scale survey. Following Mills (1969), these would be regarded as separate communities and the distribution pattern of faunal abundances would resemble model 1 (Fig. 2A). However, if a more detailed (10-m) scale SCUBA transect survey were conducted across the boundary of these two substrata, then we would expect patterns to resemble model 2 (Fig. 2B).

The explanations offered for these patterns are diverse: some may reflect biotic interactions among species (competition, coadaptation) and others may reflect shared environmental tolerances, whereas many probably reflect a combination of causes. For example, based on a series of detailed studies (Allen, 1899; Vevers, 1952; Brun, 1969; Warner, 1971; Hiscock and Mitchell, 1980), it appears that of the 22 species in the *Ophiothrix fragilis* community of offshore British waters, five species benefit by feeding on *O. fragilis*, while the others are probably present because "environmental conditions are suitable or because another food supply is present" (Hiscock and Mitchell, 1980).

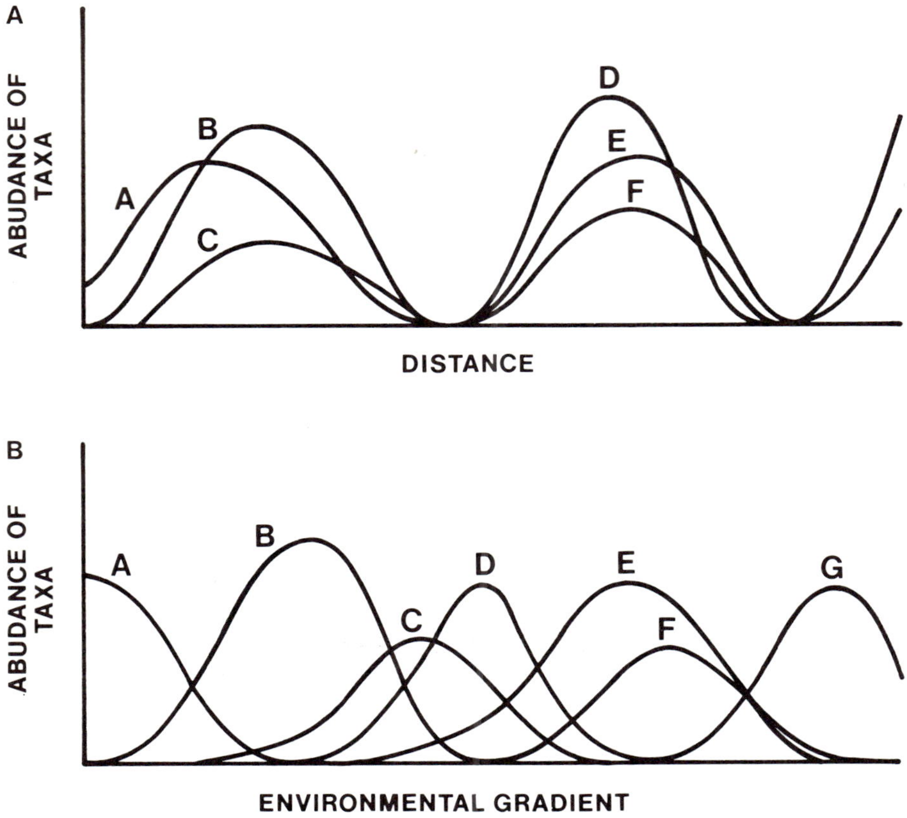

FIGURE 2. Different views of community structure: (A) Faunal abundances of different taxa occur in distinct groups A, B, C and D, E, F, possibly coinciding with environmental boundaries, (B) unrelated abundances of organisms A–G peaking separately along an environmental gradient. [After Gray (1981).]

Abundance patterns commonly are based on faunal surveys over short time scales and thus provide little insight into the temporal changes now known to characterize the structure of shelf benthic communities (Gray, 1981; Probert, 1984). Repeated sampling of shelf areas off the Northumberland coast of Britain by Buchanan *et al.* (1978), for example, indicates important seasonal and annual variation in biomass (Fig. 3A) and species abundance (Fig. 3B) attributable to larval settlement, winter mortality, and year-to-year differences in minimum winter water temperatures.

The few long-term surveys that have been carried out indicate that community structure and composition fluctuate considerably and on different time scales, and can be related to both biological disturbance (Paine, 1966) and natural cycles in the physical environment (Boesch *et al.*, 1976; Probert, 1984) (see Section 5.2.2). The data base for benthic faunas of the western English Channel, accrued over almost 100 years of sampling (Fig. 4)

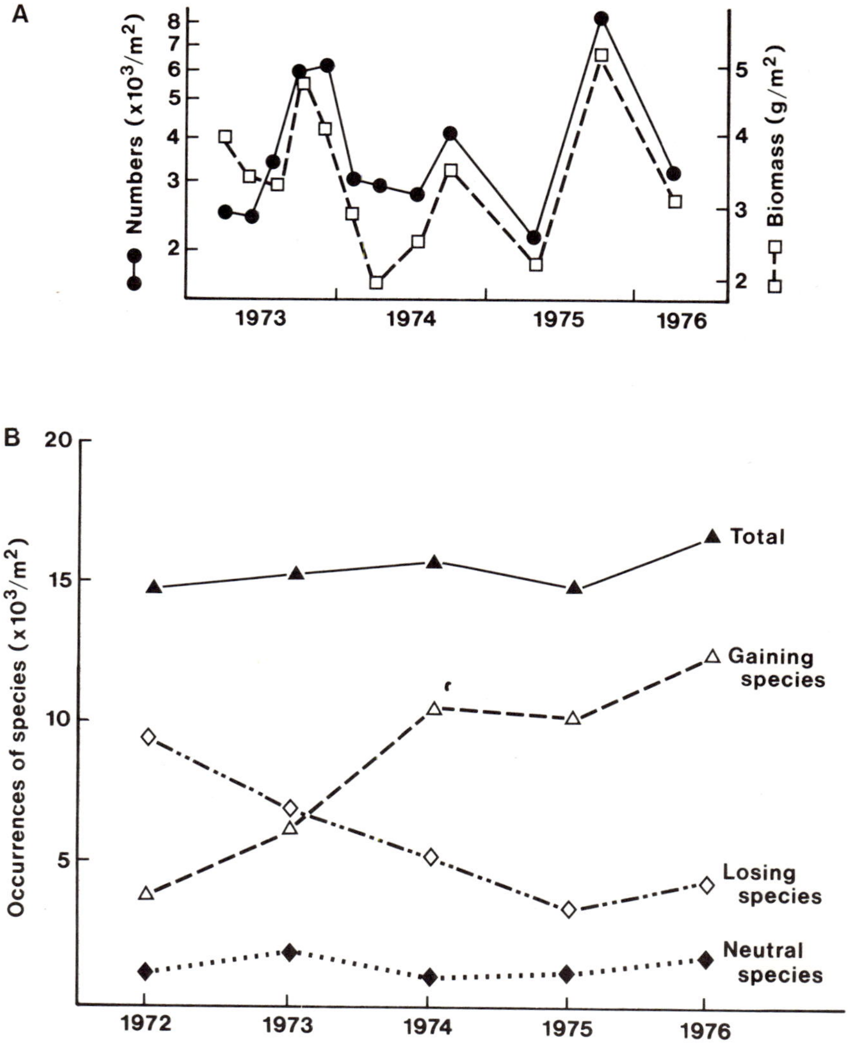

FIGURE 3. Plot of biologically long-term changes in (A) species numbers and biomass and (B) species abundances in benthos off Northumberland coast. [After Buchanan *et al.* (1978).]

(Carthew and Bosence, 1986), provides a good case in point for physical controls. Faunas are closely related to shell gravel and mud substrates (Fig. 4B) and have changed in composition over time scales of tens of years (Fig. 4C). The late 19th century faunas are clearly different from the 20th century faunas [although some of this may arise from differences in sampling techniques (Carthew and Bosence, 1986)]. The faunas of the 1920s are similar to those of the 1980s, which in turn are quite different from those of the 1930s to 1960s (Fig. 4C). These biologically long-term fluctuations (i.e., exceeding the lifespan of the organisms studied) in the composition of shell

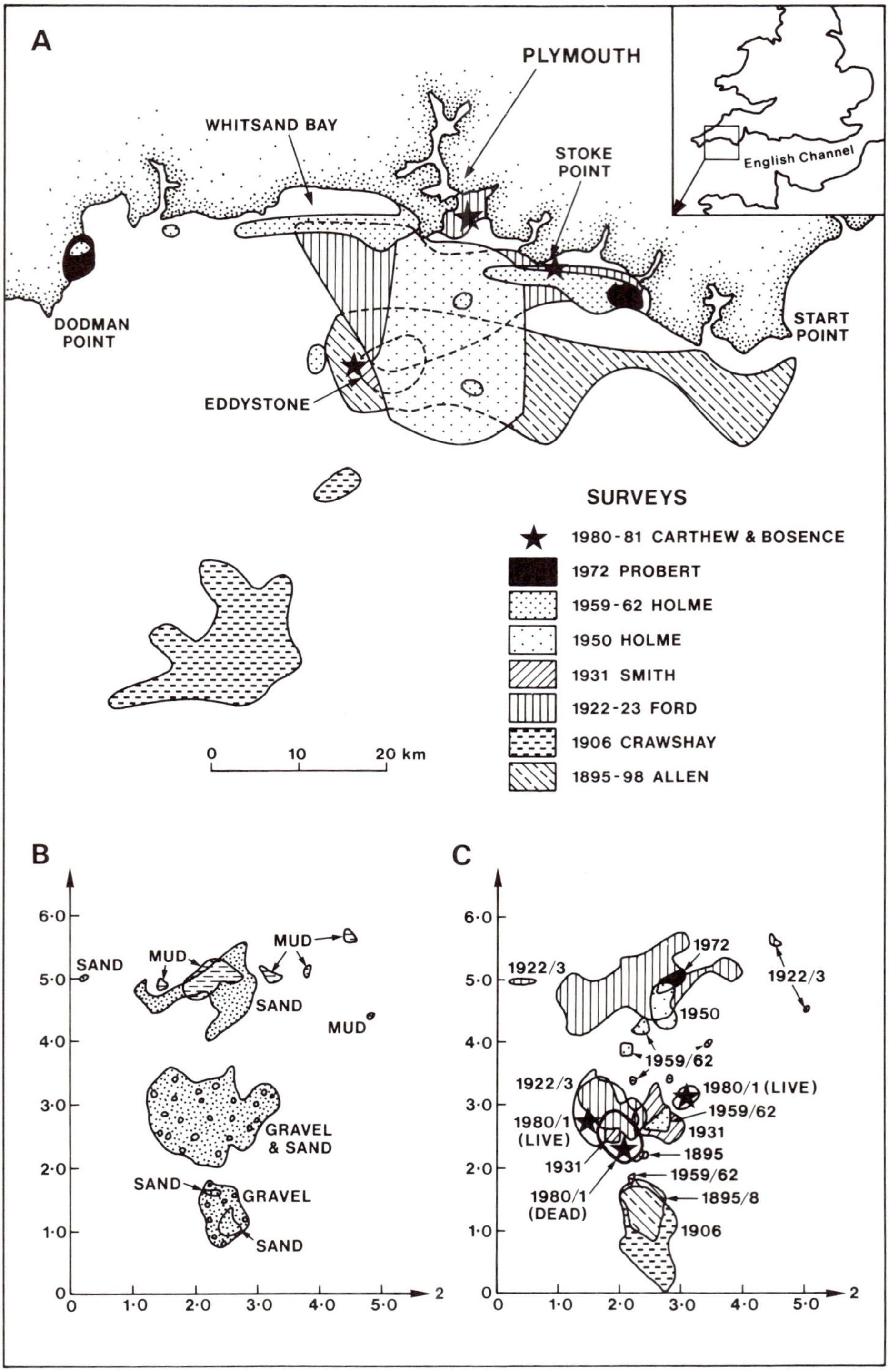

A
PLYMOUTH
WHITSAND BAY
STOKE POINT
English Channel
DODMAN POINT
START POINT
EDDYSTONE
SURVEYS
1980-81 CARTHEW & BOSENCE
1972 PROBERT
1959-62 HOLME
1950 HOLME
1931 SMITH
1922-23 FORD
1906 CRAWSHAY
1895-98 ALLEN
0 10 20 km
B
SAND
MUD
SAND
MUD
GRAVEL & SAND
SAND
GRAVEL
SAND
C
1922/3
1972
1922/3
1950
1959/62
1922/3
1980/1 (LIVE)
1959/62
1980/1 (LIVE)
1931
1931
1895
1959/62
1980/1 (DEAD)
1895/8
1906
6·0
5·0
4·0
3·0
2·0
1·0
0
1·0
2·0
3·0
4·0
5·0
2

gravel communities are also reflected in the plankton communities of the western Channel (Southward, 1962, 1980). Both correlate with long-term changes in water temperature. Long-term changes in benthic faunas following the severe winter storms of 1962–1963 in the Dutch Wadden Sea and shallow German Bight, North Sea, are discussed in Beukema and Cadée (1986) and Ziegelmeier (1963, 1964), respectively.

Gray's (1981) model of temporal and spatial fluctuations is particularly useful for framing a paleoecological concept of benthic communities. This is based on Hutchinson's (1978) niche theory that organisms normally inhabit an optimal niche within a larger habitat, which may also be colonized following environmental perturbations. Gray (1981; after Holling, 1973) depicts the stable, core state of a community as a ball within a "basin of attraction" (Fig. 5) whose form is determined by disturbance and other controlling factors. Disturbance can bring about a shift in dominance and/or structure of the community to a new successional state (B in Fig. 5); the community may either revert to state A after the disturbance is removed, or change further to another state (C or D in Fig. 5) [see Johnson (1972) for a similar view]. Because these changes can be unrelated and unpredictable, community structure cannot be understood from any single survey or area; repeated sampling is required to understand the fauna at any site.

Geologically, this means that shells can be contributed to the death assemblage from any one of a community's successive states (Fig. 5). Given that soft-bottom seafloors aggrade, on average, at rates measured in only centimeters per 10^3 years (Schindel, 1980; Sadler, 1981), and that shells can persist for decades or centuries (see Section 5 on time-averaging), different community states should rarely be recorded as discrete assemblages in shelf environments. The details of succession and temporal coexistence of species are only likely to be preserved where aggradation was episodically rapid in decimeter-thick increments, minimizing the blurring effects of later bioturbation and physical reworking. More positively, death assemblages from soft-bottom environments can provide good pictures of the long-term average composition of communities (Peterson, 1977), comparable perhaps to those produced by summing repeated benthic surveys (Fig. 4C).

Broadly speaking, patterns of community structure on hard substrata and reefs are similar to those on soft substrata, but some of the controls on

←

FIGURE 4. Evidence for decade-scale changes in shelf benthos. (A) Locations and dates of surveys of benthic communities in waters off Plymouth, England. Where surveys overlap, the more recent survey overlies the older. (B) Plot of first and second axes of Detrended Correspondence Analysis of live benthic faunas from surveys in (A). Samples have been contoured by substrate type and this illustrates the close correspondence between samples with a similar fauna inhabiting similar substrates. (C) Plot as for (B), but samples contoured by date of survey. For the shell gravel stations in the center of the plot, note the shifts in community composition through time and the averaging of these shifts by the dead shelly fauna encircled by heavy line. Surveys are displayed by date of sampling. Publication dates are: Allen, 1899; Crawshay, 1912; Ford, 1923; Smith, 1932; Holme, 1953, 1961, 1966; Probert, 1973; Carthew and Bosence, 1986. [After Carthew and Bosence (1986).]

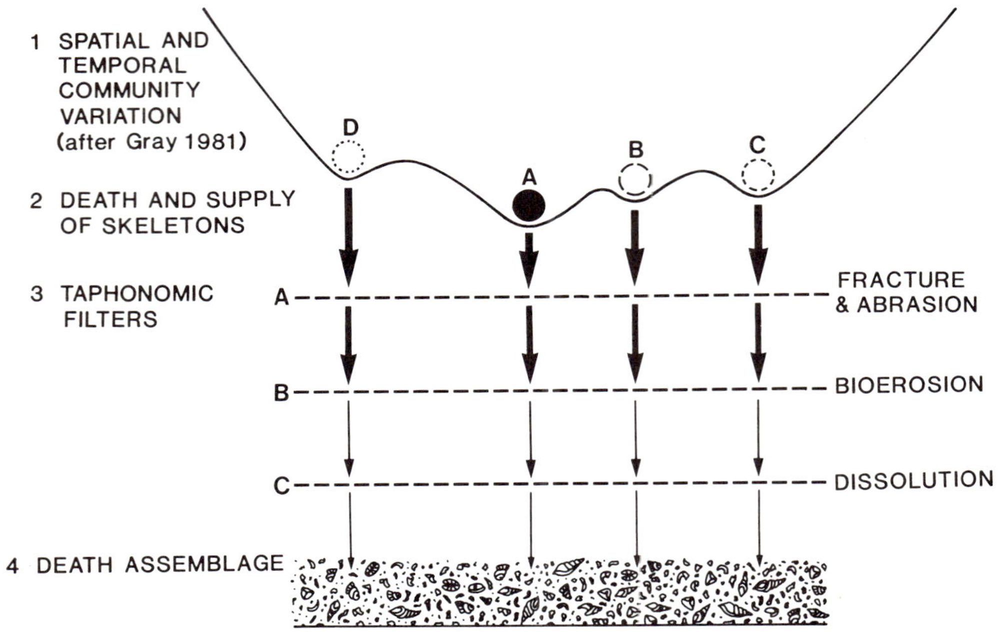

FIGURE 5. Every benthic community has a variety of locally stable states, represented by the spheres on Gray's (1981) conceptual landscape, that differ in species abundances. Perturbations cause shifts in community composition among states A–D, each of which supplies skeletons that are subjected to a series of taphonomic filters before entering the death assemblage. Therefore the composition of the death assemblage at any given locality will depend not only upon taphonomic filters, but also upon local perturbation history and community response.

species distribution have been demonstrated more successfully. Particularly well documented has been the importance of physical disturbance (Paine and Levin, 1981; Graus *et al.*, 1984; Reichelt *et al.*, 1985), grazing pressure (Steneck, 1983; Steneck and Paine, 1986; Done, 1985), and spatial competition (Jackson, 1979; Branch, 1984). Because of rapid rates of reef growth under optimal conditions, short-term biological and physical processes have some potential for preservation in ancient reefs (Jackson, 1983), notwithstanding physical and bioerosion of the reef framework (Schroeder and Zankl, 1974; Scoffin and Garrett, 1974; Bosence, 1985). Ancient patterns of ecological succession (e.g., Rowland and Gangloff, 1988; Copper, 1988) can in some instances be compared directly with experimental evidence from field surveys and settling plates in Recent environments. For example, in the Pleistocene of Kenya, Crame (1980) found that early stages of coral reef growth showed "facilitation" succession (Connell and Slatyer, 1977) whereby early colonizers "prepare the ground" for later reef builders. Subsequent reef framework successions were less orderly, so that Crame was only able to recognize early and late colonizers following disturbance within the reef, perhaps conforming to Connell and Slatyer's (1977) "inhibition model" succession [see Kidwell and Jablonski (1983) for other examples ranging through the Phanerozoic]. On a smaller scale, Bosence (1984) was able to show recurring overgrowth sequences of coralline algal and foraminiferal crusts in thin sections of preserved reef frameworks from patch reefs in St. Croix, Lesser Antilles. There the succession was the same as had been recorded by previous settlement plate experiments (Adey and Vassar, 1975). Disturbance created fresh substrates, which were first colonized by rapidly dispersing, fast-growing thin corallines, which were later overgrown by the more massive framework-forming coralline algae.

In cryptic hard-bottom environments, the main late-stage colonizers are often poorly mineralized and therefore would be underrepresented in fossil sequences (Rasmussen and Brett, 1985). However, in many instances interspecies spatial competition has been documented both on shells and on mobile pebble substrata (Liddell and Brett, 1982; Choi and Ginsburg, 1983; Taylor, 1984; M. A. Wilson, 1985, 1987). The high fidelity of preservation of these encrusting skeletal communities indicates the potential for studying the evolution of ecological dynamics in similar environments through time in a way that is probably impossible for soft-bottom environments.

3. Post-Mortem Modification of Community Structure

Benthic communities are typically composed of taxa having a variety of skeletal types and thus different preservation potentials. Macroinvertebrate groups range from those with (a) well-mineralized (= biomineralized) endo- or exoskeletons composed predominantly of calcite, aragonite, or apatite, to (b) lightly or locally mineralized proteinaceous and chitinous skeletons, (c) spiculate tissues, (d) minor mineralized hard parts, particularly feeding structures, and (e) entirely soft-tissued, hydrostatic "skeletons."

Under normal conditions of shelf sedimentation (i.e., relatively slow net aggradation of the seafloor, well-aerated overlying waters) and early diagenesis, biomineralized tissue is expected to be preserved, spiculate tissue to be dispersed, and soft tissues to decay leaving no trace (White, 1879). The first question in considering the quality of the fossil record thus must address the proportion of well-mineralized (= shelly) taxa in the benthic community: what is the maximum amount of information that a shelly assemblage can preserve from the original, total community (Section 3.1)? A second set of questions targets only the shelly portion of benthic communities: how accurately does the shelly death assemblage preserve the composition and structure of the shelled portion of the community (Section 3.2)?

3.1. Proportion of Shelled Organisms in Benthic Communities

The higher the proportion of preservable taxa and individuals, the greater the likely fidelity of the death assemblage to the live community (e.g., Lasker, 1976). Proportions of well-mineralized taxa, however, vary widely from 0 to 70% or more (e.g., Lawrence, 1968; Schopf, 1978; McCall and Tevesz, 1983) so that live : dead discrepancies in diversity can range over nearly two orders of magnitude. Based on a literature review of soft-bottom benthic faunas, Staff *et al.* (1985) found no correlation between proportion of shelly taxa and water depth or sediment grain size (Fig. 6). Even communities that contain a large number of shelly species are commonly dominated numerically by soft-bodied taxa, thus yielding poor agreement between numerical dominance-based diversity measures for the total community at a given moment and its shelly death assemblage (Peterson, 1975; Stanton, 1976; McCall and Tevesz, 1983).

Even though most individuals are not preserved, the same death assemblage can show high fidelity to the original live community if dominance is measured by biomass and if comparisons are made to a long-term picture of the benthos rather than to a single census. Because of their typically larger body sizes and temporally stable populations, shelly taxa may dominate benthic faunas in biomass even when they do not dominate in numerical abundance at any one time (Fig. 6) (Cummins *et al.*, 1986a; Staff and Powell, 1988). In Texas lagoons, Staff *et al.* (1985) found that both rank order and trophic structure of the original community could be recovered using this approach. In fact, Powell *et al.* (1989) argued that even numerical abundance data "indicate real variation in species' [ecological] success" when the compositions of both live and dead are examined over time scales exceeding a few years.

Equally reassuring is the observation that the high rates of shell dissolution documented in Texas lagoons [e.g., 50% of all new shells lost within 307 days (Powell *et al.*, 1984; Cummins *et al.*, 1986a)] are concentrated among juveniles and poorly calcified opportunists smaller than 3mm. The ecologically most persistent, and thus, it could be argued, the most characteristic species of the benthic community had immeasurably long half-lives in the same sediments, and thus dominate death assemblages much as they dominate any long-term portrait of the combined soft-bodied and shelly source community. Paleontologists therefore should not despair [*contra* M. V. H. Wilson (1988) and others] of the small numbers of shelly taxa or individuals normally censused in modern marine habitats: data indicate that high loss rates are focused on the ecologically most transient parts of communities, i.e., short-lived small or opportunistic forms, so that the temporally persistent core of the community is captured by the shelly death assemblage.

There have been few comparable studies in reef or hardground environments, but Brett (1988) found that shelled taxa usually represent less than 40% of species richness and only 15% of the area originally covered by encrusters. These figures are largely taken from cryptic biotas (e.g., Kobluk, 1988; Rasmussen and Brett, 1985; Kobluk and van Soest, 1989). By contrast, Jamaican reef surfaces yield an average of 70% mineralized taxa in shallow water (60 m); deeper waters yield progressively lower proportions, decreasing to 1.8% shelled taxa at 120 m depth (Liddell and Ohlhorst, 1988).

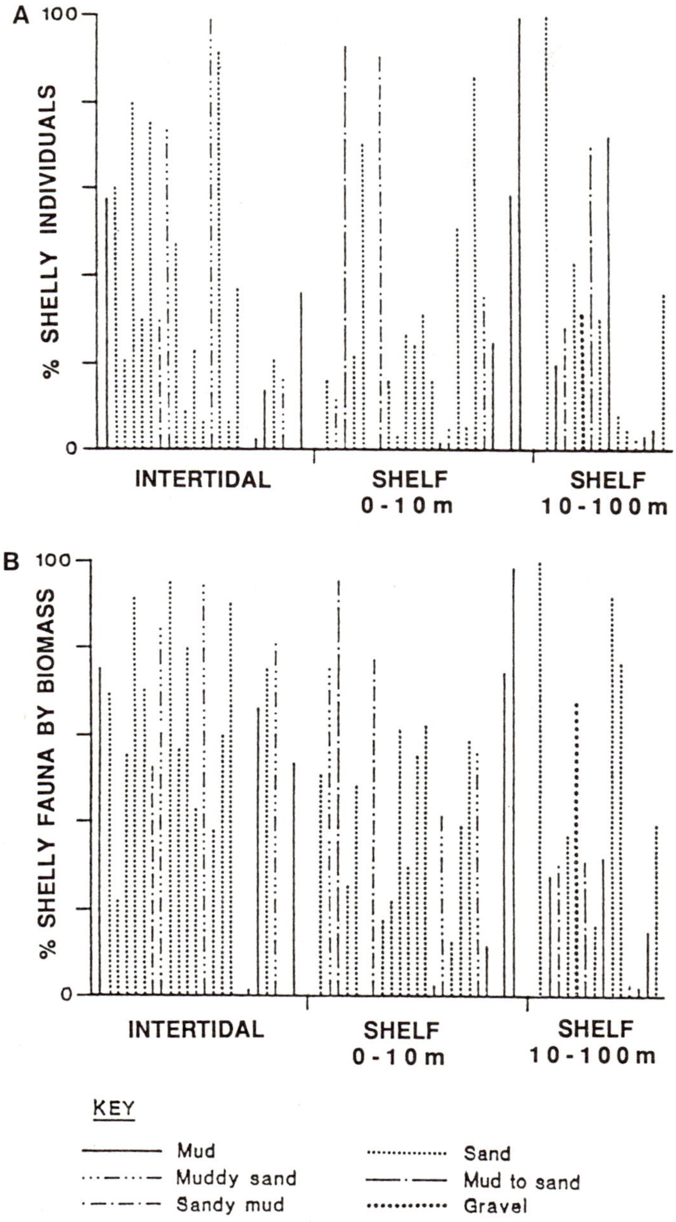

FIGURE 6. Bar charts illustrating percentages of shelly organisms in benthic communities from predominantly midlatitude soft substrata. (A) Percent of total individuals. (B) Percent of total biomass. Based on data compiled by Staff *et al.* (1985, Table 3).

Shallow-reef biotas thus have higher *potential* fidelities than deeper biotas. It remains to be tested how often this potential is actually achieved in death assemblages, given high water energies and intense bioerosion of live and dead skeletal material in shallow, photic waters.

3.1.1. Implications for Trophic Analysis

Although the temporally persistent core of the community can be captured by the shelly death assemblage, certain trophic groups and life habits will inevitably be underrepresented and thus it is unlikely that trophic structure can be satisfactorily resolved. Predators in modern shell gravel communities, for example, are mainly soft-bodied or lightly mineralized polychaetes, arthropods, echinoderms, skates, and rays (Bosence, 1979); birds are important but largely unrepresented factors in some intertidal communities (Cadée, 1989). The wide range in the proportion of low-preservation taxa even within single habitats or communities (see Section 3.1) undermines any possibility of a simple correction factor for fossil assemblages, even if benthic structures have been constant through the Phanerozoic. Reconstructing the trophic structure of the total community is further complicated by the diversity of food sources exploited by many benthic invertebrates (Cadée, 1984b).

Most trophic analyses since the late 1960s and early 1970s have not claimed to characterize the structure of the total original community (R. W. Scott, 1978; Fürsich, 1984; Fürsich and Flessa, 1987). Reconstructed trophic nuclei are used instead to describe only the shelly portion of ancient communities, and find primary application in biofacies mapping. Although in some settings such groupings may result as much from taphonomic processes as from living communities [e.g., loss of some and exotic input of others (Stanton and Dodd, 1976; Bosence, 1979)], most actualistic studies have found good agreement in the spatial distribution of live and dead shelly taxa at the biofacies scale (Section 3.2.1; Tables II–V).

3.2. Fidelity of Death Assemblages to Live Shelly Faunas

Even among shelly taxa, species differ in their susceptibility to post-mortem destruction (see Section 4 for processes and selectivity). Compared to the potential death assemblage, observed death assemblages can have lower species richness owing to preferential removal of small or fragile forms by predators and post-mortem processes. Conversely, they can have higher species richness than any single census of live shelly fauna owing to time-averaging of successive community states and the addition of allochthonous shells. In similar ways, size–frequency distributions of death assemblages may be more or less skewed than the raw input.

Many workers have found broad agreement between the most common occurrences in live and dead shelly faunas in a variety of environments (Johnson, 1965; Cadée, 1968; Warme, 1969; Warme *et al.*, 1976; Peterson, 1976; Bosence, 1976b; Ekdale, 1977; Carthew and Bosence, 1986). Shelly death assemblages often map communities and habitat types better than any single census of the living shelled community, suggesting both little out-of-habitat transport of shells and little change or lateral displacement of communities during the period of shell accumulation (Column 3 in Tables II–V).

Comparisons among studies are difficult, however, because workers measure fidelity in different ways and with varying degrees of quantification. Consequently, these data are reanalyzed here using a series of standardized metrics (Tables II–V).

3.2.1. Fidelity with Respect to Species Richness and Taxonomic Composition

In order to assess species richness and taxonomic composition, three kinds of information are necessary: number of species found live only (N_L), number of species found dead only (N_D), and number of species found both live and dead (N_S). Unfortunately, these requirements exclude several studies, most notably Johnson's (1965) seminal study of Tomales Bay, California, and Parker's (1960) data set for the Gulf of Mexico. Sixteen other studies do meet the requirements, including five from intertidal settings (Table II), five from subtidal coastal complexes (Table III), and six from open-shelf and deep-sea environments (Table IV). (Table V summarizes the data.)

Death assemblages tend to have fairly high fidelities when the question is, "what percentage of shelly species found live are also found dead at the same site?" [i.e., $(N_S \times 100)/(N_L + N_S)$; Table V). In general, most or all well-mineralized taxa contribute to the local death assemblage: mean fidelities are 83–95% if data are summed for study areas, and are 75–98% if the focus is on individual habitats (e.g., sand channels or grass beds within an intertidal study area). Such high values also characterize Pleistocene assemblages: Valentine (1989) found that 77% of living bivalves and gastropods in the Californian Province are known as Pleistocene fossils, and that the missing species are all rare today and/or minute, fragile, or from lesser known deep-shelf habitats.

The same data sets yield lower values when the question is, "what percentage of species found dead are also found live?" [i.e., $(N_S \times 100)/(N_D + N_S$; Table V]: 33–54% for entire study areas and 42–57% for individual facies, but the range is great (4–100%). The low fidelities are probably in part a consequence of the way living communities were (under)sampled: most studies relied upon a single census or single year of sampling. A hypothesis of sampling bias is supported by the observation that single study areas subjected to longer sampling programs showed improved agreement between live and dead with progressive sampling of live fauna. Peterson (1976) demonstrated this effect by pooling successive live samples from a 3-year study: the similarity between live and dead taxa increased from approximately 20% to 75% as live taxon records accrued over successive surveys in two warm, temperate lagoons. Knight (1988) found this same relationship over a 17-year study period in open-shelf habitats by

TABLE II. Fidelity of Recent Molluscan Death Assemblages in Intertidal Environments: Presence–Absence Data

Data set	What % live species are found dead within the STUDY AREA?	What % live species are found dead within the same FACIES?	Do dead shells cluster like live taxa?	What % dead species are found alive within the STUDY AREA?	What % dead species are found alive within the same FACIES?	What % dead individuals are from species found alive in the same FACIES?	Sampling procedure for live fauna
West Coast North America, salt marshes and tidal creeks, 30°–47°N (MacDonald, 1969, 1976)	93% (avg.) (28/30 spp.)	Data not available	Yes	45% (28/62 spp.)	Data not available	Very high; quantitative data not available	One sampling of 11 sites, mesh unknown
Yellow Sea, Korea, mud and sand flats, 37°N (Frey *et al.*, 1988, Appendixes 1 and 2)	100% (13/13 spp.)	100% mid flat (8/8 spp.) 100% outer flat (11/11 spp.)	No data	46% (13/28 spp.)	44% mid flat (8/18 spp.) 48% outer flat (11/23 spp.)	82% mid flat 79% outer flat	Mesh unknown; 2 samples each of 16 stations along single transect, one season
Seto Sea, Japan, mud, sand, and gravel flats and shoreline, 34°N (Tanabe *et al.*, 1986, and unpublished data)	45% (15/22 spp.)	62% sandy mud flat (10/16 spp.) 90% sand and gravel flat (9/10 spp.)	Yes for deep-burrowing spp.; no for shallow-burrowing spp.	100% (15/15 spp.)	100% sandy mud flat 75% sand and gravel flat	100% sandy mud flat 99% sand and gravel flat	5-mm mesh, 13 transects each dredged once

Mugu Lagoon, California, tidal flat and subtidal channel complex, 34°N (Warme, 1971; and unpublished data)	97% (38/39 spp.)	95% lower intertidal (22/23 spp.) 97% seagrass beds (31/32 spp.) 89% sand channels (16/18 spp.)	Yes	51% (38/75 spp.)	38% lower intertidal (25/58 spp.) 48% seagrass beds (31/64 spp.) 44% sand channels (16/36 spp.)	86% lower intertidal 94% seagrass beds 91% sand channels	3-mm mesh, replicate samples over one season; 55 samples
Cholla Bay, Gulf of California, sand and mud flats, 31°N (Fürsich and Flessa, 1987; and unpublished data)	80% (32/40 dominant spp.); total live spp. = 62	Data not available	Yes, in general	27% (32/119 dominant spp.); total dead spp. = 212	Data not available	Data not available	3-mm mesh, 77 samples from 4 transects, collected over 5 months
Mean	83%	90%	Yes	54%	57%	90%	

[a]Based on Table I and Map 2 of Warme (1971).

TABLE III. Fidelity of Recent Molluscan Death Assemblages in Coastal Subtidal Environments: Presence–Absence Data

Data set	What % live species are found dead within the STUDY AREA?	What % live species are found dead within the same FACIES?	Do dead shells cluster like live taxa?	What % dead species are found alive in the STUDY AREA?	What % dead species are found alive within the same FACIES?	What % dead individuals are from species found alive in the same FACIES?	Sampling procedure for live fauna
Ria de Arosa, Spain, sand and clay estuary with rocky shore, and open ocean zone at mouth of estuary, 42° N(Cadée, 1968, and unpublished data)	100% (172 spp.)[a]	100% inner bay 10/10 spp.) 100% mid bay (20/20 spp.) 100% outer bay (17/17 spp.) 100% bay margin (24/24 spp.) 100% oceanic zone (23/23 spp.)[b]	Yes, except for oceanic zone	58% (172/296 spp.)[a,c]	53% inner bay (10/19 spp.) 100% middle bay (20/20 spp.) 71% outer bay (17/24 spp.) 89% bay margin (24/27 spp.) 88% oceanic zone (23/26 spp.)[b,d]	100% inner bay 100% mid bay 97% outer bay 94% bay margin >80% oceanic zone[b]	2-mm mesh; live collected over 3 years; 300 stations
Sapelo Island, Georgia, tidal sound and shelf, 31°N (Henderson and Frey, 1986, Appendix I)	100% (7/7 spp.)	100% sound (4/4 spp.) 100% channel (3/3 spp.) 100% shelf (4/4 spp.)	Yes?	10% (7/70 spp.)	9% sound (4/43 spp.) 4% channel (3/75 spp.) 24% shelf (4/17 spp.)	95% sound 94% channel 97% shelf	1.5-mm mesh, single samples at 8 stations in one season

Texas, Copano Bay, sandy mud, variable salinity, 26°N (Staff *et al.*, 1985, 1986)	≥82% based on data for one facies only	82% (14/17 spp.)[e]	Not tested	≥23% based on data for one facies only	23% (20/88 spp.)	77%[e]	0.5-mm mesh, 2 samples every 6 weeks over 2 years
Texas, Laguna Madre, sand, variable salinity, 26°N (Staff *et al.*, 1985, 1986)	≥97% based on data for one facies only	97% (28/29 spp.)[f]	Not tested	≥48% based on data for one facies only	48% (31/65 spp.)	99%[f]	0.5-mm mesh, 2 samples every 6 weeks over 2 years
Yucatan, carbonate lagoon, strait and shelf, 21°N (Warme *et al.*, 1976; Ekdale, 1972, 1977)	97% (76/78 spp.)	95% backreef (40/42 spp.) 100% mangrove lagoon (54/54 spp.) 100% strait (29/29 spp.) 93% open sea (14/15 spp.)	Yes	26% (76/288 spp.)	19% backreef (40/213 spp.) 31% lagoon (54/173 spp.) 18% strait (29/158 spp.) 15% open sea (14/94 spp.)	78% backreef 91% lagoon 92% strait 54% open sea	3.1-mm mesh; 50 stations sampled once
Mean	95%	98%	yes?	33%	42%	89%	

[a]Based on Cadée (1968, Appendix).
[b]Based on G. C. Cadée, unpublished data, for numerical dominants only.
[c]84% (58/69 spp.) if based on unpublished data (G. C. Cadée) for numerical dominants only.
[d]Average 45% dead spp. if based on entire fauna in Cadée (1968, Appendix).
[e]Based on Staff *et al.* (1985, Table 1).
[f]Based on Staff et al. (1985, Table 2).

TABLE IV. Fidelity of Recent Molluscan Death Assemblages in Open-Marine Environments: Presence–Absence Data

Data set	What % live species are found dead within the STUDY AREA?	What % live species are found dead within the same FACIES?	Do dead shells cluster like live taxa?	What % dead species are found alive in the STUDY AREA?	What % dead species are found alive within the same FACIES?	What % dead individuals are from species found alive in the same FACIES?	Sampling procedure for live fauna
North Sea, sand and muds at 20–50 m depth, 54°N (Cadée, 1984, and unpublished data)	97% (32/33 spp.)	95% mud, (20/21 spp.) 93% sand (13/14 spp.)	Yes, in general	51%(32/63 spp.)	84% mud, (21/25 spp.) 76% sand (13/17 spp.)	88% mud 93% sand	1-mm mesh, single samples of 21 stations, 4 sets of 10 samples over 8 months for 3 stations
Mannin Bay, Ireland, subtidal with rocky shore, 53°N (Bosence, 1976a, 1979)	92% (46/50 spp.)	86% algal bank (19/22 spp.) 92% clean algal gravel (33/36 spp.) 83% muddy algal gravel (30/36 spp.) 88% fine sand (23/26 spp.) 100% mud (7/7 spp.)	Yes	55% (46/83 spp.)	45% algal bank (19/42 spp.) 48% clean algal gravel (33/68 sp.) 60% muddy algal gravel (30/50 spp.) 36% fine sand (23/64 spp.) 22% mud (7/31 spp.)	94% algal bank 79% clean algal gravel 97% muddy algal gravel 66% fine sand 90% mud	2-mm mesh, 77 samples taken over 1 season
English Channel, open shelf, 50°N (Carthew and Bosence, 1986; and unpublished data)	≥92% based on data for one facies only	92% subtidal shell gravel 22/24 spp.)	Yes	≥23% based on data for one facies only	23% subtidal shell gravel (22/96 spp.)	66%	2-mm mesh, 4 quarterly replicates over one year; 11 sample sites

English Channel, subtidal bay, 50°N (Knight, 1988)	97% (40/41 spp.) (40 samples during 1985/86)	43% clean sand 60% muddy sand 69% silt 67% gravel[a]	Yes, but subcommunities overlap	37% if one uses live data from 1985/86 only 47% if one sums data from 1971 and 1985/86 54% if one sums data from 1968, 1971, and 1985/86	27% clean sand 27% muddy sand 44% silt 31% gravel[a]	6% clean sand 27% muddy sand 66% silt 67% gravel[a]	2-mm mesh, one sampling over 2 years (1985/86), one sampling of live in 1971, one sampling of live in 1968; all of same areas or stations; 103 total samples
Strait of Canso, Nova Scotia, 45°N (Wagner, 1975)	54% (47/87 spp.)	62% strait (35/56 spp.) 69% inner bay (25/36 spp.) 30% open bay (3/10 spp.)	No data	64% (47/73 spp.)	56% strait (35/62 spp.) 78% inner bay (25/32 spp.) 30% open bay (3/10 spp.)	No data	190 grab samples and ~50 shallow cores, one season
Gulf of California and Pacific Mexico, rocky intertidal to abyss, 21–31°N (Parker, 1963, Table I)	70% (281/403 spp.)	Data not available	Data not available	38% (281/736 spp.)	Data not available	Data not available	1-mm mesh, single samples of 270 stations over 3 years
Mean	84%	75%	yes	45%	46%	70%	

[a]Based on data for 12 dominant spp. in 1985/86 only.

TABLE V. Summary of Results from Tables II–IV

Setting	What % live species are found dead within the STUDY AREA? Mean	Range	What % live species are found dead within the same FACIES? Mean	Range	Do dead shells cluster like live taxa?	What % dead species are found alive within the STUDY AREA? Mean	Range	What % dead species are found alive within the same FACIES? Mean	Range	What % dead individuals are from species found alive in the same FACIES? Mean	Range
Intertidal	83%	45–100%	90%	62–100%	Yes	54%	27–100%	57%	38–100%	90%	79–100%
Coastal subtidal	95%	82–100%	98%	82–100%	Yes	33%	10–58%	42%	4–100%	89%	77–100%
Open marine	84%	54–97%	75%	30–100%	Yes	45%	38–64%	46%	22–84%	70%	6–97%
Grand means	87%		88%		Yes	44%		48%		83%	

relocating sites sampled by previous workers (Table IV). Within a single sampling program (1985/86), 37% of the dead species were found alive; this figure increased to 47% when live taxa from 1971 and 1985/86 were pooled, and increased to 54% when live taxa sampled in 1968, 1971, and 1985/86 were pooled.

This direct relationship between measured fidelity and study duration suggests that many species found "dead only" are not exotic or out-of-habitat, but simply have not yet been sampled live because of patchy or impersistent populations. Johnson (1965) and others have used the term "exotic" to designate taxa found dead only, but this term should be restricted to demonstrable instances of out-of-habitat transport. Taxa found only in death assemblages are best referred to by a more neutral term such as "dead only" (Warme, 1969), because many are simply remnants of older or very patchy indigenous populations.

Cadée (1984a, and personal communication, March 1990), for example, suggested that 26 of 32 dead-only species in molluscan assemblages of the North Sea were indigenous. The remaining 6 species not sampled live were reworked from Pleistocene intertidal deposits present 50 cm below the seafloor. Such leakage from stratigraphically older deposits is one source of discrepancy between live and dead faunas (S. W. Henderson and Frey, 1986). Another influence on fidelity is environmental change during the period of shell accumulation. Where the sedimentary environment is known to have changed during the study period, such as in Knight's (1988) study of the effects of inert clay-waste pollution in a shallow marine embayment of the English Channel (Table VI), the death assemblage differs significantly from the live community and represents an average of the older and newer communities occupying the site. It is thus essential for paleontologists to use all lines of evidence (e.g., sedimentary matrix, discontinuity surfaces, stratigraphic context, quality of shell preservation, etc.) to determine the likelihood of environmental change before death assemblages are interpreted paleobiologically (see Section 5.2.3).

Although some variation in taphonomic fidelity must reflect differences in communities, environments, and histories of shell accumulation, the positive relationship between fidelity and live-study duration described above is interesting on several accounts.

(a) It pinpoints an operational bias in data collection: because most death assemblages are time-averaged, a single live census provides an inadequate basis for comparison. Estimates of death assemblage fidelity thus should be considered to be *minimum* estimates until long-term replicate surveys of living fauna yield few new species.

(b) The between-locality variation in Tables II–IV (e.g., Mugu Lagoon versus Strait of Canso) may depend more upon how extensively live faunas were sampled than upon real differences in the fidelity of the death assemblages.

(c) Most importantly from a paleontological perspective, it provides direct evidence for scales of time-averaging. The time period over which samples must be pooled in order for the composition of the living community to converge with the death assemblage indicates the *minimum* period of time-averaging recorded by modern death assemblages. For the open shelf, living benthos must be sampled over many decades before live and dead species richnesses show good agreement (Carthew and Bosence, 1986), but evidently only a few years will suffice for lagoons (Peterson, 1976). Systematic studies to determine characteristic periods of time-averaging for Recent benthic habitats could be extremely useful for extrapolation to the fossil record because these periods are not always inherently biogenic, but a consequence of the physical environment acting on both benthic ecology and taphonomy (see Section 5). Toward these ends, it is imperative (a) that samples and records from previous faunal surveys continue to be archived, and (b) that planned and current surveys of live and dead faunas accurately locate sample sites and ensure that samples are curated for future restudy.

Finally, yet another set of fidelity estimates results if the question is, "what percentage of dead *individuals* are from species found alive?" [i.e., (dead individuals of $N_S \times 100$)/(dead individuals of $N_D + N_S$)]. Variation in this measure of fidelity is large (6–100%), but the overall average is high (70–90%; Table V), meaning that species sampled live tend to dominate the local death assemblage. This is true even for habitats with large numbers of "dead-only" taxa. For example, only 22–60% of species found dead were sampled live in subhabitats of Mannin Bay, but these accounted for the majority of individuals in death assemblages [range 66–97%, average 85% (Bosence, 1979)] (Table IV). In other words, species not sampled live, either because they have patchy and ephemeral populations or because they are truly exotic, tend to be represented by few dead individuals.

Although assemblages can be composed exclusively of exotic species (e.g., storm-washovers and turbidites dominated by bulk flow; see Section 4.2.3 and Table IX), a low number of dead individuals prove to be exotic in level-bottom settings. For example, S. W. Henderson and Frey (1986, Appendix 1) found no exotic taxa in shallow-shelf assemblages and only 9% exotic species (accounting for 10% of dead individuals) in ebb-dominated tidal sound assemblages. On intertidal flats, Frey *et al.* (1988) found 8–25% exotic species in inner flat assemblages (accounting for 5–8% of all dead individuals), 38–50% exotic species in middle flat assemblages (only 6% of dead individuals), and 17–67% exotic species in outer flat assemblages (only 1–5% of dead individuals). Exotic individuals cannot always be reliably identified based on shell damage—rafted shells may arrive in pristine condition—but transport does tend to select shells of small-bodied, free-living epifauna and very shallow burrowing infauna (see Section 4.2.3).

The rather high fidelity of death assemblages to source live communities is encouraging for paleoecological interpretation. It also might be used to advantage in environmental impact studies (e.g., Samtleben, 1981).

3.2.2. Fidelity with Respect to Species Dominance

All factors being equal, the most abundant members of a living community are the ones most likely to enter the fossil record, and the most abundant fossils are the ones most likely to be sampled by paleontologists. However, taphonomic processes can alter this simple relationship by preferentially removing certain species and size classes (see Section 4). As with species richness data, comparisons among previous studies on the fidelity of relative abundances are difficult in the absence of standardized metrics, and are limited by the scarcity of appropriate data sets.

Table VI analyzes fidelity with respect to the six most abundant taxa in Recent molluscan death assemblages. This number was chosen as a basis for comparison because (a) it permitted comparison of a larger number of published data sets, and (b) in each data set the most abundant (= top) six taxa are generally confined to the peak of the skewed species-abundance distribution typical of benthic communities. For example, the top 6 *living* mollusk species in Copano Bay, Texas, comprise 94% of all live mollusk individuals (17 species total), and the top 6 species in nearby Laguna Madre comprise 70% of all live mollusk individuals [29 species total; data from Staff *et al.* (1985)].

For each data set, fidelity was calculated several ways: (a) the number of top six taxa in the death assemblage that are also among the six most abundant taxa in the live community; (b) the percentage of dead individuals that come from the top six taxa in the live community; (c) the number of top six taxa that occur in the same *rank order* in both death assemblage and live community; and (d) the percentage of dead individuals that come from taxa ranked identically both dead and live.

Results are as follows (Table VI):

(a) One-sixth to five-sixths (grand mean = one-half) of the top taxa in the death assemblage were top taxa in the source live community.

(b) The most common live taxa generally account for slightly more than half of all individuals in the death assemblage (grand mean 57%, but range of 20–99%).

(c) Very few (zero to half) of the top dead taxa occur in the same order or ranked abundance as they do in the live community.

(d) These taxa account for about one-third of all individuals in the death assemblage (grand mean 29%, but range 0–99%).

Given that most live data sets are based on only one sampling of local communities (Table VI), all of these measures of fidelity are probably minimum estimates; presumably, fidelity values would increase with study duration, much as they do for species richness (see Section 3.2.1). The low estimates nonetheless cast doubt upon biofacies discrimination based on the rank order of the numerically most abundant species. Differences in the composition of the top taxa may be biologically meaningful, but differences in the precise ordering of those top taxa are more likely to be taphonomic in origin than biologically determined.

TABLE VI. Fidelity of Recent Molluscan Death Assemblages: Relative and Rank Abundances of the Six Most Common Species in Death Assemblages[a]

Location and sampling	Facies or community	No. of samples	How many of the 6 most abundant spp. in the death assemblage also rank as among the 6 most abundant in the source community?	What % of dead individuals come from the 6 most abundant taxa in the source community?	How many of the 6 most abundant spp. in the death assemblage occur in the same rank order as they do in the source community?	What % of dead individuals come from taxa that are ranked identically both live and dead?
Intertidal						
Yellow Sea, Korea, tidal flat (37°N) (Frey *et al.*, 1988); one sampling, unknown mesh	Muddy sand middle flat	6	5/6	81%	2/6	28%
	Sandy outer flat	10	3/6	74%	2/6	65%
Mugu Lagoon; California (34°N) (Warme *et al.*, 1976); two duplicate samples 2 years apart, 2.3-mm mesh	Eelgrass bed	8	3/6	54%	1/6	27%
	Sand channel	4	3/6	74%	3/6	74%
	Tidal creek	4	3/6	41%	1/6	8%
Seto Sea, Japan (34°N) (Tanabe *et al.*, 1986, and unpublished data); one sampling, 5-mm mesh	Sandy mud flat		6/6	99%	1/6	4%
	Sand and gravel flat		5/6	98%	NA	NA
Coastal Subtidal						
Ria de Arosa, Spain (42°N) (Cadée, 1968, and unpublished data) one sampling, 2-mm mesh	Inner bay	47	4/6	63%	0	0
	Middle bay	29	4/6	47%	2/6	24%
	Outer bay	55	4/6	49%	0	0
	Bay margin	132	1/6	10%	1/6	10%
	Oceanic zone	26	2/6	17%	0	0

Texas lagoons (26°N)	Laguna Madre sand	14	2/6	36%	0	0
(Staff *et al.*, 1986);	(biomass)		(4/6)	(97%)	(3/6)	(97%)
sampling every 6 weeks over	Copano Bay mud	14	2/6	43%	0	0
2 years, 0.5-mm mesh	(biomass)		(4/6)	(98%)	(3/6)	(98%)
Yucatan coast, Mexico (21°N)	Backreef	21	3/6	20%	1/6	11%
(Ekdale, 1972);	Mangrove lagoon	15	4/6	33%	3/6	24%
one sampling, 3.1-mm mesh	High-energy strait	12	5/6	79%	2/6	70%
Open Shelf						
North Sea (54°N) (unpublished	Muddy stations:					
raw data for Cadée, 1984a);	Single-sampled	14	4/6	40%	1/6	8%
one sampling of live and dead at 21	Quarterly-sampled	4	3/6	27%	1/6	24%
stations, 4 quarterly samplings of	Sandy stations:					
3 stations, 1-mm mesh	Single-sampled	7	2/6	52%	0	0
	Quarterly-sampled	8	3/6	85%	0	0
Mannin Bay, Ireland (53°N)	Algal bank	17	3/6	95%	1/6	78%
(Bosence, 1976a,b);	Muddy algal gravel	17	2/6	91%	1/6	85%
one sampling of live and dead	Clean algal gravel	21	3/6	66%	0	0
faunas, 2-mm mesh	Fine sand	11	3/6	59%	2/6	57%
	Mud	4	3/6	91%	3/6	91%
English Channel, embayment	Clean sand[b]	(3)	(0)	(0)	(0)	(0)
(50°N) (Knight, 1988);	Muddy sand[b]	(8)	(1/6)	(23%)	(0)	(0)
one sampling of live and dead	*A. filiformis* mud	16	4/6	36%	1/6	8%
faunas over 2 years, 2-mm mesh	*T. communis* mud	9	3/6	55%	3/6	55%
	Gravel (4-mm mesh)	4	3/6	78%	2/6	68%
English Channel, open shelf	Bay	15	2/6	75%	1/6	72%
(50°N) (Carthew and Bosence,	Inshore open shelf	12	3/6	40%	2/6	29%
1986; and unpublished	Offshore open shelf	13	2/6	23%	0	0
data) quarterly sampling for 1 year, 2-mm mesh						
Grand means			3/6	57%	1/6	29%
Range			1/6–5/6	20–99%	0–3/6	0–91%

[a]Values in parentheses are based on biomass or were from communites known to have changed and, hence, were not used in calculating grand means.
[b]Substrate type or community known to have changed during 15 years prior to sampling (Knight, 1988; Knight and Bosence, 1988).

In a unique attempt to assess the effects of ranking species according to biomass rather than number of individuals, Staff *et al.* (1985) found that live and dead assemblages in opportunist- and juvenile-dominated Texas lagoons show greater agreement when measured in terms of biomass (parenthetical values in Table VI). Two-thirds of the top dead taxa were top taxa in the live community (up from one-third; accounted for 97–98% of dead biomass), and half of the top dead taxa occurred in the same rank order as when live (up from zero; accounted for 97–98% of biomass).

3.2.3. Fidelity with Respect to Age- and Size-Frequency Distributions

The extensive literature on size–frequency and survivorship analysis has been ably reviewed by Dodd and Stanton (1981). Early work in the 1950s and 1960s emphasized size–frequency data as a means of discriminating between (a) *in situ* census assemblages, thought to be characterized by polymodal or juvenile-dominated curves, and (b) allochthonous and otherwise taphonomically sorted assemblages, thought to be characterized by normal, leptokurtic curves. Work in the 1970s shifted to using size–frequency distributions as evidence for mortality patterns among different organisms (K. R. Walker and Parker, 1976; Alexander, 1977; Hutchinson, 1978) and for the same organism in different habitats (Surlyk, 1972; Richards and Bambach, 1975; Dzik, 1979).

Actualistic work in the 1980s has provided new insights into taphonomic influences on dead shells and has been largely pessimistic concerning the paleoecological value of size–frequency data. Census assemblages (*sensu* Hallam, 1972) of articulated individuals generally permit a direct reading of shelly population structure [i.e., the age–frequency distribution of living individuals (Levinton and Bambach, 1970; Tanabe and Arimura, 1987); but see Kranz (1974a) for preferential escape of juveniles], but all other assemblages are cumulative in origin, and reflect background as well as catastrophic mortalities. At best, these assemblages provide records of local mortality (i.e., age- or size-dependent survivorship) from which population structure can be back-calculated [see Dodd and Stanton (1981) for procedures].

Unfortunately for the paleoecologist, any number of different size distributions can result from a single pattern of mortality owing to taphonomic processes; moreover, any given size distribution can have several possible explanations. For example, bimodal intertidal populations can yield (a) juvenile-dominated allochthonous assemblages by current transport, (b) adult-dominated allochthonous assemblages by hermit crab and bird transport, and (c) platykurtic residual assemblages of disarticulated shells (Fig. 7).

It appears that even with little collection bias and regardless of taxonomic group and setting, juveniles are generally underrepresented in the fossil record (Olson, 1957; Kurtén, 1954; Cadée, 1982; Yochelson and Lindemann, 1986). High taphonomic loss rates for small specimens and

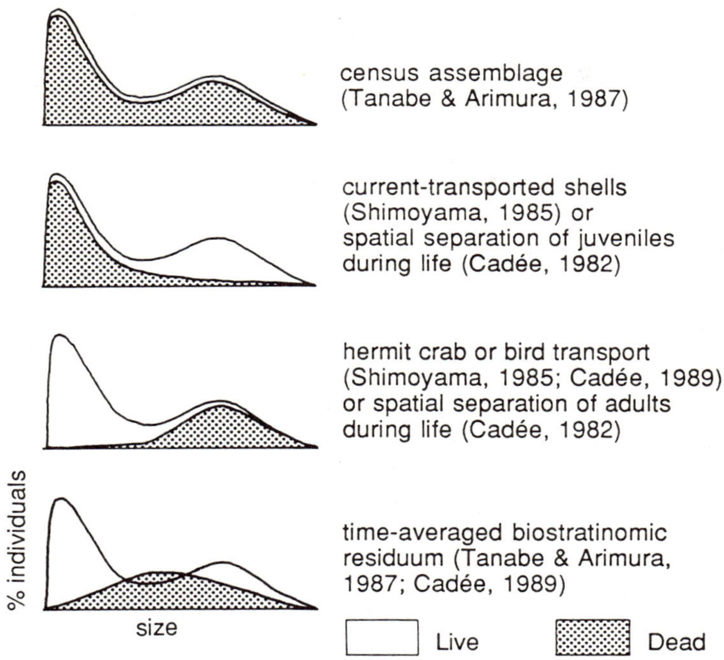

FIGURE 7. Common qualitative changes in the shape of molluscan size–frequency distributions in intertidal settings, from live (white) to dead (stippled), and documented causes.

juveniles in particular are strongly indicated by some actualistic studies (Powell *et al.*, 1982; Babcock, 1985; Cummins *et al.*, 1986a). *Adults* of small-bodied species are comparatively common in Recent and fossil death assemblages. The preferential loss of juveniles is possibly related not so much to size and surface area effects as it is to ontogenetic differences in degree and style of mineralization (e.g., high intraskeletal porosity and organic content associated with rapid juvenile growth).

In some circumstances, classes of small organisms may experience preferential preservation. Bosence (1979) and Arua (1982), for example, observed that small subtidal bivalves were less subject to borers and encrusters than were large specimens. Pip (1988) documented a similar bias against comparatively fragile large-shelled gastropods in freshwater sediments. Any species not showing an allometric increase in thickness relative to shell length might be expected to yield juvenile-dominated death assemblages owing to preferential loss of large specimens. Highly skewed frequency distributions for individual species can also reflect ecological phenomena such as spatial separation of juveniles and adults related to larval settlement (Cadée, 1982), unusually high (or low) juvenile mortality (Levinton, 1970), stunting of growth (Price, 1982), and dwarfing or paedomorphosis [reviewed by Mancini (1978)]. Multispecies assemblages dominated by small individuals generally reflect (a) hydraulic sorting and abrasive reduction, as, for example, assemblages found in beach and washover deposits, or (b) extreme diagenetic filtering by solution or mineralization, as, for example, the phosphatic steinkern-rich "pebble necrocoenoses" that characterize many Paleozoic formation boundaries in the U. S. Midcontinent (Tasch, 1955a; B. Witzke, personal communication).

Cadée (1989) suggested that in intertidal settings, size-selective predation, fragmentation, and transport are probably the rule (see also Shimoyama, 1985; Tanabe and Arimura, 1987), so that size–frequency distributions probably give little information on original population dynamics. These and other workers (Noble and Logan, 1981; Cummins *et al.*, 1986b; Westrop, 1986) have argued that size–frequency data are in general probably most revealing of post-mortem processes, except for data derived from census assemblages of articulated specimens.

Although the details of population structure are unlikely to survive taphonomic processes, data from subtidal lagoons (Cummins *et al.*, 1986b) suggest that in some instances time-averaged death assemblages provide a fairly good *qualitative* picture of molluscan mortality patterns. For example, in 10 out of 13 cases, juvenile-dominated death assemblages are associated with juvenile-dominated mortality, bimodal assemblages with bimodal mortality, and adult-dominated assemblages with adult-dominated mortality (Fig. 8). However, modes are commonly displaced by one or more size classes.

Quantitative tests of these same data yield mixed results. Based on use of the Kolmogorov–Smirnov test, a test more directly applicable to differences in the shapes of frequency distributions than the chi-square test used by Cummins *et al.* (1986b; Sokal and Rohlf, 1981), 5 of the 13 cases show no significant difference between expected and observed size distributions ($p < 0.05$) (Fig. 8). Because Cummins *et al.* (1986b) note that their estimates of juvenile mortality are probably in error by 70–95%, a second Kolmogorov–Smirnov test was run excluding the 1.0-mm size class; 6 of the 13 cases showed no significant difference (Fig. 8). In summary, approximately half of the species showed no significant difference between dead size-distributions and calculated mortalities, and virtually all show good qualitative agreement.

Cummins *et al.* (1986b) stated that their data suggested a strong preservational bias acting against juveniles. However, by their own calculations, only 4 of the 11 cases in which they tracked an "input pulse" showed any significant difference between "pre- and post-taphonomic loss" (their Table 1). Juveniles also appear to be overrepresented in death assemblages relative to mortality in 8 of their 13 graphs reprinted in our Fig. 8, but this may be a consequence of underestimated mortality, as discussed above.

The good qualitative agreement, along with quantitative agreement in half of the size–frequency distributions presented by Cummins *et al.* (1986b), suggest that such data may yet be of value to paleobiologists, although care must be taken in obtaining and interpreting these data [see, for example, the exchange between Cadée (1988) and W. Miller and DuBar (1988a,b)]. The mixed results in Texas lagoons suggest that the general conclusion that size–frequency distributions are determined primarily by taphonomy may someday prove to be overly pessimistic, even though the intertidal record does indeed appear to be strongly biased (but see Hartshorne *et al.*, 1987). Further study is warranted to determine whether time-averaged assemblages of certain taxa, life habits, or depositional environments might still yield consistently useful size–frequency data.

4. Biostratinomy: Processes and Agents of Differential Shell Preservation

Death assemblages generally do not enter the permanent record without modification on the seafloor or within surficial sediment. Throughout this period, which can include repeated cycles of burial and exhumation, individual shells and the assemblage as a whole may acquire a taphonomic overprint related to various post-mortem agents, environments, and time scales of accumulation. This overprint includes the differential loss of species and age classes (see Section 3), and the alteration of surface features, form, and facies distribution of individual shells (this section).

Biostratinomy refers to this aspect of fossilization—literally, the embedding of biological hard parts and in particular their behavior as sedimentary particles (Seilacher, 1973, 1982a) [Weigelt (1927) originally coined the term more broadly to include death and soft-tissue decay]. Most biostratinomic studies have stressed the physical, primarily hydraulic, processes that act on carcasses and hard parts during the burial interval [see recent reviews by Brett and Baird (1986), Speyer and Brett (1988), Goldring (1990)]. Because these conditions are relatively easy to manipulate experimentally, bioclast behaviors have in some instances been quantified (see below).

The post-mortem behavior of shells is usually not this simple, however, because biostratinomic histories commonly unfold in the presence of living organisms, which can bore, gnaw, swallow, break, encrust, reorient, bury, exhume, transport, trample, disarticulate, disperse, concentrate, and otherwise interfere with straightforward, physically controlled processes. In addition, the accumulation of dead shells on and just below the seafloor can alter the physical and geochemical environment and can influence subsequent benthic colonization [taphonomic feedback (Kidwell and Jablonski, 1983; Kidwell, 1991)].

A further analytical complication is that a given preservational state can reflect any number of life and post-mortem processes (Table VII), so that few unequivocal interpretations can be drawn in isolation. For example, high disarticulation might reflect high water energy, active scavenging, or general bioturbation of decayed carcasses. Future work should thus target

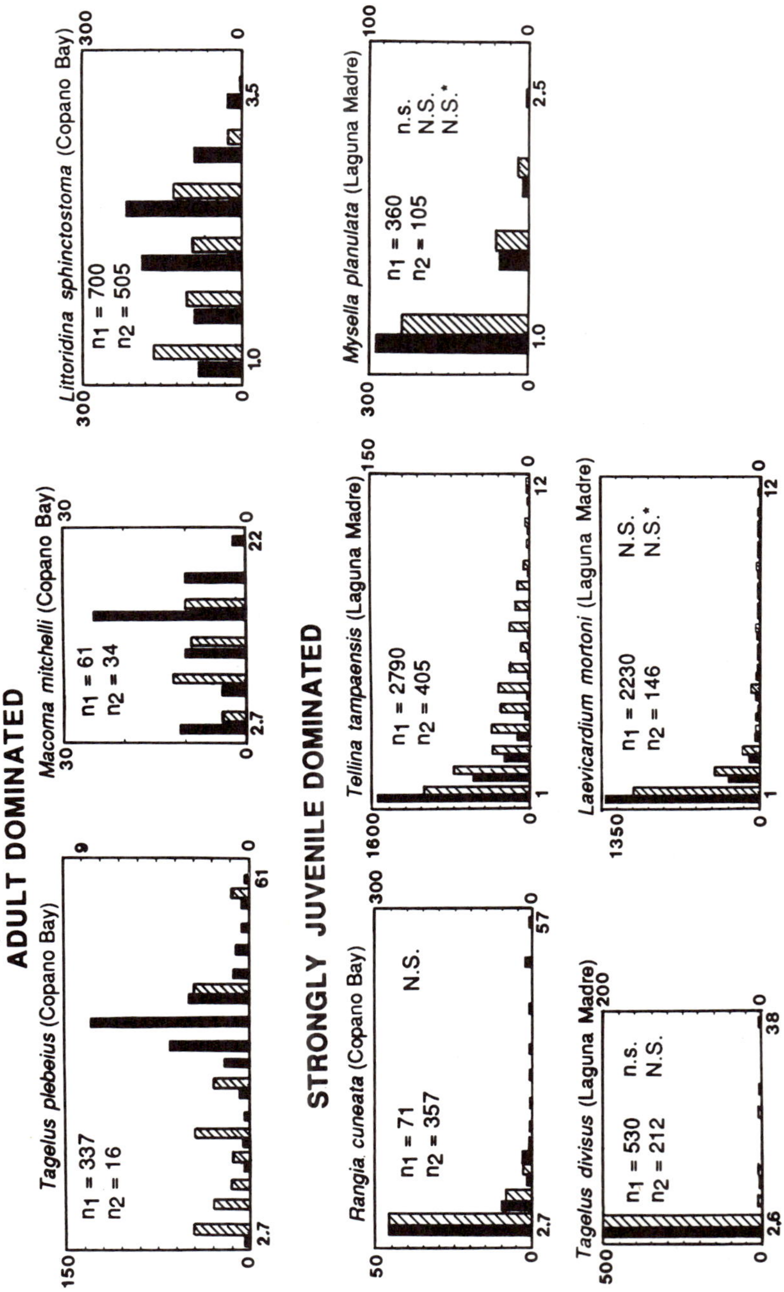

ADULT DOMINATED
Tagelus plebeius (Copano Bay)
n1 = 337
n2 = 16
Macoma mitchelli (Copano Bay)
n1 = 61
n2 = 34
Littoridina sphinctostoma (Copano Bay)
n1 = 700
n2 = 505
STRONGLY JUVENILE DOMINATED
Rangia cuneata (Copano Bay)
n1 = 71
n2 = 357
N.S.
Tellina tampaensis (Laguna Madre)
n1 = 2790
n2 = 405
Mysella planulata (Laguna Madre)
n1 = 360
n2 = 105
n.s.
N.S.
N.S.*
Tagelus divisus (Laguna Madre)
n1 = 530
n2 = 212
n.s.
N.S.
Laevicardium mortoni (Laguna Madre)
n1 = 2230
n2 = 146
N.S.
N.S.*

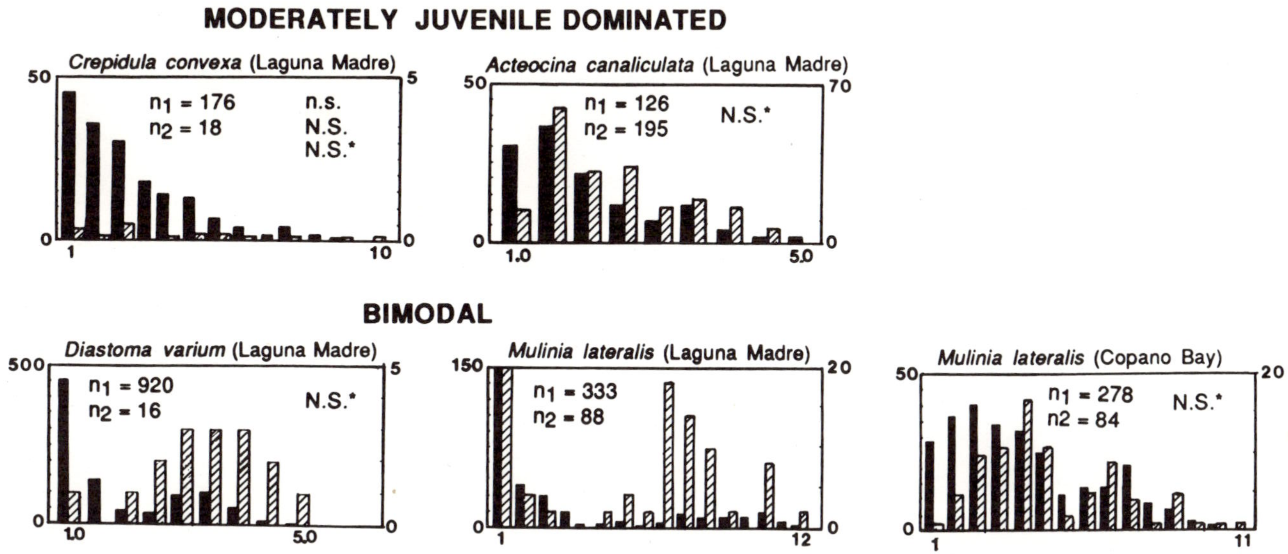

FIGURE 8. Comparison of size–frequency distributions for shells supplied by mortality (histograms with diagonal lines) and shells observed in the locally accumulating death assemblage (black histograms) for 13 molluscan species in two Texas lagoons. Based on chi-square tests, which compare expected and observed values on a size-class by size-class basis, Cummins *et al.* (1986b) concluded that death assemblages provide poor records of mortality. However, the data suggest surprisingly good *qualitative* agreement in the shapes of the distributions, and a Kolmogorov–Smirnov test sensitive to curve shape indicates no statistical difference between expected and observed frequency distributions in about half the cases. n.s., not significant based on chi-square test of Cummins *et al.* (1986b); N.S., not significant ($p > 0.05$) based on Kolmogorov–Smirnov test using all size classes; N.S.*, not significant ($p > 0.05$) based on Kolmogorov–Smirnov test excluding 1-mm size class; n_1, sample size of death assemblage; n_2, sample size of new additions (mortality during study period); ■, individuals in death assemblage (scaled on left axis); ▨, individuals/m^2 added by death during study period (right axis); mean size in mm (horizontal axis). [Adapted from Cummins *et al.* (1986b).]

TABLE VII. Origin of Biostratinomic Features

	Behavior or ecology of shell producers	Behavior or ecology of other organisms	Physical, primarily hydraulic processes	Other seafloor processes	Postburial processes	Descriptive scales
Disarticulation	Autotomy among crinoids (Emson and Wilkie, 1980); molting (Speyer and Brett, 1985; Feldman and Tshudy, 1987)	Scavengers and predators (Meyer and Meyer, 1986; Lavoir, 1956); bioturbation? of buried carcasses	Transport or *in situ* reworking, particularly of decay-loosened skeletons (Menard and Boucot, 1951; Eager, 1978; Allison, 1986; Kidwell and Baumiller, 1990)	Collapse upon microbial decomposition of ligaments (Meyer and Meyer, 1986; A. B. Smith, 1984)	Compaction?	Meyer *et al.* (1989) (crinoids); Kidwell and Baumiller (1990) (echinoids)
Sorting	Single-cohort settlement (Levinton, 1970); gregarious adults (Waage, 1964; Speyer and Brett, 1985); remnants of anastrophic burial (Kranz, 1974a)	Selective destruction of inequivalve bivalves by predators (Carter, 1974); selection by tube-building worms (Recent *Diopatra*)	According to size, right–left differences, spines, perforation during transport (Kornicker *et al.*, 1963; Martin-Kaye, 1951; Lever, 1958; Lever *et al.*, 1961; J. B. Wilson, 1967; Bosence, 1976b; Martinell and dePorta, 1982; Aigner, 1985b); according to differential durability of skeletal parts to *in situ* reworking (Holland, 1988; Velbel and Brandt, 1989; Fürsich and Werner, 1989); settling rates from suspension (Middleton, 1967; Allen, 1984)		Preferential loss of small or aragonitic shells (Fürsich, 1982); preferential preservation of small shells by phosphatization	Kidwell and Holland (1991)
Perforation	Pre-mortem infestation by boring sponges, forams, bivalves, barnacles, etc. (Highsmith *et al.*, 1983; Pisera, 1987; P. J. B. Scott, 1988; Smyth, 1988); complete abrasion of umbones and apices during life (Hollmann, 1968; Boyer, 1969; Schäfer, 1972)	Post-mortem infestation (C. M. Henderson and Styan, 1982; Akpan and Farrow, 1984, 1985; Nelson and Young 1985; K. Vogel *et al.*, 1987); hermit shell symbionts (Lambers and Boekschoten, 1986; S. E. Walker, 1988); boring and biting predators [gastropods, octopods, sharks, ichthyosaurs (Bishop, 1975; Alexander, 1986b; Mapes *et al.*, 1989)]	Mechanical "surf" breaks in bivalves, echinoids (Schäfer, 1972; Kidwell and Baumiller, 1990); interpenetration of telescoped nautiloids and hyolithids (Tasch, 1955b; Yochelson and Fraser, 1973); localized abrasion (Pratje, 1929; Hollmann, 1968; Driscoll and Weltin, 1973)	Freshwater dissolution under holes in shell periostracum (Philipon and Plaziat, 1975)		Carriker *et al.* (1969); Bromley (1970); Ekdale (1985)

Fragmentation	Asexual reproduction of corals by "autotomy" (Karlson, 1986); weakening of mollusks and corals by borers (Bergman *et al.*, 1982; Hutchings, 1986; P. J. B. Scott, 1988; Tunnicliffe, 1979; Turner, 1985)	Shell-crushing and ingestion by predatory crabs, fish, rays, etc. (Trewin and Welsh, 1976; Bertness and Cunningham, 1981; Bigelow and Schroeder, 1953; Alexander, 1986b; A. B. Smith, 1984; Hoskin, 1980); breakage on stones by otters, birds (Cadée, 1989; Hines and Pearse, 1982); bioturbation?; by-product of bioerosion	Reworking of shells weakened by bioeroders; mechanical breaks in surf and high-energy situations (Hollman, 1968); entrainment in course debris flows, crused in interstices of boulder/gravel fields (Myers, 1981; S. M. Kidwell, unpublished observtions, Recent and Pliocene)		Compaction (Brenner and Einsele, 1976)	Hollmann (1968); Davies *et al.* (1989a); Pilkey *et al.* (1967)
Surface modification by abrasion, encrustation, and dissolution	Self-inflicted abrasion of umbones of arcids, cardiids (Hollmann, 1968; Boyer, 1969; Schäfer, 1972); normal wear of apices of limpets, other gastropods (S. M. Kidsell, unpublished observations, Recent); pitting by encrusting bryozoa, algae (Boekschoten, 1966; Cutler, 1987); reduction of ornamentation, creation of rasp marks by grazing limpets, echinoids, chitons, etc., during life of epifauna; symbiotic encrustation (Alvarez and Taylor, 1987)	Reduction in ornamentation by grazers, dulling of surface by micropitting encrusters and grazers on dead shells as on live; rounding by microboring and bacterial decomposition of stroma (Ameziane-Cominardi and Roux, 1987); drag marks on underside of hermitted shells (S. E. Walker, 1988); partial to complete encrustation of dead shells by algae, bryozoans, corals, serpulids, etc. [both hermit and "rolling stone" colonies (Bishop, 1988; Ware, 1975; Balson and Taylor, 1982; Kissling, 1973; Taylor *et al.*, 1989)]	Abrasive reduction and frosting on beaches (Cutler, 1987; Driscoll, 1967) and during lab tumbling (Chave, 1964; Driscoll and Weltin, 1973); impact craters on cobble beaches (S. M. Kidwell, unpublished observations)	Corrosion by fresh and cold waters before burial (Wiedemann, 1972; Alexandersson, 1972, 1978)	Chalky textures from leaching in marsh and temperature marine sediments (Frey and Basan, 1981; Lewy, 1975; Meldahl, 1987a; Alexandersson, 1978); partial dissolution by acidic porewaters (Aller, 1982a); meteoric weathering (B. M. Walker, 1979)	Holland (1988); Pilkey *et al.* (1967, 1979); D. J. Davies *et al.* (1989a); Driscoll and Weltin (1973); Cutler (1987); criteria for pre- vs. postmortem infestations: Baird *et al.* (1989); Sando (1984)
Orientation	Life orientation, escape postures, "death throes" (Stanley, 1970; Kranz, 1974a; Seilacher, 1984; Emig, 1986; Frey *et al.*, 1987)	Modification by bioturbators, predators, scavengers (Salazar-Jimenez *et al.*, 1982; Emery, 1968; Clifton, 1971); rearrangment by tube-builders and burrow back-fillers (Schäfer, 1972)	Modification by unidirectional or reversing currents, turbulent flow, redeposition, exhumation, detachment (of sessile benthos) (Moors, 1970; Müller, 1979; Raup, 1973; Seilacher, 1973; Brett and Baird, 1986; Wulff, 1990)	Orientation determined by morphology of trap (fissues, cobble interstices)	Rotation during compaction	Kidwell *et al.* (1986)

(continued)

TABLE VII. *(Continued)*

	Behavior or ecology of shell producers	Behavior or ecology of other organisms	Physical, primarily hydraulic processes	Other seafloor processes	Postburial processes	Descriptive scales
Transportation	Deposition of epifauna from floating weeds (Cadée, 1968; Bosence, 1979)	By hermit crabs (S. E. Walker, 1988; Shimoyama, 1985), birds (Cadée, 1989; Lindberg and Kellogg, 1982; Brun, 1971; Trewin and Welsh, 1976), fish and mammals (Schäfer, 1972) in form of feces and epifauna; rafting on seaweed (Bosence, 1979; Cadée 1968); downslope "diffusion" by bioturbators (Bein and Fütterer, 1977)	Onshore, offshore, and longshore movement in coastal and nearshore marine zones; bulk transport on slopes including reef fronts (see Table IX); "wheeling" of ammonities in turbidity currents (Seilacher, 1963)	Post-mortem drift (Saunders and Spinosa, 1979; Reyment, 1986); ice-rafting (Spjeldnaes, 1978; Aitken *et al.*, 1988)		See terms in Table I
Exhumation	Epifaunal life habit; surfacing by moribund infauna (D. Nichols, personal communication in Aslin, 1968; Rosenberg, 1977; Steimle and Sinderman, 1978)	Hermits delay burial (S. E. Walker 1988); upward advection by bioturbators (McCave, 1988); excavation of infauna by predatory birds, rays (Carter, 1968; Gregory *et al.*, 1979)	Erosional reworking by storms, currents, channel migration (van Straaten, 1952; Aigner, 1985a); seafloor or subaerial winnowing, bedform migration; exposure on seafloor prolinged by sediment starvation or bypassing (Akpan and Farrow, 1984; J. B. Wilson, 1988; Nelson, 1988)			Patterns of burial-exhumation cycles: Kidwell and Aigner (1985); Kidwell (1989)

Burial	Infaunal life habit	Backfill by burrowing infauna (Seibold *et al.*, 1973); downward advection by bioturbators (Cadée, 1976; Meldahl, 1987b)	Passive infilling of burrows (Aller, 1982a; Wanless *et al.*, 1987); anastrophic burial by storm or turbidite desposition (Peterson, 1985; Aigner, 1985a; Brett and Seilacher, 1991); bedform migration, burial to base of tractive carpet (Johnson, 1957)			
Concentration	Gregarious behavior of larvae or adults, dense opportunistic settlements (biogenic event-concentrations); reef and bioherm-forming behavior (composite concentrations)	Packrat behavior by crabs, nesting birds (Cohen, 1989; Teichert and Serventy, 1947); fecal and egested masses from predators, scavengers (Trewin and Welsh, 1976); feeding pit concentration by rays, ducks (Gregory *et al.*, 1979; G. C. Cadée, personal communication); burrow construction and backfill (Hertweck, 1972; Seibold *et al.*, 1973); concentrations for camouflage and antidessication (Hart & Crowe, 1977); accretionary biogenic graded bedding (Cadée, 1976, 1979; Meldahl, 1987b)	Single-event concentration by storms, turbidites, channeled flow, bedform migration; composite shelly shoals, banks, washover fans; condensed hiatal beds produced by starvation or bypassing, associated with unconformities, flooding surfaces, condensed intervals; erosional lags (see references in Kidwell, this volume, Chapter 5)	Fissure and open burrow trapping of shell in transport (Seilacher and Westphal, 1971; Wanless *et al.*, 1987)	Sorting and concentration by earthquake liquefaction (Bradley, 1957); preferential pressure solution of micritic matrix (Kidwell *et al.*, 1986); diagenetic "extinction" of shells in intervening strata (Fürsich, 1982)	Aigner *et al.* (1978); Kidwell *et al.* (1986); Kidwell 1982, and this volume, Chapter 5)

development of diagnostic criteria for specific taphonomic processes at the level of both individual shells and whole assemblages, and (b) ranking the relative importance of different processes in a given environment and among environments. Essential to this work will be the development of standardized scales for describing shell condition (last column, Table VII).

This review stresses the implications of biostratinomic processes for the preservation of faunas. For reviews with fuller treatment of their geological applications, see Brett and Baird (1986), Speyer and Brett (1988; and this volume, Chapter 11), Goldring (1990), and Kidwell (this volume, Chapter 5). For a fuller review of the diagenesis of shelly assemblages, see Raiswell and Canfield (this volume, Chapter 9).

4.1. Basic Patterns

In general, post-mortem modification is a function of the susceptibility of shells to a particular suite of processes and the length of time that shells are exposed to attack (Fig. 1). This susceptibility is largely inherent to the organism's life position and to the shape, size, mineralogy, microarchitecture, surface roughness, organic coats, and surface area (including porosity) of its hard parts. Consequently, it varies not only among major taxonomic groups, but among species of a single group, among different elements of a single skeleton, and among individuals at different points in ontogeny. Susceptibility can also be secondarily changed by taphonomic processes before final burial (e.g., size reduction by fragmentation, removal of periostracal coverings, early syntaxial overgrowths during temporary burial) and thus, like other major factors in the taphonomic equation (Fig. 1), it is a dynamic factor.

Both laboratory and field experiments have shown that individuals and assemblages alike suffer greater modification as the period of exposure to post-mortem agents is increased (e.g., references in Table VII). However, taphonomic damage need not be a direct function of an individual shell's age since death, because its exposure to post-mortem processes may be intermittent rather than continuous. Corrosion, for example, can characterize Recent as well as relict shells in shelf death assemblages (Nelson and Hancock, 1984), and isolated old shells may be in better condition than recently-dead shells (Powell and Davies, 1989; Flessa *et al.*, 1990).

This variability owes to several phenomena. One is that taphonomic agents themselves are not always relentless: both bioerosion and pore-water chemistry, for example, can be strongly seasonal (Aller, 1982a; Hutchings, 1986; Reaves, 1986). Second, shells can gain temporary refuge by burial or encrustation, which isolate the shell from some taphonomic agents. Temporary or permanent sanctuary from one set of post-mortem agents (e.g., from physical abrasion or undersaturated seawater at the seafloor) of course may merely expose the shell to another set [e.g., acidic pore waters in the shallow subsurface (Driscoll, 1970)]. Moreover, shells can undergo several phases of modification related to burial–exhumation cycles and other environmental changes. Modification rates therefore probably vary during the history of individual shells and cannot readily be extrapolated from short-term experimental values; short-term values provide estimates of *maximum* rates. Shells do not attain true sanctuary from biostratinomic processes until final burial below the reach of bioturbation and storm reworking, at which time they enter the realm of later diagenesis.

Environments differ in their rates of shell supply and favorability for preservation, but a general correlation exists between shell condition and the short-term dynamics and *net* rate of sediment accumulation [for reviews see Kidwell, 1985, 1986a, and this volume, Chapter 5; Brett and Baird, 1986; Brandt, 1989). Both continuous, rapid sedimentation and episodic deposition of thick sedimentary increments can speed the transit of shells through the biostratinomically active zone; this rapid burial favors good preservation. Conversely, slow net rates of sedimentation and repeated exhumation retard final burial, and increase the average exposure time of shells to surficial and near-surface taphonomic agents. This should lead to poorer overall preservation of the fossil assemblage, but not necessarily to a deposit with low abundance of shells, since low dilution by sediment promotes shell concentration and positive taphonomic feedback. In these ways the time-averaging, quality of shell preservation, and paleoecology of fossil assemblages may be linked (Kidwell, 1986a, and this volume, Chapter 5).

4.2. Modification by Hydraulic Processes

4.2.1. Reworking and Reorientation of Shells

Well-documented effects of hydraulic reworking include (a) sorting by size and shape, including right–left differences in pelecypods and brachial–pedicle differences in brachiopods (Table VII), (b) abrasive size reduction according to skeletal microarchitecture (Sorby, 1879; Force, 1969; Hoskin *et al.*, 1983; Greenstein, 1989a), and (c) reorientation into a variety of highly ordered fabrics. These fabrics include bedding-plane accumulations in which elongate elements are oriented parallel, transverse, or oblique to current, and matrix- and shell-supported fabrics in which bioclasts are concordant, oblique, or perpendicular to bedding [reviewed in Müller (1979), Kidwell *et al.* (1986), Goldring (1990)] (Table VIII).

Well-ordered fabrics typically reflect steady or regularly varying flows (e.g., beach swash-backwash, tidal ebb and flood, subtidal wave oscillation), whereas poorly ordered fabrics reflect greater turbulence at the seafloor (Kellings and Williams, 1967), turbulent deposition of suspended shells (e.g., as observed in debris flows and some storm deposits and turbidites), or presumably bioturbation (Toots, 1965). Hydraulic reworking does not always result in convex-up orientations of bowl-shaped shells, and thus this is not a foolproof guide to water energy and "stratigraphic

TABLE VIII. Processes of Shell Orientation

Orientation	Hydraulic processes	Other processes
PLAN VIEW		
Unimodal 	Unidirectional current acting on shells (free or tethered) whose critical threshold velocities are lower than surrounding sediment; the more turbulent the flow, the greater the angular deviation (Kelling and Williams, 1967; Erickson, 1971; Brenner, 1976; Futterer, 1978a,b)	Rheotaxic life position (Frey *et al.*, 1987)
Bimodal 	As for unimodal patterns except reversing currents (wave oscillation, tidal ebb and flood); shell long axes aligned perpendicular to oscillatory current directions (Nagle, 1967; Kelling and Moshriff, 1977)	Free-rolling *Oliva* when alive; rheotaxic life position in oscillating currents (Frey *et al.*, 1987)
No strong trend 	Flow velocities below or near critical threshold levels for shell movement (Bayer, 1978), OR sediment moves at lower velocity than shells, leading to shell burial without azimuthal reorientation (Futterer, 1978b), OR very turbulent flow conditions, OR interference among shells or between shells and other particles (Brenchley and Newall, 1970; Futterer, 1978b)	Life position (Allmon, 1988)
CROSS-SECTIONAL VIEW		
Concordant 	Redeposition of platey fragments and shells out of suspension, OR rotation during compaction, OR complete exhumation of vertically embedded infauna by gentle winnowing (Emig, 1986)	Life position (Stanley, 1970)
Concordant and convex-down 	Redeposition out of suspension, associated with turbidity and turbulent storm currents (Middleton, 1967; Futterer, 1978c; Allen, 1984), OR redeposition on leesides of migrating bedforms (Clifton and Boggs, 1970), OR overturning of hemispheric corals, etc. (Abbott, 1974), OR burial at subthreshold velocities (Richter, 1942; Johnson, 1957; Futterer, 1978c)	Life position of concavoconvex brachiopods; bioturbation on seafloor (Emery, 1968; Clifton, 1971) and within sediments (Salazar-Jimenex *et al.*, 1982)
Concordant and convex-up 	Reorientation by currents on high-friction seafloor (Richter, 1942; Johnson, 1957; Futterer, 1978a)	Life position (Abbott, 1974)
Perpendicular 	Oscillatory wave and current action on abundant shells, interference effects for edgewise stacking (Grinnell, 1974; Sanderson and Donovan, 1974; Seilacher & Meischner, 1965), OR vertical embedding of nekton (Raup, 1973; Reyment, 1980)	Life position of solitary or clustered individuals (Papp, 1944; Frey *et al.*, 1987)

(continued)

TABLE VIII. *(Continued)*

Orientation	Hydraulic processes	Other processes
Oblique (30–60°) 	Interference effect of unidirectional or tidal current resulting in imbricate or shingled stack, with shells dipping up-current (Greensmith and Tucker, 1969; Futterer, 1978b; Seilacher, 1982b)	Life position (Alexander, 1986a)
Stacked 	Both convex-up and convex-down stacks are associated with storm-reworked concentrations; probably a result of interference among shells during transport (Kidwell *et al.*, 1986; Kidwell and Holland, 1989)	Life position of mutually attaching epifauna (e.g., anomiid bivalves)
No strong trend 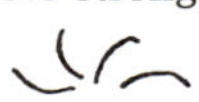	Turbulent stirring of seafloor (Greensmith and Tucker, 1969), OR rapid deposition out of high-density turbulent flows such as debris flows (Middleton, 1967), OR burial by sediment scour at subcritical velocities (Futterer, 1978b)	Life position (Allmon, 1988), OR bioturbation presumably (Elias *et al.*, 1987), OR biogenic graded bedding (van Straaten, 1956)

way-up.'' Moreover, organisms may be efficient rearrangers of shells on the seafloor and within the sediment (Tables VII and VIII), yielding both ordered and disordered fabrics. Biological influences thus must be ruled out before shell orientations are interpreted hydraulically. The association of well-ordered shells with physical sedimentary structures, textures, and/or hydraulically equivalent sediment often provides the best evidence in such situations.

Because hydraulic reorientation requires that shells be mobilized, the selective reorientation, flipping, and size sorting of shells in assemblages should permit paleocurrent velocities to be bracketed (Alexander, 1986a). For paleocurrent *direction* indicators to be reliable, the seafloor must be sufficiently firm for shells to move freely—that is, the threshold velocity for surrounding sediment must be higher than that for the shells—and so the most dependable measurements come from bedding plane surfaces where sediment dewatering has increased the internal ''bed friction'' of the sand or mud (Futterer, 1978c; Allen, 1984) and where there are few obstacles (e.g., other shells, low-amplitude bedforms). Flume studies by Futterer (1978a,b) illustrate the often surprising equilibrium orientations assumed by bioclasts, particularly when shells are in sufficient abundance to interfere with free-rolling behavior.

When the threshold velocity of surrounding sediment is *lower* than for shells, shells sink or rotate into the substratum by settling into scours formed around them. Appropriately oriented shells become traps for sediment in motion, further reducing the likelihood of shell transport and hastening shell burial to a lag position at the base of migrating bedforms (Johnson, 1957; Clifton and Boggs, 1970; Futterer, 1978a,b). Shells can thus be hydraulically buried and concentrated without necessarily acquiring a ''hydraulically stable'' orientation [and see R. L. Watson (1971) for subaerial formation of shell lags]. Presumably, most shells on seafloors above fairweather wavebase are subject to such rapid burial [e.g., burial of 85% of dead shells within 21 days observed by Conover (1975)]. It should also be important wherever bedforms migrate more episodically in response to storms. In all cases, rates of shell burial should not be limited by the typically slow *net* rates of seafloor aggradation (centimeters per thousand years) that characterize level-bottom marine settings; nor should hydraulic concentration and burial of shells be limited to the rare, high-energy events capable of causing ecologically catastrophic slumps and sediment-laden plumes.

When flow exceeds shell threshold velocities, shells are set in free motion and may themselves create bedforms (see Kidwell, this volume, Chapter 5). Such hydraulic deposits include shell megaripples and trough concentrates on tide- and wave-dominated shelves (J. B. Wilson, 1986; Norris, 1986) and sheets of storm- or turbidity-emplaced shells (Meldahl, 1987a; Aigner and Reineck, 1982; Pilkey and Curran, 1986). Bioclasts will tend to be size- or shape-sorted according to their effective diameter [a function of size, aspect ratio, and density (Ruhrmann, 1971; Seibold *et al.*, 1973; Futterer, 1978c)]. When taxa differ significantly in hydraulic properties, size sorting may be misinterpreted as having an ecological rather than taphonomic origin (Westrop, 1986).

Although convex-up orientations pervade traction-dominated flow conditions (but shell orientations can be modified by organisms during slack flow; Table VIII), concave-up orientations are common in turbidites and other mass-flow deposits. Laboratory experiments (e.g., Middleton, 1967; Futterer, 1978c; Allen, 1984; Hesselbo, 1987) have shown that resuspended shells settle concave-up through both stagnant and moving water columns, and entrain suspended sediment and thus increase in effective mass during settling. Consequently, as the effective density of the fluid increases, so does the frequency of concave-up positions [23–40% in low-density flows, 50–69% in high-density flows (Middleton, 1967)], resulting in changing proportions of concave-up orientations [and of shell size (van Tassell, 1981)] within single beds. In the absence of bioturbation and tractive movement after settling, this diversity of concave-up, oblique, and vertical positions can be preserved (Jeffry and Aigner, 1982).

Because shells are often coarser than surrounding sediment, shell-rich bedforms and lags can have higher preservation potential than nonbioclastic deposits. This is a consequence of both the high threshold velocities of individual shells and the interlocking fabrics that bioclasts can assume. Such concentrations can armor sediments against subsequent erosion (Futterer, 1978a) and become basal ''reference horizons'' for the accumulation of complexly amalgamated coquinas (Seilacher, 1985; Kidwell and Aigner, 1985; Pickrill, 1985), thus increasing the preservation potential of shelly assemblages.

4.2.2. Disarticulation and Fragmentation of Shells

These phenomena are commonly attributed to physical reworking and in particular to shell transport [e.g., the influential paper by Menard and Boucot (1951)], but until recently few systematic studies have been conducted on environmental differences and on the relative roles of physical and biological agents. In general, disarticulation should decrease with decreasing water energy, and in fact Eager (1978), S. W. Henderson and Frey (1986), and others have reported higher incidences of articulated dead bivalves in open-shelf relative to nearshore environments. Rapid burial and low-energy anoxic conditions are most frequently invoked to explain good preservation of articulated hard parts (Seilacher *et al.*, 1985).

Controlled laboratory studies indicate that disarticulation is not a simple function of water energy or distance of transport. Instead, it is strongly influenced by the life habit [e.g., epifauna are more prone than infauna (Callender *et al.*, 1990)], amount of connective tissue (Ruhrmann, 1971), and extent of organic decay before physical reworking (Schäfer, 1972; Meyer, 1971; Allison, 1986, 1990; Kidwell and Baumiller, 1990). Anoxia is relatively ineffective in preventing decay (Plotnick, 1986; Allison, 1988; Kidwell and Baumiller, 1990), but nonetheless favors preservation of articulated specimens by reducing biogenic reworking (Plotnick, 1986) and promoting early mineralization of organic connective tissues (Allison, 1986). By contrast, low temperature is very effective in retarding decay (Kidwell and Baumiller, 1990). Well-articulated specimens thus should preferentially occur in environments characterized either by frequent catastrophic burial [obrution (Brett and Seilacher, 1991)] or by low temperature and anoxia.

Susceptibility to disarticulation is also governed by how skeletal elements fit together. For example, owing to their interlocking stereom, many echinoids fracture across plates rather than separate along plate sutures even after decay of connective tissue (A. B. Smith, 1984), and the hinges of many brachiopods must break before the valves can separate (Sheehan, 1978; Alexander, 1990). Pelecypods also vary in their tendency to disarticulate owing to phylogenetic differences in ligament size and type. If kept moist, the abductin-rich types of ligament apparently remain flexible almost indefinitely. For example, oysters from subtidal middens associated with the 18th century New Haven, Connecticut, town pier were still pliable when removed from the anoxic harbor mud in the 1960s (K. M. Waage, personal communication). Butterflied scallop shells dredged from shelf fisheries, and laminated silt fills in articulated infaunal bivalves [Fig. 7 in Kidwell (this volume, Chapter 5); see Holland (1988) for brachiopod examples] indicate that individuals can survive exhumation and exposure even on current-swept seafloors without disarticulation.

For fragmentation, waves and currents are most effective when they act against anvils such as hard-packed sand, gravel, or bedrock (Hollmann,

1966). These conditions yield distinctive "surf breaks" in thin convex shells (a single hole punctured in the arc of the shell or hollow test) and impact-cratered surfaces on thick-shelled specimens. Because of the cushioning effect of water, this method of mechanical fracture is more common in beach than in subtidal settings. Alternatively, shells can be pulverized among mobile gravel (Myers, 1981) or broken by water after becoming wedged in fissures. Much fragmentation may be biogenic rather than hydraulic in origin (e.g., Cadée, 1968; J. B. Wilson, 1967; Trewin and Welsh, 1976; Bosence, 1976a; Meldahl and Flessa, 1990), but data are sparse on the proportional contribution or environmental trends (see biological effects below, Section 4.3).

Small, light shells can be entrained in turbulence and literally float on the surface or above the benthic boundary layer (D. W. J. Bosence, personal observation), and thus might suffer disproportionately less damage from hydraulic processes than will larger shells. Certainly, large, thin shells (length:thickness ≥ 10) require significantly lower stresses to fracture (S. Vogel, 1988). Older shells are more sensitive to stress than are new (recently dead) shells: bivalve shells lose mechanical strength as they dry out (Currey, 1980) and even when held in seawater [up to 50% strength loss in the bivalve *Donax variabilis* within 3–4 days (M. LaBarbera, personal communication, June 1990). Resistance to fracture also varies with skeletal microarchitecture. Among mollusks, nacre is strongest under virtually all conditions (tension, compression, impact); prismatic structures are the most flexible and cross-lamellar the least; foliated structures are least able to withstand compression (Taylor and Layman, 1972; Currey, 1980, 1988). All of these observations suggest preservational bias as a function of size, age at death (proportion of mineral and organic phases may vary ontogenetically), phylogeny, and time between death and stress.

4.2.3. Shell Transport: The Question of Spatial Fidelity

Out-of-habitat transport of shells is a major concern for paleontological interpretation. However, field and experimental studies indicate that mass (or bulk) flow is the most effective mechanism of transporting shells in large abundance or over great distances, so that allochthonous assemblages are most common in steeply sloped or episodically high-energy settings (Table IX). These deposits, such as storm- and tidal-surge washovers, turbidites, and slumps, are usually identifiable by their distinctive sedimentary structures and facies context, and the resulting predominantly allochthonous assemblages commonly occur in the form of concentrations. These may (but do not always) exhibit good size sorting of shells; shell orientations and concentrations typically are closely related to physical sedimentary structures.

In contrast, out-of-habitat transport is much less important in level-bottom sublittoral environments (Table IX). The literature suggests that: (a) most lateral movement is within-habitat [this is corroborated by the facies-level fidelity of death assemblages seen in Tables II–IV (Zenetos, 1990)]; (b) most exotic shells are derived from immediately adjacent habitats; (c) tractive transport is highly selective, affecting noncemented epifauna, small, thin shells, and shallowest-burrowing infauna in particular (van Straaten, 1956, 1960; Cadée, 1968; Shuto, 1973; Bosence, 1979; Aigner and Reineck, 1982; S. W. Henderson and Frey, 1986; Knight, 1988; Kondo, 1989); and (d) even when exotic shells account for a large percentage of species, they often comprise a minority of individuals in the assemblages that receive them (~ 10%; Tables II–IV). Data in Tables II–IV on the proportions of "dead only" taxa are maximum estimates of exotic (i.e., out-of-habitat) species: some of these "dead only" taxa are indigenous but were not sampled live because of patchy and/or episodic distributions, whereas others are mixed from local relict deposits (Sarnthein, 1971; Cadée, 1984a; S. W. Henderson and Frey, 1986).

The level-bottom settings with the largest exotic component seem to be those adjacent to habitats with abundant epifauna or epiphytic communities subject to rafting (Table IX). These exotic species will in some instances constitute distinctive trophic groups not otherwise represented in the recipient (soft-bottom) assemblage. Knowledge of the basic ecology of fossil taxa and of the regional facies context of the assemblage is thus of great help in anticipating the likely abundance and source of exotic elements.

While out-of-habitat transport is rare, it is obvious [and quantifiable (Cummins *et al.*, 1986c)] from the few shells preserved in life positions that small-scale, within-habitat movement of shells is extremely common. The critical question for paleontologists is the extent to which this small-scale movement limits the detail in which ecological variation in species abundance can be interpreted from the fossil record: how much of the original gradient in community composition (e.g., Fig. 2) is smeared by post-mortem disturbance of shells? That is, what is the precise spatial fidelity of shell material and how does this fidelity vary among environments?

As discussed in Section 3, analyses of Recent level-bottom assemblages have repeatedly demonstrated (a) the nonrandom distribution of dead shells, (b) their association with benthic communities of similar composition, and (c) the significant association of both live and dead with distinct habitats (i.e., lithofacies; Tables II–VI). The basic loyalty of death assemblages to life environments applies not only to generally low-energy, protected settings (Ekdale, 1977; Peterson, 1976; A. I. Miller, 1988) but also to many persistently high-energy settings such as macrotidal flats (Fürsich and Flessa, 1987; Meldahl and Flessa, 1990) and tide-swept shelves (Holme, 1961; J. B. Wilson, 1986; Carthew and Bosence, 1986). In fact, 75 years ago Kindle (1916) used a live:dead comparison of benthos from the macrotidal Bay of Fundy to prove that post-mortem transport does not homogenize death assemblages, and argued that fossils were reliable for paleoenvironmental interpretation even in such high-energy settings.

This correspondence indicates that net movements of shells in level-bottom settings are generally small compared with the time-averaged spatial

TABLE IX. Evidence for Out-of-Habitat Transport of Shells in Recent Marine Environments[a]

Setting	Direction, mechanism, and distance of shell transport	Selectivity of transport; net effect on assemblage at depositional site
Supratidal to nearshore		
Mesotidal estuarine sound, inter- and supratidal flats; Georgia (Wiedemann, 1972)	*Onshore* storm-surge transport and concentration of shells from intertidal flats and sound into beach ridges, cheniers, and washover fans on the supratidal marsh surface and marsh edge	Preferential transport of oysters relative to other epifauna and infauna; resulting assemblages composed predominantly of exotic species
Macrotidal flats; northern Gulf of California (M. K. Watson and Flessa, 1979; Fürsich and Flessa, 1987)	*Offshore* storm transport of rocky shore fauna onto adjacent tidal flats [distance of 3.7 km (M. K. Watson and Flessa, 1979)]	Transport of epifauna (unspecified taxa or proportion of fauna); minor effect—resulting assemblages composed predominantly of indigenous species
Macrotidal coast; northern Gulf of California (Meldahl, 1987a)	*Onshore* storm and *longshore* fair-weather transport of shells from tidal flat, formation of high-tide beach and spit	Resulting assemblage composed of 100% exotic shells
Mesotidal flats; Wadden Sea, Netherlands, and Bay of Arachon, France (van Straaten, 1956)	*Onshore* movment of shells from nearshore shelf onto open beach (2 spp.), into tidal inlet (5 spp.), and onto lower flats (3 spp.); no spp. transported as far onshore as inner flats	Of the 10 exotic species, all are small infaunal bivalves (*Spisula*, *Donax*, *Angulus*, etc.)
Mesotidal flats; Wadden Sea (Kristensen, 1957)	(1) *Onshore* from flats to concentrations along edge of salt marsh; (2) *Offshore* from flats to floor of tidal channels; few km maximum transport	Data for transport of *Cerastodoma edule* only
Macrotidal flats, strong currents; Solway Firth, Scotland (J. B. Wilson, 1965)	*Onshore* transport of shell fragments from nearshore shelf to inner? flats	Spines and broken plates of the fragile infaunal echinoid *Echinocardium*; exotic speciments are minor component of assemblages
Mesotial inlet with strong currents, associated flats, and open beach; Kyushu, Japan (Shimoyama, 1985)	(1) *Offshore* transport of small shells from sandflats to tidal channel; (2) *onshore* transport of large shells to mudflats by hermit crabs	Study limited to single species of trochid gastropod (*Umbonium*)
Mesotidal flats; Wadden Sea, Netherlands (Leopold *et al.*, 1985; Cadée, 1989)	*Onshore* transport of shells from intertidal flats to adjacent supratidal areas by birds; maximum distance 8 km	No size selection in transport of *Mytilus* or *Cerastoderma* by Oystercatchers; size-selective transport and smashing of large *Mytilus* by Herring Gulls; supratidal accumulations are 100% exotic

Supratidal; southeast Farallon Island, California (Lindberg and Kellogg, 1982)	*Onshore* transport of shelf species to supratidal rookeries by diving birds that prey upon molluscivorous fish	No data; supratidal accumulations are 100% exotic
Supratidal; Western Australia (Teichert and Serventy, 1947)	*Onshore* transport, presumably by Pacific Gulls, of shelly epifauna from exposed offshore "reefs" to rock ledges of 0–70 feet elevation	Opercula and broken shells of *Turbo* most common, plus small patellid gastropods and chitons; 100% exotic
Supratidal; Alacran Reef, Mexico (Wright and Kornicker, 1962)	*Onshore* transport of dead shells from beach to supratidal nesting sites by birds	Nesting material determined by availability of dead specimens (none are food items); selection against coral; assemblages are 100% exotic
Mesotidal flat; Essex coast of North Sea, England (Greensmith and Tucker, 1966, 1969)	*Onshore* flood-tide transport of estuarine channel species into stable banks at low-water mark, then movement onto upper flats and mobile cheniers at marsh edge; shells moved over several km	No data; bivalve *Cerastoderma* dominates (ca. 70% of individuals in subtidal banks, 50–90% in cheniers))
Mesotidal flats and adjacent shelf, to 40 m depth; Liverpool Bay, Irish Sea (Lingwood, 1976)	(1) *Onshore* transport from subtidal (0–40 m) to intertidal flats (46 spp.) (2) *Offshore* transport from intertidal (8 spp.) or from nearshore (4 spp., 0–10 m) to deeper subtidal	None apparent: exotic spp. include epifauna, interstitial forms, borers, and both shallow- and deep-burrowing infauna; exotics outnumber indigenous spp. in intertidal (46:8), but represent <10% of subtidal diversity
Normal salinity lagoon; Mauritania (Einsele *et al.*, 1974)	*Onshore* transport of shells from lower intertidal and subtidal parts of bay and of eelgrass epibionts into strandline concentrations; all transport is within bay	No data
Mesotidal estuary and open shelf, both with rocky shoreline; Ria de Arosa, Spain (Cadée, 1968)	(1) *Offshore* transport by weed drift from rocky margin to subtidal central bay; (2) *no transport* detectable within central bay; (3) *offshore* transport of shallow-water epifauna to open shelf	(1) Epiphytic gastropods; exotic individuals = 1% of bay assemblage; (2) rocky shore epifauna, especially *Mytilus*, which alone comprises 10% of shelf death assemblage
Low-energy, microtidal shoreline and shelf; Rhone Delta, France (van Straaten, 1960)	(1) *Downstream* transport of land and freshwater spp. to lagoon, less commonly to beach and to nearshore; (2) *transverse* mixing within nearshore zone above fair-weather wave base; (3) *longshore* transport within the nearshore zone; several km movement; (4) *no transport* on Rhone shelf beyond nearshore zone; little difference between live community and death assemblage	(1) small gastropods?; (2) small (<2.5 mm) bivalve *Corbula;* (3) small shells (*Gibbula, Hydrobia, Cerastoderma, Abra*)

(continued)

TABLE IX. *(Continued)*

Setting	Direction, mechanism, and distance of shell transport	Selectivity of transport; net effect on assemblage at depositional site
Mesotidal beach; Georgia (Dörjes *et al.*, 1986)	*Onshore* transport of some shells from nearshore shelf to beach; few if any shells in beach assemblage were derived from back-barrier environments, none originated from middle to outer shelf	No data
Mesotidal, ebb-dominated tidal inlet channel, estuary and adjacent shelf; Georgia (S. W. Henderson and Frey, 1986)	(1) *Onshore* transport of 4 spp. from shelf to estuarine sound via tidal inlet channel; (2) *no offshore transport* from sound or inlet channel to shelf; all spp. in shelf dead assemblages are from spp. reported living there, although some dead may be older relicts	(1) All transported spp. are small, thin-shelled infauna (*Parvilucina*, *Pleuromeris*, *Spisula*, *Tellina*); exotic shells account for 11% of spp. and 10% of individuals in sound death assemblage
Open and embayed shelf		
Nearshore; Kattegat Channel of east Denmark, and Faeroe Island off Iceland (Johansen, 1901, cited by Cadée, 1968)	Both *onshore* and *offshore* transport within the 0–15 m depth nearshore zone; the more protected the coast, the narrower the zone of transeverse mixing	Unspecified mollusks
Nearshore and offshore zone (to 15 m depth) along mesotidal, low-energy open coast; Georgia (Frey and Howard, 1986)	(1) *Onshore* movement of some offshore shells into nearshore area; (2) *no transport offshore*: offshore assemblages composed of offshore shells only	No quantitative data
Macrotidal open marine embayment, storm-influenced; Helgoland Bight, North Sea (Aigner and Reineck, 1982)	*Offshore* storm transport of shells from tidal flats and inlets: (a) common to shoreface sands; generally within 10 km of low-water mark (b) less common to 7–15 m transition zone between shoreface and self muds (proximal tempestites; within 15 km of low-water mark) (c) rare transport to shelf mud areas (distal tempestites)	(a) Mixed fauna of indigenous and allochthonous spp.; (b) most shells are from indigenous spp; exotic species are thin-shelled borers (*Barnea*, *Petricola*) from channels and small intertidal specimens (infaunal *Cerastodoma* and epifaunal *Littorina*, *Mytilus*, *Hydrobia*); (c) virtually all shells from indigenous spp.
30–75 m deep tide-swept channels in 10–20 m deep open marine embayment; Kiel Bay, North Sea (Arntz *et al.*, 1976)	*Downslope* movement of shells onto channel floors where concentrated	Transport is quantitatively "insignificant": spp. that dominate living communities are reflected in distribution of dead

Macrotidal open marine embayment; Mannin Bay, Ireland (Bosence, 1979)	*Offshore* transport by weed-drift and storm back-surge from rocky shore to sublittoral soft-sediment seafloors; 1 km maximum	Epifaunal gastropods (Patellidae, Littorinidae, Rissoidae)
Current-swept pass and open shelf; Sound of Jura and edge of Irish Sea, Scotland (Brown, 1979, reported in Akpan and Farrow, 1984)	(1) *Onshore* transport from 8 m depth onto shore; (2) *offshore* transport from intertidal to 46 m depth, both by wave-induced currents	(1) ''Sublittoral dead shells observed on beach''; (2) small epifaunal gastropods *Patella* and *Littorina*; no information on relative importance of exotic spp. in death assemblages
Meso?tidal estuaries and open shelf; Firth of Clyde, Firth of Lorne, Orkney Shelf, Scotland (Akpan and Farrow, 1984)	(1) *Transverse* transport confined to <5 m water depth in sheltered estuary but extends down to ≥ 12 m depth in less sheltered parts of shelf; (2) *no transport* on sheltered parts of shelf; (3) *ambiguous* evidence of transport on exposed parts of shelf exposed to high wave energy and strong tidal currents	(1) Epifaunal gastropods *Patella* and *Littorina*, fragments of bivalves; (3) large numbers of polished shells locally, but could be produced by passage of sandwaves
Backreef sand and mud embayment, Salt River Bay, Virgin Islands (A. I. Miller *et al.*, 1990 and A. I. Miller, personal communication, 1990)	*Onshore* transport of dead shells from shallow sandy habitats into slightly deeper ''muck pit'' mud habitat during Hurricane Hugo; lateral displacement tens of meters	(1) Prestorm assemblage had 8/14 spp. exotic, = 7% of dead individuals; (2) poststorm assemblage had 21/27 spp. exotic, accounting for 20% of dead individuals, all from nearby sands
Other settings		
Fringing reef and forereef slope to 75 m, subject to hurricanes; Discovery Bay, Jamaica (Boss and Liddell, 1987)	(1) *Minimal differential transport* of bioclasts across reef and backreef lagoon; transport from reef framework into adjacent sand channels; (2) limited *downslope* (off-reef) transport of bioclastic sand on forereef escarpment and lower forereef slope (>14 m depth); possibly storm-driven	(1) Good agreement in relative abundance of species found alive and in nearby sand-sized sediments; (2) assemblages contain minor quantities of shallow-water grain types
Carbonate bank and slope to 600 m; Bahamas (Hoskin *et al.*, 1986)	*Downslope* transport of bank-edge fauna as bedload in chutes to 45 m, then by-pass wall for deposition on mud apron below 120 m	Mollusk grains up to 1 cm diameter are transported up to 2 km offshore
Microtidal deep marine embayment, with rocky shore, narrow shelf and deep basin to 50 m; Bay of Naples, Adriatic Sea (Hollman, 1966)	(1) *Downslope* transport of surf-zone shells to 50 m depth by gravity and return flow of surf; slumps transport shells offshore to basin floor; (2) *no landward or longshore* transport; most shells are destroyed in surf zone	Fragments of epifauna broken in surf zone; *Bursa*

(*continued*)

TABLE IX. *(Continued)*

Setting	Direction, mechanism, and distance of shell transport	Selectivity of transport; net effect on assemblage at depositional site
Continental slope, to 615 m; Adriatic Sea (van Straaten, 1967)	*Downslope* transport by turbidity currents(?) of shallow-water benthos to 615 m, where mixed with indigenous but fragmented benthos and plankton	Exotic *Bittium*, *Turritella*, and *Mactra* reworked from relict Pleistocene; mixed with diverse indigenous epifaunal assemblage (no quantitative data)
Microtidal lagoon, beach, and deep borderland basin, to 2000 m depth; San Pedro and Catalina Island area, California (A. Clark, in Natland, 1957)	(1) *Onshore* transport from shelf to nearshore and beach; (2) *offshore, downslope* transport from beach and nearshore to deeper shelf	(1) Transport limited to 3 spp. (*Bursa*, *Conus*, and *Lucina*), rare individuals; (2) transport limited to 6 spp. (epifauna and shallow-burrowing infauna); total tabulated molluscan fauna = 225 spp.
Submarine canyon and continental rise; Atlantic continental margin, North Carolina (Pilkey and Curran, 1986)	*Downslope, offshore* turbidity-current transport of relict, estuarine and shallow-marine shells from middle? or outer? shelf deposits to 500 m depth on abyssal plain; 500-km lateral transport ("Black Shell Turbidite")	Exotic specimens dominated by ≤ 1.5-cm *Mulinia* and *Nassarius*; final assemblage is presumably entirely allochthonous in composition
Continental shelf, slope, and rise; northwestern Africa (Bein and Fütterer, 1977)	*Downslope* transport of relict shells from shelf to slope and basin by actions of benthic organisms, which resuspend and inject sediments into overlying water, redeposition deeper on slope	No data
Deep sea; western Pacific Ocean (Saunders and Spinosa, 1979)	*Pelagic drift* of dead *Nautilus* from Palau (Caroline Island) to northern Philippines; 1000 km transport in 138 days (contrasted with up to 150 km in 332 days when alive); beach strand deposits?	Nektonic cephalopods

[a]Most hydraulic transport removes shells only to adjacent habitats and selects for small speciments and/or epifaunal species; exotic species are usually only minor components of recipient assemblages. The major exceptions are transport by mass-flow processes associated with steep gradients (long-distance transport of entire assemblages is possible) or unusual surge conditions (large quantities of shells typically moved short distances) and transport by predators (small quantities, highly localized deposits); exotic shells typically dominate these assemblages.

range of live individuals. A. I. Miller and Cummins (1989, 1990), for example, found that physically and ecologically discrete habitat patches 10 m wide (seagrass beds) and 50–60 m wide (sandy areas) were readily detected by abundance changes in death assemblages in back-reef areas of St. Croix (also see Masse *et al.*, 1989). Smaller-scale, within-habitat ecological variability was not detectable in death assemblages, but presumably had been smeared by limited transport and temporal patchiness in benthic settlement. Spatial resolution on the low-gradient macrotidal flats of Cholla Bay, Mexico (Fürsich and Flessa, 1987) is on average coarser (five assemblage types can be discriminated across a 3-km shore-to-sea transect; habitat patches range from 10 m to hundreds of meters), but local peaks in live abundance of some species are often matched by peaks of dead individuals. On the hurricane-stirred Texas continental margin, high-fidelity death assemblages distinguish three habitats [a 15-km-wide inner shelf, 100-km-wide middle shelf, and 25-km-wide outer shelf (S. E. Walker *et al.*, 1989)]. Death assemblages thus commonly show good fidelity to life habitat, and the spatial *resolution* of these assemblages varies from very fine in very patchy shallow-water settings (on the order of tens of meters) to much coarser on more homogeneous open shelves (tens of kilometers to 100 km).

Experiments on hydraulic transport generally corroborate inferences from live:dead studies, although complex relationships abound. For example, J. B. Wilson (1967) found that under identical conditions small shells (10–14 mm) of *Cerastoderma edule* were rapidly buried, larger specimens (14–17 mm) were transported, and the largest specimens (17–24 mm) were slowly buried. Cain (1968) suggested that crinoid ossicles undergoing aerobic decomposition have different hydraulic behaviors than those under anaerobic decomposition, owing to differences in specific gravity.

Shell transport by organisms [see review by Boucot (1981, p. 333)] and passive post-mortem drift of carcasses are less well-appreciated mechanisms but highly significant for some environments and groups (Tables VII and VIII). Although rare instances of post-mortem drift have been documented for benthic organisms (Reyment, 1986), greatest drift has been documented for cephalopods. Saunders and Spinosa (1979), for example, documented 1000 km of drift among Recent *Nautilus*, underscoring the potential of taphonomic processes to increase the biostratigraphic utility of some groups (Kennedy and Cobban, 1976). Propensity for drift among cephalopods depends upon shell form, rate of epizoan growth, and mode of death (Kennedy and Cobban, 1976, and references therein; Hewitt, 1988; Donovan, 1989).

4.3. Modification by Other Organisms

Many features conventionally attributed to physical processes can be created or mimicked by biological processes (Table VII). These processes include the ecology of the shell producer, the effects of duraphagous predation and pathological infestations by endobionts, and an array of post-mortem live:dead interactions. Many live:dead interactions reflect highly selective behavior [e.g., species- and surface-selective settlement of bryozoa on dead bivalves (Bishop, 1988, 1989; Ward and Thorpe, 1989)] and possibly obligate behavior in some cases [e.g., deep-sea limpets that graze exclusively on dead squid beaks, fish, and whale bones (Hickman, 1983; Marshall, 1987)]. Other organisms select dead substrata more opportunistically on the basis of size or shape [e.g., choice of shell domiciles by hermit crabs and octopods (McClintock, 1985; Mather, 1982; Iribarne, 1990), choice of camouflage/stilt shells by *Xenophora* spp. (Ponder, 1983)], whereas still other live:dead interactions seem to be nothing more than post-mortem bad luck. This latter category would include dead shells exhumed by bioadvection (McCave, 1988; Oliver *et al.*, 1985) or inadvertently crushed by durophagous crabs (S. E. Walker, 1988) and rays (Gregory *et al.*, 1979) (Table VII). Schäfer (1962, 1972) noted many of these factors in his North Sea studies, but few have been thoroughly examined or had their importance quantified in different environments.

Recent shelly deposits commonly show both bioerosion and biofragmentation (J. B. Wilson, 1967; Cadée, 1968; Trewin and Welsh, 1976; Bosence, 1979; P. J. Davies and Hutchings, 1983), but the importance of these processes relative to physical ones is poorly known (but see Meldahl and Flessa, 1990). Wignall and Myers (1988) make a good argument for exclusively biogenic fragmentation in Jurassic mudstones where they found a consistent correlation between fragmentation and bioturbation. Compared with organic-rich laminated shales, the organic-poor bioturbated mudstones contain higher-diversity shell layers with extensive fragmentation, suggesting that fragmentation was linked with oxygenation and thus with improved access to benthic prey by predators. Extensive fragmentation is not correlated with coarser-grained layers such as would be expected had fragmentation been hydraulic or the shells exotic.

Bioturbation is a commonly invoked agent of post-mortem modification of shelly faunas and is thought to have multiple effects, but none has been adequately tested. Bulldozing species are thought to disarticulate and rotate shells within the sediment, yielding random orientations (Toots, 1965; Brett and Baird, 1986). In our separate studies of shell gravels, however, we have encountered many shallow-water accumulations with large proportions of concave-up shells, suggesting that predators and bioturbators might preferentially create concave-up rather than random fabrics within the sediment (also see Salazar-Jimenez *et al.*, 1982). Concave-up orientations on the seafloor have been attributed to "flipping" by scavengers (Emery, 1968; Clifton, 1971; J. B. Wilson, 1986) [but see Callender *et al.* (1990), who postulate that this is an artifact of human fishing activity].

Other postulated effects of bioturbators on the preservation of shelly fauna are fragmentation (Knight, 1988), dispersion of small-scale concentrations (Kidwell, 1986a), and burial +/− concentration of shells by subsurface deposit feeders (van Straaten, 1952; Rhoads and Stanley, 1965;

Trewin and Welsh, 1976; Cadée, 1976; Meldahl, 1987b). None of these effects has been tested in aquaria or under controlled field conditions. The only rigorously tested relationship between bioturbation and shell preservation is that sediment irrigation by macrobenthos can increase dissolution rates (Aller, 1982a; Raiswell and Canfield, this volume, Chapter 9).

Predators and scavengers are responsible for an undetermined amount of shell breakage, disarticulation, and dispersion. Organisms that exhume shelled infauna expose these shells to post-mortem conditions on the seafloor that they might otherwise have avoided, and some predators are capable long-distance transporters (Tables VII and IX). Digestive acids presumably accelerate destruction, as may organic coating of egested and defecated material. Such coats may make shells more attractive to microbial communities and thus to rasping grazers [see discussion in Scoffin (1987, p. 65)]. Coralline algal coats, on the other hand, seem to significantly reduce shell damage by borers (Smyth, 1989).

Algal and other organic coatings do not necessarily seal the post-mortem fates of exposed shells. For example, Cutler (1989) found that a critical density of dead shells is required to support a community of grazers; at subcritical densities, shells escape attack and thus have significantly higher likelihood of preservation. Organic coatings might actually enhance preservation potential by sealing shell carbonate from ambient, corrosive waters (Henrich and Wefer, 1986), much as periostraca protect freshwater mollusks from dissolution before death (Philippon and Plaziat, 1975; Hunter, 1990). Many encrusting organisms may actually strengthen dead shells (Stachowitsch, 1980).

Bioeroders attack live and dead shell material alike, although some are highly selective. Endobionts (boring fungi, algae, sponges, barnacles, bivalves, echinoids, various worms, bryozoa) perforate shells, thereby reducing shell strength and increasing surface area exposed to abrasion, dissolution, and maceration [see review by Young and Nelson (1988)]; raspers alter the shell surface, exposing shell layers more susceptible to dissolution (Table VII). Encrusters and other epibionts with penetrating attachment structures [e.g., lichens, byssate bivalves, foraminifera (Boekschoten, 1966; Poag, 1969)] might also hasten destruction. Measured rates of sediment production by bioeroders in carbonate habitats vary from 0.2 to 16 kg/m^2 per year (Scoffin, 1987), indicating the very large amounts of carbonate that can be removed by these processes [often equal to biogenic carbonate production (e.g., Bak, 1990)]. Some of these organisms produce distinctive shell damage (e.g., clionid and spionid borings, crab peel marks, whelk chips, bird attack), but not all fragments of such activity are diagnostic and many organisms generate as yet unrecognizable debris (especially predators that ingest and internally crush shells). Even distinctive fragments and perforations become nondescript with further breakage and bioerosion, until physical disintegration is so complete that the shell is unidentifiable (Adjas *et al.*, 1990).

The extent of shell damage by borers and encrusters and their frequency within an assemblage are usually interpreted as proportional to exposure time on the seafloor (e.g., Driscoll, 1970); this is one criterion among many commonly used to infer net rates and dynamics of sediment aggradation (Kidwell and Aigner, 1985; Brett and Baird, 1986; Brandt, 1989; Kidwell, this volume, Chapter 5, Section 3.1.1). However, not all infesters leave an identifiable skeleton or trace (some encrusters leave only vague attachment scars; Cutler, 1987), and infesting groups vary in abundance along depth and salinity gradients and over ecological time (Lawrence, 1969; Budd and Perkins, 1980; Scoffin *et al.*, 1980; P. J. Davies and Hutchings, 1983; Akpan and Farrow, 1984, 1985; Hutchings, 1986; Kobluk and van Soest, 1989; Sammarco and Risk, 1990; Günther, 1990). Inferences on relative periods of exposure thus should only be made among samples from similar environments.

Taphonomic damage must be distinguished from damage incurred while the host was alive. Infested interior surfaces of shells or of any surface thought to have been covered by soft tissues during life are the most reliable evidence. Exterior surfaces of infauna as well as epifauna may be infested during life, particularly the posterior margins of shallow-burrowers (Schäfer, 1972; Hertweck, 1979; Peterson, 1983). The distribution and orientations of infesters may in some instances be used to distinguish between life and post-mortem attack (Seilacher, 1969, 1982b; Sando, 1984; Elias, 1986; Baird *et al.*, 1989).

Some further caveats may be made on the biostratinomic influence of living organisms.

(a) Some encrusters and borers can thrive below the sediment surface, particularly within coarse rubbly deposits (Scoffin and Henry, 1984), and thus shell modification does not always cease with "burial."

(b) Many encrusters prefer cryptic areas on the undersides of shells as a means of avoiding competition with algae and rasping organisms (Bosence, 1979; Jackson, 1977; Ward and Thorpe, 1989; K. McKinney, personal communication) or avoiding sedimentation (Bishop, 1988; Cadée, 1991). Heavily encrusted surfaces thus do not provide clear-cut evidence for "way-up" (*contra* Brett and Baird, 1986); they are at least as likely to indicate "way-down."

(c) Encrusters and borers modify the hydraulic behavior of dead shells. For example, circumrotary overgrowths by bryozoans, hydroids, and corals transform plate- and stick-form fragments into spheres and add weight (Kissling, 1973; Balson and Taylor, 1982). Perforated bivalve shells have different transport characteristics than intact shells (Lever and Thijssen, 1968).

(d) Hermit crabs (a Jurassic innovation) carry dead shells, particularly from gastropods, on their backs as domiciles, thereby transporting, delaying burial, and facilitating damage and modification that create a number of distributional anomalies (Conover, 1975; Shimoyama, 1985; S. E. Walker, 1988, 1989). These shells can be identified by distinctive patterns of boring, encrustation, and physical abrasion (S. E. Walker, 1988; Smyth, 1990) and should be evaluated separately from other parts of the death assemblage.

(e) To corroborate whole-assemblage data, individual bioclasts should be examined for multiple-phase, post-mortem histories consistent with delayed burial or repeated burial–exhumation cycles. Not all individuals will exhibit such features (Frey and Howard, 1986). In fact, the hallmark of time-averaged shell accumulations having complex histories is the admixture of shells in different states of preservation, from fresh to completely altered (Kidwell and Aigner, 1985; Kidwell, this volume, Chapter 5).

4.3.1. Discrimination of Physical and Biogenic Effects

In general, shell assemblages influenced by physical processes should bear some imprint of such or have an appropriate stratigraphic context, e.g., high-energy sedimentary structures. If post-mortem processes are predominantly hydraulic, then all shells of similar or lower critical threshold velocities should be affected, whereas biological processes, if selective at all, will tend to select on other features, such as sizes and shapes most vulnerable or rewarding to infestation. With careful study, results of different processes can be distinguished. Scoffin (1987), for example, found that in situ mechanical breakdown of skeletons yields a broad range of grain sizes, whereas boring produces predominantly fine grains (also see Sammarco and Risk, 1990).

As exposure to post-mortem processes increases, taphonomic features tend to become less readily interpretable because of progressive destruction. In addition, shells can be recycled from one post-mortem environment to another through vertical reworking and time-averaging as well as by transport, and thus can accrue a succession of different taphonomic signatures. This is particularly true in concentrations of exotic shells, where taphonomic features may be more informative of shell source than of the final depositional process and thus should be interpreted with caution. For example, shells in the Recent Black Shell Turbidite (Prince *et al.*, 1987) show all stages of abrasion and have a range of radiocarbon ages, suggesting that fragmentation reflects not only downslope transport, but also initial accumulation as a time-averaged shallow-water assemblage [also see D. J. Davies *et al.* (1989a) for similar conclusions for shallow-water tempestites]. This is why stratigraphic workers have consistently used bedding features—physical scale, geometry, microstratigraphic complexity, overall bioclastic fabric, facies context—as the primary evidence for modes of shell concentration rather than relying solely upon features of component shells (see Kidwell, this volume, Chapter 5).

4.4. Geochemical Modification

The breakdown of carbonate shells into their constituent microscopic elements (prisms, needles, fibers, flakes, and granules) can proceed both on the seafloor and during shallow burial. Shells exposed on the seafloor are subject to the chemistry of overlying waters. Most open marine waters at shelf depths are saturated or oversaturated with respect to calcium carbonate and are slightly alkaline, so that aragonitic and calcitic bioclasts are chemically stable. Under such conditions, shell breakdown is largely by boring, although these pits can become micritized in warm oversaturated waters (Alexandersson, 1972; Gunatilaka, 1976). Chemical leaching or microbial decay of organic matrix can also be a factor, leading to chalky textures (Alexandersson, 1972; Lewy, 1981).

In undersaturated seawater (cold water, whether shallow or deep) and in acidic brackish and freshwater settings, carbonate shells can be etched and leached (Alexandersson, 1972; Pip, 1988) (Table VIII). Microscopic (and preferably SEM) inspection of shell surface features is almost always required in order to distinguish among the effects of chemical dissolution of crystallites, organic leaching, fungal or algal microboring, and physical abrasion [Cutler (1987); *contra* 12 × magnification used by D. J. Davies *et al.* (1989a); Staff and Powell (1990)]. Brett and Baird (1986) used the term corrasion (combining abrasion and corrosion) for dulled surfaces of uncertain origin.

Upon shallow burial, shells may pass through a series of pore-water zones whose chemistries are determined by the quantity of buried reactive organic matter and iron and by fluid exchange with overlying waters [reviews by Aller (1982b), Allison (1988), Raiswell and Canfield (this volume, Chapter 9)]. Near-surface pore waters are generally oxidizing and supersaturated owing to good exchange with saturated overlying water; the effects of this exchange generally overcompensate for acids produced by aerobic decomposition of organics. Carbonate shells thus should suffer little dissolution in this thin surficial zone unless overlying waters are undersaturated (D. E. Canfield, personal communication, March 1989).

In contrast, actively bioturbated sediments just below the surface (and extending some tens of centimeters to the depth of active burrowing) are characterized by undersaturated pore waters of high pCO_2 and low pH which promote carbonate dissolution (Aller, 1982a). Sediments are so well irrigated with sulfate-rich overlying waters that little alkalinity can build up, notwithstanding high rates of sulfate reduction and bicarbonate ion production. The accumulation of Ca^{2+} in these pore waters (Aller, 1982a) and SEM evidence of shell pitting indicate that shell dissolution occurs within the bioturbated zone. This is recorded even in sites of net carbonate accumulation such as Florida Bay, where dissolution has been reported to remove fine particles of high-Sr aragonite (Walter and Morse, 1984; Walter and Burton, 1990). If reactive detrital iron is present (as in most siliciclastic marine sediments), diagenetic pyrite forms locally by sulfate reduction; later oxidation of these reduced pockets yields sulfuric acid, which further lowers pore-water pH to levels inimical to carbonate preservation (Aller, 1982a). Where bioturbation is intense, at least seasonally, even sediment molds of dissolved shells have low likelihood of preservation.

Below the zone of sediment irrigation, anaerobic decomposition can lead to high bicarbonate levels (high alkalinity) in both carbonate and siliciclastic sediments (Raiswell and Canfield, this volume, Chapter 9). This creates supersaturated conditions that favor carbonate shell preservation. Presumably, shells must be moved quickly into this zone if they are to become part of the fossilized shelly fauna (Aller, 1982a; D. J. Davies *et al.*, 1989b). This may be accomplished by the piping of shells down open burrows (Aller, 1982a; Seibold *et al.*, 1973; Wanless *et al.*, 1987), downward advection by surface and subsurface deposit feeders (Cadée, 1976), scouring to the base of migrating bedforms (Johnson, 1957; Clifton and Boggs, 1970), and catastrophic burial by storms, slumps, and turbidites (Aigner, 1985a).

Sedimentary environments that favor carbonate shell preservation are thus those with little reactive organic matter and iron, e.g., carbonate grainstones, where any local undersaturation can be quickly buffered by dissolution of carbonate mud or can be corrected by flushing with overlying saturated seawater. Similarly, local clusters of shells in otherwise disadvantageous sediments might create their own favorable pore-water microenvironment by partial dissolution and bicarbonate buffering. There thus is potential for a positive diagenetic feedback between shell abundance and shell preservation potential [see discussion in Kidwell (1986a, 1989, and this volume, Chapter 5). Sediments with bioturbated zones that are thin, heterogeneous (i.e., characterized by discrete burrows rather than by thorough admixing), or highly seasonal in development should also favor carbonate shell preservation by providing various time and space windows for shells to escape undersaturated pore-water conditions (Kidwell, 1989).

Alternatively, shells can persist in the face of antagonistic geochemical conditions if their reactivities are modified soon after death. Although the mineralogies of shells are generally well known (Milliman, 1974; Scoffin, 1987; Lowenstam and Weiner, 1989; Saelen, 1989), the kinetics of shell dissolution is less well understood, inasmuch as species are not always lost in the order or at the rates predicted by their chemical activities (Berner and Morse, 1974; Walter and Morse, 1984, 1985; Walter, 1985; Morse, 1986). Peterson (1976), for example, estimated the differential preservability of shelly benthos from Mugu Lagoon (California) by tracking weight loss of individual shells over 7.5 months (shells ≥ 2 cm) (C. H. Peterson, personal communication, June 1990). After 50-cm-deep burial in either sand or muddy sand, high-Mg calcite echinoids and some aragonitic gastropods had suffered 10–20% weight loss, but most aragonitic species lost only 0–4%, and the only low-Mg calcite scallop lost 0.14%. Linear extrapolation suggests that the post-mortem longevities of the different taxa ranged from 4 to 894 years.

High surface area and disseminated organic material tend to speed shell destruction, and may even reverse the susceptibilities of aragonitic and calcitic bioclasts (Flessa and Brown, 1983; Walter and Morse, 1984, 1985; Ameziane-Cominardi and Roux, 1987; Schott *et al.*, 1989). High organic content, for example, appears to be an important factor in the rapid post-mortem destruction of calcitic brachiopods relative to many aragonitic bivalves (Collins, 1985; Gaspard, 1989; Curry *et al.*, 1989; Emig, 1990; Kidwell, 1990). On the other hand, organic coatings on shells (Henrich and Wefer, 1986; Morse, 1986), adsorption of less soluble ions [surface "poisoning" by Zn, Mg, PO_4, and others (Berner and Morse, 1974; Walter and Hanor, 1979; Morse, 1986)], periostracal coverings (Phillippon and Plaziat, 1975; Hunter, 1990), and conchiolin layers within shells (Lewy and Samtleben, 1979) can significantly impede dissolution. Rapid conversion of high-Mg calcite to low-Mg calcite quickly stabilizes the tests of some echinoderms and bryozoans (Nelson, 1978), but not all [e.g., rapid destruction of ophiuroids (Jensen and Thomsen, 1987)].

The importance of kinetics in determining shell loss rates is underscored by the abundance of shells that accumulate in sites where water chemistry should not permit shell preservation. Deep-sea oozes, for example, can accumulate on seafloors overlain by undersaturated water despite metastable mineralogies (Morse, 1986). Kinetics is also implicated in the persistence of shell carbonate on and shallowly buried within modern shelf sediments at low, middle, and high latitudes (Nelson, 1988), even where total carbonate production rates are low.

Work in Long Island Sound (New York) on carbonate productivity and carbonate dissolution illustrates this apparent paradox of preservation in estuarine siliciclastic muds. Macrobenthic carbonate production has a mean of 67 g $CaCO_3/m^2$ per year [ranging from 6 to 180 g/m^2 per year (Sanders, 1956)], similar to rates established from other temperate level-bottom environments (Bosence, 1980). In the same area, carbonate dissolution rates of 10–70 g $CaCO_3/m^2$ per year (Aller, 1982a) have been measured from pore waters, depending on season and bioturbation intensity. Similar budgets have been documented for Californian and Texan lagoons (Peterson, 1976; Cummins *et al.*, 1986a).

Based on measured oxygen demands in muds, D. J. Davies *et al.* (1989b) calculated that the potential for dissolution is much more extreme on continental shelves—ranging from 400 to >2000 g $CaCO_3/m^2$ per year and usually >1000 g/m^2 per year in surficial, bioturbated sediments. Because these values exceed average rates of carbonate production (300–500 g/m^2 per year), they predicted that shells should only be preserved if they are quickly buried to a depth below the bioturbated zone. Although logical, and certainly one viable scenario for shell preservation often invoked in the paleontological literature, this conclusion is belied by the presence of abundant old shells in surficial sediments across many clastic-starved shelves (Nelson, 1988; Kidwell, this volume, Chapter 5) (see data in Section 5.1) and even in the bioturbated, organic- and clastic-rich Texas shelf (S. E. Walker *et al.*, 1989). In addition, field studies demonstrate that much of early diagenetic dissolution is probably focused on fine particulate carbonate (e.g., Walter and Morse, 1985), leaving shells of several millimeters or larger intact. It is thus clear that, notwithstanding high rates of carbonate dissolution measured for single seasons or burrow systems, and the high rates that can be calculated if these short-term rates are multiplied out to their annual values (e.g., D. J. Davies *et al.*, 1989b), individual shells face a temporally and spatially heterogeneous diagenetic environment and can survive by many pathways other than immediate deep burial. These pathways include very early post-mortem modification of shell structure and composition that alter the kinetics of dissolution.

Loss rates definitely are high for many juvenile mollusks: notwithstanding high juvenile mortality following large larval settlements, juveniles are rare in most Recent and fossil death assemblages (Section 3.2.3). Shell loss is thus not a simple matter of shell size or surface area, but also of microarchitecture, including organic content. Based on a capture/recapture method (Powell *et al.*, 1982), Cummins *et al.* (1986a, Table III) estimated that small (0.8–3.12 mm) aragonitic mollusks have post-mortem "half-lives" of less than 40–307 days in two Texas lagoons (half-life = length of time for half the members of a larval cohort to disappear from the death assemblage). It should be noted, however, that variance is very high: three out of the four taxa that occur in both lagoons (*Acteocina, Caecum, Laevicardium*) showed high loss in one but no loss (i.e., immeasurably long half-lives) in the other lagoon, even after 20 months of study. Four other numerically dominant taxa in the Texas death assemblages also showed no loss (all size classes of *Anomalocardia, Chione,* and *Brachiodontes; Tagelus plebius* >16 mm). Because the shortest half-lives and highest loss rates are documented for the most juvenile and presumably most susceptible members of the shelly fauna, these values should not be taken as typical of all aragonitic taxa and all size classes (*contra* M. V. H. Wilson, 1988, and Brett and Baird, 1986).

5. Estimating Relative and Absolute Scales of Time-Averaging

Most benthic fossil assemblages are time-averaged accumulations dominated by indigenous material: instantaneous census assemblages are rare, as are assemblages composed entirely of exotic shells. One of the major challenges in taphonomic analysis, then, is to recognize and quantify degrees of time-averaging in fossil assemblages [and see Fürsich and Aberhan (1990); which appeared as this volume went into production].

Time-averaging was defined by K. R. Walker and Bambach (1971) for "fossil assemblages . . . that accumulate from the [local] live community during the time required to deposit the containing sediment." They argued that because sediment accumulation rates are slow relative to organism life spans, such assemblages do not reflect community structure at any instant, but instead are composites spanning decades, and that short-term fluctuations in populations are canceled out by the predominance of ecologically persistent, long-term trends. This concept quickly gained wide acceptance and, corroborated by actualistic studies (Peterson, 1977; Staff *et al.*, 1986; Staff and Powell, 1988), has become a cornerstone of paleoecological analysis.

5.1. Controlling Factors

Given sufficiently durable shells, the fundamental control on scales of time-averaging should be the rate of sediment accumulation, which determines the approximate stratigraphic separation of successive faunas. Because deposition is generally episodic on short time scales (e.g., Barrell, 1917; Sadler, 1981), the time span encompassed by a given deposit and its fossil assemblage cannot simply be back-calculated from longer-term sediment accumulation rates (*sensu* Schindel, 1980, 1982). Moreover, shells of several ages may be incorporated even into rapidly deposited beds [e.g., reworking of new and old shells into a single storm-bed (Kidwell, this volume, Chapter 5)], and presumably may be added later by bioadvection and other stratigraphic leakage [e.g., stratigraphic disordering *sensu* Flessa *et al.* (1989; Cutler and Flessa, 1990)]. Thus, to reconstruct the likely scale of time-averaging for a specific assemblage, the short-term dynamics of sediment accumulation must be evaluated along with shell sources and histories. This requires a variety of biostratinomic, paleoecological, sedimentological, and stratigraphic lines of evidence (e.g., Goldring, 1964; Einsele and Seilacher, 1982; Behrensmeyer, 1982; Kidwell, 1982, 1986a, this volume, Chapter 5; Kidwell and Aigner, 1985; Brett and Baird, 1986; Parsons *et al.*, 1988).

The mixed ages of shells found in surficial sediments of Recent environments indicate that many shells can endure periods of time-averaging of up to tens of thousands of years. For example, mid-shelf shell gravels from around the British Isles yield radiocarbon dates of up to ~10,000 years BP (J. B. Wilson, 1982; Sturrock, 1982), and shell gravels from the clastic-starved outer shelf of the southeastern United States yield dates as old as 10,000–15,000 years BP (Milliman *et al.*, 1972). Shells sampled from surficial carbonate-rich sediments in warm temperate and subtropical settings yield radiocarbon dates of 0–3100 years BP in south Florida, 1425–1600 years BP in the Bahamas, and 1000–4805 years BP in coastal Baja California (Taft and Harbaugh, 1964). Sealevel flooding curves for areas where Holocene sediments are still being reworked, such as Florida Bay (Bosence, 1989a,b), also indicate time-averaging over a period of 3000–4000 years, as do direct dates on beach accumulations of shells (Gould and McFarlan, 1959; Albertzart and Wilkinson, 1989; Powell and Davies, 1989) and intertidal biogenic graded beds (Meldahl, 1987b; Flessa *et al.*, 1990). In deep-sea cores, Broecker *et al.* (1988) found that even among assemblages of coexisting planktonic foraminifera, radiocarbon ages differ by 60–3100 years within sampling horizons. Assemblages composed of shells of mixed ages, formed under conditions of slow *net* aggradation, commonly record more than one community (owing to environmental change during the period of accumulation) and several different states of degradation [see

Kidwell and Aigner (1985) for fossil examples; Kidwell (this volume, Chapter 5)].

Many authors have postulated that time-averaging is also governed by the depth of sediment (and presumably faunal) reworking by bioturbators [for recent examples, see Schiffelbein (1985), Backman (1986), Meldahl (1987b)]. This model, although often invoked, has yet to be tested for macroinvertebrates; some recent analyses suggest that bioturbation can yield a far more complex record. Kershaw *et al.* (1988), for example, found that only fine carbonate particles are vertically mixed by bioturbators in areas of slow clastic sedimentation (0.02–0.3 cm/year) within the Irish Sea. Notwithstanding a deep-burrowing shrimp (*Callianassa subterranea*) which moves sediment to the surface from depths ≥140 cm and an echiuran worm (*Maxmuelleria lankesteri*) which moves surface sediment to depth, large *Turritella communis* shells increased in age down-core from 270 to 4250 years BP. In contrast, radiocarbon ages of bulk carbonate (particles <1 mm) were roughly constant at 10,000–15,000 years BP throughout each 0.6- to 2.7-m-long core. These results suggest that the large *Turritella communis* retained their stratigraphic positions, whereas fine-grained sediment and shell fragments were homogenized by advection. The preservation of a relative age stratigraphy among large molluscs in ~70-cm cores of bioturbated sediment from Texas bays [Powell *et al.* (1989), dated using free amino acids] provides a second, independent example of this "no-homogenization" model within the errors of dating. Numerical simulation of bioadvection by Flessa *et al.* (1989; Cutler and Flessa, 1990) yielded comparable, well-ordered synthetic stratigraphies.

Many of the consequences of time-averaging for shelly faunas have already been discussed (see Sections 2 and 3). These include heightened species richnesses in death assemblages, relative abundances that reflect long-term averages rather than successional states, and poor records of *short-term* ecological structure. The following discussion focuses on methods of estimating absolute scales of time-averaging and on criteria for identifying taphonomically comparable assemblages. [Contrary to Fürsich and Aberhan (1990), we have found little evidence that time-averaging lowers the species richness of death assemblages (see Tables II–IV).]

5.2. Magnitudes of Time-Averaging

Census assemblages and *biostratigraphically condensed* assemblages are end members in the spectrum of time-averaging (Fig. 9; Table I). Assemblages that accumulate over intermediate time scales may form: (a) during periods in which the physical environment remained stable (*within-habitat* assemblages); such assemblages sum within-habitat perturbations in community state or physical environment; or (b) during periods of significant physical environmental change at the site of accumulation [*environmental condensation sensu* Fürsich (1975)]; these assemblages are derived from two or more successive communities and reflect taphonomic overprints of two or more environments. [Compare with Fig. 6 of Fürsich and Aberhan (1990).]

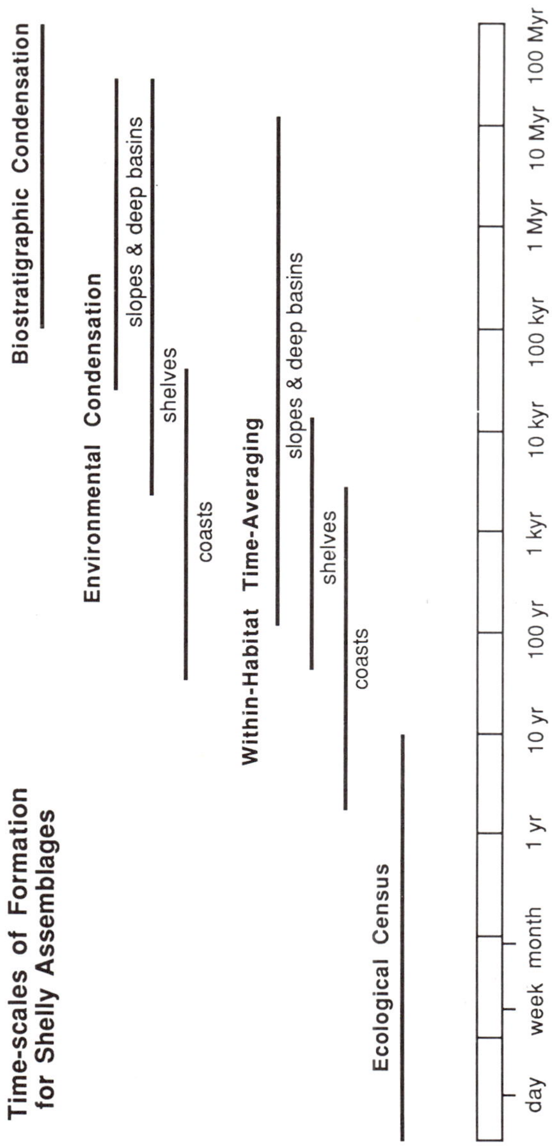

FIGURE 9. Estimates for relative and absolute scales of time-averaging in census, within-habitat, environmentally condensed, and biostratigraphically condensed assemblages. See text for explanation.

5.2.1. Census Assemblages

Sudden burial [''obrution''; for review see Brett and Seilacher (1991)] can entomb a benthic community, yielding a census assemblage that has undergone little or no time-averaging. Such assemblages provide the highest resolution records of community structure, including virtual ''snapshots'' of species relative abundances and size–frequency distributions. Other ecological features such as commensal relationships, substratum preferences for byssate or pediculate taxa, and individual spacing can also be preserved.

The hallmarks of such assemblages should be high proportions of articulated specimens, with many in life position, and sedimentary evidence of rapid permanent burial or asphyxiation (graded sediment caps, ''smothering'' black mud, early concretion formation, escape traces). Census assemblages are typically associated with single beds or bedding planes and, if shells are densely packed, these concentrations are typically thin and local in distribution [pods, pavements, and thin beds; local overgrowths on hard substrata; = simple event-concentrations in Kidwell (this volume, Chapter 5)].

Catastrophic mortality and/or burial can be an important mode of assemblage formation in environments of variable oxygenation, salinity, or temperature such as lagoons, deep basins, and inland seas (Noe-Nygaard *et al.*, 1987; Oschmann, 1988). In higher-energy and persistently aerobic soft-bottom environments, *recognizable* census assemblages are less likely to be preserved, owing to physical reworking and bioturbation, even though mass mortality may be common.

Preservation of discrete census assemblages depends upon the immediacy, thickness, and permanence of burial sediments (Kidwell, 1990). In the presence of burrowing organisms, burial increments must be at least tens of centimeters thick to sequester the death assemblage from taphonomic modification and time-averaging (Bjerstedt and Erickson, 1989). A minimum thickness of sedimentary cover [~30 cm for Recent infauna (Nichols *et al.*, 1978)] is also necessary to limit the number of organisms that can escape entombing sediments, with mobile epifauna fleeing the scene and infauna burrowing *up* to a new equilibrium position; such behaviors have considerable potential to bias census assemblages (Kranz, 1974a,b). The lack of an entombing sedimentary increment may explain in part the failure of the extraordinary 1983–1984 mass mortality of the Caribbean urchin *Diadema antillarum* to leave a signal in the record (Greenstein, 1989b). Census assemblages are probably more common on hard substrata where encrusters and largely sessile borers and cavity dwellers can be entombed by later encrustations and organic frameworks, although cross-cutting relationships must be carefully distinguished before the true census community can be recognized (Fürsich, 1979; Jackson, 1983; Brett and Brookfield, 1984; M. A. Wilson, 1985).

Biological interpretation of census assemblages is not necessarily straightforward: buried along with the catastrophically censused community will be the shelly remains of any earlier colonists of the site together with introduced shelly material. Moreover, few mass mortalities kill all individuals of a species or all species in a community (e.g., Rees *et al.*, 1977; Von Esternhagen and Dethlefsen, 1983; Hunte *et al.*, 1986; Mah and Stearn, 1986; Stebbins, 1987; Lessios, 1988), with local exceptions (Tsuchiya, 1983), and nonobrution types can be markedly diachronous or involve time lags in mortality (Knowlton *et al.*, 1981; Lessios, 1988), or even fail to leave a record (Greenstein, 1989b). Thus, strictly speaking, only some of all assemblages conventionally categorized as census types are likely to reflect complete censuses or truly zero time-averaging of shelly faunas. Absolute time resolution for census assemblages therefore ranges from zero to perhaps a few decades (Fig. 9).

5.2.2. Within-Habitat Time-Averaged Assemblages

Death assemblages that accumulate through sporadic, attritional mortality or that sum several census events are necessarily more complex, both biologically and taphonomically. The longer the period of accumulation, the greater the number of biological processes that can influence shell supply and preservation even within a single environment, and the greater the likelihood of both short- and long-term changes in the local physical environment, whether driven by external forces, by the community itself (sediment baffling, growth up into shallower waters), or by live/dead interactions (taphonomic feedback) (Figs. 1 and 5).

Time-averaged indigenous assemblages from a single environment, here termed within-habitat assemblages, should be good sources of paleoecological information, judging by analyses of Recent examples (e.g., most entries in Tables II–VI). Environmental conditions during the period of shell accumulation need not be static: the background or average state may be punctuated by short-term events and natural oscillations in water energy, emersion, salinity, temperature, and oxygenation, any of which might have ecological and taphonomic consequences for the time-averaged assemblage. Equally important, the substratum itself might change, owing to erosion, dewatering (firming), cementation, deposition of new sediment, or accumulation of shells; many of these processes can be biologically mediated.

Low-density fossil assemblages embedded within a larger homogeneous sedimentary body [e.g., Thorson's (1957) stereotypic *Venus* community in a bioturbated muddy sand facies] are likely to reflect within-habitat time-averaging. Other typical features include (a) diverse mixture of shells supplied by different community states (*sensu* Johnson, 1972, and Gray, 1981) (Fig. 5), (b) mixture of shells with differing degrees of physical, biological, or chemical damage, (c) association of apparently indigenous but ecologically disparate forms, such as soft-bottom and shell-gravel taxa in shell-rich intervals, owing to taphonomic feedback [Kidwell and Jablonski (1983); one pathway of faunal condensation *sensu* Fürsich (1978)], and (d) fine-scale heterogeneity in composition owing to incomplete mixing of successive states by physical or biogenic reworking [another pathway of faunal condensation *sensu* Fürsich (1978)].

The repetitive and distinctive associations of taxa [recurrent assemblages of Johnson (1962); Valentine and Mallory (1965)] that characterize stratigraphic facies provide strong circumstantial evidence that many fossil assemblages are time-averaged over periods that encompass the entire range of within-community variability (K. R. Walker and Bambach, 1971), i.e., over at least decades to centuries on the open shelf. If fossil assemblages typically formed over shorter time scales, then the record should be far more mosaic in its composition and less likely to yield recurrent associations in a given facies type.

The fossil record of a community or habitat can be made more complex by taphonomic feedback between accumulating dead shells and living benthos (Kidwell and Jablonski, 1983; Kidwell, 1986b; K. B. Miller *et al.*, 1988). Dead shells, whether supplied by local mortality or by transport from adjacent habitats, provide substrata for attachment and might eventually transform a soft substratum into coarser, firmer seafloor. Such a change could inhibit the initial soft-bottom fauna and facilitate colonization and survival of new taxa that tolerate or prefer shell-gravel conditions (e.g., J. B. Wilson, 1988), modifying both the lithofacies and community structure. Actualistic studies suggest that changes driven by live/dead interactions require many decades to accomplish. Knight (1988; Knight and Bosence, 1988), working in a coastal shelf embayment where the timing of clastic deposition was well known, found that while shell accumulation over several decades resulted in some changes in the benthic fauna (facilitation of encrusters and borers not otherwise present in the soft-bottom habitat), longer periods would be required to generate a community-wide change in faunal composition by live/dead interactions.

Absolute scales for within-habitat time-averaging can be estimated in a number of ways from Recent faunas. One approach (Section 3.2.1) measures the number of years over which a local fauna must be sampled before its cumulative species richness approximates that of the local death assemblage. By this method, data from the English Channel suggest at least 70 years of time-averaging in shell gravel habitats (Carthew and Bosence, 1986), whereas the less diverse faunas of Californian lagoons with higher sedimentation rates require only 3 years of pooled data (Peterson, 1976).

Another approach that provides minimum estimates of time-averaging is based on the temporal scale of physical and biological variability in modern environments. For example, the natural hydrographic period for coastal environments, over which storms and other perturbations become a steady-state process, is on the order of a few decades, whereas the natural period for the depth-buffered open shelf is on the order of a century (Moore and Curray, 1964; Swift, 1970, 1976). From a sedimentological perspective, only those sampling programs that span these long cycles can reveal the true "average condition" of the coast or shelf as a depositional system, and detect the infrequent but high-energy processes that disproportionately influence the permanent rock record. Because benthic populations commonly track physical environmental fluctuations even on seasonal scales (e.g., Nichols and Thompson, 1985), these long-term scales almost certainly influence the composition of death assemblages.

Analyses of benthic ecosystems yield virtually identical estimates for the minimum period over which a community must be studied for its entire range of states to be sampled (e.g., Livingstone, 1987). Pearson and Barnett (1987) suggested that observations along coasts of western Europe should span ≥25 years and preferably 50–70 years, and that in general the minimum length of study should equal the lifespan of the longest-lived organisms in the target community (also see Connell and Sousa, 1983). Because populations can oscillate over periods of several years or more, even benthic surveys separated by several decades cannot always differentiate between short-term "noise" and long-term unidirectional trends or stasis (Gray and Christie, 1983; Nichols and Thompson, 1985; Shillabeer and Tapp, 1990).

All of these lines of evidence estimate the *minimum* period over which communities and their within-habitat death assemblages are time-averaged: from years to decades in coastal environments and from many decades to centuries on the open shelf (Fig. 9). *Maximum* periods may be estimated from rates of habitat migration and change. Coastal environments can shift rapidly, in part because of their mosaic nature, so that a single habitat is unlikely to persist at a single site for time-averaging over more than a few thousand years. In contrast, open-shelf habitats are typically more extensive and more intergradational in nature, so that time scales of habitat migration will more closely approximate those of transgression and regression, yielding within-habitat scales of time-averaging up to tens of thousands of years (Fig. 9). These minimum and maximum estimates thus bracket time-averaging values derived from absolute dating of individual shells (see Section 5.1).

5.2.3. Environmentally Condensed Assemblages

When environmental change is rapid relative to net rates of sediment aggradation, faunas from successive habitats can accumulate *in situ* to form a single fossil assemblage. These assemblages, such as might be formed during salinity fluctuations in marine embayments (e.g., Fürsich and Kauffman, 1984) or marine transgression of a sediment-starved shelf (e.g., J. B. Wilson, 1988; Kidwell, 1989), are examples of environmental condensation as defined by Fürsich (1975; Fürsich and Kauffman, 1984, p. 505): "faunas representing different environments in time are telescoped into one stratigraphic horizon."

Environmental condensation should be suspected for assemblages that (a) lie along facies boundaries or along discontinuity surfaces of any type, (b) occur as dense concentrations, particularly if these coquinas are microstratigraphically complex and extend laterally into adjacent facies, (c) have shell infills that differ from the surrounding matrix, or (d) include shells showing a wide range of preservational states (Fürsich, 1978; Kidwell and Aigner, 1985). On both hard and soft substrata, these assemblages commonly show features consistent with multiple cycles of burial and exhumation, early diagenesis, and benthic recolonization including taphonomic feedback. In carbonate settings, the matrix can undergo early cementation and thus may also be subjected to dissolution, bioerosion, and encrustation (e.g., Schroeder and Zankl, 1974; Scoffin and Garrett, 1974; Palmer and Fürsich, 1974; Fürsich, 1978; Walker and Diehl, 1985).

Almost by definition, time scales of environmental condensation fall at the high end of those estimated for within-habitat time-averaging (Fig. 9). These processes operate on time scales that approach or exceed the limits of direct observation, so that Quaternary cores and surficial deposits are the chief evidence. Up to 15,000 years of *in situ* time-averaging during the Holocene transgression are recorded by mixed nearshore and shelf faunas (see Section 5.1). Similarly, thousands of years of shell accumulation with only microstratigraphic separation of successive faunas has been documented in a variety of Holocene coastal settings. Examples include the amalgamation of open marine and lagoonal assemblages off Mauritania (Einsele *et al.*, 1974), the reworking of ~9000-year-old intertidal shells into modern subtidal communities in the North Sea (Cadée, 1984a), and biogenic mixing of upper and lower intertidal shells whose ages range from 0 to ~4000 years in the Gulf of California (Meldahl, 1987b).

5.2.4. Biostratigraphically Condensed Assemblages

If net rates of sediment aggradation are so slow that a time-averaged assemblage includes taxa that ordinarily have mutually exclusive geological ranges (i.e., taxa from different biozones), then that assemblage is more precisely described as biostratigraphically condensed (Wendt, 1970; Jenkyns, 1971) (Fig. 9). Assemblages can also be mixed biostratigraphically in the absence of time-averaging, for example by stratigraphic leakage (exhumation, reworking) of older taxa into a young assemblage (Middlemiss, 1962; Frey *et al.*, 1975; W. Miller, 1983; Goldring, 1990; Ruffell, 1990), or by piping of young taxa into an older assemblage (Garrison *et al.*, 1987) (Table I). These phenomena, along with a number of others discussed in Sections 5.2.2 and 5.2.3, all fall under the general concept of faunal condensation *sensu* Fürsich (1978).

Given that biostratigraphically condensed deposits must encompass evolutionarily significant periods of time (perhaps 100,000 years or more), most shelf-depth and shallower examples will record environmental condensation as well (Machalski and Walaszczyk, 1987). Within-habitat biostratigraphic condensation, in contrast, is most likely to occur in deep-sea sediments, where time scales of facies change can be slow relative to evolutionary rates, although environmental condensation can occur even here (Wendt, 1970, 1988; Bernoulli and Jenkyns, 1974; Urlichs, 1975; Bergner *et al.*, 1982).

5.3. Identifying Taphonomically Comparable Assemblages

In his paleoecological analyses of Jurassic molluscan faunas, Fürsich (1977, 1984, 1986) focused on beds containing relatively few, dispersed shells and on monospecific shell pavements. By avoiding the most shell-rich beds, even though these are conspicuous, he restricted his paleobiological interpretations to the taphonomically most comparable census and within-habitat accumulations. This strategy could be applied effectively in many other settings.

When shell gravels are the predominant lithofacies or are the benthic habitat of interest, some degree of taphonomic comparability can be achieved by grouping them according to the inferred process and time scale of concentration (Kidwell, 1982, 1986a, this volume, Chapter 5). Shell assemblages can be time-averaged to varying degrees before final deposition, of course, and so there will not always be a precise correlation between the time scale over which the concentration formed and the age spectrum of constituent shells. For example, a Recent tempestite could incorporate shells from a single mass mortality (i.e., census assemblage), from a time-averaged indigenous assemblage, from exotic sources, from relict Pleistocene deposits, from eroded Miocene strata, or from any combination of these.

Nonetheless, simple event-concentrations formed by brief episodes of shell influx or reworking should generally be taphonomically more straightforward than shell concentrations formed by the amalgamation or stacking of many event-concentrations. Such multiple-event, composite concentrations commonly involve environmental condensation and other mixing processes, so that samples that are taphonomically comparable to those from simple event-concentrations will be difficult to extract, even by microstratigraphic dissection. The cumulative effects of taphonomic processes should tend to be most pronounced in stratigraphically condensed hiatal concentrations and in erosional lags (Kidwell, 1982, and this volume, Chapter 5).

6. Bathymetric Variation in the Preservation of Shelly Faunas

All factors in the basic taphonomic equation—quantity and susceptibility of shells, post-mortem environment, time scale of accumulation (Fig. 1)—are likely to vary over ecological time scales, among environments, and through the Phanerozoic. Comparative actualistic studies are rare, however, and, owing to little standardization or quantification in the description of taphonomic features, it is difficult to compare among published accounts and thus to characterize bathymetric and other gradients in much detail. Moreover, because many actualistic studies describe taphonomic end-products rather than processes, extrapolation to the fossil record is uncertain, particularly for Mesozoic and Paleozoic strata.

Figure 10 summarizes bathymetric trends in the taphonomy of benthic marine faunas, as developed in an aggradational facies tract. Long-term condensed accumulations are excluded (see Kidwell, this volume, Chapter 5, Fig. 16), as are anaerobic and dysaerobic environments. Environments within tidal range and within SCUBA depths are best known; reef work is only getting underway, and very little evidence is available for deeper-shelf environments.

In terms of shell condition, soft substrata of the outer shelf and presumably the slope evidently yield the best preservation: low frequency of fragmentation, virtually no abrasion or bioerosion, and deep infauna articulated though generally not in life positions (S. E. Walker *et al.*, 1989, and unpublished data). Nearshore, beach, and other shoal environments within the photic zone show the poorest preservation, whereas intertidal, sheltered coast, and middle-shelf environments are highly variable (Fig. 10).

Bathymetric trends in time-averaging can only be postulated on the basis of stratigraphic evidence and a handful of direct-dated assemblages. Within-habitat assemblages should occur in all settings, whereas censuses are likely to be preserved only in areas of rapid and/or substantial episodic deposition; these include lagoons, reefs, and the transitional zones of open shelves. Environmentally condensed assemblages should be most common in higher-energy and shallow-water environments where environments shift most rapidly. These trends broadly correspond with bathymetric patterns in shell concentration (Kidwell, this volume, Chapter 5, Fig. 15). [Compare with Fig. 6 of Fürsich and Aberhan (1990).]

Dissolution rates are poorly constrained, as indicated by the ranges in Fig. 10, with rates varying from low in sands to high in muds. Precipitation within micrite envelopes, between particles, and in intraskeletal pores is better documented. In tropical and subtropical environments, early precipitation is found in fully marine environments characterized by frequent wave pumping through pore spaces.

Despite poorly understood limitations on applicability to the pre-Cenozoic fossil record, owing to the evolution of shell-producing and shell-destroying organisms (Efremov, 1940; Behrensmeyer and Kidwell, 1985; Speyer and Brett, 1988; Kidwell, 1990), actualistic studies provide valuable insights into the variety of end-products that can arise under a set of relatively well-understood and tightly constrained conditions. Some of this variation is environmentally arrayed, although the broad range of preservational quality even within single environments underscores just how difficult it will be to characterize taphofacies uniquely. This does not mean that taphofacies analysis (e.g., Speyer and Brett, 1986, 1988; Norris, 1986; Kidwell *et al.*, 1986; D. J. Davies *et al.*, 1989a; Staff and Powell, 1990) is premature, but its state of development resembles that of conventional facies analysis in the early 1960s. It is also more difficult, simply because patterns and effects can be generated by more than one agent, shells can

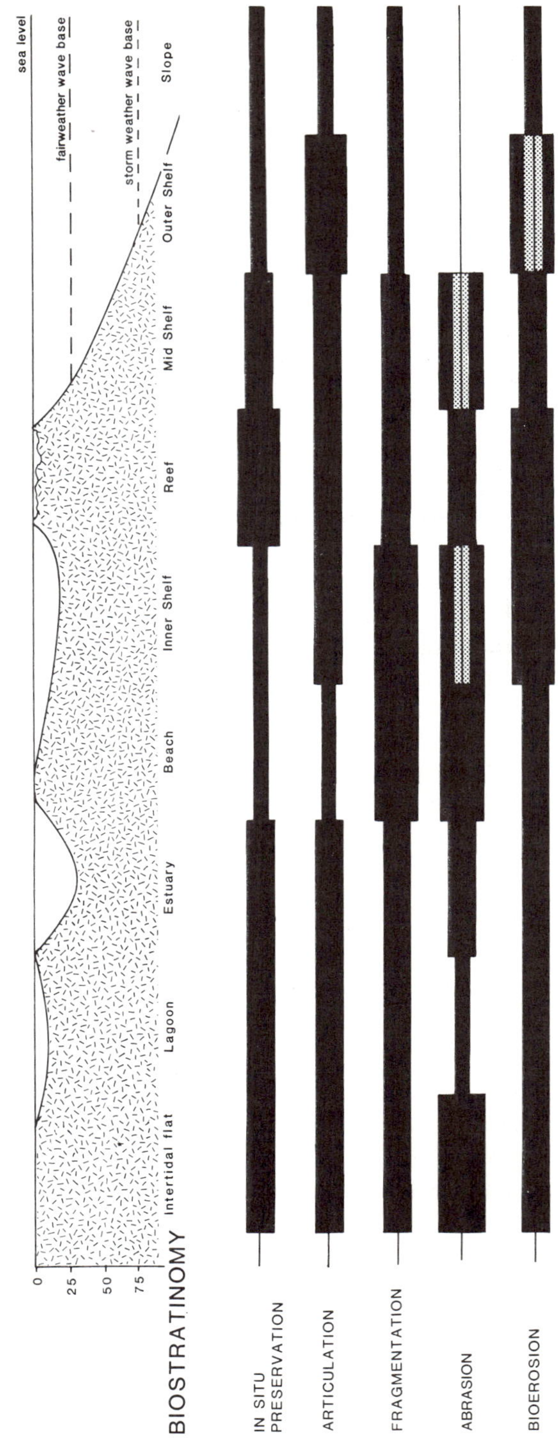
BIOSTRATINOMY
sea level
fairweather wave base
storm weather wave base
0
25
50
75
Intertidal flat
Lagoon
Estuary
Beach
Inner Shelf
Reef
Mid Shelf
Outer Shelf
Slope
IN SITU PRESERVATION
ARTICULATION
FRAGMENTATION
ABRASION
BIOEROSION

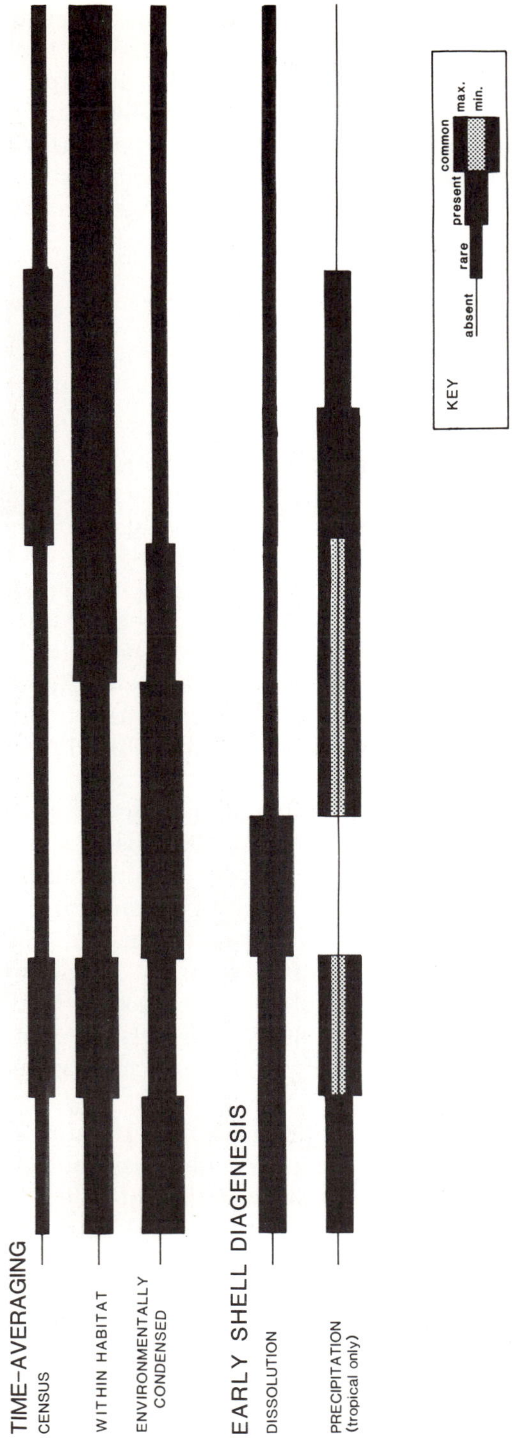

FIGURE 10. Bathymetric trends in taphonomy and time-averaging of predominantly molluscan assemblages.

experience more than one cycle of modification, and generations can be time-averaged to many different degrees. These complications make it all the more important to establish objective, preferably quantitative descriptors of post-mortem alteration (see references in Table VII), and to design studies that directly link taphonomic patterns and effects to processes. Meldahl and Flessa (1990), for example, focused on the taphonomic behavior of a single bivalve species into order to rank the relative importance of diverse factors among intertidal to shallow subtidal environments.

7. Interpretation of the Fossil Record: Conclusions and Strategy

1. After decades of stressing post-mortem loss of information, taphonomic studies provide ample evidence for the wealth of sedimentological and paleoecological information preserved in shelly assemblages. One major challenge for taphonomy in the 1990s is geological: how to improve the reliability and sophistication of taphonomic criteria bearing on depositional and diagenetic histories of sedimentary deposits. The second challenge is biological and lies in quantifying the ratio of signal to noise in the paleontological record: what is the fidelity of shelly assemblages to original biological patterns, how does fidelity vary as a function of environment and scale of time-averaging, and how are these values best estimated for different assemblages?

2. Many kinds of shell modification can be generated by more than one agent (Table VII). For example, much surface dulling and shell loss casually referred to as "dissolution" is attributable to leaching of organic matrix, microboring, and bioabrasion rather than to dissolution of mineral phases alone; SEM examination allows these processes to be distinguished. The origin of many shell fragments remains ambiguous: we cannot yet distinguish the effects of different mechanical and biological processes. These complications underscore the importance of developing objective, quantitative descriptors of taphonomic features and of improving the criteria used to recognize the effects of separate processes.

3. Although shell transport has been documented in virtually all environments, its effect on the composition of open-shelf shelly assemblages appears to be minor at the scale of facies. This justifies a host of paleontological applications, from biofacies analysis to evolutionary paleoecology. Except for settings characterized by steep gradients or bulk-flow phenomena, death assemblages show strong loyalty to life habitats (Tables II–IV and VIII). Out-of-habitat transport in level-bottom settings usually only moves shells into adjacent habitats and involves mainly small, epifaunal shells; these exotic shells generally constitute a small minority of individuals in the recipient assemblage. Shell movement *within* habitats, on the other hand, is pervasive, as evidenced by the scarcity of fossils preserved in life positions.

These processes (small-scale transport, disruption by bioturbators) smear ecological patterns, so that within-habitat details of community structure usually must be derived from exceptional, instantaneously buried communities.

4. The temporal resolution of shelly assemblages varies immensely, from high-resolution census assemblages to low-resolution biostratigraphically condensed accumulations. Intermediate scales of time-averaging characterize most of the fossil record. These assemblages can form either (a) during a period of environmental stability (within-habitat time-averaging), or (b) during a period of environmental change (environmental condensation). Absolute scales of time-averaging can be bracketed on the basis of rates of environmental and community change in Recent settings, and these estimates are corroborated by direct-dated shells in Recent death assemblages. The sparse data suggest some systematic variation in scales of time-averaging among environments (Figs. 9 and 10).

5. Actualistic investigations indicate that the compositions of death assemblages are in many respects faithful reflections of long-term community structure, further justifying certain kinds of paleobiological interpretation. Based on studies of level-bottom soft-substrata conducted over the last 25 years, it appears that (a) virtually all live shelly taxa leave some record in the local death assemblage, whether examined at the facies or larger scale, (b) species found "dead only" are usually low-abundance taxa, indicating that ecologically transient and exotic species will only dominate assemblages in unusual situations, even though they may constitute a large proportion of the species richness of an assemblage, and (c) only half of the top (most abundant) dead taxa are top taxa in the living community (Tables II–VI). The longer the period over which live data are pooled, the higher the measured fidelities of the death assemblages, underscoring the time-averaged nature of most shell accumulations in surficial marine sediments and the importance of long-term baselines of live data for comparison. Consequently, much of the variance and many of the low resemblance values may be attributable to inappropriate comparisons of time-averaged death assemblages with living communities sampled for only 1 or 2 years. Surprisingly, fidelity with respect to species presence/absence and relative abundance does not appear to vary strongly across environmental gradients (Tables II–VI). Discriminating biofacies and interpreting community structure on the basis of the *rank* abundance of the most common species (rather than of the entire assemblage) cannot be justified actualistically (Table VI). The fidelity of size–frequency data in time-averaged assemblages is highly variable; further work might yet demonstrate such data to be biologically meaningful, at least in some environments.

The results of actualistic studies are fundamentally encouraging for both geologists and paleontologists. Taphonomic criteria for paleoenvironmental interpretation show great promise, although still in an early stage of development. Moreover, when viewed at the appropriate scale, the record clearly yields more biological signal than noise, and this can be scaled quantitatively for Recent environments. The greatest difficulty in extrapolating these results to the older fossil record, whether for geological or paleontological application, will be biological factors—namely, evolutionary changes in the behavior and environmental deployment of organisms as producers, utilizers, and destroyers of dead hard parts. A strategy informed rather than ruled by actualism will thus be the most fruitful.

ACKNOWLEDGMENTS. Our grateful thanks to: G. C. Cadée, H. Cummins, K. W. Flessa, R. Goldring, and D. Jablonski for their careful reviews; to G. C. Cadée, A. A. Ekdale, K. W. Flessa, A. P. Knight, K. Tanabe, and J. E. Warme for providing unpublished data; and to D. Archer, D. E. Canfield, S. M. Holland, M. C. LaBarbera, A. I. Miller, and the 1990 University of Chicago graduate taphonomy class for their helpful feedback and advice. Research and manuscript production supported by NSF EAR85-52411-PYI (to S.M.K.)

References

Abbott, B. M., 1974, Flume studies on the stability of model corals as an aid to quantitative paleoecology, *Palaeogeogr. Palaeoclimatol. Palaeoecol.* **15**:1–27.

Adey, W. H., and Vassar, J. M., 1975, Colonization, succession and growth rates of tropical coralline algae (Rhodophyta, Cryptonemiales), *Phycologia* **14**:55–69.

Adjas, A., Masse, J.-P., and Montaggioni, L. F., 1990, Fine-grained carbonates in nearly closed reef environments: Mataiva and Takapoto atolls, central Pacific Ocean, *Sediment. Geol.* **67**:115–132.

Aigner, T., 1985a, *Storm Depositional Systems: Dynamic Stratigraphy in Modern and Ancient Shallow-Marine Sequences,* Springer-Verlag, Berlin.

Aigner, T., 1985b, Biofabrics as dynamic indicators in nummulite accumulations, *J. Sediment. Petrol.* **55**:131–134.

Aigner, T., and Reineck, H.-E., 1982, Proximality trends in modern storm sands from the Helgoland Bight (North Sea) and their implications for basin analysis, *Sencken. Marit.* **14**:183–213.

Aigner, T., Hagdorn, H., and Mundlos, R., 1978, Biohermal, biostromal and storm-generated coquinas in the Upper Muschelkalk, *Neues Jahrb. Geol. Paläontol. Abh.* **157**:42–52.

Aitken, A. E., Risk, M. J., and Howard, J. D., 1988, Animal–sediment relationships on a subarctic intertidal flat, Pangnirtung Fjord, Baffin Island, Canada, *J. Sediment. Petrol.* **58**:969–978.

Akpan, E. B., and Farrow, G. E., 1984, Shell-boring algae on the Scottish continental shelf: Identification, distribution, bathymetric zonation, *Trans. R. Soc. Ed. Earth Sci.* **75**:1–12.

Akpan, E. B., and Farrow, G. E., 1985, Shell bioerosion in high-latitude low-energy environments: Firths of Clyde and Lorne, Scotland, *Mar. Geol.* **67**:139–150.

Albertzart, L. S., and Wilkinson, B. H., 1989, Genesis of Holocene shell assemblages along barrier backbeaches of the central Texas coast, *Geol. Soc. Am. Abstr. Progr.* **21A**:A290–A291.

Alexander, R. R., 1977, Growth morphology, and ecology of Paleozoic and Mesozoic opportunistic species of brachiopods from Idaho–Utah, *J. Paleontol.* **51**:1133–1149.

Alexander, R. R., 1986a, Life orientation and post-mortem reorientation of Chesterian brachiopod shells by paleocurrents, *Palaios* **1**:303–311.

Alexander, R. R., 1986b, Resistance to and repair of shell breakage induced by durophages in Late Ordovician brachiopods, *J. Paleontol.* **60**:273–285.

Alexander, R. R., 1990, Disarticulated shells of Late Ordovician brachiopods: Inferences on strength of hinge and valve architecture, *J. Paleontol.* **64**:524–532.

Alexandersson, T., 1972, Micritization of carbonate particles: Processes of precipitation and dissolution in modern shallow-marine sediments, *Bull. Geol. Inst. Univ. Uppsala N. S.* **3**(7):201–236.

Alexandersson, T., 1978, Destructive diagenesis of carbonate sediments in the eastern Skagerrak, North Sea, *Geology* **6**:324–327.

Allen, E. J., 1899, On the fauna and bottom deposits near the thirty fathom line from the Eddystone grounds to Start Point, *J. Mar. Biol. Assoc. U. K.* **5**:365–541.

Allen, J. R. L., 1984, Experiments on the settling, overturning and entrainment of bivalve shells and related models, *Sedimentology* **31**:227–250.

Aller, R. C., 1982a, Carbonate dissolution in nearshore terrigenous muds: The role of physical and biological reworking, *J. Geol.* **90**:79–95.

Aller, R. C., 1982b, The effects of macrobenthos on chemical properties of marine sediment and overlying water, in: *Animal-Sediment Relations* (P. L. McCall and M. J. S. Tevesz, eds.), Plenum Press, New York, pp. 53–102.

Allison, P. A., 1986, Soft-bodied animals in the fossil record: The role of decay in fragmentation during transport, *Geology* **14**:979–981.

Allison, P. A., 1988, The role of anoxia in decay and mineralization of proteinaceous macrofossils, *Paleobiology* **14**:139–154.

Allison, P. A., 1990, Variation in rates of decay and disarticulation of Echinodermata: Implications for the application of actualistic data, *Palaios* **5**:432–440.

Allmon, W. D., 1988, Ecology of Recent turritellid gastropods (Prosobranchia, Turritellidae): Current knowledge and paleontological implications, *Palaios* **3**:259–284.

Alvarez, F., and Taylor, P. D., 1987, Epizoan ecology and interactions in the Devonian of Spain, *Palaeogeogr. Palaeoclimatol. Palaeoecol.* **61**:17–31.

Ameziane-Cominardi, N., and Roux, M., 1987, Biocorrosion et micritisation des ossicules d'Echinodermes en milieu bathyal au large de la Nouvelle-Calédonie, *C. R. Acad. Sci. Paris Sér. II* **305**:701–705.

Arntz, W. E., Brunswig, P., and Sannthern, M., 1976, Zonierung um Mollusken und Schill im Rinnensystem der Kieler Bucht (Westliche Ostsee), *Senckenb. Marit.* **8**:189–269.

Arua, I., 1982, Borings and shell damage in Eocene Gastropoda: Southeastern Nigeria, *Palaeogeogr. Palaeoclimatol. Palaeoecol.* **38**:269–282.

Aslin, C. J., 1968, Echinoid preservation in the Upper Estuarine Limestone of Blisworth, Northamptonshire, *Geol. Mag.* **105**:506–518.

Babcock, R. C., 1985, Growth and mortality in juvenile corals (*Goniastrea, Platygyra* and *Acropora*): The first year, in: *Proceedings 5th International Coral Reef Congress, Tahiti,* Volume 4, pp. 355–360.

Backman, J., 1986, Accumulation patterns of Tertiary calcareous nannofossils around extinctions, *Geol. Rundsch.* **75**:185–196.

Baird, G. C., Brett, C. E., and Frey, R. C., 1989, "Hitchhiking" epizoans on orthoconic cephalopods: Preliminary renewal of the evidence and its implications, *Senckenb. Lethaea* **69**:439–465.

Bak, R. P. M., 1990, Patterns of echinoid bioerosion in two Pacific coral reef lagoons, *Mar. Ecol. Progr. Ser.* **66**:267–272.

Balson, P. S., and Taylor, P. D., 1982, Palaeobiology and systematics of large cyclostome bryozoans from the Pliocene Coralline Crag of Suffolk, *Palaeontology* **25**:529–554.

Barrell, J., 1917, Rhythms and the measurement of geologic time, *Geol. Soc. Am. Bull.* **28**:745–904.

Bayer, U., 1978, Multimodale Häufigkeitsverteilung bei der Analyse von Strömungseinregelungen, *Neues Jahrb. Geol. Paläontol. Monatsh.* **1978**:747–748.

Behrensmeyer, A. K., 1982, Time resolution in fluvial vertebrate assemblages, *Paleobiology* **8**:211–227.

Behrensmeyer, A. K., 1988, The Pull of the Recent analogue, *Palaios* **3**:373.

Behrensmeyer, A. K., and Kidwell, S. M., 1985, Taphonomy's contributions to paleobiology, *Paleobiology* **11**:105–119.

Bein, A., and Fütterer, D., 1977, Texture and composition of continental shelf to rise sediments off the northwestern coast of Africa: An indication for downslope transportation, *Meteor Forschungsergeb. Reihe C* **27**:46–74.

Bergman, K. M., Elner, R. W., and Risk, M. J., 1982, The influence of *Polydora websteri* borings on the strength of the shell of the sea scallop, *Placopecten magellanicus*, *Can. J. Zool.* **60**:2551–2556.

Bergner, H. D., Gebhard, G., and Wiedmann, J., 1982, Kondensationerscheinungen in der marokkanischen und alpinen Mittelkreide (Apt, Alb), *Neues Jahrb. Geol. Paläontol. Abh.* **165**:102–124.

Berner, R. A., and Morse, J. A., 1974, Dissolution kinetics of calcium carbonate in sea water IV. Theory of calcite dissolution, *Am. J. Sci.* **274**:108–134.

Bernoulli, D., and Jenkyns, H. C., 1974, Alpine, Mediterranean, and central Atlantic Mesozoic facies in relation to the early evolution of the Tethys, *Soc. Econ. Paleontol. Mineral. Spec. Publ.* **19**:129–160.

Bertness, M. D., and Cunningham, C., 1981, Crab shell-crushing predation and gastropod architectural defense, *J. Exp. Mar. Biol. Ecol.* **50**:213–230.

Beukema, J. J., and Cadée, G. H., 1986, Zoobenthos responses to eutrophication of the Dutch Wadden Sea, *Ophelia* **26**:55–64.

Bigelow, H. B., and Schroeder, W. C., 1953, Sawfishes, guitarfishes, skates, rays, and chimaeroids, *Mem. Sears Found. Mar. Res.* **2**:1–514.

Bishop, G. A., 1975, Traces of predation, in: *The Study of Trace Fossils* (R. W. Frey, ed.), Springer-Verlag, New York, pp. 261–281.

Bishop, J. D. D., 1988, Disarticulated bivalve shells as substrates for encrustation by the Bryozoan *Cribrilina puncturata* in the Plio-Pleistocene Red Crag of Eastern England, *Palaeontology* **31**:237–253.

Bishop, J. D. D., 1989, Colony form and the exploitation of spatial refuges by encrusting bryozoa, *Biol. Rev.* **64**:197–218.

Bjerstedt, T. W., and Erickson, J. M., 1989, Trace fossils and bioturbation in peritidal facies of the Potsdam-Theresa Formations (Cambrian–Ordovician), Northwest Adirondacks, *Palaios* **4**:203–224.

Boekschoten, G. J., 1966, Shell borings of sessile epibiontic organisms as paleoecological guides (with examples from the Dutch coast), *Palaeogeogr. Palaeoclimatol. Palaeoecol.* **3**:333–379.

Boesch, D. F., Wass, M. L., and Virnstein, R. W., 1976, The dynamics of estuarine benthic communities, in: *Estuarine Processes* (M. Wiley, ed.), Academic Press, New York, Volume 1, pp. 177–196.

Bosence, D. W. J., 1976a, Ecological and sedimentological studies on some carbonate sediment producing organisms, Co. Galway, Ireland, Ph.D. thesis, University of Reading, Reading, United Kingdom.

Bosence, D. W. J., 1976b, Ecological studies on two unattached coralline algae from Western Ireland, *Palaeontology* **19**:365–395.

Bosence, D. W. J., 1979, Live and dead faunas from coralline algal gravels, Co. Galway, Eire, *Palaeontology* **22**:449–478.

Bosence, D. W. J., 1980, Sedimentary facies, production rates and facies models for Recent coralline algal gravels, Co. Galway, Ireland, *Geol. J.* **15**:91–111.

Bosence, D. W. J., 1984, Construction and preservation of two Recent algal reefs, St. Croix, Caribbean, *Palaeontology* **27**:549–574.

Bosence, D. W. J., 1985, Preservation of coralline algal reef frameworks, in: *Proceedings 5th International Coral Reef Congress, Tahiti,* Volume 6, pp. 623–628.

Bosence, D. W. J., 1989a, Surface sublittoral sediments of Florida Bay, *Bull. Mar. Sci.* **44**:434–453.

Bosence, D. W. J., 1989b, Carbonate budgets for carbonate mounds, Florida, U.S.A., in: *Proceedings 6th International Coral Reef Symposium, Townsville.*

Boss, S. K., and Liddell, W. D., 1987, Patterns of sediment composition of Jamaican fringing reef facies, *Sedimentology* **34**:77–87.

Boucot, A. J., 1953, Life and death assemblages among fossils, *Am. J. Sci.* **251**:25–40.

Boucot, A. J., 1981, *Principles of Benthic Marine Paleoecology,* Academic Press, New York.

Boyer, P. S., 1969, Pelecypod beak wear: An indicator of postmortem transport, *Texas J. Sci.* **20**:385–388.

Bradley, W. H., 1957, Physical and ecologic features of the Sagadahoc Bay tidal flat, Georgetown, Maine, *Geol. Soc. Am. Mem.* **67**(2):641–682.

Branch, G. M., 1984, Competition between marine organisms: Ecological and evolutionary implications, *Oceanogr. Mar. Biol. Annu. Rev.* **22**:429–593.

Brandt, D. S., 1989, Taphonomic grades as a classification for fossiliferous assemblages and implications for paleoecology, *Palaios* **4**:303–309.

Brandt Velbel, D., 1984, On defining limits to paleoecological interpretation in the fossil record, *Géobios Mém. Spéc.* **8**:415–418.

Brenchley, P. J., and Newall, G., 1970, Flume experiments on the orientation and transport of models and shell valves, *Palaeogeogr. Palaeoclimatol. Palaeoecol.* **7**:185–220.

Brenner, K., 1976, Ammoniten-Gehäuse als Anzeiger von Palaeo-strömungen, *Neues Jahrb. Geol. Paläontol. Abh.* **151**:101–118.

Brenner, K., and Einsele, G., 1976, Schalenbruch im Experiment, *Zentralbl. Geol. Paläontol. II* **1976**:349–354.

Brett, C. E., 1988, Palaeoecology and evolution of marine hard substrate communities: An overview, *Palaios* **3**:374–378.

Brett, C. E., and Baird, G. C., 1986, Comparative taphonomy: A key to paleoenvironmental interpretation based on fossil preservation, *Palaios* **1**:207–227.

Brett, C. E., and Brookfield, M. E., 1984. Morphology, faunas and genesis of Ordovician hardgrounds from southern Ontario, Canada, *Palaeogeogr. Palaeoclimatol. Palaeoecol.* **46**:233–290.

Brett, C. E., and Seilacher, A., 1991, Obrution deposits: A taphonomic consequence of event sedimentation, in: *Cycles and Events in Stratigraphy* (G. Einsele, W. Ricken, and A. Seilacher, eds.), Springer-Verlag, Berlin, in press.

Broecker, W. S., Andree, M., Bonani, G., Wolfli, W., Klas, M., Mix, A., and Oeschger, H., 1988, Comparison between radiocarbon ages obtained on coexisting planktonic foraminifera, *Paleoceanography* **3**:647–657.

Bromley, R. G., 1970, Borings as trace fossils and *Entobia cretacea* Portlock, as an example, in: *Trace Fossils: Geological Journal Special Issue 3* (T. P. Crimes and J. C. Harper, eds.), Liverpool, pp. 49–90.

Brouwer, A., 1959, *General Palaeontology*, University of Chicago Press, Chicago.

Brown, B. J., 1979, Shell transport and community composition in modern shelf seas, Ph. D. thesis, Glasgow University, Glasgow.

Brun, E., 1969, Aggregation of *Ophiothrix fragilis* (Abildgaard) (Echinodermata: Ophiuroidea), *Nytt Mag. Zool.* **17**:153–160.

Brun, E., 1971, Predation of *Chlamys islandica* (O. F. Müller) by eiders (*Somateria* spp.), *Astarte* **4**:23–29.

Buchanan, J. B., Sheader, M., and Kingston, P. R., 1978, Sources of variability in the benthic macrofauna off the south Northumberland coast, 1971–6, *J. Mar. Biol. Assoc. U. K.* **58**:191–210.

Budd, D. A., and Perkins, R. D., 1980, Bathymetric zonation and paleoecological significance of microborings in Puerto Rican shelf and slope sediments, *J. Sediment. Petrol.* **50**:881–903.

Cadée, G. C., 1968, Molluscan biocoenoses and thanatocoenoses in the Ria de Arosa, Galicia, Spain, *Rijksmus. Nat. Hist. Leiden Zool. Verhandl.* **95**:1–121.

Cadée, G. C., 1976, Sediment reworking by *Arenicola marina* on tidal flats in the Dutch Wadden Sea, *Neth. J. Sea Res.* **10**:440–460.

Cadée, G. C., 1979, Sediment reworking by the polychaete *Heteromastus filiformis* on a tidal flat in the Dutch Wadden Sea, *Neth. J. Sea Res.* **13**:441–456.

Cadée, G. C., 1982, Low juvenile mortality in fossil brachiopods, some comments, *Interne Versalagen Ned. Inst. Onderzoek Zee, Texel* **3**:1–29.

Cadée, G. C., 1984a, Macrobenthos and macrobenthic remains on the Oyster Ground, North Sea, *Neth. J. Sea Res.* **18**:160–178.

Cadée, G. C., 1984b, 'Opportunistic feeding', a serious pitfall in trophic structure analysis of (paleo)-faunas, *Lethaia* **17**:289–292.

Cadée, G. C., 1988, The use of size frequency distribution in paleoecology, *Lethaia* **21**:289–290.

Cadée, G. C., 1989, Size-selective transport of shells by birds and its palaeoecological implications, *Palaeontology* **32:**429–437.

Cadée, G. C., 1991, Carapaces of the shore crab *Carcinus maenas* as a habitat for encrusting organisms, in: *8th International Bryzoological Association Conference*, Paris, in press.

Cain, J. D. B., 1968, Aspects of the depositional environment and paleoecology of crinoidal limestones, *Scot. J. Geol.* **4:**191–208.

Callender, W. R., Staff, G. M., Powell, E., and MacDonald, I. R., 1990, Gulf of Mexico hydrocarbon seep communities V. Biofacies and shell orientation of autochthonous shell beds below storm wave base, *Palaios* **5:**2–14.

Carriker, M. R., Smith, E. H., and Wilce, R. T. (eds.), 1969, Penetration of calcium carbonate substrates by lower plants and invertebrates, *Am. Zool.* **9**(3)**:**629–1020.

Carter, R. M., 1968, On the biology and paleontology of some predators of bivalve molluscs, *Palaeogeogr. Palaeoclimatol. Palaeoecol.* **4:**29–65.

Carter, R. W. G., 1974, Feeding sea birds as a factor in lamellibranch valve sorting patterns, *J. Sediment. Petrol.* **44:**689–692.

Carthew, R., and Bosence, D. W. J., 1986, Community preservation in Recent shell-gravels, English Channel, *Palaeontology* **29:**243–268.

Chave, K. E., 1964, Skeletal durability and preservation, in: *Approaches to Paleoecology* (J. Imbrie and N. D. Newell, eds.), Wiley, New York, pp. 377–387.

Choi, D. R., and Ginsburg, R. N., 1983, Distribution of coelobites (cavity dwellers) in coral rubble across the Florida reef tract, *Coral Reefs* **2:**165–172.

Clifton, H. E., 1971, Orientation of empty pelecypod shells and shell fragments in quiet waters, *J. Sediment. Petrol.* **41:**671–682.

Clifton, H. E., and Boggs, S., Jr., 1970, Concave-up pelecypod (*Psephidea*) shells in shallow marine sand, Elk River Beds, southwestern Oregon, *J. Sediment. Petrol.* **40:**888–897.

Cohen, A. S., 1989, The taphonomy of gastropod shell accumulations in large lakes: An example from Lake Tanganyika, Africa, *Paleobiology* **15:**26–45.

Collins, M. J., 1985, Post mortality strength loss in shells of the Recent articulate brachiopod *Terebratulina retusa* (L.) from the west coast of Scotland, in: Les Brachiopodes fossiles et actuels (P. R. Racheboeuf and C. C. Emig, eds.), *Biostratigr. Paléozoique* **4:**209–218.

Connell, J. H., and Slatyer, R. O., 1977, Mechanisms of succession in natural communities and their roles in community stability and organization, *Am. Nat.* **111:**1119–1144.

Connell, J. H., and Sousa, W. P., 1983, On the evidence needed to judge ecological stability or persistence, *Am. Nat.* **121:**789–824.

Conover, M. R., 1975, Prevention of shell burial as a benefit hermit crabs provide their symbionts (Decapoda, Paguridea), *Crustaceana* **29:**311–313.

Copper, P., 1988, Ecological succession in Phanerozoic reef ecosystems: Is it real?, *Palaios* **3:**136–152.

Craig, G. Y., 1953, Discussion: Fossil communities and assemblages, *Am. J. Sci.* **251:**547–548.

Craig, G. Y., and Hallam, A., 1963, Size frequency and growth ring analysis of *Mytilus edulis* and *Cardium edule* and their palaeoecological significance, *Palaeontology* **6:**731–750.

Crame, J. A., 1980, Succession and diversity in the Pleistocene coral reefs of the Kenya coast, *Palaeontology* **23:**1–37.

Crawshay, L. R., 1912, On the fauna of the western outer area of the English Channel, *J. Mar. Biol. Assoc. U. K.* **9:**292–393.

Cummins, H., Powell, E. N., Stanton, R. J., and Staff, G., 1986a, The rate of taphonomic loss in modern benthic habitats: How much of the potentially preservable community is preserved?, *Palaeogeogr. Palaeoclimatol. Palaeoecol.* **52:**291–320.

Cummins, H., Powell, E. N., Stanton, R. J., and Staff, G., 1986b, The size frequency distribution in palaeoecology: Effects of taphonomic processes during formation of molluscan death assemblages in Texas bays, *Palaeontology* **29:**495–518.

Cummins, H., Powell, E. N., Newton, H. J., Stanton, R. J., Jr., and Staff, G., 1986c, Assessing transportation by the covariance of species with comments on contagious and random distributions, *Lethaia* **19:**1–22.

Currey, J. D., 1980, Mechanical properties of the mollusc shell, *Symp. Soc. Exp. Biol.* **34:**75–97.

Currey, J. D., 1988, Shell form and strength, in: *The Mollusca 11* (E. R. Trueman and M. R. Clarke, eds.), Academic Press, San Diego, pp. 183–210.

Curry, G. B., Ansell, A. D., James, M., and Peck, L., 1989, Physiological constraints on living and fossil brachiopods, *Trans. R. Soc. Ed. Earth Sci.* **80**:255–262.

Cutler, A. H., 1987, Surface textures of shells as taphonomic indicators, *Paleontol. Soc. Spec. Publ.* **2**:164–176.

Cutler, A. H., 1989, Shells survive—Loss, persistence and accumulation of hardparts in shallow marine sediments, *Geol. Soc. Am. Abstr. Progr.* **21**:A71.

Cutler, A. H., and Flessa, K. W., 1990, Fossils out of sequence: Computer simulations and strategies for dealing with stratigraphic disorder, *Palaios* **5**:227–235.

Davies, D. J., Powell, E. N., and Stanton, R. J., Jr., 1989a, Taphonomic signature as a function of environmental process: Shells and shell beds in a hurricane-influenced inlet on the Texas coast, *Palaeogeogr. Palaeoclimatol. Palaeoecol.* **72**:317–356.

Davies, D. J., Powell, E. N., and Stanton, R. J., Jr., 1989b, Relative rates of shell dissolution and net sediment accumulation—A commentary: Can shell beds form by the gradual accumulation of biogenic debris on the sea floor?, *Lethaia* **22**:207–212.

Davies, P. J., and Hutchings, P. A., 1983, Initial colonization, erosion and accretion on coral substrate. Experimental results, Lizard Island, Great Barrier Reef, *Coral Reefs* **2**:27–35.

Dodd, J. R., and Stanton, R. J., Jr., 1981, *Paleoecology, Concepts and Applications,* Wiley, New York.

Done, T. J., 1985, Effects of two *Acanthaster* outbreaks on coral community structure. The meaning of devastation, in: *Proceedings 5th International Coral Reef Congress Tahiti,* Volume 5, pp. 315–320.

Donovan, S. K., 1989, Taphonomic significance of the encrustation of the dead shell of Recent *Spirula spirula* (Linné) (Cephalopoda: Coleoidea) by *Lepas anatifera* Linné (Cirripedia: Thoracia), *J. Paleontol.* **63**:698–702.

Dörjes, J., Frey, R. W., and Howard, J. D., 1986, Origins of, and mechanisms for, mollusk shell accumulations on Georgia beaches, *Senckenb. Marit.* **18**:1–43.

Driscoll, E. G., 1967, Experimental field study of shell abrasion, *J. Sediment. Petrol.* **37**:1117–1123.

Driscoll, E. G., 1970, Selective bivalve shell destruction in marine environments, a field study, *J. Sediment. Petrol.* **40**:898–905.

Driscoll, E. G., and Weltin, T. P., 1973, Sedimentary parameters as factors in abrasive shell reduction, *Palaeogeogr. Palaeoclimatol. Palaeoecol.* **13**:275–288.

Dzik, J., 1979, Some terebratulid populations from the Lower Kimmeridgian of Poland and their relations to the biotic environment, *Acta Palaeontol. Polon.* **24**:473–492.

Eagar, R. M. C., 1978, Shape and function of the shell: A comparison of some living and fossil bivalve molluscs, *Biol. Rev.* **53**:169–210.

Efremov, J. A., 1940, Taphonomy: New branch of paleontology, *Pan Am. Geol.* **74**:81–93.

Einsele, G., and Seilacher, A. (eds.), 1982, *Cyclic and Event Stratification,* Springer-Verlag, Berlin.

Einsele, G., Herm, D., and Schwartz, H. U., 1974, Holocene eustatic(?) sea level fluctuation at the coast of Mauritania, *Meteor Forschungsergeb. Reihe C* **18**:43–62.

Ekdale, A. A., 1972, Ecology and paleoecology of marine invertebrate communities in calcareous substrates, northeast Quintana Roo, Mexico, Ph. D. thesis, Rice University, Houston, Texas.

Ekdale, A. A., 1977, Quantitative paleoecological aspects of modern marine mollusk distribution, northeast Yucatan coast, Mexico, *Am. Assoc. Petroleum Geol. Stud. Geol.* **4**:195–207.

Ekdale, A. A., 1985, Paleoecology of the marine endobenthos, *Palaeogeogr. Palaeoclimatol. Palaeoecol.* **50**:63–81.

Elias, R. J., 1986, Symbiotic relationships between worms and solitary rugose corals in the Late Ordovician, *Paleobiology* **12**:32–45.

Elias, R. J., McAuley, R. J., and Mattison, B. W., 1987, Directional orientations of solitary rugose corals, *Can. J. Earth Sci.* **24**:806–812.

Emery, K. O., 1968, Positions of empty pelecypod valves on the continental shelf, *J. Sediment. Petrol.* **38**:1264–1269.

Emig, C. C., 1986, Conditions de fossilisation de genre *Lingula* (Brachiopoda) et implications paléoécologiques, *Palaeogeogr. Palaeoclimatol. Palaeoecol.* **53**:245–253.

Emig, C. C., 1990, Examples of post-mortality alteration in Recent brachiopod shells and (paleo)ecological consequences, *Mar. Biol.* **104**:233–238.

Emson, R. H., and Wilkie, I. C., 1980, Fission and autotomy in echinoderms, *Oceanogr. Mar. Biol. Annu. Rev.* **18**:155–250.

Erickson, J. M., 1971, Wind-oriented gastropod shells as indicators of paleowind direction, *J. Sediment. Petrol.* **41**:589–593.

Erwin, D. G., 1983, The community concept, in: *Sublittoral Ecology, The Ecology of the Shallow Sublittoral Benthos* (R. Earll and D. G. Erwin, eds.), Clarendon Press, Oxford, pp. 144–164.

Fagerstrom, J. A., 1964, Fossil communities in paleoecology: Their recognition and significance, *Geol. Soc. Am. Bull.* **75**:1197–1216.

Feldmann, R. M., and Tshudy, D., 1987, Ultrastructure in cuticle from *Hoploparia stokesi* (Decapoda: Nephropidae) from the Lopez de Bertodano Fornatra (Late Cretaceous–Paleocene) of Seymour Island, Antarctica, *J. Paleontol.* **61**:1194–1203.

Flessa, K. W., and Brown, T. J., 1983, Selective solution of macro-invertebrate calcareous hard parts, *Lethaia* **16**:193–205.

Flessa, K. W., Cutler, A. H., Meldahl, K. H., and Fürsich, F. T., 1989, Taphonomic processes and stratigraphic disorder, in: *Abstracts, 28th International Geological Congress,* Volume I, Washington, D.C., pp. 493–494.

Flessa, K. W., Meldahl, K. H., and Cutler, A. H., 1990, Quantitative estimates of stratigraphic disorder and time-averaging in a shallow marine habitat, *Geol. Soc. Am. Abstr. Progr.* **22**:A83.

Force, L. M., 1969, Calcium carbonate size distribution on the West Florida shelf and experimental studies on the microarchitectural control of skeletal breakdown, *J. Sediment. Petrol.* **39**:902–934.

Ford, E., 1923, Animal communities of the level sea bottom in the waters adjacent to Plymouth, *J. Mar. Biol. Assoc. U. K.* **13**:164–224.

Frey, R. W., and Basan, P. B., 1981, Taphonomy of relict Holocene salt marsh deposits, Cabretta Island, Georgia, *Senckenb. Marit.* **13**:111–155.

Frey, R. W., and Howard, J. D., 1986, Taphonomic characteristics of offshore mollusk shells, Sapelo Island, Georgia, *Tulane Stud. Geol. Paleontol.* **19**:51–61.

Frey, R. W., Voorhies, M. R., and Howard, J. D., 1975, Estuaries of the Georgia coast, U.S.A.: Sedimentology and biology. VII. Fossil and recent skeletal remains in Georgia estuaries, *Senckenb. Marit.* **7**:257–295.

Frey, R. W., Basan, P. B., and Smith, J. M., 1987, Rheotaxis and distribution of oysters and mussels, Georgia tidal creeks and salt marshes, U.S.A., *Palaeogeogr. Palaeoclimatol. Palaeoecol.* **61**:1–16.

Frey, R. W., Hong, J.-S., and Hayes, W. B., 1988, Physical and biological aspects of shell accumulation on a modern macrotidal flat, Inchon, Korea, *Neth. J. Sea Res.* **22**: 267–278.

Fürsich, F. T., 1975, Trace fossils as environmental indicators in the Corallian of England and Normandy, *Lethaia* **8**:151–172.

Fürsich, F. T., 1977, Corallian (Upper Jurassic) marine benthic associations from England and Normandy, *Palaeontology* **20**:337–386.

Fürsich, F. T., 1978, The influence of faunal condensation and mixing on the preservation of fossil benthic communities, *Lethaia* **11**:243–250.

Fürsich, F. T., 1979, Genesis, environments and ecology of Jurassic hardgrounds, *Neues Jahrb. Geol. Paläontol. Abh.* **158**:1–63.

Fürsich, F. T., 1982, Rhythmic bedding and shell bed formation in the Upper Jurassic of east Greenland, in: *Cyclic and Event Stratification* (G. Einsele and A. Seilacher, eds.), Springer-Verlag, Berlin, pp. 208–222.

Fürsich, F. T., 1984, Palaeoecology of boreal invertebrate faunas from the Upper Jurassic of central east Greenland, *Palaeogeogr. Palaeoclimatol. Palaeoecol.* **48**:309–364.

Fürsich, F. T., 1986, Benthic associations and their environmental significance in the

Lusitanian Basin (Upper Jurassic, Portugal), *Neues Jahrb. Geol. Paläontol. Abh.* **172**:271–329.

Fürsich, F. T., 1990, Fossil concentrations and life and death assemblages, in: *Paleobiology: A Synthesis* (D. E. G. Briggs and P. R. Crowther, eds.), Blackwell Scientific Publications, Oxford, pp. 235–238.

Fürsich, F. T., and Aberhan, M., 1990, Significance of time-averaging for palaeocommunity analysis, *Lethaia* **23**:143–152.

Fürsich, F. T., and Flessa, K. W., 1987, Taphonomy of tidal flat molluscs in the northern Gulf of California: Paleoenvironmental analysis despite the perils of preservation, *Palaios* **2**:543–559.

Fürsich, F. T., and Kauffman, E. G., 1984, Palaeoecology of marginal marine sedimentary cycles in the Albian Bear River Formation of south-western Wyoming, *Palaeontology* **27**:501–536.

Fürsich, F. T., and Werner, W., 1989, Taxonomy and ecology of *Juranomia calcibyssata* gen. et sp. nov.—A widespread anomiid bivalve from the Upper Jurassic of Portugal, *Géobios* **22**:325–337.

Futterer, E., 1978a, Hydrodynamic behavior of biogenic particles, *Neues Jahrb. Geol. Paläontol. Abh.* **157**:37–42.

Futterer, E., 1978b, Studien über die Einregelung, Anlagerung und Einbettung biogener Hartteile im Strömungskanal, *Neues Jahrb. Geol. Paläontol. Abh.* **156**:87–131.

Futterer, E., 1978c, Untersuchungen über die Sink- und Transportgeschwindigkeit biogener Hartteile, *Neues Jahrb. Geol. Paläontol. Abh.* **155**:318–359.

Garrison, R. E., Kennedy, W. J., and Palmer, T. J., 1987, Early lithification and hardgrounds in Upper Albian and Cenomanian calcarenites, southwest England, *Cretaceous Res.* **8**:103–140.

Gaspard, D., 1989, Quelques aspects de la biodégradation des coquilles de brachiopods; Conséquences sur leur fossilisation, *Bull. Soc. Géol. Fr. Sér. 8*, **5**:1207–1216.

Gauch, H. G., 1982, *Multivariate Analysis in Community Ecology*, Cambridge University Press, Cambridge.

Goldring, R., 1964, Trace fossils and the sedimentary surface in shallow water marine environments, in: *Deltaic and Shallow Marine Deposits* (L. M. J. U. van Straaten, ed.), Elsevier, Amsterdam, pp. 136–143.

Goldring, R., 1990, *Fossils in the Field: Information Potential and Analogues*, Longman, London.

Gould, H. R., and McFarlan, E., Jr., 1959, Geologic history of the chenier plain, southwestern Louisiana, *Trans. Gulf Coast Assoc. Geol. Soc.* **9**:261–270.

Graus, R. R., MacIntyre, I. G., and Herchenroder, B. E., 1984, Computer simulation of reef zonation at Discovery Bay, Jamaica: Hurricane disruption and long term physical oceanographic controls, *Coral Reefs* **3**:59–68.

Gray, J. S., 1981, *The Ecology of Marine Sediments*, Cambridge University Press, Cambridge.

Gray, J. S., and Christie, H., 1983, Predicting long-term changes in marine benthic communities, *Mar. Ecol. Progr. Ser.* **13**:87–94.

Greensmith, J. T., and Tucker, E. V., 1966, Morphology and evolution of inshore shell ridges and mud-mounds on modern intertidal flats, near Bradwell, Essex, *Proc. Geol. Assoc.* **77**:329–346.

Greensmith, J. T., and Tucker, E. V., 1969, The origin of Holocene shell deposits in the chenier plain facies of Essex (Great Britain), *Mar. Geol.* **7**:403–425.

Greenstein, B. J., 1989a, The effect of taphonomic bias on the fossil record of echinoids, *Geol. Soc. Am. Abstr. Progr.* **21**:A70–A71.

Greenstein, B. J., 1989b, Mass mortality of the West-Indian echinoid *Diadema antillarum* (Echinodermata: Echinoidea): A natural experiment in taphonomy, *Palaios* **4**:487–492.

Gregory, M. R., Ballance, P. F., Gibson, G. W., and Ayling, A. M., 1979, On how some rays (Elasmobranchia) excavate feeding depressions by jetting water, *J. Sediment. Petrol.* **49**:1125–1130.

Grinnell, R. S., Jr., 1974, Vertical orientation of shells on some Florida oyster reefs, *J. Sediment. Petrol.* **41**:116–122.

Gunatilaka, A., 1976, Thallophyte boring and micritization within skeletal sands from Connemara, western Ireland, *J. Sediment. Petrol.* **46**:548–554.

Günther, A., 1990, Distribution and bathymetric zonation of shell-boring endoliths in Recent reef and shelf environments: Cozumel, Yucatan (Mexico), *Facies* **22**:233–262.

Hallam, A., 1972, Models involving population dynamics, in: *Models in Paleobiology* (T. J. M. Schopf, ed.), Freeman, Cooper and Co., San Francisco, pp. 62–80.

Hart, C. E., and Crowe, J. H., 1977, The effect of attached gravel on survival of intertidal anemones, *Trans. Am. Microsc. Soc.* **96**:28–41.

Hartnoll, R. G., 1983, Substratum, in: *Sublittoral Ecology, The Ecology of the Shallow Sublittoral Benthos* (R. Earll and D. G. Erwin, eds.), Clarendon Press, Oxford, pp. 97–124.

Hartshorne, P. M., Gillespie, W. B., and Flessa, K. W., 1987, Population structure of live and dead gastropods from Bahia la Choya, in: Paleoecology and Taphonomy of Recent to Pleistocene Intertidal Deposits, Gulf of California (K. W. Flessa, ed.), *Paleontol. Soc. Spec. Publ.* **2**:139–149.

Henderson, C. M., and Styan, W. B., 1982, Description and ecology of Recent endolithic biota from the Gulf Islands and banks in the Strait of Juan de Fuca, British Columbia, *Can. J. Earth Sci.* **19**:1382–1394.

Henderson, S. W., and Frey, R. W., 1986, Taphonomic redistribution of mollusk shells in a tidal inlet channel, Sapelo Island, Georgia, *Palaios* **1**:3–16.

Henrich, R., and Wefer, G., 1986, Dissolution of biogenic carbonates: Effects of skeletal structure, *Mar. Geol.* **71**:341–362.

Hertweck, G., 1972, Georgia coastal region, Sapelo Island, USA: Sedimentology and biology. V. Distribution and environmental significance of lebensspurren and *in-situ* skeletal remains, *Senckenb. Marit.* **4**:125–167.

Hertweck, G., 1979, Aufwuchs von See-pocken auf Hartteilen anderer Tiere, Teil I, *Nat. Mus.* **109**:305–311.

Hesselbo, S. P., 1987, The biostratinomy of *Kikelocephalus* sclerites: Implications for the use of trilobite attitude data, *Palaios* **2**:605–608.

Hewitt, R. A., 1988, Nautiloid shell taphonomy: Interpretations based on water pressure, *Palaeogeogr. Palaeoclimatol. Palaeoecol.* **63**:15–25.

Hickman, C. S., 1983, Radular patterns, systematics, diversity, and ecology of deep-sea limpets, *Veliger* **26**:73–92.

Highsmith, R. C., Lueptow, R. L., and Schonberg, S. C., 1983, Growth and bioerosion of three massive corals on the Belize barrier reef, *Mar. Ecol. Progr. Ser.* **13**:261–271.

Hines, A. H., and Pearse, J. S., 1982, Abalones, shells, and sea otters: Dynamics of prey populations in central California, *Ecology* **63**:1547–1560.

Hiscock, K., 1983, Water movement, in: *Sublittoral Ecology, The Ecology of the Shallow Sublittoral Benthos* (R. Earll and D. G. Erwin, eds.), Clarendon Press, Oxford, pp. 58–96.

Hiscock, K., and Mitchell, R., 1980, The description and classification of sublittoral epibenthic ecosystems, in: *The Shore Environment* (J. H. Price, D. E. G. Irvine, and W. F. Farnham, eds.), Academic Press, London, pp. 323–370.

Holland, S. M., 1988, Taphonomic effects of sea-floor exposure on an Ordovician brachiopod assemblage, *Palaios* **3**:588–597.

Holling, C. S., 1973, Resilience and stability of ecological systems, *Annu. Rev. Ecol. Syst.* **4**:1–23.

Hollmann, R., 1966, Tauchbeobachtungen zur Mollusken-Verbreitung und Sediment-Umlagerung im Golf von Neapel, *Neues Jahrb. Geol. Paläontol. Abh.* **125**:499–526.

Hollmann, R., 1968, Zur morphologie rezenter mollusken-bruchschille, *Paläontol. Z.* **42**:217–235.

Holme, N. A., 1953, The biomass of the bottom fauna in the English Channel off Plymouth, *J. Mar. Biol. Assoc. U. K.* **32**:1–49.

Holme, N. A., 1961, The bottom fauna of the English Channel, *J. Mar. Biol. Assoc. U. K.* **41**:397–461.

Holme, N. A., 1966, The bottom fauna of the English Channel, *J. Mar. Biol. Assoc. U. K.* **46**:401–496.

Hoskin, C. M., 1980, Flux of barnacle plate fragments and fecal pellets measured by sediment traps, *J. Sediment. Petrol.* **50:**1213–1218.

Hoskin, C. M., Geier, J. C., and Reed, J. K., 1983, Sediment produced from abrasion of the branching stony coral *Oculina varicosa, J. Sediment. Petrol.* **53:**779–786.

Hoskin, C. M., Reed, J. K., and Mook, D. H., 1986, Production and off-bank transport of carbonate sediment, Black Rock, southwest Little Bahama Bank, *Mar. Geol.* **73:**125–144.

Hughes, R. G., and Thomas, M. L., 1971, The classification and ordination of shallow water benthic samples from Prince Edward Island, Canada, *J. Exp. Mar. Biol. Ecol.* **7:**1–39.

Hunte, W., Cote, I., and Tomascik, T., 1986, On the dynamics of the mass mortality of *Diadema antillarum* in Barbados, *Coral Reefs* **4:**135–139.

Hunter, R. D., 1990, Shell dissolution rate in freshwater pulmonate snails: Influence of periostracum and water chemistry, *Nautilus* **104:**87–91.

Hutchings, P. A., 1986, Biological destruction of coral reefs, *Coral Reefs* **4:**239–252.

Hutchinson, G. E., 1978, *An Introduction to Population Ecology,* Yale University Press, New Haven, Connecticut.

Iribarne, O. O., 1990, Use of shelter by the small Patagonian octopus *Octopus tehuelchus:* Availability, selection, and effects on fecundity, *Mar. Ecol. Progr. Ser.* **66:**251–258.

Jackson, J. B. C., 1977, Habitat area, colonization, and development of epibenthic community structure, in: *Biology of Benthic Organisms* (B. F. Keegan, P. O'Ceidigh, and P. J. S. Boaden, eds.), Pergamon Press, Oxford, pp. 349–358.

Jackson, J. B. C., 1979, Morphological strategies of sessile animals, in: *Biology and Systematics of Colonial Organisms* (G. Larwood and B. R. Rosen, eds.), Academic Press, London, pp. 499–556.

Jackson, J. B. C., 1983, Biological determinants of present and past sessile animal distributions, in: *Biotic Interactions in Recent and Fossil Benthic Communities* (M. J. S. Tevesz and P. L. McCall, eds.), Plenum Press, New York, pp. 39–120.

Järvinen, O., Babin, C., Bambach, R. K., Flügel, E., Fürsich, F. T., Futuyma, D. J., Niklas, K. J., Panchen, A. L., Simberloff, D., Underwood, A. J., and Weidich, K. F., 1986, The neontologico-paleontological interface of community evolution: How do the pieces in the kaleidoscopic biosphere move?, in: *Patterns and Processes in the History of Life* (D. M. Raup and D. Jablonski, eds.), Springer, Berlin, pp. 331–350.

Jeffrey, D., and Aigner, T., 1982, Storm sedimentation in the Carboniferous limestones near Weston-Super-Mare (Dinantian, SW-England), in: *Cyclic and Event Stratification* (G. Einsele and A. Seilacher, eds.), Springer, Berlin, pp. 240–247.

Jenkyns, H. C., 1971, The genesis of condensed sequences in the Tethyan Jurassic, *Lethaia* **4:**327–352.

Jensen, M., and Thomsen, E., 1987, Ultrastructure, dissolution and "pyritization" of Late Quaternary and Recent echinoderms, *Bull. Geol. Soc. Den.* **36:**275–287.

Johansen, A. C., 1901, Om aflejungen af molluskerner skuller i indooer og i havet, *Vidensk. Medd. Dansk. Naturh. Foren.* **53:**5–46.

Johnson, R. G., 1957, Experiments on the burial of shells, *J. Geol.* **65:**527–535.

Johnson, R. G., 1960, Models and methods for analysis of the mode of formation of fossil assemblages, *Geol. Soc. Am. Bull.* **71:**1075–1086.

Johnson, R. G., 1962, Interspecific associations in Pennsylvanian fossil assemblages, *J. Geol.* **70:**32–55.

Johnson, R. G., 1965, Pelecypod death assemblages in Tomales Bay, California, *J. Paleontol.* **39:**80–85.

Johnson, R. G., 1972, Conceptual models of benthic marine communities, in: *Models in Paleobiology* (T. J. M. Schopf, ed.), Freeman, Cooper and Co., San Francisco, pp. 148–159.

Jones, N. F., 1950, Marine bottom communities, *Biol. Rev.* **25:**283–313.

Karlson, R., 1986, Disturbance, colonial fragmentation, and size-dependent life history variation in two coral reef cnidarians, *Mar. Ecol. Progr. Ser.* **28:**245–249.

Kauffman, E. G., and Scott, R. W., 1976, Basic concepts of community paleoecology, in: *Structure and Classification of Paleocommunities* (R. W. Scott and R. R. West, eds.), Dowden, Hutchinson and Ross, Stroudsburg, Pennsylvania, pp. 1–28.

Kelling, G., and Moshrif, M. A., 1977, The orientation of fossil bivalves in a pene-littoral sequence (the Rhaetian of South Wales), *J. Sediment. Petrol.* **47**:1342–1346.

Kelling, G., and Williams, P. F., 1967, Flume studies of the reorientation of pebbles and shells, *J. Geol.* **75**:243–267.

Kennedy, W. J., and Cobban, W. A., 1976, Aspects of ammonite biology, biogeography, and biostratigraphy, *Spec. Pap. Palaeontol.* **17**:1–94.

Kershaw, P. J., Swift, D. J., and Denoon, D. C., 1988, Evidence of recent sedimentation in the eastern Irish Sea, *Mar. Geol.* **85**:1–14.

Kidwell, S. M., 1982, Time scales of fossil accumulation: Patterns from Miocene benthic assemblages, in: *Proceedings 3rd North American Paleontological Convention, Montreal,* Volume 1, pp. 295–300.

Kidwell, S. M., 1985, Palaeobiological and sedimentological implications of fossil concentrations, *Nature* **318**:457–460.

Kidwell, S. M., 1986a, Models for fossil concentrations: Paleobiologic implications, *Paleobiology* **12**:6–24.

Kidwell, S. M., 1986b, Taphonomic feedback in Miocene assemblages: Testing the role of dead hardparts in benthic communities, *Palaios* **1**:239–255.

Kidwell, S. M., 1989, Stratigraphic condensation of marine transgressive records: Origin of major shell deposits in the Miocene of Maryland, *J. Geol.* **97**:1–24.

Kidwell, S. M., 1990, Phanerozoic evolution of macroinvertebrate shell accumulations: Preliminary data from the Jurassic of Great Britain, in: Paleocommunity Temporal Dynamics (W. M. Miller, III, ed.), *Paleontol. Soc. Spec. Publ.* **5**:309–327.

Kidwell, S. M., 1991, Taphonomic feedback (live/dead interactions) in the genesis of bioclastic deposits: Keys to reconstructing sedimentary dynamics, in: *Cycles and Events in Stratigraphy* (G. Einsele, W. Ricken, and A. Seilacher, eds.), Springer, Berlin, pp. 268–281.

Kidwell, S. M., and Aigner, T., 1985, Sedimentary dynamics of complex shell beds: Implications for ecological and evolutionary patterns, in: *Sedimentary and Evolutionary Cycles* (U. Bayer and A. Seilacher, eds.), Springer-Verlag, Berlin, pp. 382–395.

Kidwell, S. M., and Baumiller, T., 1990, Experimental disintegration of regular echinoids: Roles of temperature, oxygen and decay thresholds, *Paleobiology* **16**:247–271.

Kidwell, S. M., and Holland, S. M., 1991, Field description of coarse bioclastic fabrics, *Palaios* **6**(4):in press.

Kidwell, S. M., and Jablonski, D., 1983, Taphonomic feedback: Ecological consequences of shell accumulation, in: *Biotic Interactions in Recent and Fossil Benthic Communities* (M. J. S. Tevesz and P. L. McCall, eds.), Plenum Press, New York, pp. 195–248.

Kidwell, S. M., Fürsich, F. T., and Aigner, T., 1986, Conceptual framework for the analysis of fossil concentrations, *Palaios* **1**:228–238.

Kindle, E. M., 1916, Bottom control of marine faunas as illustrated by dredging in the Bay of Fundy, *Am. J. Sci.* **41**:449–461.

Kissling, D. L., 1973, Circumrotary growth form in Recent and Silurian corals, in: *Animal Colonies* (R. S. Boardman, A. H. Cheetham, and W. A. Oliver, Jr., eds.), Dowden, Hutchinson and Ross, Stroudsburg, Pennsylvania, pp. 43–58.

Knight, A. P., 1988, Ecological and sedimentological studies on china clay waste deposits in Mevagissey Bay, Cornwall, Ph. D. thesis, University of London.

Knight, A. P., and Bosence, D. W. J., 1988, China clay waste pollution: A giant palaeoecological experiment, in: *Abstracts, Society of Economic Paleontologists and Mineralogists Annual Midyear Meeting*, Volume V, p. 29.

Knowlton, N., Lang, J. C., Rooney, M. C., and Clifford, P., 1981, Evidence for delayed mortality in hurricane-damaged Jamaican staghorn corals, *Nature* **294**:251–252.

Kobluk, D. R., 1988, Cryptic faunas in reefs: Ecology and geological importance, *Palaios* **3**:379–390.

Kobluk, D. R., and van Soest, R. W. M., 1989, Cavity-dwelling sponges in a southern Caribbean coral reef and their paleontological implications, *Bull. Mar. Sci.* **44**:1207–1235.

Koch, C. F., and Sohl, N. F., 1983, Preservational effects in paleoecological studies: Cretaceous mollusc examples, *Paleobiology* **9**:26–34.

Kondo, Y., 1989, Bivalve orientation analysis as a method of biostratinomic interpretation of shallow sea fossiliferous facies, in: *28th International Geological Congress Abstracts*, Volume 2, Washington, D.C., pp. 2–210.

Kornicker, L. S., Wise, C. D., and Wise, J. M., 1963, Factors affecting the distribution of opposing mollusk valves, *J. Sediment. Petrol.* **33**:703–712.

Kranz, P. M., 1974a, The anastrophic burial of bivalves and its palaeoecological significance, *J. Geol.* **82**:237–265.

Kranz, P. M., 1974b, Computer simulation of fossil assemblage formation under conditions of anastrophic burial, *J. Paleontol.* **48**:800–808.

Kristensen, I., 1957, Differences in density and growth in a cockle population in the Dutch Wadden Sea, *Arch. Néer. Zool.* **12**:351–453.

Kurtén, B., 1954, Population dynamics—A new method in paleontology, *J. Paleontol.* **28**:286–292.

Lambers, P., and Boekschoten, G. J., 1986, On fossil and recent borings produced by acrothoracic cirripeds, *Geol. Mijn.* **65**:257–268.

Lasker, H., 1976, Effects of differential preservation on the measurement of taxonomic diversity, *Paleobiology* **2**:84–93.

Lavoir, M. E., 1956, How sea stars open bivalves, *Biol. Bull.* **111**:114–122.

Lawrence, D. R., 1968, Taphonomy and information losses in fossil communities, *Geol. Soc. Am. Bull.* **79**:1315–1330.

Lawrence, D. R., 1969, The use of clionid sponges in paleoenvironmental analysis, *J. Paleontol.* **43**:539–543.

Leopold, M. F., Marteijn, E. C. L., and Swennen, C., 1985, Long-distance transport of prey from the intertidal zone to high-tide roosts by the oystercatcher, *Ardea* **73**:76–82.

Lessios, H. A., 1988, Mass mortality of *Diadema antillarum* in the Caribbean: What have we learned?, *Annu. Rev. Ecol. Syst.* **19**:371–393.

Lever, J., 1958, Quantitative beach research. I. The "left-right phenomenon": Sorting of lamellibranch valves on sandy beaches, *Basteria* **22**:21–51.

Lever, J., and Thijssen, R., 1968, Sorting phenomena during the transport of shell valves on sandy beaches studied with the use of artificial valves, *Symp. Zool. Soc. Lond.* **22**:259–271.

Lever, J., Kessler, A., van Overbeeke, A. P., and Thijssen, R., 1961, Quantitative beach research. II. The "hole effect": A second mode of sorting of lamellibranch valves on sandy beaches, *Neth. J. Sea Res.* **6**:339–358.

Levinton, J. S., 1970, The paleoecological significance of opportunistic species, *Lethaia* **3**:69–78.

Levinton, J. S., and Bambach, R. K., 1970, Some ecological aspects of bivalve mortality patterns, *Am. J. Sci.* **268**:97–112.

Lewy, Z., 1975, Early diagenesis of calcareous skeletons in the Baltic Sea, western Germany, *Meyniana* **27**:29–33.

Lewy, Z., 1981, Maceration of calcareous skeletons, *Sedimentology* **28**:893–895.

Lewy, Z., and Samtleben, C., 1979, Functional morphology and palaeontological significance of the conchiolin layers in corbulid pelecypods, *Lethaia* **12**:341–351.

Liddell, W. D., and Brett, C. E., 1982, Skeletal overgrowths among epizoans from the Silurian (Wenlockian) Waldron Shale, *Paleobiology* **8**:67–78.

Liddell, W. D., and Ohlhorst, S. L., 1988, Hard substrata community patterns, 1–120 m, north Jamaica, *Palaios* **3**:413–423.

Lindberg, D. R., and Kellogg, M. G., 1982, Bathymetric anomalies in the Neogene fossil record: The role of diving marine birds, *Paleobiology* **8**:402–407.

Lingwood, P. F., 1976, The marine Mollusca of Liverpool Bay (Irish Sea), *J. Conchol.* **29**:51–56.

Livingstone, R. J., 1987, Field sampling in estuaries: The relationship of scale to variability, *Estuaries* **10**:194–207.

Lowenstam, H. A., and Weiner, S., 1989, *On Biomineralization*, Oxford University Press, New York.

Machalski, M., and Walaszczyk, I., 1987, Faunal condensation and mixing in the uppermost Maastrichtian/Danian Greensand (Middle Vistula Valley, central Poland), *Acta Geol. Polon.* **37**:75–91.

MacDonald, K. B., 1969, Quantitative studies of salt marsh mollusc faunas from the North American Pacific Coast, *Ecol. Monogr.* **39**:33–60.

MacDonald, K. B., 1976, Paleocommunities: Toward some confidence limits, in: *Structure and Classification of Paleocommunities* (R. W. Scott and R. R. West, eds.), Dowden, Hutchinson and Ross, Stroudsburg, Pennsylvania, pp. 87–106.

Mah, A. J., and Stearn, C. W., 1986, The effect of Hurricane Allen on the Bellaire fringing reef, Barbados, *Coral Reefs* **4**:169–176.

Mancini, E. A., 1978, Origin of micromorph faunas in the geologic record, *J. Paleontol.* **52**:321–333.

Mapes, R. H., Fahrer, T. R., and Babcock, L. B., 1989, Sublethal and lethal injuries of Pennsylvanian conulariids from Oklahoma, *J. Paleontol.* **63**:34–37.

Marshall, B. A., 1987, Osteopeltidae (Mollusca: Gastropoda): A new family of limpets associated with whale bone in the deep-sea, *J. Molluscan Stud.* **53**:121–127.

Martin-Kaye, P., 1951, Sorting of lamellibranch valves on beaches in Trinidad, *Geol. Mag.* **288**:432–434.

Martinell, J., and dePorta, J., 1982, Observations on the molluscan thanatocoenoses from Platja Llarga (Salou, Spain), *Malacologia* **22**:587–591.

Masse, J.-P., Thomassin, B. A., and Acquaviva, M., 1989, Bioclastic sedimentary environments of coral reefs and lagoon around Mayotte Island (Comoro Archipelago, Mozambique Channel, SW Indian Ocean), *J. Coastal Res.* **5**:419–432.

Mather, J. A., 1982, Choice and competition: Their effects on occupancy of shell homes by *Octopus joubini*, *Mar. Behav. Physiol.* **8**:285–293.

McCall, P. L., and Tevesz, M. J. S., 1983, Soft-bottom succession and the fossil record, in: *Biotic Interactions in Recent and Fossil Benthic Communities* (M. J. S. Tevesz and P. L. McCall, eds.), Plenum Press, New York, pp. 157–196.

McCave, I., 1988, Biological pumping upwards of the coarse fraction of deep-sea sediments, *J. Sediment. Petrol.* **58**:148–158.

McClintock, T. S., 1985, Effects of shell condition and size upon the shell choice behavior of a hermit crab, *J. Exp. Mar. Biol. Ecol.* **88**:271–285.

Meldahl, K. H., 1987a, Origin of shell beds and evolution of a shelly sand spit, Bahia la Choya, Northern Gulf of California, *Paleontol. Soc. Spec. Publ.* **2**:189–199.

Meldahl, K. H., 1987b, Sedimentologic and taphonomic implications of biogenic stratification, *Palaios* **2**:350–358.

Meldahl, K. H., and Flessa, K. W., 1990, Taphonomic pathways and comparative biofacies and taphofacies in a Recent intertidal/shallow shelf environment, *Lethaia* **23**:43–60.

Menard, H. W., and Boucot, A. J., 1951, Experiments on the movement of shells by water, *Am. J. Sci.* **249**:131–151.

Meyer, D. L., 1971, Post-mortem disarticulation of Recent crinoids and ophiuroids under natural conditions, *Geol. Soc. Am. Abstr. Progr.* **3**:645.

Meyer, D. L., and Meyer, K. B., 1986, Biostratinomy of Recent crinoids (Echinodermata) at Lizard Island, Great Barrier Reef, Australia, *Palaios* **1**:294–302.

Meyer, D. L., Ausich, W. I., and Terry, R. E., 1989, Comparative taphonomy of echinoderms in carbonate facies: Fort Payne Formation (Lower Mississippian) of Kentucky and Tennessee, *Palaios* **4**:533–552.

Middlemiss, F. A., 1962, Brachiopod ecology and lower Greens and palaeogeography, *Palaeontology* **5**:253–267.

Middleton, G. V., 1967, The orientation of concave-convex particles deposited from experimental turbidity currents, *J. Sediment. Petrol.* **37**:229–32.

Miller, A. I., 1988, Spatial resolution in subfossil molluscan remains: Implications for paleobiological analysis, *Paleobiology* **14**:91–103.

Miller, A. I., and Cummins, H., 1989, A numerical model for the formation of fossil assemblages: Estimating the degree of post-mortem transport along environmental gradients, *Geol. Soc. Am. Abstr. Progr.* **21**:A70.

Miller, A. I., and Cummins, H., 1990, A baseline numerical model for the formation of fossil assemblages: Estimating the amount of post-mortem transport along environmental gradients, *Palaios* **5**:303–316.

Miller, K. B., Brett, C. E., and Parsons, K. M., 1988, The paleoecologic significance of storm-generated disturbance within a Middle Devonian muddy epeiric sea, *Palaios* **3**:35–52.

Miller, A. I., Llewellyn, G., Parsons, K. M., Cummins, H., Greenstein, B. J., Boardman, M. R., Sloane, G. B., and Jacobs, D. K., 1990, The effect of Hurricane Hugo on molluscan skeletal distributions, Salt River Bay, St. Croix, U.S. Virgin Islands, *Geol. Soc. Am. Abstr.* **22**:A330.

Miller, W., III, 1983, Survey of the Pyramidellid gastropods in the Wassaw Sound area, coastal Georgia, *Nautilus* **97**:93–98.

Miller, W., III, and DuBar, J. R., 1988a, Community replacement of a Pleistocene *Crepidula* biostrome, *Lethaia* **21**:67–78.

Miller, W., III, and DuBar, J. R., 1988b, The use of size frequency distribution in paleoecology: A reply, *Lethaia* **21**:416.

Milliman, J. D., 1974, *Marine Carbonates*, Springer-Verlag, New York.

Milliman, J. D., Pilkey, O. H., and Ross, D. R., 1972, Sediments of the continental margin off the eastern United States, *Geol. Soc. Am. Bull.* **83**:1315–1334.

Mills, E. L., 1969, The community concept in marine zoology, with comments on continua and instability in some marine communities: A review, *J. Fish. Res. Board Can.* **26**:1415–1428.

Moore, D. G., and Curray, J. R., 1964, Wave-base, marine profile of equilibrium, and wave-built terraces: Discussion, *Geol. Soc. Am. Bull.* **75**:1267–1274.

Moors, H. T., 1970, Current orientation of graptolites: Its significance and interpretation, *Sediment. Geol.* **4**:117–134.

Morse, J. W., 1986, The surface chemistry of calcium carbonate minerals in natural waters: An overview, *Mar. Chem.* **20**:91–112.

Müller, A. H., 1979, Fossilization (taphonomy), in: *Treatise on Invertebrate Paleontology. Part A, Introduction* (R. A. Robison and C. Teichert, eds.), Geological Society of America, Boulder, Colorado, and University of Kansas Press, Lawrence, Kansas, pp. 2–78.

Myers, A. C., 1981, Shell hash in the gravel beaches on islands of the central Maine coast, *Geol. Soc. Am. Abstr.* **13**:167.

Nagle, J. S., 1967, Wave and current orientation of shells, *J. Sediment. Petrol.* **37**:1124–1138.

Natland, M. L., 1957, Paleoecology of west coast Tertiary sediments, *Geol. Soc. Am. Mem.* **67**(2):543–572.

Nelson, C. S., 1978, Temperate shelf carbonate sediments in the Cenozoic of New Zealand, *Sedimentology* **25**:737–771.

Nelson, C. S., 1988, An introductory perspective on non-tropical shelf carbonates, *Sediment. Geol.* **60**:3–12.

Nelson, C. S., and Hancock, G. E., 1984, Composition and origin of temperate skeletal carbonate sediments on South Maria Ridge, northern New Zealand, *N. Z. J. Mar. Freshwater Res.* **18**:221–239.

Nelson, C. S., and Young, H. R., 1985, Biodegradation of temperate-water skeletal carbonates by boring sponges on the Scott Shelf, British Columbia, Canada, *Mar. Geol.* **65**:33–45.

Nichols, F. H., and Thompson, J. K., 1985, Time scales of change in the San Francisco Bay benthos, *Hydrobiologia* **129**:121–138.

Nichols, J. A., Rowe, G. T., Clifford, C. H., and Young, R. A., 1978, *In situ* experiments on the burial of marine invertebrates, *J. Sediment. Petrol.* **48**:419–425.

Noble, J. P. A., and Logan, A., 1981, Size frequency distributions and taphonomy of brachiopods. A Recent model, *Palaeogeogr. Palaeoclimatol. Palaeoecol.* **36**:87–105.

Noe-Nygaard, N., Surlyk, F., and Piasecki, S., 1987, Bivalve mass mortality caused by toxic dinoflagellate blooms in a Berriasian-Valanginian lagoon, Bornholm, Denmark, *Palaios* **2**:263–273.

Norris, R. D., 1986, Taphonomic gradients in shelf fossil assemblages: Pliocene Purisima Formation, California, *Palaios* **1**:256–270.

Oliver, J. S., Kvitek, R. G., and Slattery, P. N., 1985, Walrus feeding disturbance: Scavenging habits and recolonization of the Bering Sea benthos, *J. Exp. Mar. Biol. Ecol.* **91**:233–246.

Olson, E. C., 1957, Size-frequency distribution of samples of extinct organisms, *J. Geol.* **65:**309–333.

Orlov, Yu. A., 1959, *Fundamentals of Paleontology,* Izdatel'stvo Akademii Nauk SSR, p. 482 [translation, Israel Program of Scientific Translations, Jerusalem (1962)].

Oschmann, W., 1988, Kimmeridge clay sedimentation—A new cyclic model, *Palaeogeogr. Palaeoclimatol. Palaeoecol.* **63:**109–140.

Paine, R. T., 1966, Food web complexity and species diversity, *Am. Nat.* **100:**65–75.

Paine, R. T., and Levin, S. A., 1981, Intertidal landscapes: Disturbance and the dynamics of pattern, *Ecol. Monogr.* **51:**145–178.

Palmer, T. J., and Fürsich, F. T., 1974, The ecology of a middle Jurassic hardground and crevice fauna, *Palaeontology* **17:**507–524.

Papp, A., 1944, Die senkrechte Einregelung von Gastropodengehäusen in Tertiärschichten des Wiener Beckens, *Palaeobiologica* **8:**144–153.

Parker, R. H., 1960, Ecology and distributional patterns of marine macroinvertebrates, northern Gulf of Mexico, in: *Recent Sediments, Northwest Gulf of Mexico* (F. P. Shepard, F. B. Phleger, and T. H. van Andel, eds.), American Association of Petroleum Geologists, Tulsa, Oklahoma, pp. 307–337.

Parker, R. H., 1963, Zoogeography and ecology of some macroinvertebrates, particularly mollusks, in the Gulf of California and the continental slope off Mexico, *Vidensk. Medd. Dansk. Naturh. Foren.* **126:**1–178.

Parsons, K. M., Brett, C. E., and Miller, K. B., 1988, Taphonomy and depositional dynamics of Devonian shell-rich mudstones, *Palaeogeogr. Palaeoclimatol. Palaeoecol.* **63:**109–140.

Pearson, T. H., and Barnett, P. R. O., 1987, Long-term changes in benthic populations in some west European coastal areas, *Estuaries* **10:**220–226.

Petersen, C. G. J., 1914, Valuation of the sea, II, The animal communities of the sea bottom and their importance for marine zoogeography, *Rep. Dan. Biol. Sta.* **21:**1–44.

Peterson, C. H., 1975, Stability of species and of community for the benthos of two lagoons, *Ecology* **56:**958–965.

Peterson, C. H., 1976, Relative abundance of living and dead molluscs in two California lagoons, *Lethaia* **9:**137–148.

Peterson, C. H., 1977, The paleontological significance of undetected short-term temporal variability, *J. Paleontol.* **51:**976–981.

Peterson, C. H., 1983, Interactions between two infaunal bivalves, *Chione undatella* (Sowerby) and *Protothaca staminea* (Conrad), and two potential enemies, *Crepidula onyx* Sowerby and *Cancer anthonyi* (Rathburn), *J. Exp. Mar. Biol. Ecol.* **68:**145–158.

Peterson, C. H., 1985, Patterns of lagoonal bivalve mortality after heavy sedimentation and their paleoecological significance, *Paleobiology* **11:**139–155.

Philippon, J., and Plaziat, J.-C., 1975, Roles respectifs de la corrosion et des Cryptogames perforantes dans la destruction des coquilles de Mollusques des mangroves. Conséquences sur la fossilisation, *C. R. Acad. Sci. Paris* **281:**617–620.

Pickrill, R. A., 1985, Sedimentation in an ebb-tidal delta, Ranganu Harbor, Northland, New Zealand, *N. Z. J. Geol. Geophys.* **28:**531–542.

Pilkey, O. H., and Curran, H. A., 1986, Molluscan shell transport—You ain't seen nothin' yet, *Palaios* **1:**197.

Pilkey, O. H., Martin, R. W., and Luternauer, J., 1967, The carbonate fraction of beach and dune sands, *Sedimentology* **8:**311–327.

Pilkey, O. H., Fierman, E. I., and Trumbull, J. V. A., 1979, Relationship between physical condition of the carbonate fraction and sediment environments: Northern Puerto Rico shelf, *Sed. Geol.* **24:**283–290.

Pip, E., 1988, Differential attrition of molluscan shells in freshwater sediments, *Can. J. Earth Sci.* **25:**68–73.

Pisera, A., 1987, Boring and nesting organisms from Upper Jurassic coral colonies from northern Poland, *Acta Palaeontol. Polon.* **32:**83–104.

Plotnick, R. E., 1986, Taphonomy of a modern shrimp: Implications for the arthropod fossil record, *Palaios* **1:**286–293.

Poag, C. W., 1969, Dissolution of molluscan calcite by the attached foraminifer *Vasiglobulina*, new genus (Vasiglobulinae, new subfamily), *Tulane Stud. Geol. Palentol.* **7**:45–72.

Ponder, W. F., 1983, A revision of the Recent Xenophoridae of the world and of the Australian fossil species (Mollusca: Gastropoda), *Mem. Aust. Mus.* **17**:1–126.

Powell, E. N., and Davies, D., 1989, When is an "old" shell really old? *Donax* on Texas beaches, *Geol. Soc. Am. Abstr. Progr.* **21**:A71.

Powell, E. N., Stanton, R. J., Jr., Cummins, H., and Staff, G., 1982, Temporal fluctuations in bay environments—The death assemblage as a key to the past, in: *Proceedings Symposium Recent Benthological Investigations in Texas and Adjacent States,* Academy of Science, Austin, Texas.

Powell, E. N., Cummins, H., Stanton, R. J., and Staff, G., 1984, Estimation of the size of molluscan larval settlement using the death assemblage, *Estuarine Coastal Shelf Sci.* **18**:367–384.

Powell, E. N., Logan, A., Stanton, R. J., Jr., Davies, D. J., and Hare, P. E., 1989, Estimating time-since-death from the free amino acid content of the mollusc shell: A measure of time averaging in modern death assemblages? Description of the technique, *Palaios* **4**:16–31.

Pratje, O., 1929, Fazettieren von Molluskenschalen, *Palaeontol. Z.* **11**:151–169.

Price, A. R. G., 1982, Western Arabian Gulf echinoderms in high salinity waters and the occurrence of dwarfism, *J. Nat. Hist.* **16**:519–527.

Price, J. H., Irvine, D. E. G., and Farnham, W. F. (eds.), 1980, *The Shore Environment,* Volume 2: *Ecosystems* (Systematics Association Special Volume 17), Academic Press, London.

Prince, C. M., Elmore, R. D., Ehrlich, R., and Pilkey, O. H., 1987, Areal and lateral changes in a major trailing margin turbidite—The Black Shell Turbidite, *Geo-Marine Lett.* **7**:103–112.

Probert, P. K., 1973, The bottom fauna of china clay waste deposits in Mevagissy Bay, Thesis, University of London.

Probert, P. K., 1984, Disturbance, sediment stability and trophic structure of soft bottom communities, *J. Mar. Res.* **42**:893–921.

Rasmussen, K. A., and Brett, C. E., 1985, Taphonomy of Holocene cryptic biotas from St. Croix, Virgin Islands: Information loss and preservation biases, *Geology* **13**:551–553.

Raup, D. M., 1973, Depth inferences from vertically imbedded cephalopods, *Lethaia* **6**:217–226.

Reaves, C. M., 1986, Organic matter metabolizability and calcium carbonate dissolution in nearshore marine muds, *J. Sediment. Petrol.* **56**:486–494.

Rees, E. I. S., Nicholaidou, A., and Laskaridou, P., 1977, The effects of storms on the dynamics of shallow water benthic associations, in: *Biology of Benthic Organisms. Proceedings of the 11th European Symposium on Marine Biology* (B. F. Keegan, P. O. Ceidigh, and P. J. S. Boaden, eds.), Pergamon Press, Oxford, pp. 465–474.

Reichelt, R. E., Green, D. G., and Bradbury, R. H., 1985, Discrete simulation of cyclone effects on the spatial patterns and community structure of a coral reef, in: *Proceedings 5th International Coral Reef Congress, Tahiti,* Volume 3, pp. 337–342.

Reyment, R. A., 1980, Floating orientations of cephalopod shell models, *Palaeontology* **24**:931–936.

Reyment, R. A., 1986, Nekroplanktonic dispersal of echinoid tests, *Palaeogeogr. Palaeoclimatol. Palaeoecol.* **52**:347–349.

Rhoads, D. C., and Stanley, D. J., 1965, Biogenic graded bedding, *J. Sediment. Petrol.* **35**:956–963.

Richards, R. P., and Bambach, R. K., 1975, Population dynamics of some Paleozoic brachiopods and their paleoecological significance, *J. Paleontol.* **49**:775–798.

Richter, R., 1942, Der Einkippungsregel, *Senckenbergiana* **25**:181–206.

Rosenberg, R., 1977, Benthic macrofaunal dynamics, production and dispersion in an oxygen-deficient estuary of west Sweden, *J. Exp. Mar. Biol. Ecol.* **26**:107–133.

Rowland, S. M., and Gangloff, R. A., 1988, Structure and paleoecology of Lower Cambrian reefs, *Palaios* **3**:111–135.

Ruffell, A. H., 1990, The mineralogy and petrology of the Sulphur Band phosphates (Aptian-Albian), at Folkestone, Kent, *Proc. Geol. Assoc.* **101**:79–84.

Ruhrmann, G., 1971, Riff-nahe Sedimentation paläzoischer Krinoiden-Fragmente, *Neues Jahrb. Geol. Paläontol. Abh.* **138**:56–100.
Sadler, P. M., 1981, Sediment accumulation rates and the completeness of stratigraphic sections, *J. Geol.* **89**:569–584.
Saelen, G., 1989, Diagenesis and construction of the belemnite rostrum, *Palaeontology* **32**:765–798.
Salazar-Jimenez, A., Frey, R. W., and Howard, J. D., 1982, Concavity orientations of bivalve shells in estuarine and nearshore shelf sediments, Georgia, *J. Sediment. Petrol.* **52**:565–586.
Sammarco, P. W., and Risk, M. J., 1990, Large-scale patterns in internal bioerosion of *Porites:* Cross continental shelf trends on the Great Barrier Reef, *Mar. Ecol. Progr. Ser.* **59**:145–156.
Samtleben, C., 1981, Die Muschelfauna der Schlei (Westliche Ostsee)—aktuopaläontologische Untersuchungen an einem sterbenden Gewässer, *Meyniana* **33**:6–183.
Sanders, H. L., 1956, Oceanography of Long Island Sound, 1952–1954. X. The biology of marine bottom communities, *Bull. Bingham Oceanogr. Coll.* **15**:345–414.
Sanderson, D. J., and Donovan, R. N., 1974, The vertical packing of shells and stones on some Recent beaches, *J. Sediment. Petrol.* **44**:680–688.
Sando, W. J., 1984, Significance of epibionts on horn corals from the Chainman Shale (Upper Mississippian) of Utah, *J. Paleontol.* **58**:185–196.
Sarnthein, M., 1971, Oberflächensedimente im Persischen Golf und Golf von Oman, II, Quantitative Komponentenanalyse der Grossfraktion, *Meteor Forschungsergeb. Reihe C* **5**:1–113.
Saunders, W. B., and Spinosa, C., 1979, *Nautilus* movement and distribution in Palau, Western Caroline Islands, *Science* **204**:1199–1201.
Schäfer, W., 1972, *Ecology and Palaeoecology of Marine Environments.* University of Chicago Press, Chicago.
Schiffelbein, D. E., 1985, Extracting the benthic mixing impulse response function: A constrained deconvolution technique, *Mar. Geol.* **65**:313–336.
Schindel, D. E., 1980, Microstratigraphic sampling and the limits of paleontological resolution, *Paleobiology* **6**:408–426.
Schindel, D. E., 1982, Resolution analysis: A new approach to the gaps in the fossil record, *Paleobiology* **8**:340–353.
Schopf, T. J. M., 1978, Fossilization potential of an intertidal fauna: Friday Harbor, Washington, *Paleobiology* **4**:261–270.
Schott, J., Brantley, S., Crerar, D., Guy, C., Borcsik, M., and Willaime, C., 1989, Dissolution kinetics of strained calcite, *Geochim. Cosmochim. Acta* **53**:373–382.
Schroeder, J., and Zankl, H., 1974, Dynamic reef formation: A sedimentological concept based on studies of Recent Bermuda and Bahama reefs, in: *Proceedings 2nd International Coral Reef Symposium, Brisbane,* Volume 3, pp. 413–428.
Scoffin, T. P., 1987, *An Introduction to Carbonate Sediments and Rocks,* Chapman and Hall, New York.
Scoffin, T. P., and Garrett, P., 1974, Processes in the formation and preservation of internal structure in Bermuda patch reefs, in: *Proceedings 2nd International Coral Reef Symposium, Brisbane,* Volume 2, pp. 429–448.
Scoffin, T. P., and Henry, M. D., 1984, Shallow-water sclerosponges on Jamaican reefs and a criterion for recognition of hurricane deposits, *Nature* **307**:728–729.
Scoffin, T. P., Alexandersson, E. T., Bowes, G. E., Clokie, J. J., Farrow, G. E., and Milliman, J. D., 1980, Recent, temperate, sub-photic, carbonate accumulation: Rockall Bank, northeast Atlantic, *J. Sediment. Petrol.* **50**:331–356.
Scott, P. J. B., 1988, Distribution, habitat and morphology of the Caribbean coral- and rock-boring bivalve, *Lithophaga bisulcata* (d'Orbigny), *J. Molluscan Stud.* **54**:83–95.
Scott, P. J. B., and Risk, M. J., 1988, The effect of *Lithophaga* (Bivalvia: Mytilidae) boreholes on the strength of the coral *Porites lobata, Coral Reefs* **7**:145–151.
Scott, R. W., 1970, Paleoecology and paleontology of the Lower Cretaceous Kiowa Formation, Kansas, *Univ. Kansas Paleontol. Contrib. Art.* **52**(*Cretaceous* 1):1–94.

Scott, R. W., 1978, Approaches to trophic analysis of paleocommunities, *Lethaia* **11**:1–14.

Seibold, E., Diester, L., Fütterer, D., Lange, H., Müller, P., and Werner, F., 1973, Holocene sediments and sedimentary processes in the Iranian part of the Persian Gulf, in: *The Persian Gulf* (B. H. Purser, ed.), Springer-Verlag, Berlin, pp. 57–80.

Seilacher, A., 1963, Umlagerung und Rolltransport von Cephalopoden-Gehäusen, *Neues Jahrb. Geol. Paläontol. Monatsh.* **1963**:593–615.

Seilacher, A., 1969, Paleoecology of boring barnacles, *Am. Zool.* **9**:705–719.

Seilacher, A., 1973, Biostratinomy: The sedimentology of biologically standardized particles, in: *Evolving Concepts in Sedimentology* (R. N. Ginsburg, ed.), Johns Hopkins University Press, Baltimore, pp. 159–177.

Seilacher, A., 1982a, General remarks about event deposits, in: *Cyclic and Event Stratification* (G. Einsele and A. Seilacher, eds.), Springer, Berlin, pp. 161–174.

Seilacher, A., 1982b, Posidonia shales (Toarcian, S. Germany)—Stagnant basin model revalidated, in: *Paleontology, Essential of Historical Geology* (E. M. Gallitelli, ed.), STEM Mucchi, Modena, Italy, pp. 25–55.

Seilacher, A., 1984, Constructional morphology of bivalves: Evolutionary pathways in primary versus secondary soft-bottom dwellers, *Palaeontology* **27**:207–237.

Seilacher, A., 1985, The Jeram model: Event condensation in a modern intertidal environment, in: *Sedimentary and Evolutionary Cycles* (U. Bayer and A. Seilacher, eds.), Springer-Verlag, Berlin, pp. 336–346.

Seilacher, A., and Meischner, D., 1965, Fazies-Analyse im Paläozoikum des Oslo-Gebietes, *Geol. Rundsch.* **54**:596–619.

Seilacher, A., and Westphal, F., 1971, Fossil-Lagerstätten, in: *Sedimentology of Parts of Central Europe, Guidebook 8th International Sedimentology Congress* (G. Müller, ed.), W. Kramer, Frankfurt-am-Main, pp. 327–335.

Seilacher, A., Reif, W.-E., and Westphal, F., 1985, Sedimentological, ecological and temporal patterns of fossil Lagerstätten, *Phil. Trans. R. Soc. Lond. B* **311**:5–23.

Sheehan, P. M., 1978, The hinging mechanisms of brachiopods—Taphonomic considerations, *J. Paleontol.* **52**:748.

Shillabeer, N., and Tapp, J. F., 1990, Long-term studies of the benthic biology of the Tees bay and the Tees estuary, *Hydrobiologia* **195**:63–78.

Shimoyama, S., 1985, Size-frequency distribution of living populations and dead shell assemblages in a marine intertidal sand snail, *Umbonium (Suchium) moniliferum* (Lamarck), and their palaeoecological significance, *Palaeogeogr. Palaeoclimatol. Palaeoecol.* **49**:327–353.

Shuto, T., 1973, Molluscan thanatocoenoses in Usuki Bay, East Kyushu, *Tohoku Univ. Sci. Rep. 2nd Ser. Spec. Vol.* **6**:15–27.

Smith, A. B., 1984, *Echinoid Paleobiology*, Allen and Unwin, London.

Smith, J. E., 1932, Shell gravel deposits, and the infauna of the Eddystone Grounds, *J. Mar. Biol. Assoc. U. K.* **18**:243–278.

Smyth, M. J., 1988, The foraminifer *Cymbaloporella tabellaeformis* (Brady) bores into gastropod shells, *J. Foram. Res.* **18**:277–285.

Smyth, M. J., 1989, Bioerosion of gastropod shells: With emphasis on effects of coralline algal cover and shell microstructure, *Coral Reefs* **8**:119–125.

Smyth, M. J., 1990, Incidence of boring organisms in gastropod shells on reefs around Guam, *Bull. Mar. Sci.* **46**:432–449.

Sokal, R. R., and Rohlf, F. J., 1981, *Biometry*, 2nd ed., W. H. Freeman, San Francisco.

Sorby, H. C., 1879, Anniversary address of the president: The structure and origin of limestones, *Proc. Geol. Soc. Lond.* **35**:56–95.

Southward, A. J., 1962, The distribution of some plankton animals in the English Channel and its approaches II. Surveys with the Gulf III high speed sampler, 1958–60, *J. Mar. Biol. Assoc. U. K.* **42**:275–375.

Southward, A. J., 1980, The western English Channel—An inconsistent system?, *Nature* **285**:361–366.

Speyer, S. E., and Brett, C. E., 1985, Clustered trilobite assemblages in the Middle Devonian Hamilton Group, *Lethaia* **18**:85–103.

Speyer, S. E., and Brett, C. E., 1986, Trilobite taphonomy and Middle Devonian taphofacies, *Palaios* **1**:312–327.

Speyer, S. E., and Brett, C. E., 1988, Taphofacies models for epeiric sea environments: Middle Paleozoic examples, *Palaeogeogr. Palaeoclimatol. Palaeoecol.* **63**:225–262.

Spjeldnaes, N., 1978, Ecology of selected late and post-glacial marine faunas in the Oslo Fjord area, *Geol. Fören. Stockh. Förh.* **100**:189–202.

Stachowitsch, M., 1980, The epibiotic and endolithic species associated with the gastropod shells inhabited by the hermit crabs *Paguristes oculatus* and *Pagurus cuanensis*, *Mar. Ecol. (Proc. Staz. Zool. Napoli)* **1**:73–101.

Staff, G. M., and Powell, E. N., 1988, The paleoecological significance of diversity: The effect of time averaging and differential preservation on macroinvertebrate species richness in death assemblages, *Palaeogeogr. Palaeoclimatol. Palaeoecol.* **63**:73–89.

Staff, G. M., and Powell, E. N., 1990, Taphonomic signature and the imprint of taphonomic history: Discriminating between taphofacies of the inner continental shelf and a microtidal inlet, in: *Paleocommunity Temporal Dynamics* (W. Miller, ed.), Paleont. Soc. N. Amer. Spec. Pub., pp. 370–390.

Staff, G. M., Powell, E. N., Stanton, R. J., and Cummins, H., 1985, Biomass: Is it a useful tool in paleocommunity reconstruction?, *Lethaia* **18**:209–232.

Staff, G. M., Stanton, R. J., Jr., Powell, E. N., and Cummins, H., 1986, Time averaging, taphonomy and their impact on paleocommunity reconstruction: Death assemblages in Texas bays, *Geol. Soc. Am. Bull.* **97**:428–443.

Stanley, S. M., 1970, Relation of shell form to life habits in the Bivalvia (Mollusca), *Geol. Soc. Am. Mem.* **125**:1–296.

Stanton, R. J., Jr., 1976, Relationship of fossil communities to original communities of living organisms, in: *Structure and Classification of Paleocommunities* (R. W. Scott and R. R. West, eds.), Dowden, Hutchinson and Ross, Stroudsburg, Pennsylvania, pp. 107–142.

Stanton, R. J., and Dodd, J. R., 1976, The application of trophic structure of fossil communities, *Lethaia* **9**:327–342.

Stebbins, T. D., 1987, Mass mortality of the bubble snail *Bulla gouldiana* Pilsbry, 1893 (Gastropoda: Opisthobranchia), *Veliger* **30**:95.

Steimle, F. W., and Sinderman, C. J., 1978, Review of oxygen depletion and associated mass mortality in the Middle Atlantic Bight in 1976, *Mar. Fish. Rev.* **40**:17–26.

Steneck, R. S., 1983, Escalating herbivory and resulting adaptive trends in calcareous algae, *Paleobiology* **9**:45–63.

Steneck, R. S., and Paine, R. T., 1986, Ecological and taxonomic studies of shallow water encrusting Corallinaceae (Rhodophyta) of the boreal northeastern Pacific, *Phycologia* **25**:221–240.

Stephenson, W., Williams, W. T., and Cook, S. G., 1971, Computer analysis of Petersen's original data on bottom communities, *Ecol. Monogr.* **42**:387–415.

Sturrock, S. J., 1982, Foraminifera and carbonate sediments in a temperate water high energy environment, Thesis, University of Exeter.

Surlyk, F., 1972, Morphological adaptations and population structures of the Danish Chalk brachiopods (Maastrichtian), upper Cretaceous, *K. Dan. Vidensk. Selsk. Biol. Skr.* **19**:1–57.

Swift, D. J. P., 1970, Quaternary shelves and the return to grade, *Mar. Geol.* **8**:5–30.

Swift, D. J. P., 1976, Coastal sedimentation, in: *Marine Sediment Transport and Environmental Management* (D. J. Stanley and D. J. P. Swift, eds.), Wiley, New York, pp. 255–310.

Taft, W. H., and Harbaugh, J. W., 1964, Modern carbonate sediments of south Florida, Bahamas and Espirito Santo Island, Baja California: A comparison of their mineralogy and chemistry, *Stanford Univ. Publ. Geol. Sci.* **8**:1–133.

Tanabe, K., and Arimura, E., 1987, Ecology of four infaunal bivalve species in the Recent intertidal zone, Shikoku, Japan, *Palaeogeogr. Palaeoclimatol. Palaeoecol.* **60**:219–230.

Tanabe, K., Fujiki, T., and Katsuta, T., 1986, Comparative analysis of living and death bivalve assemblages on the Kawarazu Shore, Ehime Prefecture, west Japan, *Bull. Jpn. Assoc. Benthol.* **30**:17–30.

Tasch, P., 1955a, The concept of psephonecrocoenosis applied to "dwarf" fossil faunas, *Am. Midl. Nat.* **53**:205–212.

Tasch, P., 1955b, Paleoecologic observations on the orthoceratid coquina beds of the Maquoketa at Graf, Iowa, *J. Paleontol.* **29:**510–518.

Taylor, J. D., and Layman, M., 1972, The mechanical properties of bivalve (Mollusca) shell structures, *Palaeontology* **15:**73–87.

Taylor, P. D., 1984, Adaptations for spatial competition in Silurian encrusting bryozoans, in: Autecology of Silurian Organisms (M. J. Bassett and J. D. Lawson, eds.), *Spec. Pap. Palaeontol.* **32:**197–210.

Taylor, P. D., Schembri, P. J., and Cook, P. L., 1989, Symbiotic associations between hermit crabs and bryozoans from the Otago region, southeastern New Zealand, *J. Nat. Hist.* **23:**1059–1085.

Teichert, C., and Serventy, D. L., 1947, Deposits of shells transported by birds, *Am. J. Sci.* **245:**322–328.

Thorson, G., 1957, Bottom communities (sublittoral or shallow shelf), in: Treatise on marine ecology and paleoecology (J. W. Hedgpeth, ed.), *Geol. Soc. Am. Mem.* **67:**461–534.

Toots, H., 1965, Random orientation of fossils and its significance, *Wyoming Univ. Contrib. Geol.* **4:**59–62.

Trewin, N. H., and Welsh, W., 1976, Formation and composition of a graded estuarine shell bed, *Palaeogeogr. Palaeoclimatol. Palaeoecol.* **19:**219–230.

Tsuchiya, M., 1983, Mass mortality in a population of the mussel *Mytilus edulis* L. caused by high temperature on rocky shores, *J. Exp. Mar. Biol. Ecol.* **66:**101–111.

Tunnicliffe, V., 1979, The role of boring sponges in coral fracture, *Colloq. Int. CNRS* **291:**309–315.

Turner, H. M., 1985, Parasites of eastern oysters from subtidal reefs in a Louisiana estuary with a note on their use as indicators of water quality, *Estuaries* **8:**323–325.

Urlichs, M., 1975, Über einen Kondensationshorizont im Pliensbachium (Lias) von Franken, *Geol. Blätter NO Bayern* **25:**29–38.

Valentine, J. W., 1989, How good was the fossil record? Clues from the Californian Pleistocene, *Paleobiology* **15:**83–94.

Valentine, J. W., and Mallory, B., 1965, Recurrent groups of bonded species in mixed death assemblages, *J. Geol.* **73:**683–701.

Van Straaten, L. M. J. U., 1952, Biogene textures and the formation of shell beds in the Dutch Wadden Sea, *Proc. K. Ned. Akad. Wet. Amst. B* **55:**500–516.

Van Straaten, L. M. J. U., 1956, Composition of shell beds formed in tidal flat environments in the Netherlands and in the bay of Arachon (France), *Geol. Mijn.* **18:**209–226.

Van Straaten, L. M. J. U., 1960, Marine mollusc shell assemblages of the Rhine Delta, *Geol. Mijn.* **39:**105–129.

Van Straaten, L. M. J. U., 1967, Turbidite, ash layers and shell beds in the bathyal zone of the southeastern Adriatic Sea, *Rev. Géogr. Phys. Géol. Dynam. Sér. 2* **9**(3):219–240.

Van Tassell, J., 1981, Silver Abyssal Plain carbonate turbidite: Flow characteristics, *J. Geol.* **89:**317–333.

Velbel, M. A., and Brandt, D. S., 1989, Differential preservation of brachiopod valves: Taphonomic bias in *Platystrophia ponderosa, Palaios* **4:**193–195.

Vevers, H. G., 1952, A photographic survey of certain areas of the sea floor near Plymouth, *J. Mar. Biol. Assoc. U. K.* **31:**215–222.

Vogel, K., Golubic, S., and Brett, C. E., 1987, Endolith associations and their relation to facies distribution in the Middle Devonian of New York State, U.S.A., *Lethaia* **20:**263–290.

Vogel, S., 1988, *Life's Devices,* Princeton University Press, Princeton, New Jersey.

Von Esternhagen, H., and Dethlefsen, V., 1983, North Sea oxygen deficiency and its effects on bottom faunas, *Ambio* **12:**264–266.

Waage, K. M., 1964, Origin of repeated fossiliferous concretion layers in the Fox Hills Formation, *Kansas Geol. Surv. Bull.* **169:**541–563.

Wagner, F. J. E., 1975, Mollusca of the Strait of Canso area, *Geol. Surv. Can. Pap.* **75-23:**1–22.

Walker, B. M., 1979, Shell dissolution: Destructive diagenesis in a meteoric environment, *Scanning Electron Microscopy/1979* **2:**463–468.

Walker, K. R., and Bambach, R. K., 1971, The significance of fossil assemblages from fine grained sediments: Time-averaged communities, *Geol. Soc. Am. Abstr. Progr.* **3:**783–784.

Walker, K. R., and Diehl, W. W., 1985, The role of cementation in the preservation of Lower Palaeozoic assemblages, *Phil. Trans. R. Soc. Lond. B* **311**:143–153.

Walker, K. R., and Parker, W. C., 1976, Population structure of a pioneer and a later stage species in an Ordovician ecological succession, *Paleobiology* **2**:191–201.

Walker, S. E., 1988, Taphonomic significance of hermit crabs (Anomura: Pagurida): Epifaunal hermit crab-infaunal gastropod example, *Palaeogeogr. Palaeoclimatol. Palaeoecol.* **63**:45–71.

Walker, S. E., 1989, Hermit crabs as taphonomic agents, *Palaios* **4**:439–452.

Walker, S. E., Kidwell, S. M., and Powell, E. N., 1989, Molluscan death assemblages on the Texas continental shelf are potentially local, not exotic, accumulations, *Geol. Soc. Am. Abstr. Progr.* **21**:A71.

Walter, L. M., 1985, Relative reactivity of skeletal carbonates during dissolution: Implications for diagenesis, *Soc. Econ. Paleontol. Mineral. Spec. Publ.* **36**:3–16.

Walter, L. M., and Burton, E., 1990, Dissolution of Recent platform carbonate sediments in marine pore fluids, *Am. J. Sci.* **290**:601–643.

Walter, L. M., and Hanor, J. S., 1979, Effect of orthophosphate on the dissolution kinetics of biogenic magnesian calcites, *Geochim. Cosmochim. Acta* **43**:1377–1385.

Walter, L. M., and Morse, J. W., 1984, Reactive surface area of skeletal carbonates during dissolution: Effect of grain size, *J. Sediment. Petrol.* **54**:1081–1090.

Walter, L. M., and Morse, J. W., 1985, The dissolution kinetics of shallow marine carbonates in seawater: A laboratory study, *Geochim. Cosmochim. Acta* **49**:1503–1513.

Walther, J., 1893–1894, *Einleitung in die Geologie als historische Wissenschaft,* Fischer, Jena.

Wanless, H., Tedesco, L., and Tyrrell, K., 1987, Production of subtidal tubular and surficial tempestites by Hurricane Kate, Caicos Platform British West Indies, *J. Sediment. Petrol.* **58**:739–750.

Ward, M. A., and Thorpe, J. P., 1989, Assessment of space utilisation in a subtidal temperate bryozoan community, *Mar. Biol.* **103**:215–224.

Ware, S., 1975, British Lower Greensand Serpulidae, *Palaeontology* **18**:93–116.

Warme, J. E., 1969, Live and dead molluscs in a coastal lagoon, *J. Paleontol.* **43**:141–150.

Warme, J. E., 1971, Paleoecological aspects of a modern coastal lagoon, *Univ. Calif. Publ. Geol. Sci.* **87**:1–110.

Warme, J. E., Ekdale, A. A., Ekdale, S. F., and Peterson, C. H., 1976, Raw material of the fossil record, in: *Structure and Classification of Paleocommunities* (R. W. Scott, and R. R. West, eds.), Dowden, Hutchinson and Ross, Stroudsburg, Pennsylvania, pp. 143–169.

Warner, G. F., 1971, On the ecology of a dense bed of the brittle star *Ophiothrix fragilis, J. Mar. Biol. Assoc. U. K.* **51**:267–282.

Wasmund, E., 1926, Biocoenose und Thanatocoenose, *Arch. Hydrobiol.* **17**:1–116.

Watson, M. K., and Flessa, K. W., 1979, Time-averaged molluscan assemblages from Cholla Bay, Gulf of California, *Geol. Soc. Am. Abstr. Progr.* **11**:134.

Watson, R. L., 1971, Origin of shell beaches, Padre Island, Texas, *J. Sediment. Petrol.* **41**:1105–1111.

Weigelt, J., 1927, *Rezente Wirbeltierleichen und ihre paläobiologische Bedeutung,* Wax Weg, Leipzig [English translation: *Recent Vertebrate Carcasses and their Paleobiological Implications,* University of Chicago Press, Chicago (1989)].

Wendt, J., 1970, Fossil-Lagerstätten, Nr. 9: Stratigraphische Kondensation in triadischen und jurassischen Cephalopodenkalken der Tethys, *Neues Jahrb. Geol. Paläontol. Abh.* **135**:171–189.

Wendt, J., 1988, Condensed carbonate sedimentation in the late Devonian of the eastern Anti-Atlas (Morocco), *Eclogae Geol. Helv.* **81**:155–173.

Westrop, S. R., 1986, Taphonomic versus ecologic controls on taxonomic abundance patterns in tempestites, *Lethaia* **19**:123–132.

White, C. A., 1879, Conditions of preservation of invertebrate fossils, *U.S. Geol. Surv. Territorial Bull.* **5**:133–141.

Whittaker, R. H., 1975, *Communities and Ecosystems,* 2nd ed. Macmillan, New York.

Whittington, H. B., and Conway Morris, S. (eds.), 1985, Extraordinary fossil biotas: Their ecological and evolutionary significance, *Phil. Trans. R. Soc. Lond. B* **311**:1–192.

Wiedemann, H. V., 1972, Shell deposits and shell preservation in Quaternary and Tertiary estuarine sediments in Georgia, U.S.A., *Sediment. Geol.* **7**:103–125.
Wignall, P. B., and Myers, K. J., 1988, Interpreting benthic oxygen levels in mudrocks: A new approach, *Geology* **16**:452–455.
Wilson, J. B., 1965, The palaeoecological significance of infauna and their associated sediments, Ph. D. thesis, University of Edinburgh.
Wilson, J. B., 1967, Palaeoecological studies on shellbeds and associated sediments in the Solway Firth, *Scot. J. Geol.* **3**:329–371.
Wilson, J. B., 1982, Shelly faunas associated with temperate offshore tidal deposits, in: *Offshore Tidal Sands* (A. H. Stride, ed.), Chapman and Hall, London, pp. 126–171.
Wilson, J. B., 1986, Faunas of tidal current and wave-dominated continental shelves and their use in the recognition of storm deposits, *Can. Soc. Petroleum Geol. Mem.* **11**:313–326.
Wilson, J. B., 1988, A model for temporal changes in the faunal composition of shell gravels during a transgression on the continental shelf around the British Isles, *Sediment. Geol.* **60**:95–105.
Wilson, M. A., 1985, Disturbance and ecologic succession in an Upper Ordovician cobble-dwelling hardground fauna, *Science* **228**:575–577.
Wilson, M. A., 1987, Ecological dynamics on pebbles, cobbles and boulders, *Palaios* **2**:594–599.
Wilson, M. V. H., 1988, Taphonomic processes: Information loss and information gain, *Geosci. Can.* **15**:131–148.
Wright, T., and Kornicker, L. S., 1962, Island transport of marine shells by birds on Perez Island Alacran Reef, Campeche Bank, Mexico, *J. Geol.* **70**:616–618.
Wulff, J. I., 1990, Biostratinomic utility of *Archimedes* in environmental interpretation, *Palaios* **5**:160–166.
Yanin, B. T., 1983, *Osnovy Taphonomii*, Nedra, Moscow.
Yochelson, E. L., and Fraser, G. D., 1973, Interpretation of depositional environment of the Plympton Formation (Permian), southern Pequop Mountains, Nevada, from physical stratigraphy and a faunule, *U. S. Geol. Surv. J. Res.* **1**:19–32.
Yochelson, E. L., and Lindemann, R. H., 1986, Considerations on systematic placement of the styliolines (Incertae sedis: Devonian), in: *Problematic Fossil Taxa* (A. Hoffman and M. H. Nitecki, eds.), Oxford University Press, Oxford, pp. 45–58.
Young, H. R., and Nelson, C. S., 1988, Endolithic biodegradation of cool-water skeletal carbonates on Scott shelf, northwestern Vancouver Island, Canada, *Sediment. Geol.* **60**:251–267.
Zenetos, A., 1990, Discrimination of autochthonous *vs.* allochthonous assemblages in the Eden Estuary, Fife, Scotland, U.K., *Estuarine Coastal Shelf Sci.* **30**:525–540.
Ziegelmeier, E., 1963, Das Makrobenthos im Ostteil der Deutschen Bucht nach qualitativen und quantitativen Bodengreifer Untersuchungen in der Zeit von 1949–1960, *Veröff. Inst. Meeresforsch. Bremerhaven Sonderbd.* **1**:101–114.
Ziegelmeier, E., 1964, Einwirkungen des kalten Winters 1962/63 auf das Makrobenthos im Ostteil der Deutschen Bucht, *Helgol. Wiss. Meeresunters.* **10**:276–282.

Chapter 5

The Stratigraphy of Shell Concentrations

SUSAN M. KIDWELL

SUSAN M. KIDWELL • Department of Geophysical Sciences, University of Chicago, Chicago, Illinois 60637.

Taphonomy: Releasing the Data Locked in the Fossil Record, Volume 9 of *Topics in Geobiology*, edited by Peter A. Allison and Derek E. G. Briggs, Plenum Press, New York, 1991.

1. Introduction

Notwithstanding the array of taphonomic agents poised to destroy skeletal hard parts, and the large proportion that probably succumb to this fate, shelly fossils pervade the stratigraphic record and are sufficiently abundant locally to form dense concentrations. Such "shell beds" are referred to by many terms—e.g., coquinas, lumachelles, shell gravels, and bioclastic calcirudites—and are familiar to paleontologists and sedimentologists. They have either a carbonate or terrigenous matrix, range from thin bedding plane pavements to laterally extensive bioclastic facies, and record a variety of shell-concentrating processes, ranging from the instantaneous (sudden influx of carcasses from mass mortality) to the geologically prolonged (low sedimentary dilution of death assemblages during marine transgression). This variety among coquinas—in macroscopic features, agents, time scales, and environments of accumulation—is reflected in the diverse approaches used in classifying them (Table I).

Understanding the origin of local peaks in shell abundance is important both to the paleontologist who preferentially collects from these richest parts of the record, and to the geologist who exploits coquinas as marker beds and petroleum reservoirs. To what extent do the agent and time scale of concentration alter a fossil assemblage? To what extent is stratigraphic variation in fossil abundance a chronicle of ancient biological productivity and behavior, as opposed to changing conditions of preservation, exotic input, erosional exhumation, and sedimentary dilution (Fig. 1)? What are the implications for paleoecological, paleoenvironmental, and biostratigraphic interpretation?

Coquinas are complex repositories of paleontological and paleoenvironmental data. Shells carry ecological and biostratigraphic information, but depositional environments may differ from the life habitat and multiple generations of death assemblages can be mixed within a single bed. Moreover, the accumulation of dead shells on or just below the sediment–water interface may modify the benthic habitat in ways that influence further benthic colonization [taphonomic feedback *sensu* Kidwell and Jablonski (1983); J. B. Wilson (1988)] and/or shell preservation [changes in sediment susceptibility to penetration and erosion, and in pore-water chemistry (Newell and Hidu, 1981; Futterer, 1978); for reviews, see Kidwell (1989a, 1991a) and Kidwell and Bosence (this volume Chapter 4)].

The dynamics of shell accumulation thus can be complicated given the diversity of hard-part types that might be supplied to an environment and the many opportunities for feedback among biological, physical, and diagenetic systems before final burial (Fig. 1). With care, however, many details of the accumulation history of individual coquinas can be reconstructed, as illustrated below, by combined study of taphonomic and stratigraphic features. In this way taphonomic analysis can be used to reconstruct both the detailed dynamics of sediment accumulation and various kinds of paleoenvironmental gradients.

TABLE I. Alternative Classifications of Shell-Rich Deposits

Descriptive Classifications	
Fabric/composition	
Dunham (1962)	Wackestone, grainstone/packstone (% volume of shells >20 μm)
Embry and Klovan (1972)	Floatstone, rudstone (% volume of shells >2 mm)
Kidwell *et al.* (1986)	Matrix-supported, bioclast-supported; monotypic, paucitypic, polytypic
Davies *et al.* (1989a)	Comminuted shelly sand, shell-rich or shell-dense sand, shell gravel (% weight of carbonate versus shell gravel)
Kidwell and Holland (1991)	Dispersed, loosely-packed, densely-packed; well-sorted, bimodal, poorly-sorted (shells >2 mm; any matrix)
Stratigraphic scale and complexity	
Kidwell (1982)	Minor simple shell beds, major complex shell beds; minor complex and major simple types possible
Parsons *et al.* (1988)	Pavements and stringers, simple shell beds, complex shell beds, condensed shell beds
Stratigraphic context	
Brinkmann (1929)	*Sohlbanke* (sole bed), *Dachbanke* (roof-bed)
Kidwell (1985, 1988b)	Shell bed capped by omission surface (I) or by erosion surface (II; end-cycle), or resting on omission surface (III) or on erosion surface (IV; base-of-cycle); composites of types I and III (midcycle) and of types II and IV (intercycle)
Genetic Classifications	
Depositional environment	
W. Schäfer (1972)	Vital-astrate, vital-lipostrate, lethal-lipostrate, vital-pantostrate, lethal-pantostrate biofacies
Process/agent	
R. G. Johnson (1960)	Sudden burial of life assemblage (I), gradual accumulation of indigenous shells (II), concentration of exotic shells (III)
Aepler and Reif (1971)	Placer, transgression, and condensation bone beds
Seilacher and Westphal (1971)	Concentration Lagerstätten: placer deposits, condensation deposits, concentration traps
Aigner *et al.* (1978)	Ecologically controlled (biohermal, biostromic), sedimentologically controlled (storm beds)
Fürsich (1982)	Current-influenced beds, *in situ* reworked beds, undisturbed condensed beds
Kidwell (1982)	Biogenic, sedimentologic, and overprinted shell beds (hydraulic reworking of biogenic types, or taphonomic feedback)
Kidwell *et al.* (1986)	Ternary diagram of biogenic, sedimentological, and diagenetic types
Strauch (1990)	Autochthonous shell banks; current-generated and storm-generated allochthonous shell banks
This chapter	Ternary diagram of factors: behavior of shell-producers, within-habitat biological/physical reworking, input of exotic shells
Accumulation history	
Aigner *et al.* (1978)	One-event beds, multiple-event beds
This chapter	Event-concentrations, composite concentrations, hiatal concentrations, lag-concentrations

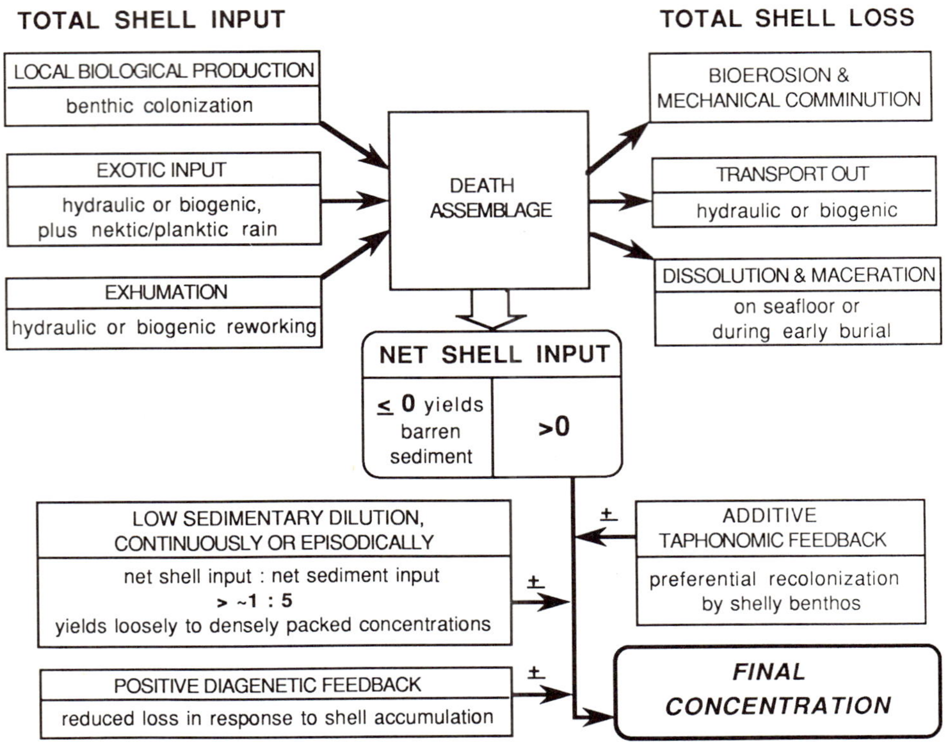

FIGURE 1. Factors in the formation of shell concentrations. The minimum requirement for any accumulation is a positive net input of shells; this may be accomplished by having a high total input, a low loss rate, or some favorable combination over the time-averaged period of accumulation. The final concentration of shells in the sedimentary deposit may be enhanced by additive taphonomic feedback (colonization of dead shells by new shell-producers, which augments initial input), by positive diagenetic feedback (e.g., self-buffering of pore water and inhibition of bioturbators by shells, which reduces loss rates), and/or by low dilution by sediment (continuous or episodic conditions of sediment starvation and winnowing).

In addition, systematic study of how different types of coquinas are distributed should eventually yield an interpretable record of larger-scale sedimentary budgets, because not all biological and geological conditions permit the accumulation of shells in great abundance. Broadly, coquinas indicate that shell supply exceeded shell loss over the period of accumulation, and that this net shell input was not overwhelmed by diluting sediment (Fig. 1) (R. G. Johnson, 1957, 1960; Kidwell, 1985, 1986a). One thus might expect, and in fact does see, the association of many coquinas with various kinds of discontinuities in the rock record.

The aims of this chapter are to summarize current understanding of processes of shell concentration, their consequences for paleontological and geological analysis of the stratigraphic record, and how best to determine the origin of particular deposits. Although the study of coquinas builds upon the general taphonomic works of W. Schäfer (1972) and R. G. Johnson (1957, 1960) and the sedimentological approach to shells advocated by Seilacher (1973), it has only gained close attention in the 1980s. Most work has focused on the stratigraphic record [see contributions in Seilacher and Westphal (1978); Einsele and Seilacher (1982); Bayer and Seilacher (1985)] rather than on actualistic studies, with the exception of intertidal coquinas (e.g., van Straaten, 1952; J. B. Wilson, 1967; Greensmith and Tucker, 1969; Farrow, 1974; Flessa, 1987). Shelf shell gravels (or temperate carbonates) have only recently become widely appreciated as depositional facies; they are a likely focus of actualistic taphonomy in future (J. B. Wilson, 1982, 1986; Nelson, 1988).

1.1. Operational Definitions

The term *shell* here denotes biomineralized remains ≥2 mm in size from any invertebrate animal. These include spicules, dermal plates, feeding structures, and articulated endo- and exoskeletons of diverse taxonomic groups (Lowenstam and Weiner, 1989) which may be preserved in their original mineralogies or as molds, casts, or replaced material. Skeletal grains smaller than 2 mm (i.e., sand-sized or finer) are considered to be *sedimentary matrix* along with all other allochems and terrigenous particles.

A strict definition of what constitutes a shell *concentration* likewise requires an arbitrary cutoff, since shells can constitute any percentage of a deposit (Fig. 2) and can be widely dispersed or densely packed (Fig. 3). This review focuses primarily on densely packed accumulations. However, because some concentrations are densely packed owing to postdepositional processes (e.g., preferential compaction and pressure solution of matrix), and because many coquinoid facies include less shelly interbeds, some loosely packed and dispersed "concentrations" are also discussed.

2. Characterizing Shell Concentrations

Many paleontological and taphonomic features can be used to describe individual shell concentrations, most obviously taxonomic and age composition, quality of preservation (evidence of abrasion, fragmentation, bioerosion, remineralization, etc.), orientation, and size or shape sorting (Table II) (R. G. Johnson, 1960; Ager, 1963; Brett and Baird, 1986; Goldring, 1990; Kidwell and Bosence, this volume Chapter 4). Taphonomic data are of greatest interpretive use if the characteristics of each taxonomic group are tallied separately, so that the condition of brachiopods, for example, can be compared against that of cooccurring trilobites or bivalves having different susceptibilities and patterns of post-mortem modification. This is the

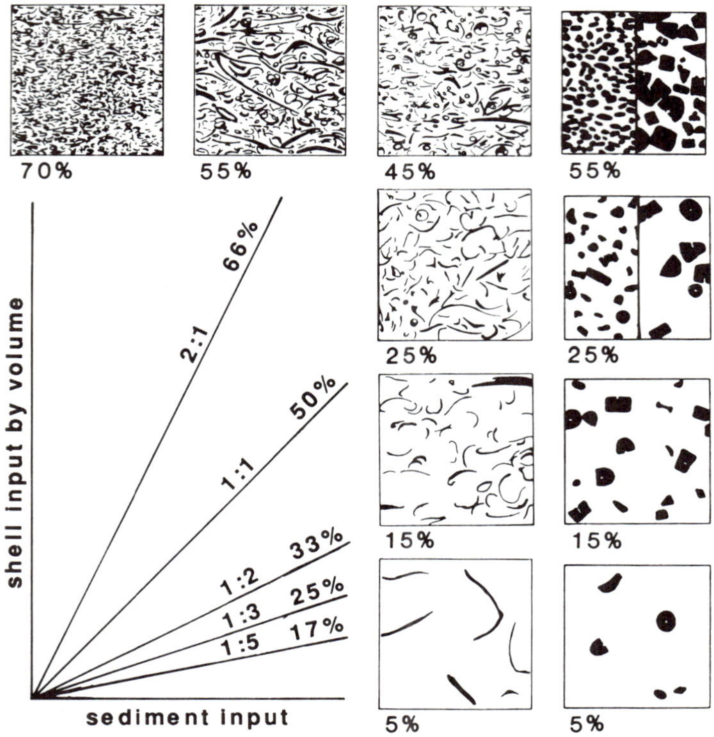

FIGURE 2. Charts for the visual estimation of %-volume abundance of shells in cross section, illustrating the effect of shell shape on close packing. Densely-packed, bioclast-supported fabrics can be produced by as little as 30% shells if bowl-shaped, but require 50% or more shells if spheroidal; these fabrics require high shell-input/sediment-input ratios of 1:2 or more. Shell-rich but loosely-packed, matrix-supported "concentrations" can be produced with as little as 10–15% bowl-shaped shells, requiring much lower 1:9 to 1:5 input ratios. [Adapted from K. Schäfer (1975).]

comparative taphonomic approach first advocated by Efremov (1940) and refined by Brett and Baird (1986); it can be applied successfully to the analysis of both spectacular Lagerstätten and more ordinary assemblages (Speyer and Brett, 1986; Feldman, 1989; Whittington and Conway Morris, 1985; Meyer *et al.*, 1990).

Valuable sedimentological observations include: the grain size of the matrix; the size (hydraulic equivalence), abundance, and close packing of shells relative to matrix; and the association of shells with sedimentary structures (Table II). Shell abundance by percent volume can be determined with great accuracy from cross-sectional views either by point counting or by visual estimation, using the comparative charts of K. Schäfer (1969; Flügel, 1982, p. 241) (Fig. 2). In contrast to percent-weight measures [e.g., Davies *et al.* (1989a); see Table I of the present chapter], volumetric measures can be used on coquinas of any age and state of preservation, including moldic faunas. Close packing of shells [genetically more informa-

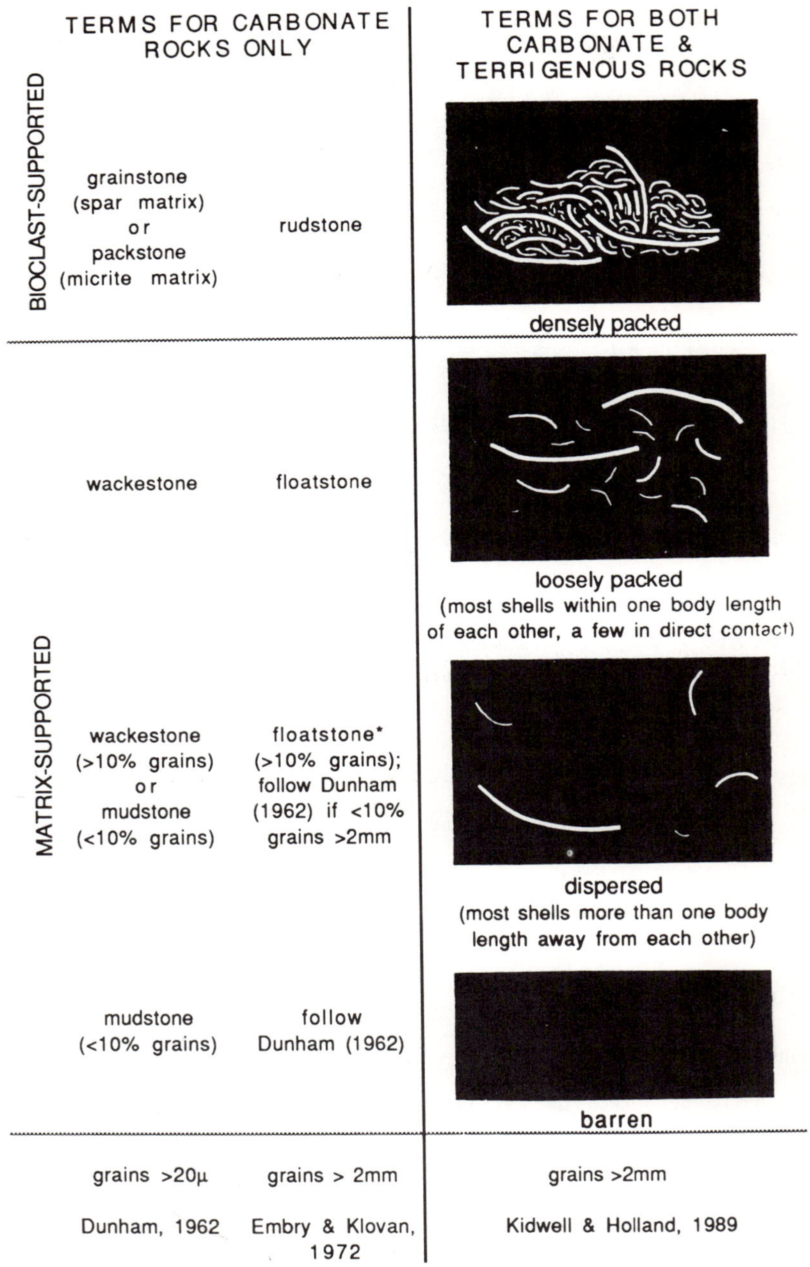

FIGURE 3. Synonymy of close-packing terms for bioclastic fabrics. [*Reid and Ginsburg (1986) redefined floatstone as containing a minimum of 20% grains larger than 2 mm.]

tive than percent abundance (Dunham, 1962)] can be described by semi-quantitative terms such as "wackestone" or "matrix-supported" (Fig. 3) or, more laboriously, by quantitative packing indices (Flügel, 1982, p. 212).

Stratigraphic features—i.e., characterization of the coquina as a three-dimensional rock body—are also useful in inferring concentration pro-

TABLE II. Attributes of Shell Concentrations for Field Description[a]

	Semiquantitative scale
Paleontologic features	
Number of species	one----------------few----------------10----------------100----------------1000
Relative abundance	{all individuals from one species}------------------------------{one individual of each species}
Taxonomic composition	dominant group(s)
Ecological spectrum (dominant life habits)	nekton, plankton, benthos, mixed; infauna, epifauna, mixed; soft-, firm-, hard-bottom dwellers, mixed; suspension-feeders, deposit-feeders, mixed
Age spectrum	juveniles only----------------------mixed----------------------adults only
Orginal mineralogy	aragonite------calcite------Mg-calcite------mixed------silica------other
Preserved mineralogy	original-recrystallized-replaced-{molds (simple, composite)}-geodal fill
Taphonomic features	
Articulation	articulated--------{disarticulated but closely associated}--------{disarticulated and dissociated}
Size sorting	very good--------{moderately good to poor}--------very poor--------bimodal
Modal size and range	−1 −2 −3 −4 −5 −6 −7 −8 φ 2 4 8 16 32 64 128 256 mm
Shape sorting	one part only------------------------sorted------------------------unsorted
Fragmentation	all whole --------------------some broken --------------------all fragments
Abrasion	{unabraded ± periostracum} -------abraded -------highly abraded or polished
Rounding	angular-----------subangular-----------rounded-----------well-rounded
Biological modification (bioerosion, encrustation)	{none (pristine)} -------{minor (post-mortem vs. pre-mortem)} -------extensive
Orientation	all in life orientation-----------------mixed-----------------all disturbed
In plan view	{preferred orientation (unimodal–bimodal)}------------------------------------high variance
In cross-section	{preferred orientation (concordant, oblique, edgewise; convex-up/down, stacked, imbricate)}----------------high variance
Sedimentological features	
Type of matrix	mud/micrite----------sand----------conglomerate----------void/cement

TABLE II. *(Continued)*

Hydraulic equivalence of matrix	same as shells ------------------more mobile------------------less mobile
Relative abundance of shells (% volume shells)	100% ---0% (visual estimation charts of K. A. Schäfer, 1975)
Close-packing of shells	densely packed (bioclast-supported) }-{ loosely packed (matrix-supported) }-{ dispersed (matrix-supported)
Associated sedimentary structures	e.g., shells at base or top of size-graded beds; within concretionary bodies; along bedset boundaries, avalanche slopes; uniform within massive beds
Stratigraphic features	
Absolute scale	
Thickness	mm -----------cm -----------dm -----------m -----------10 m -----------100 m
Lateral extent	cm---10,000 km
Relative scale	within-facies (minor)------------------within-facies (major)------------------cross-facies (major)
Geometry	pavement or bed-----------wedge-----------lens-----------pod or clump
Internal complexity, including lateral variability	none, homogeneous }--------{ simple trend, e.g., graded }--------{ complex microstratigraphy
Physical contacts (base, top, lateral)	sharp (planar scoured, burrowed, }----------------interbedded----------------gradational
Position in sequence	e.g., base or top of fining-up sequence; within a transgressive or regressive aggradational facies tract; part of lowstand or highstand systems tract
Association with significant surface	e.g., local or regional unconformity; transgressive ravinement or flooding surface; downlap horizon; transport or toplap surface

[a]Modeled on Behrensmeyes, this volume, Chapter 6.

cesses, although often omitted from descriptions. Like any other sedimentary deposit, each shell concentration has physical dimensions, a scale relative to surrounding bodies, a geometry or shape, an overall lithologic character or internal complexity, and physical contacts (sharp, gradational, intercalated) with surrounding, less fossiliferous deposits (Table II) (Kidwell *et al.*, 1986; Kidwell, 1986a). Shell concentrations also have a larger stratigraphic context with respect to facies sequences (e.g., position within a fining or coarsening cycle) and with respect to stratigraphically significant surfaces.

Stratigraphic features are as essential to interpretation as all other attributes of the concentration and, in terms of revealing the process of concentration, are commonly less ambiguous than taphonomic features. For example, a storm-concentration can have any number of *taphonomic* signatures (*sensu* Table II; see Fig. 4), depending upon shell source, and yet share diagnostic *stratigraphic* features such as a broadly tabular form, sharp erosional base, and occurrence within a hydraulically graded deposit (although the shells themselves may not be size-graded). Modern and ancient storm-concentrations in fact show a wide range of taphonomic signatures (Table III), not all of which are readily attributable to differences in storm energy. This variability, seen not only among storm-concentrations but among examples of most coquina types, is why classifications of coquinas generally rely upon sedimentological and stratigraphic rather than taphonomic criteria (Table I).

The interpretation of shell concentrations is therefore an essentially two-phase procedure:

(a) *Stratigraphic characterization* of the shell-rich sedimentary body to identify the most likely late-stage process(es) of concentration.

(b) *Taphonomic characterization* of the constituent fossil assemblage to reconstruct as much as possible of shell histories both before and during formation of the concentration. These observations, along with sedimentological and paleontological features, are used to refine or choose among rival stratigraphic hypotheses and to derive a reasonable narrative history of shell accumulation.

3. Origin and Utility of Different Types

Shell concentrations can be organized into four broad types based on their stratigraphies and inferred histories of accumulation (Fig. 5):

1. *Event-concentrations*, which have the scale of single laminae and beds, and record single, ecologically brief episodes of shell concentration.

2. *Composite or multiple-event concentrations*, which range from beds and bedsets to reservoir-scale bioclastic facies. These have more complex (and usually longer) histories than simple event-concentrations, and record the accretion or amalgamation of multiple generations and/or event-concentrations. The final, composite concentration is of average or expanded thickness relative to coeval strata.

3. *Hiatal or condensed concentrations*, which are complex accumulations like those of type 2 except that they are thin relative to coeval strata owing to slow *net* rates of accumulation (i.e., they are stratigraphically condensed).

4. *Lag-concentrations*, which are thin, highly residual concentrations associated with surfaces of significant stratigraphic truncation. The term ''significant'' is used to distinguish these from shell pavements produced by minor erosional scour. Shells are exhumed and concentrated by selective removal of sedimentary matrix, either by erosion or by corrosion.

Each of these four types has different applications to stratigraphic analysis and has different implications for paleobiological analysis of the

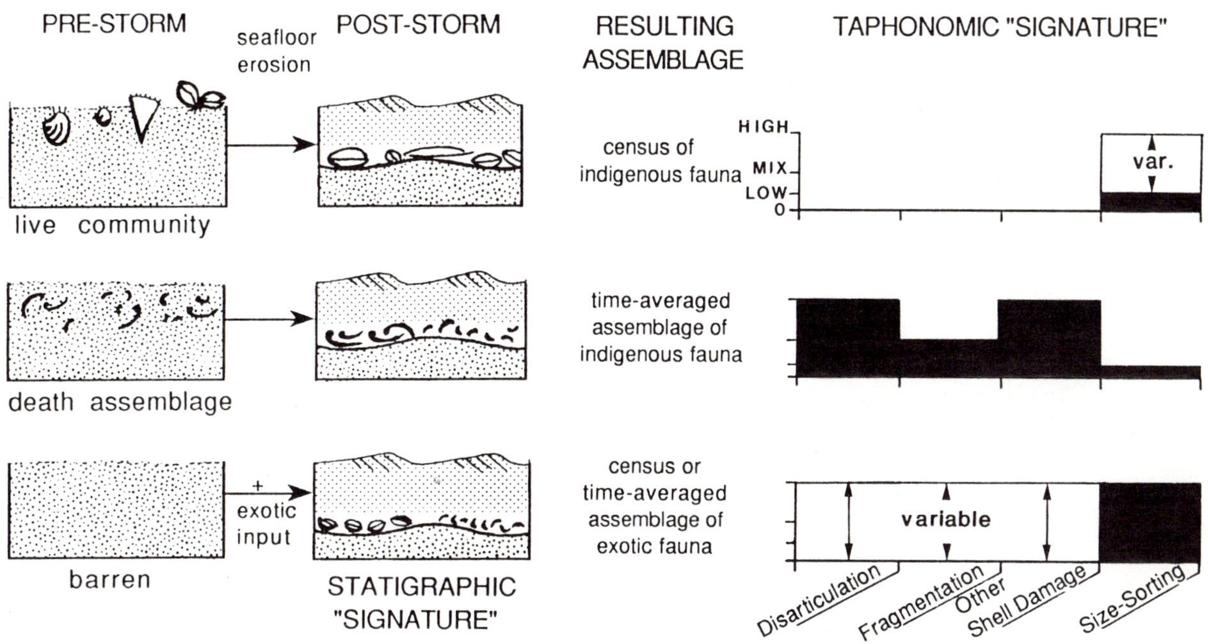

FIGURE 4. Hypothetically, the taphonomic "signature" of identical storms will be influenced by whether shells are derived locally from a live community or from a death assemblage, and whether they have been transported from other environments. The *stratigraphic* "signature" of identical storm-concentrations should be less variable: in this high-energy example, the shells occur within a graded sedimentary bed (although shells are not necessarily size-graded themselves) that is tabular or lenticular with a sharp erosive base. This is why the stratigraphic features of a shell concentration—e.g., its dimensions, geometry, internal stratification, contacts, and larger sedimentary context—are as important as taphonomic features in interpreting origins. See text for discussion.

TABLE III. Variation of Taphonomic Features of Single-Event Tempestites as a Function of Both the Preevent History of Shells ("Shell Source") and the Short-Term Dynamics of the Event Itself[a]

Example	Biostratinomic features that primarily reflect shell source				Biostratinomic features that primarily reflect storm event		
	Disarticulation	Fragmentation	Other damage	Size sorting[b]	Orientation	Close packing	Grading
Recent mollusks, nearshore Texas continental shelf (personal observations)	XXX	XXX	X	XXX 2–3 mm	Concordant	Dense	None
Recent mollusks, back-reef shelf, Australia (Gagan *et al.*, 1990)	XXX	XXX	—	XX–XXX 2–6 mm	—	Loose to dense	Fine-up
Recent mollusks (exotic in part), low-energy nearshore, Florida (Perlmutter, 1982)	XXX	XX–XXX	XX–XXX	XX	Concordant	Dense	Fine-up
Recent mollusks, shoal washovers, Florida (Aigner, 1985)	XX	XX–XXX	X-XX	X	Convex-up to chaotic	Dense to loose	Fine-up
Pliocene bivalves, open shelf California (Norris, 1986)	XX	XX	XX	—	Convex-up	Dense	None
Miocene *Glossus*, open shelf, Maryland (Kidwell, 1982) (Fig. 7, this chapter)	XX	X	O–X	XX 2–4 cm	Stacked concave-up; imbricate	Dense	None
Oligo-Miocene bivalves, open shelf, California (DeCelles, 1987)	XXX	XXX	XXX	—	Convex-up	Dispersed to loose	None
Oligocene scallops, fjord, Antarctica (Gazdzicka and Gazdzicki, 1985)	XX–XXX	O–X	X	XXX 5 cm	Convex-up and -down	Dense	None
Eocene nummulites, bank, Egypt (Aigner, 1983)	NA	O–X	O–X	XXX	Imbricate	Dense	None
Jurassic *Nanogyra*, open shelf, France (Fürsich and Oschmann, 1986)	XXX	X–XXX	O–X?	XXX	Convex-up, imbricate, or stacked	Dense	Fine-up common

Jurassic *Camptonectes*, shelf sand bar, Greenland (Fürsich and Heinberg, 1983)	XX	—	O–X	O–X	Chaotic to imbricate	Dense	None
Jurassic *Myoporella*, open shelf, Corallian, England (J. K. Wright, 1986; and personal observations)	XXX	X	X	XXX 3–4 cm	Stacked convex-up	Loose to dense	None
Triassic benthos, open shelf, Muschelkalk, Germany (Aigner, 1979, 1985)							
Thin beds or pavements	XX	O–X	X	X	Concordant	Dense	None
Thick beds with intraclasts (allochthonous shells)	XXX	XX–XXX	XXX	XX–XXX	Concordant, convex-up predominates	Dense	Fine-up
Triassic brachiopods, open shelf, Muschelkalk, Silesia (Dzulynski and Kubicz, 1975: Bodzioch, 1985):							
30-cm-thick, laterally extensive (square kilometers)	X–XXX	XX–XXX	XX–XXX	X	Concave-up predominates	Dense	Fine-up
Channelized	XX–XXX	XX	XX	XXX 3mm	—	—	Slight
Channelized	XX	XX	XX	X	—	—	Slight
Triassic bivalves, high-energy shore, New Caledonia (Freneix and Avias, 1977)	XX	XX–XXX	—	—	Predominantly convex-up, stacked and imbricated	Dense	None
Lower Carboniferous exotic brachiopods and crinoids on ramp, Wales (Faulkner, 1988)	XX–XXX	XX	—	—	Convex-up	Dense to loose	Fine-up
Lower Carboniferous lingulid brachiopods, nearshore, Scotland (Ferguson, 1963)	O–X	O–X	O–X	X	Concordant	—	—

(continued)

TABLE III. *(Continued)*

Example	Biostratinomic features that primarily reflect shell source				Biostratinomic features that primarily reflect storm event		
	Disarticulation	Fragmentation	Other damage	Size sorting[b]	Orientation	Close packing	Grading
Upper Devonian brachiopods, open shelf, New York (Bowen *et al.*, 1974; Craft and Bridge, 1987)	XXX	O–X	XX	X—XXX	Concordant, imbricated, or chaotic	Dense	None
Middle Devonian trilobites, nearshore, New York (Speyer and Brett, 1986)	XXX	XXX	—	XX–XXX	Convex-up or imbricate	Dispersed	None?
Silurian brachiopods, open shelf, New York (Eckert and Brett, 1989)	XXX	X–XX	—	X	Convex-down	Dense	None
Silurian brachiopods, open shelf, Wales (Cherns, 1988)	XXX	X–XXX	X–XX	XX–XXX?	Varies	Dense to loose	None?
Ordovician brachiopods, open shelf, Martinsburg Formation, Virginia (Kreisa, 1981)	XX–XXX	X	O–X	X–XX	Convex-up, local random	Dense	None
Cambrian trilobites, open shelf, Alberta (Westrop, 1986)	XXX	XX–XXX	XX	X–XXX	Concordant, convex-up?	Dense	Fine-up

[a]O, zero; X, low; XX, mix; XXX, high; —, no data. Shells derived entirely from indigenous fauna unless noted otherwise.
[b]Biostratinomic feature that can reflect both shell source and storm event.

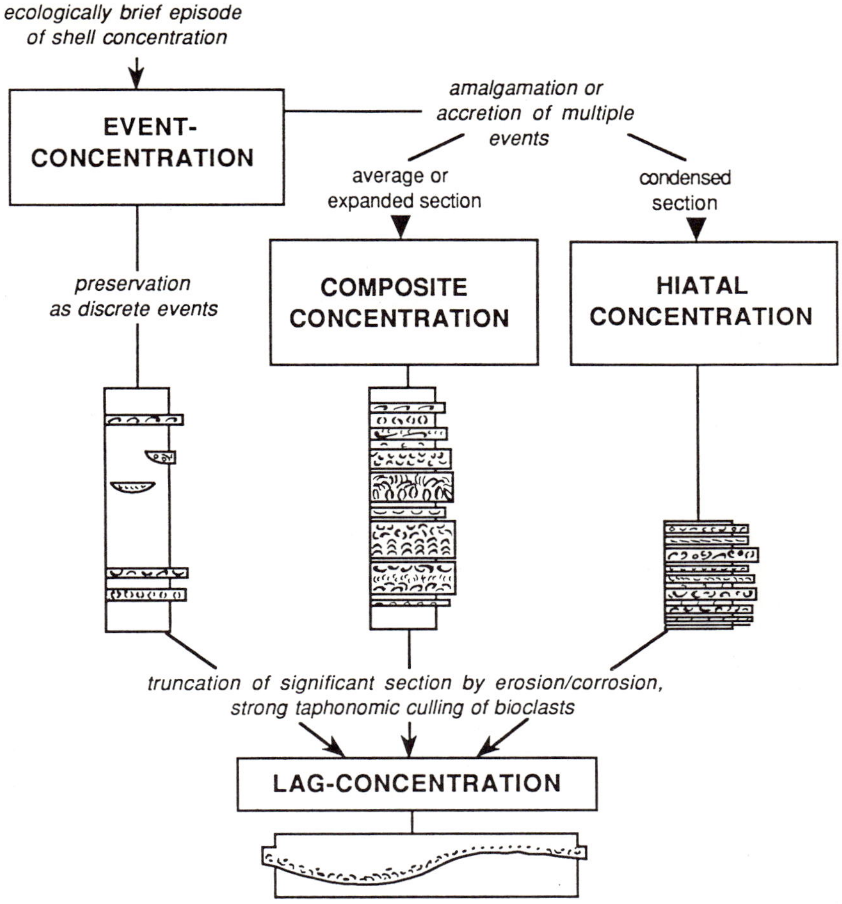

FIGURE 5. Skeletal concentrations reflect a spectrum of accumulation histories and time scales, but can be grouped into four broad categories: these range from simple *event-concentrations* to accretionary, multiple-event accumulations (both *composite concentrations* of "normal" or expanded thickness, and *hiatal concentrations* that are stratigraphically condensed), and highly derived, stratigraphically disjunct *lag-concentrations*. Each category includes biogenic, sedimentological, and mixed-origin coquinas; these may be composed of any proportion of local and exotic shells. Event-concentrations, for example, include predator and scavenger middens, *in situ* clusters created by gregarious behavior of shell-producers, shelly turbidites, and storm-concentrations that have been recolonized or reworked by benthos.

contained faunas (Tables IV–VII). The types should not be viewed as discrete categories, however, because they intergrade. For example, the boundary between erosional event-concentrations and lag-concentrations is arbitrary, and composite concentrations that are "normal" in thickness grade into hiatal concentrations, which are stratigraphically condensed. In addition, concentrations of different types can be combined—e.g., a hiatal concentration may rest directly on or terminate in a lag-concentration—and some types can be nested within others—e.g., many event-concentrations may make up a single composite or hiatal concentration. Physical scale, internal complexity, geometry, and stratigraphic context are valuable lines of evidence for the stratigraphic type of a particular coquina.

Reflecting the two-stage analysis advocated in Section 2, concentrations can be further characterized on taphonomic and other criteria. This is essential to determining the variety of biological and physical processes that may have acted as agents of concentration, and the variety of sources and past histories of contained shells. It is usually easier to reconstruct the source of preserved shells than it is to reconstruct what has been lost, even when the condition of surviving shells suggests which destructive agents may have been active. Thus, taphonomic analysis of stratigraphic types focuses on apparent shell sources—e.g., benthic colonization and behavioral aggregations of shell-producers, physical and biological reworking of local death assemblages, and exotic input by various means—rather than on possible shell "sinks"—e.g., bioerosion and fragmentation, dissolution, and transport (Fig. 1). This emphasis on preserved shells in no way implies that rates of shell loss were negligible (cf. Davies *et al.*, 1989b), only that loss apparently did not exceed the total amount of shell supplied during the time-averaged period of accumulation (Kidwell, 1986a).

In the following sections, ternary diagrams are used to display the variety of ways in which shell beds of a particular stratigraphic type can form (see Figs. 6, 9, and 12). To improve upon the ternary diagrams used by Kidwell *et al.* (1986; also see Fürsich, 1990), which used biogenic, sedimentological, and diagenetic poles, the poles in the diagrams herein were selected for their relevance to paleobiology rather than taphonomy alone. Concentrations that plot near the "behavior of shell-producers" pole reflect pulses of high shell input by local benthos, and contain largely undisturbed assemblages; coquinas near the "within-habitat reworking" pole have been created by post-mortem processes, but still comprise indigenous fauna; and those near the "exotic input" pole consist of entirely allochthonous benthic or nonbenthic assemblages. Intervening areas of the ternary diagram indicate mixtures of indigenous and exotic shells, and various degrees of modification by reworking or by shell-producers.

3.1. Event-Concentrations

These are coquinas in which the process of concentration itself is ecologically rapid and comparatively simple—e.g., a single mass mortality, a single event of storm reworking, the construction of a shell-lined burrow. Event-concentrations are typically small in scale, ranging to a few decimeters in thickness and tens of square kilometers in lateral extent (i.e., stratigraphically minor deposits compared to facies), although exceptional shell-rich turbidites can be traced over 500 km (Pilkey and Curran, 1986). They show great variation in geometry, sedimentology, and taphonomy, depending on factors such as inherited topography, shell source, whether the concentration is basically hydraulic or biogenic in origin, and the extent of later overprinting (Appendix A; Table IV).

Some event-concentrations are intrinsic to local shell-producers, for example, concentrations created by gregarious behavior of adults or by "boom-and-bust" population dynamics (behavior pole of Fig. 6). These range from small pods and clumps to very extensive pavements and beds ["intrinsically biogenic" concentrations of Kidwell *et al.* (1986)]. Other concentrations involve the rearrangement of live or dead shells and/or preferential removal of matrix (reworking pole in Fig. 6). Reworking may be by physical processes (e.g., hydraulic winnowing, or erosion and graded redeposition) or by organisms ["extrinsically biogenic" concentrations of Kidwell *et al.* (1986)], and may act upon initially dispersed or biogenically aggregated shells. Examples thus may plot anywhere along the reworking/behavior axis in the ternary diagram of Fig. 6.

Event-concentrations formed by hydraulic reworking can take many forms including burrow-fills [e.g., "tubular tempestites" (Wanless *et al.*, 1987)] and lenticular fills of scours. Pavements, stringers, beds, and coarse basal lags are the most common forms, however, as typified by shell-rich storm beds, turbidites, and debris flows (Fig. 7; Table III; Appendix A). Shells may simply be rotated in the sediment by the scour of surrounding sediment (R. G. Johnson, 1957). Alternatively, they may move with bedload or be fully resuspended and then redeposited, with or without significant (i.e., out-of-habitat) transport (Table IX in Kidwell and Bosence, this

TABLE IV. Event-Concentrations: Records of Single, Ecologically Brief Episodes of Shell Concentration

Scale	0.2 cm thick, up to single bed
Geometry	Pavement, stringer, bed, lens, pod, clump
Internal structure	Homogeneous or simple variation (e.g., vertical or lateral gradation in feature)
Context	"Float" within larger, less fossiliferous depositional body; can rest on or under minor discontinuity surface, or be unassociated with discontinuity surface
Origin	Ecologically brief episodes of hydraulic or biological reworking, mass mortality, colonization, exotic input
Taphonomic signature	Highly variable, depending on the source and history of shells, the agent of concentration, and seafloor exposure; range from "perfect" preservation (census assemblages) to highly fragmented/sorted (exotic or indigenous time-averaged death assemblages)
Diagenetic signature	Varies; census types (especiallyby obrution) may develop distinctive mineralogies related to organic decay
Paleontological use	Biological fidelity and temporal acuity varies, but some are composed of unmodified census assemblages and thus provide high-resolution data
Stratigraphic Use	Marker beds for local correlation (if laterally extensive; fossil assemblage may be census or time-averaged); keys to short-term dynamics of sediment deposition, erosion, and omission; facies differentiation and interpretation; recognition +/− calibration of bathymetric gradients; generally reliable source of biostratigraphic data

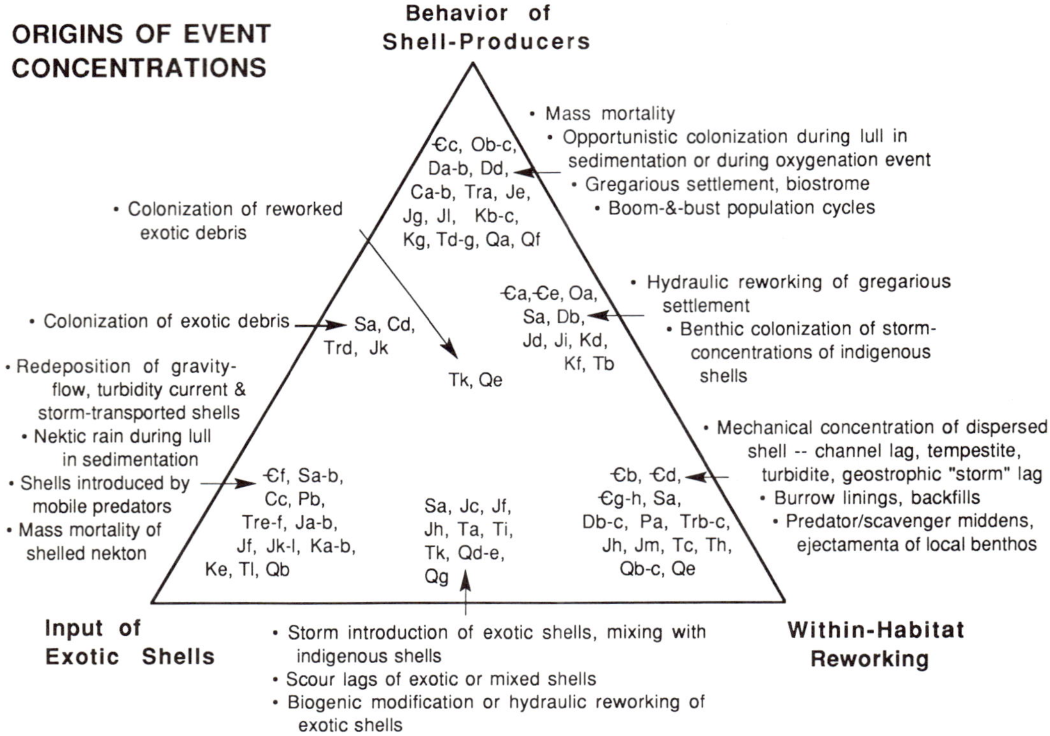

FIGURE 6. Origins of modern and ancient event-concentrations, plotted on a ternary diagram according to the relative importance of (1) the shell-producer's behavior, (2) local reworking by organisms and physical processes, and (3) input from exotic shell sources. Examples, coded by geological period (Appendix A), illustrate that event-concentrations can reflect any combination of these factors, and that this variety is apparent in Paleozoic as well as Mesozoic and Cenozoic records. Data are not based on an exhaustive or random survey, and so should not be used to infer relative frequencies of types. See Table III for more examples of storm-generated event-concentrations.

volume, Chapter 4). Hydraulic processes can be interpreted in many instances from shell sorting and orientations, including fabrics produced by the interference effects of abundant shells on the seafloor (Greensmith and Tucker, 1968; Futterer, 1978), by high- versus low-density turbidites (Allen, 1984), and by storm versus fair-weather reworking (Jeffery and Aigner, 1982; Fürisch and Heinberg, 1983; Galli, 1985) (Table VIII in Kidwell and Bosence, this volume, Chapter 4). Although exotic shells dominate many (but not all) hydraulic concentrations in supratidal and intertidal environments, out-of-habitat transport of shells >2 mm is generally minor in level-bottom subtidal environments, even above storm and fair-weather wave-bases (Fig. 6) (e.g., Aigner and Reineck, 1982; Frey and Howard, 1986; Gagan *et al.*, 1990; Table IX and Section 4.2.3 in Kidwell and Bosence, this volume, Chapter 4).

Biologically reworked examples of event-concentrations include: pred-

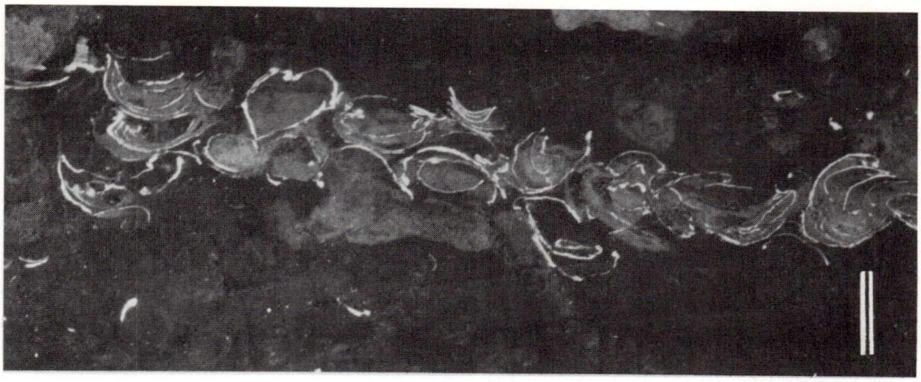

FIGURE 7. Storm-concentrated bed of the infaunal bivalve *Glossus fraterna* (Miocene Calvert Formation, Maryland). Stacked, imbricated, and articulated specimens with laminated draft-fills of well-sorted very fine sand provide the only evidence for storm reworking of the seafloor in this thoroughly bioturbated sandy-silt facies. Scale, 3 cm.

ator middens; shells concentrated by hermit crabs, nesting fish, and birds; and shells backfilled into burrows or used to construct domiciles (Fig. 6). These concentrations are usually pods, clumps, lenses, and stringers of small lateral extent (Appendix A).

A third end-member type of event-concentration is that created by the input of exotic shell material (Fig. 6). Shells may be transported out of their life habitat by hydraulic processes, by gravity (nektic/planktic rain; slumps and debris flows), by drift or rafting, and by other organisms. Admixture with indigenous death assemblages creates concentrations that plot along the exotic/reworking and exotic/behavior axes in Fig. 6.

Many event-concentrations are modified by later post-mortem conditions. Recognizing such overprinting commonly requires special attention: hydraulically reworked biogenic concentrations, for example, may look exactly like simple hydraulic event-concentrations. Overprinting can be identified, however, if the ecology and taphonomy of the shelly fauna are known from other, broadly contemporaneous deposits. For example, within bioturbated silty sands of the Maryland Miocene, the high-spired gastropod *Turritella plebeia* occurs in two modes (Kidwell, 1982): (1) in life positions within small (20 × 20 cm) clumps, suggesting life aggregations; and (2) in discrete pavements of flat-lying, bimodally-oriented shells, suggesting wave and/or tidal reworking. In light of the clumping of life-oriented individuals, the *Turritella plebeia* pavements can be interpreted as locally reworked clumps rather than hydraulic concentrations of initially dispersed individuals.

The origin of biogenically overprinted hydraulic concentrations—for example, storm-concentrations disrupted by scavengers and bioturbators, or colonized by "postevent" shelly benthos (Seilacher, 1982a)—can also be difficult to recognize (Kidwell, 1986b, 1991a). The postevent infauna may disrupt the original hydraulic fabric by burrowing through it or growing interstitially among shells of the original concentration. This latter scenario would yield an ecologically mixed assemblage of soft-ground and shell-ground taxa [one form of faunal condensation *sensu* Fürsich (1978)] in a shell concentration of indistinct origin. Alternatively, the shelly postevent fauna may simply colonize the upper surface of the shell gravel, which may lie at or just below the seafloor, yielding a comparatively clear microstratigraphic record. Through such overprinting and taphonomic feedback by shelly faunas (Figs. 1 and 8), simple event-concentrations may build into more substantial concentrations having complex ecological and taphonomic histories [see J. B. Wilson (1988) for a present-day example].

Some event-concentrations, particularly those involving rapid burial of live and dead faunas, are characterized by distinctive early diagenesis (Brett and Baird, 1986). Aigner (1979; and personal communication), for example, interpreted concretionary "underbeds" as the result of storm juxtaposition of aerobic shell carbonate upon anaerobic mud in the Triassic Muschelkalk, and Brett *et al.* (1986) attributed excellent preservation of trilobites, crinoids, and other fossils in Devonian dark shales of New York to similar early cementation. Pyritization is another diagenetic process related to event-concentration in some environments (e.g., Parsons *et al.*, 1988; Feldman, 1989).

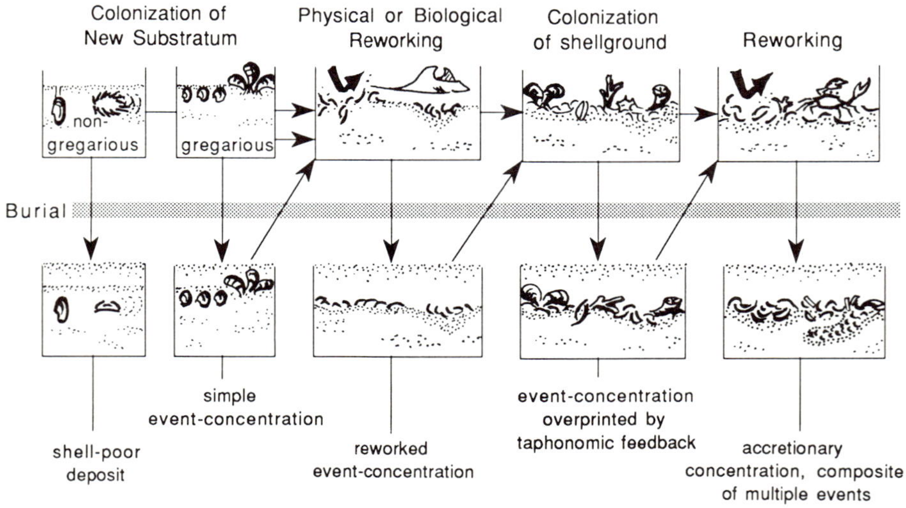

FIGURE 8. Bed-scale coquinas may have very simple accumulation histories (e.g., gregarious settlement, or a single event of reworking or of exotic input), or they may have more complex histories, in which an initial concentration is overprinted by further colonization and/or reworking. Because late-stage reworking and colonization can obliterate the evidence of earlier events, the complex origins of some small-scale shell concentrations may be underestimated. In many instances, however, the general nature of the accumulation history can be reconstructed using taphonomic and paleoecological evidence from individual shells and from the assemblage as a whole. [Figure adapted from Fursich (1982), Miller *et al.* (1988), and Kidwell (1991a).]

3.1.1. Reconstructing Sedimentary Dynamics

Microstratigraphic analysis of simple and overprinted event-concentrations can provide a powerful means of reconstructing the short-term dynamics of sediment deposition, erosion, and transport in aquatic environments (Aigner, 1979, 1985; Kidwell and Jablonski, 1983; Kidwell and Aigner, 1985; Kidwell, 1985, 1991a; Brett and Baird, 1986; K. B. Miller *et al.*, 1988), analogous to microstratigraphic dissection of hardgrounds (Fürsich, 1971).

For example, intensely bioturbated sandy silts of the Maryland Miocene are characterized by a series of *Glossus fraterna*-dominated pavements and thin beds (Fig. 7). The infaunal life habit of this bivalve indicates a minimum of 4 cm erosion to exhume shells; the high proportion of disarticulated valves suggests that most individuals were dead at the time of exhumation. The rarity of shell fragments of any size suggests a single, brief event of exhumation and exposure at the seafloor, as these large but thin shells should break easily. The absence of post-mortem borers and encrusters also suggests minimal exposure time on the seafloor after exhumation. Highly turbulent conditions are indicated by the distinctive stacking and imbrication of disarticulated valves which, judging from their laminated draft-fills, acted as seafloor traps for sand in tractive transport. These pavements are the only preserved record of turbulent, presumably storm-influenced conditions in this facies, which otherwise would be interpreted as having accumulated in much deeper waters. The sand-filled shells also provide the only evidence for sand import during storms. Although post-storm deposits were thoroughly bioturbated, the shell pavements were not disturbed, presumably owing to the large size of the shells. Shell-rich tempestites thus may have a greater preservational potential than shell-poor types.

In instances where event-concentrations are not immediately reburied or are buried under inadequate sedimentary cover, shells are vulnerable to further modification in what Davies *et al.* (1989b) referred to as the "taphonomically active zone." This provides several criteria for estimating time delays between the formation of the coquina and its final burial [i.e., permanent burial below the reach of erosional reworking and bioturbation (Kidwell, 1986a)]. Most common is the expectation of a broad, inverse correlation between shell condition and either (a) the period of exposure at or near the seafloor or (b) the number of burial/exhumation events before final burial (Driscoll, 1970). The *G. fraterna* example above provides one illustration of this style of reasoning (see Appendix A for references to many others), and Brandt (1989) has suggested a formalization in terms of time-dependent "taphonomic grades," which range from high-grade assemblages (>90% articulated skeletons, <10% shells damaged) to low (<10%; >75%).

Taphonomic grades, however scaled, are inherently difficult to calibrate to absolute time scales [Frey and Howard (1986); see review in Kidwell and Bosence (this volume, Chapter 4)]. This reflects variation (a) among environments in the vigor and styles of taphonomic attack (e.g., some variation in quality of preservation reflects environmental energy or chemistry, not time alone), and (b) among taxa in their susceptibility or response to taphonomic attack. Moreover, even if known for the Recent, absolute rates of shell modification are difficult to extrapolate confidently to past systems, owing to evolution among shell-producers and shell-destroyers or -modifiers (see Section 4.4).

Similar problems limit the usefulness of live/dead interactions (taphonomic feedback) in determining the *absolute* tempo of sedimentation. The likelihood that dead shells are colonized is a broad function of the immediacy, thickness, and permanence of sedimentary burial, but there is great variation in just how immediate and how thick a burial increment must be to sequester dead shell material from living organisms (Kidwell, 1991a). In addition, an absence of taphonomic feedback can have many explanations other than immediate burial (Kidwell, 1991a, and references therein).

Relative estimates of how long a shell is exposed before (re)burial may be possible using colonization and overgrowth relations among endo- and epibiont infesters of individual shells; a single kind of shell host should be examined. Recent infesters vary greatly in opportunism, ranging from the most opportunistic serpulids and bryozoan runners to encrusting bryozoans and sponges (Jackson, 1983); analogous sequences can be determined for ancient species (M. A. Wilson, 1985). Experiments in the Mediterranean by Bromley *et al.* (1990) suggest that shell substrata with visible clionid sponge borings record at least 2 years of exposure, and that diverse endobionts require several additional years to develop. Quantitative estimates may be derived directly from the age structure of postevent epifauna. In Ordovician strata of Norway, for example, M. E. Johnson (1989) found that the maximum age of postevent brachiopod cohorts was only 2–3 years in shallow-water facies, whereas it was 8–10 years in offshore facies. These figures can be translated into the minimum recurrence interval of "killer" storms, and thereby serve as a means of recognizing bathymetric gradients.

Combined paleoecological and taphonomic study thus can permit in some instances the interpretation of a far more detailed, bed-by-bed history of sedimentary dynamics than is possible using conventional lithologic evidence, even when some damage to shells accrued during the life of the shell-producers (Elias, 1982; K. B. Miller *et al.*, 1988; Parsons *et al.*, 1988). In heavily bioturbated facies, event-concentrations commonly preserve the only record of hydraulic events, thus contributing significantly to paleoenvironmental interpretations.

3.1.2. Facies Differentiation and Paleobathymetry

The stratigraphic close spacing and relative proportions of different shell concentrations can be used to map facies, even if the origins of the concentrations are unknown. The clearer the origins of the concentrations, the more useful they are for interpreting and contouring paleoenvironmental gradients (Fürsich and Heinberg, 1983; Fürsich and Kauffman, 1984; Aigner, 1979, 1985; Kidwell *et al.*, 1986). Norris (1986), for example, found an offshore decrease in shell concentrations of all types in the Pliocene of California (physical sedimentary structures were used for comparison). Whereas hydraulic concentrations (both storm and fair-weather types) decreased markedly from shoreface to middle-shelf facies, various biogenic concentrations were most abundant in middle- and outer-shelf settings. Storm-concentrations also vary in type bathymetrically, from erosional lags to winnowed concentrations to unreworked, smothered assemblages (Aigner, 1985; Aigner and Reineck, 1982; Speyer and Brett, 1986; K. B. Miller *et al.*, 1988; Feldman, 1989).

The precise frequency distribution of shell concentrations varies from setting to setting, of course, as does the absolute bathymetric significance of individual types. Clifton (1989), for example, suggested that hydraulic "shell lags" could be generated to depths of 100 ± 20 m on the Pleistocene California shelf, in contrast to the ~50 m estimated by Norris (1986) for the Pliocene.

3.1.3. Marker Beds

Lithologically distinctive, laterally persistent event-concentrations can be especially valuable for correlation within single basins. This application is directly analogous to that of ash beds. Waage (1964), for example, used repetitive concretionary horizons of mass-killed ammonites to correlate across diverse marginal marine facies in the Cretaceous Western Interior.

Most event-concentrations, even those formed during major storms (Kreisa, 1981; Brett *et al.*, 1986), can be traced over only part of a depositional basin, and thus many shell-rich marker beds used by stratigraphers are not single event-concentrations as defined here. Rather, they are amalgams of many events (i.e., composite or hiatal concentrations). Examples of widespread, condensed "event" horizons range from those occurring in a single basin—for example, various fish-scale beds in Cretaceous black shale basins, "epiboles" of unusual fauna in the Lower Paleozoic of North America (Brett *et al.*, 1990), and the *Spiriferina*-Bank of the German Muschelkalk (K. A. Schäfer, 1973; Schwarz, 1975)—to global "bioevents" such as the Devonian Kellwasser (Walliser, 1985). Such "event-condensation horizons" (Seilacher, 1982a) commonly coincide with maximum water depths in transgressive–regressive cycles and are essentially time-parallel. Their time acuity, however, is significantly less than that of ecological event-concentrations: in the more extreme examples, biozone boundaries may coincide with the "event."

3.1.4. Paleontological Attributes

All event-concentrations record geologically brief episodes of accumulation. In absolute terms, they record minutes (e.g., emplacement of single turbidite) to perhaps years or decades (e.g., concentrations with postevent colonists).

The fossil assemblages that make up event-concentrations, on the other hand, can range greatly in time acuity. Some are biologically instantaneous, census assemblages. Others record considerably more time, e.g., where shells were concentrated by the reworking of preexisting death assemblages (second scenario in Fig. 4). Actualistic evidence suggests that time-averaged death assemblages on modern continental shelves record a minimum of many decades to a century or more (Kidwell and Bosence, this volume, Chapter 4). Consequently, a storm-concentration of indigenous shells represents a very brief depositional event (hours to days), but might contain a paleontological assemblage whose acuity was at best several tens to hundreds of years.

Notwithstanding this variability, event-concentrations *as a class* [=minor simple shell beds of Kidwell (1982)] probably still provide the highest-resolution paleoecological and biostratigraphic data of all concentration types. This is because (a) some subset of all event-concentrations provides censuses of living communities, and (b) even those event-concentrations composed of time-averaged assemblages are likely to be superior in acuity to composite and hiatal concentrations formed by the amalgamation of multiple event-concentrations. This assessment assumes that, as a group, event-concentrations do not on average contain significantly more exotic shells than other concentration types. Among event-concentrations, examples may be ranked in relative paleobiological reliability from those that plot near the "behavior" pole to those at the "exotic" and "reworking" poles.

3.2. Composite or Multiple-Event Concentrations

These include bioclastic megaripples and dunes, tidal inlet fills, washover fans, bioherms, subtidal shoals, shelf blankets, and calciturbiditic cones (Fig. 9), and thus range in thickness from decimeters to tens of meters and in lateral extent from square meters to perhaps thousands of square kilometers (Table V; Appendix B) (Eliuk, 1989). These bedforms and facies are ordinary sedimentary bodies except for being very shell-rich; their thickness is comparable or even expanded relative to surrounding, coeval strata. This distinguishes them from hiatal concentrations, which are multiple-event accumulations that are thin (i.e., stratigraphically condensed) relative to surrounding, coeval strata. [*N.B.*: Aigner *et al.* (1978) used "multiple-event" to denote storm-beds that rested on burrowed firmgrounds rather than on scour surfaces, i.e., not as a synonym for "composite," although this is the meaning that "multiple-event" has acquired.]

Densely fossiliferous bedsets and facies can accumulate either: (a) by the accretion of multiple small-scale deposits (i.e., event-concentrations)

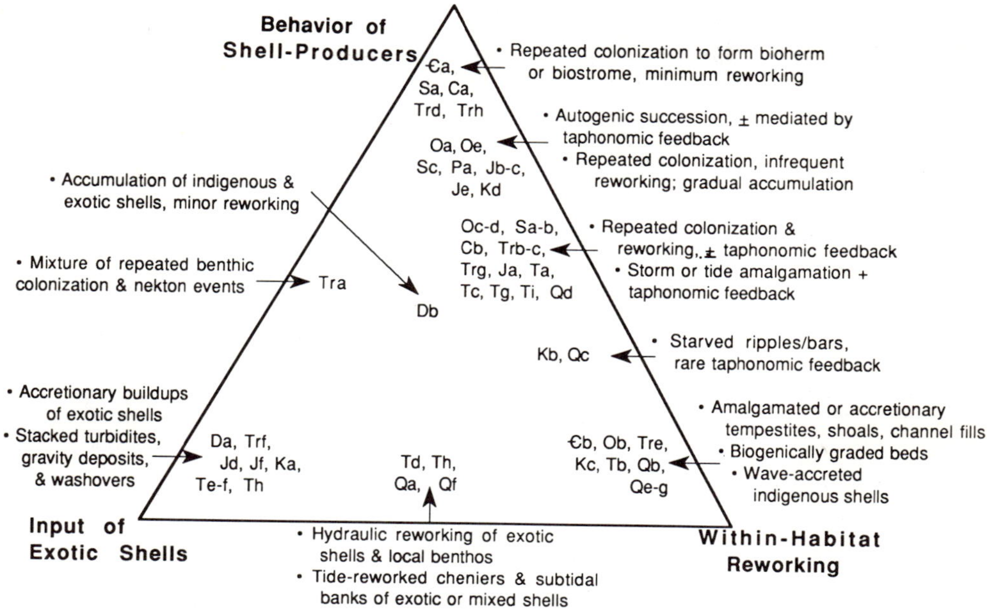

FIGURE 9. Origins of composite concentrations, plotted as in Fig. 6. Composite concentrations can reflect any combination of factors, with examples from all eras (Appendix B, same caveats as for event-concentration data set).

resulting in a larger deposit with a complex internal microstratigraphy (e.g., avalanche sheets within a megaripple, or repeated colonization horizons within a bioherm); (b) by the accretion of individual shells or complete amalgamation of event-concentrations, resulting in a deposit that may lack an obvious microstratigraphy (e.g., biogenically graded beds, quiet-water bioclastic limestones, proximal tempestite facies); or (c) by some combination of accretion and amalgamation. Like simple event-concentrations, composite concentrations vary greatly in geometry, matrix, and taphonomic attributes, and these features can be used in similar ways to determine the relative contributions of colonization, reworking, and exotic input (Fig. 9; Appendix B). Some composite concentrations are minor in scale, whereas others are major, stratigraphically significant bodies. They may accumulate rapidly or slowly, episodically or gradually, on or below the sediment–water interface.

3.2.1. Reconstructing Sedimentary Dynamics

Multiple-event accumulations must be analyzed layer by layer using the same kinds of evidence as used for single-event concentrations. Overall trends, such as the relative proportions of different kinds of event-concentrations, or up-section changes in the close spacing of discontinuity surfaces, in shell condition, and in ecological morphology, are also used

TABLE V. Composite or Multiple-Event Concentrations: Accretion or Amalgamation of Multiple Generations and/or Event-Concentrations; Average or Expanded Thickness Relative to Coeval Strata

Scale	Bed, bedset, facies
Geometry	Sheet/blanket, stringer, wedge, cone, lens, shoestring, mound, clinoform
Internal structure	Accretionary = complex microstratigraphy of discrete, subsidiary layers arranged in vertical (stacked; build up) and/or lateral succession (shingled; build up and out); amalgamated = entire concentration or large parts of it lack physical bedding
Context	May or may not be associated with discontinuity surface
Origin	Accretion or amalgamation of multiple events and populations into a single shell-rich deposit; longer-term process than the formation of any subsidiary event-concentration, ranging up to tens to hundreds of thousands of years; shells concentrated by any combination of repeated colonization, biological/hydraulic reworking, and exotic input
Taphonomic signature	Highly variable, depending on kinds of event-concentrations amalgamated, source and history of shells accreted individually, reworking subsequent to initial amalgamation/accretion, and seafloor exposure
Diagenetic signature	Varies; potential for "self-cementation" owing to quantity of soluble carbonate and primary porosity; size and geometry of some favor fluid flow, thus development of secondary porosity and/or cementation
Paleontological Use	Overall biological fidelity and temporal acuity can be no better than that of component event-concentrations, and generally should be poorer; isolated, unmodified census assemblages may be preserved within amalgamated concentrations; potential for significant complications from taphonomic feedback between dead shell and live benthos
Stratigraphic use	Fluid conduits and reservoirs; keys to short-term dynamics, but with lower acuity than event-concentrations; facies differentiation and interpretation; indicators of paleotopography (shoals, channel fills) and/or habitats and episodes of high benthic production; generally reliable sources of biostratigraphic data

(Kidwell and Aigner, 1985; Kidwell, 1985, 1986a; Miller *et al.*, 1988; Beckvar and Kidwell, 1988; Hoffman and Kobricki, 1989; Machalski, 1989). For example, Meldahl (1987a) reconstructed a history of both storm washover and longshore drift to explain a Recent well-bedded shelly spit from the Gulf of California (Fig. 10). Complex histories of repeated colonization and reworking are commonly inferred from biohermal buildups, whereas exotic input to the virtual exclusion of colonization and local reworking may characterize deep-water gravity-fed shell fans (Fig. 9; Appendix B).

An important caveat in the interpretation of multiple-event accumulations is that not all events are necessarily preserved—at least not as physically discrete events—and thus a perfect historical dissection is often not possible (Kidwell, 1982; Seilacher, 1985). The very nature of accretion,

FIGURE 10. Giant ripple (6 m maximum thickness) of oyster coquina from a tidal distributary of the ancestral Colorado River Delta (Pliocene Imperial Formation, California), providing an example of a large-scale, internally complex, composite concentration composed largely of exotic shells (scale, 1.5 m). Topsets and bottomsets contain disarticulated and densely packed whole and broken shells in tightly accreted, decimeter-thick layers (see beds of inclined and concordant shells in closeup, right; scale 5 cm); foresets are characterized by loosely-packed, better-sorted shell fragments in sand.

by which event-concentrations are plastered one upon the other, predisposes successive death assemblages to blurring by physical reworking, bioturbation, and shelly colonists [see Cutler and Flessa (1990) and Flessa *et al.* (1990) for estimates of the magnitude of "stratigraphic disorder" created by these processes]. Portions of the record can also be eroded during subsequent events. A further factor is the effect of already accumulated shell on the record of subsequent events: storms may not erode to their normal potential because of shell armoring at the seafloor or at depth, and diagnostic shell orientations may not be produced because of shell/shell interference (Futterer, 1978; Seilacher, 1985).

Because of uncertainty regarding the absolute completeness of the record at the scale of individual events, and the potential for interference effects, the dynamic histories of multiple-event concentrations can at best be *characterized,* in contrast to single-event concentrations, which commonly can be *described* with considerable confidence. Moreover, high-density and interlocking bioclastic fabrics may be overrepresented, since these have greater potential for preservation in the record. Interpretations of dynamics from composite concentrations are consequently of greatest value when the preservational styles of several accumulations are contrasted: e.g., accumulation #1 records more erosional reworking, less seafloor exposure, etc., than accumulation #2 (Fig. 11).

3.2.2. Facies Differentiation and Paleogeography

Lateral variability in shell abundance may be used as a basis for differentiating sedimentary facies and, when combined with taphonomic analysis of the concentrations themselves, improves paleoenvironmental interpretation. The advantages of such an integrated approach to basin-scale paleogeography are well illustrated by the studies of Fürsich (1981, 1982) in the Jurassic of Europe and Greenland, Aigner (1985) in the Triassic of Germany, Meyer and Ausich (1990) in the Mississippian of Kentucky, Brett (1986) in the Devonian of New York, and Cherns (1988) in the Silurian of Wales. The coquinas described in Appendix B suggest the breadth of environments in which this approach is possible: even tectonic basins of high clastic input and syndepositional deformation can be illuminated (e.g., Colella and D'Alessandro, 1988; Nebelsick, 1989). Like event-concentrations, composite coquinas occur across a lithologic spectrum from conglomerates to clays, and range bathymetrically from subaerial to abyssal settings.

Composite concentrations—particularly lenticular bodies—are also useful in reconstructing paleotopographic relief. Many facies-scale coquinas, for example, were paleohighs; these include shoals and cheniers formed of exotic, current-swept shells, and *in situ* bioherms formed by selective, gregarious colonization. Facies-scale coquinas can also record the infilling of paleolows, such as tidal inlets and distributary channels (Appendix B).

This scale of concentration can also indicate environments or episodes

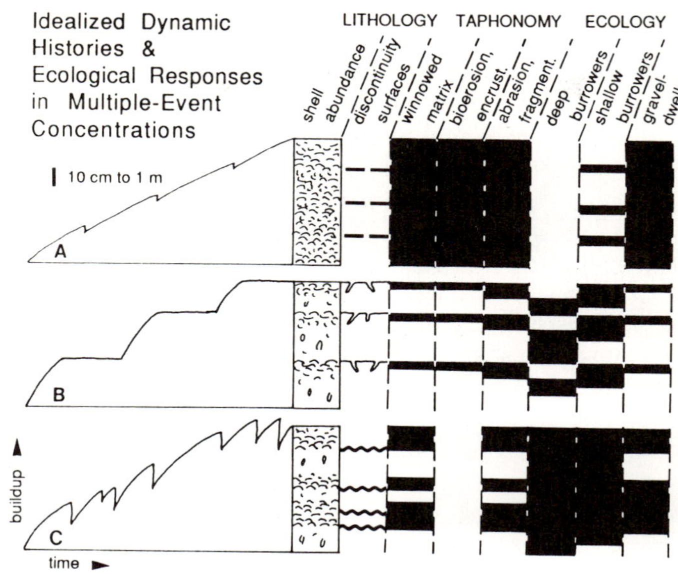

FIGURE 11. Composite concentrations may accumulate gradually or episodically, and under an array of sedimentary conditions including the idealized end-members of (A) complete sediment starvation, or repeated deposition of only temporary, thin sedimentary increments, (B) repeated deposition of thick sedimentary increments, alternating with nondeposition (omission), and (C) repeated deposition of thick sedimentary increments, alternating with partial or complete erosional reworking. The short-term sedimentary and biological dynamics of multiple-event concentrations (whether composite concentrations of normal to expanded thickness, or condensed hiatal concentrations) may be interpreted from sedimentological, taphonomic, and paleoecological features. [Adapted from Kidwell and Aigner (1985) and Kidwell (1991a).]

of high benthic production, which may or may not coincide with conditions of low sediment dilution. Sediment-starved, shallow-water areas, for example, are common sites for the accumulation of thick bioclastic deposits, especially in warm seas. These include many transgressive limestones, bioherms perched on isolated fault blocks, and level-bottom accumulations along the distal edges of coastal alluvial fans (Appendix B).

3.2.3. Fluid Reservoirs and Conduits

From the perspective of Recent facies models, coquinas have received little attention despite their being targets for petroleum exploration—coquinas (excluding true reefs) constitute 40% of productive reservoirs in some basins [proprietary data; see general review by Eliuk (1988, 1989)]. At present, taphonomic criteria for depositional environments of coquinas are still rudimentary (Kidwell and Bosence, this volume, Chapter 4): considerable empirical work will be required before coquinas can be interpreted using single cores or small samples. Nonetheless, cores may provide sufficient information on the stratigraphic context, scale, and internal stratigraphy of coquinas to be useful. For example, washover fans tend to be low-relief stratified wedges, whereas cheniers are poorly stratified elongate bodies; in general, shelly shoals tend to coarsen up and channel fills tend to fine up as with other clastic sediments.

3.2.4. Paleontological Attributes

The time acuity of these coquinas is unlikely to be finer than a sample taken from a simple event-concentration; in general, acuity can be expected to be poorer. Microstratigraphic sampling of a composite deposit might yield high-acuity samples, but not without some uncertainty owing to the likely mixing and taphonomic culling of assemblages during accretion (Kidwell, 1982).

Another cause of mixed assemblages (aside from the introduction of exotic shells, which might influence any shell concentration) is the potential for taphonomic feedback, that is, changes in the composition of the living community owing to the accumulation of dead shells in the benthic habitat (Kidwell and Jablonski, 1983; Kidwell, 1986a, 1991a) [one form of faunal condensation *sensu* Fürsich (1978)]. Shell-gravel dwelling species can live among the remains of earlier, soft-bottom dwellers, yielding an entirely indigenous but ecologically mixed fossil assemblage. The change from soft- to shell-bottom conditions can also influence the morphology and life positions of benthic species common to both (e.g., Kidwell and Aigner, 1985; Machalski, 1989). The dynamics of shell accretion in habitable environments thus can include continued benthic colonization and ecophenotypic shifts in shell morphologies that are distinctly nonrandom in nature: the resulting deposits are neither simple nor strictly taphonomic in origin.

3.3. Hiatal or Condensed Concentrations

These are complex, accretionary or amalgamated concentrations of bed to facies scale that are thin relative to nearby deposits of the same age, owing to slow *net* accumulation. Demonstration that the concentration is stratigraphically condensed usually requires regional analysis by which time planes (e.g., bedding planes, minor discontinuities, seismic reflectors, biostratigraphic datums) are shown to converge or merge. Complex microstratigraphies, faunal or event-"condensation" of distinct assemblages and sediments (*sensu* Fürsich, 1978; Seilacher, 1982a), and dense-packed bioclastic fabrics are all consistent with stratigraphic condensation, but also characterize many composite concentrations of normal or expanded stratigraphic thickness and thus do not provide diagnostic evidence. Certainly the stratigraphic term "condensation" should not be used as a synonym for "concentration."

The operational definition of condensation used here is standard: a deposit that is thin relative to coeval deposits owing to slow net accumulation [paraphrased from Jenkyns (1971); Kidwell (1991b)]. Even this precise a definition is vague: how much thinner must the deposit be than surrounding strata; how complete must it be with respect to elapsed intervals of time; over what minimum period of time must conditions of slow accumulation be maintained? The intent here is simply to discriminate qualitatively between (a) the degree of time-averaging and faunal mixing associated with normal accretionary growth of the stratigraphic record, which includes many short-term omission surfaces, and (b) the more extreme mixing and overprinting associated with geologically prolonged stillstands (or near-stillstands) in stratigraphic accumulation. The former processes result in event- and composite concentrations that constitute or are embedded within "normal" aggradational facies and facies tracts, whereas the latter yield comparably complete but thin records, and are commonly associated with stratigraphically important surfaces (e.g., sequence-bounding unconformities, flooding surfaces, downlapped surfaces).

Condensed concentrations can form because sediment was never supplied to the site (sediment starvation), was never deposited out of suspension (total passing), or was deposited only temporarily (dynamic bypassing, with alternating small-scale deposition and erosion/corrosion). Any or several of these conditions of low sedimentary dilution may influence a single condensed coquina. Net shell input, whether exotic or local in source, must also be relatively low during this period; otherwise, the coquina might build to a thickness comparable to surrounding areas, notwithstanding low sedimentary dilution. Beckvar and Kidwell (1988) applied the term hiatal to all concentrations formed during episodes of low clastic dilution, but this meaning is narrowed here to include only those concentrations that are also demonstrably condensed.

Hiatal concentrations may accumulate in a variety of water depths and oceanographic settings, including basin margins, open shelves and "swells," drowned platforms, bypassed slopes, distal or silled basins, and seamounts (Appendix C). They may reflect any combination of biological, hydraulic, and diagenetic processes (Fig. 12), and may form during any phase of a transgressive–regressive or onlap–offlap cycle. Base-of-cycle and end-cycle concentrations formed in shallow water typically have different taphonomic/ecological signatures than deeper-water midcycle concentrations (Kidwell, 1988b, 1991b; Banerjee and Kidwell, 1990, 1991); condensed coquinas also tend to vary according to the period of accumulation. In general, hiatal concentrations have more complex microstratigraphies and/or diagenetic histories than composite concentrations formed in the same environment; this probably reflects both the cumulative effects of a greater number of short-term "events" and accumulation over periods of environmental change [i.e., many hiatal coquinas are environmentally as well as stratigraphically condensed; see review in Kidwell and Bosence (this volume, Chapter 4)].

For example, shallow-water condensed deposits (e.g., transgressive

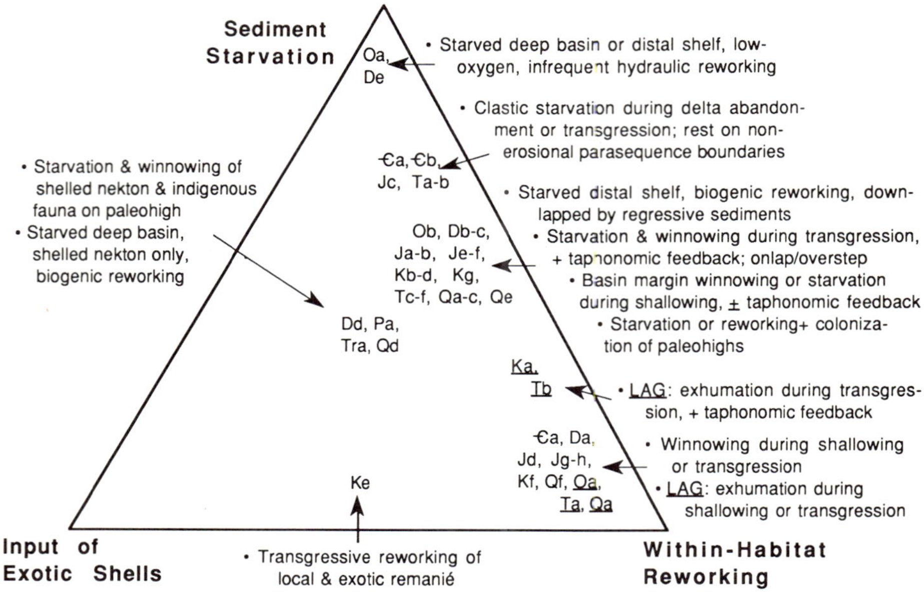

FIGURE 12. Origins of hiatal and lag-concentrations, plotted on a ternary diagram according to the relative importance of sediment starvation, biological or physical reworking, and exotic input at the accumulation site. All lag-concentrations were expected to lie within the reworking corner. Hiatal examples were expected everywhere but in the extreme exotic corner; scarcity of examples in the center of the triangle probably reflects a bias in the published literature rather than absence in the stratigraphic record. See Appendix C for description of examples.

''lags,'' seamounts) commonly contain anastomosing and complexly cross-cutting discontinuity surfaces, resulting in subsidiary ''beds'' within the coquina that are irregularly lenticular or pinch-and-swell: a few discontinuity surfaces may be laterally traceable, but in detail the microstratigraphy is highly variable [e.g., major shell deposits in Kidwell (1989a)] (Fig. 13 and Table VI). Shells may be densely packed, and generally show great preservational heterogeneity, both within and among subsidiary assemblages. A hallmark of these and other condensed coquinas, preservational heterogeneity is one of the best, single-outcrop clues that a shallow-water concentration might be condensed.

In contrast, the microstratigraphy of deep-water (below storm wave-base) hiatal concentrations is often subtle, owing to bioturbation and/or diagenetic obliteration, and in absolute terms these concentrations may be quite shell-poor [e.g., midcycle bone bed in Kidwell (1989a); midsequence biomicritic limestones in Banerjee and Kidwell (1991)] (Table VI). These, too, are characterized by shells in a range of stages of taphonomic and diagenetic alteration.

Biostratigraphically condensed concentrations (in which guide taxa from successive biozones are mixed or occur in microstratigraphic succession) may incorporate the broadest array of preservational states, ranging from reworked steinkerns to original shell material. Shells commonly are only loosely packed. Many biostratigraphically condensed beds thus are too

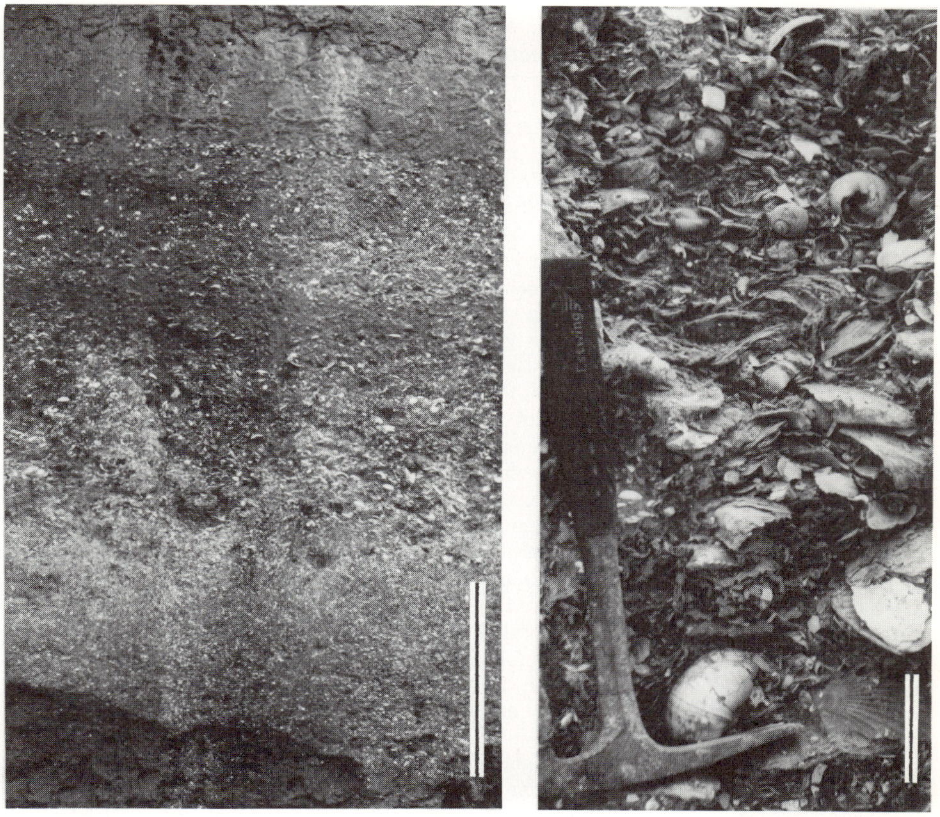

FIGURE 13. Hiatal concentration formed during marine transgression (Miocene Choptank Formation, Maryland); scale 1 m. Condensed over a paleohigh of the basal *Thalassinoides*-burrowed disconformity, the concentration has a complex internal stratigraphy of well-sorted fragmental shell hash (lowermost 1 m), poorly sorted massive coquina (middle 1 m; see closeup, right; scale, 5 cm), and incompletely amalgamated, thin event-concentrations (top 1 m).

sparsely fossiliferous to be called coquinas in a conventional sense, but may nonetheless represent local peaks of shell abundance. The frequently unimpressive, nondescript nature of these most important of all hiatal concentrations underscores the value of noting even low-density ''concentrations'' of shells when logging section.

3.3.1. Paleoenvironmental Reconstruction

Many shallow-water examples indicate dynamic histories with repeated cycles of shell burial, exhumation, winnowing, and recolonization (Kidwell and Aigner, 1985) (Fig. 12; Appendix B). Thus, although hiatal concentrations by definition form at a slow *net* rate, accumulation may be gradual or episodic with periods of omission or erosion/corrosion, just as with composite coquinas of normal thickness (Fig. 11). Similarly, shells may accumulate on or below the sediment–water interface.

TABLE VI. Condensed, Hiatal Concentrations: Complex, Amalgamated and/or Accretionary Concentrations That Are Thin Relative to Coeval Strata

Scale	Bed, bedset, facies
Geometry	Sheet/blanket, lens
Internal structure	Complex microstratigraphy of anastomosing discontinuity surfaces; discretely recognizable event-concentrations are rare; deposit may lack bedding features of any type owing to pervasive bioturbation and/or diagenesis; loosely-packed and dispersed fabrics common for some intercalations and can characterize the bulk of many biostratigraphically condensed examples
Context	Commonly but not necessarily associated with major discontinuity surface; can mark local paleotopographic highs or lows, and either base or top of shallowing-up cycle; can mantle surfaces of onlap, downlap, or flooding (i.e., transgression); can lie under surfaces of toplap; grade laterally into non-condensed deposits (shell-rich or shell-poor) or are truncated
Origin	Amalgamation or accretion of multiple events and populations into a single shell-rich deposit owing to slow net rates of both shell input and sediment accumulation; longer-term process than the formation of individual event-concentrations, up to millions of years for biostratigraphically condensed deposits; shells concentrated by any combination of colonization, biological/hydraulic reworking, and exotic input
Taphonomic signature	Typically a mixture of well-preserved and poorly preserved or highly modified shells, including reworked steinkerns and prefossilized material; post-mortem damage may be physical or biogenic, depending on setting, but repeated reworking common; hydraulic fabrics rarely preserved
Diagenetic signature	Varies; concentrations formed over subevolutionary time-scales resemble amalgamated/accretionary deposits +/− accessory authigenic minerals; biostratigraphically condensed deposits commonly are marked by more complex and pervasive diagenesis of both shells and matrix; because packing density is typically lower, biostratigraphically condensed examples tend to have lower porosities
Paleontological use	Expect lower biological fidelity and temporal acuity than amalgamated/accretionary deposits owing to slow net rates of accumulation; the more severe the condensation (i.e., the thinner the final deposit per unit time, and/or the longer the period of slow net accumulation), the less likely that microstratigraphic dissection of the deposit will yield assemblages comparable to those of event-concentrations; complex source of paleontological data owing to high potential for taphonomic feedback, overprinting/recycling of shells, and condensation of ecologically and/or evolutionarily disparate faunas
Stratigraphic use	Keys to paleotopography; keys to slowdowns in sediment accumulation, indicators of lithostratigraphic incompleteness; commonly associated with stratigraphically significant surfaces (unconformities, sequence and parasequence boundaries, flooding surfaces, horizons of downlap and backstepping); fossils may be abundant and diverse, but commonly yield complicated paleoecological and biostratigraphic signals

Condensation of shells therefore should not be stereotyped as a passive or strictly surficial process (cf. Davies *et al.*, 1989b), although shells apparently do persist directly on the seafloor in many sediment-starved settings today [e.g., temperate continental shelves (J. B. Wilson, 1986; Nelson, 1988)]. Low levels of endo- and epibiont infestation suggest that in many situations, accumulating shells were protected by a veneer of sediment despite the overall regime of slow seafloor aggradation (Kidwell, 1982, 1986b; Kidwell and Jablonski, 1983) (possible under any scenario in Fig. 11). Even under conditions of slow net sedimentation, shells may be rapidly buried by migrating ripples (R. G. Johnson, 1957) or by bioadvection (Meldahl, 1987b), and thereby escape a variety of surficial and near-surface taphonomic processes. Because episodic burial and exhumation may characterize even sediment-starved sites in shallow water, the taphonomic signature of starved hiatal concentrations in many instances may be indistinguishable from that of coquinas condensed through dynamic bypassing (Kidwell and Jablonski, 1983; Kidwell, 1986a, 1989a, 1991b).

The short-term dynamics of condensation may be evaluated using the same criteria (subject to the same pitfalls) that are applied to composite concentrations (Fig. 11). Dynamic histories are more difficult to reconstruct in detail, however, because of the even greater condensation of successive short-term events and faunas and the usually more severe near-surface diagenesis (see examples in Appendix C).

In contrast to lag-concentrations composed of material derived from older deposits, hiatal concentrations are composed of shells that are broadly contemporaneous with the period of low net sedimentation. They thus may provide a direct paleoecological and taphonomic record of paleoenvironmental conditions during the slowdown in sediment accumulation. This is illustrated by transgressive hiatal concentrations, which preserve a microstratigraphic record of successively offshore benthic assemblages and sedimentary structures (Kidwell, 1989a) (Fig. 13): had shells been supplied by the erosion of older deposits rather than by ongoing colonization during relative sea-level rise, this particular microstratigraphy could not have developed.

The distribution of hiatal concentrations themselves can be a clue to broader-scale environmental and historical conditions. Condensation is favored by: (a) certain paleotopographies that isolate parts of the seafloor from sediment supply, particularly paleohighs, silled basins, and deep (subphotic or distal) seafloors; (b) other geological conditions that either starve the seafloor (relative sea-level rise that outpaces sediment supply and/or switches sediment sinks; delta-lobe and fan switching) or cause prolonged bypassing (relative stillstand in a system at grade); (c) low to moderate rather than high rates of shell production; and (d) low to moderate rates of shell destruction, in order to keep net shell input at a reasonable level.

3.3.2. Keys to Discontinuity Surfaces

The importance of slow accumulation in the formation of hiatal coquinas tends to link them with stratigraphically important surfaces of erosion and omission. Theoretically, hiatal concentrations might be associated with any asymptotic form of stratigraphic convergence against a surface; this would include asymptotic onlap, offlap (toplap), downlap, or "backlap" [= backstepping; terminology of Van Wagoner *et al.* (1988)] (Fig. 14). Examples of each of these arrangements can be found in the rock record (Kidwell, 1991b) (Fig. 12; Appendix C). In some instances, the taphonomically mixed assemblages that characterize condensed beds provide unique evidence for hiatuses that otherwise are lithologically obscure (e.g., Baird, 1978, 1981; Cherns, 1988; Banerjee and Kidwell, 1991).

The most common associations of hiatal concentrations are: (a) in onlap or overstep relation to a low-gradient marine flooding surface; these base-of-cycle coquinas may mantle the lower boundary of third-order sequences and parasequences, with or without erosional ravinement (e.g., Kidwell, 1989a) [= "sole-beds" of Brinkmann (1929); BOP shell beds of Banerjee and Kidwell (1991)]; (b) resting on the transgressive systems tract and downlapped by the highstand tract of a third-order sequence; such midcycle coquinas approximate maximum water depth and maximum transgression within a sequence (e.g., Kidwell, 1989a; midsequence limestones of Banerjee and Kidwell, 1991) [= "condensed section" of Van Wagoner *et al.*

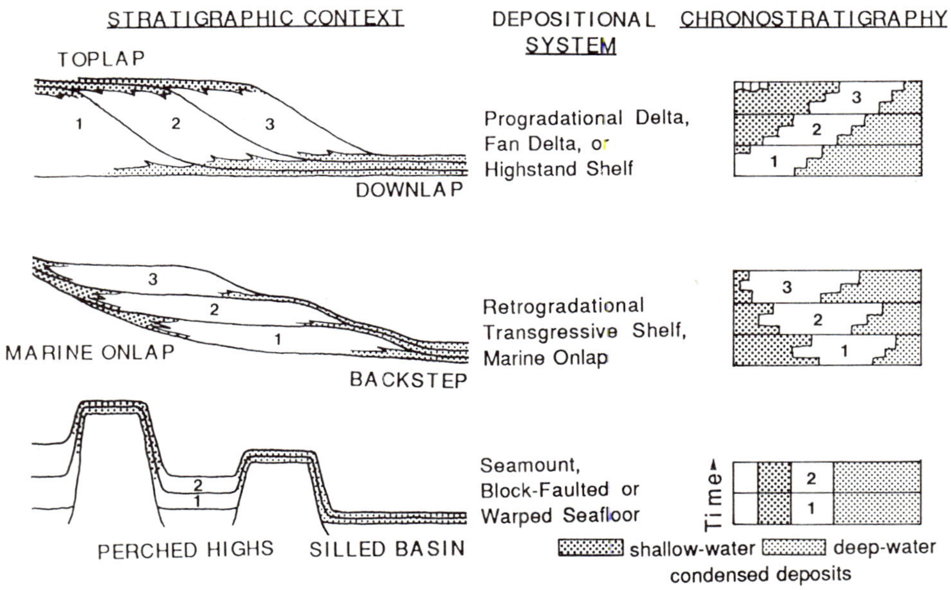

FIGURE 14. Common sites for stratigraphic condensation, and thus with potential to form hiatal shell concentrations. Shallow-water (neritic) examples are typified by physical reworking and reflect either sediment bypassing or starvation, whereas deep-water (deep neritic, bathyal, abyssal) examples generally show biogenic rather than physical reworking and reflect sediment starvation. Numbers indicate successively younger strata. Toplap, downlap, onlap, and backstep follow definitions of Van Wagoner *et al.* (1988).

(1988)]; and (c) associated with maximum marine shallowing, sediment transport, and offlap at the upper boundary of third-order sequences and parasequences (Fig. 14). These latter end-cycle or "roof beds" (Brinkmann, 1929; Klüpfel, 1917; TOP shell beds of Banerjee and Kidwell, 1991) can take several forms, depending on water depth. In deep-water slope deposits they may be winnowed horizons of concretionary material [e.g., pebble beds of Baird (1981)], but in shallow water they commonly consist of laterally accreted channelized lags or bedding plane concentrations that lie just below the sequence boundary [e.g., Rhaetic bone sand (Reif, 1982); roof-beds of Klüpfel cycles (Bayer *et al.*, 1985); bone sand of St. Marys Formation (Kidwell, 1989a, 1991b)]. End-cycle hiatal concentrations are far less common than base-of-cycle and midcycle types, possibly because they are more prone to erosion during the final stages of marine regression.

3.3.3. Chronostratigraphic Attributes

The usefulness of hiatal concentrations in time correlation varies considerably (Kidwell, 1988b, 1991b) (Fig. 14). Laterally extensive hiatal concentrations associated with onlap, for example, should be fundamentally diachronous when traced toward lap-out (updip). Thus, although they are likely to be well preserved under transgressive deposits, they are not necessarily reliable for time correlation. Hiatal concentrations associated with toplap may also be diachronous (but younger as traced downdip). They are commonly laterally discontinuous, owing to either channelized form or subsequent erosion [subaerial truncation of updip edge during later regression, or submarine reworking of downdip edge during the next transgression (e.g., Aepler, 1974)], and thus are less practical for time correlation.

Hiatal concentrations associated with backlap or downlap, on the other hand, may vary in duration when traced laterally, but should incorporate a chronostratigraphic time plane, and thus have value for correlating strata within a single basin. If eustatic in origin, they may be reliable over larger areas. Many widespread "event horizons" (see Section 3.1.3) probably are in reality condensed, mid-cycle hiatal concentrations of this type, coincident with or bracketing maximum transgression. They record longer periods of time than event-concentrations *sensu stricto*, and thus have lower chronostratigraphic acuity.

Highly localized hiatal concentrations (e.g., perched or silled types in Fig. 14) may be controlled by tectonic topography and erratic sediment transport systems rather than by broad-scale eustatic or climatic factors. They thus should not be assumed to have formed contemporaneously with other localized concentrations in the same or adjacent basins, and are of little use as chronostratigraphic markers.

Finally, although they are stratigraphically thin, hiatal concentrations provide relatively complete (but complicated!) records of elapsed time. Measures of lithostratigraphic completeness, calibrated against short-term sedimentation rates (Sadler, 1981), will therefore underestimate the completeness of the biostratigraphic record if condensed intervals are present (Kidwell, 1982; McKinney, 1985; Brett and Baird, 1986). The greater the extent to which shell concentrations are dictated by periods of slow net sedimentation, the greater the discrepancy between lithostratigraphic and biostratigraphic completeness by these measures.

3.3.4. Paleontological and Biostratigraphic Attributes

Slow net rates of accumulation have several implications for the quality of fossil data, with the net result of complicating paleobiological analysis (Kidwell and Aigner, 1985; Kidwell, 1986a). Mixing of ecologically unrelated communities and taphonomic feedback may increase species diversity in the assemblage; on the other hand, prolonged exposure to taphonomic processes and/or repeated burial–exhumation cycles may reduce diversity by removing less durable forms (Kidwell, 1986a).

Evolutionary data are also vulnerable: condensation may inflate apparent morphological variance (by mixing successive populations), drive living populations toward a particular ecophenotype (by taphonomic feedback), or skew fossil populations toward taphonomically durable morphs (by selective destruction of fragile forms) (Kidwell and Aigner, 1985; Kidwell, 1986a; Cohen, 1989; Bell *et al.*, 1989). Very prolonged condensation can obscure the stratigraphic order of first and last appearances.

Hiatal concentrations thus can yield very complicated paleoecological and evolutionary signals, a combined effect of taphonomy and ecological response. These complications are similar to those associated with noncondensed composite concentrations, but amplified by the effects of slow net accumulation.

3.4. Lag-Concentrations

Shell lags are thin (generally <1 m), biostratinomic and/or diagenetic residua, or skeletal resistates, and mark horizons of significant stratigraphic truncation. Truncation may be accomplished by mechanical erosion or by chemical corrosion, and constitutes the selective removal of matrix and of some or much of the original shelly assemblage. The criterion of "significant stratigraphic truncation" is used to differentiate lags, which are associated with various kinds of unconformities and ravinements, from concentrations associated with minor erosional events (e.g., storm scour) or with largely "within-habitat" erosion (e.g., intertidal channel migration). All of these concentrations of exhumed shells are simply differently scaled expressions of the same phenomenon.

Lags are typically thin, but this depends upon the volume of strata removed, the original quantity of shell contained in that volume, and the extent to which less durable shells are destroyed during reworking. Transport of shells out of (and into) the area can also be a factor. As a result, some lags are simple bedding-plane smears, whereas others are decimeters to 1 m thick,

TABLE VII. Lag-Concentrations: Exhumation and Concentration of Highly Resistant Shells by Erosion/Corrosion, Associated with Significant Stratigraphic Truncation

Scale	Bed, bedset
Geometry	Sheet/blanket, lens, horizon of laterally discrete or shingled lenses
Internal structure	Varies: homogeneous (structureless), size-graded, or amalgamated (e.g., microstratigraphic sequence of bedding plane pavements, or laterally shingled channel lags)
Context	Associated with major surface or with closely spaced lesser surfaces of stratigraphic truncation or stratigraphic shortening (pressure solution); can mark local paleotopographic highs or lows, and either base or top of lithologic cycles; commonly associated with sequence and parasequence boundaries, transgressive ravinements, and unconformities of varied types
Origin	Exhumation and preferential removal of matrix by erosion and/or corrosion, +/− transport, with durable shells culled from previously buried (and, in some instances, diagenetically modified) death assemblages; period of erosion/corrosion may be brief or prolonged; thickness and shelliness of lag depend upon rate of truncation, abundance and condition of shells in truncated strata, and extent of shell reworking before final burial
Taphonomic signature	Residuum of mechanically and chemically robust shells or assemblage of highly modified less robust shells; lags associated with minor scour surfaces (single storm events, channel migration) are less likely to show such modification consistently; dense and loose packing
Diagenetic signature	Varies from assemblages of exclusively original shell material to those of exclusively steinkerns and prefossilized material; preferential cementation common
Paleontological use	Low-fidelity assemblages derived from older or disparate facies
Stratigraphic use	Keys to stratigraphic truncation and age of missing strata; indicators of lithostratigraphic incompleteness; keys to stratigraphically significant surfaces (unconformities, sequence boundaries, transgressive ravinements); highly suspect sources of biostratigraphic and paleoecological data, inasmuch as fossils date from earlier (pretruncation) periods

but most are dominated by the least soluble and least transportable of the original material. Shells are typically in poor condition, with abundant disarticulation and fragmentation, plus pitting, abrasion, and/or bioinfestation in some instances. Aragonitic species may be represented by steinkerns only, and associated vertebrate material commonly shows evidence of having been reworked after some diagenetic alteration (Table VII; Appendix C).

Lag-concentrations form most commonly in subaerial and shallow-marine environments, associated with either initial transgression or late-stage regression (Fig. 12). Transgressive lags are often colonized by marine benthos, resulting in a shell concentration with mixed attributes, whereas such overprinting in less likely during regression. Lags can take several forms in siliciclastic systems, ranging from laterally continuous tabular bodies (common with rapid transgression) to laterally accreted lenses (either transgressive or regressive arrangement) and persistent horizons of discontinuous lenses (dominated by pattern of shifting shoreface or fluvial/estuarine channels).

Although truncation is usually thought of as occurring exclusively at the sediment–water (or sediment–air) interface, pressure solution can result in significant intrastratal shortening during burial [e.g., 60–70% of original thickness in Jurassic carbonates (Huber, 1987)]. Many skeletal grains are less soluble than carbonate matrix [the solubility of echinoderms < mollusks < sponges < micrite (Huber, 1987)], particularly if the hard parts have been modified during early diagenesis (e.g., high-Mg calcite transformed to low-Mg calcite). By the preferential destruction of matrix, pressure solution thus has potential to create new concentrations and to enhance preexisting concentrations. Such intrastratal diagenetic lags (Kidwell *et al.*, 1986) may be recognized by their stylonodular fabric and/or sutured contacts of skeletal grains.

3.4.1. Stratigraphic and Paleontological Implications

Lags mark stratigraphic intervals where previously deposited sediment has been removed. Because surfaces of major truncation can be lithologically obscure, particularly in marine records, concentrations of residual skeletal debris, especially bones, teeth, and phosphatic steinkerns, have become textbook indicators of unconformities.

In a hypothetical end-member example, in which the lag-concentration consists solely of fossils derived by exhumation, the paleoecological attributes and biostratigraphic age of the fossils may be used to infer the kind and age of strata removed. However, in many real examples, some part of the assemblage has been transported from adjacent areas of exhumation, or produced by organisms during the period of erosion/corrosion. For example, a transgressive lag may consist almost entirely of comminuted oysters from truncated lagoonal deposits, but include some open marine shells from benthic colonists of the lag and from marine nekton and plankton (e.g., Tl in Appendix C). The paleobiological and biostratigraphic value of fossils from true lag-concentrations therefore should be considered suspect.

4. Larger-Scale Patterns in Coquina Type and Occurrence

Although many unanswered questions remain, stratigraphic analysis of shell concentrations leads to hypotheses both of how they formed and of the likely nature of their contained paleontological assemblages. This approach also leads to testable hypotheses about basin-scale and larger patterns in the distribution of marine coquinas.

4.1. The Low-Dilution Maxim

Most of these "predictions" are simple extrapolations from the premise that sites of slow net sedimentation (and thus low dilution) should engender denser concentrations of shells than sites of higher net sedimentation. Low dilution may be an outcome of sediment starvation or of winnowing and net erosion. For bowl-shaped shells, shell-rich fabrics can be produced by shell : sediment ratios as low as 1 : 5 (Fig. 2). Of course, as rates of net sedimentation and its short-term dynamics (Fig. 11) change, other ecological and taphonomic factors may change, too, affecting the net input of shells (Kidwell, 1986a, 1989a). Any low-dilution model thus merely indicates where there is potential for concentration. Deviations in the fossil record from patterns expected from low dilution alone are a measure of more complicated sedimentological, ecological, and taphonomic factors.

For example, if shells are likely to undergo more burial–exhumation cycles, or spend longer periods on or just below the seafloor during periods of slow aggradation, then individual shells might have significantly lower preservational potentials than in settings of "normal" aggradation. The resulting deposits may be shell-poor despite low sedimentary dilution (Kidwell, 1986a; Brett and Baird, 1986; Davies *et al.*, 1989b) [also see Fürsich (1982) for late diagenetic selective destruction of shells from condensed beds]. This situation must prevail in many instances, inasmuch as not all unconformities and omission surfaces are mantled by shell-rich deposits, despite having formed in habitable marine environments (Kidwell, 1986a; Beckvar and Kidwell, 1988). The typically loosely packed fabrics of biostratigraphically condensed deposits, compared with the densely packed fabrics of many hiatal coquinas formed over shorter, subevolutionary time scales (Section 3.3), are anecdotal evidence for a threshold period of condensation beyond which shell loss either matches or exceeds the concentrating effects of low sedimentary dilution. In post-Paleozoic, mollusk-dominated shallow-water siliciclastics, this threshold period appears to lie somewhere between tens of thousands of years and a few million years.

Alternatively, net shell input might be elevated during periods of slow sedimentation, resulting in deposits that are richer than would be expected from low dilution alone (Kidwell, 1986a). This high input might reflect a direct biological response to low sedimentation rates (e.g., greater success of sessile epibenthos on stable seafloors or under less turbid water) or an indirect product of concomitant environmental changes (e.g., taphonomic feedback during progressive concentration of shells; more diverse benthos or lower shell loss because of warmer waters associated with rapid, glacioeustatic transgression). Both Tozer (1971) and Mapes (1987) suggested that aragonitic cephalopod remains have higher preservation potentials under slow sedimentation regimes than under aggradational regimes because of more favorable authigenesis.

Whatever the attributed reason for the link, the combination of low sedimentary dilution and high shell input is commonly invoked for Recent

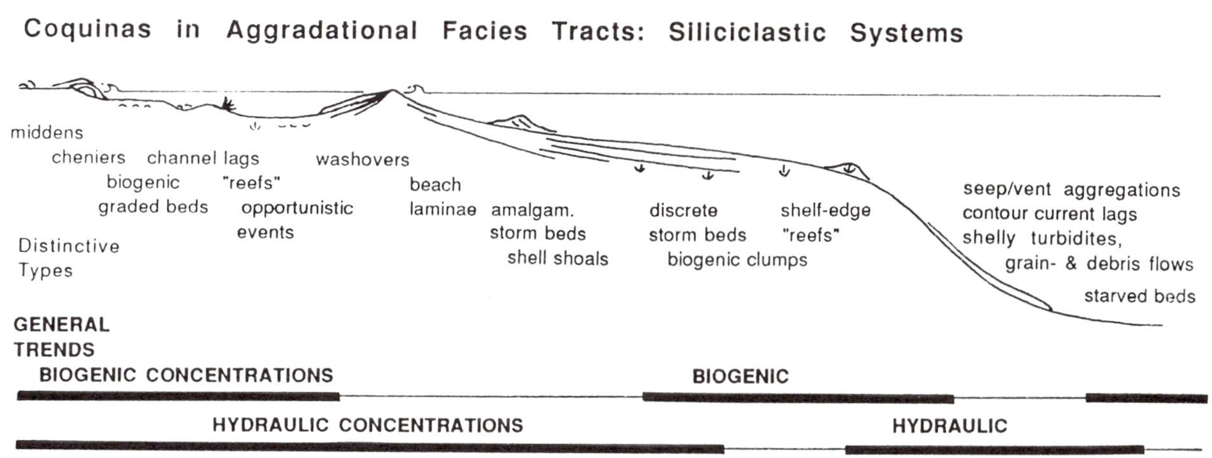

FIGURE 15. Idealized onshore–offshore trends in the distribution of event- and composite concentrations, based on sources in Appendices A and B, for an aggradational system. Biogenic concentrations can be found in virtually all environments, as can hydraulic concentrations of shells, but the relative proportions vary. [Modified from Kidwell *et al.* (1986).]

(Nelson, 1988; J. B. Wilson, 1988) and ancient concentrations alike (R. G. Johnson, 1960; Fürsich, 1990). Both theoretical models and systematic patterns in the distribution of coquinas indicate that only a few kinds of concentrations (and these are mostly event-concentrations) can be readily explained by high biological input or very low shell loss rates alone, without simultaneous low dilution (Kidwell, 1985, 1986a, 1988a) (Appendixes A–C). Independent field evidence for low dilution includes the close association of shell concentrations with erosion or omission surfaces, with condensed intervals, or in a matrix that is winnowed or rich in authigenic minerals.

The influence of sediment dilution has immediate geological interest because of the potential for ''reading'' sedimentary dynamics from patterns of shell abundance (if indeed sedimentation patterns are the primary dictator of shell accumulation in aquatic systems). This approach may be pursued at several scales, ranging from event- and composite concentrations within aggradational intervals (Aigner, 1979, 1985; Fürsich, 1982; Brett and Baird, 1986; Kidwell *et al.*, 1986; K. B. Miller *et al.*, 1988) (Fig. 15) to the distribution of hiatal and lag-concentrations within parasequences and sequences (Bayer *et al.*, 1985; Kidwell, 1985, 1988a,b, 1989a; Parsons *et al.*, 1988) (Figs. 13 and 16). In general, workers on aggradational records find that the formation of shell beds is dominated by rapid pulses of shell input, erosion, and sedimentation, whereas studies of larger stratigraphic intervals that contain diverse types of concentrations stress the importance of sedimentation slowdowns. These models are not contradictory, inasmuch as slowdowns in *net* sedimentation are one means of stacking discrete event-concentrations into composite and hiatal concentrations, and episodes of ''negative sedimentation'' (i.e., erosion) are essential to the formation of lags (Fig. 16).

4.2. Variation within and among Depositional Sequences

Most actualistic and stratigraphic studies of shell concentrations have focused on siliciclastic and mixed clastic/carbonate systems in slowly subsiding shelves and ramps, and so these are used here to illustrate the variety of ways in which coquinas can be distributed through unconformity-bound depositional sequences.

Although it is possible for an entire coastline and shelf to aggrade or prograde simultaneously over a large region (e.g., idealized transect in Fig. 15), the seafloor usually is a mosaic of sediment deposition, bypassing, starvation, and erosion at any given time, leading to a broader spectrum of types of coquinas. One common bathymetric trend consists of: (a) the onshore formation of an erosional lag, created either by shoreface erosion during transgression (cf. Liu and Zarillo, 1989) or by coastal channel migration during regression; (b) shallow subtidal hiatal concentrations

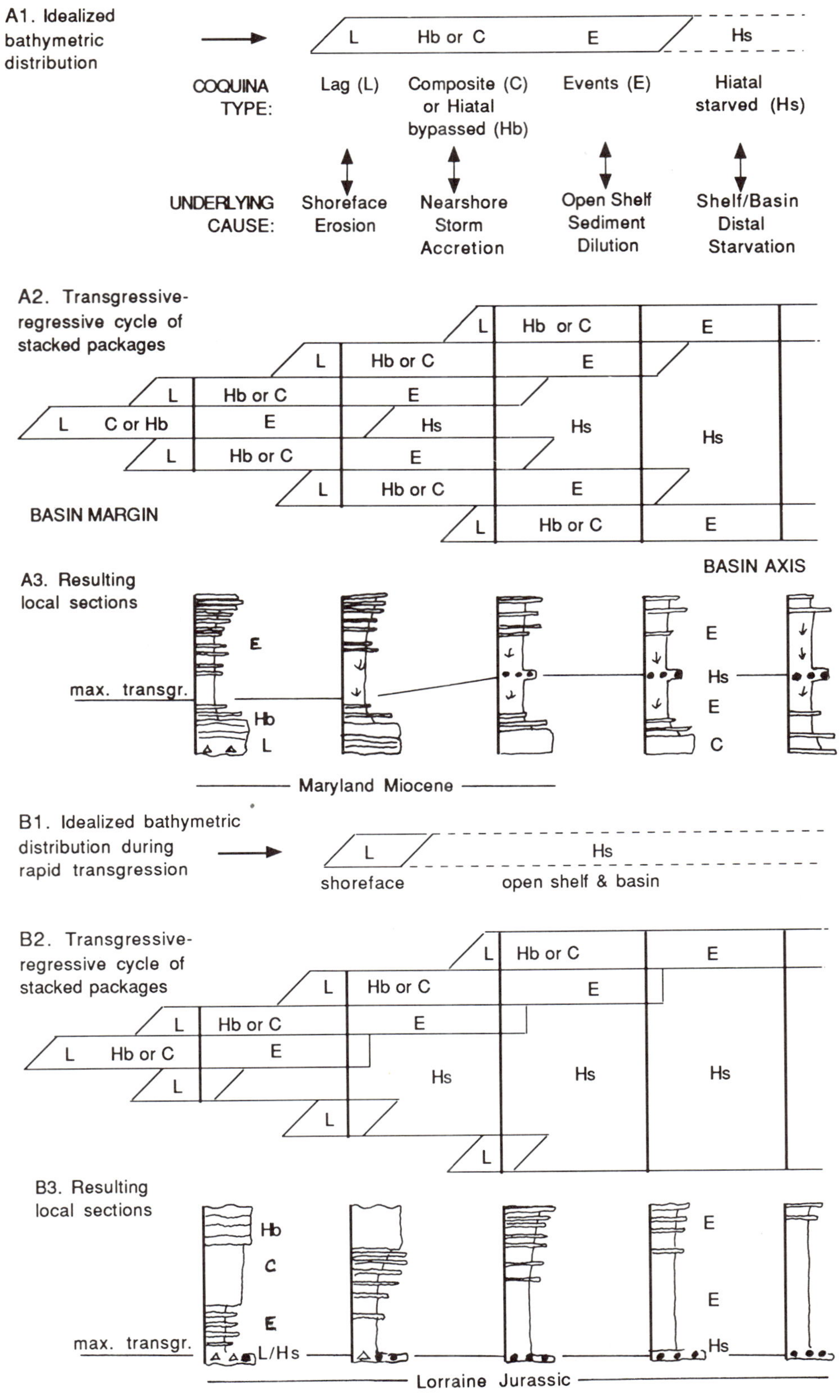

A1. Idealized bathymetric distribution
L
Hb or C
E
Hs
COQUINA TYPE:
Lag (L)
Composite (C) or Hiatal bypassed (Hb)
Events (E)
Hiatal starved (Hs)
UNDERLYING CAUSE:
Shoreface Erosion
Nearshore Storm Accretion
Open Shelf Sediment Dilution
Shelf/Basin Distal Starvation
A2. Transgressive-regressive cycle of stacked packages
BASIN MARGIN
BASIN AXIS
A3. Resulting local sections
max. transgr.
Maryland Miocene
B1. Idealized bathymetric distribution during rapid transgression
shoreface
open shelf & basin
B2. Transgressive-regressive cycle of stacked packages
B3. Resulting local sections
max. transgr.
L/Hs
Lorraine Jurassic

amalgamated by sedimentary bypassing, and/or composite coquinas created by storm accretion of event-beds or by bioherm/biostrome buildup; (c) offshore areas where discrete event-concentrations are preserved owing to permanent deposition of fine sediment; and (d) distal areas of sediment starvation with hiatal concentrations of benthic, planktic, and nektic debris (Fig. 14; A1 in Fig. 16).

When stacked in a transgressive–regressive pattern (A2 in Fig. 16), the result is a depositional sequence with several potential sites for stratigraphically significant coquinas. These include:

(a) A *base-of-cycle* coquina, which may be a transgressive lag, a condensed hiatal concentration, or a composite concentration; this rests on the transgressive surface, which may coincide with the lower sequence boundary.

(b) A *midcycle* coquina, which generally is a hiatal concentration, and occupies the deep-water core of the sequence [= "condensed section" of Van Wagoner *et al.* (1988)]; this marks the distal edge of stacked parasequences, which first backstep toward the basin margin and then prograde and downlap basinward.

(c) An *end-cycle* coquina, which may be a regressive lag, a shallow-water composite concentration, or (rarely) a condensed hiatal concentration, associated with erosional truncation, depositional offlap, or nondepositional toplap, respectively.

The distribution of coquinas through depositional sequences varies considerably, since base-of-cycle and end-cycle coquinas may be of several types and do not always develop. Moreover, end-cycle coquinas have lower preservation potential than other types, since they are often followed by conditions of subaerial erosion. The distribution of major coquinas through a third-order sequence of the Maryland Miocene [PP-1/2 (Kidwell, 1989a)] is illustrated in A3 of Fig. 16: the sequence contains a hiatal base-of-cycle coquina and a hiatal midcycle bone bed (shell-poor), and lacks an end-cycle coquina.

←

FIGURE 16. Idealized distribution of hiatal and lag concentrations in marine siliciclastic sequences, showing: (A1) common bathymetric trends in areas of sediment erosion (potential for lag formation), deposition (where event- and composite concentrations may be preserved), and nondeposition (where hiatal concentrations may be formed by sediment bypassing or starvation; (A2) the stacking of these parasequences in a transgressive-regressive cycle (vertical axis is time); and (A3) one of several possible stratigraphic records (vertical axis is thickness). Local sections in A3 are based on a third-order sequence in the Maryland Miocene [PP-1/2 (Kidwell, 1989a)], where nearshore transgression is marked by a basal bypassed hiatal coquina (similar to Fig. 13), maximum transgression by a starved hiatal bone-bed, and regressive deposits by event-coquinas (end-cycle composite and hiatal coquinas did not develop). Alternatively, the entire shelf may be starved during very rapid transgression (B1), resulting in more asymmetrical stratigraphic sequences (B2, B3). Klüpfel cycles [Jurassic of Lorraine (Klüpfel, 1917; Bayer *et al.*, 1985)] are excellent outcropping examples of such sequences, in which the entire transgressive phase is marked by a thin starved deposit, and regression comprises a full bathymetric array of event-, composite, and hiatal concentrations, capped by a lag in basin margin sections.

Transgression and regression are often characterized by different depositional systems, so that base-of-cycle and end-cycle coquinas have different features even though both are relatively shallow-water accumulations (Kidwell, 1991b). For example, during rapid marine transgression, the entire shelf may be starved even of fine clastics (B1 in Fig. 16). The resulting laterally extensive hiatal coquina mantles the basal flooding surface, and is overlain directly by a shallowing-up record. This pattern is common in parasequence-scale cycles (e.g., Ryer, 1977) and can also be observed in third-order depositional sequences, particularly in temperate latitudes where low rates of skeletal carbonate accumulation result in thin bioclastic limestones that comprise the entire transgressive record [e.g., superposition of midcycle hemipelagic sediments on clastic-starved shelf coquinas, Miocene of Italy (Simone and Carannante, 1988; Nelson, 1978)]. In other situations, clastic parasequences and sequences may have only a minor transgressive base-of-cycle coquina, but include a significant composite or condensed end-cycle coquina [e.g., Aigner (1985); Banerjee and Kidwell (1991); Jurassic Klüpfel cycles described by Bayer *et al.* (1985)] (B3 in Fig. 16).

Biostratinomic patterns in platform carbonates are less well known. Base-of-cycle concentrations in general appear to be less common. End-cycle condensed deposits are common, but may be marked by hardgrounds rather than by skeletal concentrations.

4.3. Variation among Tectonic Settings and Latitudes

The relative abundances of concentration types are expected to vary among sequences within a single basin owing to spatial and temporal variation in sediment supply, shell supply, and subsidence. For the same reasons, patterns of shell concentration may also vary among tectonic settings and during different phases of basin fill.

For example, during the early evolution of divergent margins, rapid downfaulting and flooding often create clastic-starved conditions. This can occur not only in the central basin (e.g., Recent Gulf of California) and on isolated horsts and seamounts (Wendt, 1988), but also along the shallow-water margins of the basin, which may become sites of benthic skeletal accumulation and/or reef growth (Montaggioni *et al.*, 1986; Kidwell, 1988a). Depending upon net shell input, these shell-rich deposits may be condensed, normal, or expanded in thickness. Stratigraphically similar coquinas may characterize the bedrock margins of convergent basins (Nebelsick, 1989; Braga *et al.*, 1990), but with increasing tectonic activity along the front of the allochthon, clastic-starved areas of shell concentration are shifted further out into the basin (Kamp and Nelson, 1987, 1988). Proximal areas and phases of rapid sediment infilling are marked by a minimum number of hiatal and lag-concentrations. Finally, patterns of sediment dilution and shell concentration on mature passive margins (Kidwell, 1988a, 1989a), over much of foreland basins (Parsons *et al.*, 1988; Banerjee and Kidwell, 1991), and in epicontinental seas (Hattin, 1986) appear to be most closely linked with transgressive–regressive cycles (Section 4.2).

Some of the variation in coquina styles among tectonic settings is related directly to differences in depositional systems (e.g., commonness of alluvial fan systems along shoreline; bedrock versus soft level-bottom environments), but of course these are cascading effects of the larger tectonic setting, including topographic gradients and episodicity of sediment supply.

Taphonomic variation might also be expected along paleolatitudinal gradients, owing to differences in rates of sediment supply, shell production, and shell destruction (Behrensmeyer and Kidwell, 1985). Most Recent and Late Cenozoic taphonomic studies are from temperate latitudes (Appendices A–C) [Kidwell and Bosence, this volume, Chapter 4; but see Eyles and Lagoe (1989) for a high-latitude exception], but this literature bias is balanced by carbonate-budget studies on the complete array of tropical to temperate limestones in Recent oceans [see Nelson (1988) for review; Carannante *et al.* (1988)].

As an initial hypothesis, a decrease in the average scale and stratigraphic frequency of shell concentrations might be expected with increasing distance from the highly productive, chloroalgal/coral-dominated low latitudes. On the other hand, temperate, clastic-starved carbonate shelves may be more prone to drowning than tropical equivalents because of lower rates of shell production [by mollusks, echinoderms, foraminifera, bryozoan, barnacle, and encrusting coralline algae (Simone and Carannante, 1988)], leading to condensed hiatal coquinas below the photic zone of intense shell destruction. The net latitudinal trend thus may not be striking in terms of total numbers of shell concentrations, but in relative abundance of types, with temperate records having a higher proportion of condensed coquinas (and perhaps of lags, owing to the abundance of calcite-secreting benthos), and tropical records having a higher proportion of "normal" composite coquinas.

4.4. Variation over the Phanerozoic

Given evolutionary shifts in the diversity and environmental distribution of shell-producers and of organisms that interact with shells, patterns of shell accumulation might be expected to change over the Phanerozoic. Such patterns are difficult to deduce, however, because only semiquantitative data are available, at best, for these paleobiological factors (Table VIII). An empirically determined geological history of coquinas, on the other hand, might indicate how and to what degree post-mortem bias has changed, and thus define the comparability of time-sequenced data sets used in paleobiological modeling. As products of evolutionary ecology in some guise, these

TABLE VIII. Phanerozoic Trends in Factors That Influence Post-Mortem Accumulations of Shells[a]

Shell-producers	Increase in diversity Increase in body size Increase? in shell robustness Infaunalization	Should favor bioclastic accumulations
Shell-producers	Shift to aragonite Loss of shelled nekton Loss of sessile epifauna on soft substrata Shift to mineral resorption from molts	Should reduce the likelihood of bioclastic accumulations through the Phanerozoic
Shell-destroyers	Increase bioeroders Increase duraphagous predators Increase depth and intensity of sediment irrigation by bioturbators	

[a]Adapted from Kidwell (1990).

same patterns might also represent an underexploited source of paleobiological insight.

Coquinas seem a particularly promising feature for tracking "evolutionary taphonomy" because they are the integrated product of so many different factors. However, the need to control so many variables makes the collection of an unbiased data set difficult and inhibits interpretation. For example, the North American record gives the impression that large-scale (thicker than 20–30 cm), dense-packed coquinas are far less common in Paleozoic rocks than in Cenozoic rocks (S. M. Kidwell, unpublished results, and consensus of polled paleontologists; Kidwell, 1990). Such a pattern, if true, might reflect several driving mechanisms: (a) ecological ascension of more fecund or taphonomically more durable shell-producers; (b) ecological ascension of organisms that actively concentrate dead shell material, that create conditions favoring shell preservation, or that interfere with shell-destroyers and -dispersers (or, ecological decline of organisms that limit dead shell accumulation); (c) an epiphenomenon of paleogeographic changes, including paleolatitudinal drift and proportions of tectonic basin types, especially the Cenozoic disappearance of epicontinental seas; or (d) an artifact of differential diagenesis with geological time.

Preliminary data from a Jurassic test where paleogeographic factors could largely be controlled (Kidwell, 1990), and inspection of the age distribution of coquinas reported in the literature (Appendixes A–C), suggest that thick post-mortem accumulations are at least as abundant and possibly more abundant in the younger record than in the older. Phanerozoic patterns in shell accumulation thus do not appear to be driven by externally imposed taphonomic pressures: evolutionary diversification of bioeroders, duraphagous predators, and other taphonomic agents would be expected to *reduce* the likelihood that large-scale coquinas form. Instead, the formation and preservation of large dense-packed coquinas appear to be most related to the phylogeny of certain shell-producers, most notably younger clades of mollusks, cirripeds, larger foraminifera, and some bryozoans. Although calcitic, brachiopods appear to have little post-mortem durability, with the exception of a few thick-shelled groups; pelmetazoan concentrations are a significant exception to the "thin-coquina" pattern for the Paleozoic. The organisms that dominate coquinas cannot be predicted on the basis of shell mineralogy, body size, or life habit (infauna vs. epifauna, solitary vs. gregarious, sessile vs. mobile). Instead, the most important *shared* attributes for coquina genesis seem to be (a) shell durability as imparted by skeletal microarchitecture, particularly low organic content, and (b) fecundity (Kidwell, 1990). Shell-producers with one or both of these attributes dominate coquinas because their relatively high shell input will more often overcome aggressive post-mortem taphonomic conditions. As discussed elsewhere, once shells accumulate above some threshold density or abundance, the likelihood of preservation appears to increase through several kinds of biological, physical, and biogeochemical feedbacks.

5. Conclusions

Coquinas are diverse in nature, whether considered from the perspective of composition, physical features, or processes and time scales of formation: they range from monospecific event-concentrations, to species-rich, complexly overprinted composite and hiatal concentrations and species-poor, highly residual lag-concentrations. Coquinas thus have a variety of implications for historical analysis, depending upon type. The two-stage analytic approach advocated here, based first on stratigraphic attributes and second on shell source and taphonomic features, is one way to exploit useful information from these distinctive deposits.

Paleontologic Strategies. Stratigraphic characterization of coquinas, so often omitted by paleontologists, is essential to interpreting the constituent fossil assemblages. No single taphonomic feature or suite of features can substitute for information on the dimensions, geometry, and context of shell concentrations. Stratigraphic typing immediately suggests the likely quality of paleobiological and biostratigraphical data (Tables IV–VII). The fact that complete stratigraphic information is not always attainable, for example, owing to limited outcrops and poor subsurface data, in no way downgrades its desirability or importance relative to conventional taphonomic information.

Paleontologically, coquinas may be viewed not only as variably biased repositories of specimens, but also as the record of distinctive benthic habitats where live/dead interactions might dominate community structure and dynamics. Paleoecological and paleoenvironmental interpretation of assemblages from coquinas thus is not always a simple matter of stripping the taphonomic overprint. Moreover, because shell abundance can also influence seafloor erodability and early diagenesis, fossil assemblages from coquinas may be preferentially preserved (or destroyed), thus imposing a further taphonomic structure on paleontological data.

The likelihood of taphonomic modification varies among environments, and generally increases with the period of accumulation: longer periods of formation are likely to include multiple burial–exhumation events and net shifts in the local environment. The likelihood of taphonomic feedback increases also as a function of shell abundance. Paleontological samples should thus ideally be collected from coquinas of a single type and packing density. The type of coquina to be targeted depends on the question to be asked. Simple event-concentrations that plot near the "behavior" pole are best for population-level paleoecology, and those that plot anywhere along the behavior/reworking axis are adequate for community-level characterizations. Composite concentrations from the behavior/reworking axis are also good for community-level work, with the exception of composites in which the physical environment changed markedly as indicated by change in sedimentary matrix [e.g., some forms of faunal condensation *sensu* Fürsich (1978)]. All event- and composite concentrations, including those with exotic benthos, are useful for biogeographic and biostratigraphica- nalysis. Hiatal and lag-concentrations, identified by physical stratigraphic context and by lithologic–diagenetic clues (Tables VI and VII), should be used with extreme care in all paleontological endeavors.

Geologic Strategies. Although even uninterpreted coquinas can be used to differentiate facies, information on shell condition and bioclastic fabric of coquinas permits many otherwise undetected aspects of sedimentation to be reconstructed. These details—about the intensity and frequency of storm erosion, the timing and thickness of depositional events, and the net rate and pattern of seafloor aggradation—are valuable at the very least as corroborative evidence, and commonly provide unique evidence for sedimentary dynamics. Sedimentologically, much may be gained by considering coquinas as depositional facies types and by developing explicit facies models. This has been pursued to some extent among carbonate petrologists, particularly for various shoals and reefs, but a diversity of other, underexploited coquinal types occur in carbonate and siliciclastic systems.

Stratigraphers have used coquinas, implicitly and explicitly, in sequence analysis in the past, but rarely as a cornerstone. The four types recognized here have qualitatively different applications (Tables IV–VII). These range from event-concentrations useful in high-resolution correlation within basins, to composite concentrations that track environments of high productivity or wave energy, hiatal concentrations that track environments and prolonged regimes of sediment starvation and bypassing, and lag-concentrations that mark major erosional truncation or intrastratal shortening. The link between major shell concentrations and conditions of low sediment dilution also make them valuable cues to long-term dynamics of sediment accumulation, whether controlled by tectonism, eustacy, or sediment delivery systems.

All but lag-concentrations yield faunas that are approximately contemporaneous with the concentration process, and thus are appropriate for paleoenvironmental and biostratigraphic interpretation, but the greater the exotic component, the more complex the accretionary history, and the more condensed the final concentration, then the greater the number of caveats to interpretation. On the other hand, the taphonomic nature of a coquina can indicate the significance of particular range terminations when standard lithologic evidence is ambiguous: first and last appearances that coincide with lag-concentrations are likely to reflect truncation, those that coincide with a hiatal concentration more likely reflect condensation, and those that coincide with the conformable boundary of a composite coquina most likely reflect nonevolutionary, ecological replacement.

Prognosis. Many vertebrate, microfossil, and macrofloral concentrations can be placed into these same stratigraphic types, although processes and conditions for preservation differ. Stratigraphic analysis, as opposed to taphonomic analysis in a strict sense, thus might be one way to begin more systematic comparisons across these major groups and their disparate depositional systems, as well as a means of comparing invertebrate faunas that differ radically across biogeographic boundaries and through Phanerozoic time. The variety and complexity of coquinas, on virtually all dimensions of the taphonomic equation, are sources of both their challenge and promise to historical geology and paleontology.

ACKNOWLEDGMENTS. Grateful thanks to the editors, David Jablonski, and an anonymous reviewer for their comments on the manuscript. Research and writing were supported by grants from Arco Foundation, Amoco Production Company, and the U.S. National Science Foundation (EAR 85-52411-PY1).

Appendix A: Fossil and Recent Examples of Event-Concentrations[a]

Age	Characteristics	Interpretation	Setting (Reference)
Cambrian			
Є a	Pavements of disarticulated trilobites, all ontogenetic stages, varied condition	Current reworking of possibly gregarious spp., especially of larval swarms	Nearshore shallow water, China (Zhang, 1989)
Є b	Pods, lenses, or layers of disarticulated, fragmented, well-sorted trilobite debris	Mechanically concentrated aggregates in basal scour lags	Intertidal clastics, Vermont (Speyer, 1983)
Є c	Clusters of whole trilobites	Mass burial after gregarious behavior	Subtidal, Nevada (Gunther and Gunther, 1981)
Є d	Lenses and thin beds of trilobite packstone, sorted or graded	Proximal and distal storm beds of local death assemblages	Open shelf, Alberta (Westrop, 1986)
Є e	Thin, graded beds of trilobite packstone, capped by stromatolites	Benthic colonization of storm-generated shell pavements	Open shelf, Alberta (Westrop, 1986)
Є f	< 1–10 cm trilobite grainstone beds, poor sorting, no grading, no abrasion	Gravity flow deposits of outer-shelf spp. on continental slope or basin plain	Bathyal–abyssal, Arkansas (Hohensee and Stitt, 1989)
Є g	2-cm-thick lenticular bodies of nontrilobite shell hash, in sandstone	Channel lag concentrates of hydraulically reworked shells	Intertidal?, Sonora (McMenamin, 1985)
Є h	2–8 cm lenses of non-trilobite shell hash, commonly micritized, dense-packed, in limestone	Probably storm reworking of indigenous fauna	Shallow, low-energy subtidal, Nevada (Gevirtzman and Mount, 1986)
Ordovician			
O a	Very thin beds of disarticulated brachiopods, good condition	Hydraulic reworking of unsuccessful gregarious settlements	Shallow lagoon, New South Wales (Webby and Percival, 1983)
O b	Loosely packed, complete trilobites, inverted, monospecific, poorly sorted, well-preserved, in 10–20 cm dolomitic silt bed	Mass death by burial of opportunistic holaspids	Restricted shallow marine, Missouri (Brezinski, 1986)
O c	Trilobite clusters	Mass burial after gregarious behavior	Subtidal, Oklahoma (Laudon, 1939)

Age	Characteristics	Interpretation	Setting (Reference)
Silurian			
S a	Erosion-based thin packstones and winnowed floatstones of brachiopods, high dominance	Storm concentrations of exotic, indigenous, or mixed composition, some taphonomic feedback	Open shelf, Wales, Ireland (Cherns, 1988; Nealon and Williams, 1988)
S b	Thin beds of either unsorted disarticulated brachiopods in sandstone/conglomerate or umbones and pedicle valves only with comminuted skeletal debris	Downslope transport over short or great distances of shallow-water fauna in fluxoturbidites; no mixing with autochthonous deep-water fauna	Deep basin, Ireland (Weir, 1975)
Devonian			
D a	Dense, monotypic aggregates of whole or dissociated trilobites; "body and moult clusters" of corpses and exuviae, respectively	Gregarious behavior during ecdysis and/or reproduction of specific age class, mass burial	Deep subtidal, New York (Speyer and Brett, 1985)
D b	Pavements and thin beds of soft-bottom ± shell-gravel taxa, varied fabrics	Colonization ± storm-winnowing ± post-event colonization	Marine ramp, New York (K. B. Miller *et al.*, 1988)
D c	20-cm diameter funnel-shaped pod of concave-up disarticulated brachiopods, heavy beaks down	Chance penetration of brachiopod pavement by *Arenicola*-like deposit feeder	Shallow marine, Bolivia (Barthel and Barth, 1972)
D d	Clusters of spirifers in life positions, attached to skeletal debris, few generations, in mud	Gregarious life habit in soft substratum, toxicity of fecal accumulation eventually led to mortality of new spat	Subtidal, New York (Bray, 1972)
Carboniferous			
C a	Dense-packed horizon in shale, abundant articulated blastoid crowns and crushed articulated brachiopods, rare encrusting bryozoans	Catastrophic burial, perhaps with some local transport to separate crowns and stems from holdfasts	Cratonic carbonates, Alabama (Horowitz and Waters, 1972)
C b	Clusters of articulated brachiopods in life position	Opportunistic colonization during lulls in sediment accumulation	Prodelta, Yorkshire (Masurel, 1987)
C c	10–40-cm laterally extensive, rhythmic graded beds with basal dense-packed crinoid columnels	Turbidite-resedimented debris on flanks of Waulsortian-type buildups	Shelf foreslope, toe-of-slope, and near toe-of-slope basin, Missouri (King, 1986)

(*continued*)

Age	Characteristics	Interpretation	Setting (Reference)
Carboniferous (*continued*)			
C d	Bryozoans on skeletal debris and snowshoe brachiopods in shale	Epifaunal colonization of storm-introduced exotic debris	Lagoon, Kentucky (Chesnut and Ettensohn, 1988)
Permian			
P a	13-cm bed of dense-packed bivalves, disarticulated, fragmented, some bored, poorly sorted, random orientation	Breakage, concentration, and reorientation of indigenous benthos by predaceous fish; delayed burial	Shallow marine, Wyoming (Boyd and Newell, 1973)
P b	Brachiopod shell beds, varied thickness	Storm-delivered exotic shells from above chemocline	Black shale basin, Germany (Paul, 1982)
Triassic			
Tr a	Lenses of articulated, well-preserved brachiopods in life positions	Biostromes formed by gregarious settlement	Subtidal, Poland (Bodzioch, 1985)
Tr b	Fragmented bivalves in small clumps and elongate bodies	Concentration in burrows and in tubes [biogenic? hydraulic?]	Restricted carbonate lagoon, Italy (Passeri and Pialli, 1973)
Tr c	Few-cm-thick beds of low-diversity fragmented bivalves, current-oriented, loose/dense-packed in micrite	Intertidal channel lags	Restricted carbonate lagoon, Italy (Passeri and Pialli, 1973)
Tr d	Articulated brachiopods on pavements of crinoid-bivalve reef debris	Postevent benthic colonization of storm-transported shells	Subtidal ramp, southern Germany (Aigner *et al.*, 1978)
Tr e	10–15-cm beds of disarticulated, fragmented reef fauna	Selective downslope transport of byssate and thick-shelled spp.	Slope and basin, Italy (Fürsich and Wendt, 1977)
Tr f	Thin concentrations of pelagic bivalves and ammonites	Nondilution of planktic/nektic rain	Deep basin floor, Italy; (Fürsich and Wendt, 1977)
Jurassic			
J a	Up to 1 m massive echinoderm pack/wackestone with intraclasts	Grainflows of exotic shells from nearby submarine highs	Slope, Switzerland (Eberli, 1987)
J b	0.1–1.5-m graded beds with lithoclastic base and echinoderm grain/packstone top	High-density turbidity current transport of shells from nearby submarine highs	Slope, Switzerland (Eberli, 1987)
J c	Wedge-shaped beds to 40 cm, massive calcirudite/arenite	Storm-transported and reworked shell, some related to lateral expansion and deflation of shell shoals	Subtidal, England (Holloway, 1983)

Age	Characteristics	Interpretation	Setting (Reference)
Jurassic (*continued*)			
J d	Thin beds and pavements of disarticulated and fragmented *Isognomon* bivalves	Hydraulic reworking of local banks, and channel lags	Brackish lagoon/bay, Portugal (Fürsich, 1981)
J e	Monospecific gastropod beds 10–20 cm, dense-packed, excellent preservation; 20-m lenticular bodies	Aggregation and mass mortality caused by progressive dessication; census assemblages	Back-barrier pond, Denmark (Noe-Nygaard *et al.*, 1987; Noe-Nygaard and Surlyk, 1988)
J f	Polyspecific beds with variable preservation, poorly sorted	Washover fan deposits in lagoon, mixture of live and dead fauna	Low-salinity lagoon, Denmark (Noe-Nygaard *et al.*, 1987)
J g	Small clumps of small brachiopods, life position?	Gregarious settlement	Subtidal, England (Brookfield, 1973)
J h	Lenticles of monospecific brachiopods, or of mixed indigenous–exotic assemblages	Current-swept concentration in seafloor depressions	Subtidal, England, Germany (Brookfield, 1973; Seilacher, 1982b)
J i	cm- to dm-scale beds of mollusk-crinoid debris, laterally variable fabric; some with oyster-encrusted top surfaces	Storm reworking ("swell effusion") of local biostromes; +/− post-event colonization	Subtidal, Wyoming (Specht and Brenner, 1979)
J k	Pavements of ammonites +/− encrusters in laminated black shale	Mass mortality of nekton related to anoxia events	Restricted marine basin, Germany (Kauffman, 1978)
J l	Pavements to 3-cm beds and clusters of low-diversity bivalves, nektonic ammonites, or echinoids	Colonization during oxygenated events in black shale basin	Restricted marine basin, Germany, England (Kauffman, 1978; Seilacher, 1982b; Wignall, 1989)
J m	Clusters of poorly sorted belemnites	Concentration by hybodont sharks	Marine basin, Britain (Hewitt, 1980)
Cretaceous			
K a	20–60 cm conglomerates of concretions with shelf and upper slope benthos	Downslope transport by cohesive debris flows and slumping	Base of slope or head of fan, New Zealand (Evans, 1985)
K b	Ammonite- and bivalve-rich concretionary horizons	Mass mortality of gregarious (breeding?) taxa	Subtidal, Western Interior U.S. (Waage, 1964)
K c	Decimeter-scale clusters of articulated brachiopods in life position, all sizes	Gregarious settlement of multiple cohorts on dead *Exogyra* shells	Subtidal, Isle of Wight (Middlemiss, 1962)
K d	Pockets and pavements of brachiopod/bivalve debris	Current reworking of gregarious colonies	Subtidal, Isle of Wight (Middlemiss, 1962)

(*continued*)

Age	Characteristics	Interpretation	Setting (Reference)
Cretaceous (*continued*)			
K e	1–2 m thick, 2–6 km wide lenses of bivalve wackestone; locally amalgamated to 6 m thick	High-density turbidity current transport of shallow-water assemblage	Slope and deep basin, Nigeria (Banerjee, 1981)
K f	Local pavements of inoceramid and oyster debris in carbonate mud	Storm reworking of local epifaunal biostromes	Open marine epicratonic sea, Kansas (Hattin, 1986)
K g	Clumps of heteromorph ammonoids in concretions and body chambers of large ammonites	Schooling behavior of benthic species, minimal transport	Subtidal marine, Japan (Matsumoto and Nihongi, 1979)
Tertiary			
T a	Skeletal debris at base of graded bed, patchy	Scour lags of mixed indigenous–exotic mollusks	Fluvial crevass splay, North Dakota (Hanley and Flores, 1987)
T b	Monospecific pavements and thin beds of disarticulated infaunal bivalves or gastropods	Storm reworking of gregarious settlements	Low-energy inner to middle shelf, Maryland (Kidwell, 1982)
T c	Clumps to small lenses of densely packed, disarticulated, and fragmented bivalves	Fecal masses and feeding debris of molluscivorous predators, including rays	Low-energy shelf, Maryland, and marine prodelta, California (Kidwell, 1982, 1987)
T d	Dense clumps and pavements of infaunal bivalves, articulated and disarticulated, some in life position	Gregarious colonization around hydrocarbon or cold seep	Abyssal sandstone, Japan (Niitsuma *et al.*, 1989)
T e	Beds of vermetid gastropods in life position	*In situ* biostromes	Brackish facies, Costa Rica (Seyfried *et al.*, 1987a, b)
T f	0.1–3.4 m monospecific bed of ramose corals in life position, rare debris	Thicket formed by gregarious settlement on nondepositional surface, death by catastrophic burial	Open shelf, New Zealand (Squires, 1964)
T g	Monospecific clumps of articulated? life position? bivalves (*Nemocardium*)	Gregarious settlement	Upper slope, New Zealand (Vella, 1964)
T h	Monospecific 4–6-cm bed of disarticulated well-preserved bivalves (*Nemocardium*) in sand	Erosional exhumation from mudstone	Upper bathyal, New Zealand (Vella, 1964)
T i	Channel lag of whole and disarticulated balanid barnacles, monospecific, no corrosion	Transport from space-limited shoreline habitats, local reworking in channel	Distributary channels of tide-dominated delta, California (Zullo and Buising, 1989)

Age	Characteristics	Interpretation	Setting (Reference)
Tertiary (*continued*)			
T k	Shell-lagged scour surfaces, graded beds, and sand-wave foresets of diverse mollusk–coral–echinoid debris	Hydraulic erosion, avalanching, and redeposition of exotic, shallow-water skeletal debris, mixed with indigenous benthos	High-energy floor of bathyal strait, Italy (Colella and D'Alessandro, 1988)
T l	"Bioclastic breccias" of shallow-water epibenthos	Downslope transport by gravity flows + sliding of larger blocks	Bathyal base-of-slope, Italy (Colella and D'Alessandro, 1988)
Quarternary			
Q a	Clumps of gastropods (*Crepidula*) in life position, good condition	Gregarious behavior of mutually attaching hermaphrodites	Near-normal salinity bay, North Carolina (Miller and DuBar, 1988)
Q b	Decimeter-scale lenses, typically low diversity +/− size-sorted, whole or fragmented shells	Predator middens of lobsters, crabs, birds, fish, octopi, rays; reworking +/− transport out of life habitat	Subtidal to supratidal, marine and freshwater (Randall, 1964; Shabica, 1971; Cohen, 1989; Clifton and Hunter, 1972; Gregory *et al.*, 1979; S. E. Walker, 1990)
Q c	Small clumps of gastropods, varied states of preservation	Winnowing by nesting territorial fish (cichlids)	Freshwater lake, Tanganyika (Cohen, 1989)
Q d	Shell-lagged thin sand beds, disarticulated shallow-water and intertidal mollusks	Storm reworking of local death assemblages + some admixing of exotics	Embayed shelf, North Sea (Aigner and Reineck, 1982)
Q e	Thin beds of disarticulated and fragmented bivalves, scour base	Intertidal channel lags	Intertidal, Netherlands (van Straaten, 1952)
Q f	Spikes in abundance of starfish spines and ossicles at ~20-cm intervals down-core	Normal mortality of cyclic population outbreaks of Crown-of-thorns, minimal reworking	Shallow subtidal, Great Barrier Reef (Walbran *et al.*, 1989)
Q g	Disarticulated small bivalves in backshore runnels	Backwash of shells off upper beach during ebb tide, originally derived from nearshore	Marine beach, Georgia (Dörjes *et al.*, 1986)

[a]Codes refer to data points in Fig. 6.

Appendix B: Fossil and Recent Examples of Composite Concentrations[a]

Age	Characteristics	Interpretation	Setting (Reference)
Cambrian			
Є a	Low-relief mounds of archeocyathids and algae, +/−? life position, no coarse debris	*In situ* "reefs" formed by repeated colonization, minimum reworking	Shallow marine, Siberia (Kuznetsov and Postnikova, 1988)
Є b	Small shelly fauna, fragmented and reworked, in graded sand laminae within medium to large cross-sets; coquinas up to 1 m thick	Hydraulic reworking of indigenous fauna	Intertidal flats/channels or storm-dominated subtidal, Sonora (McMenamin, 1985)
Ordovician			
O a	2–8-cm-thick shell-supported beds of brachiopods and bryozoa	Autogenic succession on soft substrata, probably mediated by taphonomic feedback	Shallow subtidal open marine, Tennessee (Walker and Alberstadt, 1975)
O b	20–100-cm sands with shell lags, interbedded with shell-rich laminated sandstone	Accretionary proximal tempestites of indigenous fauna, forming nearshore bars	Shallow subtidal marine, Wales (Hurst, 1979)
O c	2.5–5.5 m "mud-supported" sets of 8–30 cm lenticular beds of articulated brachipods in life position, alternating with beds of reoriented +/− disarticulated individuals and barren layers	Repeated gregarious colonization and storm? reworking with minimal transportation; local colonization of dead shells by coral *Tetradium*, which develops 0.2-m thickets	Transgressive'marginal marine, New South Wales (Webby and Percival, 1983)
O d	< 5-m bioclastic sand with brachiopod coquinas (mixture articulated and disarticulated) and coral-rich beds, top of channel	Colonization and physical reworking of indigenous fauna during late-stage fill tidal channel	Shallow subtidal, England (Brenchley and Cocks, 1982)
O e	10–40-cm-thick, densely/loosely packed bed of disarticulated virgianid brachiopods, pedicle-valve dominated, no size sorting, to 1 km lateral extent	Indigenous, time-averaged accumulations	Marginal marine, Greenland (Fürsich and Hurst, 1980)
Silurian			
S a	10–30-cm beds, brachiopods range from all in life position to all disarticulated	Colonization of opportunistic pavements, +/− hydraulic reworking	Subtidal, Iowa (M. E. Johnson, 1977)

Age	Characteristics	Interpretation	Setting (Reference)
Silurian (*continued*)			
S b	3–12-m massive sheets near basin edge, pelmatozoan-dominated fauna, locally articulated and cross-bedded	Fringing skeletal banks of indigenous fauna, hydraulically reworked + stabilization by colonizing pelmatozoans	Nearshore marine, Appalachians (Dorobek and Read, 1986)
S c	Think- to thick-bedded, dense-packed, poorly-sorted benthic skeletal debris, bioturbated	Gradual accumulation with infrequent storm reworking of indigenous fauna	Offshore marine, Appalachians (Dorobek and Read, 1986)
Devonian			
D a	Disarticulated, shape- and size-sorted brachiopods and bivalves in fine bryozoan mud matrix	Gradual? accumulation of shells transported from surrounding stromatoporid bioherms and coral/crinoid growths	Shallow shelf, Great Britain (Elliot, 1961)
D b	< 75-cm beds of atrypid brachiopods in micrite, low-diversity, disarticulated and articulated	Time-averaged indigenous accumulations, biogenic reworking, some exotic input from adjacent crinoid and coral/stromatoporoid shoals	Shallow-subtidal intershoal areas, Poland (Racki and Balinski, 1981)
Carboniferous			
C a	10-m coral/cryptalgal bodies, life positions with little debris	Bioherm with < 1-m relief, no reworking	Shallow subtidal, England (Adams, 1984)
C b	0.5-m-thick elongate lenses of fragmented bryozoa, parallel to megaripples, some articulated brachiopods	Wave-reworked indigenous debris formed into small shoals or trapped in seafloor irregularities, taphonomic feedback	Shallow subtidal open marine, Nevada (M. A. Wilson, 1982)
Permian			
P a	Crinoidal grainstones to 8 m thick, associated with *Eurydesma*, bored	Gradual accumulation of indigenous fauna during low terrigenous input, on fault-uplifted highs	Subtidal marine, Australia (Draper, 1988)
Triassic			
Tr a	Abundant well-preserved brachiopods in decimeter- to meter-deep fissures, filled with red and gray lime ooze; capped by tiny ammonites in oblique-bedded mud	Cavity-dwelling brachiopods, presumably multiple generations; ammonites washed in on current-swept seafloor, locally fill tops of fissures	Deep neritic, Yugoslavia (Wendt, 1973)

(*continued*)

Age	Characteristics	Interpretation	Setting (Reference)
Triassic (*continued*)			
Tr b	< 5-m-thick mounds of dense-packed disarticulated crinoids, brachiopods, bivalves, with sponges in life positions	Sponge bioherm growths on storm-reworked bioclastic banks, shallowing phase	Lagoon-shelf barrier, Poland (Bodzioch, 1989)
Tr c	1.2–1.4-m dense-packed brachiopods and bivalves, both articulated and disarticulated; postcementation encrusting of bivalves	Storm amalgamation onto exhumed hardground reference horizon or bioclastic debris, + taphonomic feedback	Shallow marine, Germany, Carpathians (Hagdorn, 1982; Michalik, 1982)
Tr d	25-m-thick accumulations of serpulids in life position, little damage or brecciation	Repeated colonization keeping up but not exceeding surrounding sediment accumulation	Deep-water carbonate ramp, Spain (Braga and Lopez-Lopez, 1989)
Tr e	Thin? sets of cross-bedded, densely packed, low-diversity bivalves, broken and size-sorted	Hydraulic reworking of indigenous fauna, concentration into laterally migrating channel bars	Intertidal flats of restricted lagoon, Italy (Passeri and Pialli, 1973)
Tr f	125-m-thick, lenticular, accretionary cross-bedded bank of large whole and fragmented bivalves, grain/packstone	Downslope transport of exotic shallow-water fauna, infilling of preexisting channels	Shelf-slope margin, British Columbia (Gibson and Hedinger, 1989; D. W. Gibson, personal communication)
Tr g	*Cochlearites* bivalves dense-packed in channels to 4 m	Repeated storm or tidal reworking of indigenous fauna in gullies	Intertidal, Morocco (Lee, 1983) [and Jurassic, Italy (Bosellini, 1972)]
Tr h	Life-positioned *Cochlearites* bivalves in mounds to 0.7 m	Bioherms formed by repeated colonization, minimal reworking	Intertidal, Morocco (Lee, 1983) [and Jurassic, Italy (Bosellini, 1972)]
Jurassic			
J a	cm–dm-thick beds, amalgamated crinoidal grain/packstone	Colonization of subtidal channels; repeated cycles of colonization, hydraulic reworking, and obrution	Subtidal open maine, Switzerland (Meyer, 1988)
J b	30–100-cm banks, hundreds of meters to 2-km lateral extent, low-diversity *Isognomon* bivalves in life position, capped by disarticulated specimens	Colonization with minor taphonomic feedback and reworking	Brackish lagoon/bay, Portugal (Fürsich, 1981)

Age	Characteristics	Interpretation	Setting (Reference)
Jurassic (*continued*)			
J c	60–80-cm-thick, 2-m-diameter patch reefs of framework *Praeexogyra* overgrown by *Nanogyra*, bored, life positions; founded on dead *Isognomon*	Largely *in situ* organic buildup initiated by taphonomic feedback	Brackish lagoon/bay, Portugal (Fürsich, 1981)
J d	Tens of meters thick massive beds of crinoidal grainstone, rare grading, lateral extent <2 km	Stacked fluxoturbidites of shallow-water spp.	Abyssal foot of slope of seamount, Hungary (Vörös, 1986)
J e	Bored, flat-lying but nonaligned nerineacean gastropods, dense-packed in 60-cm beds	Gradual accumulation of possibly gregarious species, minor winnowing	Shallow marine, Switzerland (Dauwalder and Remane, 1979)
J f	Clumps/pavements of exotic molluscan fragments and juveniles loosely packed in stacked, cross-stratified 30-cm gravelly sands	Repeated washovers with pavements on internal avalanche surfaces, mixture of shallow marine and pond faunas	Backbarrier lagoon, Denmark (Noe-Nygaard and Surlyk, 1988)
Cretaceous			
K a	0.2–1-m massive sheets of thick bivalve fragments, unabraded and randomly oriented, amalgamated into 2-m-thick bodies	Storm-generated sheets of exotic shells on lee-sides of offshore bars	Open shelf, Nigeria (Banerjee, 1981)
K b	Laminated to cross-bedded lenses and lensing grainstones of disarticulated, broken, and abraded *Inoceramus*, forams, oysters	High-energy starved ripples and bars during Cenomanian transgression and Turonian regression; rare taphonomic feedback	Shallow-subtidal open marine, Western Interior (Hattin, 1986)
K c	Dense-packed caprinid rudist shells and fragments, graded cross-beds or amalgamated grainstone	Storm- or wave-accreted shoreface deposit, indigenous fauna rarely preserved *in situ*	Marine shoreface, Texas (Scott, 1990)
K d	Rudist bafflestones several meters high, with corals and coralline algae, in micrite, terrigenous horizons	"Bioconstructed" banks on offshore highs, intermittent disturbance	Open marine subtidal, France (Masse and Philip, 1981)

(*continued*)

Age	Characteristics	Interpretation	Setting (Reference)
Tertiary			
T a	6-m-thick mounds of (a) loosely packed whole and broken bryozoa in micrite and (b) finely fragmented shell-supported bryozoa	Minor current reworking of indigenous assemblages, repeated settlement of bryozoa on dead fragments; larger upcurrent colonies trap mud	Tide-dominated shallow marine shelf, Denmark (Thomsen, 1976, 1983)
T b	Discrete beds of molluscan wacke/packstone, concave-down preferential orientation	Wave-reworked indigenous assemblages	Nearshore/shoreline lake, North Dakota (Hanley and Flores, 1987)
T c	30-m-thick nummulitic banks, thin-bedded? to massive packstone +/− wacke/grainstone, sorting varies?	High production by local colonization plus storm winnowing and minor taphonomic feedback	Shallow subtidal open marine, Egypt (Aigner, 1983)
T d	High-angle foresets of mollusk–echinoid debris in megaripples, to 35? m thick	Tidal-reworked banks of nearby/exotic? death assemblages	Terrigenous-starved shallow marine seafloor, Switzerland (Allen and Matter, 1982)
T e	0–2-m beds within 1–10-m convex-down lenses, dense-packed fragmented, unabraded bryozoans; locally worn and mixed with other benthic shells	Fragile skeletal debris transported short distances downslope from monospecific gardens; accumulated in seafloor depressions	Neritic to upper bathyal slope of fan-delta, California (Cuffey *et al.*, 1981)
T f	1–6-m mega- and giant ripples of disarticulated ± fragmented oysters, well-sorted and graded within individual layers	Exotic shells from adjacent bioherms, swept into bar forms under strong tidal currents	Tide-dominated delta front, California (Kidwell, 1987)
T g	Dense/loose-packed bioclastic sands and conglomerates, variable bedding, mollusk-dominated, mixed quality of preservation	Reworked, clastic-starved transgressive accumulations of indigenous fauna, nonconformable or at toe of alluvial fan	Shallow-subtidal, Austria, California, Sonora (Nebelsick, 1989; Kidwell, 1988a; Beckvar and Kidwell, 1988)
T h	0.2–2.5-m cross-sets of graded "extremely bioclastic" sand, amalgamated	Current-driven transport and avalanching of debris, plus tractive winnowing and gravity flows of exotic epibenthos from shallow-water habitats	High-energy bathyal strait, Italy (Colella and D'Alessandro, 1988)
T i	10–40-m cross-sets of barnacle-dominated grainstone, accreted to 100 m locally	Barnacle colonization on molluscan debris on current-swept, clastic-free ridges in forearc basin, tide-dispersed	Subtidal marine, New Zealand (Kamp *et al.*, 1988)

Age	Characteristics	Interpretation	Setting (Reference)
Quaternary			
Q a	1-m-thick, bivalve-dominated, shell-supported elongate bodies, low diversity, disarticulated, alternating beds of random and edgewise orientations	Onshore transport and reworking of lower-flat or shallow subtidal shells into mobile intertidal cheniers and low banks	Intertidal estuary, New Zealand, Britain, Georgia (Richmond *et al.*, 1984; Greensmith and Tucker, 1966, 1969; Farrow, 1974; Wiedemann, 1972)
Q b	Loose- to dense-packed bed 20–30 cm below surface, unabraded whole and broken mollusks	Subsurface concentration by infaunal deposit feeders (biogenic grading)	Intertidal flat, Netherlands, New Zealand, Mexico (Richmond *et al.*, 1984; Cadée, 1976, 1979; Meldahl, 1987b)
Q c	Dense shell layer 30-cm below surface, convex-up, encrusted, disarticulated infaunal bivalves	"Event-condensation" (amalgamation) of indigenous shells during repeated storms	Intertidal, Malaysia (Seilacher, 1985)
Q d	Shell gravels and interbedded coarse shelly sands, coral shingle ridges	Accretion of shells from high-energy inter/subtidal rock platforms	Inner shelf, northwest Australia (Dix, 1989)
Q e	30-m banks of shell sand/gravel of epifaunal and infaunal bivalves, expanded thickness	Tide-dominated reworking of indigenous fauna into mobile sand waves	Open shelf, Scotland (Farrow *et al.*, 1984)
Q f	0.3–0.6-m coarse sand with thick abraded shell fragments, discontinuous to laterally continuous irregular layer	Erosional lag in throat of tidal inlet, lateral accretion by longshore migration of inlet	Barrier island, North Carolina, New York (Moslow and Heron, 1978; Sanders and Kumar, 1975)
Q g	Asymmetrical 2-m-high sand ridges with dense/loose-packed shells, concordant with inclined bedding, some articulated	Parautochthonous marine shells in transgressive ridges similar to cheniers, but formed during glacioeustatic cycles	Tidal-influenced shelf and coast, Argentina (Aliotta and Farinati, 1990)

[a]Codes refer to data points in Fig. 9.

Appendix C. Fossil and Recent Examples of Hiatal Concentrations and Lag-Concentrations (Underlined) (Codes Refer to Data Points in Figure 12)

Age	Characteristics	Interpretation	Setting (Reference)
Cambrian			
Є a	Small shelly fossils associated with phosphates, thin beds	Physically reworked beds in condensed sections	Marine, England (Brasier, 1986)
Ordovician			
O a	Diminutive mollusks, dense-packed, associated with phosphorite, hardgrounds, and graptolitic shales; shells phosphatized or occur as phosphatic steinkerns	Sediment-starved, deep low-oxygen seafloor in cratonic basin; downlapped by shallow-water carbonates; paedomorphic species, preferential preservation of juveniles and small adults	"Pebble necrocoenosis" of Tasch (1953); recurrent type in Paleozoic cratonic seas, midcontinent U. S. (Snyder and Bretsky, 1971; Witzke, 1987)
O b	1–1.5-m diverse packstone, lenticular amalgamated bedding, abrasion varies, crinoids and phosphatic steinkerns of protoconchs	Sediment-starved condensation during transgression	Subtidal shelf, Ohio (Holland, 1990)
O a	1–2-m lenticular to cross-bedded packstone, homogeneous comminuted residuum of brachiopods and crinoids, erosional base	Reworked lag on transgressive ravinement, shallower water than Ob	Subtidal shelf, Ohio (Holland, 1990)
Devonian			
D a	Densely-packed bed, diverse worn brachiopods with ossicle-packstone cap	Condensation along basin margin by downslope winnowing of fine matrix, top of shoaling cycle	Deep water, New York (Baird and Brett, 1981)
D b	Phosphate bed with reworked shell debris, 3–10 in thick, shells worn + epizoans, bioturbated	Condensation by winnowing, + taphonomic feedback; top of shoaling cycle	Deep water, New York (Baird and Brett, 1981)
D c	Two laterally extensive, diverse shell pavements colonized by solitary corals, intercalated with shales toward clastic source; brachiopods disarticulated; bioturbated	Condensation during geologically brief period of low sedimentation and/or shallower water, + taphonomic feedback; biological reworking; basinward tongues of shallow-water fauna	Deep foreland basin, New York (Baird and Brett, 1983)

Age	Characteristics	Interpretation	Setting (Reference)
Devonian (*continued*)			
D d	Few meters thick, loose-packed and storm-layered cephalopods, reworked shells, crinoidal limestone and sandy brachiopod coquina, many internal discontinuities and hardgrounds; condensed Frasnian and Famennian Stages	Drowning and starvation of uplifted carbonate platform, flanked by thick turbiditic and nonturbiditic basin deposits; low dilution of indigenous fauna, plus corrosion + mechanical winnowing, algal encrusting	< 100-m deep shelf, Morocco; (Wendt *et al.*, 1984; Wendt, 1988); condensed cephalopod limestones also known from Upper Devonian throughout Europe (Wendt and Aigner, 1985)
D e	Shelly benthos at base of bioturbated calcareous shale	Sediment-starved, oxygen-poor seafloor in cratonic basin; downlapped by shallow-water carbonates	Cratonic sea, midcontinent U. S. (Witzke, 1987)
Carboniferous			
C a	Life-position brachiopods grading up into layers of diverse, taphonomically modified fauna; rests on ravinement at base of shallowing-up cyclothem	Low sedimentation during marine transgression, some taphonomic feedback, slow progradation of deltaic deposits	Marine delta, England (Masurel, 1987)
C b	Shell-rich limestone silicified by sponges, laterally extensive bed	Low detrital input during marine transgression	Terrigenous coast, Appalachians (Cavaroc and Ferm, 1968)
Permian			
P a	Decimeter-thick, bioturbated cephalopod limestone, loose-/dense-packed, mineral crusts	Sediment starvation at distal edge of slope apron, above carbonate compensation depth	Deep basin, Oman (Blendinger, 1988)
Triassic			
Tr a	Stratiform red micrite with fragmented, locally imbricate ammonite layers; fine shell hash; overgrown/bored microrelief hardgrounds and mineral crusts	Current reworking and accumulation of shells in seafloor depressions during period of minimum sedimentation; localized over tectonic highs	Deep neritic, Yugoslavia (Wendt, 1973)
Jurassic			
J a	Bioclastic marker beds, laterally variable in matrix and fauna, whole and fragmented shells, bioturbated	Storm-modification and amalgamation in areas of reduced net sedimentation; condensation of ammonite subzones locally	Marine basin, Germany (Bloos, 1982)

(*continued*)

Age	Characteristics	Interpretation	Setting (Reference)
Jurassic (*continued*)			
J b	Loose-/dense-packed echinoderm residue + diverse benthos and nekton in thin limestone	Condensation of two to four biozones from Middle to Upper Jurassic	Deep subtidal, Spain (Benke, 1981)
J c	Laterally persistent thin bands of shelly micrite, densely packed, high diversity, encrusted epifauna; tops of 5-m-thick coarsening-up cycles	Local accumulations during periods of sediment nondeposition following "delta abandonment"; possible marine incursion	Marginal marine low-energy coast, Portugal (Fürsich, 1981)
J d	Laterally persistent 30–60-cm dense-packed oolite, resting on erosion surface; diverse, large, and fragile bivalves	"Gradual accumulation on a bottom kept free of sediment by currents," not reworked from older deposits; transgressive?	Shallow subtidal, England; (Arkell, 1933, p. 408) (Pebble Bed at base of Berkshire Oolite)
J e	Laterally extensive, cm–dm thick nerineacean beds, dense/loose-packed, encrusted/abraded; on firmgrounds	Slow sedimentation and accumulation of indigenous fauna under favorable ecological conditions; bioturbation	Shallow carbonate bank, Poland (Wieczorek, 1979)
J f	Dense-/loose-packed diverse benthos in limestone; shells bioeroded, encrusted, + early diagenetic dissolution; on hardgrounds and in neptunian dikes	Condensation by sediment starvation on seamount; physical reworking	Seamount, Hungary (Vörös, 1986)
J g	10–30-cm bioclastic marl horizons in sponge-crust bioherm; dense-/loose-packed and dispersed fabrics	Slow accumulation due to hydraulic reworking/resedimentation during lowered sealevel; condensation and compaction of sponge colonization events	Shallow subtidal, Germany (Lang, 1989)
J h	Laterally accreted 10-cm lenticular beds of reworked concretions, epifauna	Repeated erosion, colonization, and exposure in broad channels (Klüpfel roof-beds)	Subtidal, Germany (Bayer *et al.*, 1985)
Cretaceous			
K a	Well-preserved brachiopods in phosphatic conglomerate around paleohigh, mixture of indigenous and reworked fossils	Skeletal debris eroded from island of Upper Jurassic–Lower Cretaceous bedrock, colonized by Aptian benthos	Littoral zone, Brickhill and Upware, England (Middlemiss, 1962)

Age	Characteristics	Interpretation	Setting (Reference)
Cretaceous (*continued*)			
K b	Slight to heavily abraded sponge and brachiopod bioclasts in channelized quartz sand/gravel; transgressive	Gradual accumulation of hard-bottom benthic colonists, + taphonomic feedback; physical reworking of top only	Subtidal marine, England; (Middlemiss, 1962; Krantz, 1972)
K c	Diverse, dense-/loose-packed mollusks in thin siltstone over transgressive ravinement	Low sedimentation in upper offshore environment during deepening water	Open shelf, Utah (Ryer, 1977)
K d	Phosphatized Albian fossils and piped mid/low Cenomanian fossils in glauconite, resting on densely packed silicified fossils	Repeated submarine cementation and benthic colonization during deepening phase (Glauconitic Marl/Upper Greensand junction)	Open shelf, England (Garrison *et al.*, 1987)
K e	Abraded, bored and rephosphatized steinkerns of Albian, Kimmeridgian, Oxfordian and Liassic ammonites, bivalves, brachiopods, bones	Physical reworking and boring of local and exotic fossils during transgression-related hiatus	Marine shelf, England (Ruffell, 1990)
K f	Loose-packed to dispersed bored and phosphatized Maastrichtian fossils, nonphosphatized Maastrichtian and moldic Danian fossils in Greensand	Erosional exhumation, phosphatization, recolonization, and physical reworking	Subtidal marine, Poland (Machalski and Walaszczyk, 1987)
K g	Dispersed disarticulated bivalves and diagenetic ghosts in bioturbated micrite, firmgrounds at base and top	Terrigenous sediment starvation and biological reworking on shelf below mid-depth anoxic zone, maximum transgression	Freshwater-influenced marine embayment, Alberta (Banerjee and Kidwell, 1991)
Tertiary			
T a	Loosely packed benthic shell layers in megarippled sand, poorly sorted, preferentially cemented	Mark minor and major "abandonment" surfaces formed during episodes of nondeposition	Eocene tide-influenced inner shelf, Spain (Molenaar *et al.*, 1988)
T b	1.2–3-m-thick brachiopod–scallop marker bed within a regressive facies tract of algal bioherms, back-barrier and open basin carbonates	No interpretation provided; possibly a parasequence flooding surface	Miocene shallow marine, Malta (Pedley, 1976)

(*continued*)

Age	Characteristics	Interpretation	Setting (Reference)
Tertiary (*continued*)			
T c	Very thin "biostromes" or bivalve "thanatocoenoses", many discontinuities	Transgressive onlapping deposit condensed by *in situ* reworking of indigenous fauna	Miocene shelf, Costa Rica; (Seyfried *et al.* 1987a, b)
T d	≤ 1-m amalgamated sand bed with dispersed scallops in varied states, epifaunal colonists of dense-packed tops	Sediment starvation during sea-level lowstand, storm reworking of indigenous fauna + taphonomic feedback	Miocene outer shelf high or bank, Alaska (Eyles and Lagoe, 1989)
T e	1–10-m molluscan sand, complex microstratigraphy and lateral variation in thickness, dense-packing and composition	Condensation by sand starvation and mud bypassing during marine transgression and onlap, + taphonomic feedback	Miocene inner shelf Maryland (Kidwell, 1982, 1989a)
T f	Laterally persistent, 30–100-cm fine sand with dispersed articulated bivalves, mixed quality diverse bones, authigenic minerals; maximum water depth within transgressive–regressive cycle	Sediment starvation on distal, deep shelf; intense bioturbation and diagenetic overprinting, some taphonomic feedback	Outer shelf, Pliocene California and Miocene Maryland (Norris, 1986; Kidwell, 1982, 1989a)
T a	Coarse sand lenses associated with erosional disconformity, dense-packed phosphatic steinkerns of infaunal bivalves plus comminuted bone and teeth	Erosional reworking of prefossilized material during regression, lateral amalgamation of coastal channels at top of shallowing-up trend	Shallow-marine/marginal marine, Maryland (Kidwell, 1989a)
T b	Bored concretions with associated terebratulids and glauconite	Marine transgressive lag	Muddy shelf, California (Leithold, 1989)
Quaternary			
Q a	30–100-cm bed of diverse mollusks, lateral and microstratigraphic variation in composition and dense packing	Transgressive shell accumulation on eroded/bored marine terrace, overlain by alluvium–colluvium	Pleistocene shallow subtidal open coast, California (R. H. Wright, 1972)
Q b	20–80-cm beds of shells and pebbles on Pleistocene erosional terraces	Transgressive "lags" of marine fauna, actually sediment-starved hiatal accumulations	Shallow subtidal, Oregon and Sweden (Addicott, 1964; Ahman and Königsson, 1976)

Age	Characteristics	Interpretation	Setting (Reference)
Q c	50–150-cm interbedded sand/mud with intertidal/nearshore shells, basal lag of shell fragments	Palimpsest transgressive shelly sand sheet	Shelf and upper slope, Japan (Saito *et al.*, 1989)
Q d	"Pelagic" bioclastic sand, mixture of Recent and >40% relict, on top and flanks of volcanic cone	Terrigenous starvation and local winnowing of pelagic rain and benthos; episodic encrustation and dissolution	Seamount, North Atlantic (Rad, 1974)
Q e	Sand with diverse shelf bivalves	Very low net sedimentation and high biological production during Holocene transgression	Inner continental shelf, Texas (Morton and Winker, 1979)
Q f	Laterally extensive, densely packed bed of relic gastropods	Storm-reworked "graveyard" formed during low-stand	Freshwater lake, Tanganyika (Cohen, 1989)
Q a	Reworked and fragmented bivalves (*Rangia–Ostrea* assemblage) in sand, resting on erosional ravinement	Erosional exhumation of shells from Pleistocene deltas and Holocene lagoons during marine transgression	Continental shelf, Texas (Morton and Winker, 1979)

References

Adams, A. E., 1984, Development of algal-foraminiferal-coral reefs in the Lower Carboniferous of Furness, northwest England, *Lethaia* **17**:233–249.

Addicott, W. O., 1964, A late Pleistocene invertebrate fauna from southwestern Oregon, *J. Paleontol.* **38**:650–661.

Aepler, R., 1974, Der Rhätsandstein von Tübingen—Ein kondensiertes Delta, *Neues Jahrb. Geol. Paläontol. Abh.* **147**:113–162.

Aepler, R., and Reif, W.-E., 1971, Origin of bone-beds, in: *Abstracts VIII International Sedimentology Congress*, p. 1.

Ager, D. V., 1963, *Principles of Paleoecology*, McGraw-Hill, New York.

Ahman, E., and Königsson, L.-K., 1976, A shell deposit near Akroken at Kalix in northern Sweden, *Geol. Fören. Stockh. Förh.* **98**:85–86.

Aigner, T., 1979, Schill-Tempestite im Oberen Muschelkalk (Trias, SW-Deutschland), *Neues Jahrb. Geol. Paläontol. Abh.* **157**:326–343.

Aigner, T., 1983, Fossil-Lagerstätten, Nr. 57: Facies and origin of nummulitic buildups: An example from the Giza Pyramids Plateau (Middle Eocene, Egypt), *Neues Jahrb. Geol. Paläontol. Abh.* **166**:347–368.

Aigner, T., 1985, *Storm Depositional Systems*, Springer-Verlag, Berlin.

Aigner, T., and Reineck, H.-E., 1982, Proximality trends in modern storm sands from the Helgoland Bight (North Sea) and their implications for basin analysis, *Senckenb. Marit.* **14**:183–215.

Aigner, T., Hagdorn, H., and Mundlos, R., 1978, Biohermal, biostromal and storm-generated coquinas in the Upper Muschelkalk, *Neues Jahrb. Geol. Paläontol. Abh.* **157**:42–52.

Aliotta, S., and Farinati, E., 1990, Stratigraphy of Holocene sand-shell ridges in the Bahia Blanca Estuary, Argentina, *Mar. Geol.* **94**:353–360.

Allen, J. R. L., 1984, Experiments on the settling, overturning and entrainment of bivalve shells and related models, *Sedimentology* **31**:227–250.

Allen, P., and Matter, A., 1982, Mechanics of large Miocene coquina banks, Swiss marine Molasse, in: *Abstracts XI International Sedimentology Congress*, p. 99.

Arkell, W. J., 1933, *The Jurassic System in Great Britain*, Clarendon Press, Oxford.

Baird, G. C., 1978, Pebbly phosphorites in shale: A key to recognition of a widespread submarine discontinuity, *J. Sediment. Petrol.* **48**:545–555.

Baird, G. C., 1981, Submarine erosion on a gentle paleoslope: A study of two discontinuities in the New York Devonian, *Lethaia* **14**:105–122.

Baird, G. C., and Brett, C. E., 1981, Submarine discontinuities and sedimentary condensation in the upper Hamilton Group (Middle Devonian): Examination of marine shelf and paleoslope deposits in the Cayuga Valley, in: *Guidebook for Field Trips in South-Central New York* (P. Enos, ed.), New York State Geological Association, Binghamton, New York, pp. 115–145.

Baird, G. C., and Brett, C. E., 1983, Regional variation and paleontology of two coral beds in the Middle Devonian Hamilton Group of western New York, *J. Paleontol.* **57**:417–446.

Banerjee, I., 1981, Storm lags and related facies of the bioclastic limestones of the Eze-Aku Formation (Turonian), Nigeria, *Sediment. Geol.* **30**:133–147.

Banerjee, I., and Kidwell, S. M., 1991, Significance of molluscan shell beds in sequence stratigraphy: Example from the Lower Cretaceous Manville Group of Canada, *Sedimentology*, in press.

Barthel, K. W., and Barth, W., 1972, Paleoecologic specimens from the Devonian of Bolivia, *Neues Jahrb. Geol. Paläontol. Monatsh.* **1972**:573–581.

Bayer, U., and Seilacher, A. (eds.), 1985, *Sedimentary and Evolutionary Cycles*, Springer-Verlag, Berlin.

Bayer, U., Altheimer, E., and Deutschle, W., 1985, Environmental evolution in shallow epicontinental seas: Sedimentary cycles and bed formation, in: *Sedimentary and Evolutionary Cycles* (U. Bayer and A. Seilacher, eds.), pp. 347–381, Springer-Verlag, Berlin.

Beckvar, N., and Kidwell, S. M., 1988, Hiatal shell concentrations, sequence analysis, and sealevel history of a Pleistocene coastal alluvial fan, Punta Chueca, Sonora, *Lethaia* **21**:257–270.

Behrensmeyer, A. K., and Kidwell, S. M., 1985, Taphonomy's contributions to paleobiology, *Paleobiology* **11**:105–119.

Bell, M. A., Wells, C. E., and Marshall, J. A., 1989, Mass-mortality layers of fossil stickleback fish: Catastrophic kills of polymorphic schools, *Evolution* **43**:607–619.

Benke, K., 1981, Die Dogger/Malm-Wende in den NW-Keltiberischen Ketten (Spanien) und angrenzenden Gebieten-Sedimentologie, Stratigraphie und Paläogeographie, *Facies* **4**:95–164.

Blendinger, W., 1988, Permian to Jurassic deep-water sediments of the eastern Oman Mountains: Their significance for the evolution of the Arabian margin of the south Tethys, *Facies* **19**:1–32.

Bloos, G., 1982, Shell beds in the lower Lias of south Germany—Facies and origin, in: *Cyclic and Event Stratification* (G. Einsele and A. Seilacher, eds.), pp. 223–239, Springer-Verlag, Berlin.

Bodzioch, A., 1985, Palaeoecology and sedimentary environment of the *Terebratula* beds (Lower Muschelkalk) from Upper Silesia (South Poland), *Ann. Soc. Geol. Polon.* **55**:127–138.

Bodzioch, A., 1989, Biostratinomy and sedimentary environment of the echinoderm-sponge biostromes in the Karchowice Beds, Middle Triassic of Upper Silesia, *Ann. Soc. Geol. Polon.* **59**:331–350.

Bosellini, A., 1972, Paleoecologia dei calcari a "*Lithiotis*" (Guirassico inferiore, Pre alpi Venete), *Riv. Ital. Paleontol.* **78**:441–464.

Bowen, Z. P., Rhoads, D. C., and McAlester, A. L., 1974, Marine benthic communities in the Upper Devonian of New York, *Lethaia* **7**:93–120.

Boyd, D. W., and Newell, N. D., 1972, Taphonomy and diagenesis of a Permian fossil assemblage from Wyoming, *J. Paleontol.* **46**:1–14.

Braga, J. C., and Lopez-Lopez, J. R., 1989, Serpulid bioconstructions at the Triassic–Jurassic boundary in southern Spain, *Facies* **21**:1–10.

Braga, J. C., Martin, J. M., and Alcala, B., 1990, Coral reefs in coarse-terrigenous sedimentary environments (Upper Tortonian, Granada Basin, southern Spain), *Sediment. Geol.* **66**:135–150.

Brandt, D. S., 1989, Taphonomic grades as a classification for fossiliferous assemblages and implications for paleoecology, *Palaios* **4**:303–309.

Brasier, M. D., 1986, The succession of small shelly fossils (especially conoidal microfossils) from English Precambrian–Cambrian boundary beds, *Geol. Mag.* **123**:237–256.

Bray, R. G., 1972, The paleoecology of some Ludlowville brachiopod clusters (Middle Devonian), Erie County, New York, in: 24th International Geological Congress, Section 7, pp. 66–73.

Brenchley, P. J., and Cocks, L. R. M., 1982, Ecological associations in a regressive sequence: The latest Ordovician of the Oslo-Asker District, Norway, *Palaeontology* **25**:783–815.

Brett, C. E. (ed.), 1986, Dynamic Stratigraphy and Depositional Environments of the Hamilton Group (Middle Devonian) in New York State, Part 1, *N.Y. State Mus. Bull.* **457.**

Brett, C. E., and Baird, G. C., 1986, Comparative taphonomy: A key to paleoenvironmental interpretation based on fossil preservation, *Palaios* **1**:207–227.

Brett, C. E., Speyer, S. E., and Baird, G. C., 1986, Storm-generated sedimentary units: Tempestite proximality and event stratification in the Middle Devonian Hamilton Group of New York, *New York State Museum Bull.* **457**:129–156.

Brett, C. E., Miller, K. B., and Baird, G. C., 1990, A temporal hierarchy of paleoecologic processes within a Middle Devonian epeiric sea, in: Paleocommunity Temporal Dynamics: The Long-Term Development of Multispecies Assemblages (W. Miller III, ed.), *Paleontol. Soc. Spec. Pub.* **5**:178–209.

Brezinski, D. K., 1986, An opportunistic Upper Ordovician trilobite assemblage from Missouri, *Lethaia* **19**:315–326.

Brinkmann, R., 1929, Statistisch-biostratigraphische Untersuchungen an mitteljurassichen Ammoniten über Artbegriff und Stammesentwicklung, *Abh. Ges. Wiss. Gött. Math. Phys. Kl. N.F.* **13**:1–249.

Bromley, R. G., Hanken, N.-M., and Asgaard, U., 1990, Shallow marine bioerosion: Preliminary results of an experimental study, *Bull. Geol. Soc. Den.* **38**:85–99.

Brookfield, M. E., 1973. The palaeoenvironment of the Abbotsbury Ironstone (Upper Jurassic) of Dorset, *Palaeontology* **16**:261–274.

Cadée, G. C., 1976, Sediment reworking by *Arenicola marina* on tidal flats in the Dutch Wadden Sea, *Neth. J. Sea Res.* **10**:440–460.

Cadée, G. C., 1979, Sediment reworking by the polychaete *Heteromastus filiformis* on a tidal flat in the Dutch Wadden Sea, *Neth. J. Sea Res.* **13**:441–456.

Carannante, G., Esteban, M., Milliman, J. D., and Simone, L., 1988, Carbonate lithofacies as paleoclimate indicators: Problems and limitations, *Sediment. Geol.* **60**:333–346.

Cavaroc, V. V., Jr., and Ferm, J. C., 1968, Siliceous spiculites as shoreline indicators in deltaic sequences, *Geol. Soc. Am. Bull.* **79**:263–272.

Cherns, L., 1988, Faunal and facies dynamics in the Upper Silurian of the Anglo-Welsh Basin, *Palaeontology* **31**:451–502.

Chesnut, D. R., Jr., and Ettensohn, F. R., 1988, Hombergian (Chesterian) echinoderm paleontology and paleoecology, south-central Kentucky, *Bull. Am. Paleontol.* **95**:1–102.

Clifton, H. E., 1989, Sedimentologic approaches to paleobathymetry with applications to the Merced Formation of central California, *Palaios* **3**:507–522.

Clifton, H. E., and Hunter, R. E., 1972, The sand tilefish, *Malacanthus plumieri,* and the distribution of coarse debris near West Indian coral reefs, *Nat. Hist. Mus. Los Ang. Cty. Sci. Bull.* **14**:87–92.

Cohen, A. S., 1989, The taphonomy of gastropod shell accumulations in large lakes: An example from Lake Tanganyika, Africa, *Paleobiology* **15**:26–45.

Colella, A., and D'Alessandro, A., 1988, Sand waves, *Echinocardium* traces and their bathyal depositional setting (Monte Torre Palaeostrait, Plio-Pleistocene, southern Italy), *Sedimentology* **35**:219–237.

Craft, J. H., and Bridge, J. S., 1987, Shallow-marine sedimentary processes in the Late Devonian Catskill Sea, New York State, *Geol. Soc. Am. Bull.* **98**:338–355.

Cuffey, R. J., Stadum, C. J., and Cooper, J. D., 1981, Mid-Miocene bryozoan coquinas on the Aliso Viejo Ranch, Orange County, southern California, in: *Recent and Fossil Bryozoa* (G. P. Larwood and C. Nielsen, eds.), Olsen and Olsen, Fredensborg, Denmark, pp. 65–72.

Cutler, A. H., and Flessa, K. W., 1990, Fossils out of sequence: Computer simulations and strategies for dealing with stratigraphic disorder, *Palaios* **5**:227–235.

Dauwalder, P., and Remane, J., 1979, Etude du Banc à Nérinées à la limite "Kimmeridgien-Portlandien" dans le Jura Neuchatelois méridional, *Paläontol. Z.* **53**:163–181.

Davies, D. J., Powell, E. N., and Stanton, R. J., Jr., 1989a, Taphonomic signature as a function of environmental process: Shells and shell beds in a hurricane-influenced inlet on the Texas coast, *Palaeogeogr. Palaeoclimatol. Palaeoecol.* **72**:317–356.

Davies, D. J., Powell, E. N., and Stanton, R. J., Jr., 1989b, Relative rates of shell dissolution and net sediment accumulation—A commentary: Can shell beds form by the gradual accumulation of biogenic debris on the sea floor?, *Lethaia* **22**:207–212.

DeCelles, P. G., 1987, Variable preservation of middle Tertiary, coarse-grained, nearshore to outer-shelf storm deposits in southern California, *J. Sediment. Petrol.* **57**:250–264.

Dix, G. R., 1989, High-energy, inner shelf carbonate facies along a tide-dominated non-rimmed margin, northwestern Australia, *Mar. Geol.* **89**:347–362.

Dörjes, J., Frey, R. W., and Howard, J. D., 1986, Origins of, and mechanisms for, mollusk shell accumulations on Georgia beaches, *Senckenb. Marit.* **18**:1–43.

Dorobek, S. L., and Read, J. F., 1986, Sedimentology and basin evolution of the Siluro-Devonian Helderberg Group, Central Appalachians, *J. Sediment. Petrol.* **56**:601–613.

Draper, J. J., 1988, Permian limestone in the southeastern Bowen Basin, Queensland: An example of temperate carbonate deposition, *Sediment. Geol.* **60**:155–162.

Driscoll, E. G., 1970, Selective bivalve destruction in marine environments, a field study, *J. Sediment. Petrol.* **40**:898–905.

Dunham, R. J., 1962, Classification of carbonate rocks according to depositional texture, *Mem. Am. Assoc. Petroleum Geol.* **1**:108–121.

Dzulynski, S., and Kubicz, A., 1975, Storm accumulations of brachiopod shells and sedimentary environment of the *Terebratula* beds in the Muschelkalk of Upper Silesia (Southern Poland), *Rocznik Polsk. Towarz. Geol.* **45**:157–169.

Eberli, G. P., 1987, Carbonate turbidite sequences in rift basins of the Jurassic Tethys Ocean (eastern Alps, Switzerland), *Sedimentology* **34**:363–387.

Eckert, B.-Y., and Brett, C. E., 1989, Bathymetry and paleoecology of Silurian benthic assemblages, late Llandoverian, New York State, *Palaeogeogr. Palaeoclimatol. Palaeoecol.* **74**:297–326.

Efremov, J. A., 1940, Taphonomy: New branch of paleontology, *Pan-Am. Geol.* **74**:81–93.

Einsele, G., and Seilacher, A., eds., 1982, *Cyclic and Event Stratification*, Springer-Verlag, Berlin.

Elias, R. J., 1982, Paleoecology and biostratinomy of solitary rugose corals in the Stony Mountain Formation (Upper Ordovician), Stony Mountain, Manitoba, *Can. J. Earth Sci.* **19**:1582–1598.

Eliuk, L. S., 1988a, Carbonates in clastic sequences: Mesozoic and Recent of Canada, *Can. Soc. Petrol. Geol. Mem.* **15**:576–578.

Eliuk, L. S., 1988b, Canadian Mesozoic reefs and other organic accumulations, *Can. Soc. Petroleum Geol. Mem.* **13**:695–705.

Elliott, G. F., 1961, A new British Devonian alga, *Palaeoporella lummatonensis*, and the brachiopod evidence of the age of the Lummaton Shell Bed, *Proc. Geol. Assoc.* **72**:251–260.

Embry, A. F., III, and Klovan, J. E., 1972, Absolute water depth limits of Late Devonian paleoecological zones, *Geol. Rundsch.* **61**:672–686.

Evans, R. B., 1985, Geology and paleoecology of Cretaceous rocks in northern Kaipara, New Zealand, *N. Z. J. Geol. Geophys.* **28**:609–622.

Eyles, N., and Lagoe, M. B., 1989, Sedimentology of shell-rich deposits (coquinas) in the glaciomarine upper Cenozoic Yakataga Formation, Middleton Island, Alaska, *Geol. Soc.*

Am. Bull. **101:**129–142.
Farrow, G. E., 1974, On the ecology and sedimentation of the Cardium shell sands and transgressive shellbanks of Traigh Mhor, Island of Barra, Outer Hebrides, *Trans. R. Soc. Ed.* **69:**203–229.
Farrow, G. E., Allen, N. H., and Akpan, E. B., 1984, Bioclastic carbonate sedimentation on a high-latitude tide-dominated shelf: Northeast Orkney Islands, Scotland, *J. Sediment. Petrol.* **54:**373–393.
Faulkner, T. J., 1988, The Shipway Limestone of Gower: Sedimentation on a storm-dominated early Carboniferous ramp, *Geol. J.* **23:**85–100.
Feldman, H. R., 1989, Taphonomic processes in the Waldron Shale, Silurian, southern Indiana, *Palaios* **4:**144–156.
Ferguson, L., 1963, The paleoecology of *Lingula squamiformis* Phillips during a Scottish Mississippian marine transgression, *J. Paleontol.* **37:**669–681.
Flessa, K. W., ed., 1987, Paleoecology and taphonomy of Recent to Pleistocene intertidal deposits, Gulf of California, *Paleontol. Soc. Spec. Pub.* **2:**1–237.
Flessa, K. W., Meldahl, K. H., and Cutler, A. H., 1990, Quantitative estimates of stratigraphic disorder and time-averaging in a shallow marine habitat, *Geol. Soc. Am. Abstr. Progr.* **22:**A83.
Flügel, E., 1982, *Microfacies Analysis of Limestones*, Springer-Verlag, Berlin.
Freneix, S., and Avias, J., 1977, *Caledogonia*, a new smooth trigoniacean bivalve from the Triassic of New Caledonia, *Alcheringa* **1:**279–291.
Frey, R. W., and Howard, J. D., 1986, Taphonomic characteristics of offshore mollusk shells, Sapelo Island, Georgia, *Tulane Stud. Geol. Paleontol.* **19:**51–61.
Fürsich, F. T., 1971, Hartgründe und Kondensation im Dogger von Calvados, *Neues Jahrb. Geol. Paläontol. Abh.* **138:**313–342.
Fürsich, F. T., 1978, The influence of faunal condensation and mixing on the preservation of fossil benthic communities, *Lethaia* **11:**243–250.
Fürsich, F. T., 1981, Salinity-controlled benthic associations from the Upper Jurassic of Portugal, *Lethaia* **14:**203–223.
Fürsich, F. T., 1982, Rhythmic bedding and shell bed formation in the Upper Jurassic of East Greenland, in: *Cyclic and Event Stratification* (G. Einsele and A. Seilacher, eds.), Springer-Verlag, Berlin, pp. 208–222.
Fürsich, F. T., 1990, Fossil concentrations and life and death assemblages, in: *Palaeobiology: A Synthesis* (D. E. G. Briggs and P. R. Crowther, eds.), Blackwell, Oxford, pp. 235–239.
Fürsich, F. T., and Heinberg, C., 1983, Sedimentology, biostratinomy, and palaeoecology of an Upper Jurassic offshore sand bar complex, *Bull. Geol. Soc. Denmark* **32:**67–95.
Fürsich, F. T., and Hurst, J. M., 1980, Euryhalinity of Palaeozoic articulate brachiopods, *Lethaia* **13:**303–312.
Fürsich, F. T., and Kauffman, E. G., 1984, Palaeoecology of marginal marine sedimentary cycles in the Albian Bear River Formation of south-western Wyoming, *Palaeontology* **27:**501–536.
Fürsich, F. T., and Oschmann, W., 1986, Storm shell beds of *Nanogyra virgula* in the Upper Jurassic of France, *Neues Jahrb. Geol. Paläontol. Abh.* **172:**141–161.
Fürsich, F. T., and Wendt, J., 1977, Biostratinomy and palaeoecology of the Cassian Formation (Triassic) of the Southern Alps, *Palaeogeogr. Palaeoclimatol. Palaeoecol.* **22:**257–323.
Futterer, E., 1978, Hydrodynamic behavior of biogenic particles, *Neues Jahrb. Geol. Paläontol. Abh.* **157:**37–42.
Gagan, M. K., Chivas, A. R., and Herczeg, A. L., 1990, Shelf-wide erosion, redeposition, and suspended sediment transport during Cyclone Winifred, central Great Barrier Reef, Australia, *J. Sediment. Petrol.* **60:**456–470.
Galli, G., 1985, Ecology and dispersion of the fauna of the Cima Ombladet carbonate succession, *Palaeogeogr. Palaeoclimatol. Palaeoecol.* **49:**265–275.
Garrison, R. E., Kennedy, W. J., and Palmer, T. J., 1987, Early lithification and hardgrounds in Upper Albian and Cenomanian calcarenites, southwest England, *Cretaceous Res.* **8:**103–140.

Gazdzicka, E., and Gazdzicki, A., 1985, Oligocene coccoliths of the *Pecten* Conglomerate, west Antarctica, *Neues Jahrb. Geol. Paläontol. Monatsh.* **1985**:727–735.

Gevirtzman, D. A., and Mount, J. F., 1986, Paleoenvironments of an earliest Cambrian (Tommotian) shelly fauna in the southwestern Great Basin, *J. Sediment. Petrol.* **56**:412–421.

Gibson, D. W., and Hedinger, A., 1989, Upper Triassic shell banks, Mount Laurier area, northeastern British Columbia, in: Reefs, Canada and Adjacent Area (H. H. J. Geldsetzer, N. P. James, and G. E. Tebbutt, eds.), *Can. Soc. Petroleum Geol. Mem.* **13**:721–724.

Goldring, R., 1990, *Fossils in the Field: Information Potential and Analogues*, Longman, London.

Greensmith, J. T., and Tucker, E. V., 1966, Morphology and evolution of inshore shell ridges and mud-mounds on modern intertidal flats, near Bradwell, Essex, *Proc. Geol. Assoc.* **77**:329–346.

Greensmith, J. T., and Tucker, E. V., 1968, Imbricate structure in Essex offshore shell banks, *Nature* **220**:1115–1116.

Greensmith, J. T., and Tucker, E. V., 1969, The origin of Holocene shell deposits in the chenier plain facies of Essex (Great Britain), *Mar. Geol.* **7**:403–425.

Gregory, M. R., Ballance, P. F., Gibson, G. W., and Ayling, A. M., 1979, On how some rays (Elasmobranchia) excavate feeding depressions by jetting water, *J. Sediment. Petrol.* **49**:1125–1130.

Gunther, L. F., and Gunther, V. G., 1981, Some Middle Cambrian fossils of Utah, *Brigham Young Univ. Geol. Stud.* **28**:1–87.

Hagdorn, H., 1982, The "Bank der kleinen Terebrateln" (Upper Muschelkalk, Triassic) near Schwäbisch Hall (SW-Germany)—A tempestite condensation horizon, in: *Cyclic and Event Stratification* (G. Einsele and A. Seilacher, eds.), Springer-Verlag, Berlin, pp. 261–285.

Hanley, J. H., and Flores, R. M., 1987, Taphonomy and paleoecology of nonmarine Mollusca: Indicators of alluvial plain lacustrine sedimentation, upper part of the Tongue River Member, Fort Union Formation (Paleocene), northern Powder River Basin, Wyoming and Montana, *Palaios* **2**:479–496.

Hattin, D. E., 1986, Carbonate substrates of the Late Cretaceous Sea, central Great Plains and southern Rocky Mountains, *Palaios* **1**:347–367.

Hewitt, R. A., 1980, Quantitative studies of some Jurassic belemnite assemblages, *Géobios* **13**:173–197.

Hoffman, M., and Krobicki, M., 1989, Oyster buildup within the dysaerobic-facies mudstones (Middle Jurassic, central Poland)—Example of benthic island colonization, *Ann. Soc. Geol. Polon.* **59**:299–330.

Hohensee, S. R., and Stitt, J. H., 1989, Redeposited *Elvinia* Zone (Upper Cambrian) trilobites from the Collier Shale, Ouachita Mountains, west-central Arkansas, *J. Paleontol.* **63**:857–879.

Holland, S. M., 1990, Distinguishing eustasy and tectonics in foreland basin sequences: The Upper Ordovician of the Cincinnati Arch and Appalachian Basin, Ph. D. thesis, University of Chicago, Chicago.

Holloway, S., 1983, The shell-detrital calcirudites of the Forest Marble Formation (Bathonian) of SW England, *Proc. Geol. Assoc.* **94**:259–266.

Horowitz, A. S., and Waters, J. A., 1972, A Mississippian echinoderm site in Alabama, *J. Paleontol.* **46**:660–665.

Huber, S., 1987, Drucklösungserscheinungen in Karbonaten des Oxford 1 und Kimmeridge 1 der Bohrung TB-3 Saulgau (Oberschwaben), *Facies* **17**:109–120.

Hurst, J. M., 1979, The environment of deposition of the Caradoc Alternata Limestone and contiguous deposits of Salop, *Geol. J.* **14**:15–40.

Jackson, J. B. C., 1983, Biological determinants of present and past sessile animal distributions, in: *Biotic Interactions in Recent and Fossil Benthic Communities* (M. J. S. Tevesz and P. L. McCall, eds.), pp. 39–120, Plenum Press, New York.

Jeffery, D., and Aigner, T., 1982, Storm sedimentation in the Carboniferous limestones near

Weston-Super-Mare (Dinantian, SW-England), in: *Cyclic and Event Stratification* (G. Einsele and A. Seilacher, eds.), pp. 240–247, Springer-Verlag, Berlin.

Jenkyns, H. C., 1971, The genesis of condensed sequences in the Tethyan Jurassic, *Lethaia* **4**:327–352.

Johnson, M. E., 1977, Succession and replacement in the development of Silurian brachiopod populations, *Lethaia* **10**:83–93.

Johnson, M. E., 1989, Tempestites recorded as variable *Pentamerus* layers in the Lower Silurian of southern Norway, *J. Paleontol.* **63**:195–205.

Johnson, R. G., 1957, Experiments on the burial of shells, *J. Geol.* **65**:527–535.

Johnson, R. G., 1960, Models and methods for analysis of the mode of formation of fossil assemblages, *Bull. Geol. Soc. Am.* **71**:1075–1088.

Kamp, P. J. J., and Nelson, C. S., 1987, Tectonic and sea-level controls on nontropical Neogene limestones in New Zealand, *Geology* **15**:610–613.

Kamp, P. J. J., and Nelson, C. S., 1988, Nature and occurrence of modern and Neogene active margin limestones in New Zealand, *N. Z. J. Geol. Geophys.* **31**:1–20.

Kamp, P. J. J., Harmsen, F. J., Nelson, C. S., and Boyle, S. F., 1988, Barnacle-dominated limestone with giant cross-beds in a non-tropical, tide-swept, Pliocene forearc seaway, Hawke's Bay, New Zealand, *Sediment. Geol.* **60**:173–195.

Kauffman, E. G., 1978, Benthic environments and paleoecology of the Posidonienschiefer (Torarcian), *Neues Jahrb. Geol. Paläontol. Abh.* **157**:18–36.

Kidwell, S. M., 1982, Time scales of fossil accumulation: Patterns from Miocene benthic assemblages, in: *3rd North American Paleontological Convention Proceedings,* Volume 1, pp. 295–300.

Kidwell, S. M., 1985, Palaeobiological and sedimentological implications of shell concentrations, *Nature* **318**:457–460.

Kidwell, S. M., 1986a, Models for fossil concentrations: Paleobiologic implications, *Paleobiology* **12**:6–24.

Kidwell, S. M., 1986b, Taphonomic feedback in Miocene assemblages: Testing the role of dead hardparts in benthic communities, *Palaios* **1**:239–255.

Kidwell, S. M., 1987, Origin of macroinvertebrate shell concentrations in Pliocene shallow marine environments, Gulf of California, *Geol. Soc. Am. Abstr. Prog.* **19**:726.

Kidwell, S. M., 1988a, Taphonomic comparison of passive and active continental margins: Neogene shellbeds of the Atlantic Coastal Plain and northern Gulf of California, *Palaeogeogr. Palaeoclimatol. Palaeoecol.* **63**:201–223.

Kidwell, S. M., 1988b, Significance of skeletal concentrations for the analysis of unconformities and condensed intervals: Case studies from Neogene shallow marine sequences, *Can. Soc. Petroleum Geol. Mem.* **15**:579.

Kidwell, S. M., 1989a, Stratigraphic condensation of marine trangressive records: Origin of major shell deposits in the Miocene of Maryland, *J. Geol.* **97**:1–24.

Kidwell, S. M., 1989b, Phanerozoic "evolution" in modes of shell accumulation: Preliminary data from the Jurassic, in: *Abstracts 28th International Geological Congress,* Volume 2, p. 188.

Kidwell, S. M., 1990, Phanerozoic evolution of macroinvertebrate shell accumulations: Preliminary data from the Jurassic of Britain, in: Paleocommunity Temporal Dynamics: The Long-Term Development of Multispecies Assemblages (W. Miller III, ed.), *Paleontol. Soc. Spec. Pub.* **5**:309–327.

Kidwell, S. M., 1991a, Taphonomic feedback (live/dead interactions) in the genesis of bioclastic beds: Keys to reconstructing sedimentary dynamics, in: *Cycles and Events in Stratigraphy* (G. Einsele, W. Ricken, and A. Seilacher, eds.), Springer-Verlag, Berlin, pp. 268–282.

Kidwell, S. M., 1991b, Condensed deposits in siliciclastic sequences: Expected and observed features, in: *Cycles and Events in Stratigraphy* (G. Einsele, W. Ricken, and A. Seilacher, eds.), Springer-Verlag, Berlin, pp. 682–695.

Kidwell, S. M., and Aigner, T., 1985, Sedimentary dynamics of complex shell beds: Implications for ecologic and evolutionary patterns, in: *Sedimentary and Evolutionary Cycles* (U.

Bayer and A. Seilacher, eds.), Springer-Verlag, Berlin, pp. 382–395.

Kidwell, S. M., and Holland, S. M., 1991, Field description of coarse bioclastic fabrics, *Palaios* **6**:in press.

Kidwell, S. M., and Jablonski, D., 1983, Taphonomic feedback: Ecological consequences of shell accumulation, in: *Biotic Interactions in Recent and Fossil Benthic Communities* (M. J. S. Tevesz and P. L. McCall, eds.), Plenum Press, New York, pp. 195–248.

Kidwell, S. M., Fürsich, F. T., and Aigner, T., 1986, Conceptual framework for the analysis and classification of fossil concentrations, *Palaios* **1**:228–238.

King, D. T., Jr., 1986, Waulsortian-type buildups and resedimented (carbonate-turbidite) facies, Early Mississippian Burlington shelf, central Missouri, *J. Sediment. Petrol.* **56**:471–479.

Klüpfel, W., 1917, Über die Sedimente der Flachsee im Lothringer Jura, *Geol. Rundsch.* **7**:98–109.

Krantz, R., 1972, Die Sponge-Gravels von Faringdon (England), *Neues Jahrb. Geol. Paläontol. Abh.* **140**:207–231.

Kreisa, R. D., 1981, Storm-generated sedimentary structures in subtidal marine facies with examples from the Middle and Upper Ordovician of southwestern Virginia, *J. Sediment. Petrol.* **51**:823–848.

Kuznetsov, V. G., and Postnikova, O. V., 1988, Geometry and internal structure of subsurface Lower Cambrian reefs of the Siberian Platform: Osinsky Horizon (Aldanian), Nepsko-Botoubinsky Antecline, southern central Siberia, *Facies* **19**:259–270.

Lang, B., 1989, Die Schwamm-biohermfazies der Nordlichen Frankenalb (Urspring; Oxford, Malm): Mikrofazies, Palökologie, Paläontologie, *Facies* **20**:199–274.

Laudon, L. R., 1939, Unusual occurrence of *Isotelus gigas* DeKay in the Bromide Formation (Ordovician) of southern Oklahoma, *J. Paleontol.* **13**:211–213.

Lee, C. W., 1983, Bivalve mounds and reefs of the central High Atlas, Morocco, *Palaeogeogr. Palaeoclimatol. Palaeoecol.* **43**:153–168.

Leithold, E. L., 1989, Depositional processes on an ancient and modern muddy shelf, northern California, *Sedimentology* **36**:179–202.

Liu, J. T., and Zarillo, G. A., 1989, Distribution of grain sizes across a transgressive shoreface, *Mar. Geol.* **87**:121–136.

Lowenstam, H. A., and Weiner, S., 1989, *On Biomineralization*, Oxford University Press, New York.

Machalski, M., 1989, Life position of the oyster *Deltoideum delta* (Smith) from the Kimmeridgian of Poland, and its environmental significance, *Neues Jahrb. Geol. Paläontol. Monatsh.* **1989**:603–614.

Machalski, M., and Walaszczyk, I., 1987, Faunal condensation and mixing in the uppermost Maastrichtian/Danian Greensand (Middle Vistula Valley, central Poland), *Acta Geol. Polon.* **37**:75–91.

Mapes, R., 1987, Upper Paleozoic cephalopod mandibles: Frequency of occurrence, modes of preservation, and paleoecological implications, *J. Paleontol.* **61**:521–538.

Masse, J.-P., and Philip, J., 1981, Cretaceous coral-rudist buildups of France, in: European Reef Models (D. F. Toomey, ed.), *Soc. Econ. Paleontol. Mineral. Spec. Publ.* **30**:399–426.

Masurel, H., 1987, Macrofossils and their palaeoecology in deltaic sequences of the Lower Carboniferous Yoredale Series, Yorkshire, England, *Geol. Mijn.* **66**:221–237.

Matsumoto, T., and Nihongi, M., 1979, An interesting mode of occurrence of *Polyptychoceras* (Cretaceous heteromorph ammonoid), *Proc. Jpn. Acad. B* **55**:115–119.

McKinney, M. L., 1985, Distinguishing patterns of evolution from patterns of deposition. *J. Paleontol.* **59**:561–567.

McMenamin, M. A. S., 1985, Basal Cambrian small shelly fossils from the La Cienega Formation, northwestern Sonora, Mexico, *J. Paleontol.* **59**:1414–1425.

Meldahl, K. H., 1987a, Origin of shell beds and evolution of a shelly sand spit, Bahia la Choya, northern Gulf of California, in: Paleoecology and Taphonomy of Recent to Pleistocene Intertidal Deposits, Gulf of California (K. W. Flessa, ed.), *Paleontol. Soc. Spec. Publ.* **2**:189–199.

Meldahl, K. H., 1987b, Sedimentologic and taphonomic implications of biogenic stratification,

Palaios **2**:350–358.

Meyer, C. A., 1988, Paläoökologie, Biofazies und Sedimentologie von Seeliliengemeinschaften aus dem unteren Hauptrogenstein des Nordwestschweizer Jura, *Rev. Paléobiol.* **7**:359–433.

Meyer, D. L., and Ausich, W. I., 1990, Origin and composition of carbonate buildups and associated facies in the Fort Payne Formation (Lower Mississippian, south-central Kentucky): An integrated sedimentologic and paleoecologic analysis, *Geol. Soc. Am. Bull.* **102**:129–146.

Meyer, D. L., Ausich, W. I., and Terry, R. E., 1990, Comparative taphonomy of echinoderms in carbonate facies: Fort Payne Formation (Lower Mississippian) of Kentucky and Tennessee, *Palaios* **4**:533–552.

Michalik, J., 1982, Uppermost Triassic short-lived bioherm complexes in the Tatric, western Carpathians, *Facies* **6**:129–146.

Middlemiss, F. A., 1962, Brachiopod ecology and lower Greensand palaeogeography, *Palaeontology* **5**:253–267.

Miller, K. B., Brett, C. E., and Parsons, K. M., 1988, The paleoecological significance of storm-generated disturbance within a Middle Devonian muddy epeiric sea, *Palaios* **3**:35–52.

Miller, W. M., III, and DuBar, J. R., 1988, Community replacement of a Pleistocene *Crepidula* biostrome, *Lethaia* **21**:67–78.

Molenaar, N., Van de Bilt, G. P., van den Hoek Ostende, E. R., and Nio, S. D., 1988, Early diagenetic alteration of shallow-marine mixed sandstones: An example from the Lower Eocene Roda Sandstone Member, Tremp-Graus Basin, Spain, *Sediment. Geol.* **55**:295–318.

Montaggioni, L. F., Behairy, A. K. A., El-Sayed, M. K., and Yusuf, N., 1986, The modern reef complex, Jeddah area, Red Sea: A facies model for carbonate sedimentation on embryonic passive margins, *Coral Reefs* **5**:127–150.

Morton, R. A., and Winker, C. D., 1979, Distribution and significance of coarse biogenic and clastic deposits on the Texas inner shelf, *Trans. Gulf Coast Assoc. Geol. Soc.* **24**:306–320.

Moslow, T. F., and Heron, S. D., Jr., 1978, Relict inlets: Preservation and occurrence in the Holocene stratigraphy of southern Core Banks, North Carolina, *J. Sediment. Petrol.* **48**:1275–1286.

Nealon, T., and Williams, D. M., 1988, Storm-influenced shelf deposits from the Silurian of western Ireland: A reinterpretation of deep basin sediment, *Geol. J.* **23**:311–320.

Nebelsick, J. H., 1989, Temperate water carbonate facies of the early Miocene Paratethys (Zogelsdorf Formation, Lower Austria), *Facies* **21**:11–40.

Nelson, C. S., 1978, Temperate shelf carbonate sediments in the Cenozoic of New Zealand, *Sedimentology* **25**:737–771.

Nelson, C. S., 1988, An introductory perspective on non-tropical shelf carbonates, *Sediment. Geol.* **60**:3–12.

Newell, C. R., and Hidu, H., 1981, The effects of sediment type on growth rate and shell allometry in the soft shelled clam *Mya arenaria*, *J. Exp. Mar. Biol. Ecol.* **65**:285–295.

Niitsuma, N., Matsushima, Y., and Hirata, D., 1989, Abyssal molluscan colony of *Calyptogena* in the Pliocene strata of the Miura Peninsula, central Japan, *Palaeogeogr. Palaeoclimatol. Palaeoecol.* **71**:193–203.

Noe-Nygaard, N., and Surlyk, F., 1988, Washover fan and brackish bay sedimentation in the Berriasian-Valanginian of Bornholm, Denmark, *Sedimentology* **35**:197–217.

Noe-Nygaard, N., Surlyk, F., and Piasecki, S., 1987, Bivalve mass mortality caused by toxic dinoflagellate blooms in a Berriasian-Valanginian lagoon, Bornholm, Denmark, *Palaios* **2**:263–273.

Norris, R. D., 1986, Taphonomic gradients in shelf fossil assemblages: Pliocene Purisima Formation, California, *Palaios* **1**:256–270.

Parsons, K. M., Brett, C. E., and Miller, K. B., 1988, Taphonomy and depositional dynamics of Devonian shell-rich mudstones, *Palaeogeogr. Palaeoclimatol. Palaeoecol.* **63**:109–139.

Passeri, L., and Pialli, G., 1973, L'ámbiente di sedimentazione dei Calcari a *Rhaetavicula contorta* dell'Umbria occidentale e del M. Cetona, *Geol. Rom.* **12**:177–203.

Paul, J., 1982, Types of stratification in the Kupferschiefer, in: *Cyclic and Event Stratification* (G. Einsele and A. Seilacher, eds.), Springer-Verlag, Berlin, pp. 476–481.

Pedley, H. M., 1976, A palaeoecological study of the Upper Coralline Limestone, *Terebratula-Aphelesia* Bed (Miocene, Malta) based on bryozoan growth-form studies and brachiopod distributions, *Palaeogeogr. Palaeoclimatol. Palaeoecol.* **20**:209–234.

Perlmutter, M. A., 1982, The recognition and reconstruction of storm sedimentation in the nearshore, southwest Florida, Ph. D. thesis, University of Miami, Miami, Florida.

Pilkey, O. H., and Curran, H. A., 1986, Molluscan shell transport—You ain't seen nothin' yet, *Palaios* **1**:197.

Racki, G., and Balinski, A., 1981, Environmental interpretation of the atrypid shell beds from the Middle to Upper Devonian boundary of the Holy Cross Mts and Cracow Upland, *Acta Geol. Polon.* **31**:177–211.

Rad, U. von, 1974, Great Meteor and Josephine Seamounts (eastern North Atlantic): Composition and origin of bioclastic sands, carbonate and pyroclastic rocks, *Meteor Forschungsergeb. Reihe C* **19**:1–621.

Randall, J. E., 1964, Contributions to the biology of the queen conch, *Strombus gigas*, *Bull. Mar. Sci. Gulf Carib.* **14**:246–295.

Reid, R. P., and Ginsburg, R. N., 1986, The role of framework in Upper Triassic patch reefs in the Yukon (Canada), *Palaios* **1**:590–600.

Reif, W.-E., 1982, Muschelkalk/Keuper bone-beds (Middle Triassic, SW-Germany)—Storm condensation in a regressive cycle, in: *Cyclic and Event Stratification* (G. Einsele and A. Seilacher, eds.), Springer-Verlag, Berlin, pp. 299–325.

Richmond, B. M., Nelson, C. S., and Healy, T. R., 1984, Sedimentology and evolution of Ohiwa Harbour, a barrier-impounded estaurine lagoon in Bay of Plenty, *N. Z. J. Mar. Freshwater Res.* **18**:461–478.

Ruffell, A. H., 1990, The mineralogy and petrography of the Sulphur Band phosphates (Aptian-Albian), at Folkestone, Kent, *Proc. Geol. Assoc.* **101**:79–84.

Ryer, T. A., 1977, Patterns of Cretaceous shallow-marine sedimentation, Coalville and Rockport areas, Utah, *Geol. Soc. Am. Bull.* **88**:177–188.

Sadler, P. M., 1981, Sediment accumulation rates and the completeness of stratigraphic sections, *J. Geol.* **89**:569–584.

Saito, Y., Nishimura, A., and Matsumoto, E., 1989, Transgressive sand sheet covering the shelf and upper slope off Sendai, northeast Japan, *Mar. Geol.* **89**:245–258.

Sanders, J. E., and Kumar, N., 1975, Evidence of shoreface retreat and in-place "drowning" during Holocene submergence of barriers, shelf off Fire Island, New York, *Geol. Soc. Am. Bull.* **86**:65–76.

Schäfer, K., 1969, Vergleichs-Schaubilder zur Bestimmung des Allochemgehalts bioklastischer Karbonatgesteine, *Neues Jahrb. Geol. Paläontol. Monatsh.* **1969**:173–184.

Schäfer, K. A., 1973, Zur Facies und Paläogeographie der *Spiriferina*-Bank (Hauptmuschelkalk) im nördlichen Baden-Württemberg, *Neues Jahrb. Geol. Paläontol. Abh.* **143**:56–110.

Schäfer, W., 1972, *Ecology and Palaeoecology of Marine Environments*, University of Chicago Press, Chicago.

Schwarz, H.-U., 1975, Sedimentary structures and facies analysis of shallow marine carbonates (Lower Muschelkalk, Middle Triassic, Southwestern Germany), *Contrib. Sedimentol.* **3**:1–100.

Scott, R. W., 1990, Models and stratigraphy of mid-Cretaceous reef communities, Gulf of Mexico, *Soc. Econ. Paleontol. Mineral. Concepts Sediment. Paleontol.* **2**:1–102.

Seilacher, A., 1973, Biostratinomy: The sedimentology of biologically standardized particles, in: *Evolving Concepts in Sedimentology* (R. N. Ginsburg, ed.), The Johns Hopkins University Press, Baltimore, Maryland, pp. 159–177.

Seilacher, A., 1982a, General remarks about event deposits, in: *Cyclic and Event Stratification* (G. Einsele and A. Seilacher, eds.), Springer-Verlag, Berlin, pp. 161–174.

Seilacher, A., 1982b, Posidonia Shales (Toarcian, S. Germany)—Stagnant basin model revalidated, in: *Palaeontology, Essential of Historical Geology* (E. M. Gallitelli, ed.), STEM Mucchi, Modena, Italy, pp. 25–55.

Seilacher, A., 1985, The Jeram model: Event condensation in a modern intertidal environment, in: *Sedimentary and Evolutionary Cycles* (U. Bayer and A. Seilacher, eds.), Springer-Verlag, Berlin, pp. 336–346.

Seilacher, A., and Westphal, F., 1971. Fossil-Lagerstätten, in: *Sedimentology of Parts of Central Europe, Guidebook 8th International Sedimentological Congress*, Heidelberg, pp. 327–335.

Seilacher, A., and Westphal, F. (eds.), 1978, Paleoecology: Constructions, sedimentology, diagenesis and association of fossils, *Neues Jahrb. Geol. Paläontol. Abh.* **157**:3–276.

Seyfried, H., Astorga, A., and Calvo, C., 1987a, Sequence stratigraphy of deep and shallow water deposits from an evolving island arc: The Upper Cretaceous and Tertiary of southern Central America, *Facies* **17**:203–214.

Seyfried, H., Astorga, A., Calvo, C., and Laurito, C., 1987b, Sequence response (cyclicity, biostratinomy, ichnofacies) to subsidence, sea level fluctuations, and exceptional events in Cenozoic forearc basins of southern Central America, in: *Gulf Coast Section, Society of Economic Paleontologists and Mineralogists, 8th Research Conference*, pp. 131–141.

Shabica, S., 1971, The general ecology of the Antarctic limpet *Patinigera polaris, Antarc. J. U. S.* **6**:160–162.

Simone, L., and Carannante, G., 1988, The fate of foramol (''temperate-type'') carbonate platforms, *Sediment. Geol.* **60**:347–354.

Snyder, J., and Bretsky, P. W., 1971, Life habits of diminutive bivalve molluscs in the Maquoketa Formation (Upper Ordovician), *Am. J. Sci.* **271**:227–251.

Specht, R. W., and Brenner, R. L., 1979, Storm-wave genesis of bioclastic carbonates in Upper Jurassic epicontinental mudstones, east-central Wyoming, *J. Sediment. Petrol.* **49**:1307–1322.

Speyer, S. E., 1983, Subtidal and intertidal clastic sedimentation in a lower Cambrian sequence; Monkton Quartzite, northwestern Vermont, *Northeast. Geol.* **5**:29–39.

Speyer, S. E., and Brett, C. E., 1985, Clustered trilobite assemblages in the Middle Devonian Hamilton Group, *Lethaia* **18**:85–103.

Speyer, S. E., and Brett, C. E., 1986, Trilobite taphonomy and Middle Devonian taphofacies, *Palaios* **1**:312–327.

Squires, D. F., 1964, Fossil coral thickets in Wairarapa, New Zealand, *J. Paleontol.* **38**:904–915.

Tasch, P., 1953, Causes and paleoecological significance of dwarfed fossil marine invertebrates, *J. Paleontol.* **27**:356–444.

Thomsen, E., 1976, Depositional environment and development of Danian bryozoan biomicrite mounds (Karlby Klint, Denmark), *Sedimentology* **23**:485–509.

Thomsen, E., 1983, Relation between currents and the growth of Palaeocene reef-mounds, *Lethaia* **16**:165–184.

Tozer, E. T., 1971, Triassic time and ammonoids: Problems and proposals, *Can. J. Earth Sci.* **8**:989–1012.

Van Straaten, L. M. J. U., 1952, Biogene textures and the formation of shell beds in the Dutch Wadden Sea, *Proc. K. Ned. Akad. Wet. Amst. B* **55**:500–516.

Van Wagoner, J. C., Posamentier, H. W., Mitchum, R. M., Vail, P. R., Sarg, J. F., Loutit, T. S., and Hardenbol, J., 1988, An overview of the fundamentals of sequence stratigraphy and key definitions, *Soc. Econ. Paleontol. Mineral. Spec. Publ.* **42**:39–45.

Vella, P., 1964, Foraminifera and other fossils from late Tertiary deep-water coral thickets, Wariarapa, New Zealand, *J. Paleontol.* **38**:916–928.

Vörös, A., 1986, Brachiopod palaeoecology on a Tethyan Jurassic seamount (Pliensbachian, Bakony Mountains, Hungary), *Palaeogeogr. Palaeoclimatol. Palaeoecol.* **57**:241–271.

Waage, K. M., 1964, Origin of repeated fossiliferous concretion layers in the Fox Hills Formation, *Kansas Geol. Surv. Bull.* **169**:541–563.

Walbran, P. D., Henderson, R. A., Faithful, J. W., Polach, H. A., Sparks, R. J., Wallace, G., and Lowe, D. C., 1989, Crown-of-thorns starfish outbreaks on the Great Barrier Reef: A geological perspective based upon the sediment record, *Coral Reefs* **8**:67–78.

Walker, K. R., and Alberstadt, L. P., 1975, Ecological succession as an aspect of structure in fossil communities, *Paleobiology* **1**:238–257.

Walker, S. E., 1990, Biological taphonomy and gastropod temporal dynamics, in: Paleocommunity Temporal Dynamics: The Long-Term Development of Multispecies Assemblages (W. Miller III, ed.), *Paleontol. Soc. Spec. Publ.* **5**:391–421.

Walliser, O. H., 1985, Natural boundaries and Commission boundaries in the Devonian, *Cours. Forschungsint. Senckenb.* **75**:401–408.

Wanless, H., Tedesco, L., and Tyrrell, K., 1987, Production of subtidal tubular and surficial tempestites by Hurricane Kate, Caicos Platform, British West Indies, *J. Sediment. Petrol.* **58**:739–750.

Webby, B. D., and Percival, I. G., 1983, Ordovician trimerellacean brachiopod shell beds, *Lethaia* **16**:215–232.

Weir, J. A., 1975, Palaeogeographical implications of two Silurian shelly faunas from the Arra Mountains and Cratloe Hills, Ireland, *Palaeontology* **18**:343–350.

Wendt, J., 1973, Cephalopod accumulations in the Middle Triassic Hallstatt-Limestone of Jugoslavia and Greece, *Neues Jahrb. Geol. Paläontol. Monatsh.* **1973**:624–640.

Wendt, J., 1988, Condensed carbonate sedimentation in the late Devonian of the eastern Anti-Atlas (Morocco), *Eclogae Geol. Helv.* **81**:155–173.

Wendt, J., and Aigner, T., 1985, Facies patterns and depositional environments of Palaeozoic cephalopod limestones, *Sediment. Geol.* **44**:263–300.

Wendt, J., Aigner, T., and Neugebauer, J., 1984, Cephalopod limestone deposition on a shallow pelagic ridge: The Tafilalt Platform (Upper Devonian, eastern Anti-Atlas, Morocco), *Sedimentology* **31**:601–625.

Westrop, S. R., 1986, Taphonomic versus ecologic controls on taxonomic relative abundance patterns in tempestites, *Lethaia* **19**:123–132.

Whittington, H. B., and Conway Morris, S. (eds.), 1985, Extraordinary fossil biotas: Their ecological and evolutionary significance, *Phil. Trans. R. Soc. Lond. B* **311**:1–192.

Wieczorek, J., 1979, Upper Jurassic nerineacean gastropods from the Holy Cross Mountains (Poland), *Acta Palaeontol. Polon.* **24**:299–350.

Wiedemann, H. U., 1972, Shell deposits and shell preservation in Quaternary and Tertiary estuarine sediments of Georgia, U.S.A., *Sediment. Geol.* **7**:103–125.

Wignall, P. B., 1989, Sedimentary dynamics of the Kimmeridge Clay: Tempests and earthquakes, *J. Geol. Soc. Lond.* **146**:273–284.

Wilson, J. B., 1967, Palaeoecological studies on shell-beds and associated sediments in the Solway Firth, *Scot. J. Geol.* **3**:329–371.

Wilson, J. B., 1982, Shelly faunas associated with temperate offshore tidal sands, in: *Offshore Tidal Sands, Processes and Deposits* (A. H. Stride, ed.), Chapman and Hall, London, pp. 126–171.

Wilson, J. B., 1986, Faunas of tidal current and wave-dominated continental shelves and their use in the recognition of storm deposits, *Can. Soc. Petroleum Geol. Mem.* **11**:313–326.

Wilson, J. B., 1988, A model for temporal changes in the faunal composition of shell gravels during a transgression on the continental shelf around the British Isles, *Sediment. Geol.* **60**:95–105.

Wilson, M. A., 1982, Origin of brachiopod-bryozoan assemblages in an Upper Carboniferous limestone: Importance of physical and ecological controls, *Lethaia* **15**:263–273.

Wilson, M. A., 1985, Disturbance and ecologic succession in an Upper Ordovician cobble-dwelling hardground fauna, *Science* **228**:575–577.

Witzke, B. J., 1987, Models for circulation patterns in epicontinental seas applied to Paleozoic facies of North America craton, *Paleoceanography* **2**:229–248.

Wright, J. K., 1986, A new look at the stratigraphy, sedimentology and ammonite fauna of the Corallian Group (Oxfordian) of south Dorset, *Proc. Geol. Assoc.* **97**:1–21.

Wright, R. H., 1972, Late Pleistocene marine fauna, Goleta, California, *J. Paleontol.* **46**:688–695.

Zhang, X.-G., 1989, Ontogeny of an Early Cambrian eodiscoid trilobite from Henan, China, *Lethaia* **22**:13–29.

Zullo, V. A., and Buising, A. V., 1989, An unusual species of the *Balanus amphitrite* Darwin complex (Cirripedia, Balanidae) from the ancestral Colorado River delta in western Arizona and southeastern California, *Proc. Biol. Soc. Wash.* **102**:924–932.

Chapter 6

Terrestrial Vertebrate Accumulations

ANNA K. BEHRENSMEYER

1. Introduction

The vertebrate body is made up of a large number of dissimilar, mineralized hard parts that come apart rather easily once an animal is dead. Bones are readily transported by water and also can be attractive to scavengers for long periods after death. These factors predispose vertebrate remains to consid-

ANNA K. BEHRENSMEYER • Department of Paleobiology, National Museum of Natural History, MRC NHB 121, Smithsonian Institution, Washington, D.C. 20560.

Taphonomy: Releasing the Data Locked in the Fossil Record, Volume 9 of *Topics in Geobiology*, edited by Peter A. Allison and Derek E. G. Briggs, Plenum Press, New York, 1991.

considerable physical and biological modification between the time of death and final burial. Moreover, in contrast to most shelly marine organisms, living populations of vertebrates usually have relatively low densities of individuals and range over comparatively large areas. It takes special circumstances to generate a bone accumulation, and the processes responsible for concentrating the skeletal remains into fossil localities also bias representation of species, individuals, and body parts relative to the original populations. Taphonomic biases affect paleoecological, biostratigraphic, and macroevolutionary data, including species presence/absence, relative abundance, and community structure. Taphonomic study of the vertebrate record is important because it illuminates these biases and provides guidelines for how paleontological data can be used to address evolutionary and paleoecological questions (Olson, 1980, 1985; Behrensmeyer and Kidwell, 1985).

Taphonomy is characterized by interchange between ancient and modern realms; problems encountered in the fossil record lead to research in Recent environments, generating ideas to test further in the fossil record. Johannes Weigelt's intensive study of vertebrate carcasses is a classic early example of such research (Weigelt, 1989), and this kind of conceptual interchange has become standard practice in vertebrate taphonomy (Voorhies, 1969; Schäfer, 1972; Behrensmeyer and Hill, 1980; Shipman, 1981). Experimentation and observation of recent taphonomic processes are especially important to archeology and human origins research (Gifford, 1981; Haynes, 1981; Bunn, 1982; Behrensmeyer, 1983; Blumenschine, 1986a,b). Study of modern analogues is extremely instructive, but their power to explain specific features of fossil assemblages has tended to divert attention from comprehensive treatment of taphonomic data in these assemblages. Researchers often concern themselves with sets of related variables, such as skeletal part sorting and evidence for abrasion, in order to investigate a specific taphonomic problem, such as transport. Every vertebrate accumulation provides evidence for a wide range of variables that relate to taphonomic history, and this history often is complex, involving many different phases in the transition from biosphere to lithosphere. A more deliberate multivariate approach can provide important information about these different phases, allowing taphonomic features to be analyzed in relation to one another and compared with similar data from actualistic experiments and observations.

This chapter has two major goals: (1) to review the taphonomic characteristics of terrestrial vertebrate accumulations and how these relate to interpretations of taphonomic history, and (2) to advocate and illustrate a multivariate approach to the taphonomic study of vertebrate remains. This approach emphasizes taphonomic evidence that can be derived from bone assemblages *per se*, including both eroded surface accumulations and excavated samples. Taphonomic analysis also involves study of associated lithofacies, and although this is treated in a general way in Section 4, a detailed discussion of sedimentological analysis of vertebrate-bearing deposits is not included.

In the following discussion, "accumulation" means an occurrence of vertebrate remains in a well-defined area, usually 10^0–10^4 m^2, and within a stratigraphically limited unit (a single bed or set of lithologically related beds). An accumulation is the remains of more than one individual, and includes everything from instantaneous mass death assemblages to time-averaged, dispersed occurrences of fragmentary bones and teeth.

2. Characterizing Vertebrate Accumulations

The vertebrate fossil record occurs in a wide range of sedimentary contexts. Accumulations often are described according to their environment of deposition, as in fluvial channel, lacustrine, or fissure-fill assemblages. This characterization is essentially independent of the preserved vertebrate remains and suggests taphonomic and sedimentological processes that may have interacted with them. A second approach focuses on taphonomic features of the remains themselves, which can be documented directly from the fossil accumulation. Such features include measures of frequency per unit area of bones and teeth, representation of taxonomic diversity, relative abundance, body sizes, skeletal elements, and bone damage. Thus, an assemblage might be described as a monospecific bonebed, or a fragmentary concentration of microvertebrates. Vertebrate occurrences also are characterized according to inferred taphonomic history, such as mass death, predator lair, or time-averaged attritional assemblages (Behrensmeyer, 1987a). This approach is interpretive and usually is based on a combination of evidence from sedimentary context and taphonomic features (Fig. 1). It depends to a large extent on actualistic models to identify which processes and circumstances were responsible for the accumulations.

Of these three approaches to characterizing vertebrate accumulations, inferred taphonomic history is ultimately of greatest interest to paleobiologists because (1) identification of the specific agents, processes, or circumstances responsible for an accumulation indicates what biases are likely to be present, and (2) the taphonomic features themselves provide additional information concerning the paleobiology and sedimentology of the deposit. Interpretations of taphonomic history are only as reliable as the descriptive data and primary inferences upon which they are built, however, and these are the focus of the following section.

3. Taphonomic Characteristics of Bone Assemblages

Taphonomic features listed in Table I involve identifications and counts of taxa and body parts, mapping of spatial arrays of bones *in situ*, description of bone modification (damage), and inferences concerning body size and growth stage. This information is organized according to whether it is

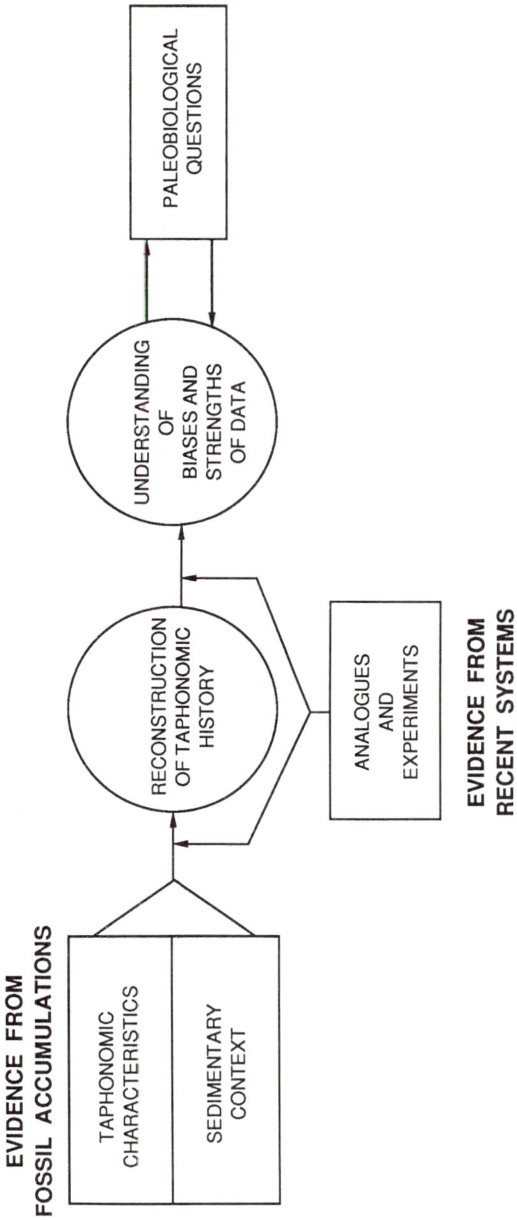

FIGURE 1. Components of the taphonomic analysis of terrestrial vertebrate accumulations.

TABLE I. Taphonomic Evidence That Can be Observed and Compared in Vertebrate Accumulations[a]

Variable	Possible manifestations at a fossil site
Assemblage data	
Sample size	--------------10 --------------100 --------------1000 --------------10,000
Number of individuals	1 ----------------10 ----------------100 ----------------1000 ----------------
Number of species	1 -----------------------10 -----------------------100 -----------------------
Relative abundance	all individuals/one species-----equal no. of individuals/species
Body size (kg)	------------5 ------------100 ------------500 ------------1000 ------------
Age spectrum	juveniles only--------------------50/50--------------------adults only
Bone articulation	articulated--{disarticulated but associated}--{associated + dispersed}--{isolated + dispersed}
Skeletal parts	unsorted ---------------------sorted ---------------------one part only
Quarry data	
Size of accumulation	---------1 m^2----------------------------100 m^2 --------1000 m^2 --------
Spatial density	---------0.1/m^2 ---------1/m^2 ---------10/m^2 ---------100/m^2 ---------
Spatial arrangement	
In plan view	random --preferred orientation
In profile	high dips --horizontal
Patchiness	even -----------------------uneven -----------------------highly patchy
Bone modification	
Breakage	complete -----------------------broken -----------------------fragments
Cracking	uncracked -------------small cracks -------------range of cracking
[Weathering, according to Behrensmeyer (1978)]	(Stage 0) --------------------(Stage 1) --------------------(Stages 0–3)
Abrasion/polish	unbraded ------------------abraded ------------------highly abraded
Surface marks	none -------------------------present -------------------------extensive (punctures, scratches, grooves, etc.)

[a]Data relating to each variable can be plotted along the horizontal scales, which are either calibrated as shown here or adjusted according to the quality of the data and the desired level of analysis. See text for details.

derived from the vertebrate assemblage itself, *in situ* evidence from quarry sites, or features of individual bones. Evidence pertaining to each variable can be arrayed along univariate scales or axes representing, for example, numerical abundance, patterns of sorting, or orientation. Thus, a given fossil accumulation can be characterized by the value of each variable, which may be scaled either qualitatively or quantitatively. The composite taphonomic evidence is then combined with sedimentological data as a basis for inferring taphonomic history (Fig. 1).

Many types of bone assemblage data are only available in carefully documented quarry samples, and at present there are few examples of assemblages with the full range of taphonomic data described below. The multivariate approach is designed so that it can be used with a limited number of variables, although its usefulness and interpretive power increase greatly with added classes of information. Specific guidelines for this method are given below, but readers are encouraged to modify it according to their own problems and assemblages. The suggested list will not be applicable to all fossil accumulations, especially those in the early part of the vertebrate record where variables such as "body size" and "age spectrum" may be difficult to determine. The most important aspect of this approach is that it allows assemblages to be characterized according to a set of taphonomically important variables and compared in a standardized manner. As long as consistency is maintained from assemblage to assemblage, variables could be added or subtracted from the suggested list in Table I, the vertical sequence of variables could be changed, and scaling values altered, while still maintaining the overall goal of the comparative multivariate approach.

3.1. Number of Specimens

The number of specimens that serve as a basis for taphonomic analysis is critically important to all other aspects of the analysis. Sample size may seem to be an obvious and easily derived feature of a bone assemblage, but complications often arise in deciding what to count and how to establish comparability of counts from different assemblages. For instance, all bone pieces may be counted, no matter how fragmentary or unidentifiable they are; alternatively, only pieces identifiable to body part and/or a particular taxon may be included in the total. In quarry samples, bones that are documented *in situ* may be counted separately from the smaller pieces that are recovered from screening. Considerable care should be taken in reporting exactly what sample size means because it affects representation of rare species and body parts, comparability of samples, and statistical tests.

The univariate scale in Table I spans the range of sample sizes for most fossil vertebrate assemblages. It refers to number of documented bones or bone fragments, not to estimates of number of remains in a deposit, although a different scale could be used for such an estimate. In general, larger samples (>100 specimens) are preferable for comparative analysis, although with appropriate caution, useful information can be derived from smaller samples.

3.2. Number of Individuals

Vertebrate fossil occurrences range from individual specimens to densely packed assemblages of thousands of animals. It is fairly common for remains of single individuals to be preserved in isolation, and some of the post-mortem processes that affected these remains can be inferred from the sedimentology of the deposit and from taphonomic features of the bones themselves. Inferences about taphonomic history for groups of organisms draw upon additional information from the associations of different individuals and taxa. Many individuals preserved together may represent instantaneous death events, longer-term attrition at a favorable site for burial, sedimentary accumulations of remains from different sources, or combinations of any of these. Between the extremes of isolated occurrences and mass accumulations are assemblages with a few to hundreds of animals, dispersed rather than packed together in a fossiliferous deposit. Many different combinations of processes can generate this pattern, which characterizes much of the vertebrate record.

Numbers of cooccurring individuals provide important clues relating to catastrophic versus attritional death and also serve as a standardized basis of comparison among different assemblages. The minimum number of individuals (MNI) is often used to describe bone accumulations, estimated on the basis of the most common unique skeletal part (e.g., a left distal humerus) (Shotwell, 1955; Shipman, 1981; Grayson, 1984; Klein and Cruz-Uribe, 1984; Badgley, 1986b). However, this method is known to underestimate the actual number of individuals (Holtzman, 1979; Badgley, 1986b). The most appropriate method for estimating MNI thus varies according to taphonomic context and type of bone assemblage (Badgley 1986a,b). In fragmentary assemblages, it can be difficult to achieve consistency in the identification of skeletal parts and the computation of MNI, and some workers prefer to use specimen counts as an estimate of number of individuals (Badgley, 1986b).

The scale in Table I utilizes number of individuals based on MNI. This scale can be divided into power-of-ten blocks if counts of MNI are imprecise; alternatively, it can be presented as a log scale to show the actual MNI. Whatever convention is used to estimate number of individuals, it must be consistent from assemblage to assemblage to allow meaningful comparative study.

3.3. Number of Species

The number of species represented in an assemblage provides evidence for taphonomic history primarily in combination with other features, such as number of individuals and body size. Mass accumulations of one or a few species may represent unusual circumstances in which herds of animals died and were buried by the same instantaneous event, but they can also reflect more diverse circumstances leading to concentrations of large groups of animals in a restricted burial situation (e.g., Colbert, 1947; Raath, 1969; Voorhies and Thomasson, 1979; Sander, 1987; Turnbull and Martill, 1988; Winkler *et al.*, 1988; Wood, 1988; Rogers, 1990; Schwartz and Gillette, in press). Preservation of animals in group aestivation burrows (Olson and Bolles, 1975; R. M. H. Smith, 1987) and recurrent trapping or other circumstances affecting single species (Agenbroad, 1984) can generate monospecific, or nearly monospecific, mass accumulations. When these include fragmentary remains of other animals, differences in preservation can be used to indicate that these represent background occurrences not related to the mass accumulation event(s) (e.g., Agenbroad, 1984).

Multispecies accumulations are common in the vertebrate record and represent a much wider range of taphonomic histories than monospecific assemblages. Such accumulations provide the primary data used to represent numbers of species within paleocommunities, regional faunas, and biochrons. Sample size plays a critical role in representations of species richness (Badgley and Gingerich, 1988). Often more fragmentary accumulations of bones record greater species richness, implying that diverse taphonomic processes acting over broader areas and longer time periods contributed to the number of species represented. Single accumulations of both macro- and microvertebrates can include 60–90 species (e.g., Solounias, 1981; Hulbert and Morgan, 1989). Assessment of species richness depends on taxonomic identifications, often of poorly preserved material. In many instances, reports of species in vertebrate assemblages also list generic or higher-level taxa not identifiable to species, and including these in diversity counts usually underestimates actual numbers of species. Moreover, species richness for particular time intervals or biochrons depends on what is considered a single accumulation and the number and types of accumulations that are combined to represent an interval.

The number of species or higher taxa is arrayed along a horizontal numerical scale in Table I. As in the scale for number of individuals, this can be divided into power-of-ten blocks or presented as a discrete $\log_{10}$ variable.

3.4. Relative Abundance of Species

The relative abundance of species, or ecological diversity, is commonly used as a descriptor of modern vertebrate communities (MacArthur, 1957; Peet, 1974; Whittaker, 1975; Magurran, 1988), with dominance by one or a few species in some communities contrasting with greater evenness in others. Ideally paleoecologists would like similar data from ancient ecosystems, and the species richness and number of individuals in fossil vertebrate accumulations have been used to estimate diversity (e.g., Rose, 1981; Grayson, 1984). The ecological validity of such estimates depends on whether the major control on preserved abundance was original abundance in a fauna. Actualistic studies indicate that biases relating to body size, habitat, mode of death, predator selectivity, and degree of water dependency, for example, affect relative abundance in bone assemblages (Behrensmeyer *et al.*, 1979; Dodson and Wexlar, 1979; Behrensmeyer and Dechant Boaz, 1980; Haynes, 1980; Badgley, 1986b; Andrews, 1990). Some of these biases undoubtedly are present in fossil assemblages, but the important question is to what degree they have altered original abundance patterns in different types of accumulations. Some workers have concluded that original relative abundances cannot be reliably reconstructed for many types of vertebrate fossil assemblage (e.g., Badgley, 1986b). Using rank order rather than counts of species abundance helps to make the data less sensitive to bias, but the possibility that rank order also was altered by taphonomic processes remains.

Multiple samples of contemporaneous paleofaunas can provide stronger evidence for original ecological dominance if one or two vertebrate species or higher taxonomic groups are consistently more abundant in assemblages with differing taphonomic histories (Rose, 1981; Badgley, 1982a, 1986a; Bown and Beard, 1990). Because such samples are derived from many different fossil localities, however, they will be time-averaged to some extent, and time-averaged dominance is not strictly analogous to more time-specific dominance in Recent communities (Rose, 1981). Damuth (1982) proposed a theoretical approach to test the ecological validity of relative abundance counts in mammal paleocommunities, based on scaling relationships of population density and body size in extant mammals. In an ideal assemblage, the regression of the logarithm of number of individuals of each species against the logarithm of body mass should have a slope of -1.05 ± 0.25 (approximate 95% confidence interval) (Damuth 1982). Assuming that the theoretical assumptions hold for extinct faunas, this method predicts the expected relative abundances of species at a given body size, thus providing a null hypothesis of "no bias" for comparison with actual fossil assemblages. Using this approach, Damuth (1982) demonstrated that Oligocene mammal assemblages show significant biases that appear to be related to differential preburial destruction of smaller species.

The horizontal scale in Table I ranges from total dominance by one species (low diversity) to an even distribution of individuals per species (high diversity). Most known communities fall between these extremes; examples approaching evenness or monospecific dominance are rare. The scale could be quantified using the Simpson index, Berger–Parker index, or other diversity indices (Simpson, 1949; Whittaker, 1975; May, 1976; Magurran, 1988), bearing in mind that these represent preserved, not necessarily original, abundance patterns.

3.5. Body Size

In practice there are two realms of vertebrate paleontology, one dealing with animals larger than about 5 kg in original body weight, the other concerned with microvertebrates weighing less than 5 kg. The division has developed along with new techniques for recovering small bones by bulk processing of sediments, using acid, water, and kerosene washes to break up matrix, fine screens to separate fossil-bearing size fractions, and heavy liquids to concentrate bones and teeth (McKenna, 1962; Estes, 1964). Larger fossils are collected using traditional methods of surface surveying, quarrying, and laboratory preparation of matrix blocks. The same number of work hours required to collect tens of specimens of larger animals can produce hundreds of small bones and teeth. Greater sample sizes and expanding interest in small vertebrates have shifted field efforts toward increased use of bulk processing strategies over the past 30 years.

Experiments demonstrate that small bones are easily transported and dispersed in water (Dodson, 1973; Korth, 1979). Rich concentrations of small vertebrates are found in fluvial deposits throughout the Phanerozoic, and these have been attributed to both biological and fluvial processes (Mellett, 1974; Korth, 1979). Such concentrations may, in part, reflect initial biological concentrating mechanisms involving predator feces or regurgitates (Mellett, 1974; Andrews, 1990). Many microvertebrate assemblages occur in poorly sorted clastic facies, suggesting at least some transport and reworking prior to burial (Eberth, 1990). Abrasion features and hydraulic equivalence between bones and associated sediment particles are further evidence for fluvial modification of the assemblage (Korth, 1979). Whether fluvial processes can concentrate small bones from dispersed sources in the absence of initial biological concentration is unclear.

Body size biases are a feature of both macro- and microvertebrate accumulations. The robust elements of larger animals are more likely to survive early destructive processes such as carnivory, trampling, and weathering (Behrensmeyer *et al.*, 1979; Behrensmeyer and Dechant Boaz, 1980; Brain, 1967a; 1980; Hill, 1978, 1980; Haynes, 1982). Smaller bones are more likely to be buried, however, and thereby protected from early destruction. Most vertebrate assemblages probably reflect some degree of size biasing relative to the body-size distribution of the original paleocommunity, but the nature of these biases varies greatly depending on the processes involved.

In Table I, the horizontal scale is divided into blocks representing size categories, and an assemblage can be represented on this scale according to its dominant body-size mode (i.e., the category with largest number of individuals). Individual size estimates are used, rather than averages for species, in order to include juveniles in the proper, taphonomically relevant, category. Given appropriate data, the scale also could use: (1) a percentage calculation of individuals larger than a minimum size, such as 5 kg, (2) a calculation of average body size using a continuous variable ($\log_{10}$), or (3) a distribution representing the number of individuals in different size groups. These require different levels of accuracy in estimates of body size, which are more or less problematic, depending on the taxonomic group and skeletal parts represented in an assemblage (Damuth, 1990).

3.6. Age Spectra

The proportions of juvenile versus adult animals can hold important clues about taphonomic history. Two basic mortality patterns are recognized; the "catastrophic age profile" representing the standing population, usually with fewer individuals in each successive age class, and the "attritional age profile" reflecting greater mortality among juveniles and old adults (Klein and Cruz-Uribe, 1984; Conybeare and Haynes, 1984; Haynes, 1985a,b, 1987, 1988). In mammal assemblages, ages of individuals can be estimated using stages of dental eruption, body size, and fusion of limb epiphyses. Age spectra can be difficult to establish for vertebrates that lack the distinct dental stages of mammals, but size groupings can serve as age classes if the taxonomy and ontogeny of a species are well known (e.g., Currie and Dodson, 1984). The age structure of monospecific fossil accumulations may resemble that of a living herd, supporting a catastropic group-death interpretation (Saunders, 1977; Voorhies and Thomasson, 1979; Turnbull and Martill, 1988). Some massed accumulations, however, have age structures implying attritional rather than catastrophic herd mortality (Agenbroad, 1984). In attritional multispecies accumulations, the proportions of juveniles and adults within particular body-size or taxonomic groupings have been used as evidence for predator activity (including human hunting) (Vrba, 1980; Klein, 1982; Klein and Cruz-Uribe, 1984; Behrensmeyer, 1987a).

A simple scale of percentage of adult individuals is used in Table I. This could be elaborated using age-range categories to indicate the age mode in an assemblage, or actual mortality profiles, given appropriate data.

3.7. Skeletal Articulation

A vertebrate skeleton consists of 100–200 separate bones and teeth of widely variable shape, size, and density (specific gravity). They are held together by tissues of different strengths, and the sequence of disarticulation of bones after death differs from species to species (Hill and Behrensmeyer, 1984). Rates of disarticulation vary with body plan and size and also with environmental conditions. Disarticulation by predators or scavengers can occur within minutes after death, although some articulated parts usually persist in the less meaty areas of the body, such as the ankle and wrist joints in ungulates (Blumenschine, 1986a,b). Mummified carcasses in arid environments or dry caves may remain articulated for centuries or longer, while carcasses in water generally fall apart in weeks to months (Weigelt, 1989; Schäfer, 1972; Hill, 1980).

Because disarticulation is relatively rapid in both subaerial and aquatic environments, preservation of articulated fossil bones indicates rapid burial or other circumstances such as anoxia that prevent scavengers and bioturbators from disturbing the skeleton (Zangerl and Richardson, 1963; Elder and Smith, 1988; Allison, 1988b; Weigelt, 1989). Often body parts are disarticulated but preserved in spatial proximity to one another, and this may indicate separation during bloating and soft-part decomposition (Schäfer, 1972; G. R. Smith and Elder, 1985; Weigelt, 1989) or localized disturbance by trampling, soil movement, plant growth, scavenging, and/or water currents too weak to move bones very far. Ingestion by predators followed by partial digestion results in associated (and some articulated) remains in pellets or scat (Dodson and Wexlar, 1979; Andrews, 1990). Transport and accumulation of articulated body segments by predators also can generate skeletal assemblages with associated parts (Brain, 1980, 1981).

Disassociated and dispersed bones and teeth are a common mode of vertebrate occurrence and represent a broad spectrum of possible histories.

Isolated bones act as sedimentary particles with widely variable transport potentials that can be roughly estimated in terms of hydraulic equivalence to quartz grains (Behrensmeyer, 1975; Korth, 1979). Once individual bones begin to move and disperse in water currents, the chance that elements from the same skeleton will be buried together decreases with distance from the site of origin (Hanson, 1980; Aslan and Behrensmeyer, 1987). Articulated bones can move as units, but information on their behavior in transport is not yet available.

States of articulation are represented on a qualitative scale in Table I. *Articulated* bones are those that retain their exact anatomical positions relative to one another. *Disarticulated but associated* bones are separated from each other but in close proximity, and they retain their identity as parts of individual animals. *Associated but dispersed* bones may be scattered over areas several times the size of the articulated skeleton, but can be related to the same individual using anatomical characteristics (e.g., matching left and right limb elements). *Isolated and dispersed* bones are widely separated from others belonging to the same skeleton and generally are derived from different individuals and taxonomic groups. The scale in Table I can be quantified as a percentage of articulations in an assemblage, or as percentages of associated elements combined with a distance measure representing dispersal.

3.8. Representation of Skeletal Parts

Bones in a single skeleton vary greatly in size, shape, and density. Consequently they are affected differentially by most taphonomic processes. Bones are mobile objects on soil surfaces, where bioturbation (chiefly trampling) can scatter them over areas of 100^2 m and scavenging can move parts for several kilometers (Gifford, 1977; Hill, 1980, 1983; Hill and Behrensmeyer, 1984). The frequencies of occurrence of different elements in bone assemblages provide evidence for taphonomic history, particularly the role of carnivore activities and water transport. In ungulates, skeletal parts such as teeth, jaws, and foot elements are more likely to survive initial carnivore or scavenger destruction because they are relatively dense and low in nutritional value (Hill, 1978, 1980). Water currents transport lighter or smaller elements such as vertebrae, ribs, and foot bones at faster rates than limb bones, cranial parts, and teeth (Voorhies, 1969; Behrensmeyer, 1975; Hanson, 1980; Aslan and Behrensmeyer, 1987). Bones and teeth of microvertebrates are easily moved by low flow velocities (<30 cm/sec), and sorting is less marked than in larger animals (Dodson, 1973; Korth, 1979).

In fluvial systems where bones are derived from multiple sources proximal to channels, skeletal parts of all types should occur at any point along the channel (Hanson, 1980; Aslan and Behrensmeyer, 1987; Behrensmeyer, 1990). Transport sorting is only well developed when there is a single point source from which bones are moved at different rates. Lighter bones move through a channel more quickly and are less likely to be buried, because of their relatively low densities. This causes a vertical (e.g., relative to the lowest point in the channel bed) as well as a horizontal (downstream) sorting effect during transport that favors longer periods of temporary burial for the denser elements as well as accelerated rates of destruction for lighter ones. Concentrations of the teeth and robust bones with large quartz-grain equivalents in the coarser facies of channel deposits are the eventual result of this process.

Sorting also occurs through surface weathering and decomposition in the soil. Weathering is effective in cracking and gradually disintegrating bones (Miller, 1975; Behrensmeyer, 1978), and this particularly affects elements with a high proportion of compact bone and natural lines of weakness, such as ribs, limbs, and jaws. These elements split longitudinally and fall apart, while podials, teeth, and vertebrae from the same skeleton may remain relatively intact after the same period of surface exposure. Following burial, biological and chemical processes in the soil may destroy bones but leave teeth relatively untouched. Porous bones are affected more quickly by subsurface agencies such as roots, fungi, and invertebrates, and by leaching. There also may be a bone size or surface-area-to-volume threshold below which soil processes are capable of destroying bones, but above which bones can generate their own microchemical environment for long enough to become mineralized (Gordon and Buikstra, 1981). This threshold would vary for different types of soils.

Differential skeletal part preservation has been of particular interest to paleoanthropologists and archeologists concerned with the role of humans in creating bone concentrations in Plio-Pleistocene to Recent deposits (Richardson, 1980; Maguire *et al.*, 1980; Binford, 1981; Isaac and Crader, 1981; Shipman, 1981; Speth, 1983; Klein and Cruz-Uribe, 1984; Bunn, 1981, 1986; Bunn and Kroll, 1986; Blumenschine, 1986a,b; Behrensmeyer, 1987a; Potts, 1988). Studies of modern bone-collectors show that humans as well as other carnivore and scavenger species transport bones from the death site and commonly prefer limb (and sometimes cranial) elements as opposed to vertebral columns (Skinner *et al.*, 1980; Hill, 1980, 1984; Brain, 1981). Thus, axial parts of the skeleton often are better represented than limb elements at the site of death. The strength of this pattern, however, depends on prey size and other factors (Vrba, 1980; O'Connell *et al.*, 1988). Skeletal part ratios in fragmentary fossil assemblages are calculated on the basis of minimum number of elements (MNE) (Holtzman, 1979; Binford, 1981; Klein and Cruz-Uribe, 1984; Badgley, 1986b), which can be affected by biases in the differential identifiability of bones as well as differential survival of the more robust parts of the skeleton. Because many of the processes affecting bone sorting tend to preserve robust elements and eliminate the more fragile ones, different histories of skeletal part sorting can result in similar assemblages (Brain 1980, 1981; Lyman 1984).

The skeletal part scale in Table I represents a gradient from unsorted carcasses with all bones present in original anatomical proportions, to the opposite extreme, where only one type of skeletal element (such as teeth) occurs. Between these endpoints are all possible manifestations of sorting caused by different sets of taphonomic processes. The scale could be quantified in terms of the frequencies of bones that reflect differential sorting potential, such as Voorhies Groups (Voorhies, 1969), ratios of teeth to vertebrae (Behrensmeyer, 1975), or an index based on ratios of observed to expected skeletal parts (Holtzman, 1979; Shipman, 1981). A scale representing hydraulic equivalents could also be used (Behrensmeyer, 1975).

3.9. Size of a Bone Accumulation

The area or the volume taken up by an accumulation of bones provides evidence for attritional versus instantaneous taphonomic processes, degree of time-averaging, influence of sedimentary processes, and general complexity of taphonomic history. In horizontally stratified deposits, the size of the accumulation refers to the lateral extent of bone occurrences along the available outcrops. Such occurrences may be limited to a few meters or tens of meters, or continue for many kilometers, such as when assemblages occur in floodplain paleosols (Bown and Kraus, 1981; Bown and Beard, 1990). Bonebeds [*Konzentrat-Lägerstätten; sensu* Seilacher *et al.* (1985)] are 10^2–10^3 m^2 (e.g., Hulbert and Morgan, 1989), although some reach square kilometers (Antia, 1979; Currie and Dodson, 1984), while dispersed assemblages cover 10^3 m^2 to many square kilometers. In fissures, caves, and burrows, accumulation size may be more appropriately recorded as volume.

Determining the original size of a bone accumulation is often problematic. Outcrop availability, topography, matrix hardness, and limited time for the recovery of spatial information may restrict quarry or surface samples to an area much smaller than the actual fossil-bearing unit. Often part of the original deposit has been lost to erosion. Because the denser or better-preserved portions of bone accumulations have been the preferred targets for paleontological collecting, much of the available taphonomic information for quarry sites is based on a limited portion of the original area occupied by a bone accumulation. The processes involved in forming the assemblage relate to the total area it occupies, however, and even rough estimates of this area are useful in interpreting taphonomic history.

Accumulation size in Table I is presented as a generalized quantitative scale for the estimated original area covered by an accumulation.

3.10. Bones per Unit Area, or Spatial Density

An excavated fossil accumulation provides direct evidence for the original spatial density and arrangement of bones and individuals at the time of lithification. These features are important to hypotheses concerning catastrophic versus attritional mortality as well as post-mortem dispersal, selective destruction of skeletal parts, and postburial disturbance.

A bonebed is a dense concentration of bones or carcasses in a single sedimentary unit, including both tabular and irregular units such as fissure fills. Bone size, proximity of bones or carcasses, and the lateral extent of the concentration are some of the features used to define a bonebed, but there is no consensus on where to draw the line between bonebeds and lower-density assemblages. A quantitative definition based on $\geq 4.5\%$ bone phosphate in a bone-bearing sedimentary unit, and a scheme for classifying bonebeds were proposed by Antia (1979), based primarily on accumulations in marine settings. A definition using percent bone phosphate is only applicable where accurate volumetric determinations are possible, however, and most nonmarine bone accumulations fall outside of Antia's system of classification. The densest Lagerstätten have tens of bones or carcasses per square meter in close juxtaposition (e.g., Peterson, 1906, 1923; Madson, 1976; Johanson *et al.*, 1976; Lawton, 1977; Norman, 1987; Hulbert and Morgan, 1989; Bell *et al.*, 1989). In cave deposits, single highly fossiliferous units can be meters thick, with irregular rather than tabular boundaries (Brain, 1958, 1981). Concentrations of small vertebrates generally have not been referred to as Lägerstätten, although numbers of bones per unit volume can be extraordinarily high. Most microvertebrate deposits are screen-washed rather than excavated, and the degree of concentration is measured in terms of number of identifiable bones or teeth per weight unit (e.g., kg) of matrix. Concentrations of 20 teeth per 30 kg of matrix are considered highly productive (W. R. Downs, personal communication, 1989).

Many accumulations of vertebrate remains are not bonebeds but laterally dispersed occurrences with spatial concentrations well below one bone per m^2. Recovery of specimens from these types of accumulation is usually by surface collection combined with local quarrying. Assemblages with a density of less than one bone per m^2 are seldom systematically excavated, and there is comparatively little information on *in situ* spatial features of such accumulations. Surface collecting often focuses on areas where fossils are secondarily concentrated by present-day erosional processes, and these may not reflect original spatial patterns or subsurface distributions (Behrensmeyer, 1985; Bown and Beard, 1990). Excavation is the only way to document precisely the original spatial densities of bones and associations of species in a sedimentary deposit.

Study of spatial densities of modern bones provides estimates of concentrations that would be expected from normal (rather than catastrophic) death over varying periods of time (Behrensmeyer, 1982, 1983; Badgley, 1982b). In a modern African game park with a relatively large population of vertebrates, initial post-mortem processes such as predation and scavenging produce a distribution of 0.0023 bone per m^2 for animals above 15 kg body weight. This represents 15–20 years of accumulation (Behrensmeyer and Dechant Boaz, 1980). Time-averaging over periods up to 10^4 years could increase concentrations to at most 0.1 bone per m^2

(Behrensmeyer, 1982), but the proportion of taxonomically identifiable elements would be much lower than this. Such densities of dispersed bones are well below those at many productive fossil sites formed by attritional processes (e.g., Johanson *et al.*, 1976; Hunt, 1978; Hunt *et al.*, 1983; Behrensmeyer *et al.*, 1989). This suggests that in order to create dense multispecies fossil accumulations, vertebrate remains must be concentrated by a factor of 10^3 or more relative to their dispersed distributions on land surfaces.

Table I provides a $\log_{10}$ scaling for spatial density. This can be adjusted to accommodate the widely different numbers of bones per unit area that occur in macro- and microvertebrate accumulations.

3.11. Patterns of Arrangement in Space

Mapping is an essential part of the taphonomic study of an *in situ* bone assemblage. A quarry map serves as a basis for interpreting the spatial patterns of different individuals, species, articulated and isolated bones, sorting, and influence of water currents. If elevations of bones are included in the mapping, then such patterns can be analyzed in three dimensions. The relationships of bones from the same carcass or conjoining pieces of the same bones can be measured precisely and used to indicate horizontal and vertical dispersion from a single source. Spatial variation in taphonomic features such as weathering, abrasion, and bone modification in excavations provides added information on processes of site formation. Work of this kind has been done in the context of archeology and human origins research (e.g., Hill and Walker, 1972; Johanson *et al.*, 1976; Bunn *et al.*, 1980; Todd, 1987) and in the paleontological record (e.g., Voorhies, 1969; Lawton, 1977; Fiorillo, 1988b).

In some fossil accumulations, bones may be so closely juxtaposed that they are very difficult to disentangle (e.g., Colbert, 1947; Madson, 1976; Maguire *et al.*, 1980). In others, they occur in variably dense and diffuse patches or evenly spaced arrays with little bone-to-bone contact. Microvertebrate (including fish) remains often are tightly packed together in ways that suggest feces or regurgitated pellets (Mellett, 1974; Dodson and Wexlar, 1979; Brain, 1981; Wilson, 1987; Andrews, 1990). Greater degrees of carcass or bone dispersion indicate originally dispersed death events and/or the influence of long periods of exposure to processes such as scavenging, bioturbation, and water transport. Patchiness versus evenness in bone distribution may indicate local variation in original carcass distributions, differential burial due to substrate texture, local effects of hydraulic sorting, or postburial diagenetic processes.

Bone orientations can reflect taphonomic processes such as water currents and trampling. Most bones are readily transported in water currents of 30–100 cm/sec, with elongate elements orienting either parallel or normal to the current direction (Voorhies, 1969; Behrensmeyer, 1975; Hanson,

1980). The behavior of each bone is determined by its shape and by water velocity and depth; elongate bones with one end larger than the other generally orient with the larger one upstream, while equidimensional bones tend to roll transverse to the current (Voorhies, 1969; Hanson, 1980). Rose diagrams or stereographic plots of bone orientations interpreted as current-influenced often show one strongly preferred direction and a second normal to this (Lawton, 1977; Voorhies 1969; Johanson *et al.*, 1976). The absence of such a pattern does not necessarily imply the absence of water transport, however, because obstructions (including other bones) and post-transport bioturbation can disrupt parallel alignment (Fiorillo, 1988a,b). Large bones that act as hydraulic barriers may accumulate smaller bones on their downstream sides and generate their own local orientation patterns.

Depending on their shapes, bones on a land surface or in fluvial transport generally have preferred stable orientations. In most cases these are horizontal, with specific sides up or down. Trampling and other forms of bioturbation can disturb this pattern and generate high angles relative to a horizontal plane (Fiorillo, 1984, 1988a,b; Gifford-Gonzalez *et al.*, 1985; Behrensmeyer *et al.*, 1989). Growth of plants (Behrensmeyer, 1978) and soil movement also can shift bones into near-vertical orientations.

The three horizontal scales referring to spatial orientation in Table I are arranged qualitatively to accommodate a range of options for significant features. The plan view orientations could be represented simply as the percentage falling into a preferred direction (e.g., NW–SE) and the in-profile orientations as a percentage greater than a specified angle of dip or plunge (e.g., >20°). Other options include measures of dispersion or plots of frequencies of different angles. Patchiness can be quantified as the coefficient of variation of the number of bones in specified units of an excavation, such as square meters.

3.12. Bone Modification

Bone modification refers to any form of alteration to individual bones, including cracking, surface marks, abrasion, polishing, and breakage (Bonnichsen and Sorg, 1989). Each type of modification bears evidence for taphonomic processes, and some of it is quite diagnostic of specific processes or agents. Experimental studies and observations of bone-modifying processes and their effects have provided evidence for features attributable to carnivore damage (Richardson, 1980; Haynes, 1982, 1983), trampling (Gifford-Gonzalez *et al.*, 1985; Behrensmeyer *et al.*, 1986; Fiorillo, 1984, 1989; Olsen and Shipman, 1988), and sediment abrasion (Behrensmeyer, 1982; Shipman, 1983) and have helped distinguish non-human from human bone modification in archeological contexts (Shipman, 1981, 1983; Behrensmeyer, 1987a; Bonnichsen and Sorg, 1989). Because modification features are expected in certain contexts, their absence also can be useful in formulating and testing hypotheses.

Fresh bones have high tensile strength, and breaking them takes considerable force, such as that exerted by carnivore mastication, stone-tool use, or trampling (Bonnichsen, 1979; Davis, 1985). Although it is commonly stated that bones are broken by fluvial transport, actualistic evidence does not support this. No breakage occurred within an experimental sample of over 100 hundred cow, horse, and sheep bones transported by natural rivers for several kilometers in sand and gravel-bed streams over periods up to 12 years (Behrensmeyer, 1982, and unpublished data). It seems unlikely that the forces required to break fresh bones (Davis, 1985) will be generated by most natural flow conditions, particularly since bones are relatively light and travel at speeds comparable to those of other particles. Much of the breakage observed in bones preserved in channel deposits probably occurred prior to transport or after they were weakened by weathering.

Breakage of fresh bones can leave characteristic patterns, such as spiral fractures, reflecting the bone structure (Bonnichsen, 1979; Myers *et al.*, 1980; Haynes, 1983; Davis, 1985). Other types of breaks, including splintered and straight patterns, are more typical of weathered or mineralized bones. Patterns of breakage in fossil accumulations can suggest whether the bones were fresh or weathered and the approximate amount of force required, but usually are not diagnostic of a specific causal agent or process (Davis, 1985; Behrensmeyer, 1987a). Recent work on human fracturing of bones with stone tools has identified features that may be diagnostic of deliberate smashing of limb elements for marrow by humans (Bunn, 1982; Davis, 1985; Blumenschine and Selvaggio, 1988). Predatory birds such as owls may impose specific breakage patterns on microvertebrate remains (Mellett, 1974; Dodson and Wexlar, 1979; Andrews, 1990).

When exposed to air, bones crack in patterns that tend to follow the original structure, and with time these multiply and lead to total disintegration. The progressive patterns of cracking and flaking are the basis for defining weathering stages that indicate the period of surface exposure prior to burial (Behrensmeyer, 1978; Lyman and Fox, 1989). Bones fully exposed on a dry soil surface develop Stage 1 cracking within months; thus, a lack of such cracks in fossils indicates that bones or carcasses were buried soon after death or were protected by water, vegetation, or tissue until burial. The absence of cracks also indicates that bones did not weather in the soil after burial. The presence of variable degrees of weathering in a fossil bone accumulation has been used as evidence for attritional accumulation (e.g., Potts, 1986). However, microenvironmental conditions such as variable vegetation cover also can result in different weathering stages for the same carcass (Behrensmeyer, 1978; Lyman and Fox, 1989). Identification of weathering stages in fossil accumulations requires excavated specimens and careful preparation. In some cases it is difficult to determine whether cracking patterns were caused by original surface weathering rather than postburial or diagenetic processes (Potts, 1986; Lyman and Fox, 1989).

Surface marks on bones are best preserved on unweathered and unabraded specimens and can be highly diagnostic of particular processes such as carnivory, human tool use, and trampling. Punctures and scoring marks with U-shaped cross sections are indicative of carnivore teeth (Haynes, 1980, 1982; Fiorillo, 1988b), and sometimes the spacing of tooth marks can be used to identify the type of carnivore (Maguire *et al.*, 1980; Brain, 1981). Bones with many bite marks are typical of denning sites where juveniles mouth and play with bones (Sutcliffe, 1970; Maguire *et al.*, 1980). Cutmarks from stone tools are usually V-shaped in cross section and occur in particular parts of the skeleton (Potts and Shipman, 1981; Bunn, 1981; Shipman and Rose, 1983; Bunn and Kroll, 1986). Trampling in sandy substrates or where hard objects can impact bones creates characteristic scratching and scoring, and some of these marks can mimic cut marks (Behrensmeyer *et al.*, 1986; Fiorillo, 1984, 1988b, 1989; Andrews and Cook, 1985; Olsen and Shipman, 1988). Multiple scratches and grooves on bone surfaces could also be caused by soil movement after burial (Wood and Johnson, 1978). Roots may cause grooves and dendritic channels on bones, both before and after mineralization. In marine settings, boring organisms can leave distinctive holes in bones (Frey *et al.*, 1975).

Bones that have lost external surfaces and projections through physical or chemical erosion are described as abraded. Abrasion is often noted as evidence for fluvial transport and reworking, but other processes may generate a similar appearance. Weathering, trampling, chewing, chemical and physical erosion during digestion, and subsurface disintegration can round bone edges and remove external surfaces. Experimental work on sediment abrasion in tumbling machines demonstrates that bones acquire polish, abrasion, and rounded edges at varying rates depending on sediment size (Shipman *et al.*, 1984). Bone surfaces easily acquire distinctive microscopic features when experimentally abraded by projected streams of particles, water, sand paper, brushing, rubbing by human hands, and dissolution in acetic acid (Bromage, 1984). In spite of this, observations in natural rivers show that unweathered bones can travel many kilometers without significant rounding (Behrensmeyer, 1982), and bones and teeth also can be reworked from older deposits with minimal evidence of abrasion (Behrensmeyer, 1982; Argast *et al.*, 1987; Eaton *et al.*, 1989). This probably reflects the fact that bones are relatively light and have a low number of forceful impacts with a channel bottom or sediment in the bed-load during transport. Sand-blasting of a bone held in a river bed is more likely to cause abrasion than impact with particles moving at approximately the same rate. Abrasion thus is an indicator of the intensity and/or time period of interaction with sediment, but not necessarily distance of transport. Bones and teeth in fossil assemblages may have been transported even if they lack evidence for abrasion, and they also may have been abraded without transport. The probabilities attached to these statements are not yet known, however, and are likely to vary depending on velocity and sediment texture. For instance, empirical evidence is lacking to estimate the probability that bones can travel 1 or 2 km or more at a given velocity over a given substrate and still remain unabraded. In a fossil accumulation, variability in abrasion features at least can serve as evidence for a complex taphonomic history, usually involving attritional input and, in river deposits, reworking and time-averaging.

Rough, irregular pitting and erosion of bone surfaces often is found in modern bones that have been exposed to algal growth and invertebrate grazing in subaquatic environments (A. K. Behrensmeyer, unpublished data). Polishing occurs when compact or dense bones acquire a smooth, shiny surface through interaction with particular materials such as fine sediment, soft tissue, or human hands. Trampling can cause polish in fine substrates (Brain, 1967b; Shipman *et al.*, 1984; Behrensmeyer *et al.*, 1989). Water-transported fine sediment also can polish bones. In archeological contexts, polish is often taken as evidence for utilization of a bone as a tool, especially if it is developed only on points or sharp edges (Brain, 1967b; Shipman *et al.*, 1984). Previously fossilized bones also can acquire a high polish through sediment abrasion during reworking.

Analysis of bone modification involves qualitative decisions about criteria for classifying them as, for example, abraded versus unabraded, followed by quantification according to this classification. The horizontal scales for bone modification features in Table I are general and qualitative, but each could be calibrated according to ratios or frequencies of bones in different states of modification. For instance, the percentage of complete bones and the percentage of unabraded bones in an assemblage provide simple ways to quantify the horizontal scales. Quantitative assessment of surface marks requires standardized observations of clean, well-preserved surfaces from a large sample (e.g., ≥ 100) of bones to establish the proportion of specimens that show this form of modification.

3.13. Chemical Alteration

Bones are chemically unstable when exposed to surface conditions and begin changing soon after death. Organic matter (approximately 20%) decomposes, and the crystallites of calcium phosphate "ripen" by changing in size and composition (Tuross *et al.*, 1989b). Trace elements may be absorbed during early crystal changes or permineralization, changing the original composition of the bone (Williams and Marlow, 1987; Williams and Potts, 1988; Tuross *et al.*, 1989a). Under favorable postburial chemical conditions (Retallack, 1984), $CaCO_3$, Fe_2O_3, and/or SiO_2 gradually penetrate bones, replacing the original minerals and/or filling spaces left by decayed organic matter (Wyckoff, 1972; Francillon-Vieillot *et al.*, 1989). In some circumstances, bones are destroyed prior to or soon after burial; biotic recycling in soils depleted in calcium or phosphorus quickly destroys them, and they are dissolved in soils with low pH (Gordon and Buikstra, 1981), unless buffered by local concentrations of calcium (White and Hannus, 1983). This is a possible cause of large-scale bias, or megabias (Behrensmeyer and Kidwell, 1985), in the land vertebrate record if preserved species preferentially represent environments with neutral to alkaline soil conditions (Retallack, 1984), e.g., the more seasonally dry, nonforested habitats.

Both organic and inorganic components of original bone contain information concerning an animal's diet and habitat. Considerable effort has gone into extracting such information from the archeological record, especially of the past few millenia, in the form of stable carbon, oxygen, nitrogen, and strontium isotopes (van der Merwe and Vogel, 1978; Sillen and Kavanagh, 1982; Price *et al.*, 1985; Nelson *et al.*, 1986; Lee-Thorp *et al.*, 1989; Koch *et al.*, 1989) and amino acids (Hare, 1980; Bada *et al.*, 1984). Diagenetic changes pose persistent problems in such studies (Kyle, 1986; Nelson *et al.*, 1986), but in special circumstances of preservation, recognizable protein signals are retained, even in well-mineralized bones (Tuross *et al.*, 1980, 1989a).

Analysis of bones in a single assemblage can reveal whether they have undergone different diagenetic histories, and thus whether they represent mixed sources (Thomas, 1986; Henderson *et al.*, 1987). Trace element chemistry can also be helpful in relating problematic surface specimens to a particular stratigraphic unit (Henderson *et al.*, 1987; Williams and Potts, 1988). Petrographic study of bone may reveal prefossilization penetration by fungi, roots, and other organisms (Hassan and Ortner, 1977), as well as the history of reworking and diagenesis (Thomas, 1986).

Although the diagenesis of a bone accumulation is an important part of its taphonomic history, it is too complex to represent on a univariate scale and therefore is not included in Table I.

4. Environmental Context and Sedimentology

Vertebrate remains occur in most types of terrestrial sedimentary deposits as well as in some marine settings. Table II provides a summary of paleoenvironmental contexts where vertebrate remains have been found, based on available documentation of fossil localities. Environmental context influences the vertebrate record in two important ways: (1) the ecological setting exerts strong controls on the types of animals available for preservation and on the causes of mortality, and (2) some taphonomic processes that affect animal remains after death vary from environment to environment, resulting in general patterns of preservation that are characteristic of different physical settings (Behrensmeyer and Hook, in press). From Table II, it is apparent that while some processes or circumstances are found in only one sedimentary setting, many occur in different environments, and some are universally present.

The strength of the environmental imprint on vertebrate preservation is determined by the degree to which particular taphonomic processes and circumstances are specific to certain environmental settings. This has yet to be systematically documented, but the number of examples of taphonomic

TABLE II. Environmentally Controlled Processes and Circumstances That Have Important Positive or Negative Effects on the Accumulation and Preservation of Vertebrate Remains[a]

Context	Processes or circumstances		
	Physical	Chemical	Biological
All environments	Water movement Temperature Rate of sediment buildup	Anoxia Water/soil pH, Eh Presence/absence of $CaCO_3$ Concentration of nutrients	Bioturbators, decomposers Bacteria Herding or schooling behavior Predator scats, regurgitates
Marine Offshore	Quiet bottom water Sediment starvation; stratigraphic condensation		Bacterial mats
Coastal: beach, lagoon, estuary	Wave action Storms Tides River input		Plant growth, roots, soil biota
Fluvial–deltaic Channel bar Channel fill Overbank: Floodplain, C-splay, levee, paleosol, swamp, pond, bay	Current action Hydraulic sorting and winnowing Slow or pulsed vertical aggradation Floods Seasonality Bank erosion	Fluctuating ground water chemistry (e.g., wet–dry) Alkaline soils	Plant growth roots, soil biota Aestivation burrows, dens Trapping–mud Trampling
Lacustrine	Thermal stratification Seasonal turnover Wave action, hydraulic sorting Quiet bottom water Fluctuations in lake margins Depth and size of lake		Benthic recyclers/bioturbators Bacterial mats Trapping–mud Trampling
Volcanic	Ash falls Mud flows Generally increased sedimentation	Increase Ca, Fl Poison gases	Temporarily reduced bioactivity
Eolian Dune, interdune loess	Rapid sedimentation Carcass dessication	Alkaline soils	Water-dependent clustering

TABLE II. *(Continued)*

Context	Processes or circumstances: Physical	Chemical	Biological
Caves, sinkholes, fissures	Protection from surface processes Stable temperature and moisture	$CaCo_3$ precipitation	Trapping Bone collecting Denning Bioturbation
Tar and amber	Adhesive texture	Low permeability to H_2O, O_2	Trapping

[a]Processes or circumstances that apply to all environments are not listed again under specific environmental categories.

histories that are associated with specific environmental contexts provides a rough measure (Koster, 1987; Behrensmeyer and Hook, in press). Thus, mass death accumulations of articulated fish are a recurring phenomenon in lake deposits because the processes that result in such taphocoenoses are relatively common in this setting (Bell *et al.*, 1989). Similar accumulations also occur in lagoons, but rarely in floodplains (where fish can be common), because a different set of processes, including extensive scavenging, prevails (G. R. Smith *et al.*, 1988).

Detailed studies of associations between sedimentary facies and taphonomic features yield further evidence for environmental controls. Concordance or discordance between hydraulic equivalence of bones and their enclosing sediment suggests differing influences of sedimentary transport and reworking (Behrensmeyer, 1975; Badgley, 1982a, 1986b). In channel deposits, there appears to be a relatively predictable (though not universal) relationship between vertebrate assemblages with different taphonomic features and coarse- versus fine-grained channel fills (Behrensmeyer, 1988a). Disarticulated and abraded teeth and bones representing the heavier and larger elements of skeletons occur in coarser channel deposits, while articulated or associated, unsorted skeletal parts occur in fine-grained fills. In floodplain paleosols, there are many examples of accumulations of relatively unsorted, disarticulated small bones with variable patterns of weathering, indicating accumulation of remains over extended periods of time (e.g., Bown and Kraus, 1981; Winkler, 1983; Maas, 1985; Badgley and Gingerich, 1988; Bown and Beard, 1990). Documentation of recurring patterns such as the above eventually could provide a foundation for defining specific taphonomic modes or "taphofacies" (*sensu* Speyer and Brett, 1986) for vertebrate assemblages based both on sedimentary context and taphonomic variables.

Although there appears to be some degree of predictability in the association of particular types of bone accumulations and depositional settings, inferring taphonomic history directly from environmental context can be misleading. For example, bones found in channels are commonly assumed to have undergone fluvial transport, but in the channel fill situations discussed above, vertebrate remains may be autochthonous (Behrensmeyer, 1987b, 1988a). Similarly, to assume that bones preserved in paleosols are untransported is to deny the role of predators and scavengers, as well as overbank flood events, in transporting whole or partial carcasses or even individual bones. Evidence contained in the bone assemblage itself, such as skeletal part sorting and bone orientation, may indicate to what extent fluvial sedimentary processes have interacted with the organic remains. Carcass transport prior to disarticulation, sorting, and weathering cannot be ruled out based on sedimentary or taphonomic evidence, however, although in some instances it might be suggested by spatial patterns of skeletal parts or lateral extent of the accumulation.

Biological processes that concentrate organic remains are associated with particular environments to varying degrees, and many are relatively independent of physical setting (Table II). Examples include herding or schooling behavior that leads to monospecific mass-death accumulations in fluvial, lacustrine, or trap situations, and predator activities that create dense concentrations of collected, fecal, or regurgitated remains in a wide variety of land-surface and cave situations. Often the taphonomic features of fluvial bone accumulations are difficult to explain based on sedimentary processes alone, implying that biological processes played a major role in their history (Lawton, 1977; Eberth, 1990). Similarities in taphonomic features of assemblages from different depositional contexts can indicate that these features were inherited from the equivalent predepositional circumstances, with minimal overprinting by processes operating during deposition and final burial. On the other hand, sedimentary processes can also impose a strong overprint on inherited features in a bone assemblage, making it difficult to reconstruct earlier parts of its taphonomic history. In any case, the pretransport role of biological agencies in shaping the features of bone concentrations in fluvial deposits probably has been underestimated.

Vertebrate faunas have long been used as evidence for paleoenvironmental reconstructions, based on ecological requirements inferred from functional morphology or relatedness to present-day species or higher-order taxonomic groups (e.g., Van Couvering, 1980; Wilson, 1988; Elder and Smith, 1988). Such paleoenvironmental reconstructions sometimes are done in the absence of taphonomic and sedimentological data. This may not affect broad-scale interpretations, for instance, when identification of cursorial or arboreal morphologies is used to indicate the existence of open savanna or forested habitats in the general vicinity of a fossil locality (Andrews and Nesbit-Evans, 1979; Van Couvering, 1980). For more detailed reconstructions, however, taphonomic mixing of taxa from different habitats can lead to erroneous conclusions concerning the composition of a community and its paleoenvironmental context, particularly where time-averaging also occurs. A stronger basis for paleoenvironmental reconstruction combines ecomorphic characters of vertebrate taxa with taphonomic and sedimentological evidence (Tables I and II) that indicate how these taxa actually represent the original communities and habitats.

5. Inferring Taphonomic History

Taphonomic characteristics of vertebrate accumulations (Table I) combined with sedimentological information (Table II) provide multiple lines of evidence with which to formulate and test hypotheses concerning taphonomic history. The first phase of inference involves deciding which taphonomic processes or circumstances have influenced the deposit, and the second involves judging their relative importance. Usually the "cause" of a fossil accumulation involves a combination of many biological, physical, and chemical processes, although these may be linked to the same overall event. It is important to determine which of these processes had the most significant effects on a bone assemblage in order to identify the major biases that affect paleobiological information.

The many different types of taphonomic data from a particular accumulation can be organized as shown in Table I, providing a visual overview that also can be used as a basis for standardized comparisons among different accumulations. Two excavated bone assemblages from the Pleistocene Koobi Fora Formation in northern Kenya (Behrensmeyer, 1985) provide examples of multivariate patterns derived from actual data (Table III, Fig. 2). These can be compared with hypothetical examples that demonstrate a range of data and graphic patterns based on different taphonomic histories (Table IV, Fig. 3). The linear patterns shown in Figs. 2 and 3 depend, of course, on the vertical order of the different variables and the scaling of each horizontal bar. The purpose of these diagrams is to demonstrate a graphic technique that helps to summarize the complex array of taphonomic information available in each bone accumulation. This serves two major goals: (1) it shows the behavior of different taphonomic variables in relation to one another within each assemblage, and (2) it encourages standardized representation of data, allowing comparative analysis of different assemblages.

5.1. Taphonomic Information from Excavated Samples

The Pleistocene examples represent relatively small-scale excavations of bone concentrations in lake margin deposits. The two sites occur in the same area and are about 13 m apart stratigraphically (Behrensmeyer, 1985). The taxa include a large and a dwarf hippopotamus, a guinea fowl, bovids, suids, crocodilians, chelonians, and fish, and many of the same taxa occur

TABLE III. Measured and Calculated Values for Taphonomic Variables in Two Excavated Bone Assemblages from the Plio-Pleistocene Koobi Fora Formation of Northern Kenya[a]

Variable	Excavation PE-1	Excavation PE-4
Assemblage data		
Number of specimens (*N*)	48	117
Number of individuals	13	17
Number of species	12	12
Relative abundance (Simpson 1/*D*)	78.13	26.32
Body size (mode, kg)	100–500	5–100
Age spectrum (% adult)	100	88
Bone articulation (mode)	Isolated–dispersed	Associated–dispersed
Skeletal parts (mode, Voorhies Group)	III	I
Quarry data		
Size of accumulation (m^2)	~20	~10
Spatial density ($1/m^2$)	2.4	21
Spatial arrangement		
In plan view	78% NW–SE	No preferred direction
In profile	15% > 20°	25% > 20°
Patchiness	Patchy ($0–8/m^2$)	Patchy ($5–34/m^2$)
Bone modification		
Breakage (% complete)	54	13
Cracking (Weathering)	Stage 0–1 (78%) (range 0–2)	Stage 0–1 (81%) (range 0–4)
Abrasion/polish (% unabraded)	42	74
Surface marks	3% (1/40)	7% (6/90)

[a]Number of specimens includes all *in situ* excavated material (mammal, reptile, bird, fish), but excludes small bones recovered from screening. Representation of skeletal parts is based on the dominant Voorhies Group (VG). Accumulation size is an estimate of original area covered by the bone assemblage, but spatial density is calculated using excavated area only. See text for details. (PE, paleontological excavation.)

in both excavations. Most species are represented by a minimum of only one individual in these small samples (Table III). The Simpson index (Magurran, 1988) is used as a measure of relative abundance and indicates somewhat greater evenness (higher diversity) for PE-1. The modal category of body size was derived from estimates for each individual, based on comparisons between the preserved body parts and those of analogous living species. In PE-1, there were very few bones or teeth that could be associated with the same individual, while in PE-4 over half of the assemblage consists of axial skeletal parts of two hippopotami (Table III). Bones that were originally articulated are separated by decimeters or, at most, a few meters. Vertebrae and vertebral epiphyses are the most common elements in PE-4, whereas relatively dense cranial parts, teeth, and dermal scutes are more abundant in PE-1. Skeletal parts are represented in terms of the dominant Voorhies Group in Table III, with VG III being the least-transportable or "lag"

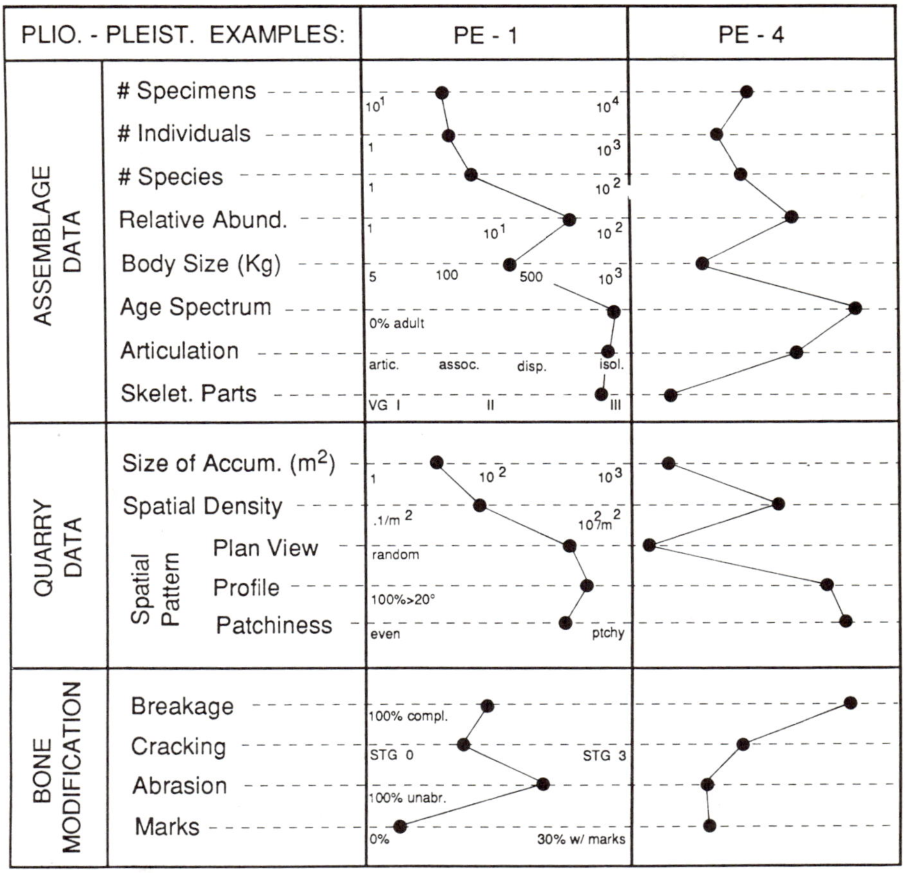

FIGURE 2. Graphic representation of data from Table III, allowing a visual comparison of patterns created by 17 different taphonomic variables in the assemblages from PE-1 and PE-4. Each horizontal dashed line represents the univariate spectrum defined in Table I, with numerical values increasing to the right, except as noted below. Relative abundance is based on calculated values for the Simpson index of diversity (Magurran, 1988), with the range in the figure being from 1 (left) to 100 (right). Spatial pattern: plan view is based on random (left) to 100% oriented in the same compass sector (NW–SE or NE–SW). Spatial arrangement: profile is based on 100% bones dipping >20° at the left, as opposed to 100% dipping <20° to the right. Breakage is based on 100% complete to the left, 0% complete to the right, and abrasion is similar, with 100% unabraded (fresh) to the left. Cracking (weathering) is more qualitative and combines both the dominant mode of weathering and an indication of the range of weathering stages in the assemblage. Surface marks range from 0% bones with these features on the left to 30% on the right. Note: connecting lines are only to aid visual comparisons of patterns among the examples and to highlight the relative positions of neighboring variables. See text for further explanation.

assemblage, VG II intermediate, and VG I the most transportable. Voorhies Groups are based on the relative transport potential of different skeletal parts of sheep and coyote in an artificial flume (Voorhies, 1969; Behrensmeyer, 1975).

The actual size of the excavations is 20 m^2 for PE-1 and 8.3 m^2 for PE-4;

estimated size of the original bone accumulation is somewhat larger than excavated area for the latter (Table III). PE-1 has a lower spatial density of bones, many of which are oriented NW–SE and most of which are within 20° of horizontal (i.e., approximately parallel to bedding). In contrast, the bones of PE-4 show no preferred compass orientation, and many of them are tilted at high angles to the bedding. Both have relatively patchy distributions across the excavated areas, with coefficients of variation in bones per square meter of $V = 35$ for PE-1 and $V = 58$ for PE-4. There is a much higher proportion of broken bones in PE-4, and some pieces of ribs, vertebrae, and other elements that were found several meters apart can be fitted back together. Most of the bones in both assemblages are in weathering Stage 0–1 (fresh to minor surface cracking), but PE-4 includes 11 bones with more advanced weathering up to Stage 4. The percent of unabraded versus abraded bones is based on rounding of break edges only, since an objective assessment of abrasion is much more difficult on natural bone surfaces. PE-4 has a much higher proportion of unabraded bones, and also a slightly higher number with surface marks, which include parallel scratches and tooth marks.

5.2. Comparative Taphonomic Analysis

The ability to observe the behavior of variables relative to one another and compare this from assemblage to assemblage is an important advantage of the graphic method. The comparison in Fig. 2, for instance, reveals that PE-1 and PE-2 are somewhat similar in overall assemblage features, but markedly different in their spatial characteristics and bone modification features. The two fossil assemblages have similar numbers of species and individuals, and they cover broadly comparable areas with a relatively patchy distribution of vertebrate remains. The most striking difference involves the presence of a large number of associated, primarily axial skeletal parts at PE-4. These belong to two individuals, and without them the assemblage and quarry data would be much more like those of PE-1. In spite of this, there are still important differences between the assemblages, including the marked compass alignment, orientations parallel to the bedding surface, and the higher proportion of bones with rounded break edges in PE-1. Greater incidence of breakage, a wider weathering spectrum, and higher angles of orientation in profile are characteristic of PE-4.

The combined information indicates that the assemblage at PE-1 was subjected to processes that differentially removed or excluded most of the lighter skeletal elements in Voorhies Groups I and II, aligned the remaining bones, and rounded many of the break edges. This is strong evidence for the influence of water currents and interaction with moving sediment immediately prior to burial and probably also for some time before burial. No such processes are indicated at PE-4; rather, the taphonomic features indicate *in situ* breakage and bone orientation by trampling or plant growth at the site where parts of two hippopotamus carcasses decomposed. The assemblage at PE-1 thus is more likely to be allochthonous, and that at PE-4 autochthonous. The range of weathering stages as well as the juxtaposition of fragmentary remains of many different species indicate attritional accumulation and time-averaging at PE-4. A lower range of weathering stages, a higher percentage of complete bones, and fewer surface marks might be used as evidence for a shorter interval of time-averaging at PE-1. On the other hand, sediment abrasion and transport may have affected these variables by smoothing bone surfaces and destroying or sorting out fragile bones in the more advanced weathering stages. The source for the bones preserved at PE-1 could have been an autochthonous assemblage similar to that of PE-4, but such inferences are limited by the extent to which current activity has overprinted pretransport features of the PE-1 assemblage.

Information on sedimentary context has been excluded from the above discussion in order to demonstrate how the taphonomic features can be used independently to build hypotheses concerning taphonomic history. Sedimentological data from the excavation sites reveal that both PE-1 and PE-4 occur on subaerial lake margin surfaces immediately underlying transgressive deposits. At PE-1, the bones rest on a siltstone with clearly preserved footprints of dwarf hippopotamus, and some of the bones are actually trapped within the footprints (Laporte and Behrensmeyer, 1980). The fossiliferous unit itself consists of a tuffaceous sand with ripple-and-trough cross-lamination indicating a NW to SE flow direction, which parallels the dominant bone orientation. This unit is part of a complex of shallow distributary channels and interdistributary mudflats bordering a lake. In contrast, the bone-bearing unit at PE-4 is a homogeneous siltstone with pedogenic features, including lack of primary stratification, carbonate glaebules, and abundant root traces. This is overlain by a thin shell coquina representing lake transgression. The excavated bone concentration occurs where the paleosol thickens from <40 cm to >50 cm, indicating a local topographic depression. This suggests a water hole or a relatively lower and damper area where plant growth, trampling, burial rates, and carcass survival may have been somewhat higher than in the surrounding landscape. The sedimentological evidence thus independently supports and also enhances interpretations of taphonomic processes that operated at the two sites.

The examples given above demonstrate how the taphonomic variables can be organized and used to develop hypotheses concerning taphonomic history. Interpretations of the taphonomic histories of PE-1 and PE-4 could be further enhanced by additional comparative data from assemblages with known taphonomic histories.

5.3. Hypothetical Examples of Taphonomic Data

Ideally, comparative models would be based on actual data resulting from known taphonomic histories, but this is not yet possible for all of the

TABLE IV. Hypothetical Examples of Taphonomic Evidence in Vertebrate Accumulations Resulting from Four Distinct Taphonomic Histories[a]

	Assemblage data	Quarry data	Bone modification
Example 1: Mass accumulation of one species (herd)	5180 specimens 17 individual 1 species Body size 150–300 kg 53% adult Articulated + associated Unsorted	Total area 150 m^2 34.5 bones/m^2 Random in plan view Horizontal + few high dips Uneven distribution	95% complete No primary cracks No abrasion No marks
Example 2: Collection of prey bones by predator	4870 specimens 460 individuals 12 species 80% individuals/ 2 species Body size 5 g–1 kg 25% adult 10% articulated, 70% associated, 20% disassociated Minor sorting: loss of small bones	Total area 5 m^2 974 bones/m^2 Random in plan view Random in profile Uneven distribution	40% complete All Stage 0–1 55% unabraded 8% with scratches
Example 3: Attritional large tetrapod assemblage	650 specimens 80 individuals 16 species 30%–25%–10% individuals in top 3 species Body size 80–1000 kg (mode 100–500 kg) 60% adult Mode associated–dispersed (85%) Limb parts, jaws, teeth, vertebrae; few ribs, skulls	Total area 8000 cm^2 0.081 bone/m^2 Random in plan view 27% high dips (>20°) Highly patchy	10% complete Mode Stage 1 (65%), range Stage 1—4 70% unabraded 10% with scratches, punctures, pits, grooves
Example 4: Winnowed, sorted assemblage	340 specimens 80 individuals 20 species 14%–8%–7% individuals in top 3 taxa Body size 1–2000 kg (mode 50–100 kg) 90% adult All isolated Durable + heavy parts only	Total area 180 m^2 1.9 bones/m^2 Long pieces oriented 15% high dips (>20°) Uneven distribution	10% complete (85% fragmentary) Mode Stage 1, range Stage 0–2 10% unabraded 3% with minor scratches, pitting

[a]The data represent reasonable estimates based on Recent and fossil accumulations resulting from these processes and circumstances, but they are not meant as definitive models for the different taphonomic histories. The four examples are portrayed graphically in Fig. 3.

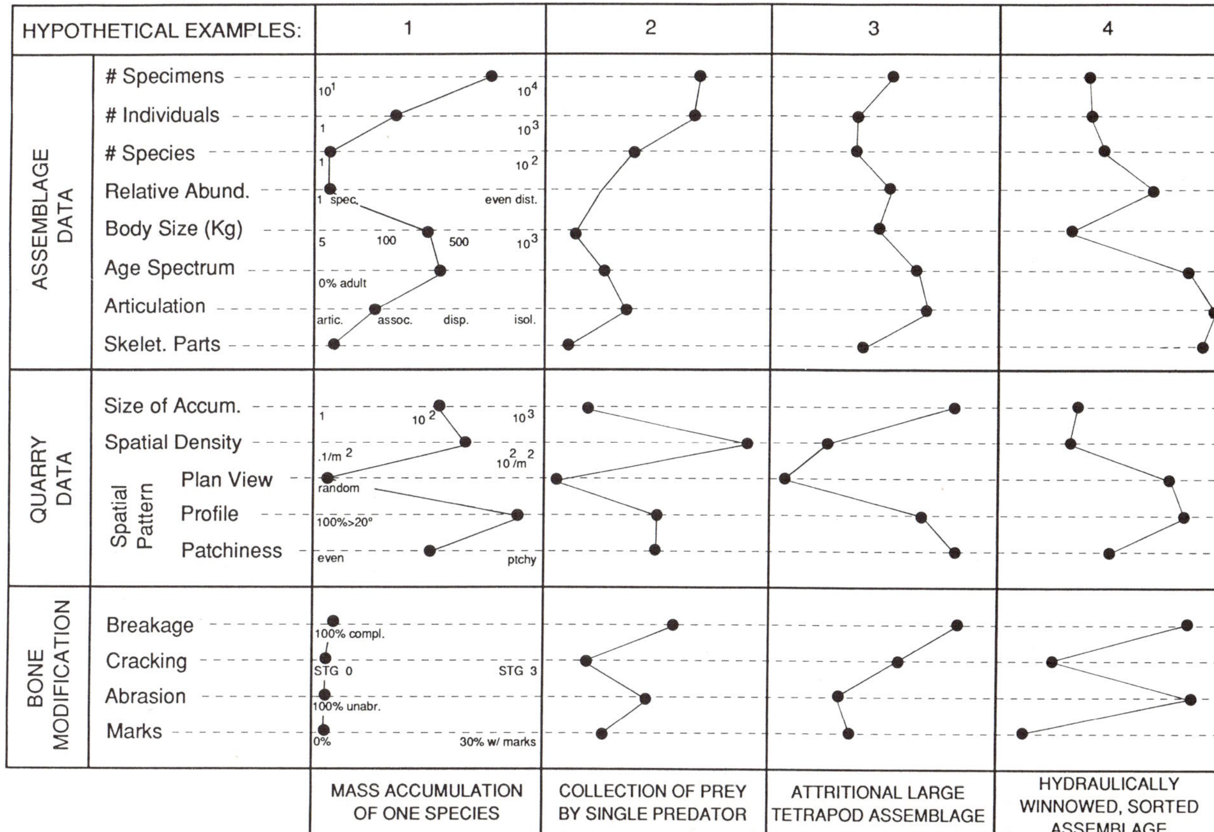

FIGURE 3. Plots of data from hypothetical examples given in Table IV, using the same univariate spectra as in Fig. 2. In these examples, however, relative abundance is plotted qualitatively using evidence for species dominance or evenness given in Table IV, and skeletal parts are represented qualitatively according to descriptive evidence for sorting. This illustrates how plots of variables can accommodate minor adjustments for different kinds of data.

variables listed in Table I. Examples of multivariate patterns shown in Fig. 3 are based on four hypothetical bone accumulations with taphonomic features specified in Table IV. These features are predictions, based on a range of published examples and the author's experience with actualistic data on taphonomic processes. The diagrams are not meant as explicit models for the four taphonomic histories at this stage in the development of the method. With real data and more precisely calibrated horizontal scales, it will be possible to characterize and compare many different types of Recent and fossil bone accumulations in order to establish consistent patterns of correlation between the different variables, using standard multivariate techniques. It should also be possible to determine with greater confidence what overall graphic patterns reflect particular taphonomic histories.

Example 1 of Table IV represents the expected features of a mass death event, such as a flood, affecting a monospecific herd (or part of herd), followed by rapid burial of whole carcasses. Judging from actualistic evidence for mass death (Weigelt, 1989), this is likely to involve species of large body size and a significant proportion of juveniles. Because in flood-related mass death, predator and scavenger populations typically are either overwhelmed by the sudden abundance of carcasses or denied access to them by rapid burial, remains generally will be articulated and unsorted. Assuming that carcasses were transported whole and deposited in a restricted backwater area by a flood, the bones should be densely packed, with an uneven spatial distribution and individual elements randomly oriented in plan view and mostly horizontal. Rapid burial would result in a lack of trampling and minimal root disturbance, preserving a high degree of articulation of complete bones, which would lack evidence for weathering and other types of bone modification.

The multivariate patterns in Examples 2–4 (Table IV) contrast with Example 1 according to expected differences in the hypothetical taphonomic histories. The taphonomic features of Example 2 represent a fissure-fill or cave accumulation of bones from a single species of predatory bird, such as a large owl (Dodson and Wexlar, 1979). Fluvial reworking is absent, and sorting and bone modification are caused by the predator itself, local bioturbators, and physical conditions within the cave. Example 3 is a widely distributed, time-averaged land surface accumulation of large vertebrates comparable to that of the present-day assemblage in Amboseli Park, Kenya (Behrensmeyer and Dechant Boaz, 1980; Behrensmeyer, 1982). This assemblage is unaffected by fluvial processes, but carnivory, scavenging, trampling, and physical weathering have played major roles in creating its taphonomic features. In contrast, fluvial processes were predominant in the taphonomic history of Example 4, which represents an attritional assemblage of winnowed, reworked, and sorted remains of both small and large vertebrates. Taphonomic features that indicate the influence of moving water include absence of articulated or associated parts, relatively greater representation of denser skeletal elements and absence of easily winnowed ones, horizontal and current-aligned orientations, and abrasion.

The overall graphic pattern for assemblage data in Example 3, which is untransported, is broadly similar to that in Example 4; the former represents pretransport features that may be inherited by the latter and not strongly altered except by accentuated sorting. Quarry and bone modification features of these two examples are more distinct and support an interpretation of autochthony for Example 3 and allochthony for Example 4. It is difficult, however, to assess the absolute degree of allochthony from such evidence. Statements that assemblages comparable to Example 4 represent "drainage basins" (Behrensmeyer, 1988a) probably overestimate the area involved, but at present there is little actualistic evidence that bears on this problem.

The data in Table IV were listed and graphed prior to analysis of PE-1 and PE-4, to avoid incorporating features of these assemblages in the hypothetical examples. Comparison of the graphic patterns in the excavated and the hypothetical assemblages shows that PE-1 is similar to Example 4 in assemblage, quarry, and bone modification data. It is relatively more patchy, perhaps because of the influence of substrate irregularities (footprints), and has a lower degree of breakage than the hypothetical winnowed and sorted assemblage. PE-4 is also similar to Example 4 in assemblage data, except that it has a higher degree of articulation and lacks skeletal part sorting, which is more like Example 3 (the attritional, untransported assemblage) or Example 1 (the mass death accumulation). The quarry data in PE-4 are similar to Example 1 or 2 because it is a relatively small, randomly oriented, dense and patchy bone accumulation. In bone modification features, PE-4 is very similar to the hypothetical untransported, attritional assemblage (Example 3) and unlike any of the other examples.

These comparisons emphasize how assemblage, quarry, and bone modification data from a real example might correspond to different parts of the graphic patterns for other (e.g., hypothetical or actualistic) examples, thereby revealing different combinations of processes affecting taxonomic and skeletal features, spatial relationships of the remains, and characteristics of bone modification.

5.4. The Problem of Autochthony or Allochthony

A perennial issue in vertebrate paleoecology is the question of allochthony or autochthony, or how fossil remains relate to the original death sites and ultimately to the habitats and communities of the living animals. Taphonomic and sedimentological features can demonstrate whether a burial site corresponds to the place where a carcass came to rest, but such evidence may not be sufficient to distinguish between an actual death site and the place where a transported carcass was grounded. Paleoecological interpretations often go beyond this, however, and may include the assumption that the place of death is representative of the preferred habitat of the living animal. Death sites for many present-day herbivores in Africa, however, occur differentially in habitats that are occupied seasonally or during parts of the daily cycle (Behrensmeyer and Dechant Boaz, 1980; Conybeare and Haynes, 1984; Haynes, 1985a). Ultimately the problem of allochthony and autochthony depends on the degree of spatial resolution that is required for paleoecological reconstructions. These terms should be clearly defined by researchers who use them, taking into account both the data at hand and the questions being addressed. In comparative studies, taphonomic features can indicate the relative degrees of allochthony or autochthony for fossil assemblages and modern analogues without requiring more specific calibration of these terms.

6. Assessing Biases in Fossil Samples

Establishing the taphonomic history of a bone accumulation leads to the next level of inference, which concerns the biases inherent in the accumulation. The significance of bias in fossil samples depends on the goals of the paleobiologist. The taphonomic loss of small species, for instance, only matters if measures of the total diversity of a fauna are important. Such biases are often taken for granted in studies of faunal diversity, and if small species are unreported, it is assumed that the numerical data apply only to large species (and *vice versa*). A companion assumption, however, is that the diversity counts for the large-body-size category are not themselves taphonomically biased. The possibility that this assumption may be incorrect underscores the need to understand taphonomic history before proceeding with paleobiological interpretations.

Understanding of taphonomic processes involved in the formation of a fossil accumulation leads to predictions concerning biases. For example, a collection of bones from the prey of a specific predator (Fig. 3, Example 2) may differentially sample particular species because of the predator's prey preferences. This type of accumulation is suspect as an estimate of standing diversity. An attritional assemblage of macrovertebrates (Example 3) may combine remains of large species over tens to hundreds of years. A time-averaged accumulation of this kind may represent several different habitat types that succeed one another within the same area (Behrensmeyer and Dechant Boaz, 1980). The resulting fossil fauna could have a higher diversity than any one of these habitats and mix species that did not interact in the original communities. A transported assemblage (Example 4) may represent a mixture of many other types of accumulations, overprinted with sorting and destruction of particular bone sizes and skeletal parts. It can include a wide range of habitats and be time-averaged over periods of 10^3–10^5 years or more (Behrensmeyer, 1982). Such accumulations obviously are biased in their representation of vertebrate ecology since the samples are not comparable to those drawn from modern, time- and site-specific communities, but they may provide the least biased estimate of diversity within the ecosystem as a whole.

Understanding the information contained in the samples also helps to define how they can be used to address specific paleobiological problems. An attritional assemblage of macrovertebrates (Example 3) can be used to estimate time-averaged diversity for a specified range of body size that is not taphonomically biased, and compared with recent faunal diversity modeled over similar time intervals (Badgley, 1982b). To examine herding behavior in extinct taxa, the age structure of a mass-death, monospecific group (Example 1) can be compared to other mass-death fossil examples or to modern ones. The predator accumulation (Example 2) can provide information on predator behavior and the presence (but not the absence) of species in the local fauna. The transported, mixed assemblage (Example 4) offers an overview of species from a wide range of taphonomic circumstances and thus may be the best source of data for first occurrences and for monitoring ecologically driven change in community composition through time (e.g., Eberth, 1990; Brinkman, 1990).

There is little doubt that taphonomic biases affect the fossil record of species diversity, paleocommunity associations, and extinction and immigration events (e.g., Bown and Beard, 1990). What typically concerns paleobiologists is how to assess the overall impact of such biases, and whether it is possible to "correct for" them or otherwise retrieve credible biological information from fossil samples. One strategy involves comparing multiple fossil samples with varying taphonomic histories within a given stratigraphic sequence. Comparative taphonomy can be used to examine recurring patterns and anomalies in faunal composition, and how these correspond to environmental context and taphonomic features. If similar faunal lists occur in assemblages with widely different taphonomic histories, then the case for a persistent, spatially homogeneous paleocommunity is strengthened. If, however, faunal composition is highly correlated with taphonomic mode, then this indicates a mosaic of faunal associations or mortality patterns and underlines the importance of sampling different modes to obtain overall diversity counts (e.g., Bown and Beard, 1990). In either case, deliberate use of fossil assemblages with different taphonomic histories should increase the probability that more preserved species in a fauna will be sampled, and it also will indicate which body sizes or ecomorphs are consistently absent.

If only a limited number of taphonomic modes is available in a rock sequence, it may not be possible to assess the impact of taphonomic biases on faunal representation, and it is difficult or impossible to "correct for" the missing information. In such cases questions can be modified to fit the taphonomic limitations of the available evidence, such as diversity for a specified body size only, or reconstructions of faunal associations that accept time-averaging and mixed habitats. There may be theoretical grounds for assuming that certain components of the fauna are absent or mixed together due to taphonomic biases (e.g., Damuth, 1982), but such assumptions should be treated cautiously because they can obscure real ecological differences in ancient communities (Behrensmeyer, 1988b) as well as unique paleotaphonomic processes.

7. Conclusion

Taphonomic history is the basis for assessing the quality of biological and ecological information contained in vertebrate accumulations. This chapter has outlined the kinds of evidence that can be used to reconstruct taphonomic history and has proposed a standardized method for comparative taphonomic study of vertebrate assemblages. As advocated here, taphonomy involves multivariate study of features in fossil assemblages combined with contextual sedimentology, followed by inference concerning taphonomic processes and circumstances, reconstruction of taphonomic history, and finally, assessment of the quality of biological and ecological information based on known biases in the assemblage (Fig. 1).

A standardized multivariate approach to taphonomic features allows the fossil record to contribute more effectively to understanding of taphonomic processes and circumstances that may not be represented in modern ecosystems. Real data are needed to establish the relationships among multivariate patterns (Fig. 3), taphonomic processes, and sedimentary context, but hypothetical patterns based on a range of examples can be useful in a heuristic sense. Perhaps the most important role of the "taphograms" is to encourage the observer to think of the different variables in relation to each other and to take a comparative approach to analyzing taphonomic data. This can help hypothesis formulation and testing in the study of fossil assemblages, while also enhancing the dialogue between modern and ancient evidence for taphonomic processes. The multivariate approach also provides a method for exploring the evolutionary history of taphonomic processes, for example, by investigation of taphonomic variables through time within well-defined sedimentary contexts, and for establishing isotaphonomic (i.e., "having the same taphonomic features and biases") contexts in which to examine faunas across major extinction events.

Ultimately it is the combination of taphonomic features with detailed sedimentological data that provides the soundest basis for inferring the history of preservation. To the extent that probabilities of different histories vary with environment, environments have exerted large-scale controls on the occurrences of different types of vertebrate fossil assemblages. It follows that the frequencies of different depositional environments through the geological record control, to some extent, the availability of different types of vertebrate samples, including the unusually well-preserved Fossil-*Lagerstätten* (Seilacher *et al.*, 1985; Brett, 1988; Allison, 1988a). Weigelt

(1989) was probably the first person to be concerned with the overall impact of *Lagerstätten* on our view of vertebrate history. It appears that the time is approaching when the data and methods of vertebrate taphonomy can seriously address the question of how much of the vertebrate record consists of samples resulting from catastrophic death and burial, and how this affects our perspective on vertebrate paleoecology and evolution.

Recurring patterns of preservation indicate that the vertebrate record is more structured than random in its sampling of faunas through time. Future research can focus on organizing the available data and establishing the characteristics of recurring taphonomic modes in the vertebrate record. This will provide a more reliable basis for assessing the effects of taphonomic processes on data ranging from species presence or absence in individual fossil assemblages to diversity patterns and paleocommunity structures throughout the Phanerozoic.

ACKNOWLEDGMENTS. Much of the background for this paper has been influenced and inspired by members of the Evolution of Terrestrial Ecosystems Program at the Smithsonian Institution. Discussions with Bob Hook, Catherine Badgley, Robyn Burnham, Richard Bateman, Bill DiMichele, Will Downs, Larry Flynn, Tom Plummer, Rick Potts, Hans Sues, and Scott Wing helped to generate references and ideas. Special thanks are due to Catherine Badgley for her careful reading of an early draft of this manuscript, to Tom Jorstad for editorial advice, and to Mary Parrish for help with the illustrations.

References

Agenbroad, L. D., 1984, Hot springs, South Dakota: Entrapment and taphonomy of Columbian Mammoth, in: *Quaternary Extinctions* (P. S. Martin and R. G. Klein, eds.), University of Arizona Press, Tucson, Arizona, pp. 113–127.

Allison, P. A., 1988a, *Konservat-Lagerstätten:* Cause and classification, *Paleobiology* **14**:331–344.

Allison, P. A., 1988b, The role of anoxia in the decay and mineralization of proteinaceous macro-fossils, *Paleobiology* **14**:139–154.

Andrews, P., 1990, *Owls, Caves and Fossils*, British Museum (Natural History), London.

Andrews, P., and Cook, J., 1985, Natural modifications to bones in a temperate setting, *Man* **20**:675–691.

Andrews, P., and Nesbit Evans, E. M., 1979, The environment of *Ramapithecus* in Africa, *Paleobiology* **5**:22–30.

Antia, D. D. J., 1979, Bone-beds: A review of their classification, occurrence, genesis, diagenesis, geochemistry, palaeoecology, weathering, and microbiotas, *Mercian Geol.* **7**:93–174.

Argast, S., Farlow, J. O., Gabet, R. M., and Brinkman, D. L., 1987, Transport-induced abrasion of fossil reptilian teeth: Implications for the existence of Tertiary dinosaurs in Hell Creek Formation, Montana, *Geology* **15**:927–930.

Aslan, A., and Behrensmeyer, A. K., 1987, Vertebrate taphonomy in the East Fork River, Wyoming, *Geol. Soc. Am. Abstr. Progr.* **1987**:575.

Bada, J. L., Gillespie, R., Gowlett, A. J., and Hedges, R. E. M., 1984, Accelerator mass spectrometry radiocarbon ages of amino acid extracts from Californian paleo-indian skeletons, *Nature* **312**:442–444.

Badgley, C. E., 1982a, Community reconstruction of a Siwalik Mammalian assemblage, Ph.D. thesis, Yale University, New Haven, Connecticut.

Badgley, C. E., 1982b, How much time is represented in the present?: The development of time-averaged modern assemblages as models for the fossil record, in: *Proceedings of the Third North American Paleontological Convention*, Volume 1, pp. 23–28.

Badgley, C. E., 1986a, Taphonomy of mammalian fossil remains from Siwalik rocks of Pakistan, *Paleobiology* **12**:119–142.

Badgley, C. E., 1986b, Counting individuals in mammalian fossil assemblages from fluvial environments, *Palaios* **1**:328–338.

Badgley, C. E., and Gingerich, P. D., 1988, Sampling and faunal turnover in Early Eocene mammals, *Palaeogeogr. Palaeoclimatol. Palaeoecol.* **63**:141–158.

Behrensmeyer, A. K., 1975, The taphonomy and paleoecology of Plio-Pleistocene vertebrate assemblages east of Lake Rudolf, Kenya, *Bull. Mus. Comp. Zool. Harv.* **146**:473–578.

Behrensmeyer, A. K., 1978, Taphonomic and ecological information from bone weathering, *Paleobiology* **4**:150–162.

Behrensmeyer, A. K., 1982, Time resolution in fluvial vertebrate assemblages, *Paleobiology* **8**:211–227.

Behrensmeyer, A. K., 1983, Patterns of natural bone distribution on recent land surfaces: Implications for archeological site formation, in: *Animals and Archeology*: Volume 1, *Hunters and Their Prey* (J. Clutton-Brock and C. Grigson, eds.), British Archeological Report No. 163, London, pp. 93–106.

Behrensmeyer, A. K., 1985, Taphonomy and the paleoecologic reconstruction of hominid habitats in the Koobi Fora Formation, in: *L'environment des Hominides au Plio-Pleistocene* (Y. Coppens, ed.), Foundation Singer-Polignac/Masson, Paris, pp. 309–324.

Behrensmeyer, A. K., 1987a, Taphonomy and hunting, in: *The Evolution of Human Hunting* (M. H. Nitecki and D. V. Nitecki, eds.), Plenum Press, New York, pp. 423–450.

Behrensmeyer, A. K., 1987b, Miocene fluvial facies and vertebrate taphonomy in northern Pakistan, *Soc. Econ. Paleontol. Mineral. Spec. Publ.* **39**:169–176.

Behrensmeyer, A. K., 1988a, The pull of the recent analogue, *Palaios* **3**:1.

Behrensmeyer, A. K., 1988b, Vertebrate preservation in fluvial channels, *Palaeogeogr. Palaeoclimatol. Palaeoecol.* **63**:183–199.

Behrensmeyer, A. K., 1990, Transport/hydrodynamics of bones, in: *Palaeobiology: A Synthesis* (D. E. G. Briggs and P. R. Crowther, eds.), Blackwell, Oxford, pp. 232–235.

Behrensmeyer, A. K., and Dechant Boaz, D. E., 1980, The Recent bones of Amboseli Park, Kenya, in: *Fossils in the Making* (A. K. Behrensmeyer and A. P. Hill, eds.), University of Chicago Press, Chicago, pp. 72–92.

Behrensmeyer, A. K., and Hill, A. P., 1980, *Fossils in the Making*, University of Chicago Press, Chicago.

Behrensmeyer, A. K., and Hook, R. S., in press, Paleoenvironments and taphonomy, in: *Terrestrial Ecosystems through Time* (Evolution of Terrestrial Ecosystems Consortium, eds.), University of Chicago Press, Chicago.

Behrensmeyer, A. K., and Kidwell, S. M., 1985, Taphonomy's contributions to paleobiology, *Paleobiology* **11**:105–119.

Behrensmeyer, A. K., Western, D., and Dechant Boaz, D. E., 1979, New perspectives in vertebrate paleoecology from a recent bone assemblage, *Paleobiology* **5**:12–21.

Behrensmeyer, A. K., Gordon, K. D., and Yanagi, G. T., 1986, Trampling as a cause of bone surface damage and pseudo-cutmarks, *Nature* **319**:768–771.

Behrensmeyer, A. K., Gordon, K. D., and Yanagi, G. T., 1989, Nonhuman bone modification in Miocene fossils from Pakistan, in: *Bone Modification* (R. Bonnichsen and M. H. Sorg, eds.), Center for the Study of the First Americans, Orono, Maine, pp. 99–120.

Bell, M. A., Wells, C. E., and Marshall, J. A., 1989, Mass-mortality layers of fossil stickleback fish: Catastrophic kills of polymorphic schools, *Evolution* **43**:607–619.

Binford, L. R., 1981, *Bones: Ancient Men and Modern Myths*, Academic Press, New York.

Blumenschine, R. J., 1986a, Early hominid scavenging opportunities: Implications of carcass availability in the Serengeti and Ngorongoro ecosystems, *Br. Archaeol. Rep. Int. Ser.* **283**:1–163.

Blumenschine, R. J., 1986b, Carcass consumption sequences and the archaeological distinction of scavenging and hunting, *J. Hum. Evol.* **15**:639–659.

Blumenschine, R. J., and Selvaggio, M. M., 1988, Percussion marks on bone surfaces as a new diagnostic of hominid behavior, *Nature* **333**:783–765.

Bonnichsen, R., 1979, Pleistocene bone technology in the Beringian Refugium, *Archaeol. Surv. Can. Pap.* **89**:1–297.

Bonnichsen, R., and Sorg, M. (eds.), 1989, *Bone Modification*, Center for the Study of the First Americans, Orono, Maine.

Bown, T. M., and Beard, K. C., 1990, Systematic lateral variation in the distribution of fossil mammals in alluvial paleosols, lower Eocene Willwood Formation, Wyoming, *Geol. Soc. Am. Spec. Pap.* **243**:135–151.

Bown, T. M., and M. J. Kraus, 1981, Vertebrate fossil-bearing paleosol units (Willwood Formation, Lower Eocene, northwest Wyoming, U.S.A.): Implications for taphonomy, biostratigraphy, and assemblage analysis, *Palaeogeogr. Palaeoclimatol. Palaeoecol.* **34**:31–56.

Brain, C. K., 1958, The Transvaal ape-man-bearing cave deposits, *Transvaal Mus. Mem.* **11**:1–131.

Brain, C. K., 1967a, Hottentot food remain and their bearing on the interpretation of fossil bone assemblages, *Sci. Pap. Namib Desert Res. Sta.* **32**:1–11.

Brain, C. K., 1967b, Bone weathering and the problem of bone pseudo-tools, *S. Afr. J. Sci.* **63**:97–99.

Brain, C. K., 1980, Some criteria for the recognition of bone-collecting agencies in African caves, in: *Fossils in the Making* (A. K. Behrensmeyer and A. P. Hill, eds.), University of Chicago Press, Chicago, pp. 108–130.

Brain, C. K., 1981, *The Hunters or the Hunted?*, University of Chicago Press, Chicago.

Brett, C. E., 1988, Comparative taphonomy and ecology of fossil "Mother Lodes", *Paleobiology* **14**:214–220.

Brinkman, D., 1990, Paleoecology of the Judith River Formation (Campanian) of Dinosaur Provincial Park, Alberta, Canada: Evidence from vertebrate microfossil localities, *Palaeogeogr. Palaeoclimatol. Palaeoecol.* **78**:37–54.

Bromage, T. G., 1984, Interpretation of scanning electron microscopic images of abraded forming bone surfaces, *Am. J. Phys. Anthropol.* **64**:161–178.

Bunn, H. T., 1981, Archaeological evidence for meat-eating by Plio-Pleistocene hominids from Koobi Fora and Olduvai Gorge, *Nature* **291**:574–577.

Bunn, H. T., 1982, Meat-eating and human evolution: Studies on the diet and subsistence patterns of Plio-Pleistocene hominids in East Africa, Ph. D. thesis, University of California, Berkeley.

Bunn, H. T., 1986, Patterns of skeletal representation and hominid subsistence activities at Olduvai Gorge, Tanzania and Koobi Fora, Kenya, *J. Hum. Evol.* **15**:673–690.

Bunn, H. T., and Kroll, E., 1986, Systematic butchery by Plio/Pleistocene hominids at Olduvai Gorge, Tanzania, *Curr. Anthropol.* **27**:431–452.

Bunn, H. Y., Harris, J. W. K., Isaac, G. L., Kaufulu, Z., Kroll, E., Schick, K., Toth, N., and Behrensmeyer, A. K., 1980, FxJj50: an Early Pleistocene site in northern Kenya, *World Archaeol.* **12**:109–144.

Colbert, E. H., 1947, Little dinosaurs of Ghost Ranch, *Nat. Hist.* **56**:392–399.

Conybeare, A., and Haynes, G., 1984, Observations on elephant mortality and bones in water holes, *Quat. Res.* **22**:189–200.

Currie, P. J., and Dodson, P., 1984, Mass death of a herd of Ceratopsian dinosaurs, in: *Third Symposium on Mesozoic Terrestrial Ecosystems, Short Papers* (W. E. Reif and F. Westphal, eds.), Attempto Verlag, Tubingen, pp. 61–66.

Damuth, J. D., 1982, Analysis of the preservation of community structure in assemblages of fossil mammals, *Paleobiology* **8**:434–446.

Damuth, J. D., 1990, Problems in estimating body masses of archaic fossil mammals, in: *Body Size in Mammalian Paleobiology* (J. D. Damuth and B. J. MacFadden, eds.), Cambridge University Press, Cambridge, pp. 229–254.

Davis, K. L., 1985, A taphonomic approach to experimental bone fracturing and applications to several South African Pleistocene sites, Ph. D. thesis, Department of Anthropology, State University of New York, Binghamton, New York.

Dodson, P., 1973, The significance of small bones in paleoecological interpretation, *Contrib. Geol. Univ. Wyoming* **12:**15–19.

Dodson, P., and Wexlar, D., 1979, Taphonomic investigations of owl pellets, *Paleobiology* **5:**275–284.

Eaton, J. G., Kirkland, J. I., and Dio, K., 1989, Evidence of reworked Cretaceous fossils and their bearing on the existence of Tertiary dinosaurs, *Palaios* **4:**281–286.

Eberth, D. A., 1990, Stratigraphy and sedimentology of vertebrate microfossil localities in uppermost Judith River Formation (Campanian) of Dinosaur Provincial Park, *Palaeogeogr. Palaeoclimatol. Palaeoecol.* **78:**1–36.

Elder, R. L., and Smith, G. R., 1988, Fish taphonomy and environmental inference in paleolimnology, *Palaeogeogr. Palaeoclimatol. Palaeoecol.* **62:**577–592.

Estes, R., 1964, Fossil vertebrates from the Late Cretaceous Lance Formation, eastern Wyoming, *Univ. Calif. Publ. Geol. Sci.* **49:**1–180.

Fiorillo, A. R., 1984, An introduction to the identification of trample marks, *Curr. Res.* **1:**47–48.

Fiorillo, A. R., 1988a, A proposal for graphic presentation of orientation data from fossils, *Contrib. Geol. Univ. Wyoming* **26:**1–4.

Fiorillo, A. R., 1988b, Taphonomy of Hazard homestead Quarry (Ogallala Group), Hitchcock County, Nebraska, *Contrib. Geol. Univ. Wyoming* **26:**57–97.

Fiorillo, A. R., 1989, An experimental study of trampling: Implications for the fossil record, in: *Bone Modification* (R. Bonnichsen and M. H. Sorg, eds.), Center for the Study of the First Americans, Orono, Maine, pp. 61–72.

Francillon-Vieillot, H., Buffrenil, V. D. E., Castanet, J., Geraudie, J., Meunier, F. J., Sire, J. Y., L. Zylberberg, L., and Ricqles, A. D. E., 1989, Microstructure and mineralization of vertebrate skeletal tissues, in: *Skeletal Biomineralization: Patterns, Processes and Evolutionary Trends* (J. G. Carter, ed.), American Geophysical Union, Washington, D. C., pp. 175–234.

Frey, R. W., Voorhies, M. R., and Howard, J. D., 1975, Estuaries of the Georgia Coast, U.S.A.: Sedimentology and biology. VIII. Fossil and Recent skeletal remains in Georgia estuaries, *Senckenb. Marit.* **7:**257–295.

Gifford, D. P., 1977, Observation of modern human settlements as an aid to archaeological interpretation, Ph. D. thesis, University of California, Berkeley.

Gifford, D. P., 1981, Taphonomy and paleoecology: A critical review of archaeology's sister disciplines, in: *Advances in Archaeological Method and Theory,* Volume 4 (M. B. Schiffer, ed.), Academic Press, New York, pp. 365–438.

Gifford-Gonzalez, D. P., Damrosch, D. B., Damrosch, D. R., Pryor, J., and Thunen, R. L., 1985, The third dimension in site structure: An experiment in trampling and vertical dispersal, *Am. Antiquity* **50:**803–818.

Gordon, C. C., and Buikstra, J. E., 1981. Soil pH, bone preservation, and sampling bias at mortuary sites, *Am. Antiquity* **46:**566–571.

Grayson, D. K., 1984, *Quantitative Zooarchaeology,* Academic Press, Orlando, Florida.

Hanson, C. B., 1980, Fluvial taphonomic processes: Models and experiments, in: *Fossils in the Making* (A. K. Behrensmeyer and A. P. Hill, eds.), University of Chicago Press, Chicago, pp. 156–181.

Hare, P. E., 1980, Organic geochemistry of bone and its relation to the survival of bone in the natural environment, in: *Fossils in the Making* (A. K. Behrensmeyer and A. P. Hill, eds.), University of Chicago Press, Chicago, pp. 208–219.

Hassan, A. A., and Ortner, D. J., 1977, Inclusions in bone material as a source of error in radiocarbon dating, *Archaeometry* **19:**131–135.

Haynes, G., 1980, Prey bones and predators: Potential ecologic information from analyses of bone sites, *Ossa* **7:**75–97.

Haynes, G., 1981, Bone modifications and skeletal disturbances by natural agencies: Studies in North America, Ph. D. thesis, Catholic University, Washington, D. C.

Haynes, G., 1982, Utilization and skeletal disturbances of North American prey carcasses, *Arctic* **35:**266–281.

Haynes, G., 1983, Frequencies of spiral and green-bone fractures on ungulate limb bones in modern surface assemblages, *Am. Antiquity* **48:**102–114.

Haynes, G., 1985a, Taphonomic studies of elephant mortality in Zimbabwe, *Elephant* **2:**57–59.

Haynes, G., 1985b, Age profiles in elephant and mammoth bone assemblages, *Quat. Res.* **24:**333–345.

Haynes, G., 1987, Proboscidean die-offs and die-outs: Age profiles in fossil collections, *J. Archaeol. Sci.* **14:**659–668.

Haynes, G., 1988, Mass deaths and serial predation: Comparative taphonomic studies of modern large mammal death sites, *J. Archaeol. Sci.* **15:**219–235.

Henderson, P., Pickford, M., and Williams, C. T., 1987, A geochemical study of rocks and spring waters at Kanam and Kanjera, Kenya, and the implications concerning element mobility and uptake, *J. Afr. Earth Sci.* **6:**221–227.

Hill, A. P., 1978, Taphonomical background to fossil man—Problems in palaeoecology, in: *Geological Background to Fossil Man* (W. W. Bishop, ed.), Scottish Academic Press, Edinburgh, pp. 87–101.

Hill, A. P., 1980, Early postmortem damage to the remains of some contemporary East African mammals, in: *Fossils in the Making* (A. K. Behrensmeyer and A. P. Hill, eds.), University of Chicago Press, Chicago, pp. 131–152.

Hill, A. P., 1983, Hyaenas and early hominids, in: Animals and Archaeology: Hunters and Their Prey (J. Clutton-Brock and C. Grigson, eds.), *Br. Archaeol. Rep. Int. Ser. Oxford* **163:**87–92.

Hill, A. P., 1984, Hyaenas and hominids: Taphonomy and hypothesis testing, in: *Hominid Evolution and Community Ecology* (R. Foley, ed.), Academic Press, New York, pp. 111–124.

Hill, A. P., and Behrensmeyer, A. K., 1984, Disarticulation patterns of some modern East African mammals, *Paleobiology* **10:**366–376.

Hill, A. P., and Walker, A., 1972, Procedures in vertebrate taphonomy, *J. Geol. Soc. Lond.* **128:**399–406.

Holtzman, R. C., 1979, Maximum likelihood estimation of fossil assemblage composition, *Paleobiology* **5:**77–89.

Hulbert, R. D., Jr., and Morgan, G. S., 1989, Stratigraphy, paleoecology, and vertebrate fauna of the Leisey Shell Pit Local Fauna, early Pleistocene (Irvingtonian) of southwestern Florida, *Pap. Fla. Paleontol.* **2:**1–19.

Hunt, R. M., Jr., 1978, Depositional setting of a Miocene mammal assemblage, Sioux County, Nebraska (U.S.A.), *Palaeogeogr. Palaeoclimatol. Palaeoecol.* **24:**1–52.

Hunt, R. M., Jr., Xue, X. X., and Kaufman, J., 1983, Miocene burrows of extinct beardogs: Indication of early denning behavior of large mammalian carnivores, *Science* **221:**364–366.

Isaac, G. L., and D. C. Crader, D. C., 1981, To what extent were early hominids carnivorous? An archaeological perspective, in: *Omnivorous Primates: Gathering and Hunting in Human Evolution* (R. Harding and G. Telecki, eds.), Columbia University Press, New York, pp. 37–103.

Johanson, D. C., Splingaer, M., and Boaz, N. T., 1976, Paleontological excavations in the Shungura Formation, Lower Omo Basin, 1969–73, in: *Earliest Man and Environments in the Lake Rudolf Basin: Stratigraphy, Paleoecology, and Evolution* (Y. Coppens, F. C. Howell, G. Ll. Isaac, and R. E. Leakey, eds.), University of Chicago Press, Chicago, pp. 402–420.

Klein, R. G., 1982, Age (mortality) profiles as a means of distinguishing hunted species from scavenged ones in Stone Age archaeological sites, *Paleobiology* **8:**151–158.

Klein, R. G., and Cruz-Uribe, K., 1984, *The Analysis of Animal Bones from Archeological Sites,* University of Chicago Press, Chicago.

Koch, P. L., Fisher, D. C., and Dettman, D. L., 1989, Oxygen isotopes in the dentition of extinct proboscideans: A measure of season of death and seasonality, *Geology* **17**:515–519.

Korth, W. W., 1979, Taphonomy of microvertebrate fossil assemblages, *Ann. Carnegie Mus.* **48**:235–285.

Koster, E. H., 1987, Vertebrate taphonomy applied to the analysis of ancient fluvial systems, *Soc. Econ. Paleontol. Mineral. Spec. Publ.* **39**:159–168.

Kyle, J. H., 1986, Effect of post-burial contamination on the concentrations of major and minor elements in human bones and teeth—The implications for palaeodietary research, *J. Archaeol. Sci.* **13**:403–416.

Laporte, L. F., and Behrensmeyer, A. K., 1980, Sediment reworking by terrestrial vertebrates, *J. Sediment. Petrol.* **50**:1337–1346.

Lawton, R., 1977, Taphonomy of the dinosaur quarry, Dinosaur National Monument, *Contrib. Geol. Univ. Wyoming* **15**:119–126.

Lee-Thorp, J. A., Sealy, J. C., and van der Merwe, N. J., 1989, Stable carbon isotope ratio differences between bone collagen and bone apatite, and their relationship to diet, *J. Archaeol. Sci.* **16**:585–599.

Lyman, R. L., 1984, Bone density and differential survivorship of fossil classes, *J. Archaeol. Sci.* **3**:259–299.

Lyman, R. L., and Fox, G. L., 1989, A critical evaluation of bone weathering as an indication of bone assemblage formation, *J. Archaeol. Sci.* **16**:293–317.

Maas, M. C., 1985, Taphonomy of a Late Eocene microvertebrate locality, Wind River Basin, Wyoming (U.S.A.), *Palaeogeogr. Palaeoclimatol. Palaeoecol.* **52**:123–142.

MacArthur, R. H., 1957, On the relative abundance of bird species, *Proc. Natl. Acad. Sci. USA* **43**:283–295.

Madson, J. H., 1976, *Allosaurus fragilis:* A revised osteology, *Utah Geol. Mineral. Surv. Bull.* **109**:1–163.

Maguire, J. M., Pemberton, D., and Collett, M. H., 1980, The Makapansgat limeworks grey breccia: Hominids, hyaenas, hystricids or hillwash?, *Palaeontol. Afr.* **23**:75–98.

Magurran, A. E., 1988, *Ecological Diversity and Its Measurement,* Princeton University Press, Princeton, New Jersey.

May, R. M., 1976, Patterns in multi-species communities, in: *Theoretical Ecology* (R. M. May, ed.), Blackwell, Oxford, pp. 142–162.

McKenna, M. C., 1962, Collecting small fossil by washing and screening, *Curator* **3**:221–235.

Mellett, J. S., 1974, Scatological origins of microvertebrate fossils, *Science* **185**:349–350.

Miller, G. J., 1975, A study of cuts, grooves and other marks on recent and fossil bone: II. Weathering cracks, fractures, splinters and other similar phenomena, in: *Lithic Technology: Making and Using Stone Tools* (E. H. Swanson, ed.), Mouton, The Hague, pp. 212–226.

Myers, T. P., Voorhies, M. R., and Corner, R. G., 1980, Spiral fractures and bone pseudotools at paleontological sites, *Am. Antiquity* **45**:483–490.

Nelson, B. K., DeNiro, J. M., Schoeninger, M. J., DePaolo, D. J., and Hare, P. E., 1986, Effects of diagenesis on strontium, carbon, nitrogen and oxygen concentration and isotopic composition of bone, *Geochim. Cosmochim. Acta* **50**:194–199.

Norman, D. B., 1987, A mass-accumulation of vertebrates from the Lower Cretaceous of Nehden (Sauerland), West Germany, *Proc. R. Soc. Lond. B* **230**:215–255.

O'Connell, J. F., Hawkes, K., and Jones, N. B., 1988, Hadza scavenging: Implications for Plio-Pleistocene hominid subsistence, *Cur. Anthropol.* **29**:356–363.

Olsen, S. L., and Shipman, P., 1988, Surface modification on bone: Trampling versus butchery, *J. Archaeol. Res.* **15**:535–553.

Olson, E. C., 1980, Taphonomy: Its history and role in community evolution, in: *Fossils in the Making* (A. K. Behrensmeyer and A. P. Hill, eds.), University of Chicago Press, Chicago, pp. 5–19.

Olson, E. C., 1985, Vertebrate paleoecology: A current perspective, *Palaeogeogr. Palaeoclimatol. Palaeoecol.* **50**:83–106.

Olson, E. C., and Bolles, K., 1975, Permo-Carboniferous fresh water burrows, *Fieldiana: Geol.* **33**:271–290.

Peet, R. K., 1974, The measurement of species diversity, *Annu. Rev. Ecol. Syst.* **5**:285–307.

Peterson, O. A., 1906, The Agate Spring Fossil Quarry, *Ann. Carnegie Mus.* **3**:487–494.
Peterson, O. A., 1923, A fossil-bearing slab of sandstone from the Agate Spring Quarries of western Nebraska exhibited in the Carnegie Museum, *Ann. Carnegie Mus.* **7**:267–273.
Potts, R., 1986, Temporal span of bone accumulations at Olduvai Gorge and implications for early hominid foraging behavior, *Paleobiology* **12**:25–31.
Potts, R., 1988, *Early Hominid Activities at Olduvai Gorge*, Aldine de Gruyter, New York.
Potts, R., and Shipman, P., 1981, Cutmarks made by stone tools on bones from Olduvai Gorge, Tanzania, *Nature* **291**:577–580.
Price, T. D., Schoeninger, M. J., and Armelagos, G. J., 1985, Bone chemistry and past behavior: An overview, *J. Hum. Evol.* **14**:419–447.
Raath, M. A., 1969, A new coelurosaurian dinosaur from the Forest sandstone of Rhodesia, *Arnoldia* **4**:1–25.
Retallack, G. J., 1984, Completeness of the rock and fossil record: Some estimates using fossil soils, *Paleobiology* **10**:59–78.
Richardson, P. R. K., 1980, Carnivore damage to antelope bones and its archaeological implications, *Palaeontol. Afr.* **23**:109–125.
Rogers, R. R., 1990, Taphonomy of three dinosaur bone beds in the Upper Cretaceous Two Medicine Formation of northwestern Montana: evidence for drought-related mortality, *Palaios* **5**:394–413.
Rose, K. D., 1981, Composition and species diversity in Paleocene and Eocene mammal assemblages: An empirical study, *J. Vertebr. Paleontol.* **1**:367–388.
Sander, P. M., 1987, Taphonomy of the Lower Permian Geraldine Bonebed in Archer County, Texas. *Palaeogeogr. Palaeoclimatol. Palaeoecol.* **61**:221–236.
Saunders, J. J., 1977, Late Pleistocene vertebrates of the western Ozark Highland, Missouri, *Ill. State Mus. Rep. Inv.* **33**:1–118.
Schäfer, W., 1972, *Ecology and Paleoecology of Marine Environments*, University of Chicago Press, Chicago.
Schwartz, H. L., and Gillette, D. D., in press, Geology and taphonomy of the *Coelophysis* quarry, Upper Triassic Chinle Fm., Ghost Ranch, New Mexico, *Bull. Geo. Soc. Am.*
Seilacher, A., Reif, W. E., and Westphal, F., 1985, Sedimentological, ecological and temporal patterns of fossil Lagerstätten, *Phil. Trans. R. Soc. Lond. B* **311**:5–23.
Shipman, P., 1981, *Life History of a Fossil*, Harvard University Press, Cambridge, Massachusetts.
Shipman, P., 1983, Early hominid lifestyle: Hunting and gathering or foraging and scavenging?, in: Animals and Archaeology: 1. Hunters and Their Prey (J. Clutton-Brock and C. Grigson, eds.), *Br. Archaeol. Rep. Int. Ser. Oxford* **163**:31–49.
Shipman, P., and Rose, J., 1983, Early hominid hunting, butchering, and carcass-processing behaviors: Approaches to the fossil record, *J. Anthropol. Archaeol.* **2**:57–98.
Shipman, P., Fisher, D. C., and Rose, J. J., 1984, Mastodon butchery: Microscopic evidence of carcass processing and bone tool use, *Paleobiology* **10**:358–365.
Shotwell, J. A., 1955, An approach to the paleoecology of mammals, *Ecology* **36**:327–337.
Sillen, A., and Kavanagh, M., 1982, Strontium and paleodietary research: A review, *Yearb. Phys. Anthropol.* **25**:67–90.
Simpson, G. G., 1949, Measurement of diversity, *Nature* **163**:688.
Skinner, J. D., Davis, S., and Ilani, G., 1980, Bone collecting by striped hyaenas, *Hyaena Hyaena*, in Israel, *Palaeontol. Afr.* **23**:99–104.
Smith, G. R., and Elder, R. L., 1985, Environmental interpretation of Clarkia fishes, in: *Late Cenozoic History of the Pacific Northwest* (C. J. Smiley, ed.), AAAS, California Academy of Sciences, San Francisco, pp. 85–93.
Smith, G. R., Stearley, R. F., and Badgley, C. E., 1988, Taphonomic bias in fish diversity from Cenozoic floodplain environments, *Palaeogeogr. Palaeoclimatol. Palaeoecol.* **63**:263–273.
Smith, R. M. H., 1987, Helical burrow casts of therapsid origin from the Beaufort Group (Permian) of South Africa, *Palaeogeogr. Palaeoclimatol. Palaeoecol.* **60**:155–170.
Solounias, N., 1981, The Turolian Fauna from the Island of Samos, Greece, in: *Contributions to Vertebrate Evolution*, Volume 6 (M. K. Hecht and F. S. Szalay, eds.), S. Karger, Basel, pp. 1–232.

Speth, J. D., 1983, *Bison Kills and Bone Counts*, University of Chicago Press, Chicago.

Speyer, S. E., and Brett, C. E., 1986, Trilobite taphonomy and Middle Devonian taphofacies, *Palaios* **1**:312–327.

Sutcliffe, A. J., 1970, Spotted Hyena: Crusher, gnawer, digester and collector of bones, *Nature* **227**:1110–1113.

Thomas, R. G., 1986, Cementation histories of Upper Cretaceous dinosaur bones, southern Alberta, and their relevance to taphonomic studies, in: *Abstracts with Programs of the 4th North American Paleontological Convention*, p. A46.

Todd, L. C., 1987, Taphonomy of the Horner II bone bed, in: *The Horner Site* (G. C. Frison and L. C. Todd, eds.), Academic Press, Orlando, Florida, pp. 107–198.

Turnbull, W. D., and Martill, D. M., 1988, Taphonomy and preservation of a monospecific titanothere assemblage from the Washakie Formation (Late Eocene), southern Wyoming, *Palaeogeogr. Palaeoclimatol. Palaeoecol.* **63**:91–108.

Tuross, N., Eyre, D. R., Holtrop, M. E., Glimcher, M. J., and Hare, P. E., 1980, Collagen in fossil bones, in: *Biogeochemistry of Amino Acids* (P. E. Hare, T. C. Hoering, and K. King, eds.), Wiley, New York, pp. 50–64.

Tuross, N., Behrensmeyer, A. K., and Eans, E. D., 1989a, Strontium increases and crystallinity changes in taphonomic and archaeological bone, *J. Archaeol. Sci.* **16**:661–672.

Tuross, N., Behrensmeyer, A. K., Eans, E. D., Fisher, L. W., and Hare, P. E., 1989b, Molecular preservation and crystallographic alterations in a weathering sequence of wildebeest bones, *Appl. Geochem.* **4**:261–270.

Van Couvering, J. H., 1980, Community evolution in East Africa during the Late Cenozoic, in: *Fossils in the Making* (A. K. Behrensmeyer and A. P. Hill, eds.), University of Chicago Press, Chicago, pp. 272–298.

Van der Merwe, N. J., and Vogel, J. C., 1978, ^{13}C content of human collagen as a measure of prehistoric diet in woodland North America, *Nature* **276**:815–816.

Voorhies, M. R., 1969, Taphonomy and population dynamics of an early Pliocene vertebrate fauna, Knox County, Nebraska, *Contrib. Geol. Univ. Wyoming Spec. Pap.* **1**:1–69.

Voorhies, M. R., and Thomasson, J. R., 1979, Fossil grass anthoecia within Miocene rhinoceros skeletons: diet in an extinct species, *Science* **206**:331–333.

Vrba, E. S., 1980, The significance of bovid remains as indicators of environment and predation patterns, in: *Fossils in the Making* (A. K. Behrensmeyer and A. P. Hill, eds.), University of Chicago Press, Chicago, pp. 247–271.

Weigelt, J., 1989, *Recent Vertebrate Carcasses and Their Paleobiological Implications*, University of Chicago Press, Chicago.

White, E. M., and Hannus, L. A., 1983, Chemical weathering of bone in archaeological soils, *Am. Antiquity* **48**:316–322.

Whittaker, R. H., 1975, *Communities and Ecosystems*, Macmillan, New York.

Williams, C. T., and Marlow, C. A., 1987, Uranium and thorium distributions in fossil bones from Olduvai Gorge, Tanzania and Kanam, Kenya, *J. Archaeol. Sci.* **14**:297–309.

Williams, C. T., and Potts, P. J., 1988, Element distribution maps in fossil bones, *Archaeometry* **30**:237–247.

Wilson, M. V. H., 1987, Predation as a source of fish fossils in Eocene lake sediments, *Palaios* **2**:497–504.

Wilson, M. V. H., 1988, Reconstruction of ancient lake environments using both autochthonous and allochthonous fossils, *Palaeogeogr. Palaeoclimatol. Palaeoecol.* **62**:609–623.

Winkler, D. A., 1983, Paleoecology of an early Eocene mammalian fauna from paleosols in the Clark's Fork Basin, northwestern Wyoming (U.S.A.), *Palaeogeogr. Palaeoclimatol. Palaeoecol.* **43**:261–298.

Winkler, D. A., Murry, P. A., Jacobs, L. L., Downs, W. R., Branch, J. R., and Trudel, P., 1988, The Proctor Lake Dinosaur locality, Lower Cretaceous of Texas, *Hunteria* **2**:1–8.

Wood, R. C., 1988, A monospecific death assemblage of fossil side-necked turtles from the Cretaceous of Brazil, in: *Abstracts, International Symposium on Vertebrate Behavior as Derived from the Fossil Record*, Museum of the Rockies, Bozeman, Montana, p. 17.

Wood, W. R., and Johnson, D. L., 1978, A survey of disturbance processes in archaeological site

formation, in: *Advances in Archaeological Method and Theory*, Volume 1 (M. B. Schiffer, ed.), Academic Press, New York, pp. 315–381.
Wyckoff, R. W. G., 1972, *The Biochemistry of Animal Fossils*, Scientechnica, Bristol.
Zangerl, R., and Richardson, E. S., 1963, The paleoecological history of two Pennsylvanian black shales, *Fieldiana Geol. Mem.* **4**:1–352.

Chapter 7

Pyrite Formation and Fossil Preservation

DONALD E. CANFIELD and ROBERT RAISWELL

DONALD E. CANFIELD • NASA–Ames Research Center, MS 239–4 Moffett Field, California 94035; *present address*: Georgia Institute of Technology, School of Earth and Atmospheric Science, Atlanta, Georgia 30332-0340. ROBERT RAISWELL • Department of Earth Sciences, University of Leeds, Leeds LS2 9JT, United Kingdom.

Taphonomy: Releasing the Data Locked in the Fossil Record, Volume 9 of *Topics in Geobiology*, edited by Peter A. Allison and Derek E. G. Briggs, Plenum Press, New York, 1991.

1. Introduction

Pyrite is the most common sulfide mineral found in marine argillaceous sedimentary rocks, where it can occur in a variety of crystallographic and textural forms. The major features of the pyrite formation process are reasonably well understood as a result of geochemical studies of modern sediments over more than 20 years (e.g., Kaplan *et al.*, 1963; Berner, 1970, 1984). However, the influence of these features on crystal and textural form is much less clear. Furthermore, the absence of any rigorous attempt to put the pyritization of fossils in the context of the work on modern sediments is a particularly striking omission, given that fossils preserved in pyrite are greatly valued for the wealth of paleobiological information that they can preserve.

This chapter endorses a more holistic approach to the study of pyrite formation in ancient sediments, encompassing scales which range from the microscopic to the macroscopic. At the microscopic level textural and petrographic information for individual pyrite phases must be integrated with the macroscopic features of sediment depositional environment, biofacies, and composition. Furthermore, no single pyrite phase should be considered in isolation. In modern sediments, pore-water geochemistry can show considerable variations during pyrite formation, both between different depositional environments and with sediment depth (as diagenesis progresses) in a single location. These variations exert considerable influence in how pyrite formation becomes localized within sediments, to be preserved as recognizable sediment features.

2. Iron Sulfide Minerals in Modern and Ancient Sediments

The iron sulfide pyrite, and its polymorph marcasite, the pyrite precursors mackinawite and greigite, as well as pyrrhotite are the principle metal sulfides found in marine sediments and display a variety of crystallographic, crystallochemical, and textural features.

2.1. Pyrite

The crystal structure of pyrite has iron and sulfur atoms arranged in a pattern with twofold symmetry which is based on a face-centered cubic array of ions with the NaCl structure. Cations are located in the Na positions and the center of each S–S unit is located on one Cl position. Although pyrite can exist as a variety of crystal forms, most common are the cube, pyritohedron, octahedron, and diploid. Combined forms (see below) are also common.

The cube has six faces consisting of three sets of parallel pairs, which are at right angles to each other (Fig. 1A). Cubes normally have fourfold symmetry, but pyrite cubes almost invariably have striated faces which diminish their symmetry to twofold. Striations result from oscillatory growth between the cube and the pyritohedron.

The pyritohedron has 12 faces arranged as six pairs of pentagonal faces, and is sometimes termed the pentagonal dodecahedron (Fig. 1B). The most common pyritohedron has {210} faces; others are {310} or {410}. These forms have angles of 43°, 36°, and 28°, respectively, between pairs of faces on either side of one of the long edges.

The octahedron has eight faces (Fig. 1C), which are equilateral triangles arranged as two four-sided pyramids placed base to base. The faces are {111}, but the octahedron is rare as a sole form and most commonly modifies cubes or pyritohedra. The octahedron may also have striated faces which reduce the symmetry to that of the diploid class.

The diploid has 24 faces which are arranged as 12 pairs of four-sided

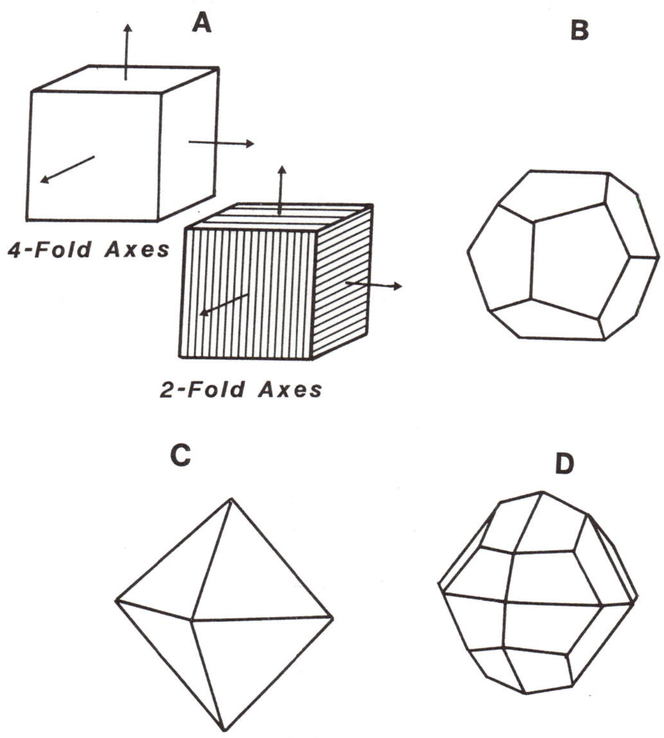

FIGURE 1. Pyrite crystal forms. (A) The cube with two-fold and four-fold symmetry axes. (B) The pyritohedron. (C) The octahedron. (D) The diploid.

faces (Fig. 1D). The most common variety is {321}; others are {421} or {532}; these have angles of approximately 64°, 51°, and 58°, respectively, between adjacent faces sharing the same long edge. As with the octahedron, the diploid is rarely seen as a sole form.

2.2. Mackinawite

Mackinawite is a sulfur-deficient tetragonal iron sulfide with a composition of $FeS_{0.9}$ to $FeS_{0.95}$ (Goldhaber and Kaplan, 1974). The crystal structure consists of a distorted cubic close-packed assemblage of sulfur atoms with iron in some tetrahedral interstitial sites, but the larger octahedral sites are vacant. Mackinawite is considered to be an important intermediate in the formation of sedimentary pyrite (Berner, 1970), and forms by the reaction between H_2S and Fe^{2+} at near neutral pH. In fact, the initial product of the reaction between Fe^{2+} and H_2S only in some cases shows X-ray diffraction peaks of mackinawite (Rickard, 1969). Berner (1964, 1967) found the product to be amorphous to X-ray diffraction and to be more soluble than mackinawite. More recently Rickard (1989a,b) has used a fast reaction mixing apparatus to show that the initial precipitate is a highly disordered gel with a composition approaching $Fe(HS)_2$ and which includes varying amounts of water and H_2S. The phase has a variable solubility which is consistently one to two orders of magnitude greater than that of amorphous FeS or mackinawite. After a few tens of milliseconds the $Fe(HS)_2$ loses sulfide (and possibly water) and is converted to an amorphous FeS which adopts a mackinawite structure after several hours. Despite the formation of short-lived intermediates, in surficial marine sediments it appears that either mackinawite or amorphous FeS controls pore-water iron and sulfide solubility (see below). The presence of mackinawite and greigite in Recent sediments is demonstrated by their black coloration.

2.3. Greigite

Greigite has the composition Fe_3S_4 and is a cubic iron sulfide with an inverse spinel structure. It can be considered to be the sulfur analogue of magnetite, and is likewise highly ferrimagnetic. Vaughan and Ridout (1971) have shown that the composition is $Fe^{3+}(Fe^{3+}, Fe^{2+})(S°, 3S^{2-})$. The presence of oxidized iron and sulfur indicates that the formation of greigite from mackinawite requires the presence of an oxidizing agent, e.g., O_2, S°, or iron oxides (Goldhaber and Kaplan, 1974). Greigite is most abundant in modern fresh- and brackish-water sediments where dissolved sulfide is absent (Dell, 1972; Berner, 1974), and in some marsh sediments where excess iron reacts sulfide from solution (Pye, 1981). Greigite may be a common trace constituent in anoxic marine sediments (Cutter and Velinski, 1988; Leslie *et al.*, 1990; Morse and Cornwell, 1987), but more work is required to establish its distribution and concentrations.

2.4. Other Iron Sulfides

In addition to the phases described above, marcasite (orthorhombic FeS_2) is present in some modern and ancient sediments (e.g., Van Andel and Postma, 1954; Maynard and Lauffenburger, 1978; Fisher, 1986). The structure of marcasite shows two pairs of S atoms with their midpoints at the centers of four edges and two faces, giving octahedral coordination of S atoms about the centrally-located Fe atoms. Marcasite is generally thought to require a low pH for its formation (Allen *et al.*, 1912; Rickard, 1969; Murowchick and Barnes, 1986), and is sometimes associated with reworked bottom sediments (Maynard and Lauffenburger, 1978; Fisher, 1986) whose oxygenation might have promoted pyrite oxidation, and hence generated the necessary acidity. In general, marcasite formation is rare during early sediment diagenesis, and to our knowledge there are no reports of marcasite formation in modern sediments, apart from in the Gulf of Paria (Van Andel and Postma, 1954).

Pyrrhotite (hexagonal FeS) is also uncommon as a product of diagenesis. Its structure is complex; sulfur atoms are arranged in hexagonal close-packing such that each iron has three sulfur atoms above it and three below (which are rotated 60° with respect to the higher set). Pyrrhotite has been reported from modern sediments in the Sea of Japan (Kobayashi and Nomura, 1972), where it occurs as veinlets within magnetite grains. Although pyrrhotite is thermodynamically stable relative to other iron monosulfides (Berner, 1967), its relatively complex structure suggests that formation is kinetically inhibited except under unusual circumstances. The sulfidization of magnetite proceeds slowly relative to other iron oxide phases (Canfield and Berner, 1987) and may provide a possible pathway. However, the sulfidized magnetites studied by Canfield and Berner (1987) contained pyrite as the only recognizable iron sulfide phase.

3. Pyrite Textures in Modern and Ancient Sediments

The term texture is here used strictly to define the characteristics (size, shape, crystallinity) and contact relationships of pyrite crystals; the external form or the shape of any aggregate of pyrite crystals is referred to as morphology. In this sense there are relatively few textures described for pyrite in the literature, but these can occur in numerous morphologies.

3.1. Framboidal and Clustered Pyrite

Framboidal pyrite (Fig. 2A) consists of discrete equigranular pyrite microcrysts (usually cubes or pyritohedra <5 μm diameter) which can be packed with variable degrees of ordering into nearly spheroidal aggregates (usually <500 μm diameter). More irregular pyrite aggregates, termed clusters (Love and Amstutz, 1966), are less strictly defined in that microcrysts of a more variable size are packed (without ordering) into aggregates which are only crudely spherical (Fig. 2B). Rickard (1970) suggested that clusters and framboids can be separated by an arbitrary ratio of microcryst diameter to aggregate diameter of 1:10.

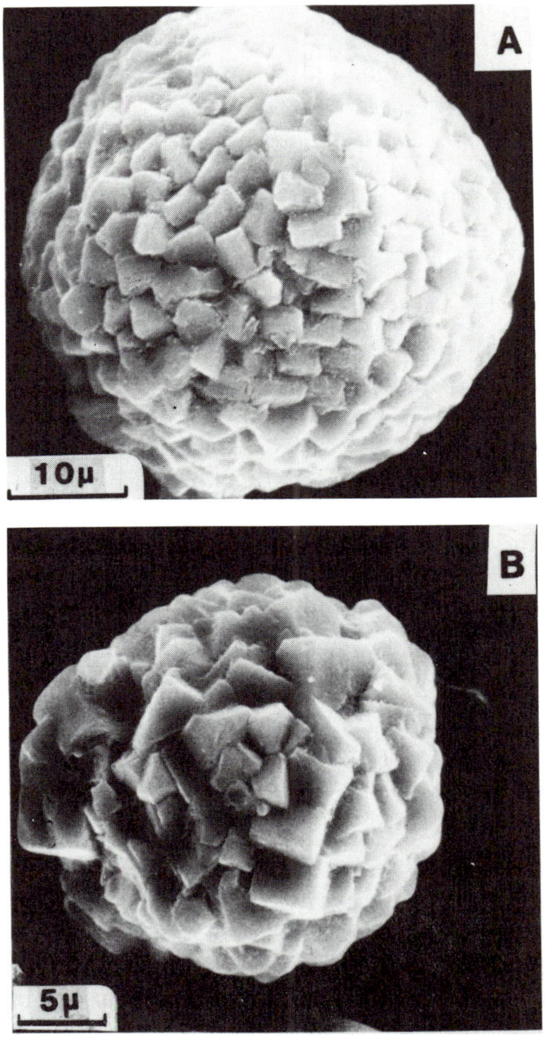

FIGURE 2. Pyrite textures. (A) Framboid. (B) Cluster.

The reactions involved in framboidal pyrite formation (Berner, 1970, 1984) are most likely

$$18CH_2O + 9SO_4^{2-} \rightarrow 18HCO_3^- + 9H_2S \quad (1)$$

$$6FeOOH + 9H_2S \rightarrow 6FeS + 3S° + 12H_2O \quad (2)$$

$$3FeS + S° \rightarrow Fe_3S_4 \quad (3)$$

$$Fe_3S_4 + 2S° \rightarrow 3FeS_2 \quad (4)$$

$$18CH_2O + 9SO_4^{2-} + 6FeOOH \rightarrow 18HCO_3^- + 3FeS + 3FeS_2 + 12H_2O \quad (5)$$

Under anoxic conditions sulfide is produced via the biologically mediated reaction (sulfate reduction) between organic compounds and dissolved pore-water sulfate [equation (1)]. Sulfide reacts with sediment-contained iron minerals [mostly iron oxides, equation (2)] to produce iron monosulfide phases (mackinawite and amorphous FeS) and elemental sulfur (Fig. 3A). The elemental sulfur becomes the oxidant required to oxidize FeS to greigite [equation (3)] and pyrite [equation (4); Berner, 1970; Roberts *et al.*, 1969; Sweeney and Kaplan, 1973; Goldhaber and Kaplan, 1974]. However, the sum of all reactions [equation (5)] shows that not enough elemental sulfur is produced to completely convert FeS to pyrite. Some elemental sulfur may be produced from the oxidation of H_2S promoted by the burrowing activities of organisms (see below), but in sediments below the zone of bioturbation the extra oxidant required for complete FeS conversion to pyrite is not known.

Experimental work has shown that the framboidal texture probably develops during the recrystallization of precursor iron sulfide phases to pyrite (Sweeney and Kaplan, 1973). The spheroidal form is associated with the transformation of mackinawite to greigite [Fig. 3B, equation (3)] and the microcrysts develop during the greigite-to-pyrite transformation [equation (4)]. That the two main features of framboids, their sphericity and their microcrysts, can be produced synthetically (Berner, 1969a) is a powerful argument that in sediments these features result from inorganic reactions and do not involve biogenic processes apart from sulfide production. Because equations (1)–(5) all occur in anoxic surficial marine sediments, framboidal pyrite starts to form in the earliest stages of diagenesis, as soon as sulfide production begins. During later diagenesis sulfide may become available through nonbiologically mediated reactions and a second generation of framboidal pyrite may form (Sassano and Schrijver, 1989).

3.2. Aggregated Pyrite

Hudson (1982) describes this texture as comprising small pyrite crystals forming irregular rounded clots, which may partly coalesce. The cores of these aggregates generally consist of fine-grained crystals, and are sur-

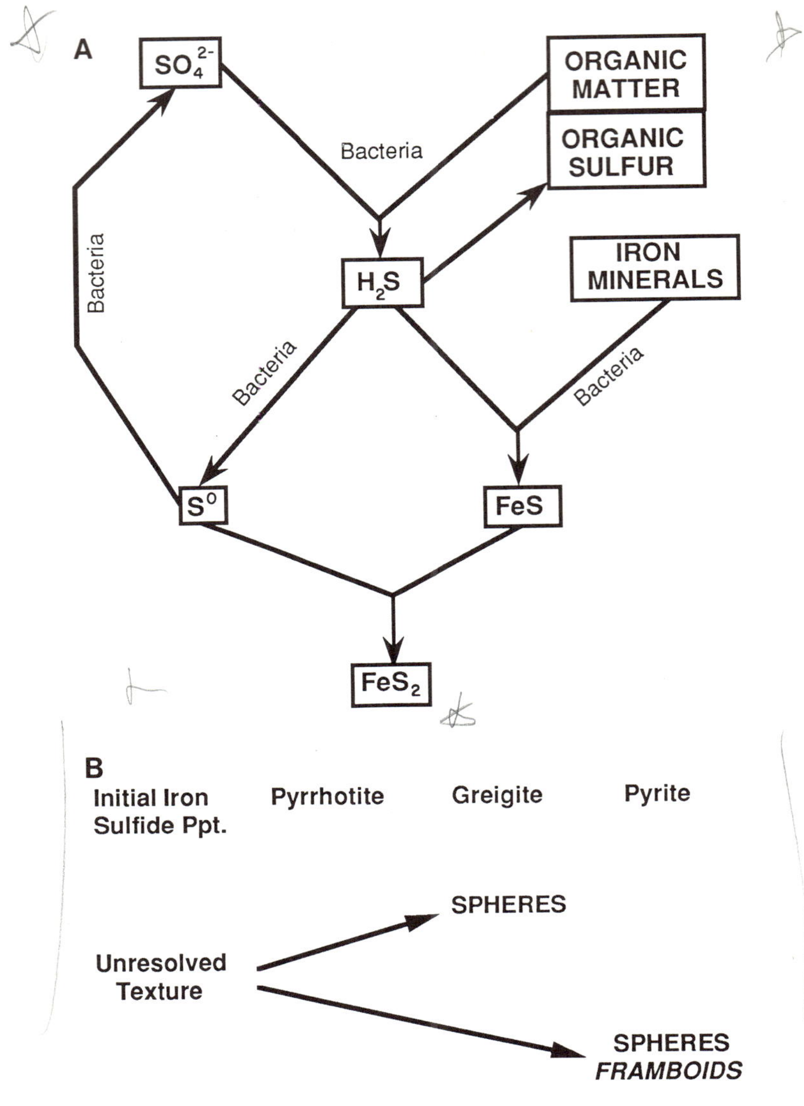

FIGURE 3. (A) Schematic representation of the pathways and processes of sulfate reduction and pyrite formation. (B) Schematic representation of the intermediate phases and morphologies involved in pyrite formation. [After Sweeney and Kaplan (1973).]

rounded by coarser crystals (Fig. 4A). Aggregates are less regular in size and shape than framboids and never display microcryst ordering. Sometimes aggregates envelope framboids. Rather similar textures have been described by Love *et al.* (1983) as irregular aggregated grains and irregular rounded bodies, both of which can be observed to envelope framboids. These textures are also illustrated by other authors (e.g., Price and Shieh, 1979; Parratt and Kullerud, 1979).

Al-Biatty and Raiswell (in preparation) consider that this texture in

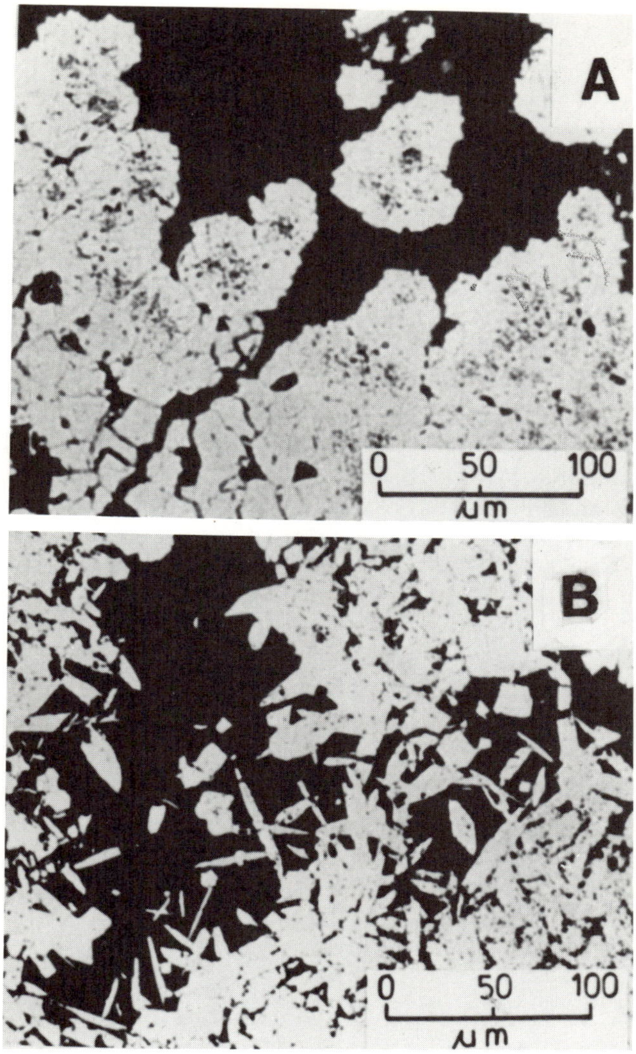

FIGURE 4. Pyrite textures. (A) Aggregated pyrite. (B) Bladed pyrite. [Photographs from Hudson (1982).]

some respects resembles a continuation of the changes found in passing from ordered framboids to disordered framboids and then to clusters. This is probably a result of changes in rates of microcryst nucleation and growth which produced more irregular outlines (due to variable microcryst size and orientation) and a less coherent product. However, the diagenetic environment required to produce aggregated pyrite is not well understood.

3.3. Bladed Pyrite

Bladed textures have been found in the linings of chambered organisms and in their sediment fills (Hudson, 1982). In thin section the crystals are elongate and acutely angled, but SEM photographs show that the crystals are bladed and not acicular, and may be randomly or radially oriented (Fig. 4B). It is possible that this texture was inherited from a marcasite precursor. Bladed pyrite is unknown in modern sediments.

3.4. Equant Pyrite

Many studies report the occurrence of individual pyrite crystals of equant habit, showing well-developed crystal faces (e.g., Goldhaber and Kaplan, 1974; Raiswell and Plant, 1980; Hudson, 1982; Love *et al.*, 1983; Allison, 1988a; Schallreuter, 1984; Morse and Cornwell, 1987). Nearly all the crystal forms discussed earlier have been found as equant crystals, although the cube and the pyritohedron are the most common. Equant crystals associated with fossils are usually relatively coarse (in contrast to those in host sediments) and, although they coalesce, they do not form subspherical aggregates. These characteristics suggest growth occurred into a void space (Hudson, 1982). Equant pyrite appears to form during early diagenesis, before compaction (Hudson, 1982).

4. Pyrite Formation in Modern Sediments

The major steps in early diagenetic sedimentary pyrite formation have been outlined in equations (1)–(5). The necessary "ingredients" for pyrite formation are presented in Fig. 3A and include organic material, sulfate, and detrital iron minerals (Berner, 1970, 1984).

4.1. Sulfate Reduction

Oxidation by oxygen is the first step in organic matter decomposition in marine sediments (Froelich *et al.*, 1979; Reimers and Suess, 1983; Bender and Heggie, 1984). However, in many sediments degradable organic matter becomes buried below the zone of oxic respiration (Canfield and Raiswell, this volume, Chapter 9, Section 3.1.1), where the bacterially mediated reduction of seawater sulfate occurs. The rate of sulfate reduction depends on the rate with which organic matter falls to the sediment surface, the availability of sulfate, and the extent of prior decomposition by oxic respiration. The latter is important because the greater the amount of decomposition by oxic respiration, the less organic matter is available for sulfate reduction, and the more refractory the residual material becomes (Westrich and Berner, 1984). Sulfate availability will also affect sulfate reduction rates, and is most important in limiting sulfate reduction in nonmarine systems where sulfate is present in low concentrations.

For marine sediments rates of sulfate reduction correlate strongly with sediment deposition rate (Fig. 5), and are seen to drop as sediment deposition rate falls. This trend in part reflects the fact that in marine sediments, the deposition rate of degradable organic matter is proportional to sedimentation rate (Canfield, 1989a). Also, the proportion of the total sedimentary organic matter flux which is decomposed by oxic respiration is a strong function of sediment deposition rate (Canfield, 1991). The relationship is such that at higher rates of sediment deposition (see Fig. 5) less of the degradable organic matter is decomposed by oxic respiration, leaving more for sulfate reduction; at lower sedimentation rates more of the organic matter is oxidized by oxic respiration (because more time is available for oxidation), leaving less for sulfate reduction. Euxinic sediments (overlain by sulfide-rich water; e.g., Black Sea) are unusual in that sulfate reduction rates are relatively high because no oxic decomposition can occur in the sediment.

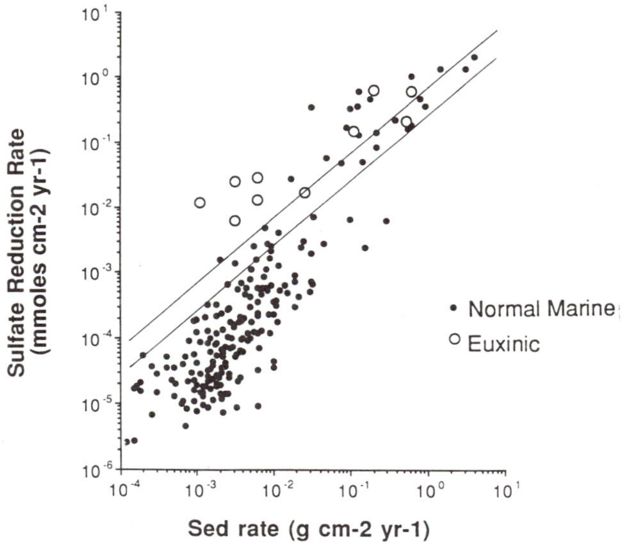

FIGURE 5. Sulfate reduction rate as a function of sediment deposition rate in marine sediments. All sedimentary environments are represented, including deep-sea red-clay and carbonate sediments, hemipelagic sediments, shelf sediments, deltaic sediments, and numerous sediments overlain by sulfide-containing waters (euxinic). Sulfate reduction rates are expressed as depth-integrated rates, which gives the total amount of sulfate reduction per area of sediment. Sedimentation rates are given as mass accumulation rates, and for porosities typical of marine sediments (60–80%), sedimentation rates in these units are similar to rates expressed in the more familiar units of cm/year. Overlain is a pair of parallel lines which define the maximum burial rate of pyrite sulfur assuming that pyrite formation becomes limited by the availability of reactive iron (see text). Where sulfate reduction rates fall below the "iron band," pyrite formation is limited by the production of sulfide, and complete pyritization is not to be expected. Where points fall above the "iron band," more sulfide is produced than iron delivered to the sediment, and pyrite formation should be limited by the availability of reactive iron. [Figure from Canfield (1988).]

4.2. Iron Availability

In many marine sediments the amount of pyrite that forms is restricted by the availability of iron minerals reactive toward sulfide (Jørgensen, 1977; Raiswell and Berner, 1985; Raiswell *et al.*, 1988); the most important of which are the iron oxides and oxyhydroxides (Berner, 1970; Canfield, 1988). In a study of the FOAM (Friends of Anoxic Mud) site in Long Island Sound, Canfield (1988) showed that iron oxides (e.g., hematite) and oxyhydroxides (e.g., goethite, lepidocrocite, and ferrihydrite) contributed about 75–80% of the iron for pyrite formation. These minerals were also the most reactive toward sulfide and, after their consumption, very little additional pyrite formation took place. In fact, the remaining iron-containing silicate minerals were found to react with sulfide with half-lives of about 50,000–100,000 years (Canfield and Raiswell, in preparation). In freshly deposited, fine-grained, siliciclastic marine sediment (collected from around the North American continental margin and the deep sea) iron oxide contents range from about 0.6 to 1.5 wt % (Canfield, 1988). This means that maximum pyrite-sulfur contents of between about 0.8 and 2.0 wt % may be expected in marine sediments whose exposure to sulfide does not exceed several thousand years. These results are consistent with published compilations of pyrite-sulfur content (Goldhaber and Kaplan, 1974; Volkov and Rosanov, 1983; Berner and Raiswell, 1983).

Maximum pyrite-sulfur contents, as estimated above, may be used to constrain the maximum burial rates of sulfur, assuming pyrite formation becomes iron-limited. These results may then be compared with sulfate reduction rate data (Fig. 5) to demonstrate the environments in which pyrite formation will be either sulfur- or iron-limited. For sediment of low deposition rate (e.g., the deep sea), sulfide production rates are lower than iron oxide delivery rates. This means that not all of the iron will react with sulfide, and iron oxides will be preserved. For sediments of intermediate sedimentation rates (e.g., some continental margins), sulfide production rates exceed iron oxide burial rates, and complete pyritization of iron oxides is to be expected. The same is true for euxinic sediments, and complete consumption of the most reactive sedimentary iron minerals is probably required to maintain sulfidic bottom-water conditions. At high sedimentation rates typical of submarine deltas, sulfate reduction rate data are limited, but these suggest that iron delivery rates and sulfide production rates are nearly balanced.

In Fig. 6, the trend of maximum pyrite burial presented in Fig. 5 is compared to actual sulfur burial rates as compiled from the literature (Berner and Canfield, 1989). Consistent with the generalizations above, at low sedimentation rates, sulfur burial rates are lower than iron oxide delivery rates and iron oxides should be preserved. For intermediate sedimentation rates and for euxinic sediments (at all sedimentation rates) sulfur burial rates compare with iron oxide delivery rates, implying that complete iron oxide consumption has taken place. At high sedimentation rates, particularly in deltas, sulfur burial rates are lower than expected if complete iron oxide

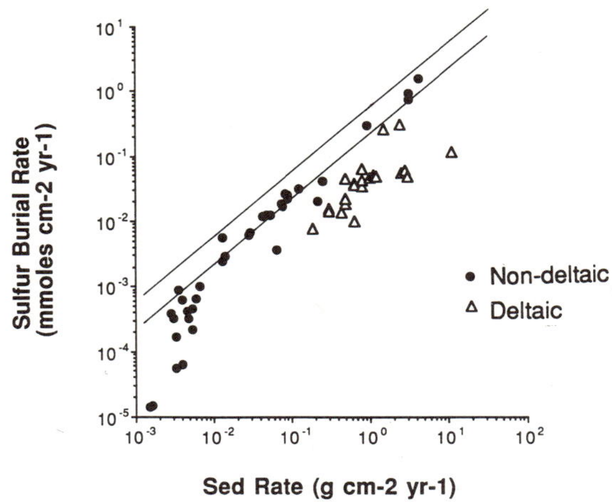

FIGURE 6. As in Fig. 5, except that actual pyrite sulfur burial rates as compiled for numerous marine sediments are compared to maximum sulfur burial rates as predicted by the "iron band." See text for discussion. [After Canfield (1988), with sulfur burial rates from Berner and Canfield (1989), with permission.]

sulfidization had taken place, demonstrating that pyrite formation is limited and iron oxides are preserved (see also Canfield, 1989b). This would seem to contrast with the results in Fig. 5, but in deltas (as with other marine sediments) considerable oxidation of sulfide may take place due to sediment disturbance by bioturbation, transport, and resuspension (Canfield, 1989b). Because iron oxide delivery rates and sulfide production rates in deltas are nearly matched (Fig. 6), sulfide oxidation will lower sulfide burial rates, allowing iron oxide preservation.

5. Localized Pyrite Formation in Modern Marine Sediments

Many types of localized pyrite formation may be found in modern and ancient sediments. Some examples are pyritized organic material, pyritized burrow tubes, shell infillings of pyrite, carbonate shell replacement by pyrite, pyrite concretions and layers, and pyrite replacement of iron minerals. Some of these will be expected in sediments undergoing steady-state diagenesis (constant deposition of similar detrital material). Others require non-steady-state conditions such as a sedimentation hiatus or a change in depositional environment, such as where organic-rich anoxic sediments are deposited over organic-poor, more oxidized sediments. Still others require conditions found in later stages of diagenesis such as basinal fluid generation. The purpose here is to develop a general model which may be used to explain most of the major types of localized pyritization, although a special emphasis will be placed on those sedimentary circumstances which favor pyritization of fossils. A full appreciation of the mechanisms responsible for such pyritization requires a solid understanding of the processes which control the sediment pore-water chemistries of iron and sulfide.

5.1. Controls on Dissolved Sulfide

A common misconception is that sediments supporting rapid rates of sulfate reduction, such as are typical of continental margin sediments (tens to hundreds of mmole/liter per year) are necessarily rich in dissolved sulfide. Though this is often true, some sediments with active sulfate reduction never accumulate dissolved sulfide (an example is the Mississippi Delta). Also, in most continental margin sediments supporting sulfate reduction, the most rapid rates (which are usually found within several centimeters of the sediment surface) are rarely associated with elevated levels of dissolved sulfide (Westrich, 1983; Jørgensen, 1977; Crill and Martens, 1987; Canfield, 1989b). These points are illustrated in Fig. 7, where sulfate reduction rates, sediment pore-water chemistry, and solid-phase iron speciation are presented for three contrasting marine sediments. The sediment sites are the well-studied FOAM site in Long Island Sound (Goldhaber *et al.*, 1977; Aller, 1980a,b; Krishnaswami *et al.*, 1984; Berner and Westrich, 1985), a rapidly depositing sediment from the subaqueous Mississippi Delta (Station 18; Canfield, 1989b), and sediment from the central depression in the low-oxygen (0.05–0.1 ml/liter) Santa Barbara Basin (K. O. Emery, 1960; K. O. Emery and Hülsemann, 1962; Hülsemann and Emery, 1961; Sholkovitz and Gieskies, 1971; Sholkovitz, 1973; D. A. Canfield and C. E. Reimers, unpublished data). At all sites dissolved sulfide was generally very low in the presence of solid iron oxide and oxyhydroxide mineral phases, regardless of sulfate reduction rate. It is important that at the FOAM site and the Santa Barbara Basin, sulfide begins to accumulate in sediment pore waters at almost the same depth where solid-phase iron oxides were exhausted in the formation of iron sulfide minerals.

These results demonstrate that rapid reaction occurs between sulfide and fine-grained sedimentary iron oxide/oxyhydroxide minerals, buffering sulfide to low levels, even with rapid rates of sulfate reduction. Experimental studies also show rapid reaction between fine-grained iron oxides and oxyhydroxides (particularly the latter) and sulfide (Rickard, 1974; Pyzik and Sommer, 1981; Canfield, 1989b). When iron oxides become depleted, the remaining iron-containing minerals [typically accounting for a majority of the iron in the sediment (Canfield, 1988)], including silicates and minor iron oxides such as magnetite and iron-containing spinels, are far less reactive (see Section 4.2), allowing sulfide to accumulate in sediment pore waters.

Recent study has shown that Mn oxide phases may also readily react with dissolved sulfide (Burdige and Nealson, 1986). However, Mn oxides may not be very important in regulating the concentration of pore-water

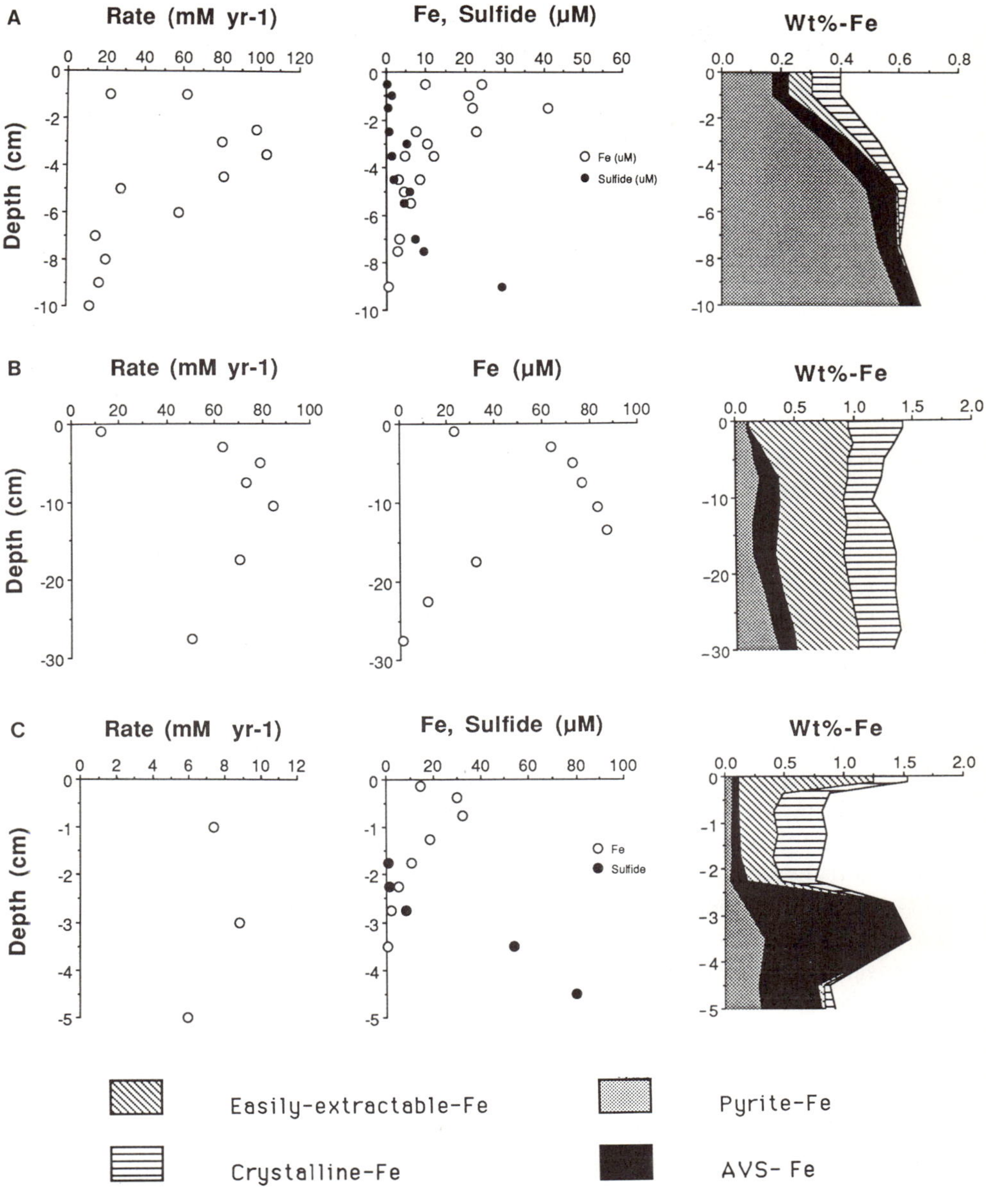

FIGURE 7. Sulfate reduction rates, pore-water iron and sulfide concentrations, and solid-phase iron speciation for three contrasting sediments from (A) Long Island Sound (FOAM), (B) the Mississippi Delta (Sta 18), and (C) the Santa Barbara Basin. Data from the FOAM site and Sta 18 are reproduced with permission from Canfield (1989b). AVS-Fe represents all acid-volatile sulfide. Easily-extractable-Fe represents Lepidocrocite and ferrihydrite which are those nonsulfidic phases removed by ammonium oxalate; crystalline Fe represents goethite and hematite, as being the remaining iron phases removed by dithionate.

sulfide because typically Mn oxides are present in much smaller concentration than iron oxides (Trefrey and Presley, 1982; Canfield, 1989b). Also, for the Santa Barbara Basin sediment, Mn oxides are virtually absent at all depths (C. E. Reimers, personal communication), and sulfide is still effectively reacted from solution.

5.2. Controls on Dissolved Iron

When sulfide is absent, or present in low concentrations in sediment pore waters, dissolved iron typically accumulates. This is seen for all three sites discussed above (Fig. 7). The source of dissolved iron is varied and may include the reduction of iron oxides by sulfide [see below, equation (6)], the reduction of iron oxides by organic compounds [equation (7)], probably mediated by bacteria (Broomfield, 1954; Jones *et al.*, 1983; Sørensen, 1982; Lovley and Phillips, 1986a,b; Tugel *et al.*, 1986; Canfield, 1989b), and the partial oxidation of iron sulfide minerals [equation (8)] (Aller, 1980b; Giblin and Howarth, 1984; Lord, 1980):

$$H_2S + 2FeOOH + 4H^+ \rightarrow S^\circ + 2Fe^{2+} + 4H_2O \quad (6)$$

$$CH_2O + 4FeOOH + 8H^+ \rightarrow CO_2 + 4Fe^{2+} + 7H_2O \quad (7)$$

$$7O_2 + 2FeS_2 + 2H_2O \rightarrow 2Fe^{2+} + 4SO_4{}^{2-} + 4H^+ \quad (8)$$

All of these mechanisms may be important. For example, high concentrations of dissolved iron, with associated low pH, tend to accompany near-surface iron sulfide oxidation in salt marsh sediments (Giblin and Howarth, 1984; Lord and Church, 1983). Also, in nearshore sediments, storm mixing of sediment and seasonal changes in the redox boundary (Aller, 1980a; Canfield, 1989b) may also result in partial iron sulfide oxidation and the accumulation of dissolved iron. When abundant reactive iron oxides are present and the sediments are at a "steady state," experimental results (Sørensen, 1982; Tugel *et al.*, 1986; Canfield, 1989b) and diagenetic modeling (Canfield, 1989b) show that the iron which migrates into solution may be from the bacterial reduction of iron oxides. The iron generated by reaction between sulfide and iron oxides [equation (6)] rapidly forms solid sulfides and does not seem to be an important source for dissolved iron (Sørensen, 1982; Tugel *et al.*, 1986; Canfield, 1989b), even though far more iron is reduced by this mechanism than by organic-matter-mediated iron reduction [equation (7)] (Canfield, 1989b).

When abundant reactive iron is present to depth in a sediment, sulfide is kept low and iron may become saturated (and precipitated) with other anionic metabolic products of anaerobic decomposition such as bicarbonate (as siderite, $FeCO_3$) and phosphate (as vivianite, $Fe_3(PO_4)_2 \cdot 8H_2O$) (Martens *et al.*, 1978; Aller *et al.*, 1986). These nonsulfide iron phases may be responsible for the removal of dissolved iron at Station 18 (Fig. 7), and in the iron oxide-rich sediments of the Amazon inner shelf (Aller *et al.*, 1986), as well as many brackish and nonmarine sediments (Emerson and Widmer, 1978; Postma, 1982).

5.3. Iron Sulfide Saturation

In general, then, when iron oxides and oxyhydroxides are present in sediments supporting sulfate reduction, dissolved sulfide is effectively reacted from solution, and dissolved iron accumulates. As iron oxides become consumed, sulfide levels rise, and the concentration of dissolved iron drops in response (see examples in Fig. 7). Throughout this transition from "iron-dominated" to "sulfide-dominated" pore waters, dissolved iron and sulfide usually remain at saturation with an iron sulfide phase, typically mackinawite (FeS) or greigite (Fe_3S_4) (Boulegue *et al.*, 1982; Elderfield *et al.*, 1981; Lord and Church, 1983; Canfield, 1989b). An example of an iron sulfide saturation state at the FOAM site is shown in Fig. 8. It is important that the solubility of even the more soluble iron sulfide minerals is low (Berner, 1967); it is this low solubility that allows for distinct zones of iron-rich and sulfide-rich pore water. This concept is illustrated in Fig. 9, where equilibrium concentrations of iron and sulfide, assuming saturation with mackinawite, are plotted at pH values ranging between 7 and 8, which are typical for marine sediments (Baas Becking *et al.*, 1960; Ben Yaakov, 1973; Gardner, 1973; Thorstenson, 1970; Boudreau and Canfield, 1988). As pH drops, saturation concentrations of sulfide and iron rise, but even at a pH of 7, iron and sulfide can coexist at a maximum of only 12 μM. From here, as the concentration of one rises, the other falls off quickly. If pore waters are saturated with greigite instead of mackinawite, saturation concentrations of both sulfide and iron drop by a factor of 2. Pyrite is even less soluble than greigite (Berner, 1967), but despite its abundance as a sedimentary mineral,

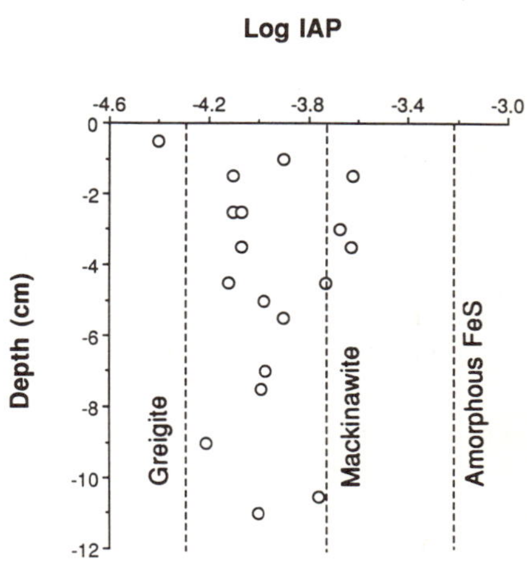

FIGURE 8. Iron sulfide saturation for pore water from the FOAM site in Long Island Sound. Log IAP is the base-10 logarithm of the ion activity products of the pore-water species iron, sulfide, and hydrogen ion assuming the following equilibrium relations: For amorphous FeS and mackinawite, $Fe^{2+} + HS^- \leftrightarrow FeS + H^+$, and for greigite, $\frac{1}{3}S^\circ + Fe^{2+} + HS^- \leftrightarrow \frac{1}{3}Fe_3S_4 + 3H^+$. The saturation IAPs for greigite, mackinawite, and amorphous FeS are shown by the dashed lines with values from Emerson *et al.* (1983). The higher the IAP, the more soluble the mineral, so amorphous FeS is more soluble that mackinawite, which is more soluble than greigite. For the FOAM site, pore waters are intermediate between mackinawite and greigite solubility. [Figure redrawn with permission from Canfield (1989b).]

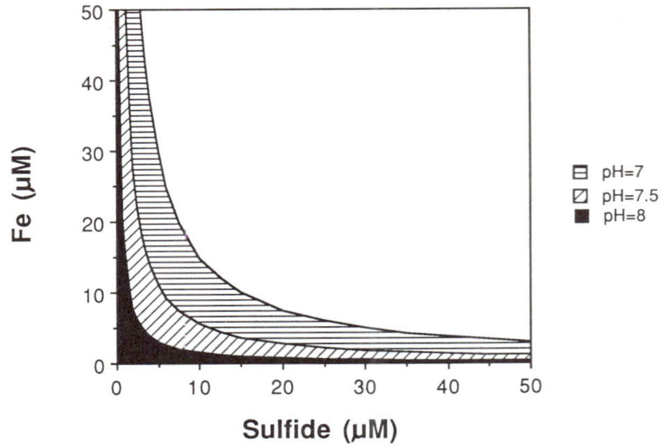

FIGURE 9. Equilibrium concentrations of Fe^{2+} and sulfide (total dissolved sulfide) at several different pH values assuming a seawater solution at saturation with mackinawite. Note that solubility drops as pH rises, and than when concentrations of either iron or sulfide are high, the other is necessarily low.

it is rarely a controlling factor in iron sulfide saturation (Berner, 1967; Elderfield *et al.*, 1981; Boulegue *et al.*, 1982; Lord and Church, 1983). An exception may be surficial, low-pH, relatively oxidizing salt-marsh sediments (Giblin and Howarth, 1984), but more work is required to substantiate this.

5.4. Mechanisms of Localized Pyritization in Recent Sediments

Many possible mechanisms of localized pyritization exist in marine sediments and much of the conceptual framework to understand these processes has been developed and discussed by Berner (1969b, 1980). Five possible mechanisms are described below, each giving rise to morphologically distinct pyrite accumulations as coatings, layers, nodules, and pyrite replacement of carbonate.

5.4.1. Pyritization of Organic Matter

The reduction of sulfate in anoxic marine sediments [above the zone of anaerobic methane oxidation] (Canfield and Raiswell, this volume, Chapter 9, Section 3.1.4) occurs mostly at the site of the decomposing organic carbon (Berner, 1980). However, some more dispersed sulfate reduction is likely to occur from dissolved organic fermentation products which are able, in low concentrations (Sansone, 1986), to diffuse away from the particulate organic carbon source. From the previous discussion, when the concentration of pore-water iron is high, pore-water sulfide is necessarily low, and elevated sulfide levels at the site of sulfate reduction will result in local iron sulfide supersaturation. Because there is no significant kinetic inhibition to the formation of iron sulfide phases such as mackinawite and amorphous FeS (R. A. Berner, personal communication; Rickard, 1989b), precipitation of solid-phase iron sulfide occurs immediately at the site of supersaturation. The high levels of dissolved iron in the ambient pore fluid ensure a ready source of iron for diffusion into the site of sulfate reduction, hence fixing most of the hydrogen sulfide at the site of decay by the precipitation of iron sulfide. In this way, iron sulfide (which will eventually convert to pyrite) will accumulate in and around the decomposing organic matter (see Fig. 10 and Section 6.1.2). This scenario for pyrite formation has also been discussed by Berner (1969b).

Examples of pyrite formation in association with decomposing organic matter are common in Quaternary sediments and include shell infillings of pyrite (Kaplan *et al.*, 1963; Honjo *et al.*, 1965; Sweeney and Kaplan, 1973; Morse and Cornwell, 1987; Schallreuter, 1984), pyritized burrow tubes (Hein and Griggs, 1972), where sulfate reduction rates are enhanced in the mucus layer lining the inner tube wall (Aller and Yingst, 1978), pyritized vascular land plant debris, and other organic accumulations (Kaplan *et al.*, 1963; Morse and Cornwell, 1987). The morphologies of these occurrences include nodular pyrite and pyrite coatings, which preserve little of the original organic substrate if this were completely metabolized.

Unfortunately, most observations of organic-matter pyritization have been made in the absence of supporting pore-water chemistry. However, in the surface few centimeters at both the FOAM site and at Station 18, pore-water iron is abundant (Fig. 7), as are pyritized vascular land plant material and pyrite infillings of foraminiferal tests (Figs. 11A–11D). These observations support the proposed ''iron-dominated'' pore-water model for organic-matter pyritization outlined above. Note that this phase of pyritization is the earliest which can be associated with fossils, and also that it may occur quite quickly. The time scale over which organic-matter pyritization occurs can be estimated from the work of Westrich (1983), who found that bulk sedimentary organic matter could be divided into various fractions with half-lives ranging from 77 days to 2×10^6 years with respect to oxidation by sulfate reduction. The organic matter typical of near-surface continental margin sediments, where dissolved iron accumulates, decomposes with half-lives ranging between 2 and 50 years. Hence, organic-matter pyritization should occur over the same time scales. However, in more slowly deposited hemipelagic sediments, dissolved iron may be present at much greater sediment depths (for example, Froelich *et al.*, 1979), where organic matter decomposes more slowly. Under these circumstances organic-matter pyritization may occur over much longer time scales, perhaps hundreds to thousands of years, as long as sulfate reduction occurs in the zone of iron accumulation, and is not inhibited by other metabolic organic-matter decomposition reactions (Froelich *et al.*, 1979).

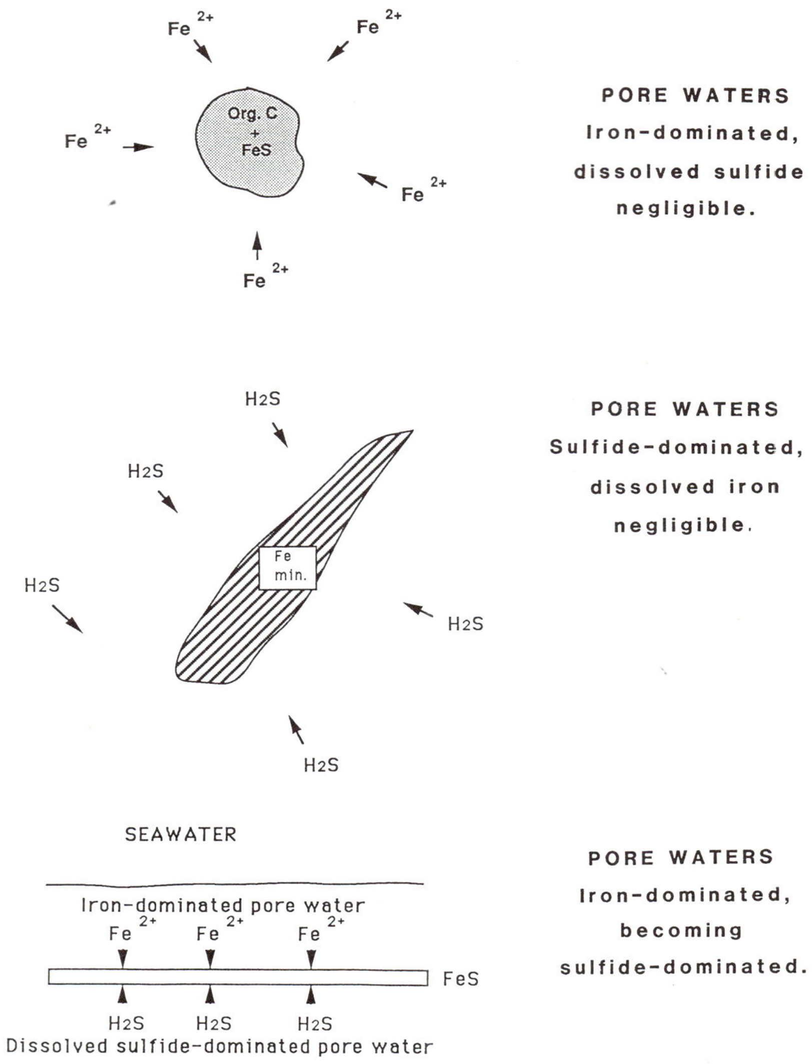

FIGURE 10. (A) pyritization of organic matter. Organic matter fuels the production of H_2S which is locally fixed as FeS in iron-dominated pore waters. (B) Pyritization of iron minerals. Sulfide-dominated pore waters react *in situ* with iron minerals. (C) Pyrite layer formation at the interface of iron-dominated and sulfide-dominated pore waters.

5.4.2. Pyritization of Iron Minerals

When consumption of the major iron oxides [goethite, hematite, lepidocrocite, and ferrihydrite (Canfield, 1989b)] is almost complete, sulfide levels rise and dominate pore-water chemistry (Fig. 7). In this case, the concentration of dissolved iron becomes too low to allow effective diffusion into regions of locally produced sulfide, and the sulfide produced by sulfate reduction migrates into the pore solution. With such a pore-water chemistry, all pyrite formation should be localized at the only source of iron, which is the solid phase. The scenario for pyrite formation in sulfide-rich water is shown in Fig. 10B. Examples are found at depth at the FOAM site, where sulfide levels are high (Fig. 7), and include rims of pyrite on the iron oxide magnetite (Figs. 11E–11G) and pyrite growths on, and dispersed within, iron-containing sheet silicate minerals (Figs. 11H and 11I). It is important that where sulfide levels are low, such as in surficial FOAM sediment and at Station 18, no pyritization of magnetite and sheet silicates is observed. The sulfide levels required for magnetite and mica mineral pyritization are relatively high and probably exceed 1 mM (Canfield and Berner, 1987; and D. E. Canfield, unpublished results). With the millimolar concentrations of sulfide at FOAM (Goldhaber *et al.*, 1977; Canfield and Berner, 1987; Canfield, 1988), substantial pyritization of detrital magnetite takes several hundred years (Canfield and Berner, 1987). The pyritization of iron-containing sheet silicates is also observable on a time scale of several hundred years, but appreciable pyritization of sheet silicate-bound iron may take at least 100,000 years (D. E. Canfield and R. Raiswell, unpublished data). This type of pyritization must postdate pyrite associated with organic matter (see above).

5.4.3. Pyrite Layer Formation

A third type of localized pyrite formation is that of pyrite layers and may occur during times of hiatuses in sediment deposition. In this case, with no sediment burial, dissolved iron and sulfide will diffuse from opposite directions into a common zone (see pore-water profiles in Fig. 7 and schematic in Fig. 10C) of pyrite accumulation. For example, the interface between the iron-rich and sulfide-rich pore waters occurs at 6 cm depth at the FOAM site and at 2–3 cm depth in Santa Barbara Basin sediment (Fig. 7). If sediment deposition were to cease, accumulations of pyrite in layers would be expected at these depths. In some types of sediment, such a mechanism may operate very slowly. For example, with the present pore-water gradients of iron and sulfide at FOAM and the Santa Barbara Basin, a 1-mm-thick layer of pore-filling pyrite would take 1300 and 840 years, respectively, to form. In fact, the actual time would probably be longer because the ability of the sediment bacteria to liberate reduced iron and to produce sulfide should decrease with time. By contrast, in salt marsh sediments, chemical gradients of dissolved iron and sulfide become much steeper (Lord, 1980; Reaves, 1984), giving similar accumulations of pyrite in as little as 50 years (Pye, 1981).

A pyrite layer could also form if suboxic sediment containing iron oxide-rich layers, such as found in the Gulf of Bothnia, Baltic Sea (Ingri and Ponter, 1986), were to become overlain by anoxic sediment containing dissolved sulfide. Such a change in depositional conditions should result in a migrating sulfide-rich "reduction front" that would pyritize the iron oxide-rich layer. This scenario could account for the pyrite "layers"

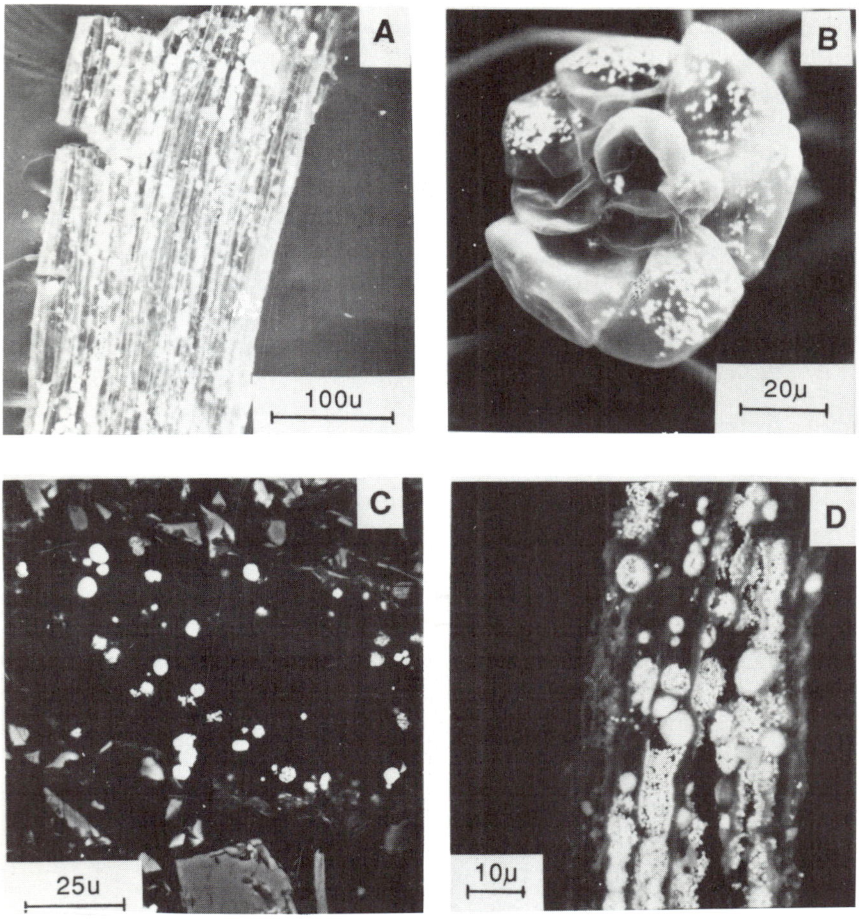

FIGURE 11. SEM micrographs (all save panels E–G are backscattered images) of pyrite/organic and pyrite/mineral associations. In all cases pyrite was verified by elemental analysis using EDS and, when feasible, X-ray diffraction. (A) Pyritized woody material from the 1–2 cm depth interval at the FOAM site in Long Island Sound. (B) Pyrite lining the inner organic coating of a benthic foramanifer from the 0–1 cm depth interval at FOAM. The carbonate shell had been removed with a buffered acetate solution. (C) Pyrite framboids forming within an organic maceral from the 1–2 cm depth interval at FOAM. (D) Various pyrite morphologies forming in vascular plant material from Sta 18 in the Mississippi Delta. (E) Pyrite replacing magnetite at 40–50 cm depth at the FOAM site. (F) Magnetites from 60–70 cm depth are completely covered with pyrite. (G) EDS sulfur map of pyrite coating magnetite at FOAM (100–200 cm depth). (H) Biotite in thin section from 250–260 cm depth at FOAM, showing pyrite forming between and within biotite layers. (I) Pyrite forming on the surface of a biotite grain from the 210–220 cm depth interval at FOAM.

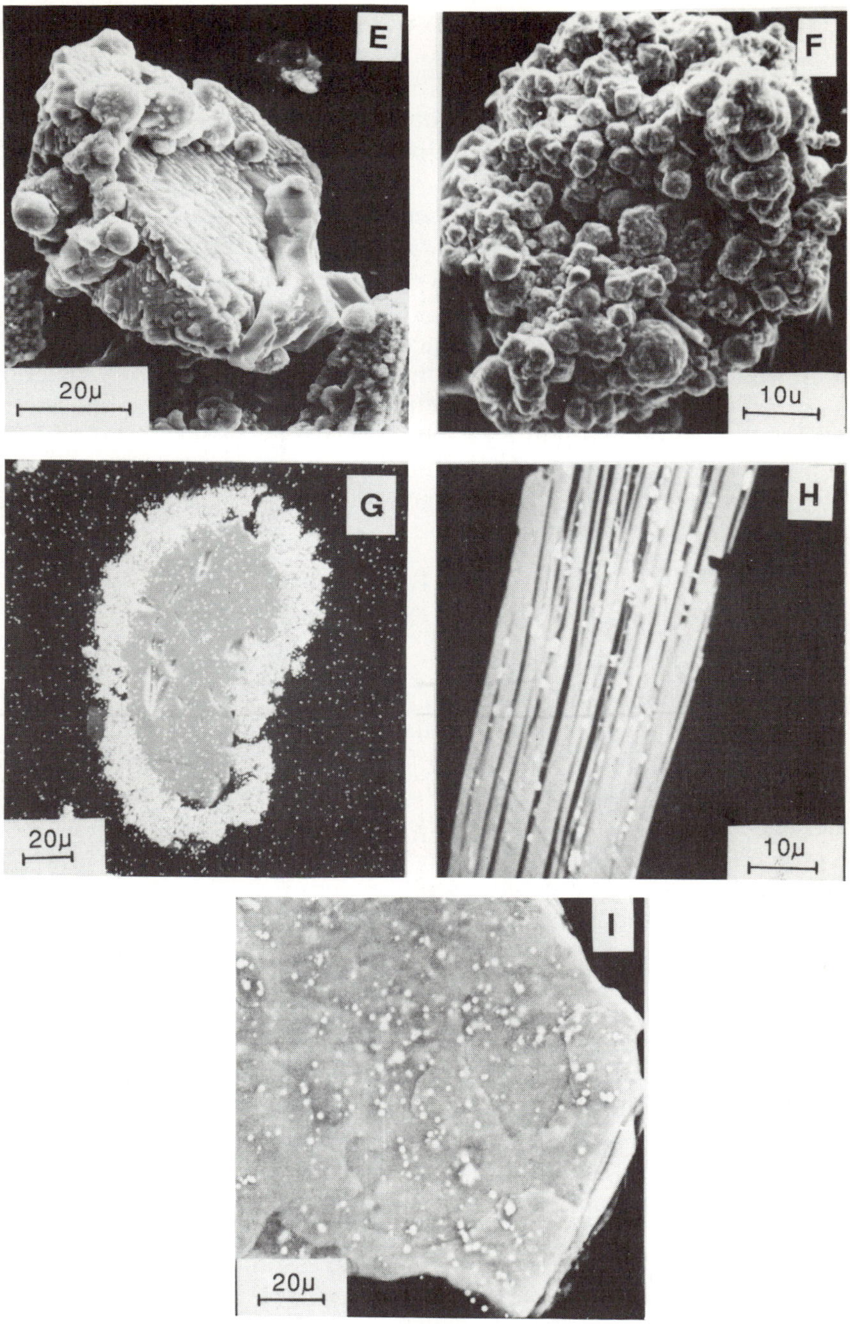

FIGURE 11. *(continued)*

reported by Ross and Degens (1974) in unit 3 (lacustrine) Black Sea sediments, and by Field and Pilkey (1970) in lower continental rise sediments off Cape Hatteras. Iron oxides may also be enriched at the oxic/anoxic interface in active sediment burrows. Hence, both localized sulfate reduction at the burrow lining (as discussed previously) and the burial of iron oxide-rich relict burrows into sulfide-containing pore water may conspire to pyritize burrow tubes. Despite these two potential pyritization processes, reports of pyritization of burrow tubes in modern sediments are very uncommon.

Another mechanism for pyrite layer formation is the winnowing of sedimentary fines and subsequent accumulation of heavier pyrite on the seafloor (Baird and Brett, 1986). It is interesting that for pyrite to accumulate as a winnowed lag deposit, the seabed must be shallow enough to intersect wind waves, yet be overlain by suboxic to anoxic water (Baird and Brett, 1986). Similarly, the winnowing and accumulation of iron oxides observed by Aller *et al.* (1986) and their subsequent reaction with dissolved sulfide could also produce pyrite layers.

5.4.4. Pyrite Replacement of Shelly Carbonate

A final mechanism for localized pyrite formation in Recent sediments involves the local elevation of pH in pore waters saturated with dissolved iron and sulfide. Because iron sulfide saturation is pH sensitive (Fig. 9), solid-phase iron sulfide will precipitate in order to maintain pore-water saturation when solution pH rises. In sediments the dissolution of shelly carbonate could provide the necessary acid consumption to raise pH locally and precipitate iron sulfides. This mechanism can potentially explain many of the occurrences of completely pyritized shell material encountered in the sedimentary record (e.g., Fisher, 1986; Hudson, 1982):

$$CaCO_3 + Fe^{2+} + H_2S \rightarrow FeS + Ca^{2+} + CO_2 + H_2O \qquad (9)$$

The prerequisites for this mode of pyrite precipitation are pore fluids which are both undersaturated with respect to calcium carbonate (calcite, aragonite, or both, depending on the mineralogy of the shell carbonate) and saturated with dissolved iron sulfides. A range of pore-water chemistries fulfills these criteria, including variable pore-water pH values (and subsequent degrees of carbonate saturation) and variable ratios of dissolved iron to sulfide.

To explore the circumstances by which dissolving carbonate may be most effectively replaced by pyrite, a theoretical model was developed which predicts the amount of iron sulfide that may precipitate as calcium carbonate dissolves. The model assumes that pore waters always remain at saturation with the iron sulfide mineral mackinawite, and that as carbonate dissolves and pH rises, just enough iron sulfide is precipitated to maintain saturation. The assumption that a precursor iron sulfide phase and not pyrite controls pore-water saturation is consistent with observations of iron sulfide saturation in anoxic sediments (see previous discussion and Fig. 8). As initial conditions, the model assumes a dissolved calcium concentration of 10 mM (seawater value) and a total dissolved inorganic carbon content of 2 mM (also a reasonable seawater value, but low for pore waters with extensive sulfate depletion by sulfate reduction). Initial pH values were varied between 4 and 7, and initial iron-to-sulfide ratios (Fe/S) ranged from 1 to 10,000. A more complete discussion of the model, including mass balance equations, equilibrium constants, activity coefficients, etc., is outside of the scope of the present review and will be presented elsewhere (Canfield and Raiswell, in preparation).

Model results (Fig. 12) show, for a number of different solution compositions, how much iron sulfide will replace calcium carbonate (as the ratio of Fe precipitated to Ca dissolved) as calcium carbonate dissolves. The model allows that the initial solution composition may variably change due to reaction with the carbonate. If the dissolution rate of the carbonate is slow compared to the rate with which fresh fluid may be reintroduced to the site of the dissolving carbonate, the solution remains close to its starting composition (the lowest values of omega: defined in Fig. 12). When carbonate dissolution is faster relative to fluid reintroduction, dissolution causes the fluid to evolve from its original state and more closely approach carbonate saturation. In all cases (Fig. 12), the further the fluid evolves from its original starting composition, the less effective the replacement of carbonate by FeS. In general, the most complete replacement of carbonate by iron sulfide occurs when the initial solution pH is below about 5. Effective replacement results, in part, because at these lower pH values, the concentrations of dissolved iron and sulfide are relatively high (Fig. 9) and available for precipitation. Model results also show that most FeS replaces dissolving carbonate when the ratio of dissolved iron to sulfide in the reacting fluid is unity. It is important that for fluid pH values >6, no combination of water chemistries will give effective replacement of calcium carbonate by iron sulfide.

Other initial conditions will change model results slightly, but not the main conclusions. For example, if greigite is chosen as the phase controlling pore-water saturation, the model results remain about the same at low pH values, but the ability of iron sulfide to replace carbonate becomes even less at higher initial pH values. If higher initial dissolved inorganic carbon concentrations are used [as may be found deeper in anoxic sediments (Canfield and Raiswell, this volume, Chapter 9, Sections 3.1.3–3.1.5)] the amount of carbonate replacement by iron sulfide drops off more steeply as pore water approaches carbonate saturation. A similar result is obtained if dissolved sulfide, rather than iron, dominates pore-water chemistry. This is because of the buffering capacity of H_2S.

To apply this model to modern marine sediments, we must pinpoint where pore waters rich in dissolved iron and sulfide, but undersaturated with calcium carbonate, are found. Such conditions are not found deep in

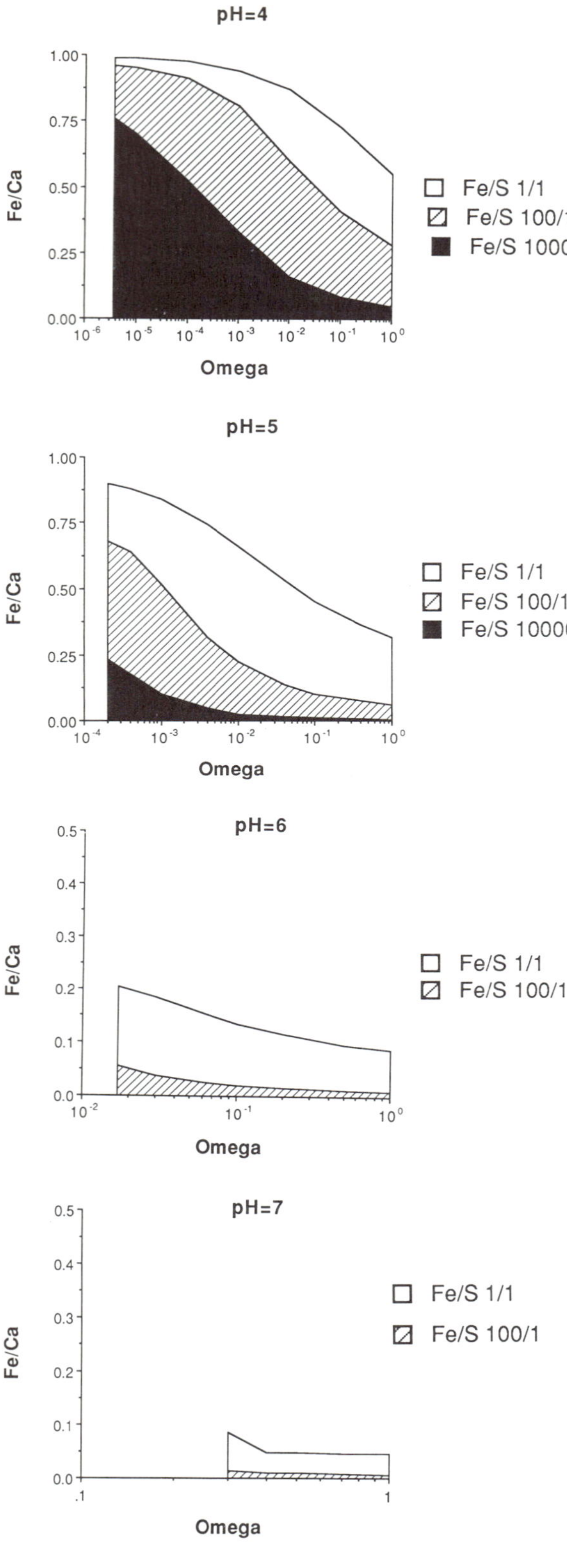

FIGURE 12. Results from model calculating the amount of iron sulfide that may form from the dissolution of carbonate by a solution that is undersaturated with respect to calcite, but saturated with respect to the iron sulfide mineral mackinawite. The model was run for several initial pH values and for various initial ratios of dissolved iron and sulfide, always maintaining saturation with mackinawite. Omega is the saturation state of the dissolving fluid with respect to calcite and is defined as the ratio of the ion activity product (Ca^{2+}) $(CO_3{}^{2-})$ of the fluid with the equilibrium ion activity product of calcite in seawater. The ratio Fe/Ca expresses the amount of iron (forming as iron sulfide FeS) that will replace calcium as the carbonate dissolves.

anoxic sediments (Canfield and Raiswell, this volume, Chapter 9, Sections 3.1.3–3.1.5), where pore waters become supersaturated with respect to carbonate (Sholkovitz, 1973; Berner *et al.*, 1970, 1978; Aller, 1982). By contrast, carbonate undersaturation is found in many near-surface continental margin sediments (e.g., Sholkovitz, 1973; Aller, 1982), particularly in salt marshes where low sediment pH and carbonate undersaturation can become extreme (see below). Focusing on permanently submerged near-surface sediments, undersaturation results both from sulfide oxidation (Aller, 1982) and as a consequence of the early stages of sediment diagenesis (Canfield and Raiswell, this volume, Chapter 9, Sections 3.1.1–3.1.3). The region where carbonate undersaturation is encountered will typically be enriched in dissolved iron and depleted in dissolved sulfide (Fig. 7) (also see Aller, 1980b). Also, pore-water pH values are nearly always above 6, and usually above 7. Both of these conditions make pyrite replacement of dissolving carbonate unlikely. Consistent with this, to our knowledge there are no reports of pyrite replacing carbonate in surficial normal marine sediments.

Heavily irrigated anoxic sediments may also be undersaturated with calcium carbonate to depths of 1 m or more. Specific examples are the NWC site in Long Island Sound (Aller, 1982) and sediments on the flanks of the Santa Barbara Basin, where bottom-water oxygen levels are relatively high (Sholkovitz, 1973). Carbonate dissolution in these sediments is well documented (Aller, 1982; Sholkovitz, 1973; Berger and Soutar, 1970); however, at both locations, as for most permanently submerged marine sediments, pore-water pH values are above 7 (Aller, 1982; Kaplan *et al.*, 1963). From model predictions, pyrite replacement of dissolving carbonate would not be expected. Consistent with this, no pyrite replacement of carbonate has been observed at NWC (D. E. Canfield unpublished data) or reported in the Santa Barbara Basin.

Srivastava (1975) reports pyrite forming within surface borings of mollusk shells in modern sediments from New Zealand. In these shells, pyrite probably formed locally at sites of decomposing organic matter within the borings (see Fig. 10A) rather than as a result of carbonate dissolution. Carbonate dissolution can be ruled out as a mechanism of pyrite formation because dissolution was observed only within the bore holes and not on the main body of the shells. This means that the ambient pore waters were saturated with respect to carbonate, and that the local activity of borers reduced the solution pH (and hence the carbonate saturation state). This is precisely the opposite of what is required to precipitate iron sulfide from a saturated solution.

The combination of model results with observations of pore-water chemistry in permanently submerged marine sediments indicates that carbonate replacement by pyrite is unlikely to occur during early diagenesis. The generation of the appropriate combination of low pH with Fe- and H_2S-rich pore solutions is not observed. We will consider the case for shell pyritization in salt marsh sediments below.

5.4.5. Pyrite Overgrowths on Shells in Salt Marsh Sediments

Pyritization of living mollusk shells, including *Mercenaria mercenaria* and *Geukensia demissa* (Clark and Lutz, 1980; Reaves, 1984), is particularly interesting, as it may serve as a modern analogue of shell pyritization in ancient sediments. Pyrite coating modern mollusks was first reported by Clark and Lutz (1980) for specimens recovered from salt marsh sediment located on St. Catherines Island, Georgia. These authors reported that the brass-yellow color of pyrite was in many instances apparent in the field and that "the greatest concentrations [of pyrite] were near the surface, but in some areas the pyrite had penetrated as much as 1 mm into the shell. Initial penetration seemed to follow fractures, borings, and concentrations of organic matter along growth lines."

Clark and Lutz (1980) suggested that the pyritization of mollusks at St. Catherines Island may accompany aragonite dissolution in the low-pH surficial salt marsh sediments, a mechanism similar to that discussed above. Unfortunately, without supporting data on pore-water chemistry, these authors could only suggest, but not prove pyrite formation mechanisms.

To determine the mechanisms of shell pyritization, Reaves (1984) studied salt marsh sediments on Sapelo Island, Georgia [analogous to the sediment studied by Clark and Lutz (1980)] and marsh sediment from Branford Harbor, Connecticut. At both locations the mollusks *M. mercenaria* and *G. demissa* were found, and *Spartina alterniflora* was the dominant marsh grass. At Sapelo Island, living specimens of both mollusks were found with pyrite overgrowths, whereas at Branford Harbor no pyritized mollusks were present. Replacement of dissolving carbonate by pyrite was considered a possible mechanism for shell pyritization, but dismissed because at Sapelo Island active shell pyritization was observed despite pore-water supersaturation with both calcite and aragonite (Fig. 13). By contrast, at Branford Harbor, tidal creek sediment showed marked undersaturation with calcium carbonate phases (Fig. 13), yet no shell pyritization was observed. Further, for specimens of pyritized *G. demissa* from Sapelo Island, pyrite was almost always absent in regions of the shell where extensive dissolution was observed.

These observations are also consistent with the model results outlined previously (Section 5.4.4). In the study of Reaves, sediment pH was restricted to values greater than 6 and dissolved iron-to-sulfide ratios ranged from 5/1 to 100/1. The combination of these factors makes pyrite replacement of dissolving carbonate an unlikely mechanism of shell pyritization (Fig. 12).

The mechanism of shell pyritization was revealed by careful study of pore-water iron and sulfide chemistry. At Sapelo Island surficial pore waters were enriched in dissolved iron and depleted in sulfide; conversely, at Branford Harbor high concentrations of dissolved sulfide were found with low concentrations of dissolved iron (Fig. 14). With an iron-enriched pore solution such as at Sapelo Island, pyrite formation should be localized on shell material if the shell surface was also enriched in metabolizable organic

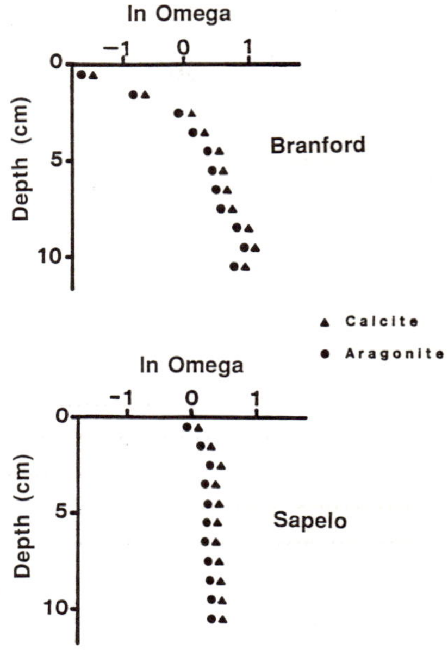

FIGURE 13. Carbonate saturation state at the Branford Harbor, Connecticut and Sapelo Island, Georgia tidal creek sampling sites of Reaves (1984). Note that omega expresses the saturation state with respect to calcite and aragonite, and when ln[Omega] = 0, the solutions are saturated with the respective carbonate phases. [Reproduced with permission.]

carbon (Fig. 10A). Shell periostracum is a potential source of decomposing organic matter; however, Clark and Lutz (1980) noted that at their study site, *M. mercenaria* had a less extensive periostracum, but a more extensive pyrite covering than *G. demissa*. Also, Reaves (1984) observed that at Sapelo Island disarticulated shells with intact periostracum and with no pyrite overgrowths could be found in the immediate vicinity of living mollusks undergoing active pyritization.

Reaves suggested that the excretion of organic-rich substances into the sediment by the mollusks provided the necessary source of organic compounds. Evidence for this mechanism came from a laboratory experiment where live *M. mercenaria* were housed in an artificial organic carbon-poor (but anoxic) sediment rich in dissolved iron. Sediment at and near the shell surface became enriched with organic carbon, and iron sulfides formed on the shell surface. The organic enrichment resulted in high rates of sulfate reduction and the formation of iron sulfides on the shell, completely analogous to the shell pyritization observed at Sapelo Island. At Branford Harbor organic carbon is probably also enriched at the surface of living mollusk shells, but because the surrounding sediment is rich in dissolved sulfide and not iron, sulfide produced from localized sulfate reduction migrates freely into solution and does not precipitate at the shell surface.

In nearshore sediments, shelly fauna typically live in surface sediment enriched in dissolved iron (Fig. 7). If these organisms also secrete organic substances, it becomes difficult to understand why pyrite overgrowths on shells are not more widespread. Perhaps the mobility of most sediment infauna (*M. mercenaria* and *G. demissa* are relatively sedentary) does not allow for organic accumulations to persist at the shell surface.

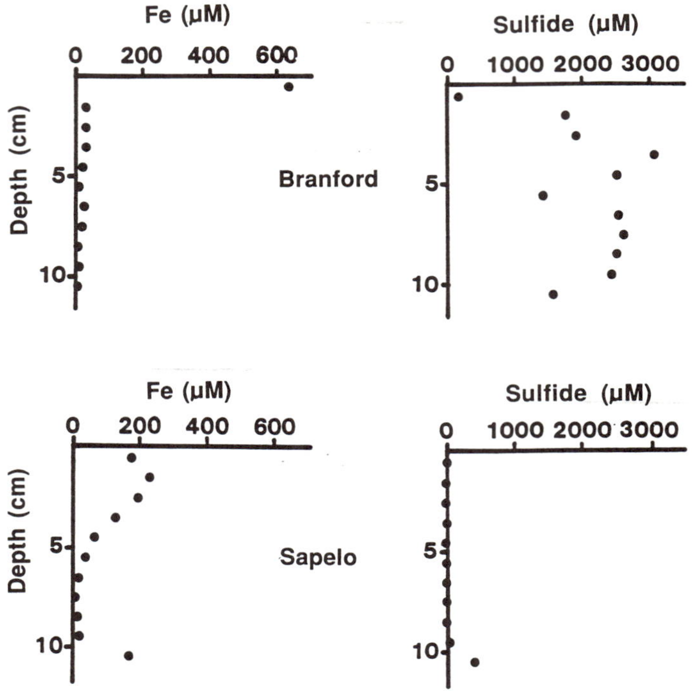

FIGURE 14. Dissolved iron and sulfide at the Branford Harbor and Sapelo Island tidal creek sampling sites of Reaves (1984). [Reproduced with permission.]

6. Pyritization of Fossils in Ancient Sediments

Surprisingly little detailed work has been published on pyrite in fossils and the following discussion is based mainly on the studies of Hudson and Palframan (1969), Love (1971), Seilacher *et al.* (1976), Hudson (1982), Love *et al.* (1983), Fisher (1986), and Allison (1988a,b). Here we attempt to place the morphologies recognized by these workers in the context of the processes of pyritization observed in modern sediments (Section 5.4). In considering the literature, a first distinction that emerges is that different and less varied morphologies are associated with soft-part mineralization than with mineralization of skeletal materials.

6.1. Pyritization of Soft Parts

6.1.1. Observations

The preservation of soft-bodied material in ancient sediments requires exceptional physical and chemical conditions (Seilacher *et al.*, 1985), believed to include rapid burial, stagnation, and cyanobacterial coatings. Recent study has shown, however, that although rapid burial leads to the preservation of articulated hard parts, it is rarely important in soft-part preservation (Allison, 1988b,c). Similarly, stagnation does not act to preserve soft-bodied material, but merely prevents disarticulation by inhibiting bioturbation. Subsequent decomposition of organic material can occur just as rapidly under anoxic as under oxic conditions (Westrich and Berner, 1984; Allison, 1988b, Canfield, 1989a). Soft-part preservation by cyanobacterial coatings (Seilacher *et al.*, 1985) can only occur in shallow, well-lit environments, and has only been reported from a few localities (see below). In general, preservation of soft-bodied material is best achieved by early diagenetic mineral formation (Allison, 1988c), with pyrite the least common replacement mineral.

Pyritized soft parts have been observed in the Hunsrück Slate of Budenbach, Germany (Stürmer, 1970; Stürmer and Bergström, 1973; Stürmer, 1985), in the Beecher's Trilobite Bed of New York State (Cisne, 1973), and sparsely in the Burgess Shale of British Columbia (Conway Morris, 1986). All of these studies concentrate on paleobiological detail and observations of pyrite texture and morphology are scarce.

In general, three modes of soft-part preservation by pyrite can be recognized (Allison, 1988c).

(a) Permineralized tissue. Refractory tissues such as cellulose and chitin may be preserved by the precipitation of pyrite in their pore spaces [e.g., by infilling cellular cavities (Boctor *et al.*, 1976; Price and Shieh, 1979; Love *et al.*, 1983; Allison, 1988a); see Figs. 11A and 11D]. Also, pyrite may preferentially replace more degradable plant components such as cell walls (Love *et al.*, 1983) leaving more refractory lignin-type material unreplaced as a carbonaceous residue (Allison, 1988a; Kenrick and Edwards, 1988). The preservation of very degradable soft parts (e.g., muscle fibers) has not been documented. Pyrite textures include framboids and clustered and aggregated pyrite, the last of which envelopes framboids (Love *et al.*, 1983; Allison, 1988a).

(b) Mineral coats. The preservation of very degradable soft parts most commonly occurs by outline pyritization, which rarely preserves internal structure. Here we define mineral coats as being essentially monomineralic and of a limited and clearly defined thickness (compare with casts and molds, below). Pyrite coats around soft parts occur in Beecher's Trilobite Bed, the Hunsrück Slate, and the Burgess Shale. These coats may be formed by the pyritization (i.e., permineralization) of a layer of bacteria which pseudomorph the original soft parts, as has been suggested for siderite mineral coats (Wüttke, 1983). Few textural observations have been made of soft-bodied pyrite mineral coats, although both framboidal and aggregated pyrite have been recorded (Stürmer, 1970, 1985).

(c) Mineral casts or molds. This style of preservation involves the greatest degree of information loss, since only the fossil outline is preserved. Here casts and molds are defined as resulting from diffuse early diagenetic cementation in the surrounding sediments (to which pyrite formation may be only contributory). Other diagenetic phases, e.g., carbonates, are probably more important than pyrite in forming casts and molds because they can be abundant and because they tend to form more dispersed textures which lithify more effectively.

6.1.2. The Model

There are significant differences in the three modes of preservation, which are essentially distinguished by the extent to which the locus of mineral precipitation is separated from the boundary between the decaying organism and the surrounding sediment: In permineralized tissue, pyrite occurs inside the boundary, in pore spaces, and replacing organic matter. In mineral coats, pyrite occurs in a thin, well-defined zone immediately outside the boundary. In mineral casts or molds, pyrite occurs in a diffuse zone outside the boundary.

The following model [based on Helfferich and Katchalsky (1970)] illustrates how the size, shape, and position of the steady-state precipitation flux of iron sulfide varies with the pore-fluid composition and rates of organism decay by sulfate reduction. The model (Coleman and Raiswell, in preparation) envisages the existence of two infinite reservoirs which, in the case of pyrite precipitation, would contain solutions of dissolved iron and sulfide (Fig. 15). It is assumed that (a) the reservoirs are joined by a permeable plug of uniform thickness and cross section which represents the boundary between the decaying organism and the surrounding sediment, (b) mass transfer occurs only by diffusion and is unaffected by electrostatic interactions between ions, (c) precipitation does not impede diffusion, (d) the reservoirs each remain uniform, (e) no supersaturation can be maintained in the zone of precipitation, that is, precipitation is instantaneous and only diffusion-limited, (f) the size of the decaying organism is large relative to the mean size of the precipitating sulfide.

Model results show that precipitation occurs at the plug, or boundary between the two reservoirs, only when the product of the reservoir concentrations (moles/cm^3) of dissolved sulfide (C_S) and dissolved iron (C_{Fe}) exceed four times the apparent equilibrium constant of iron sulfide (for mackinawite $K = 4.17 \times 10^{-17}$, see Fig. 15). If this condition is only barely exceeded, then precipitation is restricted to a narrow zone around the center of the plug (Figs. 15A and 15B). However, the precipitation zone widens and develops an increasingly sharp maximum as the concentration product increasingly exceeds $4K$ (compare Figs. 15A and 15C). Additionally, as the ratio of the reservoir concentrations increases, so the zone of precipitation shifts toward the reservoir of lower concentration (compare Fig. 15A with Fig. 15B; Fig. 15C with Fig. 15D), although in the case where the reservoir concentrations are equal the precipitation maximum is offset toward the dissolved iron reservoir (Fig. 15D) because the diffusion coefficient for dissolved sulfide (376 cm^2/year) is larger than that of dissolved iron (149

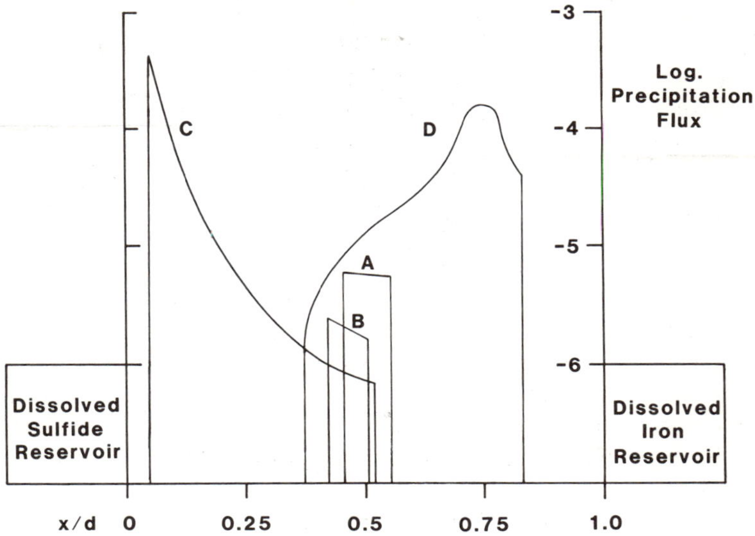

FIGURE 15. Model results showing relative size, shape, and position of precipitation fluxes arising from the diffusion of dissolved iron and sulfide from reservoirs of different concentrations. Saturation with respect to mackinawite (apparent equilibrium constant, $\log K = -16.38$ at pH 7.5) is assumed. Concentrations of dissolved sulfide and dissolved iron are C_S and C_{Fe}, respectively, in mole/cm^3:
(A) $C_{Fe} = 25 \times 10^{-9}$, $C_S = 8 \times 10^{-9}$, $C_S \times C_{Fe} = 2 \times 10^{-16}$ (just exceeds 4K);
(B) $C_{Fe} = 100 \times 10^{-9}$, $C_S = 2 \times 10^{-9}$, $C_S \times C_{Fe} = 2 \times 10^{-16}$;
(C) $C_{Fe} = 500 \times 10^{-9}$, $C_S = 5 \times 10^{-9}$, $C_S \times C_{Fe} = 2.5 \times 10^{-15}$ ($\gg 4K$);
(D) $C_{Fe} = C_S = 50 \times 10^{-9}$, $C_S \times C_F = 2.5 \times 10^{-15}$.

cm^2/year). Note that all conditions give some precipitation on both sides of the midpoint between the reservoirs.

Clearly the position of sulfide precipitation relative to the decaying soft parts will depend on the concentration of dissolved sulfide (and thus the rate of sulfate reduction) at the site of the decomposing organic matter, and the concentration of dissolved iron in the pore waters. Optimum conditions for iron sulfide mineralization at the organic site occur (a) where the sulfate reduction process generates dissolved sulfide comparatively slowly, maintaining low concentrations of dissolved sulfide (relative to dissolved iron) in the vicinity of the decaying organics, and (b) the more the product $C_S \times C_{Fe}$ exceeds 4K. Under these conditions replacement of organic matter (Section 5.4.1) represents a likely possibility. Where rates of sulfate reduction are rather fast, raising the concentrations of sulfide at the organic site (or the pore-water iron concentrations are correspondingly lower), the locus of iron sulfide precipitation will move outward to form a coating around the soft-body material. Finally, in the case where rates of sulfate reduction at the organic site are very fast, and concentrations of dissolved sulfide high at the organic site (and/or pore-water dissolved iron concentrations are low), the zone of sulfide precipitation will occur in the host sediment (not necessarily even adjacent to the organic material). *In extremis* iron sulfide precipitation may be too diffuse to be attributed to decay of the soft material, occurring only as a weak corona in the host sediment.

This model clearly shows how the concentrations of dissolved iron and sulfide may determine the style of pyritization. For any given iron concentration, concentrations of dissolved sulfide around a decaying organism will be lowest where the soft parts are poorly metabolized. Thus, pyritization will only occur at the site of soft parts where these are poorly metabolizable and/or where dissolved iron is high. This prediction is supported by Allison (1988c), who observes that permineralized tissue most frequently occurs in association with poorly metabolizable organic matter, e.g., cellulose and chitin. The same arguments also suggest that pyrite mineral coats should represent more readily metabolizable soft parts. However, although mineral coats are the most common mode of preservation of soft-bodied biotas, this form of pyritization is rare. Moreover, the origin of mineral coats has usually been attributed to the presence of decaying bacteria which pseudomorph original parts [e.g., Wüttke (1983) for siderite]. In these circumstances, where the organic matter has been completely consumed by diagenesis, it may be difficult to distinguish a coating of pyrite from replacement/permineralization.

Most soft parts are more readily metabolizable than cellulose and chitin (Fig. 6 of Allison, 1988c); hence, mineral coats of pyrite (as opposed to pyrite permineralization of bacterial coats) should result from the higher concentrations of dissolved sulfide which can be generated in the vicinity of the decaying soft parts. Still more readily metabolizable organic material would simply generate sulfide which would be fixed as iron sulfide in the surrounding sediment (and would aid mold or cast formation) unless pore-water dissolved iron is unusually high.

Clearly decay cannot produce pyritization in sulfide-dominated pore waters (Section 5.3); thus, the window for organic-matter pyritization lies only within the upper part of the sulfate reduction zone. Furthermore, the position of pyritization with respect to the decomposing organism depends on the iron concentrations in the pore-water system, which show significant changes with depth (Fig. 7). Highest concentrations of iron are found near the sediment surface; here sulfate reduction of poorly metabolizable material will produce permineralization, while more readily metabolizable organics will usually form pyrite mineral coats. However, as iron concentrations decrease with depth, the potential for fossil pyritization depends on whether residual soft-body material is broken down by sulfate reduction relatively slowly (such that the concentration of dissolved iron > dissolved sulfide, and pyritization is possible) or relatively quickly (such that sulfide > iron and no pyritization can occur). It is anticipated that different sediments may show considerable variations in this respect.

The circumstances most favorable to pyritization of soft parts will clearly be those in which the bulk sediment is rich in dissolved iron, while sulfate reduction is limited to discrete microenvironments where more reducing conditions can be maintained. Such conditions are optimized (a) at moderate sedimentation rates (~0.001–0.05 cm/year; Figs. 5 and 6), where sulfate reduction occurs but is insufficient to titrate sediment reactive iron, and (b) where high sedimentation rates (~ >1 cm/year) are associated with sediment reworking, such as in deltas (Fig. 6). However, a scarcity of pyritized fossils may not reflect inappropriate geochemical conditions, but rather, a lack of sufficient time for pyritization due to rapid sediment burial through the zone of sulfate reduction (Brett and Baird, 1986). We must add, however, that in sediments from the Mississippi Delta, we have observed abundant pyritized plant material (Fig. 10D) and pyrite-filled forams.

Non-steady-state conditions, such as turbidite deposition, may also generate appropriate conditions for organic-matter pyritization by overlying otherwise oxidized sediment with organic-rich reduced sediment (Wilson *et al.*, 1985). An advantage to this mode of pyritization is that bioturbation and feeding by sediment infauna will be suppressed, the activity of infauna otherwise acts to disturb or destroy preservable pyritization of organic material.

Soft-part pyritization is not to be expected under euxinic conditions. Hence the presence or absence of early diagenetic soft-body pyritization may be used to distinguish environments with poorly-oxygenated bottom waters, such as the Santa Barbara Basin, from euxinic basins. In both cases finely-laminated sediments may accumulate, but in sediments underlying poorly-oxygenated waters the pore fluids may still contain elevated levels of dissolved iron (Fig. 7), allowing organic-matter pyritization.

Model-predicted pore-water conditions responsible for the pyritization of organic matter and the pyritization of iron minerals are illustrated in Fig. 16. The reservoir concentrations (in moles/cm^3) on the axes correspond to pore-water concentrations of dissolved iron and sulfide, and the horizontal and vertical dashed lines show typically high (but not maximum) concentrations of 100 μM for dissolved iron (near the sediment surface; see Fig. 7) and 5 mM for dissolved sulfide [at depth in the sediment (Canfield and Berner, 1987)] as found in FOAM sediments. The diagonal lines (dashed) with negative slopes define the criteria for precipitation, i.e., that $C_S \times C_{Fe} > 4K$, drawn for pH values of 7.0, 7.5, and 8.0. The solid diagonal lines with positive slopes show the position of maximum precipitation, on a relative scale ranging from 0 (at the dissolved sulfide reservoir) to 1 (at the dissolved iron reservoir). Values <0.1 are arbitrarily considered to favor organic-matter pyritization, while values >0.9 favor iron mineral pyritization.

Model predictions show only a small field for organic-matter pyritization, but a much larger field for iron mineral pyritization. FOAM pore waters in the region of organic-matter pyritization (see Section 5.4.1) have dissolved iron concentrations of 10–25 μM approximately, with bulk pore-water dissolved sulfide concentrations below detection (0.5 μM) (Canfield, 1989b). These measured values plot in the model-predicted region of organic-matter pyritization, although below the boundary for precipitation from pore waters with the pH range found at FOAM (approximately 7.5–8.0). However, since standard analytical techniques measure bulk (average) pore-water concentrations, the concentration of sulfide at the site

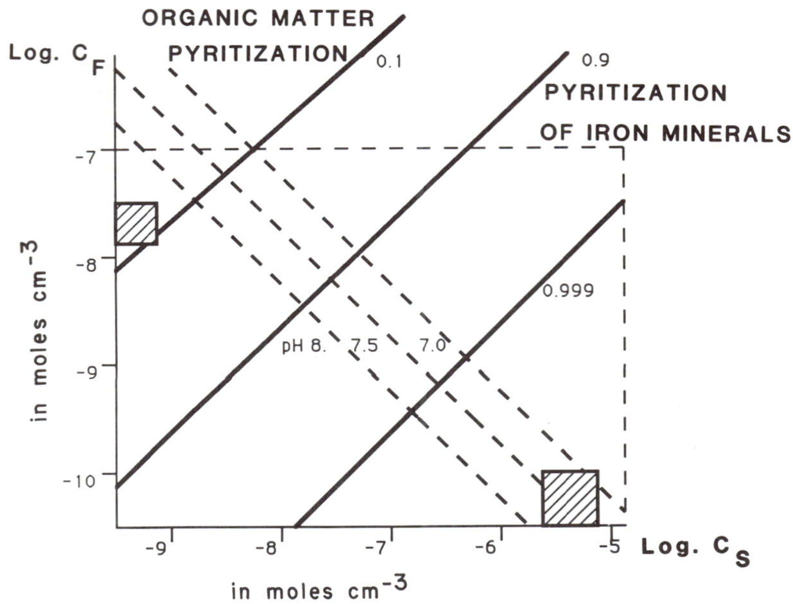

FIGURE 16. Model predictions of pore-water compositions favoring the pyritization of organic matter and iron minerals. The heavy, positive-slope diagonal lines represent the position of maximum pyrite formation on a distance scale where 0 corresponds to the dissolved sulfide reservoir and 1 to the dissolved iron reservoir. The values 0.1 and 0.9 are arbitrarily chosen to indicate pyritization associated with organic matter and iron minerals, respectively. The broken lines with negative slopes represent the constraint that the product of the dissolved concentrations of sulfide and iron must exceed four times the solubility product of mackinawite for precipitation to occur (plotted for pH 7.0, 7.5, 8.0). The dashed vertical and horizontal lines represent reasonable arbitrary maximum concentrations of dissolved sulfide and iron in pore waters. The upper shaded area shows the composition of near-surface pore waters in FOAM sediments where organic-matter pyritization occurs; the lower shaded area shows compositions associated with iron mineral pyritization.

of organic decomposition, and the concentration of iron away from this site, may both have been higher than measured. Somewhat higher concentrations of iron and sulfide would push the FOAM field closer to model predictions. Also, the boundaries marking the criteria for precipitation (diagonal dashed lines) are defined by the choice of an iron sulfide solubility product, for which there are variable values (Davison, 1980), so model and measurements appear in reasonable agreement.

The model-predicted region of iron mineral pyritization is also in good agreement with the deep-pore-water compositions at FOAM. Magnetite pyritization is observed at pore-water concentrations of approximately 3 mM dissolved sulfide (depths of 40–50 cm), with other iron minerals being pyritized at concentrations of up to 5 mM (dissolved iron below detection limit of 0.1 μM). The increase in the position scale from 0.9 to 0.999 indicates increasingly intimate associations of iron sulfide with sediment iron minerals; the values >0.999 suggested by the FOAM pore waters are entirely consistent with the pyritization of iron in the interlayer sites of sheet silicates (see Fig. 11H).

Overall the model gives a very encouraging summary of the conditions under which organic matter and iron minerals may become pyritized. It will be most successful where the site of pyritization is physically separated from the surrounding pore waters (for example, by permeable skeletal material), or where the approach to steady-state precipitation creates a diffusion-limited zone around the decaying organism. Such a zone would then behave as the permeable plug envisaged as part of the basis of the model. The clear agreement between the model and the measurements at FOAM suggests that the model assumptions, although idealized, are at least reasonable approximations to the sediment system.

6.2. Skeletal Carbonate/Pyrite Associations

6.2.1. Observations

Hudson (1982) has recognized the following categories of pyrite associated with carbonate macrofossils: internal sediment, chamber linings, overpyrite, replacement pyrite, stalactitic pyrite, and pyrite nodules or concretions. All except replacement pyrite and nodular/concretionary pyrite (e.g., Gad *et al.*, 1969; Hudson, 1982; Allison, 1988a) require that pyrite formation takes place close enough to carbonate saturation to prevent carbonate dissolution, at least during their initial stages of formation.

Studies of modern sediments suggest that the process of sulfate reduction is normally accompanied by an increasing degree of carbonate supersaturation (Sholkovitz, 1973; Berner *et al.*, 1970, 1978; Aller, 1982; Canfield and Raiswell, this volume, Chapter 9, Sections 3.1.3–3.1.4). Supersaturation results from a number of factors, including the alkalinity generation accompanying sulfate reduction (Berner *et al.*, 1970; Sholkovitz, 1973) and the buffering of pore-water pH values to between 7 and 8 (Ben Yaakov, 1973; Thorstenson, 1970; Gardner, 1973; Boudreau and Canfield, 1988). Undersaturation can arise in the earlier stages of diagenesis (Canfield and Raiswell, this volume, Chapter 9, Sections 3.1.1–3.1.3) and from partial sulfide oxidation (Aller, 1982), but high degrees of oversaturation are the norm in any permanently submerged, near-surface anoxic sediments.

Sediment found inside macrofossils (here termed internal sediment) frequently contains framboidal and/or aggregated pyrite. In some cases pyrite, in association with surrounding sediment, has penetrated breaks in the shell, while in other cases part or all of the framboidal pyrite has formed within the chamber (Hudson and Palframan, 1969; Seilacher *et al.*, 1976; Hudson, 1982). In addition to breakage, sediment may enter the shell as soft parts decay (Allison, 1988b). Internal sediments may be layered and may have a near horizontal upper surface if the chamber is incompletely filled. They are always the earliest chamber filling.

Chamber linings are one of the best-documented pyrite morphologies (Hudson and Palframan, 1969; Seilacher *et al.*, 1976; Hudson, 1982), because of their importance in preservation. Pyrite generally lines all of the available internal surfaces, including both chamber walls and internal sediment surfaces. Chamber linings clearly postdate internal sediment but predate shell dissolution (where this occurs) and compactional crushing of the shell. Hudson (1982) observes that the outer parts of ammonite chamber linings in the Upper Oxford Clay were composed of well-crystallized pyrite, generally of an equant form. Combinations of the cube and pyramid are common, but octahedral habits also occur (Allison, 1988a).

Stalactitic pyrite typically attaches to chamber lining pyrite, and extends downward into cylindrical protuberances with rounded ends (Hudson and Palframan, 1969; Seilacher *et al.*, 1976; Hudson, 1982). Stalagtites occur in dense clusters, but only rarely coalesce. They are orientated geopetally and usually grow from the chamber roof, but sometimes from side wall protuberances. The pyrite is equant, but has a complex form with some crystal faces having curved edges. Only cube-octahedron and octahedron crystals occur (Hudson, 1982; Allison, 1988a).

Overpyrite is found on or near the outer shell surface, and is formed during compaction, since cracked surfaces may be pyritized (Hudson, 1982) and there may be only two-dimensional preservation (Seilacher *et al.*, 1976). Texturally, overpyrite occurs as equant crystals, mainly octahedra, which may coalesce into aggregates a few millimeters in size (Hudson, 1982). More varied textures were observed by Allison (1988c) in the London Clay (Eocene), including octahedra and cauliform textures of cubic pyrite.

6.2.2. The Model

Hudson (1982) discussed a possible sequence of events leading to the varied morphologies often observed with shell pyritization. Pyrite formation in internal sediment may be viewed in the conventional way with the sediment itself supplying the requisite organic carbon and iron. Seawater sulfate is also readily available for reduction either prior to or after entry into the chamber. The other pyrite morphologies—overpyrite, chamber-lining, and stalactitic pyrite—have little in common in terms of magnitude, morphology, and timing with potential modern analogues such as pyrite coatings (Section 5.4.5) and shell pyrite infillings (Section 5.4.1). Hudson (1982) explains these other morphologies by drawing on the experimental work of Rickard (1969), who observed (during anaerobic organic matter decomposition) the formation of an oil which was a mixture of dead bacteria and their waste products with precipitating iron sulfides. The mixture was immiscible with water, and Hudson (1982) suggests that pyrite forming in the oil would preferentially crystallize on preexisting pyrite due to surfactant forces. Al-Biatty and Raiswell (in preparation) also suggest that, in addition to surfactant forces, conversion of an iron sulfide to greigite can be accompanied by the development of ferrimagnetism, which could encourage colloidal FeS to gather on preexisting sulfide surfaces.

Chamber-lining, stalactitic, and overpyrite may be *initiated* by the decay of the organism (see Section 5.4.5) or the organic matrix of the shell, by sulfate reduction (Hudson, 1967). This is followed by the precipitation of amorphous iron sulfide phases, and these become ferrimagnetic on recrystallization to greigite. Surfactant forces cause liquid organic breakdown products to deemulsify on the surfaces and act as a focus for further sulfate reduction and sulfide precipitation. The ferrimagnetic properties of greigite further assist in scavenging colloidal iron sulfides and associated organics from the surrounding pore waters.

This mechanism would allow coatings of pyrite to grow on exposed skeletal surfaces during early diagenesis, although it is difficult to envisage thicknesses of several millimeters being achieved. Reasonably thick (millimeter scale) pyrite layers require substantial amounts of organic carbon, iron, and sulfur for their formation. For example, organic carbon in marine benthic animals comprises about 5–10% of the total wet weight of animal tissue (Parsons *et al.*, 1977). With a wet density of about 1.1 g/cm^3 and assuming all carbon is decomposed by sulfate reduction and all sulfide is precipitated as pyrite at the site of organic decomposition, at maximum 1 cm^3 of animal tissue can form about 0.025–0.05 cm^3 of pyrite, a volume reduction of 95–97.5%. Far more pyrite than this is often observed in pyritized fossils (for example, Fisher, 1986), requiring an external source of sulfur (not derived from sulfate reduction using the organism) as well a source of iron to precipitate it.

A potential source is the surrounding pore water, but, at normal sediment pH values, dissolved sulfide and iron can coexist at only very low concentrations (see Fig. 9). This is an important limiting factor on iron and sulfur availability for pyritization, because the further removed from the sediment–water interface and the zone of sulfate reduction (recall that much of the pyrite contained in fossils is relatively "late" stage), the more closed the sediment becomes, and the less able are chemical species to diffuse to depth to replace those lost to reaction. In this way, pore waters can supply only a minimal amount of iron and sulfur for relatively "late" stage pyritization. However, for pyritization occurring in the zone of sulfate reduction, pore waters may be a viable source of iron and sulfur beyond that which the decomposition of the organism could support.

The absence of modern examples of morphologies such as chamber lining pyrite and stalactitic pyrite suggests that these do not form during early diagenesis. Multiple stages of pyrite growth have been observed in modern sediments as pyritohedra, octahedra, and "granules" formed on framboids (Schallreuter, 1984; Morse and Cornwell, 1987), as well as pyrite layers encrusting earlier generations of pyrite (Section 5.4.3). However, except for layers (whose formation requires non-steady-state conditions) multiple generations of pyrite growth, as observed in modern sediments, are of an extremely small magnitude when compared with examples from the

Upper Oxford Clay and the London Clay. Together these observations suggest that a deeper, possibly external, mineralizing fluid may be involved in delivering dissolved iron and sulfide for the later stages of pyrite growth in some sediments (see below). Such mineralizing fluids would need to have been supersaturated with respect to $CaCO_3$ in addition to having suitable mineralization capacity. With fluids undersaturated with respect to $CaCO_3$ (and again assuming suitable mineralizing capacity) pyrite replacement of $CaCO_3$ can occur.

6.3. Replacement of Carbonate by Pyrite

6.3.1. Observations

There have been a variety of well-documented studies of pyrite replacing skeletal carbonates, although the literature needs to be evaluated carefully to separate those cases where pyrite replaces organic matter associated with skeletal material from those where pyrite truly replaces carbonate. Hudson and Palframan (1969) observed pyrite replacing nacreous structures in a crushed kosmoceratid ammonite. Duff (1975, 1978) pointed out that bivalve shells replaced by pyrite showed good external morphology but poor preservation of internal structure. Hudson (1982) observed many kinds of ammonite shell replacement, including pyrite replacing septa and shell walls as rods (normal to the shell surface) and sheets (parallel to the surfaces). In the case of one ammonite, pyritization was shown to predate crushing of the shell. Pyrite which truly replaces carbonate requires pore waters to be simultaneously undersaturated with respect to carbonates, but saturated to oversaturated with respect to iron sulfides (see Section 5.4.4).

By far the most detailed study of pyrite replacement has been made by Fisher (1986), who found a range of textures in crushed ammonites and in small gastropods and bivalves in the Lower Oxford Clay. In the ammonites, pyrite replaced the organic shell matrix. No carbonate was removed during this process, which occurred prior to crushing. At some point, still prior to crushing, diffuse ovoid pyrite aggregates formed, which encased large masses of carbonate. These ovoids were probably located at discrete concentrations of organic matter, and may have replaced an earlier marcasite phase. The ovoids consist of aggregated pyrite that range up to about 200 μm in diameter (average 20–30 μm). These pyritization events all represent replacement of organic matter and are attributed to bacterial sulfate reduction. However, at a later stage large numbers of ovoids coalesced and replaced substantial areas of shell material. Coalescence of the aggregates appears to be contemporaneous with the formation of a pyrite phase free of carbonate inclusions, which was associated with cracking and fracturing of the shell. Shell replacement thus spanned a considerable period of sediment burial history.

The gastropods and bivalves often show more faithful replacement of original shell structures. Three types of replacement were observed. Type I replacement of organic matrix by pyrite preserves good microstructure, whereas type II preserves only the gross features of shell microstructure (e.g., growth lines), although the outer margins and their morphology are well-preserved. Type III pyrite represents complete replacement of shell material, leaving only lines of carbonate inclusions at the shell margins. Many of the smaller shells have resisted crushing better than the ammonites; hence, timing the different types of pyrite replacement relative to compaction is difficult.

These detailed descriptions indicate that replacement is initially confined to the organic shell matrix, and that this pyrite formed early before microstructure could be fractured or distorted by compaction. Although this pyrite morphology is referred to as replacement, its origin is essentially the same as that proposed for the pyritization of soft parts (Section 6.1.2). The proteinaceous material comprising the organic matrix (Hudson, 1967) is readily decomposed, but insufficiently abundant to maintain high local concentrations of sulfide. Hence, initiation of pyritization would occur in the iron-dominated, early phases of diagenesis.

6.3.2. The Model

Except for the early pyrite which replaces the organic matrix of the shell, the remaining true replacement pyrites postdate crushing of ammonite, bivalve, or gastropod shells (Fisher, 1986). To our knowledge there are no data to indicate how much burial is necessary to fracture different types of shell; thus, these petrographic observations only poorly constrain the timing of pyritization. The only other constraint could be supplied by sulfur isotope studies of the pyrite. Fisher (1983) reports that the pyrite replacing the above skeletal carbonates ranges from $\partial^{34}S = -6$ to $+25‰$, and Hudson (1978a) has one value of $+44‰$!

These isotope data are sparse and provide only limited clues as to the timing of pyritization and the origins of the pyritizing fluids. The lightest pyrite values are still consistent with an early diagenetic origin for the H_2S, produced by microbial sulfate reduction, but in a closed (or partially closed) system where the formation of earlier, isotopically lighter pyrites have left a residual, isotopically heavy sulfate phase for late-stage reduction. The heaviest values might also represent reduction from the last vestiges of heavy sulfate, but fractionation to this extent implies reduction of the last few percent of the original sulfate. Thus, a continuous resupply of heavy sulfate would be required to produce significant amounts of heavy pyrite. The source and transport of this sulfate are difficult to reconcile with mass balance criteria. There is, furthermore, still the difficulty that alkalinity, and hence degree of carbonate supersaturation, reach their maximum values toward the end of sulfate reduction (Canfield and Raiswell, this volume, Chapter 9, Sections 3.1.3 and 3.1.4), which indicates carbonate precipitation rather than the dissolution necessary for pyrite replacement.

One possibility is that shell replacement uses the dissolved sulfide generated by sulfate reduction, but actually occurs in the underlying zone of methanogenesis. After sulfate reduction the residual pore fluids may be low in dissolved metals and rich in dissolved sulfide (Fig. 7). The addition of carbon dioxide from methanogenesis and the thermal breakdown of organic matter (Canfield and Raiswell, this volume, Chapter 9, Section 3.1.5) and/or the addition of meteoric water could now drive the pore fluids toward carbonate undersaturation. The resulting acidity also enables reactive iron to be leached from the surrounding sediments (causing iron sulfides to become supersaturated). Different styles of pyritization could then result according to the relative timing of the passage from (a) carbonate over- to undersaturation and (b) iron sulfide saturation to oversaturation. A fluid which is carbonate undersaturated and sulfide oversaturated may cause shell replacement, whereas a fluid which becomes sulfide oversaturated while still remaining carbonate oversaturated will produce pyrite coatings on shell material. This explanation is attractive in that shell replacement by pyrite (or pyrite coatings on shells) could occur at relatively shallow burial depths and would not require access by external fluids [although meteoric waters might assist the process (Hudson, 1978b; Marshall, 1982; D. Emery *et al.*, 1988)]. However, from the preceding discussion (Section 6.2.2), the mass balance of dissolved sulfide availability versus the amount of pyrite formed must be carefully considered.

An alternative explanation rests on a possible analogy with Mississippi Valley Type (MVT) ores, which are also formed by interactions between carbonates and basinal fluids. The timing of MVT ore formation can range from syngenetic to late diagenetic and cover temperatures of ~50–200°C (Ohmoto and Rye, 1979). Reduced sulfur may be derived from a number of sources with variable $\partial^{34}S$ values, depending on the sulfur source and the mechanisms of H_2S production. Possibilities include sulfur derived by thermochemical sulfate reduction of seawater or anhydrite sulfur (Machel, 1987), either at depth or in the host sediment. Sulfate reduction in the Pine Point MVT ores (Michigan) occurred along with sulfide deposition and was made possible by the presence of bitumen and elemental sulfur. This process was not mediated by bacteria (with their resulting isotopic fractionation) and produced "heavy" pyrite $\partial^{34}S$ values of approximately +15 to +25‰. Very light pyrite $\partial^{34}S$ values (approximately −15 to −20‰ at Hansonberg, New Mexico) are attributed to the leaching of sulfur from sedimentary sulfides or to the production of sulfide by bacterial sulfate reduction, with little sulfate depletion occurring because of the low organic carbon content of the host rocks (Ohmoto and Rye, 1979). Intermediate values reflect precipitation by cooling and mixing with near-surface fluids.

Sulfide may also be generated from the thermal decomposition of organic sulfur compounds, whose isotopic composition in marine sediments is variable, but may be as low as −30‰ (Francois, 1987; Aizenshtat *et al.*, 1981). Higher values, however, may also be observed, with Nissenbaum and Kaplan (1972) reporting values as high as −2‰ for organic sulfur in marine humics. Also, the isotopic composition of organic sulfur from oil seeps off the California coast range from −3 to +16‰ (Hartman and Hammond, 1981), showing that organic sulfur may be quite heavy, with much of the isotopically heavy sulfur likely incorporated during later diagenesis.

There is additional support for carbonate replacement by pyrite from studies of basinal brine chemistry in relation to the formation of MVT ores (Kharaka *et al.*, 1987). This study demonstrates that basinal brines normally carry only low concentrations of metals (well below those believed to be necessary to form ore bodies), but may have high concentrations of dissolved sulfide. These fluids are close to equilibrium with base metal sulfides, but are undersaturated with respect to carbonates. Kharaka *et al.* (1987) argue that these fluids could evolve to become metal-rich and sulfur-deficient by interactions with red beds or shale (see also Raiswell and Al-Biatty, 1991). Mixing of the metal-rich, sulfur-deficient and metal-poor, sulfur-rich fluids is then proposed as the mechanism by which MVT Pb and Zn ores are formed.

Metal-poor, sulfur-rich basinal brines may, on their own, be capable of causing shell pyritization. First, the brines studied by Kharaka *et al.* (1987) contain substantially more Fe than Pb or Zn, and hence have a greater capacity to form pyrite than base metal sulfides. This capacity could be further augmented by passage of the fluids through shales containing reactive iron (Raiswell and Al-Biatty, 1991). Second, the brines are saturated with iron sulfides, but undersaturated with carbonates, and hence, in contrast to the pore fluids generated by sulfate reduction, have the necessary characteristics for shell pyritization (see Section 5.4.4).

In another study, Sassano and Schrijver (1989) argued that "hydrothermal" fluids have transported iron and sulfur and resulted in late-stage pyrite formation in carbonates from the Acton-Vale limestone quarry (Cambro-Ordovician) in Quebec, Canada. Evidence for late-stage pyrite includes anomalously high pyrite concentrations in relatively pure carbonates (suggesting that the source of iron must have been external), unusual "columnar" pyrite morphologies, somewhat reminiscent of the stalactitic pyrite discussed earlier, and the observation that pyrite seems concentrated in "discontinuous zones sub-parallel to bedding (that)may indicate former channels of greater than average permeability, now sealed by quartz and carbonate." The origin and composition of the fluids are not clear, but one may envisage acidic iron- and sulfide-containing fluids as capable of replacing carbonate with pyrite as discussed previously (Section 5.4.4). Finally, basinal fluids capable of precipitating sulfide (e.g., through cooling) are not necessarily undersaturated with carbonates. If in fact a basinal fluid became saturated with carbonates and sulfides, then pyrite coatings rather than replacement pyrite would result.

7. Conclusions

Studies of sulfate reduction and pyrite formation in modern sediments place valuable constraints on the mechanisms, timing, and style of fossil pyritization. It is found that the nature of the organic material and, to a lesser extent, the concentration of dissolved sulfate only control sulfate reduction rates (and hence H_2S production), whereas it is the mineralogy of the detrital iron phases that determines how much pyrite can form and whether formation is localized or dispersed. Iron oxides or hydroxides (hematite, goethite, lepidocrocite, and ferrihydrite) react rapidly with dissolved H_2S and are therefore effective at buffering pore waters at low levels of dissolved H_2S. Once these iron phases have been consumed, sulfide accumulates in the pore water and dissolved iron becomes negligible, because the solubility products of iron sulfides are very small. Sediments which contain reactive iron at burial thus pass from "iron-dominated" to "sulfide-dominated" pore waters. This transition is one of the principal controls on the style of fossil pyritization because it determines that the mobilities of iron and sulfide are mutually exclusive during early diagenesis.

Hence, in an iron-dominated pore water, sulfate reduction generates H_2S, which can be fixed as iron sulfide in, or around, decomposing organic matter, by drawing on a diffusive supply of pore-water iron. In modern sediments this style of pyritization includes shell infillings of pyrite, pyritized burrow tubes, and pyritized plant debris. Conversely, in a sulfide-dominated pore water, pyrite formation will be located at any reactive iron source (usually residual, poorly reactive phases such as magnetite or iron silicates). This style of pyritization postdates the pyrite associated with organic matter, but does not produce volumetrically significant amounts of pyrite in association with skeletal material.

Ancient sediments display a greater variety of associations between organic matter and pyrite (for example, biogenic tissue may be replaced, coated, or even preserved as a cast or mold), but these are here considered to differ only in degree. Thus, the site of pyrite precipitation is separated to a variable extent from the site of tissue decomposition, and these styles can be viewed as a complex analogue of the modern sediment case where organic matter is pyritized in an iron-dominated pore water. A simple mathematical model shows that the site of sulfide precipitation depends on the relative concentrations of dissolved sulfide (generated at the organic-matter site) and pore-water-dissolved iron. Relatively high concentrations of H_2S allow dissolved sulfide to penetrate further away from the organic matter before sulfide precipitation occurs. The concentrations of H_2S and dissolved iron can both vary; the former decreases as the more readily decomposable organic matter is progressively used up, while the latter decreases as the sediment passes from iron-dominated to sulfide-dominated pore waters. Thus, the optimum conditions for a close association of pyrite with organic matter will occur where poorly metabolizable material generates H_2S in pore water with high concentrations of dissolved iron. Thus, comparatively refractory tissues are most likely to be permineralized.

Ancient sediments also display a far greater range of pyrites associated with skeletal material (e.g., internal sediment, chamber linings, overpyrite, and stalactitic pyrite) than is the case in modern sediments, where only thin pyrite coatings have been found on shells in salt marshes. The latter owe their origin, however, to the generation of H_2S from an organic coating on the shell in contact with an iron-dominated pore water (as described above). These are not an appropriate analogue for the ancient pyrite/shell associations, because the latter would require unreasonably thick coatings of organics. Furthermore, at least some of the ancient pyrites postdate sediment compaction. We therefore argue that deep basinal fluids have been responsible for the formation of these pyrites, as for the replacement pyrite discussed below.

We attach particular importance to the observation that the modern sediment studies have never unequivocally demonstrated the replacement of skeletal carbonate by pyrite. In fact, theoretical models suggest that the necessary conditions of low-pH, carbonate-undersaturated fluids are difficult to combine with conditions where significant volumes of metal sulfides can be formed (i.e., where fluids contain high concentrations of dissolved iron and sulfide). Carbonate undersaturation can occur through pyrite oxidation (due to subaerial exposure or bioturbation) and also, briefly, during the early stages of progressive burial diagenesis, but in none of these cases is the pH low enough or the ratio of dissolved sulfide to iron appropriate for pyrite to replace carbonate.

However, basinal brines do appear to have the required chemical characteristics; they are low in pH and hence can carry enough iron in the presence of high concentrations of dissolved sulfide to form volumetrically significant amounts of pyrite. Such brines are also often undersaturated with respect to carbonates. The sparse data on timing and sulfur isotope compositions of fossil pyritization are consistent with an origin from basinal brines, whose interactions with carbonates caused pyrite precipitation. In many ways Mississippi Valley Type ores can be considered to represent a more extreme development of the same processes.

An extension of this argument would consider the thick pyrite coatings on shells as also originating from basinal brines, in this case from fluids in equilibrium with carbonates. Precipitation here would be induced probably by cooling or dilution, rather than interaction with carbonates to increase pH. These mechanisms involving basinal brines can only be considered as conjectural at present; nevertheless, the modern sediment data are abundant and consistent enough to indicate that early diagenetic processes are unable to explain many features of fossil pyritization.

Acknowledgments. We would like to acknowledge the comments and suggestions of Marty Goldhaber, John Hudson, Tim Lyons, Derek Briggs, and Peter Allison. In numerous ways Carlton Brett has provided inspiration for this work. One of us (DEC) would like to acknowledge the invitation of Claire Reimers and the support of Bernie Boudreau and Bob Berner for a cruise to the Santa Barbara Basin. DEC would like to acknowledge the National Research Council (NRC) of the United States for support during the writing of this manuscript. Pergamon Press and the American Journal of Science are thanked for permission to reproduce material. Both authors thank Bob Berner for supporting and encouraging their interests in pyrite formation.

References

Aizenshtat, Z., Stoler, A., Cohen, Y., and Nielsen, H., 1981, The geochemical sulfur enrichment of recent organic matter by polysulfides in the Solar-lake, in: *Advances in Organic Geochemistry,* Volume 18 (M. Bjoroy, ed.), Wiley, New York, pp. 279–288.

Al-Biatty, H. J., and Raiswell, R., in preparation, The internal structure of framboidal and clustered pyrite.

Allen, E. T., Crenshaw, J. L., Johnston, J., and Larsen, E. S., 1912, Mineral sulfides of iron, *Am. J. Sci. Ser. 4* **33:**169–236.

Aller, R. C., 1980a, Diagenetic processes near the sediment–water interface of Long Island Sound. I Decomposition and nutrient element geochemistry, *Adv. Geophys.* **22:**237–350.

Aller, R. C., 1980b, Diagenetic processes near the sediment–water interface of Long Island Sound. II Fe and Mn, *Adv. Geophys.* **22:**351–415.

Aller, R. C., 1982, Carbonate dissolution in near-shore terrigenous muds: Role of physical and biological re-working, *J. Geol.* **90:**79–95.

Aller, R. C., and Yingst, J. Y., 1978, Biochemistry of tube dwellings: A study of the sedentary polychaete *Amphitrite ornata* (Leidy), *J. Mar. Res.* **36:**201–254.

Aller, R. C., Mackin, J. E., and Cox, R. T., Jr., 1986, Diagenesis of Fe and S in Amazon Inner Shelf muds: Apparent dominance of Fe reduction and implications for the genesis of ironstones, *Cont. Shelf Res.* **6:**263–289.

Allison, P. A., 1988a, Taphonomy of the Eocene London Clay biota, *Paleontology* **31:**1079–1100.

Allison, P. A., 1988b, The role of anoxia in the decay and mineralization of proteinaceous macrofossilis, *Paleobiology* **14:**139–154.

Allison, P. A., 1988c, *Konservat-Lagerstätten:* Cause and classification, *Paleobiology* **14:**331–344.

Baird, G. C., and Brett, C. E., 1986, Erosion of an anaerobic seafloor: Significance of reworked pyrite deposits from the Devonian of New York State, *Paleogeogr. Paleoclimatol. Paleoecol.* **57:**157–193.

Baas Becking, L. G. M., Kaplan, I. R., and Mare, D., 1960, Limits of the natural environment in terms of pH and oxidation-reduction potentials, *J. Geol.* **68:**243–284.

Ben Yaakov, S., 1973, pH buffering of pore water of Recent anoxic marine sediments, *Limnol. Oceanogr.* **18:**86–94.

Bender, M. L., and Heggie, D. T., 1984, Fate of organic carbon reaching the deep-sea floor: A status report, *Geochim. Cosmochim. Acta.* **48:**977–986.

Berger, W. H., and Soutar, A., 1970, Preservation of plankton shells in an anaerobic basin off California, *Geol. Soc. Am. Bull.* **81:**275–282.

Berner, R. A., 1964, Distribution and diagenesis of sulfur in some sediments from the Gulf of California, *Mar. Geol.* **1:**117–140.

Berner, R. A., 1967, Thermodynamic stability of sedimentary iron sulfides, *Am. J. Sci.* **265:**773–785.

Berner, R. A., 1969a, The synthesis of framboidal pyrite, *Econ. Geol.* **64:**383–384.

Berner, R. A., 1969b, Migration of iron and sulfur within anaerobic sediments during early diagenesis, *Am. J. Sci.* **267:**19–42.

Berner, R. A., 1970, Sedimentary pyrite formation, *Am. J. Sci.* **268:**1–23.

Berner, R. A., 1974, Iron sulfides in the Pleistocene deep Black Sea sediments and their paleoceanographic significance, *Am. Assoc. Petroleum Geol. Mem.* **20:**523–531.

Berner, R. A., 1980, *Early Diagenesis: A Theoretical Approach,* Princeton University Press, Princeton, New Jersey.

Berner, R. A., 1984, Sedimentary pyrite formation: An update, *Geochim. Cosmochim. Acta.* **48:**605–615.

Berner, R. A., and Canfield, D. E., 1989, A new model for atmospheric oxygen over Phanerozoic time, *Am. J. Sci.* **289:**333–361.

Berner, R. A., and Raiswell, R., 1983, Burial of organic carbon and pyrite sulfur over Phanerozoic time: A new theory, *Geochim. Cosmochim. Acta.* **47:**855–862.

Berner, R. A., and Westrich, J. T., 1985, Bioturbation and the early diagenesis of carbon and sulfur, *Am. J. Sci.* **285:**193–206.

Berner, R. A., Scott, M. R., and Thomlinson, C., 1970, Carbonate alkalinity in the pore waters of anoxic marine sediments, *Limnol. Oceanogr.* **15:**544–549.

Berner, R. A., Westrich, J. T., Graber, R., Smith, J., and Martens, C. S., 1978, Inhibition of aragonite precipitation from supersaturated seawater: A laboratory and field study, *Am. J. Sci.* **278:**816–837.

Boctor, N. Z., Kullerud, G., and Sweany, J. L., 1976, Sulfide minerals in the Seelyville Coal, III. Chinook Mine, Indiana, *Mineral. Deposita.* **11:**249–266.

Boudreau, B. P., and Canfield, D. E., 1988, A provisional diagenetic model for pH in anoxic porewaters, *J. Mar. Res.* **46:**429–455.

Boulegue, J., Lord III, C. J., and Church, T. M., 1982, Sulfur speciation and associated trace metals (Fe, Cu) in the pore waters of Great Marsh, Delaware, *Geochim. Cosmochim. Acta.* **46:**453–464.

Brett, C. E., and Baird, G. C., 1986, Comparative taphonomy: A key to paleoenvironmental interpretation based on fossil preservation, *Palaios.* **1:**207–227.

Broomfield, S. M., 1954, Reduction of ferric compounds by soil bacteria, *J. Gen. Microbiol.* **10:**1–6.

Burdige, D. J., and Nealson, K. H., 1986, Chemical studies of sulfide-mediated manganese reduction, *Geomicrobiol. J.* **4:**361–397.

Canfield, D. E., 1988, Sulfate reduction and the diagenesis of iron in anoxic marine sediments, Ph. D. thesis, Yale University, New Haven, Connecticut.

Canfield, D. E., 1989a, Sulfate reduction and oxic respiration in marine sediments: Implications for organic carbon preservation in euxinic environments, *Deep Sea Res.* **36:**121–138.

Canfield, D. E., 1989b, Reactive iron in marine sediments, *Geochim. Cosmochim. Acta* **53:**619–632.

Canfield, D. E., 1991, Sulfate reduction in deep-sea sediments, *Am. J. Sci.* **291:**177–188.

Canfield, D. E., and Berner, R. A., 1987, Dissolution and pyritization of magnetite in anoxic marine sediments, *Geochim. Cosmochim. Acta* **51:**645–659.

Canfield, D. E., and Raiswell, R., in preparation, A model for the replacement of calcium carbonate by pyrite.

Cisne, J. L., 1973, Anatomy of *Triarthrus* and the relationships of the Trilobita, *Fossils Strata* **4:**45–64.

Clark, G. R., and Lutz, R. A., 1980, Pyritization in the shells of living bivalves, *Geology* **8:**268–271.

Coleman, M. L., and Raiswell, R., in preparation, Sources of carbonate and origin of zonation in pyritiferous carbonate concretions.

Conway Morris, S., 1986, The community structure of the Middle Cambrian phyllopod bed (Burgess Shale), *Paleontology* **29:**423–458.

Crill, P. M., and Martens, C. S., 1987, Biogeochemical cycling in an organic-rich coastal basin. 6. Temporal and spatial variations in sulfate reduction rates, *Geochim. Cosmochim. Acta* **51:**1175–1186.

Cutter, G. A., and Velinsky, D. J., 1988, Temporal variations of sedimentary sulfur in a Delaware salt-marsh, *Mar. Chem.* **23:**311–327.

Davison, W., 1980, A critical comparison of the measured solubilities of ferrous sulphide in natural waters, *Geochim. Cosmochim. Acta* **44:**803–808.

Dell, C. I., 1972, An occurrence of greigite in Lake Superior sediments, *Am. Mineral.* **57:**1303–1304.

Duff, K. L., 1975, Paleoecology of a bituminous shale—The Lower Oxford Clay of central England, *Paleontology* **18:**433–482.

Duff, K. L., 1978, Bivalvia from the English Lower Oxford Clay (Middle Jurassic), *Palaeontgr. Soc. Mem.* **1978:**1–137.

Elderfield, H., McCaffery, R. J., Luedtke, N., Bender, M., and Truesdell, V. W., 1981, Chemical diagenesis in Narragansett Bay sediments, *Am. J. Sci.* **281:**1021–1055.

Emerson, S., and Widmer, G., 1978, Early diagenesis in anaerobic lake sediments. II Thermodynamic and kinetic factors controlling the formation of iron phosphate, *Geochim. Cosmochim. Acta* **42:**1307–1316.

Emerson, S., Jacobs, L., and Tebo, B., 1983, The behaviour of trace metals in marine anoxic waters: Solubilities at the oxygen–hydrogen sulfide interface, in: *Trace Metals in Sea Water* (C. S. Wong, E. Boyle, K. W. Bruland, J. D. Burton, and E. D. Goldberg, eds.) pp. 599–609, Academic Press, New York.

Emery, D., Hudson, J. D., Marshall, J. D., and Dickson, J. A. D., 1988, The origin of late spar cements in the Lincolnshire Limestone, Jurassic of central England, *J. Geol. Soc. Lond.* **145:**621–633.

Emery, K. O., 1960, *The Sea off Southern California: A Modern Habitat of Petroleum*, Wiley, New York.

Emery, K. O., and Hülsemann, J., 1962, The relationship of sediments, life and water in a marine basin, *Deep Sea Res.* **8:**165–180.

Field, M. E., and Pilkey, O. H., 1970, Lithification of deep sea sediments by pyrite, *Nature* **226:**836–837.

Fisher, I. St. J., 1983, Studies on the formation of pyrite in Jurassic shales, Ph. D thesis, University of Leicester.

Fisher, I. St. J., 1986, Pyrite replacement of mollusc shells from the Lower Oxford Clay (Jurassic) of England, *Sedimentology* **33:**575–585.

Francois, R., 1987, A study of sulfur enrichment in the humic fraction of marine sediments during early diagenesis, *Geochim. Cosmochim. Acta* **5:**17–27.

Froelich, P. N., Klinkhammer, G. P., Nender, M. L., Luedtke, N. A., Heath, G. R., Cullen, D., Dauphin, P., Hammond, D., Hartmann, D., and Maynard, V., 1979, Early oxidation of organic matter in pelagic sediments of the Eastern equatorial Atlantic: Suboxic diagenesis, *Geochim. Cosmochim. Acta* **43:**1075–1090.

Gad, M. A., Catt, J. A., and Le Riche, H. H., 1969, Geochemistry of the Whitbian (Upper Lias) sediments of the Yorkshire coast, *Proc. Yorks. Geol. Soc.* **37:**105–136.

Gardner, L. R., 1973, Chemical models for sulfate reduction in closed anaerobic marine environments, *Geochim. Cosmochim. Acta* **37:**53–68.

Giblin, A. E., and Howarth, R. W., 1984, Porewater evidence for a dynamic sedimentary iron cycle, *Limnol. Oceanogr.* **29:**47–63.

Goldhaber, M. B., and Kaplan, I. R., 1974, The sulfur cycle, in: *The Sea* (E. D. Goldberg, ed.), Wiley-Interscience, New York, Volume 5, pp. 569–655.

Goldhaber, M. B., Aller, R. C., Cochran, J. K., Rosenfeld, J. K., Martens, C. S., and Berner, R. A., 1977, Sulfate reduction, diffusion and bioturbation in Long Island Sound sediments: Report from the FOAM group, *Am. J. Sci.* **277:**193–237.

Hartman, B., and Hammond, D. E., 1980, The use of carbon and sulfur isotopes as correlation parameters for the source identification of beach tar in the southern California borderland, *Geochim. Cosmochim. Acta* **45:**309–319.

Hein, J. R., and Griggs, G. B., 1972, Distribution and scanning electron microscope (SEM) observations of authigenic pyrite from a deep-sea core, *Deep Sea Res.* **19:**133–138.

Helfferich, F., and Katchalsky, A., 1970, A simple model of interdiffusion with precipitation, *J. Phys. Chem.* **74:**308–314.

Honjo, S., Fischer, A. G., and Garrison, R., 1965, Geopetal pyrite in fine grained limestones, *J. Sediment. Petrol.* **35:**231–240.

Hudson, J. D., 1967, The elemental composition of the organic fraction, and the water content of some Recent and fossil mollusc shells, *Geochim. Cosmochim. Acta* **31:**2361–2378.

Hudson, J. D., 1978a, Pyrite in ammonite shells and in shales, *Neues Jahrb. Geol. Palaontol. Abh.* **157:**190–193.

Hudson, J. D., 1978b, Concretions, isotopes and the diagenetic history of the Oxford Clay (Jurassic) of central England, *Sedimentology* **25:**339–369.

Hudson, J. D., 1982, Pyrite in ammonite-bearing shales from the Jurassic of England and Germany, *Sedimentology* **29:**639–667.

Hudson, J. D., and Palframan, D. F. B., 1969, The ecology and preservation of the Oxford Clay fauna at Woodham, Buckinghamshire, *Q. J. Geol. Soc. Lond.* **124:**387–418.

Hülsemann, J., and Emery, K. O., 1961, Stratification in Recent sediments of Santa Barbara as controlled by organisms and water character, *J. Geol.* **69:**279–290.

Ingri, J., and Ponter, C., 1986, Iron and manganese layering in Recent sediments in the Gulf of Bothnia, *Chem. Geol.* **56:**105–116.

Jones, J. G., Gardner, S., and Simon, B. M., 1983, Bacterial reduction of ferric iron in a stratified eutrophic lake, *J. Gen. Microbiol.* **192:**131–139.

Jørgensen, B. B., 1977, The sulfur cycle of a coastal marine sediment (Limfjorden, Denmark), *Limnol. Oceanogr.* **5:**814–832.

Kaplan, I. R., Emery, K. O., and Rittenberg, S. C., 1963, The distribution and isotopic abundance of sulfur in Recent sediment off Southern California, *Geochim. Cosmochim. Acta* **27:**297–331.

Kenrick, P., and Edwards, D., 1988, The anatomy of Lower Devonian *Gosslingia breconensis* Heard based on pyritized axes, with some comments on the permineralisation process, *Bot. J. Linn. Soc.* **97:**95–123.

Kharaka, Y. K., Maest, A. S., Carothers, W. W., Law, L. M., Lamothe, P. J., and Fries, T. L., 1987, Geochemistry of metal-rich brines from central Mississippi Salt Dome basin, U.S.A., *Appl. Geochem.* **2:**543–561.

Kobayashi, K., and Nomura, M., 1972, Iron sulfides in sediment cores from the sea of Japan and their geophysical implications, *Earth Planet. Sci. Lett.* **16:**200–208.

Krishnaswami, S., Monaghan, M. C., Westrich, J. T., and Turekian, K. K., 1984, Chronologies of sedimentary processes in sediments of the FOAM site, Long Island Sound, Connecticut, *Am. J. Sci.* **284:**706–733.

Leslie, B. W., Lund, S. P., and Hammond, D. E., 1990, Rock magnetic evidence for dissolution and authigenic growth of magnetic minerals within anoxic marine sediments of the California Continental Borderland, *J. Geophys. Res.* **95:**4437–4452.

Lord, C. J., III, 1980, The chemistry and cycling of iron, manganese, and sulfur in salt marsh sediments, Ph. D thesis, University of Delaware, Newark, Delaware.

Lord, C. J., III, and Church, T. M., 1983, The geochemistry of salt marshes: Sedimentary ion diffusion, sulfate reduction and pyritization, *Geochim. Cosmochim. Acta* **47:**1381–1391.

Love, L. G., 1971, Pyrite and marcasite in fossilisation, in: Studies in Earth Sciences, *West Commemoration Volume,* (T. V. V. G. R. K. Murty and S. S. Rao, eds.), pp. 439–469, *Saugar University,* Sagar, India.

Love, L. G., and Amstutz, G. C., 1966, Review of microscopic pyrite from the Devonian Chattanooga Shale and Rammelsberg Banderz, *Fortschr. Mineral.* **43:**273–309.

Love, L. G., Coleman, M. L., and Curtis, C. D., 1983, Diagenetic pyrite formation and sulphur isotope fractionation associated with a Westphalian marine incursion, northern England, *Trans. R. Soc. Ed.* **74:**165–182.

Lovely, D. R., and Phillips, E. J., 1986a, Organic matter mineralisation with reduction of ferric iron in anaerobic sediments, *Appl. Environ. Microbiol.* **51:**683–689.

Lovely, D. R., and Phillips, E. J., 1986b, Availability of ferric iron for microbial reduction in bottom sediments of the freshwater tidal Potomac River, *Appl. Environ. Microbiol.* **52:**7511–7517.

Machel, H. G., 1987, Some aspects of diagenetic sulfate–hydrocarbon redox reaction, in: Diagenesis of Sedimentary Sequences (J. D. Marshall, ed.), *Geol. Soc. Lond. Spec. Publ.* **36:**15–28.

Marshall, J. D., 1982, Isotopic composition of displacive fibrous calcite veins: Reversals in pore-water composition trends during burial diagenesis, *J. Sediment. Petrol.* **52:**615–630.

Martens, C. S., Berner, R. A., and Rosenfeld, J. K., 1978, Interstitial water chemistry of anoxic Long Island Sound sediments, 2. Nutrient regeneration and phosphate removal, *Limnol. Oceanogr.* **22:**10–25.

Maynard, J. B., and Lauffenburger, S. K., 1978, A marcasite layer in prodelta turbidites of the Borden Formation (Mississippian) in Eastern Kentucky, *Southeast. Geol.* **20:**47–58.

Morse, J. W., and Cornwell, J. C., 1987, Analysis and distribution of iron sulfide minerals in Recent anoxic marine sediments, *Mar. Chem.* **22:**55–69.

Murowchick, J. B., and Barnes, H. L., 1986, Marcasite precipitation from hydrothermal solutions, *Geochim. Cosmochim. Acta* **50**:2615–2629.

Nissenbaum, A., and Kaplan, I. R., 1972, Chemical and isotopic evidence for *in situ* origin of marine humic substances, *Limnol. Oceanogr.* **17**:570–582.

Ohmoto, H., and Rye, R. O., 1979, Isotopes of sulfur and carbon, in: *Geochemistry of Hydrothermal Ore Deposits* (H. L. Barnes, ed.), Wiley, New York, pp. 509–567.

Parratt, R. L., and Kullerud, G., 1979, Sulfide minerals in Coal Bed V, Minnehaha Mine, Sullivan County, Indiana, *Mineral. Deposita* **14**:195–206.

Parsons, T. R., Takahashi, M., and Hargrave, B., 1977, *Biological Oceanographic Processes*, 2nd ed., Pergamon Press, New York.

Postma, D., 1982, Pyrite and siderite formation in brackish and freshwater swamp sediments, *Am. J. Sci.* **282**:1151–1183.

Price, F. T., and Shieh, Y. N., 1979, The distribution and isotopic composition of sulfur in coals from the Illinois Basin, *Econ. Geol.* **74**:1445–1461.

Pye, K., 1981, Marshrock formed by iron sulphide and siderite cementation in saltmarsh sediments, *Nature* **294**:650–652.

Pyzik, A. J., and Sommer, S. E., 1981, Sedimentary iron monosulfides: Kinetics and mechanism of formation, *Geochim. Cosmochim. Acta* **45**:687–698.

Raiswell, R., and Al-Biatty, H. J., 1991, Depositional and diagenetic C-S-Fe signatures and the potential of shales to generate metal-rich formation water, in: *Early Organic Evolution: Implications for Mineral and Energy Resources* (M. Schidlowski ed.), Springer-Verlag, Berlin.

Raiswell, R., and Berner, R. A., 1985, Pyrite formation in euxinic and semi-euxinic sediment, *Am. J. Sci.* **285**:710–724.

Raiswell, R., and Plant, J. A., 1980, The incorporation of trace elements into pyrite during the diagenesis of black shales, Yorkshire, England, *Econ. Geol.* **75**:684–699.

Raiswell, R., Buckley, F., Berner, R. A., and Anderson, T. F., 1988, Degree of pyritization of iron as a paleoenvironmental indicator of bottom water oxygenation, *J. Sediment. Petrol.* **58**:812–819.

Reaves, C. M., 1984, The migration of iron and sulfur during the early diagenesis of marine sediments, Ph. D. thesis, Yale University, New Haven, Connecticut.

Reimers, C. E., and Suess, E., 1983, The partitioning of organic carbon fluxes and sedimentary organic matter decomposition rates in the oceans, *Mar. Chem.* **13**:141–168.

Rickard, D. T., 1969, The chemistry of iron sulphide formation at low temperatures, *Stockh. Contrib. Geol.* **20**:67–95.

Rickard, D. T., 1970, The origin of framboids, *Lithos* **3**:269–273.

Rickard, D. T., 1974, The kinetics and mechanism of the sulfidation of geothite, *Am. J. Sci.* **274**:941–952.

Rickard, D. T., 1989a, An apparatus for the study of fast precipitation kinetics, *Min. Mag.* **50**:527–530.

Rickard, D. T., 1989b, Experimental concentration–time curves for the iron (II) sulphide precipitation process in aqueous solutions and their interpretation, *Chem. Geol.* **78**:315–324.

Roberts, W. M. B., Walker, A. L., and Buchanan, A. S., 1969, The chemistry of pyrite formation in aqueous solution and its relation to the depositional environment, *Mineral. Deposita* **4**:18–29.

Ross, D. A., and Degens, E. T., 1974, Recent sediments of the Black Sea, in: The Black Sea—Geology, Chemistry and Biology (E. T. Degens and D. A. Ross, eds.), *Am. Assoc. Petroleum Geol. Mem.* **20**:183–199.

Sansone, F. J., 1986, Depth distribution of short-chain organic acid turnover in Cape Lookout Bight sediments, *Geochim. Cosmochim. Acta* **50**:99–105.

Sassano, G. P., and Schrijver, K., 1989, Framboidal pyrite: Early-diagenetic, late-diagenetic, and hydrothermal occurrences from the Acton Vale quarry, Cambro-Ordovician, Quebec, *Am. J. Sci.* **289**:167–179.

Schallreuter, R., 1984, Framboidal pyrite in deep-sea sediments, in: *Initial Reports of the Deep*

Sea Drilling Project (W. W. Hay and J. C. Sibuet, eds.), U.S. Government Printing Office, Washington, D.C., Volume 75, pp. 875–891.

Seilacher, A., Andalib, F., Dietl, G., and Gocht, H., 1976, Preservational history of compressed Jurassic ammonites from Southern Germany, *Neues Jahrb. Geol. Paläontol. Abh.* **152**:307–356.

Seilacher, A., Reif, W. E., and Westphal, F., 1985, Sedimentological, ecological and temporal patterns of fossil Lagerstätten, *Phil. Trans. R. Soc. Lond. B* **311**:5–23.

Sholkovitz, E., 1973, Interstitial water chemistry of the Santa Barbara basin sediments, *Geochim. Cosmochim. Acta* **37**:2043–2073.

Sholkovitz, E., and Gieskes, J. M., 1971, A physical study of the flushing of the Santa Barbara basin, *Limnol. Oceanogr.* **16**:479–489.

Sørensen, J., 1982, Reduction of ferric iron in anaerobic, marine sediment and interaction with reduction of nitrate and sulfate, *Appl. Environ. Microbiol.* **43**:319–324.

Srivastava, N. K., 1975, Early diagenetic changes in recent molluscan shells., *Neues Jahrb. Geol. Paläontol. Abh.* **148**:380–403.

Stürmer, W., 1970, Soft parts of Cephalopods and Trilobites: Some surprising results of X-ray examination of Devonian slates, *Science* **170**:1300–1302.

Stürmer, W., 1985, A small coleoid cephalopod with soft-parts from the Lower Devonian discovered using X-ray radiography, *Nature* **318**:53–54.

Stürmer, W., and Bergström, J., 1973, New discoveries on trilobites using X-rays, *Paläontol. Z.* **47**:104–141.

Sweeney, R. E., and Kaplan, I. R., 1973, Pyrite framboid formation: Laboratory synthesis and marine sediments, *Econ. Geol.* **68**:618–634.

Thorstenson, D. C., 1970, Equilibrium distribution of small organic molecules in natural waters, *Geochim. Cosmochim. Acta* **34**:745–770.

Trefrey, J. H., and Presley, B. J., 1982, Manganese fluxes from Mississippi Delta sediments, *Geochim. Cosmochim. Acta* **46**:1715–1726.

Tugel, J. B., Hines, M. E., and Jones, G. E., 1986, Microbial iron reduction by enrichment cultures isolated from estuarine sediments, *Appl. Environ. Microbiol.* **52**:1167–1172.

Van Andel, T. J., and Postma, H., 1954, Recent sediments of Gulf of Paria, Verland, *K. Ned. Akad. Wet.* **20**:1–244.

Vaughan, D. J., and Ridout, M. S., 1971, Mössbauer studies of some sulfide minerals, *J. Inorg. Nucl. Chem.* **33**:741–746.

Volkov, I. I., and Rosanov, A. G., 1983, The sulfur cycle in the oceans, in: *The Global Biogeochemical Sulfur Cycle* (M. V. Ivanov and J. R. Freney, eds.), Scope Volume 19, pp. 357–423, Wiley, New York.

Westrich, J. T., 1983, The consequences and controls of bacterial sulfate reduction in marine sediments, Ph. D. thesis, Yale University, New Haven, Connecticut.

Westrich, J. T., and Berner, R. A., 1984, The role of sedimentary organic matter in bacterial sulfate reduction: The G model tested, *Limnol. Oceanogr.* **29**:236–249.

Wilson, T. R. S., Thomson, J., Cooley, S., Hydes, D. J., Higgs, N. C., and Sørensen, J., 1985, Early organic diagenesis: The significance of progressive subsurface oxidation fronts in pelagic sediments, *Geochim. Cosmochim. Acta* **49**:811–822.

Wüttke, M., 1983, Weilteil-Ehraltung durch lithifizierte Microorganismen bei mitteleozanen Vertebraten aus den Ölschiefern der Gube Messel bei Darmstadt, *Senckenb. Lethaea* **64**:509–527.

Chapter 8

Phosphates and Fossil Preservation

JACQUES LUCAS and LILIAN E. PRÉVÔT

1. Introduction

1.1. Phosphorus in Nature

Phosphorus is an essential constituent of living cells and, as such, a limiting element, its concentration in waters being one of the factors which regulate biomass. Concentrations in natural waters are generally less than 1 mg/liter, except in some lakes (Table I). In oceans, deep water is richer in phosphorus than surface water; thus, upwelling zones are areas of high productivity.

The concentration of phosphorus in rocks is generally about 100 times that in water. It is irregularly distributed in the earth's crust, which averages about 0.27% P_2O_5, but may occur in concentrations 100 times greater than

JACQUES LUCAS and LILIAN E. PRÉVÔT • Centre de Géochimie de la Surface, Centre National de la Recherche Scientifique, Institut des Sciences de la Terre, Université Louis Pasteur, 67084 Strasbourg Cedex, France.

Taphonomy: Releasing the Data Locked in the Fossil Record, Volume 9 of *Topics in Geobiology,* edited by Peter A. Allison and Derek E. G. Briggs, Plenum Press, New York, 1991.

TABLE I. Phosphorus Content of Some Natural Environments[a]

Atmosphere dust (% P_2O_5)	0.2
Igneous rocks (% P_2O_5)	0–2 (0.2)[b]
Sedimentary rocks (% P_2O_5)	
Deep-sea clays	0.28
Shelf clays and shales	0.15
Graywackes	0.13
Quartzites	0.06
Limestone	0.07
Marls	0.12
Evaporites	0.002
Phosphorites	>18
Water (mg PO_4/liter)	
Rain	0.001–0.52
Rivers	0.01–1.00 (0.02–0.35)[c]
Lakes	0–3.43 (0.004–1.07)[c]
Subsurface	< 1
	2.5 (associated with volcanism)
Sea	0.27 (average)

Variation	North Atlantic	South Atlantic	North Pacific	South Pacific
Surface	0.03	0.18	0.02	0.12
Deep	0.18	0.25	0.27	0.27

[a]After Koritnig (1978).
[b]Extreme known values; most common value in brackets.
[c]Extreme known values; range of average values in brackets.

this in carbonatites and sedimentary phosphorite deposits. The distribution of sedimentary phosphorites through time and space reflects global paleogeography and regional plate movements (Cook and McElhinny, 1979). These tremendous accumulations of phosphorus require extraordinary mechanisms of formation.

1.2. Phosphate Minerals

There are more than 300 phosphate-bearing minerals (Nriagu, 1984), most of which belong to the apatite group, the general formula of which is $Ca_{10}(PO_4, CO_3)_6(F,OH,Cl)_{\geq 2}$. Various cations (e.g., Mg, Na) may substitute for Ca; substitution of PO_4^{3-} by CO_3^{2-} is usually balanced by excess F^- (McClellan, 1980). Fluorapatite $Ca_{10}(PO_4)_6F_2$ and the rare chlorapatite $Ca_{10}(PO_4)_6Cl_2$ are the thermodynamically stable apatites; they also have the highest formation energies.

In nonweathered sediments of marine origin, the most common phosphate mineral is carbonate-fluorapatite (CARFAP). Substitution of CO_3^{2-} for PO_4^{3-} is such that the CO_2 concentrations can be as high as 6% for a P_2O_5 content of 35%, which implies a correlative increase in the fluorine content.

Several ways to reach electrostatic equilibrium are possible, including: substitution of Ca^{2+} by a monovalent cation, Na^+; two substitutions of PO_4^{3-} by CO_3^{2-} balanced by Ca^{2+} vacancy; introduction of one F^- for one substitution. The last is the most common. The CO_2 content of apatites decreases with geological time (McClellan, 1980), but this is presumably the result of diagenetic alteration, as the original composition of apatite is considered to have remained constant (Flicoteaux and Lucas, 1984). CARFAP is not the most stable apatite in all surface conditions, however; during meteoric weathering, for example, CO_2 is leached out prior to the formation of aluminum or iron phosphates. In freshwater, higher concentrations of Fe lead to the formation of vivianite (Emerson and Widmer, 1978). In general, all nonweathered phosphatic grains, teeth, bones, or diagenetically phosphatized fossils are made of CARFAP.

Sedimentary phosphates were once thought to be amorphous, but XRD and SEM studies have shown the great majority to be cryptocrystalline. Amorphous phases are known to occur, however, in living organisms or in *fresh* biogenic skeletal remains.

1.3. Phosphate in Organisms

Biogenic phosphorus occurs in both soft and mineralized tissues. The average PO_4 content of seawater is around 0.27 mg/liter, but that of animal soft tissue is between two and ten times this value. Soft tissues may contain both organic and inorganic phosphorus phases. The former accounts for almost 90% of the total phosphorus content (Table II) and includes compounds used in energy transfer, such as ATP and ADP, as well as DNA and RNA.

The skeletons of invertebrates today are principally calcareous or siliceous, and low in phosphorus. However, this has not always been so. Biomineralization may have originated in the Precambrian as an extracellular secretion, initially through sulfate reduction (Lowenstam, 1980); intracellular mineralization evolved later. Few examples of these early hard tissues are preserved, and those that are may have been diagenetically altered. However, it is clear that phosphate was the dominant hard-part-forming mineral in the Cambrian (Lowenstam, 1981; Lowenstam and Margulis, 1980; Lowenstam and Weiner, 1983; Cook and Shergold, 1984; Nyers, 1987).

Phosphatic hard tissues are known in several groups of marine invertebrates, including brachiopods, annelids, mollusks, arthropods, echinoderms, and conodonts (Lowenstam, 1972; Lowenstam and Weiner, 1983). Some taxa contain up to 36% P_2O_5. Sixteen biogenic phosphate minerals are known, containing various amounts of Ca, Mg, NH_4, Fe, and Al (McConnell, 1973). The most common are the calcium phosphates: dahllite (carbonate hydroxyapatite), francolite (carbonate fluorapatite), brushite ($CaHPO_4 \cdot 2H_2O$), carbonate apatite, and octocalcic phosphate.

TABLE II. Phosphorus Content of Some Organisms[a]

Organism	% P dry weight
Terrestrial mammals	
Soft tissues	0.1–0.4
Bones	12
Terrestrial dry plant tissues	0.1–1
Bacteria and viruses	1.5–5
Marine phytoplankton	0.5 (average)
Diatoms	0.4–2.0
Peridineans	0.6–1.3
Green algae	0.9–3.3
Golden algae	1.2–3.0
Calcareous blue, brown, red algae	trace–0.6
Marine animals	
Soft tissues	0.5–3.2
Skeletons:	
Polychaetes	2.8–9.4
Echinoderms	trace–0.8
Mollusk shells	trace–1.59
Lobster shells	2.34–6.15
Carbonatic brachiopods	trace–0.12
Phosphatic brachiopods	16–17
Lingulid brachiopods	15–19
Fish soft tissues	0.1–1.1
Fish scales	6.5–8.3
Fish bones	up to 16

[a]Koritnig (1978); Baturin (1982).

Transitory amorphous phosphate has also been recorded (Legeros *et al.*, 1975; Lowenstam and Margulis, 1980; Lowenstam and Weiner, 1985; Posner *et al.*, 1984). Some skeletons, such as those of inarticulate brachiopods or trilobites, are composed of chitinophosphate (Speyer and Brett, 1988).

Vertebrate skeletons are particularly rich in mineral phosphorus. An adult human, for example, contains about 650 g of phosphorus, about 85% of which is present in bones and teeth (Machlin, 1973). A precise determination of the bone mineral is difficult, because of its close association with organic matter. Treatments to remove this organic matter may alter the mineral. Nevertheless there is general agreement that bone is made mainly of dahllite-carbohydroxyapatite. The unbalanced chemical composition given by McConnell (1973) (see Table III) is due to the presence of organic matter. Posner *et al.* (1980, 1984) and Newesely (1989) gave further reasons

TABLE III. Composition of Bovine Bone, in Weight Percentages of Oxides[a]

CaO	MgO	Na_2O	K_2O	SrO	P_2O_5	CO_2	F	Cl	Sum[b]
37.56	0.72	0.99	0.07	0.04	28.58	3.48	0.07	0.08	71.54

[a]After McConnell (1973).
[b]The sum has been corrected for F and Cl by −0.05.

why the chemistry and structure of biological apatite are difficult to define: (1) The primary mineral is a poorly crystallized [crystals less than 500 × 350 × 50Å (Posner *et al.*, 1980)], nonstoichiometric, Ca- and hydroxy-deficient, F- and CO_2-enriched hydroxyapatite; (2) the crystals remain submicroscopic in size because of their structural imperfection, even though they become larger and increase in stoichiometry during the organism's life; and (3) the protein component of bones and teeth is predominantly composed of collagen, which occurs as a dense network of fibers which orient the apatite crystals (Fig. 1). Only tooth enamel is noncollagenous.

2. The Destruction of Phosphate-Bearing Tissues

2.1. Soft Parts

Biogenic tissues, in comparison with surface waters, are rich in phosphorus and constitute a significant reservoir. Upon death, organisms are quickly degraded by autolysis (Hooper, 1973) and microbes, and phosphorus is liberated to pore-water solution (Scharpf, 1974). Microbes respire aerobically and are capable of completely degrading all sedimentary organic carbon in the presence of oxygen. In the absence of oxygen, bacteria utilize a series of alternative oxidants (manganese oxides, nitrates, iron hydroxide, and finally sulfates) to degrade organic carbon [see Redfield (1958) and Berner (1981) for review]. Eventually these oxidants are consumed and fermentative microbes dominate, producing methane. The ordering of these bacterial reduction zones is dependent upon the free-energy yield of the particular reaction; those providing most energy (e.g., manganese reduction) occur highest in the pile, while those that provide less (methanogenesis) occur lowest (see this volume, Chapter 2). At each stage of this process biogenic phosphorus is released into solution (see Fig. 2, this volume, Chapter 2).

2.2. Hard Parts

The large surface area (due to small crystal size) and poorly developed crystal form of biogenic hydroxyapatites (Nriagu, 1983) result in apparent solubilities 10^4 times that of the true solubility of stoichiometric hydroxyapatite (Posner *et al.*, 1984). This supports the suggestion (Suess, 1981) that biogenic remains such as fish skeletons may be a major regenerative source of dissolved phosphate in the sediments of the Peru–Chile upwelling (Van Cappellen and Berner, 1988). Biogenic hard parts are composed of both mineral and organic matter. The intergrowth of these two phases may promote their preservation, the mineral encapsulation inhibiting decay and the collagen inhibiting phosphate dissolution. Degradation does develop, however, and filamentous schizophytes, for example, can transform bone

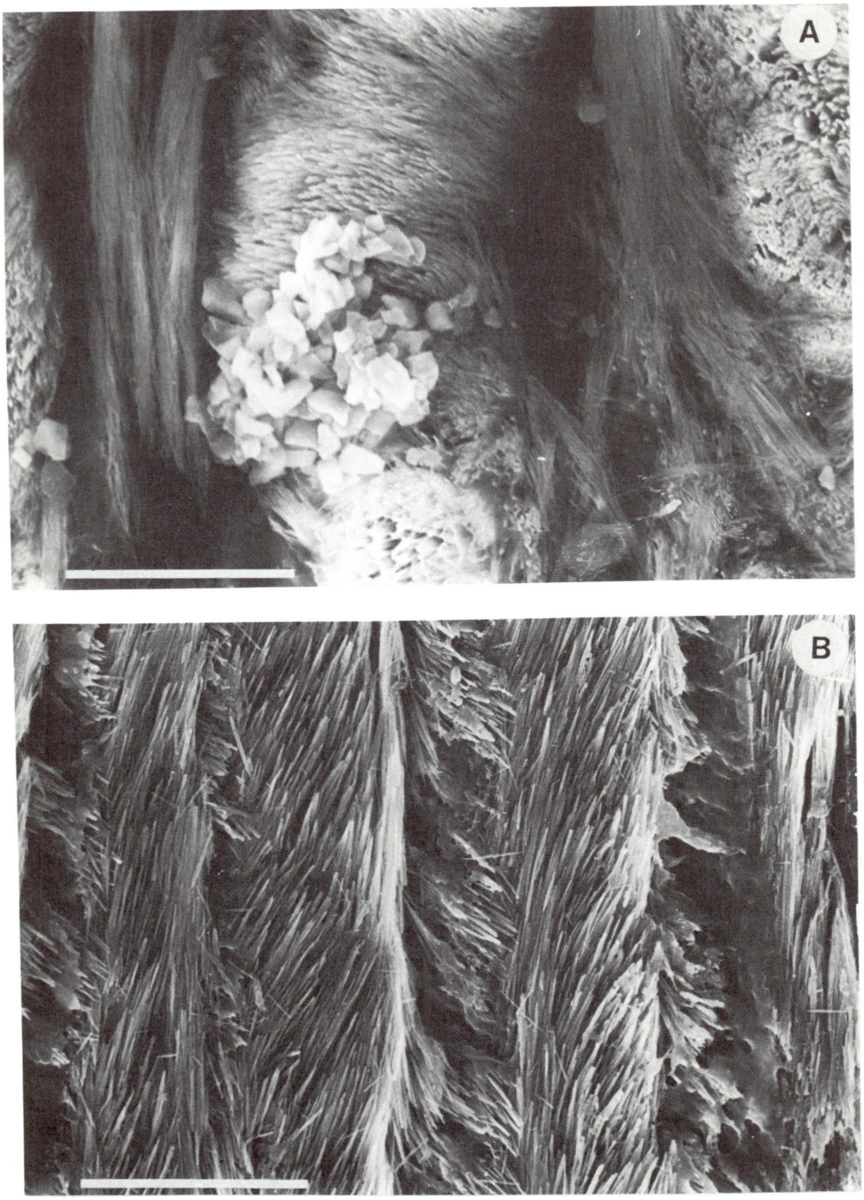

FIGURE 1. Structure of the enameloid covered zone of shark teeth: (A) modern tooth; (B) fossil tooth from the Cretaceous–Tertiary phosphorites of Morocco. SEM photographs. Scale, 10 μm. The structure of the modern and the fossil tooth is very similar. However, the crystalline fibers of the fossil tooth are sharper and thicker, probably because of removal of the collagenous organic matter and of slight crystal growth. [Photographs by B. Kaess.]

FIGURE 2. Fossil bone tunneled by microorganisms (Bengrir phosphate deposit, Morocco, Paleocene, Montian). Thin section photographed in plane-polorized light. Scale, 10 μm.

into a phosphomicrite which is either dispersed or aggregated as undiffe-rentiated peloids (Bathurst, 1971; Soudry, 1979) (Fig. 2). Porous bones are more susceptible to such degradative transformation than teeth, since the latter are denser and are protected by noncollagenous enamel (Fig. 3). Shark teeth from phosphorite levels in North Africa, for example, preserve the upper enamel, but the bony base is commonly absent (Prévôt, 1988). Some organisms break down bone directly, by eating it, e.g., mollusks and worms (Gibbs, 1987; Haszprunar, 1988).

Although phosphorus occurs in all organisms, it is present in very low concentrations (C:N:P = 106:16:1) (Redfield, 1958). Its accumulation in significant quantities in sediment requires special conditions leading to concentration in interstitial waters by repeated microbial cycles. In well-oxygenated environments, organisms with calcareous shells may be abundant, and clastic or calcareous sediments are poor in organic matter. Phosphorus supply to the sediment is low, and apatite is dissolved in the interstitial waters which are saturated in calcium and undersaturated in phosphate [models of Berner (1980), Van Cappellen and Berner (1988)]. Phosphatic fossils are rare in calcium carbonate sediments, not because fishes were absent, but because their bones were dissolved.

Conversely, in anoxic basins, the development of calcareous shells is inhibited, and algae and other soft-bodied organisms dominate (Rhoads and Morse, 1971). Fine-grained organic-rich black shales are deposited, in which organic phosphate is not liberated. In conditions intermediate between these two extremes, i.e., depleted in oxygen, but not anoxic, the

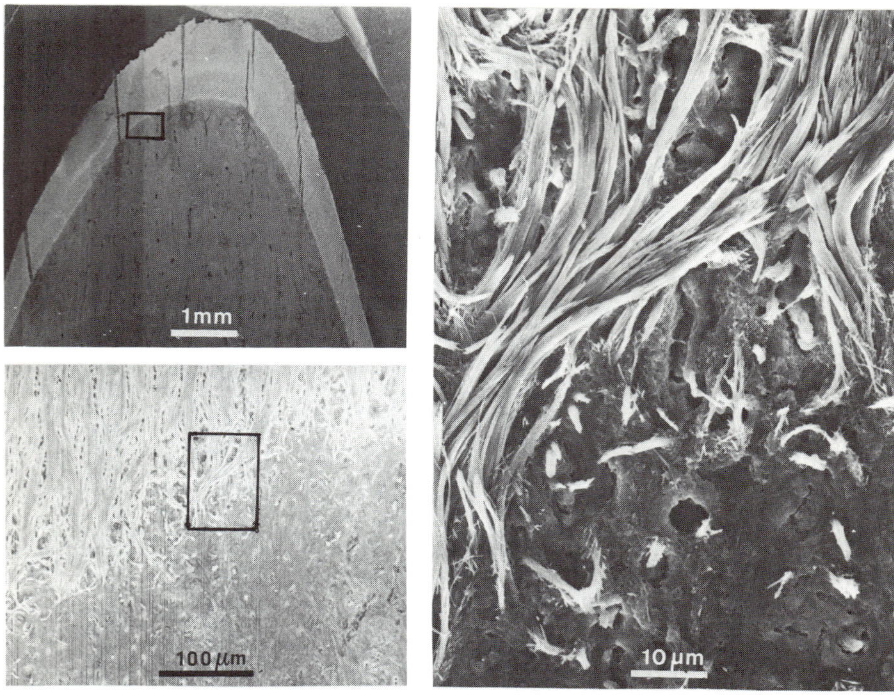

FIGURE 3. Structural change at the transition between dentine and enameloid in a fossil tooth; fibers are more and more dense toward the external edge of the tooth. SEM of polished section. [Photographs by B. Kaess.]

sediment is high in organic matter, and moderate levels of calcium carbonate are present. Degradation of the organic matter releases phosphorus as phosphoric acid, which promotes dissolution of the calcium carbonate and precipitation of apatite. Where phosphorus saturation occurs, organic matter may be partially replaced by apatite, whereas skeletal phosphate is transformed into CARFAP, which is more stable in such an environment than primary hydroxyapatite.

In summary, all the dissolved phosphorus in surface waters is used by organisms. It is only temporarily trapped, to be recycled on the death and decay of the organism. During each cycle, however, a small fraction of mineralized phosphorus escapes and is concentrated in the sediment. Apatite may precipitate from this mineralized phosphorus to form fossils.

3. Apatitic Fossil Record

3.1. Fossilization of Primarily Phosphatic Remains

Due to the intimate association between the organic and inorganic components of bone, both fractions undergo diagenesis in parallel: the organic matter is degraded and hydroxyapatite, which is thermodynamically unstable, is transformed (Tuross *et al.*, 1989). In terrestrial environments bone seldom survives more than 10,000 years (Newesely, 1989), because it easily breaks down in soils (Behrensmeyer, 1978; Retallack, 1984; Brothwell, 1987). In a marine environment most of the phosphate minerals in skeletons are altered ultimately to full carbonate fluorapatite or dissolve (Manheim and Gulbrandsen, 1979). Dissolution may be promoted by the activity of tunneling microorganisms whose acid metabolites dissolve both the mineral and the organic matrix, starting around the osteon canals (Piepenbrink, 1989). The original mineralogy and the histological microstructure may be completely destroyed before a new mineral recrystallizes. In terrestrial conditions the new mineral which precipitates depends on soil pH. Different phosphate minerals are stable at different pH values, e.g., hydroxyapatite (pH 7–7.5), octocalcium phosphate (pH 6–7), brushite (pH 4.5–6). Where iron is available, vivianite is also formed in acidic conditions. Thus, the very soluble brushite, or vivianite, can occur in association with archeological bone (Piepenbrink, 1989). In terrestrial conditions diagenetic recrystallization to apatite is unlikely, since many soils are starved of phosphorus (Pate *et al.*, 1989). In marine conditions the stable authigenic mineral is CARFAP.

The crystallographic maturation of hydroxyapatite, initiated *in vivo*, continues during diagenesis. The low-F ($<1\%$) apatite species which occur *in vivo* are transformed to a high-F ($>1\%$) species during diagenesis (Lozinski, 1973; Newesely, 1989). Fluorine not only replaces the hydroxyls, which may disappear completely, but also enters the structure to balance CO_3^{2-} substitutions for PO_4^{3-} as the apatite becomes a carbonate-fluorapatite. These chemical changes occur by simple ion exchange because the initial imperfect hydroxyapatite belongs to the same hexagonal system as CARFAP. Maltese crosses are frequently formed as interference figures in thin sections of older fossil bone (Fig. 4) where the concentric arrangement of apatite around the Haversian canals is preserved. The increase in fluorine during diagenesis leads to greater crystallinity (Schoeninger *et al.*, 1989). CARFAP is also more stable and less soluble than fresh bone.

Other chemical changes can occur without altering the crystallographic structure of fossilized bone (Piepenbrink, 1989). The concentrations of trace elements—Sr, Zn, Ba, Cu, rare earths (RE)—which are low *in vivo* (but vary according to the animal's diet) may increase. Increased concentrations may also occur in other types of phosphate grains, depending on the composition of sea and interstitial water (Altschuler *et al.*, 1967; Darlet, 1984; McArthur and Walsh, 1984; Prévôt, 1988; Grandjean, 1989; Williams, 1989) and may even rise by several orders of magnitude (Lozinski, 1973). Some elements, especially strontium, which is very similar in valency and ionic radius to calcium, are incorporated into the Ca sites of the CARFAP lattice (Lucas *et al.*, 1990), although most are simply adsorbed onto the crystals or trapped between them (Prévôt, 1988) [e.g., uranium (Lucas and Abbas, 1989)].

Microbial activity influences the isotopic fractionation of carbon and nitrogen in bone during early diagenesis (Grupe *et al.*, 1989), but, as in the

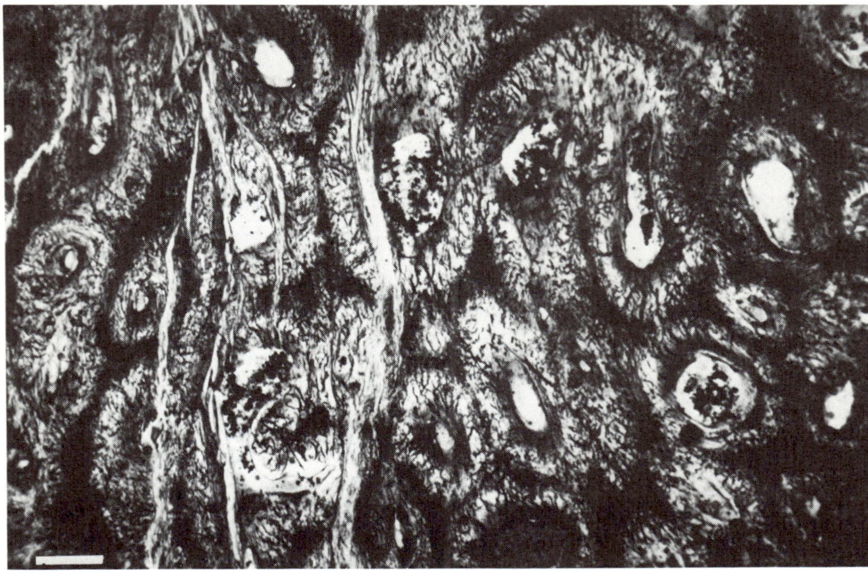

FIGURE 4. Fossil bone showing the apatite oriented around the Haversian canals by the organization of collagen fibrils (Ganntour phosphorite deposit, Morocco, Paleocene, Thanetian). Some canals are also filled with apatite. Thin section photographed in plane-polarized light. Scale, 100 μm.

case of trace elements, it is difficult to separate the effect of diagenetic environment from that of diet. Oxygen isotope fractionation in mammal bone reflects the composition of fluid intake rather than environmental factors (Luz and Kolodny, 1985). A relationship exists between water temperature and $\partial^{18}O$, however, in fish bone and teeth (Luz *et al.*, 1984).

3.2. Phosphatization of Nonphosphatic Hard Tissues

Phosphate may also preserve calcareous and siliceous skeletons, either by replacing the primary mineral or by forming internal or external molds (Dietz *et al.*, 1942; Baturin *et al.*, 1975; Manheim *et al.*, 1975; Pasho, 1976; Prévôt, 1988). Such preservations are rare because phosphate concentrations are usually low in natural environments. Hence, secondarily phosphatized skeletons are predominantly associated with phosphate beds, or with hard grounds where conditions favorable for apatite precipitation prevail (see Section 6). Predictably, as biological activity is a primary mechanism for concentrating phosphate, apatite precipitation is usually associated with an abundance of organic matter. This realization has prompted a program of experiments to synthesize apatite in near natural conditions, using an organic phosphorus donor (Lucas and Prévôt, 1981). In these experiments (El Faleh, 1988; Prévôt *et al.*, 1989) ribonucleic acid (RNA) was dissolved in natural freshwater or seawater; calcium carbonate and sodium fluoride were added. Microorganisms colonized the solution using the RNA as an energy source.

Several observations regarding the genesis of apatite and its role in fossilization were made on the basis of these experiments. (1) Microbial activity is critical to apatite formation; no apatite formed under sterile conditions, even after several years. (2) The apatite synthesized is CARFAP, identical mineralogically and chemically to natural apatite. The most common substitutions in nature (Sr, U, Ba, F) were obtained from solutions enriched in these elements (Lucas *et al.*, 1990; Prévôt *et al.*, 1989). (3) Apatite can form either by replacement of calcium carbonate in the solid state or by precipitation from dissolved calcium. (4) Magnesium, which inhibits apatite formation (Martens and Harris, 1970; Posner *et al.*, 1984), can be removed temporarily from a closed environment by precipitation of a transitory mineral such as struvite ($NH_4MgPO_4 \cdot 6H_2O$).

Further experiments (Hirschler, 1990; Hirschler *et al.*, 1990) demonstrated that bacteria mediate apatite formation through the action of their enzymes; apatite was obtained by adding these enzymes directly to the solution. The activity of phosphatase and other nucleases (e.g., 5'-nucleotidase, cyclic phosphodiesterase), which are enhanced by the bacteria, cut the RNA, producing nucleotides and liberating inorganic phosphorus. This inorganic phosphorus, which is in solution as H_3PO_4, attacks calcium carbonate and precipitates calcium phosphate. In some bacteria, e.g., *Escherichia coli,* enzymes are released on death, and apatite crystallizes externally. In other cases, e.g., *Providencia rettgeri,* enzymes are liberated during life. Almost all bacteria and other enzyme-producing microorganisms are able to mediate apatite crystallization, but the details of the process are not clearly known. In experiments apatite replaces the preexisting calcium carbonate by a mechanism of dissolution–recrystallization. Whether apatite forms within the bacterial cell (Ennever *et al.*, 1981) or outside but very close to it, calcium is used to precipitate apatite as soon as it is released from calcite. Hence, phosphoric acid does not accumulate in an alkaline environment (pH 8–8.5) even though calcite is dissolved in ephemeral acidic conditions in microenvironments.

Apatite commonly occurs as agglomerates of microspherules approximately 1 μm in diameter, each of which is an assemblage of cryptocrystals. This diagenetic apatite faithfully replicates the external shape of the primary object. However, the detailed microstructure is only retained in phosphatic skeletons such as bone. Where nonphosphatic material is replaced the microstructure is obscured by the microspherules at the micrometer scale (Fig. 5).

These results can be used as a basis for a model for the genesis of apatite and resulting fossilization. The use of RNA as a phosphorus donor in experiments is not as artificial as might appear; nucleic acids are abundant

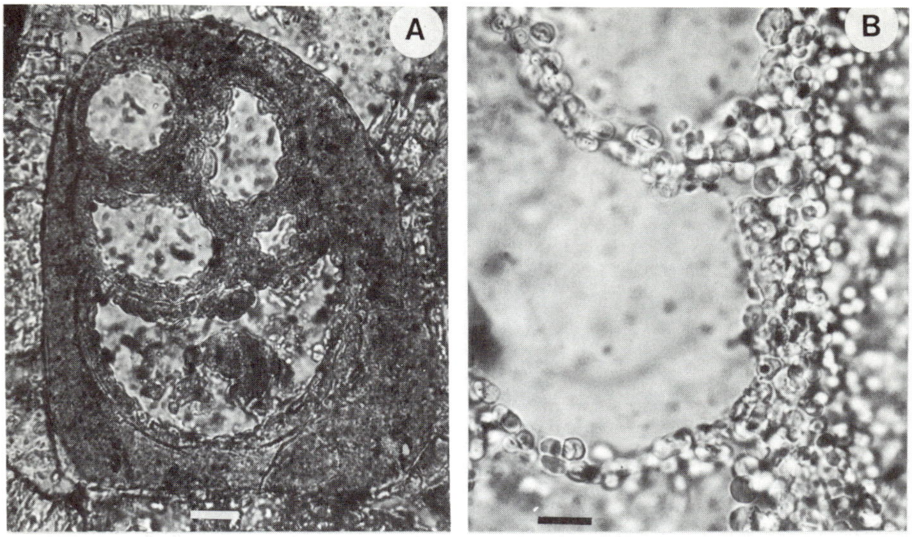

FIGURE 5. Phosphatization of primarily calcareous organisms. (A) Foraminiferan in a natural phosphorite (Ganntour deposit, Morocco, Late Cretaceous, Maastrichtian); (B) Cuttle bone transformed in the laboratory. Thin sections photographed in plane-polarized light. Scale, 20 μm. In both cases, walls are preserved, but with their microstructure changed. Calcium carbonate crystals are substituted by apatite spheres with a diameter of around 1 μm, evoking bacteria. In the natural example, the foram is coated with apatite, forming a coated grain.

in the degradation of organic matter in seawater (Solorzano and Strickland, 1968). Apatite has also been precipitated experimentally using mussels as a source of organic phosphorus (Hirschler, 1990); thus, microorganisms can apparently produce apatite based on any natural phosphorus-containing organic matter. The similarity between the microstructures produced experimentally and those described from a range of phosphatized fossils (Fig. 5) (Krajewski, 1984; Prévôt and Lucas, 1986; Soudry and Lewy, 1988) is an additional argument for a similar mechanism of formation.

3.3. Phosphatization of Nonmineralized Organic Material

Nonmineralized tissues, including microorganisms which lack shells (algae, fungi, bacteria), the soft tissues of certain plants and animals, and coprolites, may be preserved by diagenetic mineralization in apatite.

Algae, fungi, and bacteria are preserved in apatite in natural phosphorites, where they form rolls, fibers, or sheaths. Examples of many ages are known [Campanian phosphorites from Negev, Israel (Soudry and Champetier, 1983); Neogene examples from the Atlantic coastal shelf system of the United States (Riggs, 1982); Cretaceous–Tertiary phosphorites of Ganntour, Morocco (Prévôt, 1988)]. These organisms probably colonized organic matter within the sediment immediately after deposition. They promoted mineralization of the organic phosphorus, and were themselves phosphatized. Where stromatolites are preserved in phosphorites, the cell walls are mineralized in phosphate rather than in calcium carbonate (Krumbein and Cohen, 1977; Monty, 1986).

Fecal pellets are often regarded as one of the major constituents of organic deposits. Those of zooplankton, generally encased in thin membranes, are resistant enough to sink at rates ranging from less than 10 to up to 100 m/day (Turner and Ferrante, 1979). Fecal pellets have an average content of 20% organic carbon, 4.5% nitrogen, and 1.7% phosphorus (Porter and Robbins, 1981). They are important purveyors of organic matter to the sea floor, providing 14% of the daily energy requirement of the benthos through fluxes of 4.7–6.4 mg C/m^2 per day (Wiebe *et al.*, 1976). Pellets are in large part recycled by coprophagous organisms, but some reach the sea floor to be preserved and buried, especially in shallow water. They are colonized in the water column by microorganisms which begin to decay the membrane and transform the organic matter within. The mechanism of mineralization is poorly understood. Enrichment in NH_4, for example, occurs (Turner and Ferrante, 1979), increasing the content of nitrogen 48–85 times in 3 days. This rise in NH_4 provokes the precipitation of struvite, which is regarded as a metastable precursor of apatite (Pacey, 1985), implying a correlative enrichment in phosphorus.

Apatite must have formed very early to preserve the external aspect of fresh fecal pellets which is evident on many coprolites. Experiments have shown that fecal bacteria, such as *Providencia rettgeri*, promote rapid apatite synthesis in experiments (Hirschler *et al.*, 1990). The mechanism is clearly the same as the general one proposed to account for the formation of apatite in seawater (Section 4). The presence of magnesium inhibits the precipitation of apatite in seawater and it has to be removed through the agency of a transitory mineral; struvite has been shown to fulfil this role in experimental conditions (Lucas and Prévôt, 1984).

Replication of organic soft tissues by an early diagenetic mineral is rare and requires exceptional conditions (Allison, 1988a). Some of the most remarkable soft-bodied fossils are those preserved in phosphate (e.g., Müller and Walossek, 1985; Martill, 1990; Allison, 1988a,b). Such fossils normally occur in argillaceous sediments deposited in more or less anoxic conditions, but the precise circumstances of fossilization are unknown. Early diagenetic mineralization may be a major factor in the preservation of soft tissues (Allison, 1988c,d); anoxia and rapid burial alone are not sufficient. The mechanism is likely to be similar to that proposed for coprolites (Hirschler *et al.*, 1990), although these may mineralize more readily than muscle tissue, for example, because they already contain the necessary bacteria.

4. Conditions for Phosphate Enrichment

Experimental results and field observations allow some of the conditions necessary for the formation and preservation of apatite to be defined.

1. The first requirement is a concentration of phosphorus, which can only be produced by biological activity in surface environments, e.g., by microorganism blooms.

2. The phosphorus, which occurs in organic compounds, must accumulate in the sediment rather than being recycled. Blooms due to pollution in modern environments, which result in mass mortality, show that all organisms may be incorporated into the sediment in extreme conditions. Such blooms lead to anoxia, one condition considered necessary for apatite formation in the marine environment (Burnett, 1977).

3. High concentrations of calcium carbonate inhibit apatite precipitation. Hence, apatite forms where organisms using other skeletal materials dominate, e.g., dinoflagellates, diatoms, and radiolarians. Such forms are abundant in sedimentary sequences rich in phosphate (Fauconnier and Slansky, 1978; Prévôt *et al.*, 1979), where poorly oxygenated conditions largely eliminate foraminiferans.

4. The presence of fluorine enhances apatite precipitation in experimental conditions, and most phosphatic fossils, including bone, are enriched in this element. In view of this requirement for fluorine, a marine environment is likely to be most favorable for apatite formation.

5. Magnesium must be removed (by precipitation in a transitory mineral, for example) because it inhibits apatite formation.

The requirements for apatite formation—removal of magnesium, concentration of phosphorus, acid conditions to destabilize calcium carbonate—occur in the closed environment of the interstitial water of an organic-rich fresh sediment. Apatite forms in the pores of the sediment, either by direct precipitation or by replacing carbonate, in response to the local disequilibrium resulting from bacterial activity. Apatite is thus formed in microenvironments, which added together and juxtaposed finally coalesce into extensive volumes. If natural sedimentary apatite is also formed through bacterial mediation, as we believe, it must likewise be the result of microreactions, even in the case of deposits containing millions of tons. Environments propitious for apatite formation can occur at all scales.

5. Large-Scale Phosphate Deposits

Major phosphate deposits have attracted attention due to their economic importance; they are also the most promising environment for the preservation of fossils in apatite. Their formation is the result of a number of conditions. High biological productivity reflects an exceptional nutrient supply, usually due to upwelling. The organic phosphorus is concentrated as a result of rapid sedimentation and preservation of the organic matter, usually in calm shallow water on a wide continental margin, for example, with restricted connection with the deep ocean. Detrital input is minimal due to tectonic stability and desert conditions on land. In this setting high levels of microbiological activity recycle organic matter, calcite is dissolved due to the acid conditions produced by bacterial reactions, and even bones are degraded, contributing phosphorus to the sediment (Froelich *et al.*, 1988). Once concentrations of soluble mineral phosphorus are sufficiently high, apatite precipitation begins either directly or by replacement of nonphosphatic skeletons. The remaining bones, now in equilibrium with the environment and resistant due to fluoration, have a high preservation potential. Not only bone fragments, but all kinds of skeletal remains provide a locus for apatite precipitation. The microorganisms involved in the process—bacteria, fungi, algae—are also fossilized and form a significant part of the phosphatic material (Soudry and Champetier, 1983; Krajewski, 1984; Soudry and Southgate, 1989). Sedimentation is condensed. Hardgrounds and infilled burrows represent numerous nondepositional periods [e.g., in the Cambrian of Australia (Southgate, 1986)].

This "sequential" model of phosphorite formation is generalized. Events in the microenvironments represented by individual pores in the sediment are not in phase. It is the repetition of dissolution and precipitation in time and space which results in the formation of extensive beds of decimeter thickness. These beds differ in details, such as the abundance of preserved fossils, for reasons which are not yet understood. The representation of skeletal remains in these deposits is limited and biased. Many bones, teeth, large shells, and carapaces are dissolved; only small fragments and the tests of planktonic microorganisms are normally preserved by diagenesis in the sediment.

6. Small-Scale Occurrences of Phosphate

Phosphatic fossils also occur where phosphatization is less extensive. Most examples are associated with hiatuses in condensed series. Hardgrounds in calcareous sequences, for example, which are known from rocks as old as Cambrian (Lindström 1963), often include crusts or nodules formed by glauconite and phosphate mineralization. Fossilization in the nodules takes a range of forms: calcareous tests preserved in a phosphatic matrix; internal molds due to infilling of chambers with phosphatic mud; or epigenic phosphatization of the skeleton itself. Most described occurrences are Cretaceous (Kennedy and Garrison, 1975a; Jarvis, 1980; Krajewski, 1984; Delamette, 1988; Föllmi, 1989), Tertiary (Pedley and Bennett, 1985), and Recent (Manheim *et al.*, 1975; Pasho, 1976; Baturin, 1982). Several hypotheses have been proposed to explain the formation of these phosphate nodules. Kennedy and Garrison (1975a,b) believed that mineralization occurred in the sediment while "omission surfaces" were exposed for long periods during low rates of sedimentation. In the Cretaceous of southern England, nodules which were originally calcareous were altered to glauconite and then phosphatized. Internal molds, together with some residual skeletons, were secondarily phosphatized by phosphorus supplied by circulating seawater.

Krajewski (1984) studied the condensed series of glauconitic limestones in the Albian of the Tatras where phosphate occurs at many horizons within semipelagic fossiliferous marine sediments and associated hardgrounds. Apatite occurs in many forms (rim envelopes, infilling intraparticle porosity, rim cement, palissade fabrics, cluster cement) which coat the tests or infill chambers. Sometimes shell fragments, such as ammonite phragmocones, are phosphatized. Krajewski attributed this kind of phosphatization, which is accompanied by the formation of glauconite and often associated with microstromatolites, to bacterial mediation in semiclosed microenvironments within soft sediment. Mineralization appears to be related to rapid sea-level rises. Föllmi (1989), however, did not consider that concentrations of phosphorus in soft sediment were likely to be high enough. He suggested the following process for the Middle Cretaceous along the northern margin of Tethys: settling of a benthic fauna where sedimentation rate is low; sudden burial of the biotic community by a migrating sandbody; phosphatization of the organisms by enrichment of PO_4^{3-} through a "pump-and-shuttle" mechanism (Shaffer, 1986); winnowing and erosion, concentrating phosphatized particles in a gravel; repetitions of this complete cycle.

These mechanisms, although very different, all require bacterial intervention. This is frequently evident as bacterial films, which are believed to concentrate the phosphorus which is subsequently mineralized (Lucas *et al.*, 1978). Phosphatic stromatolites, which occur in a range of sizes from the Precambrian to the present day, are also formed by microbial populations, characteristically in shallow-water environments. Stromatolites may constitute major phosphate deposits (Choudhuri *et al.*, 1980).

Preservation of phosphatic skeletal debris, and even phosphatization of calcareous tests, also occurs in argillaceous sediments, especially where they are organic-rich. Examples include complete ichthyosaurs from the Upper Lias Posidonia Shale of Holzmaden, West Germany, phosphatized ammonites in the Gault Formation of the Paris Basin, and skeletons of *Iguanodon bernissartensis* and other vertebrates preserved in the lacustrine sediments of the Wealden. Finally, originally phosphatic skeletons may be preserved by burial in an inert medium where scavenging and bacterial action are inhibited. Examples include the vertebrate skeletons of the Jurassic Solnhofen Limestone, which are buried in inorganically-precipitated calcium carbonate, and the Pleistocene vertebrates preserved in the La Brea tar pits of southern California.

7. Conclusions

Apatite can preserve a complete range of marine and terrestrial organisms, uni- and multicellular, including all kinds of biological material, from plant and animal soft tissues to skeletons—whether originally phosphatic, or not (e.g., calcareous, siliceous). Phosphatization, in common with any other mode of fossilization, is selective, even in the favorable setting of phosphorites. Preservation in apatite requires transformation; no biological material, even hydroxyapatite, is stable in the natural environment.

A remarkable feature of preservation in apatite is that all fossils acquire the same composition, CARFAP, whether through simple fluoration, in the case of bone, or complete transformation of soft parts. In all cases the alteration appears to involve very similar processes, occurring in microenvironments, as a result of the mediation of microorganisms which transfer the energy required for the chemical reactions.

Given the ubiquity of biological activity, any environment is potentially favorable for phosphatization provided that sufficient nutrients, including phosphorus, are available to stimulate microbiological activity. Here is the unsolved problem; experimental and field observations have illuminated the mechanism of phosphatization, but how the concentrations of the requisite ions are achieved remains to be resolved in detail.

ACKNOWLEDGMENT. We thank very much Prof. B. Kaess, Institut d'Odontologie, Université Louis Pasteur, Strasbourg, who kindly allowed us to use some of his unpublished SEM photographs. This chapter is a contribution to IGCP 325.

References

Allison, P. A., 1988a, Phosphatized soft-bodied squids from the Jurassic Oxford Clay, *Lethaia* **21**:403–410.

Allison, P. A., 1988b, Taphonomy of the Eocene London Clay biota, *Palaeontology* **31**:1079–1100.

Allison, P. A., 1988c, The role of anoxia in the decay and mineralization of proteinaceous macro-fossils, *Paleobiology* **14**:139–154.

Allison, P. A., 1988d, Konservat-Lagerstätten: Cause and classification, *Paleobiology* **14**:331–344.

Altschuler, Z. S., Berman, S., and Cuttila, F., 1967, Rare earths in phosphorites—Geochemistry and potential recovery, *U. S. Geol. Surv. Prof. Pap.* **575B**:1–9.

Bathurst, R. G. C., 1971, *Carbonate Sediments and Their Diagenesis*, Elsevier, Amsterdam.

Baturin, G. N., 1982, *Phosphorites on the Sea Floor. Origin, Composition and Distribution*, Volume 33 of *Developments in Sedimentology*, Elsevier, Amsterdam.

Baturin, G. N., Bliskovskiy, V. Z., and Lisitsyn, A. P., 1975, Phosphorite concretions off the coast of Peru, *Okeanologiya* **3**:331–334.

Behrensmeyer, A. K., 1978, Taphonomic and ecological information from bone weathering, *Paleobiology* **4**:150–162.

Berner, R. A., 1980, *Early diagenesis: A Theoretical Approach*, Princeton University Press, Princeton, New Jersey.

Berner, R. A., 1981, Authigenic mineral formation resulting from organic matter decomposition in modern sediments, *Fortschr. Mineral.* **59**:117–135.

Brothwell, D., 1987, *The Bog Man and the Archaeology of People*, Harvard University Press, Cambridge, Massachusetts.

Burnett, W. C., 1977, Geochemistry and origin of phosphorite deposits from off Peru and Chile, *Geol. Soc. Am. Bull.* **88**:813–823.

Choudhuri, R., Balasubramanian, N., Kapoor, G. P. S., Sharma, D. S., and Dhabai, D. R., 1980, Stromatolites as parameter in grade control on Jhamar Kotra phosphorites, in: Stromatolites, characteristics and utility, *Geol. Surv. India Misc. Publ.* **44**:330–339.

Cook, P. J., and McElhinny, M. W., 1979, A reevaluation of the spatial and temporal distribution of sedimentary phosphate deposits in the light of plate tectonics, *Econ. Geol.* **74**:315–330.

Cook, P. J., and Shergold, J. H., 1984, Phosphorus, phosphorites and skeletal evolution at the Precambrian–Cambrian boundary, *Nature* **308**:231–236.

Darlet, P., 1984, Les terres rares comme traceurs de la voie biogène et indicateurs de diagenèse dans la phosphatogenèse, Thèse 3ème cycle, Nancy.

Delamette, M., 1988, L'évolution du Domaine Helvétique (entre Bauges et Morcles) de l'Aptien Supérieur au Turonien: Séries Condensées, Phosphorites et Circulations Océaniques, *Publication du Département de Géologie et de Paléontologie de l'Université de Genève*, Volume 5.

Dietz, R. S., Emery, K. O., and Shephard, F. P., 1942, Phosphorite deposits on the sea floor off southern California, *Geol. Soc. Am. Bull.* **53**:815–848.

El Faleh, E. M., 1988, Les mécanismes de synthèse de l'apatite par activité bactérienne; rôle et comportement de quelques éléments minéraux—Application aux phosphates sédimentaires, Thèse Doctorat, University of Strasbourg, Strasbourg, France.

Emerson, S., and Widmer, G., 1978, Early diagenesis in anaerobic lake sediments—II. Thermodynamic and kinetic factors controlling the formation of iron phosphate, *Geochim. Cosmochim. Acta* **42**:1307–1316.

Ennever, J., Streckfuss, J. L., and Goldschmidt, M. C., 1981, Calcifiability comparison among selected micro-organisms, *J. Dent. Res.* **60**:1793–1796.

Fauconnier, D., and Slansky, M., 1978, Rôle possible des dinoflagellés dans la sédimentation phosphatée, *Bull. Bureau de Recherches Géologiques et Minières* **4**:191–200.

Flicoteaux, R., and Lucas, J., 1984, Weathering of phosphate minerals, in: *Phosphate Minerals* (J. O. Nriagu and P. B. Moore, eds.), Springer-Verlag, Berlin, pp. 292–317.

Föllmi, K. B., 1989, Evolution of the Mid-Cretaceous Triad-Platform Carbonates, Phosphatic Sediments and Pelagic Carbonates along the Northern Tethys Margin, No. 23 in *Lecture Notes in Earth Sciences*, Springer-Verlag, Berlin.

Froelich, P. N., Arthur, M. A., Burnett, W. C., Deakin, M., Hensley, V., Jahnke, R., Kaul, L., Kim, K. H., Soutar, A., and Vathakanon, C., 1988, Early diagenesis of organic matter in Peru continental margin sediments: Phosphorite precipitation, *Mar. Geol.* **80**:309–343.

Gibbs, P. E., 1987, A new species of *Phascolosoma* (Sipuncula) associated with a decaying whale's skull trawled at 880 m depth in the S.W. Pacific, *N. Z. J. Zool.* **4**:135–137.

Grandjean, P., 1989, Les terres rares et la composition isotopique du néodyme dans les phosphates biogènes: Traceurs des processus paléo-océanographiques et sédimentaires, Thèse Doctorat, University of Nancy.

Grupe, G., Piepenbrink, H., and Schoeninger, M. J., 1989, Note on microbial influence on stable carbon and nitrogen isotopes in bone, *Appl. Geochem.* **4**:299.

Haszprunar, G., 1988, Anatomy and relationships of the bone feeding limpets *Cocculinella minutissima* (Smith) and *Osteopelta mirabilis* (Marshall) (Archaeogastropoda), *J. Molluscan Stud.* **54**:1–20.

Hirschler, A., 1990, Etude de l'intervention des microorganismes dans la formation de l'apatite, Thèse Doctorat, University of Strasbourg, Strasbourg, France.

Hirschler, A., Lucas, J., and Hubert, J. C., 1990, Bacterial involvement in apatite genesis, *FEMS Microbiol. Ecol.* **73**:211–220.

Hooper, F. F., 1973, Origin and fate of organic phosphorus compounds in aquatic systems, in: *Environmental Phosphorus Handbook* (E. J. Griffith, A. Beeton, J. M. Spencer, and D. T. Mitchell, eds.), Wiley, New York, pp. 179–201.

Jarvis, I., 1980, The initiation of phosphatic chalk sedimentation—the Senonian (Cretaceous) of the Anglo-Paris Basin, *Soc. Econ. Paleontol. Mineral. Spec. Publ.* **29**:167–192.

Kennedy, W. J., and Garrison, R. E., 1975a, Morphology and genesis of nodular phosphates in the Cenomanian glauconitic marl of south-east England, *Lethaia* **8**:339–360.

Kennedy, W. J., and Garrison, R. E., 1975b, Morphology and genesis of nodular chalks and hardgrounds in the Upper Cretaceous of southern England, *Sedimentology* **22**:311–386.

Koritnig, S., 1978, Phosphorus, in: *Handbook of Geochemistry* (K. H. Wedepohl, ed.), Springer-Verlag, Berlin, pp. 15-B-1–15-O-5.

Krajewski, K. P., 1984, Early diagenetic phosphate cements in the Albian condensed glauconitic limestone of the Tatra Mountains, Western Carpathians, *Sedimentology* **31**:443–470.

Krumbein, W. E., and Cohen, C., 1977, Primary production, mat formation and lithification chances of oxygenic and facultative anoxygenic cyanophytes (cyanobacteria), in: *Fossil Algae* (E. Flügel, ed.), Springer-Verlag, Berlin, pp. 37–57.

Legeros, R. Z., Shirra, W. P., Miravite, M. A., and Legeros, J. P., 1975, Amorphous calcium phosphates: Synthetic and biological, in: Physico-Chimie et Cristallographie des Apatites d'Intérêt Biologique, *Coll. Int. CNRS Paris* **230**:105–116.

Lindström, M., 1963, Sedimentary folds and development of limestone in an early Ordovician sea, *Sedimentology*, **2**:243–275.

Lowenstam, H. A., 1972, Phosphatic hard tissues of marine invertebrates: Their nature and mechanical function, and some fossil implications, *Chem. Geol.* **9**:153–166.

Lowenstam, H. A., 1980, Bioinorganic constituents of hard parts, in: *Biochemistry of Amino-Acids* (P. E. Hare, ed.), John Wiley & Sons, New York, pp. 3–14.

Lowenstam, H. A., 1981, Minerals formed by organisms, *Science* **211**:1126–1131.

Lowenstam, H. A., and Margulis, L., 1980, Calcium regulation and the appearance or calcareous skeletons in the fossil record, in: *The Mechanisms of Biomineralization in Animals and Plants* (M. Omori and N. Watabe, eds.), Tokai University Press, Tokyo, pp. 289–300.

Lowenstam, H. A., and Weiner, S., 1983, Mineralization by organisms and the evolution of biomineralization, in: *Biomineralisation and Biological Metal Accumulation* (M. Westbrock and E. W. de Jong, eds.), Reidel, Dordrecht, pp. 191–203.

Lowenstam, H. A., and Weiner, S., 1985, Transformation of amorphous calcium phosphate to crystalline dahllite in the radular teeth of chitons, *Science* **227**:51–53.

Lozinski, J., 1973, Rare earth elements in fossil bone, *Rocs. Polsk. Tow. Geol. Polska* **43**:407–435.

Lucas, J., and Abbas, M., 1989, Uranium in natural phosphorites: The Syrian example, *Bull. Sci. Géol. Strasbourg* **42**:223–236.

Lucas, J., and Prévôt, L., 1981, Synthèse d'apatite à partir de matière organique phosphorée (ARN) et de calcite par voie bactérienne, *C. R. Acad. Sci. Paris II* **292**:1203–1208.

Lucas, J., and Prévôt, L., 1984, Les synthèses de l'apatite. Données nouvelles pour un modèle de genèse des phosphorites sédimentaires, in: *Non-Metallic Mineral Ores* (Proceedings of the 27th International Geological Congress, Volume 15), VNU Science Press, Utrecht, pp. 173–186.

Lucas, J., Prévôt, L., and Lamboy, M., 1978, Les phosphorites de la marge nord de l'Espagne—Chimie, minéralogie, genèse, *Oceanol. Acta* **1**:55–72.

Lucas, J., El Faleh, E. M., and Prévôt, L., 1990, Experimental study of the substitution of Ca by Sr and Ba in synthetic apatites, in: Phosphorite Research and Development (A. J. G. Notholt and I. Jarvis, eds.), *Geol. Soc. Lond. Spec. Publ.* **52**:33–47.

Luz, B., and Kolodny, Y., 1985, Oxygen isotope variations in phosphate of biogenic apatites, IV. Mammal teeth and bones, *Earth Planet. Sci. Lett.* **75**:29–36.

Luz, B., Kolodny, Y., and Kovach, J., 1984, Oxygen isotope variations in phosphate of biogenic apatites, III. Conodonts, *Earth Planet. Sci. Lett.* **69**:255–262.

Machlin, L. J., 1973, Phosphorus in human nutrition, in: *Environmental Phosphorus Handbook* (E. J. Griffith, A. Beeton, J. M. Spencer, and D. T. Mitchell, eds.), Wiley, New York, pp. 413–424.

Manheim, F. T., and Gulbrandsen, R. A., 1979, Marine phosphorites, in: *Reviews in Mineralogy*, Volume 6, *Marine Minerals*, Mineralogical Society of America, pp. 151–173.

Manheim, F., Rowe, G. T., and Jipa, D., 1975, Marine phosphorite formation off Peru, *J. Sediment. Petrol.* **45**:243–251.

Martens, C. S., and Harris, R., 1970, Inhibition of apatite precipitation in the marine environment by magnesium ions, *Geochim. Coschim. Acta* **34**:621–625.

Martill, D. M., 1990, Macromolecular resolution of fossilized muscle tissue from an elopomorph fish, *Nature* **346**:171–172.

McArthur, J. M., and Walsh, J. N., 1984, Rare-earth geochemistry of phosphorites, *Chem. Geol.* **47**:191–220.

McClellan, G. H., 1980, Mineralogy of carbonate fluorapatites, *J. Geol. Soc. Lond.* **137**:675–681.

McConnell, D., 1973, Biomineralogy of phosphates and physiological mineralization, in: *Environmental Phosphorus Handbook* (E. J. Griffith, A. Beeton, J. M. Spencer, and D. T. Mitchell, eds.), Wiley, New York, pp. 425–442.

Monty, C., 1986, Interactions événements géologiques—Stromatolithes, *Bull. Cent. Rech. Explor. Prod. Elf Aquitaine* **10**:537–545.

Müller, K. J., and Walossek, D., 1985, A remarkable arthropod fauna from the Upper Cambrian "Orsten" of Sweden, *Trans. R. Soc. Ed. Earth Sci.* **76**:161–172.

Newesely, H., 1989, Fossil bone apatite, *Appl. Geochem.* **4**:233–246.

Nriagu, J. O., 1983, Rapid decomposition of fish bones in Lake Erie sediments, *Hydrobiologia* **106**:217–222.

Nriagu, J. O., 1984, Phosphate minerals: Their properties and general modes of occurrence, in: *Phosphate Minerals* (J. O. Nriagu and P. B. Moore, eds.), Springer-Verlag, Berlin, pp. 1–136.

Nyers, A. J., 1987, *Thorslundella:* A proposed early Cambrian protogastropod that secreted a phosphatic shell due to environmental constraints, *Neues Jahrb. Geol. Paläontol. Abh.* **174**:171–192.

Pacey, N. G., 1985, The mineralogy, geochemistry and origin of pelletal phosphates in the English chalk, *Chem. Geol.* **48**:243–256.

Pasho, D. W., 1976, Distribution and morphology of Chatham Rise phosphorites, *N. Z. Oceanogr. Inst. Mem.* **77**:5–22.

Pate, F. D., Hutton, J. T., and Norrish, K., 1989, Ionic exchange between soil solution and bone: Towards a predictive model, *Appl. Geochem.* **4**:303–316.

Pedley, H. M., and Bennett, S. M., 1985, Phosphorites, hardgrounds and syndepositional solution subsidence: A palaeoenvironmental model from the Miocene of the Maltese islands, *Sediment. Geol.* **45**:1–34.

Piepenbrink, H., 1989, Examples of chemical changes during fossilization, *Appl. Geochem.* **4**:273–280.

Porter, K. G., and Robbins, E. I., 1981, Zooplankton fecal pellets link fossil fuel and phosphate deposits, *Science* **212**:931–933.

Posner, A. S., Blumenthal, N. C., and Betts, F., 1980, Formation and structure of chemically precipitated hydroxyapatites, in: *Impurity Elements in Phosphate Rock and Phosphoric Acid—Characteristics, Elimination, Recovery* (2nd International Congress on Phosphorus Compounds Proceedings), IMPHOS, Paris, pp. 25–39.

Posner, A. S., Blumenthal, N. C., and Betts, F., 1984, Chemistry and structure of precipitated hydroxyapatites, in: *Phosphate Minerals* (J. O. Nriagu and P. B. Moore, eds.), Springer-Verlag, Berlin, pp. 330–350.

Prévôt, L., 1988, Géochimie et pétrographie de la formation à phosphate des Ganntour (Maroc). Utilisation pour une explication de la genèse des phosphorites crétacé-eocènes, Thèse Dr. ès Sciences, Strasbourg [English translation: *Mém. Soc. Géol. Fr.* **158**:1–230 (1990)].

Prévôt, L., and Lucas, J., 1986, Microstructure of apatite replacing carbonate in synthesized and natural samples, *J. Sediment. Petrol.* **56**:153–159.

Prévôt, L., Lucas, J., and Doubinger, J., 1979, Une correspondance entre le contenu palynologique et la composition minéralogique et chimique d'une série phosphatée sédimentaire (Ganntour, Maroc), *Bull. Sci. Géol. Strasbourg* **32**:69–90.

Prévôt, L., El Faleh, E. M., and Lucas, J., 1989, Details on synthetic apatites formed through bacterial mediation—Mineralogy and chemistry of the products, *Bull. Sci. Géol. Strasbourg* **42**:237–254.

Redfield, A. C., 1958, The biological control of chemical factors in the environment, *Am. Sci.* **48**:206–226.

Retallack, G. J., 1984, Completeness of the rock and fossil record; some estimates using fossil soils, *Paleobiology* **10**:59–78.

Rhoads, D. C., and Morse, J. W., 1971, Evolutionary and ecologic significance of oxygen-deficient marine basins, *Lethaia* **4**:413–428.

Riggs, S. R., 1982, Phosphatic bacteria in the Neogene phosphorites of the Atlantic coastal plain-continental shelf system, *IGCP Project 156 "Phosphorites" Newsl.* **11**:34.

Scharpf, L. G., Jr, 1974, Transformations of naturally occurring organophosphorus compounds in the environment, in: *Environmental Phosphorus Handbook* (E. J. Griffith, A. Beeton, J. M. Spencer, and D. T. Mitchell, eds.), Wiley, New York, pp. 393–412.

Schoeninger, M. J., Moore, K. M., Murray, M. L., and Kingston, J. D., 1989, Detection of bone preservation in archaeological and fossil samples, *Appl. Geochem.* **4**:281–292.

Shaffer, G., 1986, Phosphate pumps and shuttles in the Black Sea, *Nature* **321**:515–517.

Solorzano, L., and Strickland, J. D. H., 1968, Notes and comments—Polyphosphate in seawater, *Limnol. Oceanogr.* **13**:515–518.

Soudry, D., 1979, Intervention de schizophytes dans la phosphomicritisation des débris osseux, *C. R. Acad. Sci. Paris* **288D**:669–671.

Soudry, D., and Champetier, Y., 1983, Microbial processes in the Negev phosphorites (southern Israel), *Sedimentology* **30**:411–423.

Soudry, D., and Lewy, Z., 1988, Microbially influenced formation of phosphate nodules and megafossil moulds (Negev, Southern Israel), *Palaeogeogr. Palaeoclimatol. Palaeoecol.* **64**:15–34.

Soudry, D., and Southgate, P., 1989, Ultrastructure of a Middle Cambrian primary nonpelletal phosphorite and its early transformation into phosphate vadoids: Georgina Basin, Australia, *J. Sediment. Petrol.* **59**:53–64.

Southgate, P. N., 1986, Middle Cambrian phosphatic hardgrounds, phoscretes profiles and stromatolites, and their implications for phosphogenesis, in: *Phosphate Deposits of the World*, Volume 1, *Proterozoic and Cambrian Phosphorites* (P. J. Cook and J. H. Shergold, eds.), Cambridge University Press, Cambridge.

Speyer, S. E., and Brett, C. E., 1988, Taphofacies models for epeiric sea environments: Middle Paleozoic examples, *Palaeogeogr. Palaeoclimatol. Palaeoecol.* **63**:225–262.

Suess, E., 1981, Phosphate regeneration from sediments of the Peru continental margin by dissolution of fish debris, *Geochim. Cosmochim. Acta* **302**:603–606.

Turner, J. T., and Ferrante, J. G., 1979, Zooplankton fecal pellets in aquatic ecosystems, *Bioscience* **29**:670–677.

Tuross, N., Behrensmeyer, A. K., Eanes, E. D., Fisher, L. W., and Hare, P. E., 1989, Molecular preservation and crystallographic alterations in a weathering sequence of wildebeest bones, *Appl. Geochem.* **4**:261–270.

Van Cappellen, P., and Berner, R. A., 1988, A mathematical model for the early diagenesis of phosphorus and fluorine in marine sediments: Apatite precipitation, *Am. J. Sci.* **288**:289–333.

Wiebe, P. H., Boyd, S. H., and Winget, C., 1976, Particulate matter sinking to the deep-sea floor at 2000 m in the Tongue of the Ocean, with a description of a new sedimentation trap, *J. Mar. Res.* **34**:341–354.

Williams, C. T., 1989, Trace elements in fossil bone, *Appl. Geochem.* **4**:247–248.

Chapter 9

Carbonate Precipitation and Dissolution

Its Relevance to Fossil Preservation

DONALD E. CANFIELD and ROBERT RAISWELL

DONALD E. CANFIELD • NASA–Ames Research Center, MS 239–4 Moffett Field, California 94035; *present address*: Georgia Institute of Technology, School of Earth and Atmospheric Science, Atlanta, Georgia 30332-0340. ROBERT RAISWELL • Department of Earth Sciences, University of Leeds, Leeds LS2 9JT, United Kingdom.

Taphonomy: Releasing the Data Locked in the Fossil Record, Volume 9 of *Topics in Geobiology*, edited by Peter A. Allison and Derek E. G. Briggs, Plenum Press, New York, 1991.

1. Introduction

Quite extraordinary levels of paleobiological detail can occur as a result of the three-dimensional preservation of soft-bodied organisms by early diagenetic carbonate cementation. For example, the Mazon Creek siderite nodules contain superbly preserved jellyfish and siphonophores, yolk sacs on larval fish, color markings on bivalves, and delicate radulae on chitons and cephalopods (Baird *et al.*, 1985a,b). Such concretionary carbonates are *usually* found in organic carbon-bearing sediments and must clearly form early, close to the sediment–water interface, before any compaction occurred. This near-surface environment is readily accessible in modern sediments and hence has been the subject of detailed geochemical study for more than 30 years (e.g., Rittenberg *et al.*, 1955; Berner, 1971; Nissenbaum *et al.*, 1972; Froelich *et al.*, 1979; Westrich, 1983; Canfield, 1988). Considerable variations in the state of carbonate saturation have been observed both at the sediment surface (within the same basin) and with depth beneath the sediment–water interface (e.g., Berger and Soutar, 1970; Sholkovitz, 1973) and it is now possible to identify the processes capable of causing carbonate precipitation and dissolution. This suggests that it should be a relatively simple matter to predict the preservation potential of ancient sediments; however, caution is needed because kinetic effects (e.g., rate of burial) can either enhance or diminish the influence of equilibrium factors (e.g., degree of undersaturation), sometimes with dramatic consequences. For example, it is an intriguing fact that there are simply no modern analogues for the concretionary carbonates found in ancient sediments, although these clearly originate by early diagenetic microbial processes. The microbial decay of organic matter is generally believed to be responsible for the formation of those diagenetic carbonates associated with unusually well-preserved faunas. Chemical reactions resulting from organic-matter degradation also govern the degree of preservation/dissolution of skeletal carbonate. It is the purpose here to present a quantitative synthesis of geochemical information which will facilitate predictions on how, where, and when carbonates are dissolved/precipitated. Synthesis and modelling will therefore concentrate on organic carbon-bearing elastic sediments. This approach will be generally relevant to organic carbon-bearing carbonate muds, but not otherwise to carbonate sediments [the cementation of which involves altogether different processes (Bathurst, 1975)]. Particular, but not exclusive attention will be paid to marine sediments because of their predominance in the geological record.

2. Carbonate Diagenesis in Modern Clastic Sediments

2.1. Carbonate Equilibria

The carbonate species found dissolved in natural waters are carbon dioxide (CO_2), carbonic acid (H_2CO_3), bicarbonate (HCO_3^-) and carbonate

(CO_3^{2-}), which are related to one another through the following reactions:

$$CO_2 + H_2O \leftrightarrow H_2CO_3 \tag{1}$$

$$H_2CO_3 \leftrightarrow H^+ + HCO_3^- \tag{2}$$

$$HCO_3^- \leftrightarrow H^+ + CO_3^{2-} \tag{3}$$

The distribution of species in seawater is quantified through the following equilibrium relations:

$$K'CO_2 = \frac{[H_2CO_3]}{{}_{\rho}CO_2} \tag{4}$$

$$K'_1 = \frac{[HCO_3^-]\,\{H^+\}}{[H_2CO_3]} \tag{5}$$

$$K'_2 = \frac{[CO_3^{2-}]\,\{H^+\}}{[HCO_3]} \tag{6}$$

where $K'CO_2$ is the apparent equilibrium constant for the hydration of CO_2 to carbonic acid [equation (1)], and K_1' and K_2' are the apparent equilibrium constants for the first and second dissociations of carbonic acid [equations (2), (3)]. For seawater the convention is to use apparent equilibrium constants, which incorporate concentrations for the carbonate species and activities for H^+. This is opposed to the "normal" approach used in calculations of equilibria, where activities (the products of concentration and activity coefficient) are used for all dissolved species. The apparent equilibrium constant convention has been adopted because seawater is an essentially constant ionic medium, with activity coefficients that do not vary from place to place in either seawater or marine sediments. Apparent equilibrium constants are a function of temperature and pressure (Table I lists values at 25°C and 1 atm pressure), and it is clear that as pH ($=-\log_{10}\{H^+\}$) increases, the dominant dissolved carbonate species shifts from H_2CO_3 (at pH < 6.0), to HCO_3^- (6.0 < pH < 9.1), to CO_3^{2-} (pH > 9.1) (Fig. 1).

Calcium carbonate dissolves to give a calcium and a carbonate ion,

$$CaCO_3 \leftrightarrow Ca^{2+} + CO_3^{2-} \tag{7}$$

with the appropriate equilibrium expression defined as follows:

$$K'CaCO_3 = [Ca^{2+}][CO_3^{2-}] \tag{8}$$

As with dissolved carbonate species in seawater, an apparent equilibrium constant is used ($K'CaCO_3$), with the square brackets denoting concentrations. Values of $K'CaCO_3$ for calcite and aragonite are given in Table I. From equation (8), any process which increases the carbonate ion concentration

TABLE I. Values for Apparent Equilibrium Constants at 25°C, 1 atm Pressure, and 35% Salinity

$K'CO_2$	2.84×10^{-2}	Weiss (1974)
K'_1	9.99×10^{-7}	Mehrbach *et al.* (1973)
K'_2	7.68×10^{-10}	Mehrbach *et al.* (1973)
$K'CaCO_3$ (calcite)	4.30×10^{-7}	Mucci (1983)
$K'CaCO_3$ (aragonite)	6.65×10^{-7}	Mucci (1983)

$[CO_3^{2-}]$ will increase the value of $[Ca^{2+}][CO_3^{2-}]$, which is termed the ion molarity product (IMP). Variations in $[CO_3^{2-}]$ are more common than in $[Ca^{2+}]$, as the latter is essentially constant in seawater, though some variability may be found in pore waters. When IMP $> K'CaCO_3$ the solution is supersaturated and $CaCO_3$ precipitation may occur; when IMP $= K'CaCO_3$ the solution is just at saturation, and when IMP $< K'CaCO_3$ the solution is undersaturated and carbonate dissolution may occur.

Whether calcium carbonate dissolves or is stabilized and preserved in marine waters and sediments depends to a large degree on the carbonate saturation state of the water surrounding the carbonate. Also, numerous other factors control carbonate preservation, including carbonate mineralogy (e.g., aragonite is a more soluble mineral than calcite; Table I), predation by fauna and colonization by flora, and the depth of water where the carbonate is deposited.

2.2. Biological Influences on Carbonate Dissolution

In both siliciclastic and carbonate-rich continental margin marine sediments, carbonate dissolution may result directly from the activities of

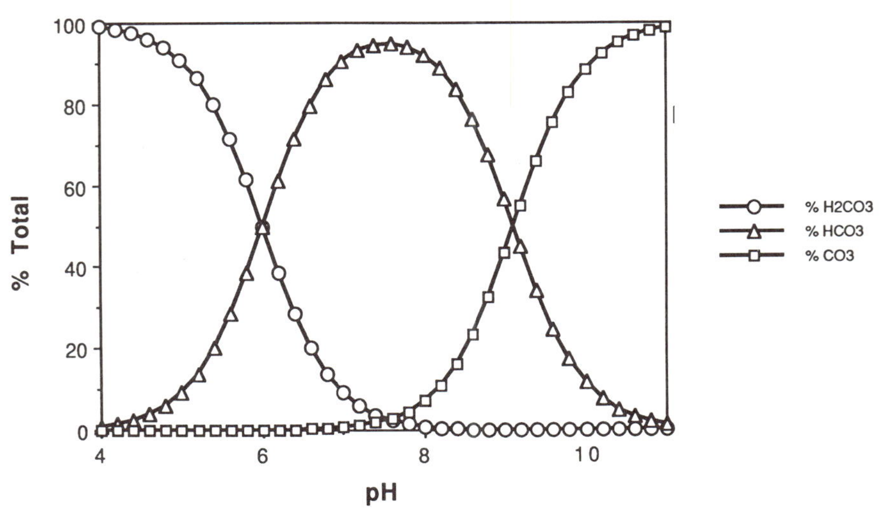

FIGURE 1. The relative importance of each of the carbonate species—carbonic acid (which is defined by convention to include both carbonic acid, H_2CO_3, and dissolved carbon dioxide, CO_2), bicarbonate (HCO_3^-), and carbonate (CO_3^{2-})—as a function of pH. At low pH nearly 100% of the dissolved inorganic carbon is present as carbonic acid. As pH rises, bicarbonate becomes more important, and at high pH all inorganic carbon exists as carbonate. Figure is drawn from the equilibrium values in Table I.

both micro- and macroorganisms (Milliman, 1974; Bathhurst, 1975), though the discussion here will focus on dissolution in siliciclastic sediments only. It is important that dissolution by organisms is independent of the carbonate saturation state of the surrounding waters. Some mollusks and sponges bore through shells to expose the biomass within, the former for feeding purposes, the latter to create a dwelling (Cobb, 1969; Day, 1968; Carriker *et al.*, 1967; Carriker, 1969; Chétail and Fournié, 1969; and references therein). For example, gastropods secrete acidic substances to soften the shell of their prey, scraping away the softened shell with their radula (Day, 1968; Carriker *et al.*, 1967; Carriker, 1969; Chétail and Fournié, 1969). Sponges bore by dissolving the shell with a chemical secretion (Cobb, 1969), although there is also evidence that some may excavate by purely mechanical means (Ansell and Nair, 1969). Some macroorganisms (including fishes and some benthic fauna) ingest whole organisms during feeding, while others (e.g., starfish) prise shells open. These processes may affect carbonate preservation by breaking up the shell material (Ginsberg, 1957; Hobson, 1965). Acid dissolution of shelly carbonate in the digestive tract may also occur (Emery *et al.*, 1954; Mayor, 1924), although the amount of carbonate so dissolved may be minimal (Emery *et al.*, 1954; Milliman, 1974).

Microorganisms (algae, fungi, and bacteria) may also promote carbonate dissolution by boring into shells (Kobluk and Risk, 1977; Alexandersson, 1972; Golubic, 1969; Schroeder, 1972; Perkins and Halsey, 1971; Kohlmeyer, 1969; Srivastava, 1975; Milliman, 1974). Algal activity may be quite extensive, creating vast networks of channels that may penetrate up to 1 mm or more (Golubic, 1969; Alexandersson, 1972); however, their boring is restricted to the photic zone. Fungi may bore into shelly carbonate in water depths of up to 800 m (Perkins and Halsey, 1971), apparently to obtain energy from oxidizing the organic material within the shell matrix (Milliman, 1974). In a comprehensive report on carbonate preservation in several marine basins, Srivastava (1975) concluded that microorganisms accounted for only a small amount of calcium carbonate dissolution. In other environments boring microorganisms may cause more extensive dissolution, but, in many instances, the dissolving carbonate may also be replaced by another calcium carbonate phase, preserving the original shell morphology (Alexandersson, 1972).

2.3. Mineralogical Influences on Dissolution

An important factor influencing carbonate preservation is the mineralogy of carbonate phase. As an example, in sediments from the North Sea, the aragonite shells of *Nucula* may be extensively dissolved, whereas calcite molluscan shells show no evidence of corrosion (Srivastava, 1975). Also, in sediments of the Agean Sea (Srivastava, 1975) the proportion of aragonite in shells of the pelecypod *Pteria*, which contain both aragonite and calcite, decreases with sediment depth, providing evidence of preferential aragonite dissolution. In the deep sea, aragonite becomes undersaturated and subsequently lost from sediments at considerably shallower water depths than calcite (Murray and Renard, 1891; Chen, 1964; Berner *et al.*, 1976; Broecker and Peng, 1982). In fact, in some surface ocean waters aragonite production by pteropods may be quite high, with no aragonite found buried in the underlying sediment, though calcite may be abundant (Berner and Honjo, 1981).

The amount of Mg which precipitates in the crystal lattice of skeletal calcite is another factor which affects calcite solubility (Chave, 1952). The Mg content of calcite varies between species and is latitude dependent, presumably responding to rates of calcite precipitation (Chave, 1954; Chave *et al.*, 1962). In all experimental studies, calcite solubility increases with its magnesium content [see Mackenzie *et al.* (1983) for a review], making high-Mg calcite metastable relative to low-Mg calcite. Despite this metastability, transformations between high- and low-Mg calcite under early diagenetic marine conditions have not been reported, though such transformations readily occur during later diagenesis and subaerial exposure (Turekian and Armstrong, 1961; Gavish and Friedman, 1969; Berner, 1971).

Shell material of the same mineralogy produced by different organisms may be variably resistant to carbonate dissolution, further affecting carbonate preservation (Arrhenius, 1952; Berger, 1971; Hsü and Andrews, 1970; A. McIntyre and McIntyre, 1971; Berger and von Rad, 1972; Berger, 1976; Walter and Morse, 1984; Wu and Berger, 1989, and references therein). Arrhenius (1952), for example, observed that as planktonic foraminifers of the same mineralogy undergo progressively more extensive dissolution in deep-sea sediments, the shells become fragmented and only a few dissolution-resistant species are left intact. Resistance to dissolution results from a number of factors, including shell wall thickness, distribution of organic matter (such as a periostracum and organic matrix), and shell porosity. More porous shells, with more specific surface area for attack by corrosive fluids, dissolve faster than less porous shells (Berger, 1971; Walter and Morse, 1984; Wu and Berger, 1989).

3. Microbially Mediated Carbonate Saturation

3.1. Model Development

The preceding observations demonstrate that carbonate preservation in marine sediments is complex, being affected by a number of variables, including predation, pore-water and water-column chemistry, shell mineralogy, shell morphology, and shell dissolution kinetics. However, despite these numerous variables, the saturation state of the water contacting carbonate is of fundamental importance in controlling carbonate preservation. In what follows, a model is developed which predicts how carbonate saturation state responds to the progression of organic carbon oxidation reactions that occur during early diagenesis. As such, different diagenetic environments are shown to have different carbonate preservation potential.

The model starts with a parcel of marine water, saturated with oxygen, which undergoes a series of carbon oxidation reactions beginning with oxic respiration, proceeding to denitrification, sulfate reduction, methane oxidation by sulfate reduction, and finally methane production. Organic matter is assumed to decompose with a stoichiometry characterized by the Redfield ratio of C/N/P equal to 106/16/1. This is a typical composition for marine plankton (Redfield, 1942; Redfield *et al.*, 1963; Richards *et al.*, 1965), but an obvious oversimplification, as variations from Redfield behavior are observed in the release of carbon, nitrogen, and phosphorus from decomposing organic matter (e.g., Richards *et al.*, 1965; Nissenbaum *et al.*, 1972; Sholkovitz, 1973; Berner, 1977; Aller, 1977; Martens *et al.*, 1978; Broecker and Peng, 1982). Model results, however, are not particularly sensitive to reasonable variations in the release ratio of C/N/P. Also, iron and manganese reduction are not included in the model because the overall importance of these processes in marine sedimentary organic-matter oxidation is not yet clear, and may not be large (Bender and Heggie, 1984; Henrichs and Reeburgh, 1987). Iron and manganese reduction may, however, be important in freshwater sediments (see Section 3.7).

Following the approach of Ben Yaakov (1973) and those of Gardner (1973) and Thorstenson (1970), seawater is assumed to be composed of a number of different ionic species, some of which are protolytic (e.g., ammonia, carbonate, borate, and sulfide) and exist in seawater in different chemical forms which are related to one another by association/dissociation reactions involving protons (H^+) and hydroxide ions (OH^-). Others are nonprotolytic (e.g., calcium, magnesium, sulfate, and chloride) and are not involved in acid/base reactions. The dominant chemical form, and hence charge, held by any protolytic species is pH dependent. An example is the carbonate system (Fig. 1), where dissolved inorganic carbon may exist as the doubly charged carbonate ion (CO_3^{2-}), the singly charged bicarbonate ion (HCO_3^-), and an uncharged species (H_2CO_3).

Solution pH and its subsequent effect on carbonate saturation state are controlled by the stoichiometries of the various organic-matter oxidation reactions (see below) and the necessary constraint of charge balance. As an example, during sulfate reduction, the negative charge lost by the reduction of sulfate ions must be balanced by the charge added from the products of sulfate reduction, including carbonate, sulfide, ammonia, and phosphate. In modeling, the products of the various carbon oxidation reactions are added to the solution as the respective oxidants become consumed. Assuming equilibrium among acid/base pairs, the pH of the solution is adjusted until charge balance is achieved. Once solution pH is obtained, the concentration of carbonate ion (CO_3^{2-}) is determined and the ion molarity product (IMP) is calculated. The relative sizes of the IMP with respect to the apparent equilibrium constants for calcite and aragonite ($K'CaCO_3$; see Table I) determine the saturation state of the solution.

The model developed here assumes a closed system and does not account for the differential transport of solutes as may occur by molecular diffusion in marine sediments [see Boudreau and Canfield (1988) for a more rigorous approach]. This is, however, probably not a serious difficulty, and the results should be at least qualitatively correct, particularly as the effects of differential diffusion can be approximated by altering the stoichiometries of the oxidation reactions to fit pore-water profiles of oxidation products (Section 3.4). Model results should be directly applicable to ocean waters because large-scale mixing by eddy diffusion, as occurs in the ocean (Sverdrup *et al.*, 1942), overwhelms transport by molecular diffusion.

Finally, the effects of carbonate dissolution and precipitation, as driven by the saturation state of the water, are not included. However, carbonate dissolution and precipitation only act to reduce the magnitudes of under- or oversaturation, and the model should still predict whether under- or oversaturation occurs.

3.1.1. Oxic Respiration

The following reaction stoichiometry (Richards *et al.*, 1965) describes the oxic decomposition of organic matter:

$$(CH_2O)_{106}(NH_3)_{16}H_3PO_4 + 138O_2 \rightarrow 106CO_2 + 122H_2O + 16NO_3^- + 16H^+ + H_3PO_4 \qquad (9)$$

A "standard" initial seawater composition is assumed with total dissolved carbonate equal to 2.2 mM, phosphate equal to 1 μM, oxygen equal to 300 μM, and borate equal to 420 μM (Broecker and Peng, 1982) (borate must be considered because, although not involved in organic-matter diagenesis, it is a protolytic species and contributes to the pH buffering of seawater). With initial pH values of 7.85, 8, and 8.15 [a reasonable range for surface seawater (Horne, 1969)], IMP is plotted in Fig. 2 as a function of progressive oxygen consumption [see also results of Gardner (1973)]. These compositions produce an initial oversaturation with respect to both aragonite and calcite, but, not surprisingly, as oxygen becomes consumed, solution pH and carbonate saturation drop due to CO_2 addition and the oxidation of ammonia to nitrate [equation (9)]. As oxygen consumption becomes complete, with the whole range of initial pH values, the solution becomes undersaturated with respect to aragonite. Calcite undersaturation is observed for initial pHs of 8 and below.

3.1.1.a. Field Observations. Model results suggest that in marine waters and pore waters with starting values of pH, oxygen, and total dissolved carbonate similar to those used here, oxygen consumption will lead to carbonate undersaturation and dissolution. It is well known that in deep ocean waters and sediments, both calcite and aragonite undersaturation occurs. Undersaturation results, in a large part, from organic-matter

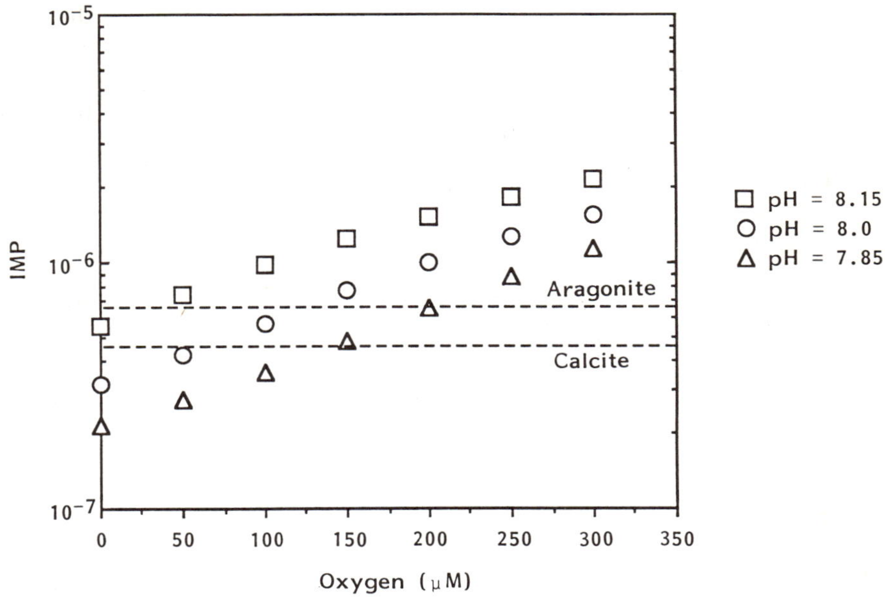

FIGURE 2. Carbonate saturation expressed by ion molarity product (IMP) as a function of progressive oxygen utilization with different starting values of solution pH.

oxidation by oxygen and subsequent CO_2 buildup (Sverdrup *et al.*, 1942; Peterson, 1966; Berger, 1967, 1970; Ben Yaakov *et al.*, 1974; Emerson *et al.*, 1980, 1982; Broecker and Peng, 1982; Boudreau, 1987), and also, to a lesser extent, by the effects of decreasing temperature and increasing pressure on calcite solubility (W. G. McIntyre, 1965; Millero and Berner, 1972; Ingle, 1975; Broecker and Takahashi, 1978). Carbonate undersaturation leads to dissolution, and at a rate which is a rather complex function of the degree of undersaturation [see Morse (1983) for a review]. Where the rate of carbonate dissolution is greater than the delivery rate from settling particles, dissolution is complete and sediments are left barren of calcium carbonate (e.g., Biscaye *et al.*, 1976; Berger, 1976). In the present-day oceans carbonate-free sediments are found at depths greater than about 4–6 km, depending on location [see Broecker and Peng (1982) for a review], with the boundary between carbonate-containing and carbonate-free sediments termed the carbonate compensation depth (CCD). An example of the change in sedimentary carbonate contents that occurs with water depth in the Atlantic Ocean is shown in Fig. 3.

Another example where oxic organic-matter decomposition leads to carbonate undersaturation is Lake Nitinat, a permanently euxinic (lacking bottom-water oxygen) fjord located on Vancouver Island, British Columbia, Canada. Data from Richards *et al.* (1965) were used to calculate IMP values as a function of progressive oxygen consumption and are shown along with

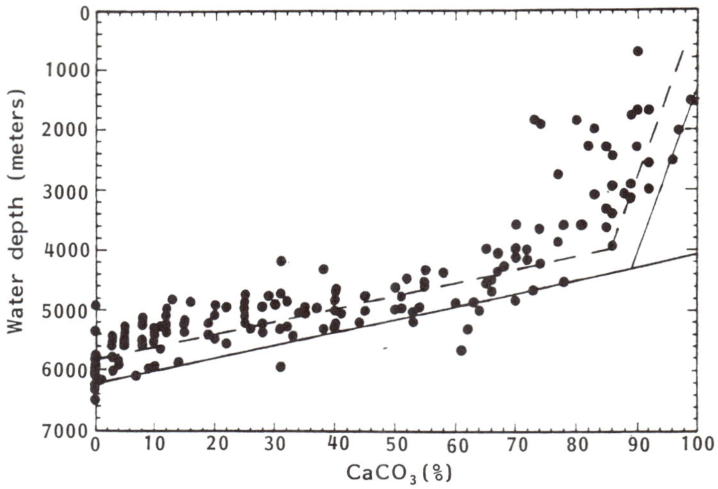

FIGURE 3. Calcium carbonate content as a function of water depth for sediment from the North American Basin in the Atlantic Ocean. [From Biscaye *et al.* (1976). Reproduced with permission of the publisher.]

the model results in Fig. 4. It is obvious that Lake Nitinat becomes undersaturated with respect to both calcite and aragonite as a result of oxic organic-matter decomposition, and by an extent more or less predicted by the model. It is cautioned, however, that not all dysaerobic and euxinic water bodies will become undersaturated with calcium carbonate. For

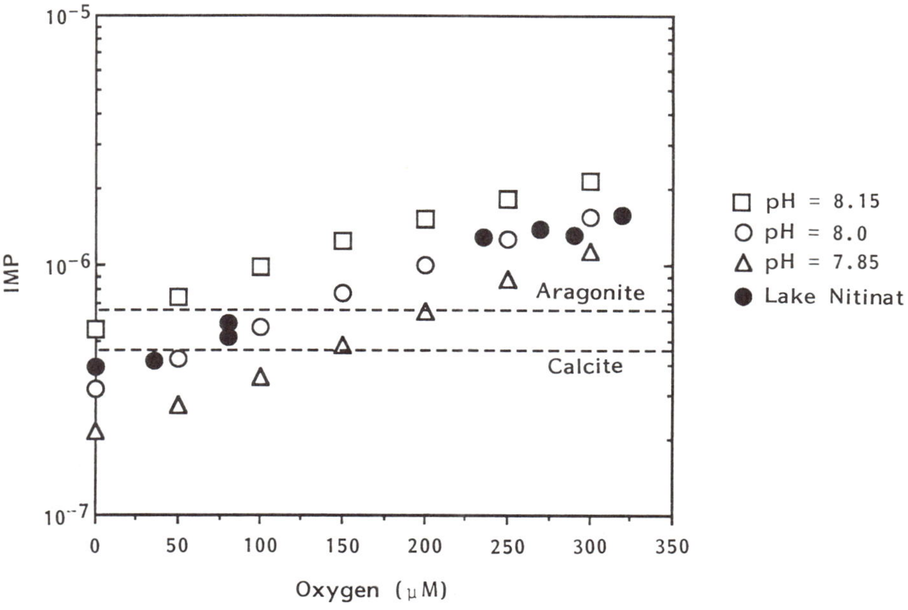

FIGURE 4. Carbonate saturation expressed by ion molarity product (IMP) from Lake Nitinat compared to model results from Fig. 2.

example, rivers draining into the Black Sea are relatively alkaline, resulting in high surface-water pH values and total inorganic carbon concentrations (Grasshoff, 1976). Hence, Black Sea water is better buffered than "normal" seawater and oxic decomposition does not lower pH to the extent of calcium carbonate undersaturation (Alekin and Moricheva, 1966).

3.1.2. Denitrification

Denitrification is the next important process of organic matter decomposition (e.g., Froelich *et al.*, 1979), responsible in some sediments for up to 10% or more of the organic carbon oxidation (Bender and Heggie, 1984; Henrichs and Reeburgh, 1987; Jørgensen, 1983). Assuming Redfield stoichiometry for the decomposing organic matter, one may write the following reaction (Richards *et al.*, 1965):

$$(CH_2O)_{106}(NH_3)_{16}H_3PO_4 + 84.8HNO_3 \rightarrow 106CO_2 + 42.4N_2 + 148.4H_2O + 16NH_3 + H_3PO_4 \qquad (10)$$

To model denitrification, initial starting conditions (pH, total inorganic carbon, etc.) are chosen to be the same as those determined at the point of oxygen depletion in the oxic decomposition model. An initial nitrate concentration of 51 μM is also assumed, which is the value predicted from equation (9) assuming an initial O_2 concentration of 300 μM and a nitrate concentration of 16 μM. This falls within the range of maximum nitrate concentrations (about 20–60 μM) normally found in marine sediments (Froelich *et al.*, 1979; Klinkhammer, 1980; Jahnke *et al.*, 1982; Aller *et al.*, 1985; Heggie *et al.*, 1987), but is higher than maximum values measured in euxinic basins (Richards *et al.*, 1965; Grasshoff, 1976). Model results (Fig. 5) show that denitrification more or less maintains the carbonate saturation state inherited at the completion of oxic decomposition.

3.1.3. Sulfate Reduction

Sulfate reduction accounts for about one-half of the carbon oxidation in nearshore sediments, becoming progressively less important as sedimentation rate falls (Jørgensen, 1982; Canfield, 1989). The decomposition of organic matter by sulfate reduction is expressed by

$$(CH_2O)_{106}(NH_3)_{16}H_3PO_4 + 53SO_4^{2-} \rightarrow 106HCO_3^- + 53H_2S + 16NH_3 + H_3PO_4 \qquad (11)$$

The sulfate reduction model begins with the same water chemistry as at the completion of the pH = 8.0 denitrification model (Fig. 5). Chemical species are added to solution by equation (11), with the additional consideration that one of the species, sulfide, may react to form solid iron sulfide phases, which also affects solution pH and carbonate saturation (Ben Yaakov, 1973; Gardner, 1973; Boudreau and Canfield, 1988). Model results including the effects of sulfide reaction on solution pH are shown in Fig. 6. Consistent

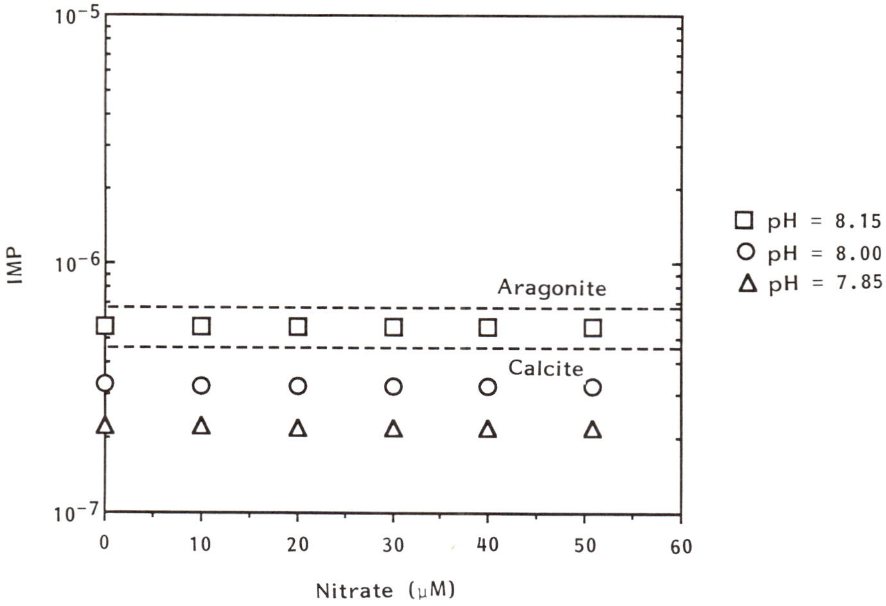

FIGURE 5. Carbonate saturation expressed by ion molarity product (IMP) as a function of nitrate utilization by denitrification. pH values are those used at the beginning of the oxygen utilization model (Fig. 2; see text for details).

with field observations (Sholkovitz, 1973; Berger and Soutar, 1970; Murray *et al.*, 1978; Aller, 1982), bicarbonate addition from sulfate reduction causes supersaturation with respect to both calcite and aragonite. However, carbonate undersaturation is observed in the initial stages of sulfate reduction, and by an extent which is dependent on the degree to which sulfide reacts from solution to form solid sulfide phases. For example, if sulfide does not react from solution and just accumulates, the solution becomes fairly acidic (the model predicts pH values of just below 7) and carbonate undersaturation may persist until 5–7 mM of sulfate has been reduced (supplying 10–14 mM of carbonate alkalinity). When sulfide reacts completely from solution (Canfield and Raiswell, this volume, Chapter 7, Section 5.1) the solution is less acidic and supersaturation occurs after only about 0.2–0.3 mM of sulfate has been reduced.

An intermediate case is also considered where the sulfide produced from the first 2 mM of reduced sulfate completely reacts from solution. After the initial 2 mM of sulfate is reduced, 25% of the remaining sulfide produced builds into solution. The 75% sulfide that does not accumulate may be viewed as lost from solution. Alternately, even without reaction, sulfide will accumulate more slowly in sediments than sulfate concentrations decrease, because the diffusion coefficient for sulfide is greater than for sulfate (Boudreau and Canfield, 1988). Model results will fit with either scenario for sulfide loss. The "intermediate" model is designed to mimic

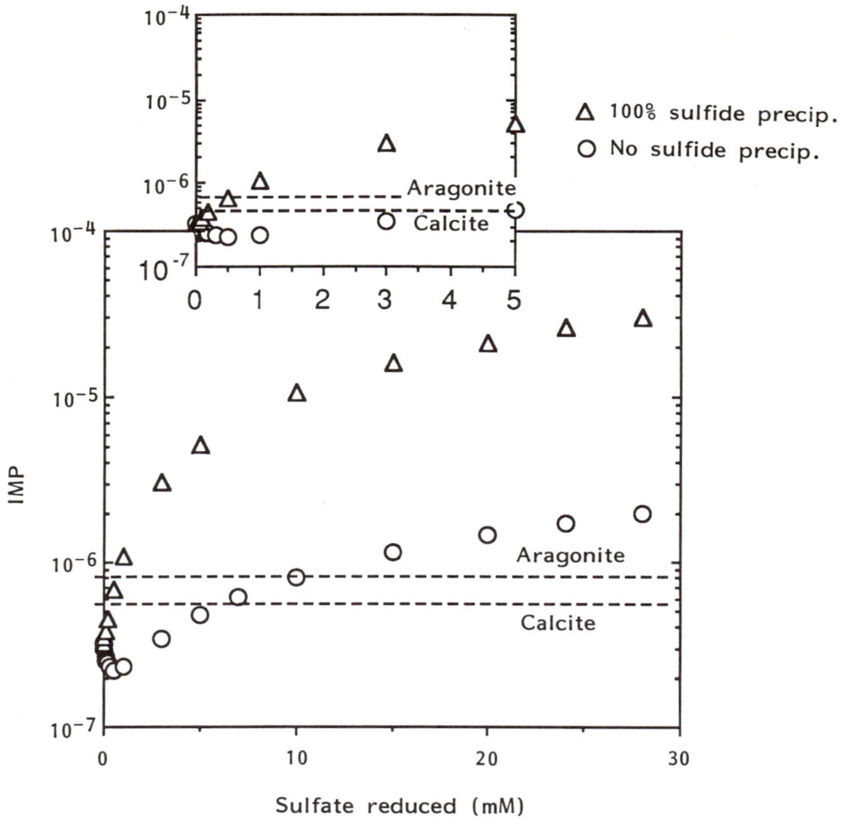

FIGURE 6. Results from the sulfate reduction model considering two cases, one where all of the sulfide produced accumulates into solution, and one where all of the sulfide precipitates. The inset is an expanded scale of the first few mM of sulfate reduced. Carbonate saturation expressed by ion molarity product (IMP).

continental margin sediments where, in the near surface, iron oxides are present and react sulfide from solution (Canfield, 1989; Canfield and Raiswell, this volume, Chapter 7, Fig. 7). When iron oxides become consumed with sediment depth, sulfide builds into solution (Berner, 1969; Canfield, 1988). With the "intermediate" model, dissolved sulfide is at a concentration of about 6.5 mM when sulfate is depleted; a sulfide level that is fairly typical for anoxic, organic-rich continental margin sediments.

Results from the "intermediate model" (Fig. 7) show that carbonate supersaturation is quickly established as sulfate is depleted, just as for the "100% sulfide precipitation" model. However, because 25% of the subsequent sulfide produced builds into solution, final supersaturation is intermediate between the "100% sulfide precipitation" model and the "no sulfide precipitation" model.

The process of sulfide oxidation, which occurs to a great extent in continental margin sediments (Jørgensen, 1977), may be a further factor

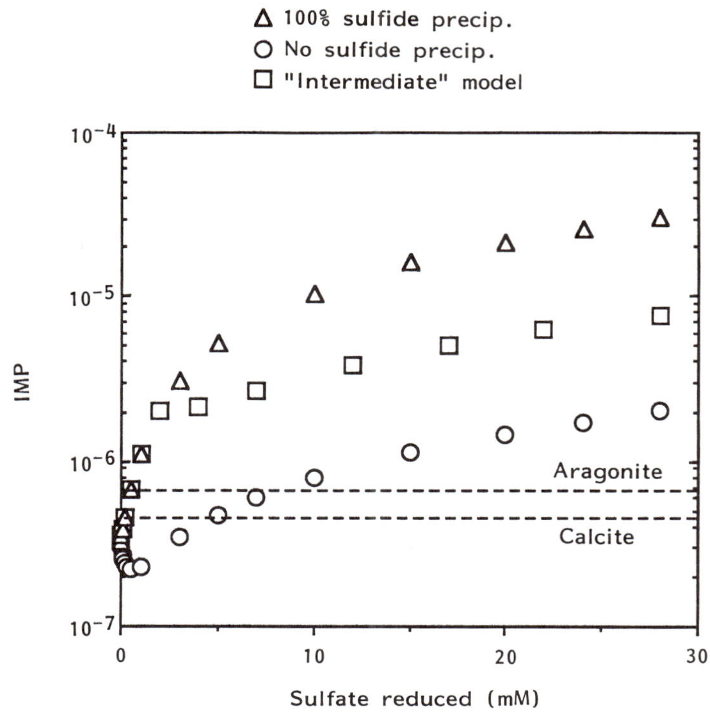

FIGURE 7. An "intermediate" model is added to the two models in Fig. 6. For the first 2 mM sulfate reduced all of the sulfide is assumed to react from solution. For the remaining sulfate reduced, 25% of the produced sulfide accumulates. This scenario is similar to many anoxic continental margin marine sediments. Carbonate saturation expressed by ion molarity product (IMP).

affecting carbonate saturation because acidity is generated that is potentially corrosive to carbonate minerals (Berger and Soutar, 1970; Aller, 1982; Reaves, 1986). In a steady state, that is, where chemical gradients, rates of decomposition processes, and sedimentation rates have been constant for an extended period, the net effect of sulfate reduction followed by sulfide oxidation is to generate CO_2, as in oxic decomposition. This is demonstrated by summing the sulfate reduction (in its simplified form) and sulfide oxidation reactions [equations (12) and (13)] to yield equation (14):

$$SO_4^{2-} + 2CH_2O \rightarrow 2HCO_3^- + H_2S \qquad (12)$$

$$H_2S + 2O_2 \rightarrow SO_4^{2-} + 2H^+ \qquad (13)$$

$$2CH_2O + 2O_2 \rightarrow 2CO_2 + 2H_2O \qquad (14)$$

The overall effect of steady-state sulfide oxidation on carbonate saturation is difficult to evaluate, however, because bioturbation, which mixes oxygen into the sediment (thus promoting sulfide oxidation), also mixes out the CO_2

generated by sulfide oxidation. Also, the oxygen used to oxidize sulfide will not be used to oxidize organic matter and, as both processes produce CO_2, it may not matter which process in fact consumes oxygen in steady state conditions. However, in non-steady-state situations, such as occur seasonally in salt marsh sediments and some nearshore sediments (Aller, 1982; Lord, 1980; Reaves 1986), net sulfide oxidation may occur [equation (12)], generating considerable amounts of acid and resulting in carbonate dissolution (Reaves, 1986). Because of the difficulty in quantifying the effects of sulfide oxidation on carbonate saturation in sediments, this process has not been included in the model, although it may be important in some instances.

3.1.3.a. Field Observations. Model results are compared with carbonate saturation for two sites in Long Island Sound (Connecticut): the FOAM (Friends of Anoxic Mud) site (Goldhaber *et al.*, 1977), which is located in 10 m of water and is moderately bioturbated in the top 10 cm, and the Northwest Control (NWC) site, which is in 15 m of water and is heavily irrigated to below 1 m depth (Aller, 1977, 1982; Goldhaber *et al.*, 1977; Krishnaswami *et al.*, 1984; Canfield, 1988). Beginning with FOAM, in the surficial 8–10 cm, sulfide is nearly absent from sediment pore waters (Canfield, 1989; Canfield and Raiswell, this volume, Chapter 7, Section 5.1), and sulfate is depleted by 1–2 mM compared to overlying water. Below, sulfide accumulates to a maximum of about 6 mM at the depth of sulfate depletion (Goldhaber *et al.*, 1977; Canfield, 1988), making FOAM porewater chemistry similar to the "intermediate" model discussed above.

In Fig. 8, results from the "intermediate" model are compared to carbonate saturation for FOAM, calculated using data from Aller (1977) and Canfield (1988). The agreement between model results and FOAM porewater chemistry is fairly good, with both showing initial undersaturation with aragonite (and possibly calcite, but to a lesser degree), and subsequent supersaturation with progressive sulfate depletion. The major discrepancy between model results and field measurements occurs in the initial stages of sulfate reduction, where the model overestimates the carbonate saturation state compared to actual sediment chemistry. In this zone the additional process of sulfide oxidation, as discussed above, may be contributing to a greater range of aragonite undersaturation (Fig. 8) than predicted by the model.

NWC differs from FOAM in that there is a large population of irrigating infauna (Aller, 1977) which continually exchange, to depth, overlying Long Island Sound water with diagenetically enriched pore water. As a result, even though sulfate reduction occurs at a similar rate to FOAM (Aller and Yingst, 1980; Berner and Westrich, 1985), sulfate concentrations are only reduced by about 0.5 mM in the top 30 cm (Aller, 1977; Westrich, 1983). Through this zone dissolved sulfide is reacted from solution to a very low concentration [about 20 μM or less (Canfield and Berner, 1987)], and so chemical diagenesis is similar to the "100% sulfide precipitation" model (Fig. 6). Sulfate depletion at NWC was estimated from the alkalinity data in

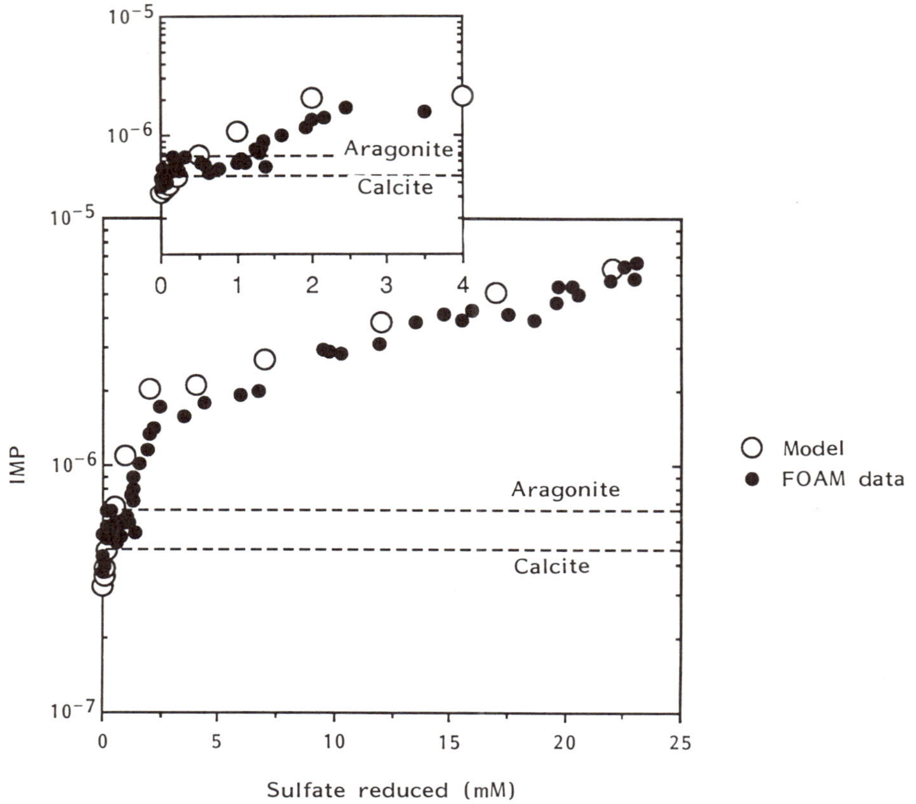

FIGURE 8. Results from the FOAM site are compared to the "intermediate" model. The inset shows an expanded scale where, during the early stages of sulfate reduction, more extensive undersaturation is observed at FOAM than is predicted by the model. Carbonate saturation expressed by ion molarity product (IMP).

Aller (1977). Alkalinity data were used because the measurement is much more precise than direct sulfate measurements, though sulfate depletion estimates from the alkalinity data are broadly comparable to those obtained directly from the sulfate data.

Carbonate saturation at NWC is reasonably close to results from the "100% sulfide precipitation" model (Fig. 9). Because the model considers only the effects of organic-matter diagenesis on carbonate saturation, the implication is that carbonate undersaturation at NWC is a natural consequence of the organic-matter decomposition process. Undersaturation, then, is caused by organisms whose rapid pore-water exchange allows only small amounts of sulfate depletion, and not enough carbonate alkalinity is accumulated to cause carbonate supersaturation (Figs. 6 and 7, also see Walter and Burton, 1990). This conclusion is somewhat different from that of Aller (1982), who favored sulfide oxidation as an important factor in creating undersaturation at NWC. Indeed, considerable sulfide oxidation

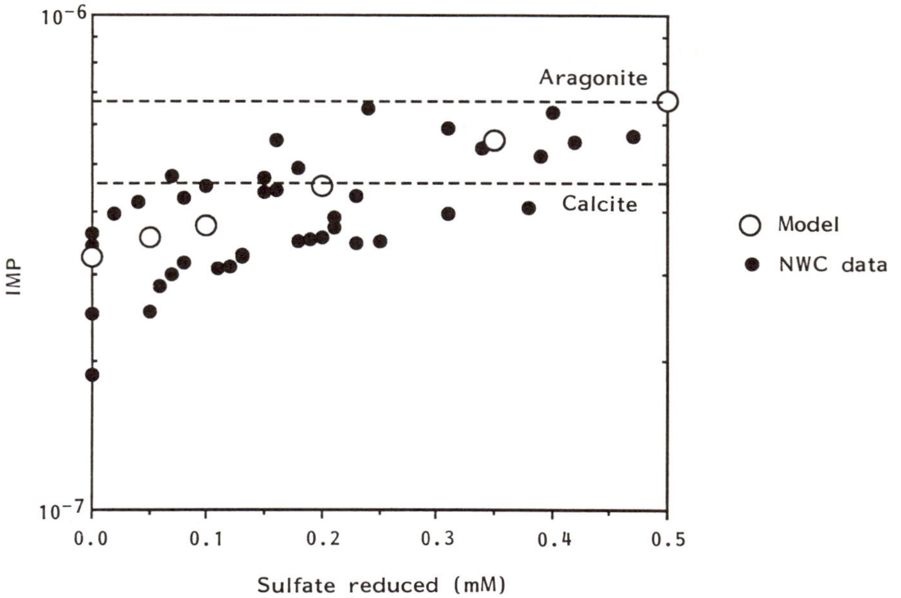

FIGURE 9. Results from Northwest Control (NWC) are comapred to results from the initial stages of the "100% sulfide precipitation" model. Carbonate saturation expressed by ion molarity product (IMP).

occurs at this site (Aller, 1982; Berner and Westrich, 1985) and may be a contributing factor in carbonate undersaturation. However, the real importance of sulfide oxidation is difficult to evaluate because the rapid pore-water exchange (as occurs here) will ventilate the CO_2 generated by sulfide oxidation [equations (12)–(14)].

Both NWC and FOAM show differing styles of chemical diagenesis and differing extents of carbonate undersaturation, which profoundly affect carbonate preservation. At NWC, where minimal sulfate depletion occurs and consequent carbonate undersaturation is observed to 30–60 cm depth, only about 10% of the carbonate produced is preserved (Aller, 1982). At FOAM, sulfate is readily depleted with sediment depth and carbonate supersaturation is established by 1–6 cm depth (Aller, 1977, 1982). Carbonate preservation is better here, with >50% of the carbonate production preserved [from data in (Aller, 1982)]. Carbonate preservation may in fact be quite good at FOAM since there is a general lack of etching and dissolution features on the buried shell material (D. A. Canfield, unpublished SEM observations), and because fairly fresh-looking, unfragmented benthic foraminiferal tests are preserved to make up a large fraction of the total carbonate. It is important that NWC and FOAM both have similar sedimentation rates of about 0.1 cm/year (Benoit *et al.*, 1979; Krishnaswami *et al.*, 1984), which allows roughly 300–600 years for dissolution to occur at NWC and only about 10–70 years at FOAM.

The two contrasting Long Island Sound sites are analogous to various locations in the Santa Barbara Basin (Berger and Soutar, 1970). For sediment from the central basin depression, underlying waters with low oxygen content, bioturbation is minimal, and pore-water chemistry is similar to the FOAM site (Canfield and Raiswell, this volume, Chapter 7, Fig. 7). In the initial stages of sulfate reduction, sulfide is completely reacted from solution, and sulfate becomes quickly depleted from sediment pore waters with depth (Sholkovitz, 1973; D. E. Canfield and C. E. Reimers, unpublished data). Also, surface sediments accumulate at the rapid rate of about 0.8 cm/year (Reimers *et al.*, 1990), allowing rapid transport of carbonate to sulfate-depleted (and alkalinity-enriched) pore waters. With this scenario of pore-water chemistry, good carbonate preservation might be expected, and in fact, excellent preservation is reported (Berger and Soutar, 1970). For sediment from basin margins, bottom-water oxygen levels are higher and extensive benthic irrigation allows very little sulfate depletion (<1 mM) to depths greater than 30 cm (Sholkovitz, 1973). By analogy with NWC, carbonate undersaturation would be expected, and indeed, pervasive carbonate dissolution does occur (Berger and Soutar, 1970).

3.1.4. Methane Oxidation by Sulfate Reduction

In anoxic marine sediments, particularly those with rapid sediment deposition, methane is produced by fermentation reactions at depths below sulfate depletion (Rittenberg *et al.*, 1955; Nissenbaum *et al.*, 1972; Claypool and Kaplan, 1974). The methane diffuses upward to become the main source of organic carbon for sulfate-reducing bacteria at the sulfate/methane interface, where a peak in sulfate reduction rate often results (Fig. 10) (Devol, 1983; Iversen and Jørgensen, 1985; Alperin, 1988; Alperin *et al.*, 1988). In some cases, methane may contribute to up to 20% of the total organic carbon used by sulfate-reducing bacteria (Devol, 1983; Alperin, 1988; Henrichs and Reeburgh, 1987), though in other sediments the proportion may be less (Iversen and Jørgensen, 1985).

The anaerobic oxidation of methane by sulfate reduction does not involve the release of nutrients, and the reaction may be written as follows:

$$CH_4 + 2H^+ + SO_4^{2-} \rightarrow H_2S + 2H_2O + CO_2 \qquad (15)$$

Model results were generated by modeling sulfate reduction as discussed above and then switching to methane as an organic source. Two cases were explored; one where 50% of the sulfate depletion occurred by anaerobic methane oxidation, and in the other, 10%. These values were chosen to reflect maximum and typical contributions of methane for sulfate reduction in anoxic sediments. Results (Fig. 11) demonstrate that methane oxidation causes a rather abrupt increase in carbonate supersaturation. The extent of the increase is sensitive to the preexisting carbonate saturation state, being more substantial the more supersaturated the solution. It is important that model results may reflect maximal effects. This is because, in sediments,

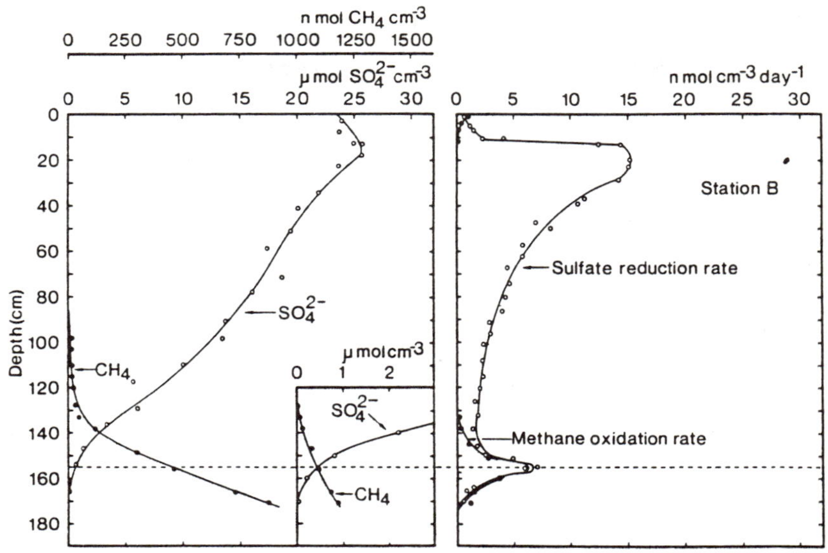

FIGURE 10. Anaerobic methane oxidation is observed at the interface between sulfate and methane in sediments from coastal Denmark (Iversen and Jørgensen, 1985). Enhanced sulfate reduction is also observed in the same region as high methane oxidation rates. Figure reproduced by permission of the publisher.

diffusion of nutrients and CO_2 generated below the zone of methane oxidation (see following discussion) should act to decrease the extent of supersaturation in Fig. 11. Nevertheless, enhanced supersaturation caused by methane oxidation may, in some instances, be enough to cause carbonate to precipitate. Such a mechanism, combined with sedimentation hiatus events, has been proposed by Raiswell (1987, 1988a) to explain the precipitation of carbonate as concretions and beds in carbonate-shale sequences. Unfortunately, we are unaware of any modern sediments supporting anaerobic methane oxidation where enough pore-water data exist to allow a test of the model predictions.

3.1.5. Methane Production

Below the zone of sulfate depletion in marine sediments, methane accumulates (see Fig. 10) (Emery and Hoggan, 1958; Claypool and Kaplan, 1974; Martens and Berner, 1977; Reeburgh, 1980). Though methane production involves many steps (Berner, 1980), an overall reaction may be written as follows:

$$2(CH_2O)_{106}(NH_3)_{16}H_3PO_4 \rightarrow 106CO_2 + 106CH_4 + 32NH_3 + 2H_3PO_4 \quad (16)$$

Adopting this stoichiometry, carbonate saturation was calculated assuming in one case that enough methane is produced to supply 10% of the organic

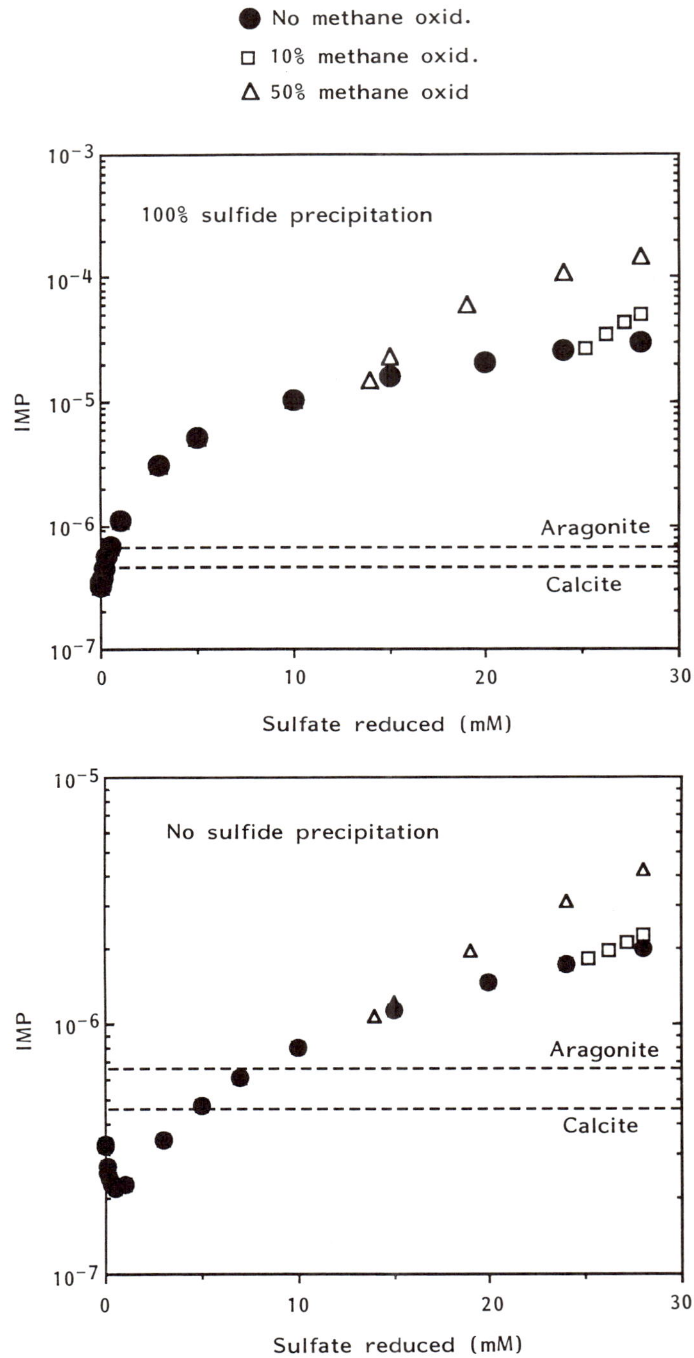

FIGURE 11. Model results including methane oxidation are compared to those where methane oxidation does not occur. In all cases, methane oxidation causes an increase in carbonate supersaturation. Methane oxidation is added to both (A) the "100% sulfide precipitation" model and (B) the "no sulfide precipitation model." Carbonate saturation expressed by ion molarity product (IMP).

carbon for sulfate reduction, and in another case 50%, making these amounts comparable to the methane oxidation results discussed above. In the modeling, as in marine sediments, methane production follows methane oxidation by sulfate reduction. In general (results in Table II), methane production lowers carbonate saturation (by lowering sediment pH); however, in no case does carbonate undersaturation occur. Hence, in marine sediments undergoing early diagenesis methane production should do little to affect carbonate preservation, unless little alkalinity accumulates during sulfate reduction and initial IMP values are lower than in Table II (see Section 3.3).

3.2. Synthesis of Results for Marine Sediments

Overall, carbonate preservation is maximized in oxic marine sediments located above the CCD (carbonate compensation depth) and in anoxic sediments where sulfate becomes rapidly depleted with sediment depth, and where, during the initial stages of sulfate depletion, sulfide is reacted from solution. These conditions are common in organic-rich shelf sediments. Also important is rapid burial through the initial stages of carbonate undersaturation, allowing minimal contact between carbonate and corrosive pore solutions. As sediment is buried into the zone of anaerobic methane oxidation, enhanced carbonate supersaturation may encourage carbonate precipitation. Carbonate dissolution is maximized in oxic sediments deposited below the CCD, and in bioturbated anoxic sediments where irrigation keeps the products of sulfate reduction from accumulating (e.g., NWC and the flanks of the Santa Barbara Basin). Deep irrigation seems to persist in organic-rich sediments below mean wave base, where relatively stable conditions allow irrigating organisms to construct and maintain burrow networks (Rhoads, 1974). This trend is exacerbated in carbonate sediments, where sulfide does not react from solution, but builds into the pore solution (Walter and Burton, 1990). Carbonate undersaturation and subsequent dissolution may also be expected in euxinic water bodies when sulfate is not

TABLE II. Carbonate Saturation Resulting from Methane Production[a]

No H_2S precipitation	$IMP_{beginning}$ [b,c]	IMP_{end} [d]
10%	2.27×10^{-6}	2.10×10^{-6}
50%	4.20×10^{-6}	2.44×10^{-6}
100% H_2S precipitation		
10%	5.14×10^{-5}	3.41×10^{-5}
50%	1.47×10^{-4}	4.99×10^{-5}

[a]The methane model was grafted onto both the no-precipitation and 100%-precipitation models of Fig. 11. Model is run for two cases, one case where enough methane is generated to fuel 10% of the sulfate reduction, and for another case, 50%.
[b]Compare to $K'CaCO_3$ values in Table I.
[c]Carbonate saturation at the beginning of the model run.
[d]Carbonate saturation at the end of the model run.

greatly depleted from ambient seawater levels, and where sulfide accumulates. It is important that water-column sulfate need not be much depleted under euxinic conditions, as is the case in the best modern examples of euxinic basins, the Black Sea and the Cariaco Trench. However, regionally important factors such as the high alkalinities seen in the Black Sea may allow supersaturation to persist through all organic-matter decomposition reactions.

3.3. Carbonate Preservation in Nonmarine Sediments

Freshwaters vary widely in their state of saturation with respect to $CaCO_3$ (and also in their calcium content), hence sediment diagenesis begins from initial conditions which differ from those in marine sediments. The presence of organic carbon fuels the same sequence of microbial processes as described in Sections 3.1.1–3.1.5, but with the important difference that sulfate reduction is *much* less significant (because freshwater sulfate concentrations are typically two orders of magnitude lower than those in seawater). Hence, little alkalinity can be generated by this process. Instead, larger concentrations of organic matter are degraded by methanogenesis, which is initiated as soon as anaerobic conditions are established (Lovely and Klug, 1986). Since methanogenesis generates CO_2 (Section 3.1.5), nonmarine sediments can become quickly undersaturated with respect to $CaCO_3$ (e.g., Postma, 1981). However, because of the small amounts of sulfate reduction and subsequent sulfide reaction with iron minerals, large amounts of iron oxides may survive to be reduced microbially:

$$4FeOOH + CH_2O + H_2O \rightarrow 4Fe^{2+} + HCO_3^- + 7OH^- \quad (17)$$

This process directly adds alkalinity and tends to neutralize the acidic CO_2 formed by methanogenesis (Coleman, 1985). Thus, the preservation potential of skeletal $CaCO_3$ is highly variable, depending on initial water composition and the relative significance of methanogenesis and iron reduction. As a result of iron reduction [equation (17)] pore waters in nonmarine sediments may accumulate high concentrations of Fe^{2+} and become supersaturated and precipitate siderite (e.g., Postma, 1981, 1982).

4. Carbonate Preservation in Ancient Clastic Sediments

The modern sediments discussed in Sections 2 and 3 show that carbonate dissolution can occur by a variety of processes in near-surface sediments. Deep subsurface environments are less well studied. However, during burial diagenesis carbonate dissolution can arise from the evolution of CO_2 by methanogenesis and decarboxylation reactions (e.g., Schmidt and McDonald, 1979; Tissot and Welte, 1984) or the formation of carboxylic acids (e.g., Surdam *et al.*, 1984). Carbonate dissolution can also result from acidity generated by clay mineral transformations during deep burial (e.g., Bjørlykke, 1983) or introduced through the invasion of meteoric water.

In general, fluids derived from greater depths and introduced into fine-grained clastic sediments can be extremely variable in their state of saturation with respect to carbonates, depending on their origin and subsequent evolution [see discussion in Holland and Malenin (1979); Butler (1982)]. At any given temperature the solubility of calcite in equilibrium with a vapor phase increases with increasing CO_2 pressure and, at constant pressure, increases with increasing CO_2 concentration (up to ~1 mole/kg). Furthermore, the addition of NaCl causes increased calcite solubility, especially at temperatures greater than 200°C. Temperature effects are complex. The solubility product of calcite increases with decreasing temperature, but is accompanied by a still larger increase in CO_2 solubility. The combination of these two effects on an ascending fluid saturated with CO_2 and calcite will often cause undersaturation (due to cooling and pressure reduction) unless hydrogen ions are buffered by fluid–rock interactions.

However, even comparatively small changes in fluid chemistry and/or pressure and temperature conditions may be sufficient to produce a highly selective loss of the least stable carbonate skeletal mineralogies and hence produce significant changes in the fossil record. Carbonate preservation can be envisaged as a transient state because carbonates can be destroyed by a single dissolution event occurring at any time in the deposition–burial–uplift cycle. In these circumstances it may seem surprising that skeletal material is at all commonly preserved; three factors are mainly responsible. First, the kinetics of carbonate dissolution is so rapid that saturation by dissolution will be commonly achieved in much less time than the replacement time of deep fluids, provided these have access to carbonate phases. Thus, fluids in sedimentary basins may be more commonly carbonate-saturated than undersaturated. Second, many fine-grained shales (and especially those enriched in organic matter and/or bitumen) will not be readily permeable to fluids derived externally—hence, only *in situ* processes may affect the saturation state. Third, preservation can be enhanced by cementation of the enclosing sediments, especially the early diagenetic cementation which is characteristic of concretionary growth.

4.1. Fossil Preservation in the Absence of Concretionary Growth

Few attempts have been made to link preservation to primary depositional parameters, e.g., bottom-water oxygenation, sedimentation rate, and sediment composition, although the modern sediment studies (Section 3.1) do suggest that the opportunities for preservation are maximized under conditions where bottom waters are poorly oxygenated (hence suppressing bioturbation), and diagenesis produces enhanced pore-water alkalinity. Thus, calcareous shells deposited in modern oxygenated, well-bioturbated sediments show extensive dissolution, in contrast to those found in poorly-oxygenated and weakly bioturbated sediments on the basin floor (see Section 3.1.3). Unfortunately, this simple pattern does not persist through the ancient record. Paleoenvironmental studies of both the Jurassic shales of southern Germany (Seilacher *et al.*, 1976) and the Middle Devonian of New York state (Brett *et al.*, in press) show that calcitic fossils in the bioturbated gray shales demonstrate less evidence of dissolution than in the poorly oxygenated black shales. Reaves (1986) has attempted to resolve this dilemma by arguing that the reactivity of the organic matter is more important than bottom-water oxygenation levels. He maintains that sediments with relatively low inputs of poorly metabolizable organic matter can host a diverse benthic fauna and are hence deeply bioturbated. However, below the zone of bioturbation chemical conditions are comparatively stable and the enhanced alkalinity levels promote carbonate preservation. Conversely, where organic-matter contents are higher (and potentially variable in composition) there is less benthic activity (and bioturbation), but fluctuations in microbial activity allow periods of oxygenation (seasonally) when sulfide minerals can be oxidized and acidity can be generated to cause carbonate dissolution.

Note, however, that even when inputs of organic matter are low and poorly metabolizable, dissolution can still occur due to bioturbation before more stable conditions are established, as is shown at the NWC site (Long Island Sound) and on the flanks of the Santa Barbara Basin (Section 3.1.3). Clearly, dissolution can only be less extensive in these more bioturbated sediments if less undersaturation results from irrigation than can result from acidity generated by sulfide oxidation in the organic-rich sediments. This is not necessarily true, as is shown by Hudson (in preparation), who points out that European Jurassic shales show a range in preservation of aragonitic fossils which cannot be explained on this basis. The Upper Oxford Clay is a typical bioturbated gray shale, low in organic carbon and pyrite, yet aragonite is only rarely preserved. By contrast, the Lower Oxford clay has a restricted fauna, due to poorly-oxygenated bottom waters, and is richer in both organic carbon and pyrite, but has a perfectly preserved aragonite fauna. Richest of all in organic carbon and pyrite is the Posidonienschiefer, a classic black shale, probably deposited under euxinic conditions at times (Seilacher and Westphal, 1971; but see Kauffman, 1981). Here ammonites are preserved as flattened films of periostracum and the aragonite part of the shell has dissolved completely.

Hudson (in preparation) emphasizes that these differences in preservation cannot be considered to arise solely as a result of differences in degree of bioturbation and bottom-water oxygenation, and instead the different styles of diagenesis undergone by each sediment must be considered. The major conclusions of Hudson (in preparation) are integrated below with the model results of Section 3.1.

The Upper Oxford Clay [see also above and Hudson and Palframan (1969), Hudson (1982), and Fisher and Hudson (1987)] contains pyrite internal molds of the inner phragmacones of ammonites from which the aragonite shell has been dissolved. Thus, pyrite formation preceded the dissolution of aragonite, which in turn only occurred after some compaction in the sediment. These sediments would appear to have lost aragonite through a similar mix of processes as occurred at NWC and on the flanks of the Santa Barbara Basin (Section 3.1.3), where bioturbation prevents alkalinity buildup and produces acidity through respiration and sulfide oxidation. The analogy is strengthened by the survival of calcitic skeletal debris in the Upper Oxford Clay, even though aragonitic material is dissolved, as would be expected from the relative solubilities of the two minerals. Model results (see Section 3.1.3) are also consistent in indicating that low amounts of sulfate depletion (which results from pervasive irrigation) can produce carbonate undersaturation.

In the Lower Oxford Clay, aragonitic shells are generally preserved and the fauna is dominated by infaunal deposit feeders and epifaunal suspension feeders (Duff, 1975), which produced an intense but shallow zone of bioturbation. Hence the sediment shows evidence of more extensive sulfate reduction (Fisher and Hudson, 1987) and alkalinity loss by bioturbation was suppressed. Carbonate supersaturation and aragonite preservation would be expected, assuming that rapid sediment burial below the near-surface zone of possible carbonate undersaturation occurred (see Section 3.1.3).

In the Posidonienschiefer, aragonitic shell material has been completely dissolved away, apparently at very shallow burial depths within reach of the rare bioturbation events (Savrda and Bottjer, 1989) and probably before pyrite formation ceased. Hudson (in preparation) suggests several ways in which aragonite could have been dissolved, of which the most attractive is to infer the development of H_2S-bearing bottom waters aggressive to carbonates. Our model results suggest that H_2S buildup without removal by iron leads initially to undersaturation (Section 3.1.3), before alkalinity becomes sufficiently high to cause carbonate supersaturation. Such a scenario could also account for the poor preservation of calcitic fossils in the Devonian black shales of New York state (Brett *et al.*, in press).

A comparison of the Posidonienschiefer with the Jet Rock [equivalent sediments exposed in northeastern England (see Jenkyns, 1988)] is revealing (Raiswell and Berner, 1985). Both sediments are rich in organic carbon, but the Jet Rock contains considerably more pyrite because the initial content of iron reactive toward H_2S (4–8%) was considerably higher than in the Posidonienschiefer (~2.5%). The Jet Rock also still contains carbonate which was probably aragonitic. The low iron contents of the Posidonienschiefer suggest the possibility that sulfide removal from the water column was minimal; hence the bottom waters may have been poised at the initial stages of the "no sulfide precipitation" model (Section 3.1.3) where carbonate undersaturation is maintained. Conversely, in the Jet Rock, enough reactive iron may have been available to drive the system past the point where acid generation by H_2S could exceed alkalinity generation (see Fig. 6) approaching the "100% sulfide precipitation" model (Section 3.1.3). However, it is also possible that sulfate reduction in the bottom waters was extensive and hence sufficient alkalinity was generated to preserve aragonite.

A superb example of dissolution due to water column accumulation of hydrogen sulfide occurs with the "half" ammonites described by Seilacher *et al.* (1976) and Tanabe *et al.* (1983). In the latter case two types of ammonite preservation are found in Lower Jurassic shales (similar in age to the Jet Rock and the Posidonienschiefer); horizontally bedded ammonites are found either complete, or with their upper flanks dissolved away. Tanabe *et al.* (1983) envisaged that the upper flanks were exposed to seawater, whilst the lower flanks were buried in sediment. Complete preservation was attributed to the presence of poorly oxygenated bottom waters, in contrast to "half" preservation which occurred in sulfidic bottom waters. These differences correspond well with model results. The "half" ammonites occur partly in sulfidic bottom waters (which represent the early, carbonate under-saturated stages of the "no-precipitation" model) and partly in pyritic sediments (where sulfide precipitation and continued sulfate reduction has generated sufficient alkalinity to cause carbonate over-saturation). Conversely, complete preservation represents exposure either to poorly oxygenated bottom waters (where water column sulfate reduction did not occur) or to highly reduced, carbonate-saturated sediment porewaters.

These case studies should be taken to indicate that a comprehensive approach is required to understand how different diagenetic events affect preservation in marine, organic carbon-bearing sediments. The approach of correlating preservation with degree of bottom water oxygenation is clearly inadequate, since the models (and the modern sediment data) show that all the following factors are important:

(a) Extent of bioturbation. Low bottom water oxygenation levels suppress bioturbation, which otherwise causes carbonate dissolution through the addition of carbon dioxide by respiration and through the generation of acidity by the oxidation of iron sulfides. In the absence of bioturbation, carbonate supersaturation results from the alkalinity produced by sulfate reduction (see below).

(b) Rate of sulfate reduction. Rates of sulfate reduction which are rapid enough to completely remove pore-water sulfate (rates of addition/removal exceed opposing rates of mass transfer, e.g., by diffusion) will cause the pore system to pass rapidly through the early stages of carbonate undersaturation, to the point where alkalinity addition causes supersaturation.

(c) Reactivity of sediment iron phases. The addition of hydrogen sulfide by sulfate reduction produces acidity, unless reactive iron phases are available to be transformed to iron sulfides, and hence maintain low concentrations of pore-water-dissolved sulfide.

Finally, it is stressed that dissolution events can occur throughout sediment history and are not just an early diagenetic phenomenon.

4.2. Fossil Preservation in Concretionary Carbonate

In this section the term "concretion" is used to embrace all forms of early diagenetic carbonate cementation, whether these have concretionary or a bedded morphology. In all cases early carbonate cementation provides excellent preservation of a wide variety of organisms, including fish (e.g., Weeks, 1957), plants and associated cell materials (Allison, 1988; Scott and Rex, 1988), insects and jellyfish (Woodland and Stenstrom, 1979) and even assemblages of flora and/or fauna (Baird *et al.*, 1985a; Allison, 1988).

It might be thought that such detailed preservation would place some important constraints on the timing and nature of the concretionary growth process. Thus, good preservation clearly indicates that carbonate cementation began early, close to the sediment surface, and was completed quickly before decay of the occluded organic parts occurred. In fact, however, the paleobiological studies have tended to generate additional questions regarding the causes, timing, and duration of concretionary growth for which the geochemical data is only indicative. It is also unfortunate that almost all of the isotopic/geochemical studies of carbonate concretions have been made on material from which significant macrofossils are absent, making it difficult to compare the fossiliferous and nonfossiliferous end-products of concretionary growth. This has been a deliberate choice on the part of geochemists to simplify their sampling strategy, but it could be a major weakness in the following discussion.

4.2.1. The Timing and Duration of Concretionary Growth

Carbonate cementation as concretions may take place over a wide span of diagenetic history; only the earliest of these events will preserve uncompacted fossil material and successively later events will generally be associated with greater information loss (although this itself may be useful in providing a record of the degradation process). A time framework is therefore required to date the inception of concretionary growth relative to the burial history of the enclosing sediments and relative to other concretions in the same sediments. Ideally this time framework should be based on, or correlated with, field observations which allow early concretions to be distinguished from those formed later.

A variety of approaches will give relative time scales (Raiswell, 1971) of which the most useful have proved to be (a) the presence of uncrushed skeletal material or soft body matter; (b) the degree of compaction of sedimentary fabrics inherited by the concretion; and (c) mineralogical/geochemical evidence, essentially timing concretionary growth relative to the diagenetic history of the enclosing sediments. Unfortunately, none of these criteria can be applied with precision in the field. The most precise relative dating evidence is provided by estimates of the degree of compaction of the host sediment during concretion growth (Raiswell, 1971).

Lippmann (1955) and Raiswell (1971) suggested that carbonate cement passively fills the pore spaces of the host sediment, hence the volume percentage cement provides an estimate of the host sediment porosity (and degree of compaction) at the time of concretionary growth. Oertel and Curtis (1972) confirmed that the clay fabric inherited by a Carboniferous siderite concretion did indeed show the fabric changes associated with the gradual compaction of a highly porous sediment during concretionary growth.

Estimates of host sediment porosity clearly demonstrate that concretionary growth is possible throughout a substantial period of diagenesis (Raiswell, 1971), spanning porosities from near-surface values of approximately 85% to less than 40%. It is possible to relate porosity estimates to burial depth (Curtis *et al.*, 1986) using the clay compaction relationships of Baldwin and Butler (1985). However, care needs to be exercised because modern sediments can vary greatly in the porosity of their surficial layers (for example, FOAM, 67%; NWC, 84%; Santa Barbara Basin, 95%) depending on grain-size, composition, bioturbation, sedimentation rate, etc. Such porosity estimates are clearly best used for relative dating only, between concretions in similar sediments. Despite this, dating by porosity estimates has proven to be useful in numerous studies of concretionary growth (Curtis *et al.*, 1972; Raiswell, 1976; Dickson and Barber, 1976; Irwin *et al.*, 1977; Hudson, 1978; Gautier, 1982; Gluyas, 1984; Curtis *et al.*, 1986) but should now be supported by petrographic studies since Feistner (1989) has been able to demonstrate the existence of replacive/displacive textures, which would render porosity estimates as only approximate.

A variety of mineralogical/geochemical criteria can also be used to time concretionary growth relative to the progress of diagenesis in the surrounding sediments. For example, Raiswell (1976, 1982) demonstrated that growth of carbonate concretions in the Jet Rock (Lower Jurassic) post-dated the formation of an early diagenetic, isotopically light pyrite phase in the host sediment, but was contemporaneous with the formation of later, isotopically heavy pyrite (and hence still in the zone of sulfate reduction). Similarly, Hudson and Andrews (1987) dated concretionary growth in Jurassic sandstones as occurring after the pressure solution of shells against quartz and feldspar, but prior to final compaction. In addition, isotopic evidence can be used in some circumstances to determine the relative age of concretions, either through the use of the oxygen isotope paleotemperature scale (Hudson, 1978) or the zonal sequence of diagenetic carbon isotope changes defined by Irwin *et al.* (1977; but see Raiswell, 1987).

Estimating the duration of concretionary growth is much more problematic. Berner (1968a) presented a theoretical model based on the assumption that growth is transport limited. If transport is by diffusion only, the time-span of growth (t) is given by:

$$t = \frac{R^2}{2\nu D(C_\infty - C_R)} \tag{18}$$

where ν is the molar volume of the concretion cement ($cm^3\ mole^{-1}$), D is the diffusion coefficient ($cm^2 s^{-1}$), R is the concretion radius (cm) and ($C_\infty - C_R$) is the degree of supersaturation ($mol\ cm^{-3}$). In using these equations, most investigators (e.g., Hudson and Andrews, 1987; Thyne and Boles, 1989) have emphasized the difficulty of obtaining realistic values for the degree of oversaturation. The problem is that natural and experimental systems can sustain considerable degrees of oversaturation with precipitation occurring only slowly, if at all (Berner *et al.*, 1970; Berner *et al.*, 1978). This suggests that growth is controlled by surface reactions and not by transport processes (Raiswell, 1988b). Growth may therefore be much slower than predicted by equation (18) and may vary greatly with differing concentrations of those aqueous species which inhibit $CaCO_3$ precipitation (e.g., magnesium, phosphate and dissolved organics; see Morse, 1983).

Also, given that growth must occur over a relatively short period, in rocks that are often comparatively old, it is unlikely that any dating technique could be developed which would be sufficiently precise to measure differences across a radial center/margin traverse. At present, routine porosity estimates may be the most useful measure of the timing of concretionary growth. There is a tendency for earlier concretions to be more spherical (Raiswell, 1971), which might prove to be sufficiently widespread to aid collection of the earliest cemented concretions.

4.2.2. Stratigraphic Constraints on Concretionary Growth

There is now general agreement that concretionary growth occurs in pore waters where carbonate supersaturation is brought about by the alkalinity generated from the microbial decay of organic matter. The evidence for this is quite compelling; the depths inferred from porosity estimates are consistent with the depths in modern sediments where microbial activity is well-documented; furthermore many concretions are associated with diagenetic phases whose microbial origin can be demonstrated in modern sediments (e.g., pyrite, Berner, 1970, 1984; siderite, Postma, 1981). Finally, carbon isotope studies of the carbonates show the expected fractionations of different microbial processes (e.g., Curtis *et al.*, 1972; Irwin *et al.*, 1977; Hudson, 1978; Coleman and Raiswell, 1981; Gauthier, 1982).

However, despite the wealth of evidence for the involvement of microbial processes in generating carbonate supersaturation, the nature of the processes which determine the siting of cementation have proved elusive. Until these can be identified it cannot be known whether fossil preservation is adventitious or selective. The first problem is to identify what controls the stratigraphic position of prolific concretionary horizons. It seems likely for the following reasons that episodic sedimentation encourages the formation of concretionary horizons, which arise during a pause in deposition (or a period of greatly reduced sedimentation) at a shallow depth above the horizon of growth (Raiswell, 1971, 1987, 1988a).

(a) Our model results (Figs. 6, 7, and 11) show a gradual increase in the degree of carbonate supersaturation with depth, as do many modern anaerobic sediments (Fig. 8 and 9; see also Sholkovitz, 1973). However, with steady-state diagenesis (uniform sedimentation rates, particle phases of constant composition), the saturation state depth profile always remains the same and the horizon of maximum supersaturation (the most probable location of carbonate precipitation) moves continuously up through the sediment. Under steady-state conditions only a diffuse phase of cementation can result; a pause in deposition is required to locate the probable zone of precipitation at one depth to produce localized, stratigraphically confined precipitation.

(b) There are no quantitative data on rates of carbonate precipitation in modern marine anaerobic sediments, but inferred rates are simply too slow relative to steady-state sedimentation rates to produce localized cementation. For example, calcium concentrations at FOAM decrease from about 8 mM in the surface layer of sediment to about 5.5 mM one meter lower. Assuming this were all due to $CaCO_3$ precipitation (in fact a small proportion of the calcium may be removed by phosphate precipitation; Aller, 1982), then, using the maximum concentration gradient for Ca^{2+} in the surface sediment and the appropriate diffusion coefficient, Ca^{2+} is removed at a rate of only 7×10^{-12} mol cm^{-2} yr^{-1}. Although, with no sediment depositing, this could form a 3–4 cm layer in 11,000 years, the current burial rate (0.1 cm yr^{-1}) would add approximately 10m of sediment over this time. Continuous sedimentation is clearly much too rapid to produce localized cementation. FOAM may not represent the most extreme supersaturation that can be achieved during anaerobic diagenesis, but its organic carbon contents (1–2%) and alkalinity concentrations (up to 35 meq. $liter^{-1}$; Aller, 1982) are high enough to be typical of many modern and ancient sediments. Precipitation is generally slow to negligible in anaerobic pore waters (see earlier) and would need to be more than two orders of magnitude faster to produce obvious cementation. A pause in deposition seems the only viable alternative.

Some concretionary horizons (Hallam, 1969) are in fact observed to be associated with a pause in deposition. Savrda and Bottjer (1988) also suggested that the formation of one concretionary horizon was related to temporary oxygenation of the seafloor. They conjectured that oxygenation was associated with increased sulfate availability and hence also increased alkalinity for carbonate precipitation. In fact, oxygenation would cause sulfide oxidation and carbonate dissolution (Section 3.1.3) and it is more likely that the changes in bottom water conditions which promoted oxygenation at the sediment surface were also associated with decreased sedimentation rates, which allowed precipitation at depth.

The model (Section 3.1.4) shows that, in marine sediments, maximum degrees of carbonate supersaturation are reached at the interface between the sulfate reduction and methanogenic zones, where anaerobic methane oxidation (AMO) occurs. There are also several other reasons for associating the process of AMO and concretionary growth (Raiswell, 1987, 1988a). However, in general terms the formation of a concretionary horizon appears to require that a supersaturation maximum is maintained in the same position for long enough (see above) to allow growth within a stratigraphically confined interval. An association between concretionary horizons and pauses in deposition suggests that the concretions may occur near (but below) horizons with distinct faunal differences indicative of depositional breaks.

4.2.3. The Siting of Concretionary Growth

Localized cementation as carbonate concretions may result either from isolated nucleation (presumably onto skeletal $CaCO_3$ grains in most cases) within an environment which is uniformly supersaturated with respect to carbonates, or from the generation of localized supersaturation by the decay of high concentrations of organic matter at the concretionary site (Raiswell, 1976). The difference between these two cases for fossiliferous concretions resides in whether $CaCO_3$ skeletal material simply promotes nucleation of authigenic carbonate, or whether there is instead (or as well) alkalinity generated (from organic matter decay by sulfate reduction) faster than it can be removed by diffusion, so maintaining local supersaturation. The excellent preservation of skeletal macroorganisms, associated with soft body material, inside concretions could support either case. However, it is also true that some concretions only preserve soft body material and others only skeletal material; also many concretions contain neither organic nor skeletal material. This seems to be true of both calcite (compare Weeks, 1957 and Müller, 1985 with Raiswell, 1987) and siderite concretions (compare Baird *et al.*, 1985b and Carpenter *et al.*, 1988 with Pearson, 1979).

Siting control by nucleation alone is also difficult to reconcile with the observations of Weeks (1957), who found extreme variability in the relative proportions of barren and ammonite-bearing concretions, and Waage (1964), who found barren concretions coexisting with fossiliferous (mainly mollusc-bearing) concretions and with fossils not enveloped in concretions. Raiswell (1987) points out that in many cases potential macrofossil nuclei appear far more abundant than concretions, such that the distribution of concretions is essentially unrelated to macrofossils. Clearly the availability of fossil nuclei cannot solely dictate the siting of individual concretions.

In some cases it has been argued that siting does appear to be controlled by alkalinity generated locally from microbial decay of soft-body material (Weeks, 1957; Zangerl *et al.*, 1969; Sass and Kolodny, 1972; Dickson and Barber, 1976). One example is in the Mississippian Fayetteville Shale (Zangerl *et al.*, 1969). Here concretionary growth took place around coprolite material and shells containing organic carcasses. The presence of organic material was recorded texturally in the carbonate; also, the shells had been broken and it was argued that this occurred explosively, due to the pressure of gases generated by microbial decay (whose escape was prevented by cementation). Even here, however, there are a number of difficulties in accepting the decay of the soft-bodied material as the sole cause of enveloping cementation. For example, it is difficult to envisage how cementation could have allowed the continual escape of the alkalinityne-cessary for growth, whilst also trapping gases with enough efficiency to result in explosive breakage. It seems unlikely that locally elevated alkalinity levels alone (if these occurred) could have caused growth of these concretions. Instead, it seems more likely that local decay only initiated growth by producing favorable surfaces for precipitation, when the incipient concretion entered the zone of maximum carbonate supersaturation.

This argument is supported by studies of calcite precipitation kinetics in the laboratory (Reddy and Nancollas, 1971; Nancollas and Reddy, 1971) and in the field (Pye *et al.*, 1990), which clearly show growth to be surface-reaction controlled. Raiswell (1988b) has extended these arguments to the diagenetic environment of concretion growth. Two experimental observations (Berner, 1975; Berner *et al.*, 1978) seem particularly important. First, calcite precipitation from normal seawater appears only to occur above a "threshold" supersaturation, due to the inhibiting effects of dissolved magnesium. Phosphate and dissolved organics also adsorb on the growing crystal surfaces and hence inhibit aragonite and calcite precipitation (Burton and Walter, 1990; see also review in Morse, 1983). Second, even when growth has been aided by the addition of calcite seeds, fresh surfaces soon become poisoned and growth then ceases. Further growth can only occur either when fresh seeds are added or when fresh surfaces can be created by precipitation. The picture that emerges is that high pore-water supersaturations need to be maintained to trigger precipitation of calcite, which forms small crystals, because continued growth is inhibited by poisoning. Difficult though it is to induce precipitation on clean surfaces, this is still easier than sustaining precipitation on the poisoned growing surfaces; although, clearly the new surfaces age and alter with time to become less favorable for carbonate precipitation. A further possibility is that the first-formed precipitate is not calcium carbonate, instead a calcium soap is formed which is subsequently transformed to calcium carbonate (as suggested by the laboratory experiments of Berner, 1968b). However, metastable intermediates such as these may result from the much more rapid decomposition of organic matter in laboratory systems, since recent work (Pye *et al.*, 1990) clearly shows that calcium carbonate can precipitate from anoxic pore waters.

Taken together this evidence suggests that where concretions contain readily metabolizable organic matter, its rapid decay may locally enhance supersaturation to the point where kinetic barriers to precipitation can be overcome and cementation can occur around, and in the vicinity of, the soft-body material. However, because $CaCO_3$ precipitation is surface-reaction controlled, a considerable proportion of the alkalinity will escape precipitation locally and contribute to raising overall $CaCO_3$ saturation levels. These may accordingly reach levels where precipitation then becomes possible on skeletal material not associated with soft-body material. Equally, once decay of soft-bodied material has been completed the concretion so formed may continue to grow by virtue of the overall pore-water saturation levels, even though local enhancement of supersaturation is no longer possible.

Mass balance calculations (Raiswell, 1976) show that most concretions contain more authigenic $CaCO_3$ cement than could be derived from the decay of organic matter at the concretionary site, especially since many organisms may (for their size) contain a comparatively small mass of metabolizable organic matter. This suggests that organisms containing large amounts of readily metabolizable body material will be initially favored for preservation by concretionary growth. Growth at these sites however may continue even after the body material has been degraded and, if growth at these sites can maintain supersaturation below the levels necessary to utilize less favorable nuclei, then concretions will only be found associated with soft-body preservation. However, if this is not the case, successively less favorable nuclei will be utilized until the overall rate of $CaCO_3$ precipitation is such as to prevent supersaturation rising further. This model suggests that selective preservation is to be expected where levels of supersaturation are comparatively low but, with increasing levels of supersaturation, a wider range of nuclei may be used. This view is supported by the observations of Pye *et al.* (1990), who conducted field experiments demonstrating that authigenic carbonate and sulfide precipitation were possible on a range of different nuclei buried in anoxic intertidal sediments, but that rates of precipitation varied widely on the different nuclei.

5. Conclusions

The dissolution of carbonate skeletal debris generally results from contact with surrounding waters which are undersaturated with respect to carbonates. A notable exception occurs with the activity of those macro- and microorganisms which can abrade, bore, or locally dissolve carbonate skeletal debris. The quantitative significance of these processes on carbonate preservation (and their effects on the fossil record) is unclear. However, the nonbiogenic dissolution of carbonate skeletons in contact with undersaturated waters is affected by mineralogy (aragonite is thermodynamically more soluble than calcite) and Mg content (solubility increases with increasing Mg content) and by physical parameters such as shell wall thickness, organic coatings, and shell porosity.

During the diagenesis of organic carbon-bearing sediments, the organic carbon can be microbially oxidized either to CO_2 (producing carbonate undersaturation) or HCO_3^- (producing carbonate oversaturation). Additionally, the decomposition of organic matter with a Redfield stoichiometry (C : N : P = 106 : 16 : 1) produces other dissolved species which range from negligible to significant in their effects on the state of carbonate saturation (e.g., NO_3^-, H_3PO_4, N_2, NH_3, H_2S).

A simple closed-system model evaluates the progressive changes in the carbonate saturation state of seawater, which successively undergoes oxic respiration, denitrification, sulfate reduction, methane oxidation, and methane production, and which precipitates solid sulfides by reaction with sediment iron phases. The model predicts that oxic respiration leads to carbonate undersaturation and dissolution (as is in fact observed in deep-sea sediments, and some euxinic basins). Subsequent denitrification only maintains the saturation state existing at the end of oxic respiration.

Sulfate reduction is extremely important in producing early diagenetic cementation in organic carbon-bearing sediments. Interestingly, the initial stages of the models predict undersaturation as a result of acidity from increasing concentrations of dissolved sulfide, which are later overwhelmed by the production of alkalinity. Carbonate preservation is thus maximized in sediments where sulfate is rapidly exhausted (producing large amounts of alkalinity) and where sulfide can be reacted from solution by iron minerals during the early stages of sulfate reduction. These model conditions are approximate to those at the FOAM site in Long Island Sound, and the deepest Santa Barbara Basin, where there is indeed good carbonate preservation. Conversely, where bioturbation allows only small amounts of sulfate reduction, not enough alkalinity is produced to cause carbonate supersaturation and poor preservation is predicted by the model. This situation is observed in the well-bioturbated sediments at NWC, Long Island Sound, and on the flanks of the Santa Barbara Basin.

The oxidation of upward-diffusing methane by sulfate reduction causes a further abrupt increase in supersaturation, and this increase is larger for higher degrees of supersaturation at the start of methane oxidation. Beneath the zone of methane oxidation, methane production acts to decrease the existing degree of oversaturation, although not generally by enough to produce undersaturation with respect to carbonates.

Model results also provide satisfactory explanations for the different styles of preservation in Jurassic organic carbon-bearing sediments. The Upper Oxford Clay is a well-bioturbated gray shale in which benthic activity would have suppressed alkalinity build-up and would also have produced acidity through sulfide oxidation and oxic respiration. Poor preservation is predicted and, in fact, aragonite has been dissolved although calcite still survives. By contrast, in the weakly bioturbated Lower Oxford Clay, extensive sulfate reduction and the consequent alkalinity build-up allowed the preservation of both aragonite and calcite, as the modelling predicts. Differences in the preservation of fossils in the Posidonienschiefer and the Jet Rock (two synchronous Jurassic shales deposited under anoxic conditions in Germany and the United Kingdom respectively) can also be explained by models which differ in the extent of dissolved sulfide accumulation in the water column, as a result of sulfate reduction. The Posidonienschiefer contains low concentrations of reactive iron and hence has little capacity to remove dissolved sulfide from the water column. Carbonate undersaturation is to be expected and is observed in the quantitative dissolution of aragonite. By contrast, in the Jet Rock, higher reactive iron contents allowed dissolved sulfide to be removed from the water column, hence preventing acid conditions from establishing carbonate undersaturation before alkalinity had accumulated to a sufficient level to preserve carbonates through oversaturation. The Jet Rock consequently contains skeletal debris which was probably originally aragonitic.

In some cases diagenesis can produce sufficient oversaturation to cement fossils within concretionary carbonate promoting exceptional preservation of organic soft parts. Concretionary cementation can occur very early in the burial history in highly porous sediment (perhaps within a meter of the sediment surface) and also later, after some compaction and perhaps at burial depths of several hundred meters. No realistic estimates can be made for the duration of concretionary growth in shales, although rates of Ca^{2+} removal from modern sediment pore waters implies that precipitation is so slow that the amounts of cement necessary to form a concretionary bed could only accumulate during a pause in deposition. A pause in deposition serves to locate a zone of carbonate supersaturation at a fixed distance below the sediment/water interface, whereas continuous sedimentation would only produce a diffuse phase of cementation as the zone of precipitation passed continuously through the sediment. Model results suggest that the zone of maximum supersaturation is reached where anaerobic methane oxidation occurs.

Within the zone of anaerobic methane oxidation it seems likely that organisms with readily metabolizable body material will be favored for preservation by concretionary growth, because decay of the organic matter produces locally enhanced supersaturation and leads to carbonate precipitation. Subsequently, these fresh surfaces are favored for continued concretionary growth even after decay of the associated organic matter has ceased. Cementation is restricted to these sites as long as precipitation is able to prevent pore-water saturation from rising sufficiently high to utilize other, less favorable sites.

ACKNOWLEDGMENT. We would like to express our thanks to Bernie Boudreau, John Hudson, Lynn Walter, Susan Kidwell, Derek Briggs, Tom Anderson, and Peter Allison for their helpful comments. DEC would like to acknowledge the National Research Council (NRC) of the United States for support during the writing of this manuscript. The authors wish to thank Bob Bener for developing their interests in diagenesis and encouraging its application to the rock record.

References

Alekin, A. O., and Moricheva, N. P., 1966, On the saturation of Black Sea water with $CaCO_3$, *Dokl. Akad. Nauk SSSR* **167:**423–425.

Alexandersson, E. T., 1972, Intergranular growth of marine aragonite and Mg calcite: Evidence of precipitation from supersaturated seawater, *J. Sediment. Petrol.* **42:**441–460.

Aller, R. C., 1977, The influence of macrobenthos on chemical diagenesis of marine sediments, Ph. D. thesis, Yale University, New Haven, Connecticut.

Aller, R. C., 1982, Carbonate dissolution in near-shore muds: The role of physical and biological re-working, *J. Geol.* **90:**79–95.

Aller, R. C., and Yingst, J. Y., 1980, Relationships between microbial distributions and the anaerobic decomposition of organic matter in surface sediments of Long Island Sound, U.S.A. *Mar. Biol.* **56:**29–42.

Aller, R. C., Mackin, J. E., Ullman, W. J., Chen-Hou, W., Shing-Min, T., Yong-Nian, S., and Jia-Zhen, H., 1985, Early chemical diagenesis, sediment–water solute exchange, and storage of reactive organic matter near the mouth of the Changjiang, East China Sea, *Cont. Shelf Res.* **4:**227–251.

Allison, P. A., 1988, Taphonomy of the Eocene London Clay biota, *Paleontology* **31:**1079–1110.

Alperin, M. J., 1988, The carbon cycle in an anoxic marine sediment: Concentrations, rates, isotopic ratios and diagenetic models, Ph. D. thesis, University of Alaska, Fairbanks.

Alperin, M. J., Reeburgh, W. S., and Whiticar, M. J., 1988, Carbon and hydrogen isotope fractionations resulting from anaerobic methane oxidation, *Global Biogeochem. Cycles* **2:**279–288.

Ansell, A. D., and Nair, N. B., 1969, A comparative study of bivalves which bore mainly by mechanical means, Am. *Zoologist* **9:**851–858.

Arrhenius, G., 1952, Sediment cores from the East Pacific, Rept. Swed. Deep-sea Exped., 5, 1–227.

Baird, G. C., Shabica, C. W., Anderson, J. L., and Richardson, E. S. Jr, 1985a, Biota of a Pennsylvanian muddy coast: habitats within the Mazonian Delta Complex, northeast Illinois. *J. Paleontology,* **59:**253–281.

Baird, G. C., Sroka, D., Shabica, C. W., and Beard, T. L., 1985b, Mazon Creek-type fossil assemblages in the U.S. midcontinent Pennsylvanian: their recurrent character and paleoenvironmental significance. *Phil. Trans. R. Soc. Lond.* **B311:**87–99.

Baldwin, B., and Butler, C. O., 1985, Compaction curves, *Am. Assoc. Petrol. Geol. Bull.,* **69:**622–626.

Bathurst, R. G. C., 1975, *Carbonate Sediments and their Diagenesis,* Developments in Sedimentology, Volume 12, Elsevier, Amsterdam.

Ben Yaakov, S., 1973, pH buffering of pore water of Recent anoxic marine sediments, *Limnol. Oceanogr.,* **18:**86–94.

Ben Yaakov, S., Ruth, S. E., and Kaplan, I. R., 1974, Carbonate compensation depth: relation to carbonate solubility in ocean waters, *Science,* **184:**982–984.

Bender, M. L., and Heggie, D. T., 1984, Fate of organic carbon reaching the deep sea floor: a status report. *Geochim. Cosmochim. Acta* **48:**977–986.

Benoit, G. T., Turekian, K. K., and Benninger, L. K., 1979, Radiocarbon dating of a core from Long Island Sound, *Estuarine Coastal Mar. Sci.,* **21:**325–336.

Berger, W. H., 1967, Foraminiferal ooze: Solution at depths, *Science,* **156:**383–385.

Berger, W. H., 1970, Planktonic foraminifera: selective solution and the lysocline, *Mar. Geol.,* **8:**11–138.

Berger, W. H., 1971, Sedimentation of planktonic foraminifera, *Mar. Geol.,* **11:**325–358.

Berger, W. H., 1976, Biogenous deep-sea sediments: production, preservation and interpretation, in: *The Sea,* Volume 5, (J. P. Riley and R. Chester, eds.), 265–388. Academic Press, New York.

Berger, W. H., and Soutar, A., 1970, Preservation of plankton shells in an anaerobic basin off California, *Geol. Soc. Am. Bull.,* **81:**275–282.

Berger, W. H., and von Rad, U., 1972, Cretaceous and Cenozoic sediments from the Atlantic Ocean, in: *Initial Reports of the Deep-Sea Drilling Project* (D. E. Hayes, A. C. Pimm, J. P. Bachmann, W. E. Benson, W. H. Berger, P. H. Roth, P. R. Supko and U. von Rad, eds.) **14:**787–954, U.S. Government Printing Office.

Berner, R. A., 1968a, Rate of concretion growth, *Geochim. Cosmochim. Acta* **32:**477–484.

Berner, R. A., 1968b, Calcium carbonate concretions formed by the decomposition of organic matter, *Science* **159:**195–197.

Berner, R. A., 1969, Migration of iron and sulfur within anaerobic sediments during early diagenesis, *Am. J. Sci.* **265:**773–785.
Berner, R. A., 1970, Sedimentary pyrite formation, *Am. J. Sci.* **268:**1–23.
Berner, R. A., 1971, *Principles of Chemical Sedimentology* McGraw-Hill, New York.
Berner, R. A., 1975, The role of magnesium in the crystal growth of aragonite and calcite from seawater, *Geochim. Cosmochim. Acta* **39:**489–504.
Berner, R. A., 1977, Sedimentation and dissolution of pteropods in the ocean, in: *Fate of Fossil Fuel Carbonate* (N. R. Anderson and A. Malhoff, eds.), pp. 243–260. Plenum Press, New York.
Berner, R. A., 1980, *Early Diagenesis: A Theoretical Approach*, Princeton, New Jersey.
Berner, R. A., 1984, Sedimentary pyrite formation: an update, *Geochim. Cosmochim. Acta* **48:**605–615.
Berner, R. A., and Honjo, S., 1981, Pelagic sedimentation of aragonite: its geochemical significance, *Science* **211:**940–942.
Berner, R. A., and Westrich, J. T., 1985, Bioturbation and the early diagenesis of carbon and sulfur, *Am. J. Sci.* **285:**193–206.
Berner, R. A., Scott, M. R., and Thomlinson, C., 1970, Carbonate alkalinity in the pore waters of anoxic marine sediments, *Limnol. Oceanogr.* **15:**544–549.
Berner, R. A., Berner, E. K., and Keir, R. S., 1976, Aragonite dissolution on the Bermuda pedestal: its depth and geochemical significance, *Earth. Planet. Sci. Lett.,* **30:**169–178.
Berner, R. A., Westrich, J. T., Graber, R., Smith J., and Martens, C. S., 1978, Inhibition of aragonite precipitation from supersaturated seawater: a laboratory and field study, *Am. J. Sci.* **278:**816–837.
Biscaye, P. E., Kolla, V., and Turekian, K. K., 1976, Distribution of calcium carbonate in surface sediments of the Atlantic Ocean, *J. Geophys. Res.* **81:**2595–2603.
Bjórlykke, K., 1983, Diagenetic reactions in sandstones, in: *Sediment Diagenesis*, (A. Parker and B. W. Sellwood, eds.), pp. 169–213. D. Reidel, Dordrecht.
Boudreau, B. P., 1987, A steady-state diagenetic model for dissolved carbonate species and pH in the pore waters of oxic and suboxic sediments, *Geochim. Cosmochim. Acta* **51:**1985–1996.
Boudreau, B. P., and Canfield, D. E., 1988, A provisional diagenetic model for pH in anoxic porewaters: Application to the FOAM site, *J. Mar. Res.,* **46:**429–455.
Brett, C. E., Dick, V. B., and Baird, G. C., in press, Taphonomy and paleoecology of Middle Devonian (Hamilton Group) dark gray and black shales from Western New York.
Broecker, W. S., and Peng, T.-H., 1982, *Tracers in the Sea*, Eldigio Press, Palisades, New York.
Broecker, W. S., and Takahashi, T., 1978, The solubility of calcite in seawater, in: *Thermodynamics in Geology* (D. G. Fraser, ed.), Reidel, Dordrecht, pp. 365–379.
Burton, E. A., and Walter, L. M., 1990, The role of pH in phosphate inhibition of calcite and aragonite precipitation rates in seawater, *Geochim. Cosmochim. Acta* **54:**797–808.
Butler, J. N., 1982, *Carbon Dioxide Equilibria and Their Applications*, Addison-Wesley, Reading, Mass.
Canfield, D. E., 1988, Sulfate reduction and the diagenesis of iron in anoxic marine sediments, Ph. D. thesis, Yale University, New Haven, Connecticut.
Canfield, D. E., 1989, Reactive iron in marine sediments, *Geochim. Cosmochim. Acta* **53:**619–632.
Canfield, D. E., and Berner, R. A., 1987, Dissolution and pyritization of magnetite in anoxic marine sediments, *Geochim. Cosmochim. Acta* **51:**645–659.
Carpenter, S. J., Erickson, J. M., Lohmann, K. C., and Owen, M. R., 1988, Diagenesis of fossiliferous concretions from the Upper Cretaceous Fox Hills Formation, North Dakota, *J. Sediment. Petrol.* **58:**706–723.
Carriker, M. R., 1969, Excavation of bore holes by the gastropod *Urosalpinx:* An analysis by light and scanning electron microscopy, *Am. Zool.* **9:**917–933.
Carriker, M. R., Charlton, G., and van Zandt, D., 1967, Gastropod *Urosalpinx:* pH of accessory boring organ while boring, *Science* **158:**920–922.
Chave, K. E., 1952, A solid solution between calcite and dolomite, *J. Geol.* **60:**190–192.

Chave, K. E., 1954, Aspects of the biochemistry of magnesium: Calcareous and marine organisms, *J. Geol.* **62:**266–283.

Chave, K. E., Deffeyes, K. S., Weyl, P. K., Garrels, R. M., and Thompson, M. E., 1962, Observations on the solubility of skeletal carbonates in aqueous solutions, *Science* **137:**33–34.

Chen, C., 1964, Pteropod ooze from Bermuda pedastal, *Science* **144:**60–62.

Chétail, M., and Fournié, J., 1969, Shell-boring mechanism of the gastropod *Purpura (Thais-lapillus)*: a physiological demonstration of the role of carbonic acid anyldiase in the dissolution of $CaCO_3$, *Am. Zoologist,* **9:**983–990.

Claypool, G. E., and Kaplan, I. R., 1974, The origin and distribution of methane in marine sediments, in: *Natural Gases in Marine Sediments* (I. R. Kaplan, ed.), pp. 99–139. Plenum Press, New York.

Cobb, W. R., 1969, Penetration of calcium carbonate substrates by the boring sponge, *Clione, Am. Zoologist,* **9:**783–790.

Coleman, M. L., 1985, Geochemistry of diagenetic non-silicate minerals: kinetic considerations, *Phil. Trans. R. Soc. Lond.* **A315:**39–56.

Coleman, M. L., and Raiswell, R., 1981, Carbon, oxygen and sulphur isotope variations in concretions from the Upper Lias of N.E. England, *Geochim. Cosmochim. Acta* **45:**329–340.

Curtis, C. D., Petrowski, C., and Oertel, G., 1972, Stable carbon isotope ratios within carbonate concretions: a clue to time and place of origin, *Nature,* **235:**98–100.

Curtis, C. D., Coleman, M. L., and Love, L. G., 1986, Pore water evolution during sediment burial from isotopic and mineral chemistry of calcite, dolomite and siderite concretions, *Geochim. Cosmochim. Acta* **50:**2321–2334.

Day, J., 1968, The proboscis and probosic gland of the cymatiial megogostropod *Argobuccinum argus, Am. Zoologist* **8:**801.

Devol, A. H., 1983, Methane oxidation rates in the anaerobic sediments of Saanich Inlet, *Limnol. Oceanogr.,* **28:**738–742.

Dickson, J. A. D., and Barber, C., 1976, Petrography, chemistry and origin of early diagenetic concretions in the Lower Carboniferous of the Isle of Man, *Sedimentology,* **23:**189–211.

Duff, K. A., 1975, Paleoecology of a bituminous shale—the Lower Oxford Clay of central England, *Palaeontology,* **18:**443–482.

Emerson, S., Jahnke, R., Bender, M., Froelich, P., Klinkhammer, G., Bowser, C., and Setlock, G., 1980, Early diagenesis in sediments from the eastern equatorial Pacific, I. Pore water nutrient and carbonate results, *Earth. Planet. Sci. Lett.,* **49:**57–80.

Emerson, S., Grundmanis, V., and Graham, D., 1982, Carbonate chemistry in marine pore waters: MANOP sites C and S, *Earth Planet. Sci. Lett.,* **61:**220–232.

Emery, K. O., and Hoggan, D., 1958, Gases in marine sediments, *Am. Assoc. Petrol. Geol. Bull.,* **42:**2174–2188.

Emery, K. O., Tracey, J. I. Jr., and Lodd, H. S., 1954, Geology of Bikini and nearby atolls, *U.S. Geol. Surv. Prof. Paper, 260A, 265p.*

Feistner, K. W. A., 1989, Petrographic examination and re-interpretation of concretionary carbonate horizons from the Kimmeridge Clay, Dorset, *J. Geol. Soc. Lond.* **146:**345–350.

Fisher, I. St. J., and Hudson, J. D., 1987, Pyrite formation in Jurassic shales of contrasting biofacies, in: *Marine Petroleum Source Rocks* (J. Brooks and A. J. Fleet, eds.), pp. 69–78. Spec. Publ. Geol. Soc. London, Volume 26, Blackwell Scientific Publications, Oxford.

Froelich, P. N., Klinkhammer, G. P., Bender, M. L., Luedtke, N. A., Heath, G. R., Cullen, D., Dauphin, P., Hammond, D., Hartman, B., and Maynard, V., 1979, Early oxidation of organic matter in pelagic sediments of the eastern equatorial Atlantic: suboxic diagenesis, *Geochim. Cosmochim. Acta* **43:**1075–1090.

Gardner, L. R., 1973, Chemical models for sulfate reduction in closed anaerobic marine environments, *Geochim. Cosmochim. Acta* **37:**53–68.

Gautier, D. L., 1982, Siderite concretions—indications of early diagenesis in the Gammon Shale (Cretaceous), *J. Sediment. Petrol.,* **52:**859–871.

Gavish, E., and Friedman, G. M., 1969, Progressive diagenesis in Quatenary to Late Tertiary carbonate sediments: sequence and time scale, *J. Sediment. Petrol.,* **39:**980–1006.

Giles, M. R., and Marshall, J. D., 1986, Constraints on the development of secondary porosity in the subsurface: re-evaluation of processes, *Mar. Petrol. Geol.*, **3**:243–256.

Ginsberg, R. N., 1957, Early diagenesis and lithification of shallow-water carbonate sediments in South Florida, in: *Regional Aspects of Carbonate Deposition*, S.E.P.M. Spec. Publ. Volume 5, (R. J. LeBlanc and J. G. Breeding, eds.), pp. 80–90.

Gluyas, J. G., 1984, Early carbonate diagenesis within Phanerozoic shales and sandstones of the N.W. European Shelf, *Clay Minerals*, **19**:309–321.

Goldhaber, M. B., Aller, R. C., Cochran, J. K., Rosenfeld, J. K., Martens, C. S., and Berner, R. A., 1977, Sulfate reduction, diffusion, and bioturbation in Long Island Sound sediments: report of the FOAM Group, *Am. J. Sci.*, **277**:193–237.

Golubic, S., 1969, Distribution, taxonomy and boring patterns of marine endolithic algae, *Am. Zoologist*, **9**:747–751.

Grasshoff, K., 1976, The hydrogeochemistry of landlocked basins and fjords, in: *Chemical Oceanography*, Volume 2, 2nd edition, (J. P. Riley and G. Skirrow, eds.), pp. 456–597. Academic Press, New York.

Hallam, A., 1969, A pyritised limestone hardground in the Lower Jurassic of Dorset, England, *Sedimentology*, **12**:231–240.

Heggie, D., Maris, C., Hudson, A., Dymond, J., Beach, R., and Cullen, J., 1987, Organic carbon oxidation and preservation in N.W. Atlantic continental margin sediments, in: Geology and Geochemistry of Abyssal Plains (P. P. E. Weaver and J. Thompson, eds.), *Geol. Soc. Lond. Spec. Publ.* **31**:215–236.

Henrichs, S. M., and Reeburgh, W. S., 1987, Anaerobic mineralization of marine organic matter: Rates and the role of anaerobic processes in the ocean carbon economy, *Geomicrobiol. J.* **5**:191–237.

Hobson, E. S., 1965, Diurnal-nocturnal activity of some inshore fishes in the Gulf of California, *Copeia* **3**:291–302.

Holland, H. D., and Malenin, S. D., 1979, The solubility and occurrence of non-ore minerals, in: *Geochemistry of Hydrothermal Ore Deposits*, 2nd ed. (H. L. Barnes, ed.), Wiley, New York, pp. 461–508.

Horne, R. A., 1969, *Marine Chemistry*, Wiley-Interscience, New York.

Hsü, K. J., and Andrews, J. E., 1970, Summary and conclusions, in: *Initial Reports of the Deep Sea Drilling Project* (A. E. Maxwell, R. P. Von Herzon, J. E. Andrews, R. E. Boyce, E. D. Milow, K. J. Hsu, S. F. Percival, and T. Saito, eds.), U. S. Government Printing Office, Washington, D. C., Volume 3, pp. 445–453.

Hudson, J. D., 1978, Concretions, isotopes, and the diagenetic history of the Oxford Clay (Jurassic) of central England, *Sedimentology* **25**:339–370.

Hudson, J. D., 1982, Pyrite in ammonite-bearing shales from the Jurassic of England and Germany, *Sedimentology* **29**:639–667.

Hudson, J. D., in preparation, Preservation, dissolution and pyritization of aragonitic fossils in organic-rich sediments: Geochemical considerations and empirical evidence.

Hudson, J. D., and Andrews, J. E., 1987, The diagenesis of the Great Estuarine Group, Middle Jurassic, Inner Hebrides, Scotland, in: *Diagenesis of Sedimentary Sequences*, Geol. Soc. London Spec. Publ. Volume 18 (J. D. Marshall, ed.), pp. 259–276. Blackwell Scientific, Oxford.

Hudson, J. D., and Palframan, D. F. B., 1969, The ecology and preservation of the Oxford Clay fauna at Woodham, Buckinghanshire, *Q. J. Geol. Soc. Lond.* **124**:387–418.

Ingle, S. E., 1975, The solubility of calcite in the oceans, *Mar. Chem.*, **3**:301–319.

Irwin, H., Coleman, M. L., and Curtis, C. D., 1977, Isotopic evidence for source of diagenetic carbonates formed during burial of organic-rich sediments, *Nature*, **269**:209–213.

Iversen, N., and Jórgensen, B. B., 1985, Anaerobic methane oxidation rates at the sulfate-methane transition in marine sediments from Kattegat and Skagerrak (Denmark), *Limnol. Oceanogr.*, **30**:944–955.

Jahnke, R., Emerson, S. R., and Murray, J. W., 1982, A model of oxygen reduction, denitrification and organic matter mineralization in marine sediments, *Limnol. Oceanogr.* **27**:610–623.

Jenkyns, H. C., 1988, The early Toarcian (Jurassic) anoxic event: Stratigraphic, sedimentary and geochemical evidence, *Am. J. Sci.*, **288**:101–151.
Jørgensen, B. B., 1977, Bacterial sulfate reduction within reduced microenvironments of oxidised marine sediments, *Mar. Biol.*, **41**:7–17.
Jørgensen, B. B., 1982, Mineralization of organic matter in the sea bed—the role of sulfate reduction, *Nature*, **296**:643–645.
Jørgensen, B. B., 1983, Processes at the sediment-water interface, in: *The Major Biogeochemical Cycles and their Interactions* (B. Bolin and R. B. Cook, eds.), pp. 477–509. Scope 21, Wiley, New York.
Kauffman, E. G., 1981, Ecological reappraisal of the German Posidonienschiefer, in: *Communities of the Past* (J. Gray, A. J. Boucot and W. B. N. Berry eds.), pp. 311–381. Hutchinson Ross, Stroudsberg, Pa.
Klinkhammer, G. P., 1980, Early diagenesis in sediments from the eastern equatorial Pacific, II Pore water metal results, *Earth Planet. Sci. Lett.*, **49**:81–101.
Kobluk, D. R., and Risk, M. J., 1977, Algal borings and framboidal pyrite in Upper Ordovician brachiopods, *Lethia*, **10**:135–143.
Kohlmeyer, J., 1969, The role of marine fungi in the penetration of calcareous substrates, *Am. Zoologist*, **9**:741–746.
Krishnaswami, S., Monaghan, M. C., Westrich, J. T., Bennett, J. T., and Turekian, K. K., 1984, Chronologies of sedimentary processes in sediments of the FOAM site, Long Island Sound, Connecticut, *Am. J. Sci.*, **284**:706–733.
Lippmann, F., 1955, Ton, Geoden und Minerale des Barreme von Hoheneggelsen, *Geol. Rundschau*, **43**:475–503.
Lord, C. J. III, 1980, The Chemistry and Cycling of Iron, Manganese and Sulfur in Salt Marsh Sediments, Ph.D. thesis, University of Delaware.
Lovely, D. R., and Klug, M. J., 1986, Model for the distribution of sulfate reduction and methanogenesis in freshwater sediments, *Geochim. Cosmochim. Acta*. **50**:11–18.
Mackenzie, F. T., Bischoff, W. D., Bishop, F. C., Loyens, M., Schoonmaker, J., and Wollast, R., 1983, Magnesian calcites: Low temperature occurrence, solubility and solid solution behavior, in: *Reviews in Mineralogy* Volume 11 (R. J. Reeder, ed.), Min. Soc. America, Washington, pp. 97–144.
Martens, C. S., and Berner, R. A., 1977, Interstitial water chemistry of anoxic Long Island Sound sediments. I. Dissolved gases, *Limnol. Oceanogr.*, **22**:10–25.
Martens, C. S., Berner, R. A., and Rosenfeld, J. K., 1978, Interstitial water chemistry of anoxic Long Island Sound sediments. II. Nutrient regeneration and phosphate removal, *Limnol. Oceanogr.*, **23**:605–617.
Mayor, A. G., 1924, Causes which produce stable conditions in the depth of the floors of Pacific fringing reef-flats, *Carnegie Inst. Washington Publ.*, **340**:27–36.
McIntyre, A., and McIntyre, R., 1971, Coccolith concentration and differential solution in oceanic sediments, in: *The Micropaleontology of the Oceans* (B. M. Funnell and W. R. Reidel, eds.), Cambridge University Press, Cambridge, pp. 253–261.
McIntyre, W. G., 1965, The temperature variation of the solubility product of calcium carbonate in seawater, *Fish. Res. Bull. Can. Ms. Rep. Ser.* **200**:1–153.
Mehrbach, C., Culberson, C. H., Hawley, J. E., and Pytkowicz, R. M., 1973, Measurements of the apparent dissociation constants of carbonic acid in seawater at atmospheric pressure, *Limnol. Oceanogr.* **18**:897–907.
Millero, F. J., and Berner, R. A., 1972, Effect of pressure on carbonate equilibria in seawater, *Geochim. Cosmochim. Acta* **36**:92–98.
Milliman, J. D., 1974, *Marine Carbonates*, Springer-Verlag, Berlin.
Morse, J. W., 1983, The kinetics of calcium carbonate dissolution and precipitation, in: *Reviews in Mineralogy*, Volume 11 (R. J. Reeder ed.), Mineralogical Society of America, Washington, D. C., pp. 227–264.
Mucci, A., 1983, The solubility of calcite and aragonite in seawater of various salinities, temperatures and one atmosphere total pressure, *Am. J. Sci.* **283**:780–799.
Müller, K. J., 1985, Exceptional preservation in calcareous nodules, *Phil. Trans. R. Soc. Lond. B* **311**:67–73.

Murray, J., and Renard, A. F., 1891, Report on deep-sea deposits based on the specimens collected during the voyage of H.M.S. Challenger in the years 1872 to 1876, Reports Scientific Results Exploration Voyage H.M.S. Challenger, Her Majesty's Stationery Office.

Murray, J. W., Grundmanis, V., and Smethie, W. M., 1978, Interstitial water chemistry in the sediments of Saanich Inlet, *Geochim. Cosmochim. Acta* **42:**1011–1026.

Nancollas, G. H., and Reddy, M. M., 1971, The crystallization of calcium carbonate. II. Calcite growth mechanism, *J. Colloid Interface Sci.* **37:**824–829.

Nissenbaum, A., Presley, B. J., and Kaplan, I. R., 1972, Early diagenesis in a reducing fjord, Saanich Inlet, British Columbia. I. Chemical and isotopic changes in major components of interstitial water, *Geochim. Cosmochim. Acta* **36:**1007–1027.

Oertel, G., and Curtis, C. D., 1972, Clay-ironstone preserving fabrics due to progressive compaction, *Geol. Soc. Am. Bull.* **83:**2597–2606.

Pearson, R., 1979, Geochemistry of the Hepworth Carboniferous sediment sequence and origin of the diagenetic iron minerals and concretions, *Geochim. Cosmochim. Acta* **43:**927–941.

Perkins, R. D., and Halsey, S. D., 1971, Geologic significance of microboring fungi and algae in Carolina shelf sediments, *J. Sediment. Petrol.* **41:**843–853.

Peterson, M. N. A., 1966, Calcite: Rates of dissolution in a vertical profile in the central Pacific, *Science* **154:**1542–1544.

Postma, D., 1981, Formation of siderite and vivianite and the pore water composition of a recent bog sediment in Denmark, *Chem. Geol.* **31:**225–244.

Postma, D., 1982, Pyrite and siderite formation in brackish and freshwater swamp sediments, *Am. J. Sci.* **282:**1151–1183.

Pye, K., Dickson, J. A. D., Schiavon, N., Coleman, M. L. and Cox, M., 1990, Formation of siderite-Mg calcite-iron sulphide concretions in intertidal marsh and sandflat sediments, north Norfolk, England, *Sedimentology* **37:**325–343.

Raiswell, R., 1971, The growth of Cambrian and Liassic concretions, *Sedimentology* **17:**147–171.

Raiswell, R., 1976, The microbiological formation of carbonate concretions in the Upper Lias of N.E. England, *Chem. Geol.* **18:**227–244.

Raiswell, R., 1982, Pyrite texture, isotopic composition and the availability of iron, *Am. J. Sci.* **282:**1244–1263.

Raiswell, R., 1987, Non-steady state microbiological diagenesis and the origin of concretions and nodular limestones, in: Diagenesis of Sedimentary Sequences (J. D. Marshall, ed.), *Geol. Soc. Lond. Spec. Publ.* **36:**41–54.

Raiswell, R., 1988a, A chemical model for the origin of minor limestone-shale cycles by anaerobic methane oxidation, *Geology* **16:**641–644.

Raiswell, R., 1988b, Evidence for surface-reaction controlled growth of carbonate concretions in shales, *Sedimentology* **35:**571–575.

Raiswell, R., and Berner, R. A., 1985, Pyrite formation in euxinic and semi-euxinic sediment, *Am. J. Sci.* **285:**710–724.

Reaves, C. M., 1986, Organic matter metabolizability and calcium carbonate dissolution in nearshore marine muds, *J. Sediment. Petrol.* **56:**486–494.

Reddy, M. M., and Nancollas, G. H., 1971, The crystallization of calcium carbonate. I. Isotopic exchange and kinetics, *J. Colloid Interface Sci.* **36:**166–172.

Redfield, A. C., 1942, The processes determining the concentration of oxygen, phosphate and other organic derivatives within the depth of the Atlantic Ocean, *Pap. Phys. Oceanogr. Meteorol.* **9:**1–22.

Redfield, A. C., Ketchum, B. J., and Richards, F. A., 1963, The influence of organisms on the composition of seawater, in: *The Sea,* Volume 2 (M. N. Hill, ed.), Wiley-Interscience, New York, pp. 26–77.

Reeburgh, W. S., 1980, Anaerobic methane oxidation rate depth distribution in Skan Bay sediments, *Earth Planet. Sci. Lett.* **47:**345–352.

Reimers, C. E., Lange, C. B., Tabak, M., and Bernhard, J. M., 1990, Seasonal spillover and varve formation in the Santa Barbara Basin, California. *Limnol. Oceanogr.* **35:**1577–1585.

Rhoads, D. C., 1974, Organism–sediment relation on the muddy seafloor, *Oceanogr. Mar. Biol. Annu. Rev.* **12:**263–300.

Richards, F. A., Cline, J. D., Broenkow, W. W., and Atkinson, L. P., 1965, Some consequences of the decomposition of organic matter in Lake Nitinat, an anoxic fjord, *Limnol. Oceanogr.* **10:**R185–R200.

Rittenberg, S. C., Emery, K. O., and Orr, W. L., 1955, Regeneration of nutrients in sediments of marine basins, *Deep Sea Res.* **3:**23–45.

Sass, E., and Kolodny, Y., 1972, Stable isotopes, chemistry and petrology of carbonate concretions (Mishash Formation, Israel), *Chem. Geol.* **10:**261–286.

Savrda, C. E., and Bottjer, D. J., 1988, Limestone concretion growth documented by trace-fossil relations, *Geology* **16:**908–911.

Savrda, C. E., and Bottjer, D. J., 1989, Anatomy and implications of bioturbated beds in "black shale" sequences: Examples from the Jurassic Posidonienschiefer (Southern Germany), *Palaios* **4:**330–342.

Schmidt, V., and McDonald, D. A., 1979, The role of secondary porosity in the course of sandstone diagenesis, *Soc. Econ. Paleontol. Mineral. Spec. Publ.* **26:**175–207.

Schroeder, J. H., 1972, Calcified algal filaments of an endolithic algae in Recent Bermuda reefs, *Neues Jahrb. Geol. Paläontol. Abh.* **1:**16–33.

Scott, A. C., and Rex, G., 1988, The formation and significance of Carboniferous Coal Balls, *Phil. Trans. R. Soc. Lond. B* **311:**123–137.

Seilacher, A., 1970, Begriff and bedeutung der Fossil-Lagerstätten, *Neues Jahrb. Geol. Paläontol. Abh.* **1970:**34–39.

Seilacher, A., and Westphal, F., 1971, *Fossil-Lagerstätten*, in: *Sedimentology of Parts of Central Europe, Guidebook Eighth International Sedimentological Congress* (G. Muller, ed.), pp. 327–335.

Seilacher, A., Andalib, F., Dietl, G., and Gocht, H., 1976, Preservational history of compressed Jurassic ammonites from southern Germany, *Neues Jahrb. Geol. Paläontol. Abh.* **152:**307–356.

Sholkovitz, E., 1973, Interstitial water chemistry of the Santa Barbara Basin sediments, *Geochim. Cosmochim. Acta* **37:**2043–2073.

Srivastava, N. K., 1975, Early diagenetic changes in Recent molluscan shells, *Neues Jahrb. Geol. Paläontol. Abh.* **148:**380–403.

Surdam, R. C., Boese, S. W., and Crossey, L. J., 1984, The chemistry of secondary porosity, *Am. Assoc. Petrol. Geol. Mem.* **37:**127–151.

Sverdrup, H. V., Johnson, M. W., and Fleming, R. H., 1942, *The Oceans: Their Physics, Chemistry and General Biology*, Prentice Hall, New York.

Thorstenson, D. C., 1970, Equilibrium distribution of small organic molecules in natural waters, *Geochim. Cosmochim. Acta* **34:**745–770.

Tanabe, K., Inazumi, I., Tamahama, K., and Katsuta, T., 1983, Taphonomy of half and compressed ammonites from the Lower Jurassic black shales of the Toyora Area, West Japan, *Paleogeogr. Paleoclimatol. Paleoecol.* **47:**329–347.

Thyne, G., and Boles, J. R., 1989, Isotopic evidence for the origin of the Moerki septarian concretions, *J. Sediment. Petrol.* **59:**272–279.

Tissot, B. P., and Welte, D. H., 1984, *Petroleum Formation and Occurrence*, 2nd ed., Springer-Verlag, Berlin.

Turekian, K. K., and Armstrong, R. L., 1961, Chemical and mineralogical composition of fossil molluscan shells from the Fox Hills Formation, South Dakota, *Geol. Soc. Am. Bull.* **72:**1817–1828.

Waage, K. M., 1964, Origin of repeated concretion layers in the Fox Hills Formation, *Kansas Geol. Surv. Bull.* **169:**541–563.

Walter, L. M., and Burton, E. A., 1990, Dissolution of recent platform carbonate sediments in marine pore fluids, *Am. J. Sci.* **290:**601–643.

Walter, L. M., and Morse, J. W., 1984, Reactive surface area of skeletal carbonates during dissolution: Effect of grain size, *J. Sediment. Petrol.* **54:**1081–1090.

Weeks, L. G., 1957, Origin of carbonate concretions in shales, Magdelena Valley, Columbia, *Geol. Surv. Am. Bull.* **68:**95–102.

Weiss, R. F., 1974, Carbon dioxide in water and seawater: The solubility of a nonideal gas, *Mar. Chem.* **2:**203–215.

Westrich, J. T., 1983, The consequences and controls of bacterial sulfate reduction in marine sediments, Ph. D. thesis, Yale University, New Haven, Connecticut.

Woodland, B. A., and Stenstrom, R. C., 1979, The occurrence and origin of siderite concretions in the Francis Creek Shale (Pennsylvanian) of northeast Illinois, in: *Mazon Creek Fossils* (M. H. Nitecki, ed.), Academic Press, New York, pp. 69–103.

Wu, E., and Berger, W. H., 1989, Planktonic foraminifera: Differential dissolution and the Quatenary stable isotope record in the west equatorial Pacific, *Paleoceanography* **4**:181–198.

Zangerl, R., Woodland, B. G., Richardson, E. S. Jr., and Zachry, D. L. Jr., 1969, Early diagenetic phenomena in the Fayetteville Black Shale (Mississippian) of Arkansas, *Sediment. Geol.* **3**:87–119.

Chapter 10

Silicification of Fossils

GREG A. CARSON

GREG A. CARSON • Geochem Group Ltd., Chester CH4 8RD, United Kingdom.

Taphonomy: Releasing the Data Locked in the Fossil Record, Volume 9 of *Topics in Geobiology,* edited by Peter A. Allison and Derek E. G. Briggs, Plenum Press, New York, 1991.

1. Introduction

The level of fossil preservation is a function of the composition of the living organism and the diagenetic realms through which it passes during its burial history. Generally, the more recent the specimen, the more it is an accurate representation of the original organism. The older the fossil, the more likely it is that the original fabric has been altered by diagenetic processes. It is less common to find well-preserved soft tissue than well-preserved skeletal components. Likewise, the preservation potential of skeletal matter may be ranked according to its mineralogy. For instance, biogenic phosphatic material tends to be better preserved than low-Mg-calcite material, which in turn is less soluble than aragonite (Walter, 1985).

If destruction of body fossils occurs prior to burial and lithification of the surrounding sediment, all record of the organism (save for ichnofabrics) will be lost. When dissolution and decay of the organism postdates lithification, then a mold results. Subsequent mineral precipitation in the mold, from diagenetic fluids, may be unrelated to the original mineralogy. As well as wholesale degradation of the original organism, skeletal replacement commonly occurs by neomorphic processes, through alteration by a thin-film dissolution process (Pingitore, 1982; Sandberg and Hudson, 1983; James and Choquette, 1984).

The resulting replacement, either void-fill or neomorphic, may also be susceptible to further diagenetic modification. As with the original mineralogy, the diagenetic potential of the replacement will be affected by its mineralogy. Silica in the form of quartz is a highly stable mineral at earth surface temperatures and pressures. Once a fossil is silicified, it tends to resist further morphological change, although some diagenetic alteration may occur. Consequently, silicified fossils may be most prominent in successions where aggressive diagenesis has occurred.

2. Morphology/Petrography of Sedimentary Silica

2.1. Hydration States

In saturated or near-saturated solutions, silicon exists as hydrated molecules, H_4SiO_4 (Jones and Segnit, 1972; Williams *et al.*, 1985). These link up to form silica by the elimination of water. Thus, the principal crystalline forms of silica exist predominantly as tetrahedra (Deer *et al.*, 1966). In the solid state, SiO_4 groups are linked as each oxygen is shared between two tetrahedra (Deer *et al.*, 1966). Hypothetically, this leaves free oxygen (O^-) at the surface of each crystallographic unit. These may be strongly bonded by hydroxyl groups to form anhydrous vicinal bonding (Iler, 1979), where each tetrahedron shares a hydrogen atom. This is usually overlain by two or more layers of molecular water, the hydrogen bonded to the hydroxyl groups [hydrated vicinal bonding (Iler, 1979)]. In addition to these structurally adsorbed waters, absorbed waters may be present as free molecules on the silica surface. Combined isotopic and differential thermal analysis experiments performed on biogenic silica from the Monterey Formation (Knauth and Epstein, 1982) identified five different forms of water within the silica with respect to hydration state and isotopic exchange properties. Only a small proportion of these waters (bonded hydroxyl groups) were resistant to isotopic exchange with ground waters at temperatures greater than 100°C. Water may also be present as fluid inclusions (Folk and Weaver, 1952).

2.2. Minerals and Fabrics

Types of sedimentary silica may be subdivided initially into two groups according to the crystal structure, as determined by X-ray diffraction (XRD): quartz and opal. Quartz may be identified by prominent peaks at 3.34, 4.26, and 1.82 Å (Berry, 1974), whereas different varieties of opal are identified by peaks around 4 Å.

2.2.1. Quartz

The most useful classification of chemically precipitated quartz in sedimentary rocks is based on petrography (Folk and Pittman, 1971). Two principal forms are defined, megaquartz (crystal size >20 μm) and microquartz (crystal size <20 μm). Microquartz may be divided into two morphologies: (1) microcrystalline quartz, where the grains are equant (usually 1–4 μm across), and (2) chalcedonic quartz, where crystals have a bladed or fibrous nature (length:width ratio >4). Depending upon the optical orientation of the fibers, chalcedonic quartz (or chalcedony) may be classified as chalcedonite (length-fast, *c* axis perpendicular to the fibers), quartzine (length-slow, *c* axis parallel to the fibers), and lutecite (length-slow, *c* axis ~30° to the fibers).

There is little structural variation among the various types of quartz (as determined by XRD) within the range of 4.44–1.82 Å. There is, however, variation within a set of five peaks between 1.40 and 1.36 Å. This variation allows a classification of quartz to be developed on the basis of crystal structure. By measuring the intensity of the (212) peak at 1.38 Å relative to the background, Murata and Norman (1976) established an index of crystallinity for quartz. This varies from less than one, e.g., a silicified *Tempskya* trunk from the Cretaceous of Oregon (Read and Brown, 1937), up to ~10.0, e.g., clear euhedral vein quartz of Paleozoic age. The highest value obtained for a silicified fossil was 4.6 (Pennsylvanian silicified wood from Kentucky). Further XRD analyses of silicified wood by Stein (1982) demonstrated that the crystallinity index increases proportionally with age. The large spread of the data (coupled with petrographic evidence) indicated that chalcedony may be present alongside euhedral quartz in samples as old as the Carboniferous.

Chalcedony usually forms as a true replacement of the bioclast structure, whereas megaquartz and microcrystalline quartz infill pore space, forming a mold or cast. An exception is the direct replacement of bioclasts by euhedral megaquartz, as documented by Orme (1974), Henderson (1984), and Maliva and Siever (1988). Also, chalcedony replacement of a fossil may transform (Section 2.3) to a micro- or megaquartz fabric (e.g., Eley and Jull, 1982; Scurfield and Segnit, 1984).

2.2.2. Opal

Types of natural opal may be distinguished by gross morphological variation. However, structural variation of natural hydrous silicates allows them to be divided into four well-defined groups: opal-C, opal-A, opal-CT (Jones and Segnit, 1971), and disordered tridymite (Tada and Iijima, 1983). The most ordered of these is opal-C (α-cristobalite); it is mostly associated with lava flows (Jones and Segnit, 1971) and thus will not be considered here. All biogenic silica is opal-A, amorphous silica characterized by a broad XRD peak (centered around 4.1 Å) representing the (101) reflection of cristobalite. Opal-CT (disordered α-cristobalite, α-tridymite) shows a pair of better-defined (but nevertheless still broad) XRD peaks. The strongest of these lie between 4.05 and 4.12 Å (Iijima and Tada, 1981), with the lower $d(101)$ spacing representing the most deeply buried samples (Murata and Nakata, 1974). A weak peak around 4.3 Å [and thus frequently masked by the (100) reflection of associated quartz, (Tada and Iijima, 1983)] is attributed to the (200) reflection of tridymite. Opal-CT usually precipitates as small ($\sim$3–12 μm, depending upon location) spherulitic masses of $\sim$300- to 500-μm-thick bladed crystals (Wise and Kelts, 1972) known as lepispheres. Clayton (1982) suggested that aggregated masses of lepispheres may have functioned as precursors for Upper Cretaceous flints. The fourth phase of opal, disordered tridymite (Tada and Iijima, 1983), is characterized by a pair of strong peaks at 4.11 and 4.32 Å, a moderate peak at 2.51 Å, and a broad peak around 3.9 Å. It occurs as pore cement and veins (Iijima and Tada, 1981) in opaline chert and porcelanite, and is likely to be the phase observed in many wood opals (Mitchell and Tufts, 1973; cf. Scurfield and Segnit, 1984).

2.3. Mineral Transformations

Diagenetic modification most commonly proceeds in the direction opal-A to opal-CT to quartz (Fig. 1) (Kastner, 1981; Stein, 1982). The opal-A to opal-CT (e.g., cherts) or opal-A to quartz (e.g., bioclast replacement, void fills) transformation is a dissolution–reprecipitation reaction. This reaction occurs both within pore waters [e.g., remobilization of sponge spicule silica (Jarvis *et al.*, 1982)] and in oceanic water; the general dissolution of siliceous plankton results in a higher silica concentration at depth (Calvert, 1974,

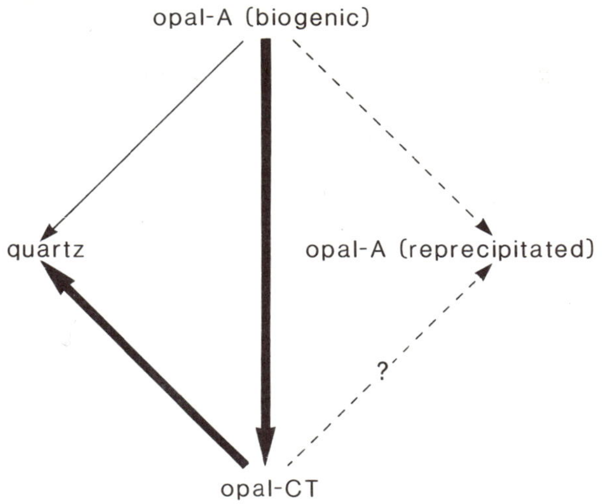

FIGURE 1. Phase transformations of silica in deep-sea sediments. The thicker the arrow, the more common the transformation. [After Kastner (1981).]

Fig. 1). Recent work on chert formation (Clayton, 1984; Williams *et al.*, 1985) suggests that the opal-CT to quartz transformation is also a solution reprecipitation reaction. The opal-A to opal-CT to quartz transformation may occur due to increases in temperature and pressure (i.e., depth of burial) (Murata *et al.*, 1977; cf. Land, 1979, p. 230). However, since sediment pore waters may be undersaturated with respect to opal-A yet at the same time supersaturated with respect to quartz, the transformation may be due to a kinetic effect involving a mineralizing agent such as organic matter or carbonate ions (Clayton, 1984, p. 45).

The actual paths by which these transformations take place are complex and depend largely upon surface area affecting solubility (Williams *et al.*, 1985). The total dissolved silica concentration in equilibrium with any silica phase (opal-A, opal-CT, or quartz) depends upon the surface area of the phase in question. Thus, a solution in equilibrium with a phase of higher surface area will be supersaturated with respect to one in equilibrium with the same phase of lower surface area. So dissolution of opal-A with a higher surface area may theoretically lead to the precipitation of an opal-A with a lower surface area (Fig. 2, arrow "a"). Where a point is reached where further changes in specific surface area of the opal-A produce negligible effects on its solubility, the solution will still be supersaturated with respect to opal-CT and quartz. Since for a total dissolved silica concentration, opal-CT will have a lower specific surface area than quartz (Fig. 2), opal-CT will be the next phase to precipitate. Depending upon the speed of opal-CT nucleation relative to growth, opal-CT of high (nucleation rapid relative to growth; Fig. 2, arrow "b") or low specific surface area (nucleation slow relative to growth; Fig. 2, arrow "c") may form. The opal-CT produced will then evolve to a configuration of lower specific surface area (Fig. 2, arrow "d"). The transformation of opal-CT to quartz will undergo similar diage-

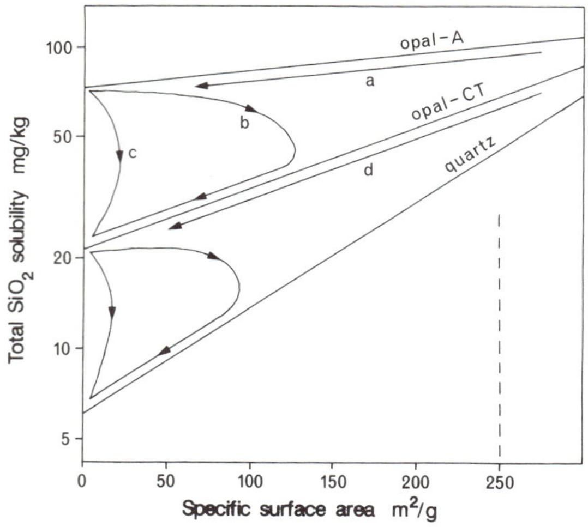

FIGURE 2. The relationship between total SiO_2 solubility and specific surface area of opal-A, opal-CT, and quartz. The vertical dashed line indicates the approximate surface area of the diatom *Thalassiosira decipiens*. See text for details. [After Williams *et al.* (1985).]

netic processes. It may also be possible for opal-A to transform directly into quartz (Kastner, 1981; Williams *et al.*, 1985).

In addition to burial and kinetic effects, chemical variations may be responsible for the transformation of silica minerals. Folk and Pittman (1971) stressed the association between length-slow chalcedony and evaporites. Kastner (1980) showed that the formation of length-slow chalcedony was dependent on the presence of Mg^{2+} and SO_4^{2-} ions. Using hydrothermal experiments, she also demonstrated that the transformation of opal-CT to chalcedony (either length-slow or length-fast), as opposed to microcrystalline quartz, is dependent upon the concentration of SO_4^{2-} and the temperature. Silica solutions rich in SO_4^{2-} form relatively more chalcedony than those rich in Cl^-, since the latter are more supersaturated with respect to quartz. Under certain circumstances, the reverse transformation may occur. Khoury (1987) described Late Cretaceous chert from Jordan composed predominantly of cristobalite and/or euhedral quartz. He suggested that the cherts were composed initially of a dense mass of microcrystalline quartz (as in the adjacent Silicified Limestone Unit). The composition of the chert was then altered to cristobalite by fracturing and subsequent leaching by alkaline solutions. Since it is a metastable phase, paramorphism of the cristobalite occurred, forming euhedral quartz crystals.

3. Fossil Preservation

3.1. Skeletal Preservation

3.1.1. Previous Research

The silicification of bioclasts has preserved data of great significance to paleontology (e.g., Brunton, 1966, 1968, 1984; Curry, 1986). Only recently, however, has attention been paid to processes of replacement (Schmitt and Boyd, 1981; Holdaway and Clayton, 1982; Carson, 1987; Maliva and Siever, 1988).

In a study of the silica diagenesis in Upper Jurassic limestones of southern England, Wilson (1966) erected a classification of silica morphologies based on general grain shape, intergrain relationships, and optical properties. He differentiated five morphological variations, four of which (mosaic quartz, spherulitic chalcedony, lutecite, and microgranular quartz) were present as bioclast replacement fabrics. Apart from the general suggestion that spherulitic chalcedony replaced aragonitic skeletons, while lutecite preferentially replaced calcitic shells, no attempt was made to relate the style of silicification to the type of fossil.

Such a relationship was ascertained by Brown *et al.* (1969), in their study of partially silicified Cretaceous brachiopods, echinoids, and bivalves from England. They established that, while rhynchonellids were completely silicified, the silicification of terebratulids tended to be restricted to the posterior part of the shell. They also determined that the quartz crystals in the rhynchonellids were lath-shaped and elongated in a posterior–anterior direction with the flat side oriented parallel to the shell surface. The quartz crystals in the terebratulids, as well as being smaller, were more rod-shaped, the long axis running perpendicular to the shell surface. Siedlecka (1972) documented the selective silicification of similar faunas from the Permian of Svalbard. She noted that although the silica fibers usually paralleled the lamellar structure of the brachiopods, length-slow chalcedony spherulites transecting the fabric of the shell were also developed.

Although Brown *et al.* (1969) established the relationship between silica morphology and fossil taxa, no geochemical considerations of the replacement process were put forward until the study by Jacka (1974) of silicified fossils from the Middle Permian turbidites of the Delaware Basin. He suggested that the association of dolomite rhombs with the silica replacement was dependent on the original mineralogy of the skeleton (high-Mg calcite vs. low-Mg calcite). He also stressed the importance of organic-matter decomposition in producing a microenvironment conducive to silicification.

3.1.2. Classification and Mechanisms of Skeletal Silicification

Holdaway and Clayton (1982) divided the bioclast silicification fabrics found in the Cenomanian of southwest England into three types: a granular white crust, beekite rings [concentric rings of silica formed by a "liesegang" diffusion mechanism (Hedges, 1932)], and a fine-scale textural replacement. These different styles reflect the rate of silica supply relative to carbonate dissolution. The rate of carbonate dissolution is in turn linked to the size of the individual crystals within the original skeleton (as well as the presence of organic layers/sheaths; Section 5.2.2). For example, the large crystal size in echinoids caused a granular crust (cryptocrystalline quartz) to form, dissolution of carbonate being slow compared to the precipitation of silica.

Beekite was formed in bivalves (especially *Exogyra*) due to the rapid dissolution of the smaller calcite crystals. A fine-scale replacement occurred in brachiopods due to the intermediate size of the shell prisms. Holdaway and Clayton (1982) concluded that the different styles of silica replacement are thus restricted to different fossil taxa. Henderson (1984) also suggested a rate control on gross silica morphology within a silicified Devonian stromatoporoid from northeast Australia. He noted that megaquartz replaced the interior of the skeleton. However, shorter diffusion pathways adjacent to the external part of the skeleton caused faster replacement and the restricted formation of quartzine.

There is little doubt that primary skeletal structure does play an important role in determining sites for silicification. Elorza (1987) documented preferential silicification (aggregates of quartzine-lutecite) of nummulites and bivalve fragments in the Middle Eocene turbidites of northeast Spain. Associated coralline algae, bryozoans, worm tubes, and echinoderm fragments remain unaffected. Fabric selectivity may also occur within individual bioclasts. In gigantoproductids from central England, replacement is essentially restricted to individual lamellar layers, while the adjacent prismatic layers remain largely unaltered (Fig. 3) (Crowley, 1988).

However, the fabrics are by no means mutually exclusive (Carson, 1987). In brachiopods where concentric beekite structures were visible in

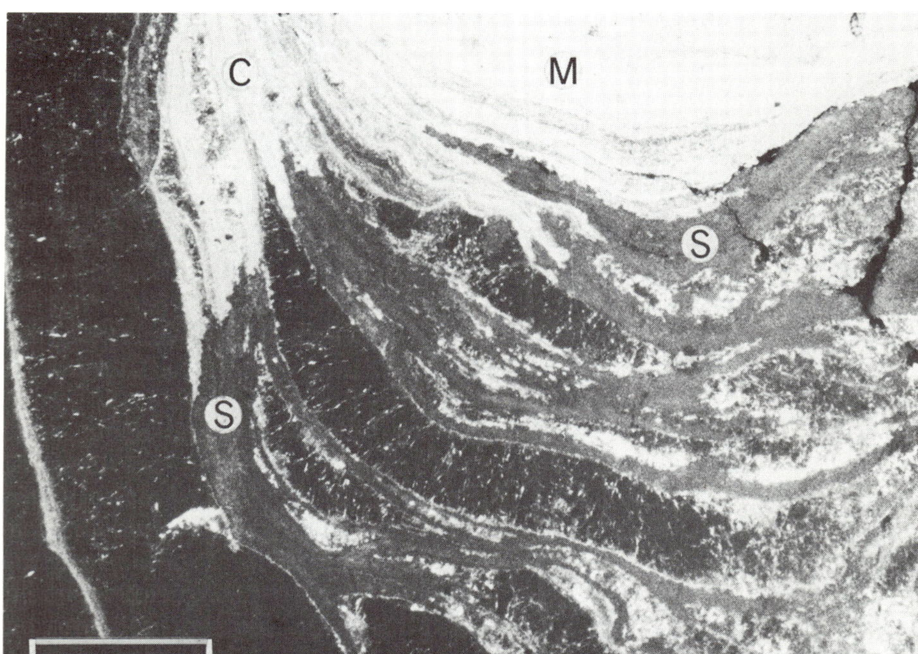

FIGURE 3. Gigantoproductid, Carboniferous, central England. Note that silicification (S) preferentially occurs along the lamellar shell layers. Where silicification has not taken place (C), the lamellar shell exhibits luminescence similar to that of the surrounding sediment (M). Thin section, cathodoluminescence. Scale, 500 μm. [Photograph from Crowley (1988).]

hand specimen, scanning electron microscopy (SEM) revealed fine-scale replacement. Even when observed at the same magnification, completely different fabrics may be evident within a single specimen. At a macroscopic scale, one fragment of *Inoceramus* displayed two completely different styles of silicification: replacement along prism boundaries (Fig. 4), and formation of chalcedony botryoids that completely transect prism boundaries (Fig. 5). Similarly, Carson (1987) noted that one valve of a pectinid comprised at least three texturally different forms of silica (Fig. 6): a very soft (chalklike) pure white replacement occurring as small (~2 mm) blebs sporadically in the shell, a white beekite replacement (from which individual zones were sampled), and a gray fine-scale replacement. Some types of fabrics also appeared to be gradational within single specimens. In the pectinid valve described above, the white beekite frequently passes into the gray replacement (i.e., the latter may possess concentric rings), although the boundary may also be abrupt. Beekite was also observed grading into fine-grained quartz in *Exogyra* specimens (Carson, 1987), and the fine-scale replacement in *Gigantoproductus* observed by Crowley (1988) grades into euhedral microcrytalline quartz (Fig. 7).

From the above, it is obvious that the relationship between the type of silica replacement and the host bioclast is extremely complex. Although the size of the individual $CaCO_3$ crystals and rate of silica supply play a role in

FIGURE 4. Silica following prism boundaries (arrows) in *Inoceramus*. Note the corroded prisms where silicification has occurred (right) compared to the unaltered shell (left). Mid-Cretaceous, southern England. SEM photograph. Scale, 0.1 mm. [Reproduced from Carson (1987) with permission of Blackwell Scientific Publications Ltd. (Oxford).]

FIGURE 5. Chalcedony botryoid cutting across prisms of *Inoceramus*. Same specimen as in Fig. 4. SEM photograph. Scale, 0.1 mm. [Reproduced from Carson (1987) with permission of Blackwell Scientific Publications Ltd. (Oxford).]

determining the type of fabric that develops, there are clearly other factors (perhaps the presence of microorganisms within the structure of the shell) that induce the formation of one fabric preferentially over another.

In interpreting replacement fabrics, relating the same phenomena observed at different scales is also a problem. For example, faint concentric beekite observed in hand specimen may not be visible under the SEM. Also, where beekite is prominent in hand specimen, a transverse cut under thin section may appear as radiating chalcedony rosettes. It would be incorrect, however, to assume that the presence of chalcedony rosettes in thin section automatically indicates the development of beekite.

Thus it seems that only a broad morphological classification of replacement fabrics may be useful, based on the morphology of the silica types (Wilson, 1966) as well as the scale at which the fabric is considered. The different forms are beekite, fine-scale replacement [the exact form dependent upon the fossil taxon (Carson, 1987)], granular structure, megaquartz, botryoids/rosettes/spherulites, and cryptocrystalline quartz (Fig. 8).

3.1.3. Pyritization

Although pyritization of fossils is discussed elsewhere (Raiswell and Canfield, this volume, Chapter 9), the frequent intimate association of

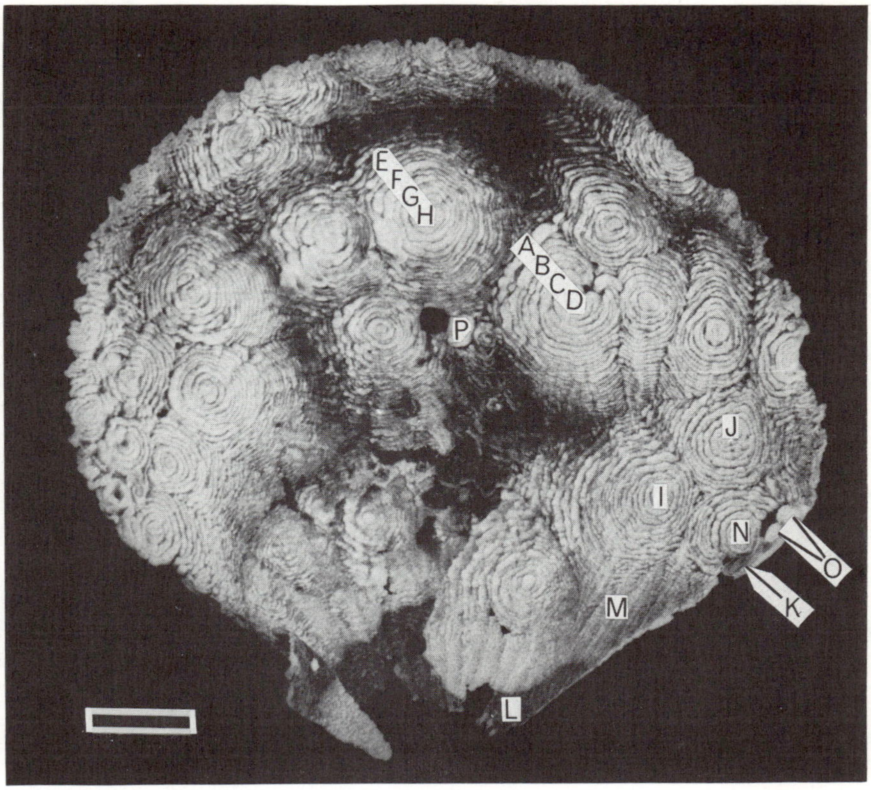

FIGURE 6. Silicified pectinid (etched in dilute HCl). Most of the shell consists of prominent beekite (A–J) and a dark fine-scale replacement (K–M) with occasional blebs of soft, white silica (N–P) (N is scraped from the surface of a beekite ring). Mid-Cretaceous, southern England. Scale, 1 cm. Lettering corresponds to analyses shown in Table I. [Reproduced from Carson (1987) with permission of Blackwell Scientific Publications Ltd. (Oxford).]

opaque minerals with silicified bioclasts must be stressed. The opaque minerals may be pyrite (Holdaway and Clayton, 1982), its alteration products (e.g., goethite), hematite (Loope and Watkins, 1989), or occasionally a Mn-bearing phase [probably pyrolusite) (Carson, 1988)]. The gross structure of the pyrite may closely follow the skeletal fabric or occur in a more irregular form (Fig. 9); both may occur within individual bioclasts. Study of a silicified *Chlamys aspersa*, for example, revealed a thick ($\sim 150\ \mu m$) layer of pyrite separating the silicified area of the inner part of the shell from the unsilicified outer part. The latter, however, was also mineralized, the pyrite following the foliated shell fabric (Carson, 1988).

3.2. Soft-Tissue Preservation

3.2.1. Wood and Other Plant Material

The most striking examples of fossil silicification are petrified forests such as that in Arizona, where *in situ* trunks cover an area of over 100 km^2

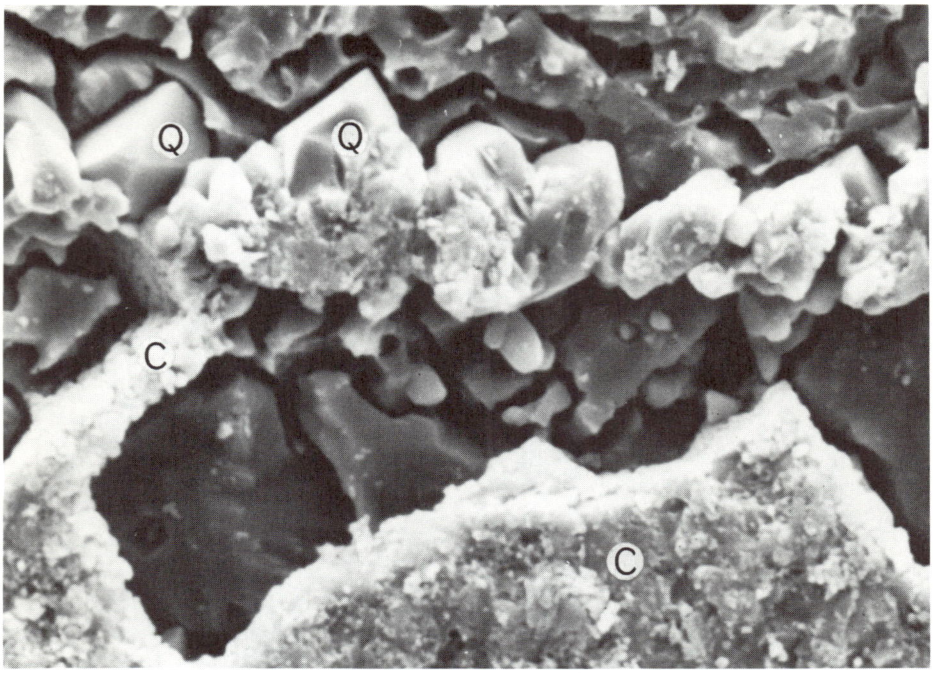

FIGURE 7. Gigantoproductid, Carboniferous, central England. Replacement chalcedony (C) grades into euhedral quartz (Q), indicating a decrease in silica supply and the generation of porosity prior to silica replacement. SEM photograph. Scale, 1 μm. [Photograph courtesy of S. F. Crowley.]

(Dorf, 1964). Silicified wood has been documented from numerous other localities (Murata, 1940; Francis, 1983; Astin, 1987). Silicified plants range from the Devonian (Leo and Barghoorn, 1976) to Recent (Buurman *et al.*, 1973).

Cenozoic silicified plant material is in the form of opal which may show characteristics of tridymite. With time/burial the concentration of silica tends to increase (Karowe and Jefferson, 1987) and the concentration of water diminishes (Scurfield and Segnit, 1984), and eventually the opal transforms to quartz (see below), so that ancient specimens are composed of microcrystalline quartz or chalcedony (Leo and Barghoorn, 1976). At a macroscopic scale, silicified plant material does not show different surficial morphological variation (i.e., beekite, granular crusts, etc.). At higher magnifications, the silica appears to mimic the cellular microstructure, and lepispheric structures (opal-CT) may be visible in young samples (Stein, 1982). Scurfield and Segnit (1984) observed botryoidal opal-CT and chalcedony within silicified softwoods and hardwoods, similar to structures found in skeletal silica replacements.

Silicification of plant material is not a replacement, but a permineralization (Buurman *et al.*, 1973; Leo and Barghoorn, 1976; Scurfield and Segnit, 1984; Knoll, 1985; Francis, 1982), where the organic structure acts

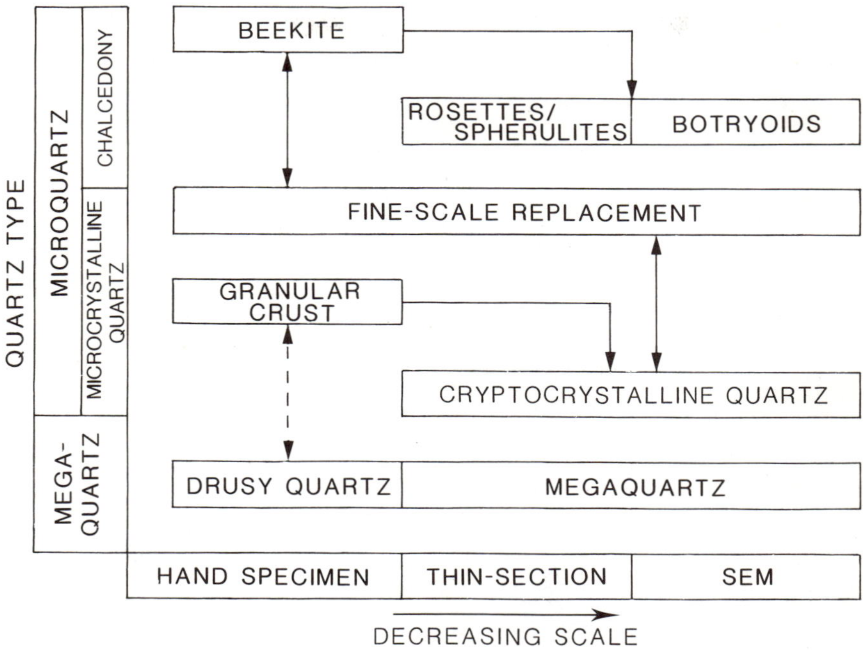

FIGURE 8. Schematic diagram illustrating the relationship between the various morphological forms of quartz in silicified bioclasts.

as a template for silica deposition (Fig. 10). The preservation of cell walls and bordered pits in many examples of fossil wood provides evidence for silicification prior to significant decay (Karowe and Jefferson, 1987). Leo and Barghoorn (1976) emphasized that the information preserved in a silicified plant is limited by the state of the organic structure immediately prior to silicification. This produces a preservational bias since certain components of vascular tissue are more resistant to decay than others [lignin is more resistant than holocellulose (Kenrick and Edwards, 1988)]. However, Leo and Barghoorn (1976) stressed that a certain amount of decay is necessary for effective mineralization, in order to increase the permeability of cell walls, create void space for the deposition of silica, and create active sites for bonding between the organic and mineral components (Section 5.2.1).

3.2.2. Microorganisms

The study of silicified prokaryotes has been most valuable in elucidating the earliest forms of life (Hofmann and Schopf, 1983). Most "silicified" Proterozoic organisms have not been replaced by silica, but are in the form of organic structures embedded in silica [although the organic matter may have been replaced by other minerals (Barghoorn and Tyler, 1965)].

The chert within Banded Iron Formations (e.g., the Gunflint Formation)

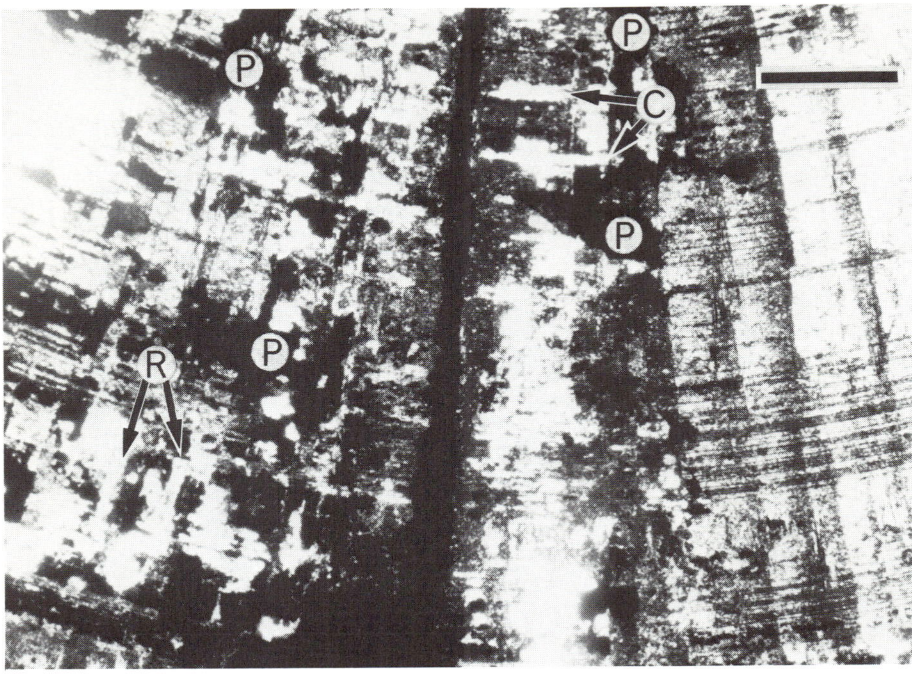

FIGURE 9. Silicification of a belemnite following the concentric (C) and radial (R) structure of the guard. Note the association of pyrite (P) with the silica. Cenomanian, southern England. Scale, 500 μm.

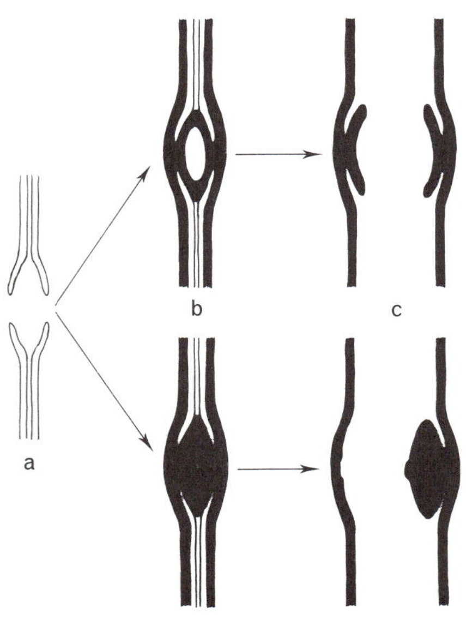

FIGURE 10. The emplacement of silica within vascular tissue. (a) Unmineralized wood, showing two adjacent tracheid walls and the interconnection of bordered pits. (b) Early stages of silica impregnation (black). Silica may partially (top) or completely (bottom) fill the pit chamber. (c) Sample preparation: Following removal of the organic component and maceration, the bordered pits may be represented by either convex or concave deposits of silica. [After Leo and Barghoorn (1976).]

is believed to have originated from a primary precipitate in seawater (Barghoorn and Tyler, 1965; Knoll and Simonson, 1981), but most other Proterozoic cherts are considered to be secondary, either after carbonates and evaporites, or as permineralizations of microbial peats (Knoll, 1985). Some microbial structures are composed of decay-resistant compounds which possess a higher preservation potential than others (Knoll, 1985); e.g., the sheaths of cyanobacteria (Hofmann and Schopf, 1983) and the reproductive cysts of protists.

Although early diagenetic silicification is a prerequisite for the exceptional preservation of microorganisms, Knoll (1985) stressed the importance of the short-term preservation of the organic matter prior to silicification. Microbial degradation in organic-rich sediments has to be arrested at an early stage [probably by cells binding metallic iron (Ferris *et al.*, 1988)] to ensure structural preservation. For this reason the information gained from Proterozoic microfossil assemblages may be environmentally limited (Knoll, 1985), producing a biased view of early microbial life (Section 6).

Silicified microorganisms are not restricted to the Proterozoic. Knoll (1985) illustrated fungi embedded within silicified peat of Devonian and Miocene–Pliocene ages. Small ($\sim$15–30 μm in diameter) spherical to hemispherical bundles of chalcedony are found in clumps on detrital quartz in mid-Cretaceous silicified nodules in southwest England (Carson, 1988). They are found as annular rings (presumably an intermediate stage of development; Fig. 11) and leave a circular "residue" when plucked off during sample preparation, indicating that they are hollow structures which formed around nucleii $\sim$10 μm in diameter. Cyanophytes vary between 1 and 25 μm in diameter, may be spherical or cylindrical in outline (Brasier, 1980), and thus could possibly have been the nuclei. Of more problematic affinity are the spherical to elongate chalcedony botryoids found on the interior surface where the silicified nodules are hollow (Section 3.3). These are $\sim$300–400 μm in diameter, with the chalcedony fibers radiating out from a hollow central "canal" of $\sim$100 μm diameter. This suggests that silica nucleated initially on elongate organic substrates present on the burrow walls.

3.3. Ichnofabric Preservation

Silicified ichnofaunas are common in parts of the Phanerozoic, conspicuously in the Chalk (Upper Cretaceous) of northwest Europe. Although the average concentration of silica in Chalk sections as a whole amounts to $\sim$2–5% (Håkansson *et al.*, 1974; Clayton, 1984), it is mainly concentrated at discrete chert horizons, intimately linked with omission surfaces (Kennedy and Garrison, 1975). It was during periods of nondeposition that significant biological activity occurred. Trace fossils were first recognized as exerting an influence on flint formation by Voigt and Häntzschel (1956), who noted that spiral flints had developed in *Zoophycos* traces. Bromley (1967)

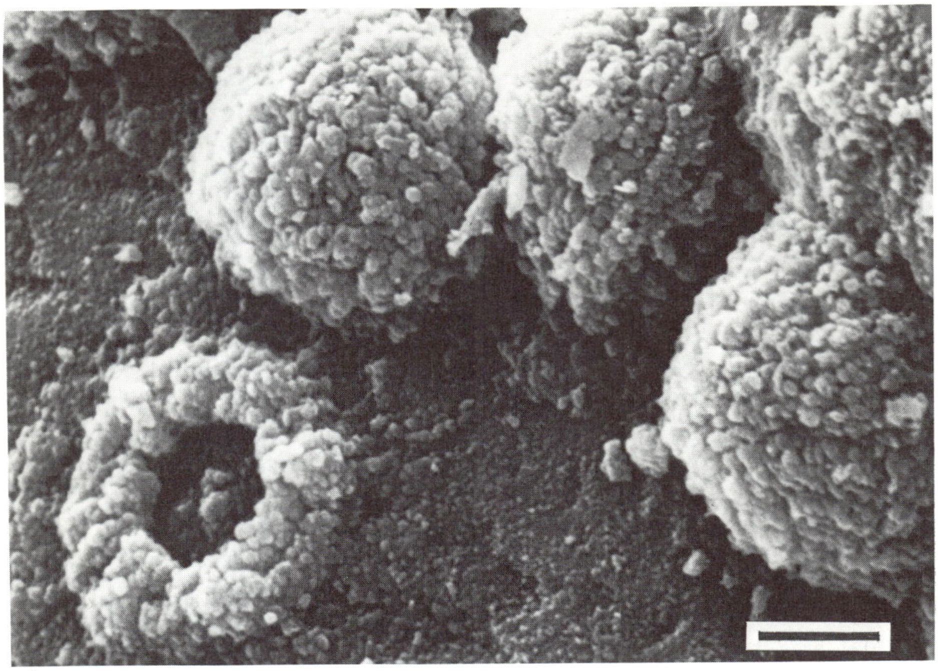

FIGURE 11. Chalcedony botryoids on the surface of a detrital quartz grain in a siliceous nodule. The development of an annular structure suggests that these botryoids are hollow and nucleated around an object of ~10 μm diameter. Cenomanian, southern England. SEM photograph. Scale, 1 μm.

suggested that many flints reflected the presence of *Thalassinoides* burrows. The role of bioturbation in influencing flint morphology is widely accepted (e.g., Håkansson *et al.*, 1974; Bromley and Gale, 1982; Bromley and Ekdale, 1986). The silicification mechanism is linked to the presence of organic material (Bromley *et al.*, 1975; Clayton, 1984), such as fecal matter and burrow linings.

Flints display the sediment fabric in three independent ways: (1) in their external gross morphology, (2) their surface sculpture; and (3) their internal inhomogeneities (Bromley and Ekdale, 1986). *Thalassinoides* burrows are commonly recognized on the basis of their gross morphology. Slender fingerlike projections, about 2 cm in diameter, reflect the morphology of the burrow-fill in which the flint has formed. Larger and more irregular cylindrical bodies represent silicification beyond the walls of the burrow. Digitate burrows are preserved both as solid forms and as hollow tubes within silicified nodules (Carson, 1987). Much ichnological information can be deduced from the surface sculpture and internal inhomogeneities of flints, as well as from morphology (Bromley and Ekdale, 1986).

Silicification may indicate the presence of trace fossils indirectly, as in the formation of paramoudras. These vertical, cylindrical flints may reach a length of several meters and are usually in the form of an annulus surrounding a core of cemented chalk. The formation of a paramoudra is associated with a narrow (about 1 cm) *Bathichnus* burrow in the center of the chalk core.

3.4. Timing of Silicification

Silicification prior to deep-burial diagenesis is a common phenomenon in sedimentary carbonates. Geeslin and Chafetz (1982) demonstrated that silicification of Ordovician anhydrite nodules from southwest New Mexico commenced on the exterior. Frequently, the interior of the nodules was dissolved, and then geopetally infilled with dolomite, indistinguishable from the surrounding matrix. Fractures within the chert truncate sedimentary structures and allochems, and are also infilled with matrix dolomite. Silicification of associated bioclasts may have occurred prior to carbonate lithification.

Similar evidence was recorded by Eley and Jull (1982) in Middle Silurian replacive cherts and silicified fossils from Ontario. Prelithification replacement is indicated by: (1) occurrence of cherts in what were unlithified interbioherm rather than biohermal sediments; (2) differential compaction of laminae around chert nodules and lenses; (3) nodules elongated parallel to bedding planes, indicating formation in unconsolidated sediments; (4) silicification preceding dolomitization; and (5) preservation of organic matter and microorganisms.

Meyers (1977) noted that chert predates both ferroan calcite cementation and intense compaction of Carboniferous sediments in New Mexico, indicating silicification prior to significant burial.

Silicification of fossils in a partially lithified sediment has been postulated (Jacka, 1974; Holdaway and Clayton, 1982). Carson (1988) noted that silicification postdated boring (Fig. 12) and that silica follows the boundaries of sparry calcite cement crystals and intercrystalline boundaries within calcitic neomorphosed shells (Carson, 1987). Bioclast silicification postdated aragonite dissolution and the initial stages of syntaxial cementation, yet was contemporaneous with, or predated, pyrite formation (Holdaway and Clayton, 1982). This suggests that silicification occurred within the zone of sulfate reduction (Section 5.2.2). Clayton (1984) postulated the same environment for the silicification of ichnofaunas within the Chalk (forming flint). He suggested that this occurred at a depth of 5–10 m below the sediment surface on the basis of the height of paramoudras continuous with horizontal flint beds at the base.

Aragonitic faunas preserved in flint are rare. This indicates that aragonite is generally dissolved at depths in excess of 5–10 m. However, large pachydiscid ammonites have been documented as molds in flint from the Upper Cretaceous of eastern England (P. S. Whittlesea and N. Peake, personal communication, 1989). Strong bacterial reduction of organic matter locally within the shells may have induced the conditions necessary for early carbonate dissolution and silica precipitation. The ammonites

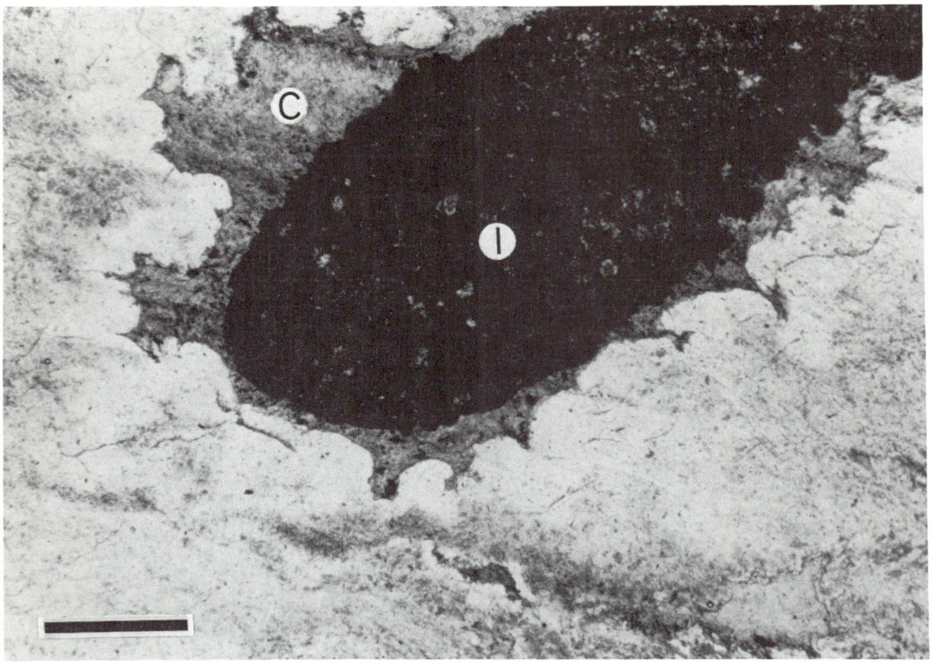

FIGURE 12. Silicification postdating a boring in a large bivalve. The infill of the boring (I) and the shell calcite immediately surrounding it (C) remain unmineralized. Cenomanian, southern England. Scale, 400 μm.

occur in association with hardgrounds. Hardground formation may have been associated with periods of reduced bottom-water oxygenation when sulfate reduction occurred near the sediment surface (Clayton, 1984). Thus, silicified ammonites may also reflect sulfate reduction near the sediment–water interface, prior to aragonite dissolution.

Matrix immediately adjacent to the inner surface of the silicified outer ostracum of pectinids demonstrates how the (nonsilicified) inner ostracum was diagenetically removed following burial but prior to sediment lithification (Schmitt and Boyd, 1981; see also Newell and Boyd, 1970). However, not all skeletal silicification may have occurred during such early burial. Henderson (1984) noted quartzine and megaquartz replacement of a Devonian stromatoporoid from northeast Australia. Silicification, postdating skeletal recrystallization and associated with microstylolite formation, is believed to have taken place during deep burial.

In the case of the silicification of organic material such as wood, a certain amount of structural degradation is necessary to allow silicification to take place, but where excess decay occurs no structure will be preserved (Section 3.2.1). Assuming that no exceptional environment was involved [such as within peat (Knoll, 1985) or under volcanic debris], the preservation of structures associated with soft tissues per se indicates that silicification took place prior to significant burial.

Knoll (1985) implied that time scales of 1–10 years were likely for the silicification of most Proterozoic fossils. He did, however, note that Golubic (1976) documented 8000-year-old nonmineralized microbial peats from Abu Dhabi. Since the cyanobacteria within these were structurally intact, it is reasonable to surmise that ancient microbes could avoid decay for as long as this prior to mineralization.

During the Proterozoic, inorganic silica cementation took place close to the sediment–water interface (Cloud *et al.*, 1979; Simonson, 1987) or even prior to burial (Simonson, 1985b). This probably reflects higher concentrations of dissolved silica within the oceans (Knoll and Simonson, 1981; Simonson, 1985a). Thus, it is likely that silicification of microbes occurred soon after sediment deposition. Indeed, Ferris *et al.* (1986) noted that bacterial cells within Recent geothermal sediments are silicified *at* the sediment–water interface.

4. Sources of Silica through Geologic Time

In modern marine sediments silica may accumulate as biogenic debris (opal-A) in areas of high primary productivity (Calvert, 1974) or from the alteration of clay minerals (Altschuler *et al.*, 1963; Buurman *et al.*, 1973; Leo and Barghoorn, 1976). It may also be released during periods of submarine volcanism (or ocean ridge spreading) and by the submarine breakdown of volcanic rocks. These processes may be interlinked (Garrison, 1974)—the silica released during the alteration of volcanic rocks may enhance the productivity and preservation of siliceous microplankton.

Various authors have identified silica phases as part of volcanic alteration products (Calvert, 1974; Kastner, 1981). Wise *et al.* (1973) observed the conversion of volcanic glass to smectite and opal-CT, apparently by a dissolution–reprecipitation reaction. Berner (1971) suggested that the alteration of volcanic glass in seawater leads to the crystallization of montmorillonite and zeolites. Some silica may be derived subsequently from the transformation of the clay minerals themselves (McBride and Folk, 1979). Altschuler *et al.* (1963) suggested that regional conversion of montmorillonite to kaolinite during subaerial weathering provided the silica for the silicification of bivalve shells in the Pliocene of central Florida. Likewise, Buurman *et al.* (1973) identified the transformation of smectite to kandite (kaolinite) during acid sulfate soil formation as the source of silica in Recent plant fossils from Thailand and the Netherlands. However, Wise and Kelts (1972) indicated that the limited quantity of clinoptilolite (a diagenetic alteration product of volcanic glass with a lower silica content than the original material) present in Tertiary deep-sea chalks is insufficient to account for the large volumes of chert present in the associated lithologies. They concluded that siliceous organisms were the major source of silica.

A biogenic origin for silica from Mesozoic times to present is now widely accepted (Wise *et al.*, 1973; Chester, 1990; cf. Orme, 1974). Calvert (1974) showed that in deep waters around latitude 60°S and in several nearshore areas such as the Gulf of California and the southwest African continental shelf, biogenic opal may reach concentrations of over 70% by weight of the total sediment. Regional variations in the distribution of biogenic opal within sediments can be directly ascribed to a high nutrient flux at regions of upwelling causing high primary productivity. The type of siliceous plankton depends upon position in the water column: 70–90% of suspended silica in present-day surface waters may consist of diatoms, while Radiolaria form the larger portion in deeper waters (Zijlstra, 1987). Regions of upwelling are in turn controlled by oceanic circulation patterns (Calvert, 1966; Garrison, 1974).

In Tethyan Jurassic radiolarites, McBride and Folk (1979) found that the concentration of Radiolaria within chert nodules (preserving the composition of the original sediment) was up to 70% by volume. They also noticed poor preservation of Radiolaria in the adjacent shales: microquartz casts of the central capsules indicated that dissolution of the Radiolaria could not have occurred on the sea floor. In the shale beds, the Radiolaria comprise only 10% of the sediment. Thus, it may be inferred that a large proportion of the biogenic silica was mobilized during early diagenesis. In shallower water conditions, diatoms may have been the primary source of silica [e.g., the Maastrichtian chalk of the Netherlands (Zijlstra, 1987)].

In addition to planktonic siliceous organisms, sponges may have provided an important source of biogenic silica. Siliceous spicules, now replaced by calcite, have been noted in pelagic carbonates throughout the Phanerozoic. The abundance of siliceous sponge spicules in the Aleman Formation (Ordovician) of southwest New Mexico suggests that they provided silica for the formation of ribbon cherts (Geeslin and Chafetz, 1982). Such a source is substantiated by the paucity of clays and other terrigenous material within the formation. Sponge spicules are common in the Fossil Hill Formation (Silurian) of Manitoulin Island, Ontario. Eley and Jull (1982) suggested that this may reflect the proximity of Precambrian quartzite (La Cloche Mountains) enriching the ocean with silica and stimulating "blooms" of siliceous organisms. These are believed to have contributed to the formation of chert. Meyers (1977) believed that the dissolution of clays (even under optimum conditions where pore fluids were initially free of dissolved silica) would not yield enough silica for the chertification of the Carboniferous skeletal packstones and grainstones of New Mexico. He calculated the calcitized spicule/chert volume ratios and estimated that, depending upon the member, sponge spicules produced about 0.4–4.4 times the amount of silica necessary to produce the chert nodules. Wilson (1966) noted that formerly siliceous (*Rhaxella*) sponge spicules (now calcite) formed over 25% of the sediment within the Corallian and Portland Beds of southern England. These would have provided enough silica to form all the observed silicification fabrics in the associated rocks.

Clearly a biogenic source of silica is favored for bioclast silicification during the Phanerozoic, with Porifera (sponges) contributing since the Cambrian, siliceous Sarcodina (Radiolaria) since the Ordovician, and siliceous Chrysophyta (diatoms) since the Cretaceous. Although it has been suggested that organisms may have been responsible for silica deposition during the Precambrian (Barghoorn and Tyler, 1965), it is unlikely that such a source would have been sufficient to account for the ancient chert formations (Simonson, 1985a) and associated silicification of microfossils. An alternative inorganic source of the silica must have been available.

Probably the most widely accepted source for Precambrian silica is volcanic. Murata (1940) first noted a systematic relationship between silicified wood and volcanic ash. He stated that fine-grained ash is highly unstable and easily and rapidly altered (to bentonite), providing a rich source of silica for the petrification of wood embedded in volcanic ash or lava. There are many Phanerozoic locations where the presence of a silicified flora may be related to proximal volcanic lithologies [e.g., the preservation of Tertiary silicified plants in freshwater sediments interbedded with lavas of the Deccan Plateau (Sahni, 1941) and the Eocene petrified forests of Wyoming (Dorf, 1964)]. More recently, a similar volcanic source has been postulated for the silicification of trees engulfed by the eruptions of Mount Saint Helens, Washington (Karowe and Jefferson, 1987). The dominance of volcanic-igneous terrains during the Precambrian (Eugster, 1969) suggests that volcanic or volcanic-associated sources may have provided silica for the silicification of Proterozoic fossils.

The possibility of other sources of silica during the Precambrian cannot be excluded. Many Proterozoic sequences with abundant silicified fossils contain no evidence for nearby volcanism. Knoll (1985) suggested that the silica may have originated by evaporation at ocean margins, along broad-tidal flats, or in restricted lagoons. Although the amount produced would have been small, it would dissolve readily in percolating meteoric groundwaters. A similar evaporative source for silica has been identified in Pleistocene lake deposits in East Africa. Chert forms by leaching of a sodium silicate precursor, which precipitates in sodium carbonate-rich lakes (Eugster, 1967; Hay, 1968). The occurrence of these alkaline lakes is related to the tectonic setting of the region (i.e., rift valley). Kempe and Degens (1985) presented thermodynamic, mass-balance, and kinetic arguments for an early ocean enriched in sodium carbonate and bicarbonate, which would have existed until the end of the Proterozoic. Thus, it is likely that a similar source of silica contributed to the formation of ancient siliceous deposits and the preservation of early biotas.

Sea-floor hot spring effluents are highly supersaturated with silica and iron, suggesting a hydrothermal source for Precambrian iron formations

(Simonson, 1985a,b). Concentrations of silica in the Precambrian may have been sufficiently high to allow the solutes to disperse over the required distances. The lack of silica-secreting organisms may have contributed to the maintenance of a higher silica concentration.

Another source of silica is detrital quartz. Thiry *et al.* (1988) suggested that the silica for the local formation of quartzites in the Oligocene sands of the Paris Basin was derived initially from the dissolution of the quartz grains, flint pebbles, and clay minerals. However, as quartz equilibrium is reached, detrital quartz is preserved and silica is derived only from more poorly crystalline phases. Henderson (1984) postulated a nonbiogenic source of silica for the silicification of a Middle Devonian stromatoporoid, derived from the dissolution of detrital quartz and other silicates during late diagenesis.

In summary, it seems that nearly all the silica involved in fossil silicification during the Phanerozoic was biogenic in origin. The silica source during the Proterozoic was almost certainly inorganic. Volcanism, hydrothermal activity, and evaporation in the shallower settings are all possible major contributors. However, the exact sources of silica at that time remain unknown (Barghoorn and Tyler, 1965).

5. Geochemical Considerations

5.1. The Silica Gel Theory

Tresise (1961) summarized the earliest ideas for the formation of chert beds: (1) sediment was directly replaced by silica solutions derived from an inorganically precipitated silica gel, initially precipitated on the sea floor; (2) dense concentrations of siliceous sponges were diagenetically transformed into chert; and (3) chert formed directly from colloidal silica precipitated on the sea floor. He considered the last the most likely because of the botryoidal nature of the silica which is characteristic of solidified gels (Rogers, 1917). Such a mechanism was also proposed to explain the spherulitic nature of the silica observed in Upper Jurassic silicified faunas (Wilson, 1966).

Oehler and Schopf (1971) synthesized an organic chert by embedding filamentous blue-green algae in a silica gel and subjecting the mixture to pressures between 2000 and 4000 bars at temperatures of about 150°C for 2–4 weeks. It is unlikely, however, that silica precipitated in this fashion during the Phanerozoic. The structure of fossils and the distribution of organic matter within silicified rocks also preclude precipitation of a silica gel within a void resulting from the dissolution of carbonate during diagenesis (Buurman and Van der Plas, 1971). Colloidal particles do not exist in solutions dilute in silica (Leo and Barghoorn, 1976) and, in any event, this form of silica would be unable to penetrate small cellular openings within plant tissue (cf. Scurfield and Segnit, 1984). Leo and Barghoorn (1976) therefore considered a (poly)silicic acid to be a more feasible mechanism for silicification of wood.

5.2. Soft-Tissue Silicification

One of the most important landmarks in understanding silicification processes was the experimental work of Leo and Barghoorn (1976). They used water to hydrolyze ethyl silicate to monosilicic acid, the principal soluble form of silicon in nature. They proposed that, as the concentration of the solution increased beyond saturation with respect to amorphous silica, polymerization of the silicic acid took place via the formation of siloxane bonds and the elimination of water:

$$Si(OH)_4 + Si(OH)_4 \rightarrow Si(OH)_3OSi(OH)_3 + H_2O \quad (1)$$

As polymer growth continued, amorphous silica precipitated from solution. With the initiation of structural ordering and further loss of water, opaline silica formed. The silicified wood formed in this fashion bears a striking resemblance to fossil material (Leo and Barghoorn, 1976, Plate 2, Figs. 1 and 2).

Leo and Barghoorn (1976) proposed a model for the petrification of wood involving the process outlined above, together with the interaction of degraded organic matter with the silicic acid prior to its polymerization. The degradation of holocellulose and lignin components of the vascular tissue produces abundant functional groups (especially hydroxyl groups) suitable for the formation of hydrogen bonds (Ferris *et al.*, 1988). The interaction of these groups with dilute solutions of silicic acid causes the oriented bonding of silicic acid to the organic framework:

$$C_2H_5(OH) + Si(OH)_4 \rightarrow C_2H_5\text{–}(OH)_2\text{–}Si(OH)_3 \quad (2)$$

Subsequent polymerization and dehydration of silicic acid eventually produce a thin film of silica along cellular surfaces which replicates the histological character of the wood. Although monosilicic acid is the more stable form, it is likely that polysilicic acid (released from weathering of volcanic ash) is the agent involved in petrifaction, forming multiple hydrogen bonds (R. K. Iler, in Leo and Barghoorn, 1976). Leo and Barghoorn (1976) cited four observations, in addition to their experiments, that substantiate their hypothesis that hydrogen bonding is involved in the silicification of wood: (1) The lack of silicification of coalified wood, which may be due to the loss of available functional groups (capable of forming bonds with the silicic acid) during the coalification process. (2) The quality of replication of woody structure in carbonate or sulfide mineralizations is inferior to that in silicified examples. Since neither carbonates nor sulfides are able to establish hydrogen bonds with holocellulose or lignin, deposition is not oriented with respect to the ultrastructure of the vascular tissue. (3) The sediment surrounding the silicified wood is itself rarely silicified. This probably reflects the potential of the wood to bond with the silicic acid and thus act as a "seed," nucleating further silica precipitation. (4) The deposition of dense silica noted in living plants [frequently up to 3% dry wood weight (Amos, 1952)] may represent the earliest stages of the silicification of nonliving wood (Scurfield *et al.*, 1974). The association of silica particles with starch grains and polyphenolic material in the same cell suggests a causal relationship which may be explained by hydrogen bond formation. The model for the silicification of wood proposed by Leo and Barghoorn (1976) has been upheld in subsequent studies of silicified wood (e.g., Karowe and Jefferson, 1987; Sigleo, 1978; Stein, 1982).

The behavior of minor, trace and rare earth elements during silicification was studied by Sigleo (1979). She noted that the concentration of Al, Mg, Ca, and Na in silicified wood from the Triassic of Arizona increases from the core outward. The variation of Al is the most dramatic, increasing from 0.87 to 4.39 wt %. The Al content is representative of the clay mineral fraction incorporated within the silicified wood, which may comprise up to 15% of the sample. The clay is unlikely to have been detrital, due to particle-size exclusion; hence the silica and clay components must have been in equilibrium and precipitated simultaneously (Sigleo, 1979). Trace element abundances are generally higher in the associated sediments and clay mineral separates than in the silicified wood. An important exception is Sb, which is greatly enriched in the wood. Since Sb is most soluble in the oxidized state (SbO_3^-), precipitates when reduced (Sb_2O_3), and is commonly enriched in coal (Onishi, 1972), Sigleo (1979) suggested that it underwent reduction with the organic matter. The highest Mn value was obtained from the sediment adhering to a silicified branch. This possibly indicates the redox boundary for silicification, since Mn in its reduced, mobile state will tend to migrate to more oxic conditions where precipitation may take place. The same sort of process is believed to control Mn distribution in flints (Section 5.2.2), although Mn may be in a detrital mineral rather than a precipitated oxide (Sigleo, 1979). The highest Fe concentration is in red silicified wood, where it is in the oxidized state. The iron may have been transported as a fine-grained colloid or adsorbed on detrital minerals. Low total lanthanide-group concentrations may suggest a slightly acidic setting, typical of decaying organic matter in fluvial swamp environments, while little variation in Ce indicates a reducing environment. Sigleo (1979) concluded that silicification of the wood took place in an anoxic environment, probably in a swamp. This is compatible with paleoenvironmental reconstructions of the area.

Leo and Barghoorn's (1976) hypothesis for the silicification of wood has also been used to explain the silicification of early prokaryotic material (e.g., Knoll, 1985). Until recently, it has generally been accepted that it is because of this early silicification (preventing the total degradation of cellular material) that microorganisms have been preserved. Ferris *et al.* (1988), however, provided experimental evidence that it is not the silicification, but the prior reaction of iron that is responsible for the preservation of cellular material. *Bacillus* was cultured under different aqueous conditions involving the addition of $FeCl_3$, colloidal silica, or both. Resulting cells were aged for up to 150 days at 70°C. Bacteria which were untreated displayed a complete loss of cell shape. Bacteria treated only with silica also suffered in the same manner, but the remains of the cell walls showed the development of silica crystallites. The bacteria treated with the iron, however, retained their cell shape, as did those treated with iron and silica, which also showed the development of silica crystallites. Dead microbial cells are normally decomposed by autolysins (enzymes capable of degrading the cell wall). Since autolytic activity is inhibited by metal cations (Herbold and Glaser, 1975; Leduc *et al.*, 1982), Ferris *et al.* (1988) concluded that it was the retention of iron that led to preservation of microorganisms prior to the silicification of intact cellular structures. They also suggested that the inhibition of decay in previous investigations of the silicification of microbial cells (e.g., Oehler, 1976; Francis *et al.*, 1978) was the result of experimental conditions which denatured the autolysins within the microorganisms.

5.3. Skeletal and Ichnofabric Silicification

The exclusion of processes involving silica gels (Section 5.1) leaves three possible explanations for the replacement of carbonate by silica: (1) a calcium silicate precursor, (2) environmental conditions created by mixing marine and meteoric water, and (3) microenvironmental conditions associated with decaying organic matter.

5.3.1. Calcium Silicate Precursor

The involvement of a calcium silicate precursor in the genesis of flints and cherts (of Carboniferous–Tertiary age) was proposed by Buurman and Van der Plas (1971). Based largely on thermodynamic arguments, they suggested that replacement was a two-stage process: (1) transformation of calcite to calcium silicate with the removal of CO_2, and (2) transformation of this calcium silicate into silica, with the removal of Ca^{2+}. They considered that the formation of submarine chert thus depends upon changes in P_{CO_2}, pH, local concentrations of Ca^{2+}, and temperature.

5.3.2. Marine–Meteoric Mixing

Knauth (1979) recognized that diagenetic waters must be simultaneously supersaturated with silica and undersaturated with calcite. He suggested that such conditions are fulfilled at coastal locations where marine and meteoric waters mix. Modern surface seawater is saturated with calcite. As meteoric water passes through a carbonate sequence, it, too, becomes saturated. Where the P_{CO_2} and/or temperature of the two solutions differ (James and Choquette, 1984), mixing results in an undersaturated solution. In this way, corrosion of the carbonate is possible. At the same time, as meteoric water passes through the carbonate sequence, it becomes increasingly concentrated with silica due to dissolution of biogenic opal. Since the silica concentration of modern surface seawater is low, the maximum silica saturation will occur at the zone of mixing between meteoric and marine waters. Thus, at the mixing zone, waters are saturated with respect to silica while being undersaturated with respect to calcite, allowing replacement of carbonate by silica.

The chemical arguments for this process are supported by petrographic observations. The mixing zone has also been identified as a location of dolomitization (Badiozamani, 1973; cf. Hardie, 1987), and it is not uncommon for dolomite rhombs to be enclosed in chert (e.g., Pittman, 1959; Dietrich *et al.*, 1963). In a study of silicified Upper Cretaceous *Gryphaea*, Elorza and Orue-Etxebarria (1985) noted quartzine-lutecite spherulites associated with exsolved dolomite rhombs in void-filling sparite. They suggested that at least this initial phase of bioclast silicification took place in a mixing-zone environment. Jacka (1974) observed that dolomite in echinoderm debris from the Middle Permian of the Delaware basin is optically continuous with the unreplaced calcite plates, suggesting that it exsolved from the high-Mg calcite. He proposed that this exsolution occurred during replacement of high-Mg calcite by silica in areas enriched in Mg^{2+} ions. Such a process may be enhanced in the mixing zone, where waters are supersaturated with dolomite (Knauth, 1979). There is some doubt concerning the effectiveness of the mixing zone in the dolomitization process. Hardie (1987) pointed out that it is disordered dolomite (the more soluble, Ca-rich type) that forms at surface temperatures. Solubility values of ordered dolomite have been used previously in calculating the theoretical range of marine–meteoric mixtures available for dolomitization. If, however, the solubility values for disordered dolomite are used, then the range of mixtures conducive to dolomitization is considerably reduced. Thus, it is the mineralogy of dolomites that casts doubt on the mixing-zone dolomitization model, not the chemistry of marine–meteoric fluids. So, even if the dolomite–silica associations cited above are fortuitous, the mixing zone cannot be precluded as a possible site for silicification.

Astin (1987) suggested a modification of Knauth's (1979) model for the formation of cherts in discussing Upper Jurassic cherts and silicified biotas from southern England. He proposed that silicification occurred during emergence of the sequence, under the influence of meteoric water *alone* (see also Meyers and James, 1978). The meteoric water was mixed initially with remnant marine water saturated with carbonate and amorphous silica. This resulted in the precipitation of opal-CT cement. As the meteoric component increased, the pore fluid became undersaturated with carbonate, causing dissolution and silica precipitation. As the source of biogenic opal was depleted, silica concentrations were reduced, and subsequent silica cements formed only during deeper burial.

5.3.3. Organic-Matter Oxidation

In many examples of bioclast silicification there is no relationship to dolomitization or terrestrial emergence, even in marginal environments where such situations would have been most likely to develop (e.g., Carson, 1987). Such circumstances imply a different mechanism for silicification.

Silicification of uncompacted shell microstructure demonstrates an intimate relationship between carbonate dissolution and silica precipitation (Eley and Jull, 1982; Holdaway and Clayton, 1982). Silica may exist in metastable solution in a highly supersaturated condition with respect to crystalline silica (Holdaway and Clayton, 1982). Hence, although a drop in pH would cause pore waters to become simultaneously undersaturated with calcite and supersaturated with quartz, this is an unlikely mechanism to explain bioclast silicification. In addition, anaerobic decay may result in an increase in pH, depending upon the nature of the organic source (Berner, 1971).

Silicification may extend into surrounding syntaxial calcite cement (Carson, 1987), yet the restriction of silica to bioclasts or burrows suggests that organic matter played a role, presumably via bacterial breakdown products (Holdaway and Clayton, 1982). Calcite dissolution and resulting silicification may have occurred as a result of CO_2 buildup produced by bacterial oxidation of organic matter under aerobic conditions (Holdaway and Clayton, 1982). Reduction of Mn and Fe is rejected as a mechanism since it results in a net precipitation of calcite. Nitrate reduction may result in carbonate dissolution (Froelich *et al.*, 1979), but is unimportant in shelf sediments and thus an unlikely mechanism. Bacterial fermentation reactions are excluded on the grounds that pyrite would predate silicification; most of the associated pyrite postdates the silicification. Sulfate reduction is also excluded since it may lead to calcite precipitation (Irwin *et al.*, 1977; Coleman and Raiswell, 1981). However, where CO_2 is generated during the oxidation of organic matter, its hydrolysis to H_2CO_3 will invariably decrease pH and promote carbonate dissolution (Coleman, 1985). Also, the precipitation/ dissolution of carbonate in anoxic pore waters is dependent on the availability of iron. Where iron is absent, the S^{2-} is lost as free hydrogen sulfide, causing carbonate dissolution (Coleman, 1985). This conflicts with the frequent observation of pyrite intimately associated with the silica of a replaced fossil. However, simultaneous reduction of iron and sulfate liberates protons, which promote carbonate dissolution [equation (3)], and further reduction of sulfate converts FeS to FeS_2 and increases pH, causing carbonate precipitation [equation (4)]:

$$2Fe_2O_3 + 4SO_4^{2-} + 9CH_2O \rightarrow 4FeS + 9CHO_3^- + 4H_2O + H^+ \quad (3)$$

$$SO_4^{2-} + CH_2O \rightarrow 4S^\circ + 6HCO_3^- + 2H_2O + 2OH^- \quad (4)$$

This hypothesis (Coleman, 1985) may explain why pyritization appears to predate, to be contemporaneous with (Holdaway and Clayton, 1982), and also to postdate (Carson, 1988) silicification. Individual crystal size within a shell may control rate of carbonate dissolution, but it is the presence of organic matter (both in the sediment and skeleton) which "seeds" the silicification process.

The limited quantity of organic material available for oxidation in calcite shells may have been insufficient to account for their dissolution by CO_2 buildup. Replacement may have been controlled by the force of crystallization of the silica phase itself, causing an increase in calcite solubility at a thin (>1 nm) solution–film contact between the quartz and calcite (Maliva and Siever, 1988; Dewers and Ortoleva, 1990). A zone of recrystallized shell, which is frequently associated with the silica–carbonate contact, may have been formed by pressure solution in pore fluids nearly saturated with respect to low-Mg calcite. The released calcium and carbonate would have caused the intrashell fluids adjacent to the silicification front to become supersaturated. Low-Mg calcite would have then precipitated and grown at the expense of the less stable shell calcite. Further evidence in support of this hypothesis is provided by euhedral quartz crystals present as a replacement within a shell. Such a feature could have only been produced by the growth of the quartz (Maliva and Siever, 1988). The dissolution of calcite would not by itself result in a void the shape of euhedral quartz. Such quartz crystals have been documented from a Devonian trilobite (Maliva and Siever, 1988) and stromatoporoid (Henderson, 1984), and from Carboniferous brachiopod debris (Orme, 1974, Plate 3, Fig. 1), but are by no means a common feature of silicified skeletal debris.

Stable carbon and oxygen isotope data from the Upper Cretaceous Chalk show that flints formed in the upper 5–10 m of sediment, at the top of the sulfate reduction zone (Clayton, 1984). Decay of organic matter by sulfate-reducing bacteria released carboxyl and sulfide ions (Fig. 13). Some sulfide is precipitated as pyrite. However, due to the paucity of iron in the chalk, the bulk of the sulfide hydrolyzed to H_2S. This then diffused toward more oxic conditions, where it was reoxidized to sulfate and hydrogen ions. The sulfate then diffused back toward the sulfate reduction zone, while the increased acidity caused the local dissolution of carbonate. The high concentration of carbonate ions "catalyzed" the precipitation of silica from the surrounding pore fluids (Clayton, 1986).

This model is supported by trends in the concentration of Ca, Mg, Al, K, Fe, and Mn within the flint and associated cemented cores of paramoudras (Clayton, 1986). Highest concentrations of Ca and Mg occur in the core of the flint where diffusion was inhibited. The concentration of clays in the flint relative to the surrounding sediment is reduced during compaction, and this

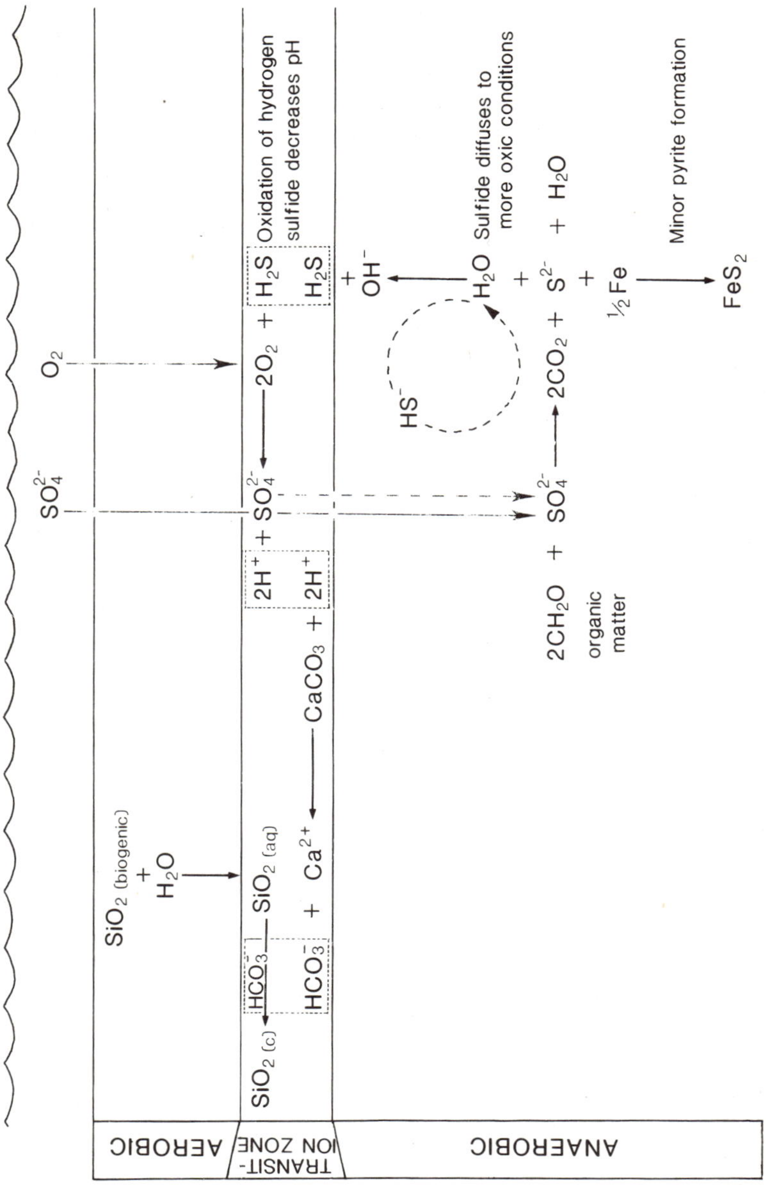

FIGURE 13. Schematic diagram of the diagenetic reactions leading to silicification in the Chalk (aq, in solution; c, solid). See text for details.

may be reflected in relatively low Al values. These values indicate a porosity at the time of flint formation of about 80–85% (Clayton, 1986). The concentration of K increases away from the core of the paramoudra. Clayton (1986) ascribed this trend to a glauconite component in a mixed layer illite–smectite which formed prior to silicification and sediment lithification. Glauconite formation took place at the anoxic–oxic interface (Ireland *et al.*, 1983); hence the onset of anoxic conditions within the central burrow would have eradicated the glauconite and reduced the K concentration. Fe and Mn are highly mobile in the reduced state (Fe^{2+} and Mn^{2+}), and highly insoluble in the oxidized state (Fe^{3+} and Mn^{4+}). Hence there is a tendency for these elements to migrate from reducing areas, precipitating at the edge of oxidizing areas (Clayton, 1986). Their relatively high concentration within the flint thus indicates the boundary between oxidizing and reducing conditions. Although the data cited above refer to paramoudra flints, the general model is applicable to bedded flints, where similar chemical analysis revealed comparable patterns of trace element concentrations (Clayton, 1986).

Zijlstra (1987) presented a similar model for the formation of Upper Cretaceous flints from the Netherlands. He also suggested that the precipitation of the silica occurred at the transition zone between oxic and anoxic sediment. However, instead of a concentration of dissolved bicarbonate, he postulated that a prerequisite for silica precipitation is a low hydrogen ion concentration. This caused supersaturation of negative silica complexes (H_3SiO_4, $Si_4O_6(OH)_6{}^{2-}$), which eventually polymerized and precipitated. Zijlsrta (1987) suggested that this was achieved by the presence of archaeobacteria which utilized the hydrogen ions in the decomposition of organic matter.

5.4. Isotopic Evidence

Most stable isotope work on natural silica has been carried out on cherts in relation to the problem of secular oxygen isotope variation since the Precambrian (e.g., Perry, 1967; Perry and Tan, 1972; Becker and Clayton, 1976; Oskvarek and Perry, 1976; Knauth and Lowe, 1978; Perry *et al.*, 1978). Isotopic studies devoted to the genesis and diagenesis of cherts have been limited (e.g., Knauth and Epstein, 1975, 1976; Kolodony *et al.*, 1980; Clayton, 1984), as has been the contribution of stable isotopes to the understanding of the silicification of fossils (Carson, 1987).

Knauth and Epstein (1975) illustrated how the $\delta^{18}O$ values of opal-CT were established prior to or during shallow burial in Upper Cretaceous and Eocene cherts from the Pacific Ocean, and were preserved to depths of at least 1000 m. Granular microcrytalline quartz (chert) invariably had more negative $\delta^{18}O$ values than coexisting opal-CT, suggesting that the chert formed at a greater depth. The same general trend of $\delta^{18}O$ values decreasing with burial depth during the transformation of silica minerals occurs in the Monterey Shale of California (Murata *et al.*, 1977).

Using boron α-track mapping in combination with stable oxygen isotope ratios, Kolodony *et al.* (1980) differentiated two stages of silicification in Upper Cretaceous chert breccias of southern Israel. Chert fragments within the breccia were invariably depleted in ^{16}O relative to the matrix (mean $\delta^{18}O$ of 32‰ vs. 27.5‰). This cannot be accounted for by a difference in temperature due to burial. Kolodony *et al.* (1980) postulated that the difference was the result of an initial stage of equilibration of the chert with seawater and a second stage of equilibration with freshwater, following brecciation. Since boron tends to remain in solution during evaporation retaining a constant ratio to chlorinity, the relative enrichment of boron in the fragments confirms that the silica equilibrated in hypersaline water.

The above studies demonstrate the environment of silicification, but do not account for the genesis of the chert. Clayton (1984) used a combination of petrography, trace element geochemistry, and stable oxygen and hydrogen isotope analyses to demonstrate the subtle variations in genesis and recrystallization of the different components of Upper Cretaceous chalk flints. Although some of his data were tentative, he ascribed the consistently more negative $\delta^{18}O$ values of "white bodies" within black flint to closed-system buffering of the oxygen isotopes. This was combined with a kinetic fractionation of the hydrogen isotopes during white-body formation, which was promoted by the exsolution of structurally-bonded hydroxyl groups into water-filled fluid inclusions. Furthermore, Clayton (1984) noted that there was no correlation between the $\delta^{18}O$ and the lepisphere content of black flint. This suggests that the lepispheres and chalcedony comprising the flint recrystallized at the same time, and that the opal-CT–quartz transition is independent of the original precipitated morphology. Clayton (1984) also correlated the δD and $\delta^{18}O$ of both the white bodies and the black flint (Fig. 14), and noted that all of the flint samples fall on or below the line defined by the white bodies. This suggests that the current isotopic composition of the black flint was acquired after the recrystallization which formed the white bodies. Since isotopic exchange has been documented in hydrated, disordered forms of silica (Labeyrie and Juillet, 1982; Matheney and Knauth, 1989), Clayton (1984) suggested that the more ordered yet still hydrated silica in flint would have been susceptible to isotopic alteration by meteoric waters. He concluded that water content is a measure of the structural disorder of the silica, which is controlled in turn by the rate of inversion of opal-CT to quartz. Thus, a lower H_2O content suggests slower recrystallization. This may result in recrystallization continuing to a greater depth, and/or greater exchange with pore fluids external to the flint nodule. Both these effects would cause a decrease in the $\delta^{18}O$ of the black flint.

Clayton (1984) suggested that flint formation in the chalk is initiated by lepisphere growth. This results in a diffusion gradient with more closely spaced silica nucleation centers toward the center of a flint. Carson (1987), on the other hand, believed that chalcedony-cemented chert nodules from the mid-Cretaceous of southern England precipitated initially around the

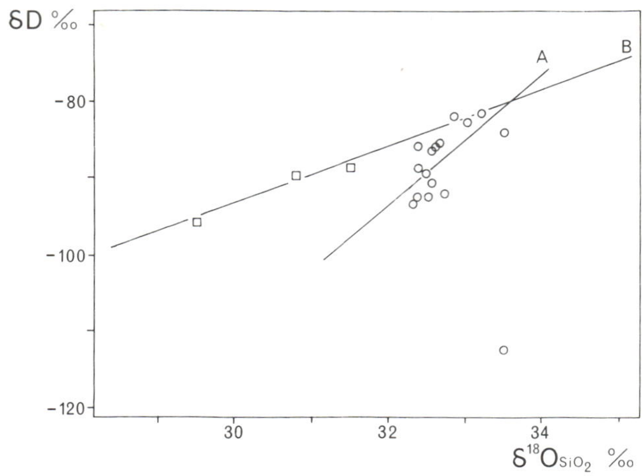

FIGURE 14. Isotopic analysis of the components of flint from the Upper Cretaceous. Circles, black flint; squares, white bodies. Regression lines: (A) $\delta D = 8.6\delta^{18}O - 368.3$, for black flint (excludes sample with $\delta^{18}O$ of 33.5 and δD of −112.4‰); (B) $\delta D = 3.6\delta^{18}O - 201.1$ for white bodies. See text for details. [Modified from Clayton (1984).]

circumference of burrows (such as *Thalassinoides*). Despite petrographic differences between the interior and exterior of these nodules, the isotopic compositions are similar. Recrystallization within a closed system (Pingitore, 1982) was considered an unlikely mechanism to explain these similar values, since the sandy facies in which the nodules occur would have been too permeable to restrict pore-fluid circulation. An alternative possibility is that the initial precipitation rate of the exterior of the nodule was more rapid than the interior. The relative structural disorder of the exterior zone would render it more susceptible to later exsolution of structural water, producing the observed petrographic fabrics.

The models discussed above for the formation of chert are relevant to the silicification of carbonate skeletons, although some aspects are not applicable. For example, it is believed that bioclast silicification does not necessarily involve a maturation sequence from opal to quartz, the observed mineral variations being primary (Holdaway and Clayton, 1982). However, primary fabrics may undergo some form of recrystallization with respect to isotopic composition. This has been demonstrated in silicified bivalves and brachiopods from mid-Cretaceous sands of southern England (Carson, 1987). Comparison of $\delta^{18}O$ values from different phases within individual specimens (Table I) and from groups of specimens (Fig. 15) revealed the lowest $\delta^{18}O$ values in later replacements. Field and petrographic evidence precluded recrystallization during deeper burial and formation within a mixing zone (respectively) as mechanisms responsible for the observed variations. It was concluded that recrystallization preferentially affected the later silicification fabrics during uplift, when the sands were subjected to isotopically light meteoric waters.

TABLE I. Details of Isotopic Results from the Pectinid Shown in Fig. 6: $\delta^{18}O$ Values Relative to Standard Mean Ocean Water[a]

		$\delta^{18}O$	Average	1σ
Individual beekite rings—from exterior to interior	A	28.6		
	B	30.7	29.7	0.96
	C	29.3		
	D	30.3		
	E	29.3		
	F	30.2	28.1	1.86
	G	26.6		
	H	26.5		
Whole beekite "Disks"	I		29.9	
	J		29.3	
Beekite average			29.3	(±0.79)
Late gray replacement	K	27.7		
	L	27.9	27.6	0.36
	M	27.2		
Soft, white beekite	N	30.8		
	O	30.0	30.5	0.42
	P	30.5		

[a]Mean deviation for duplicate sample analysis ±45‰ (Carson, 1987). Reproduced from Carson (1987) with permission of Blackwell Scientific Publications Ltd (Oxford).

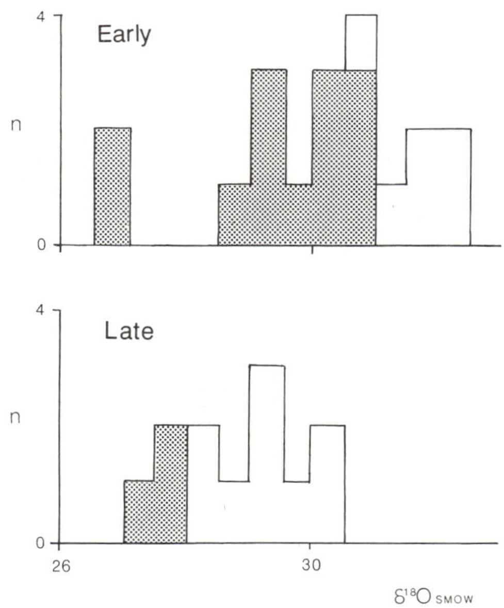

FIGURE 15. Histograms of $\delta^{18}O$ values obtained for early and late skeletal mineralization. Where duplicate analyses were made from one specimen, the mean value was used. Stippled area indicates values obtained from pectinid shown in Fig. 6 (n = number of analyses; SMOW = Standard Mean Ocean Water). [Reproduced from Carson (1987) with permission of Blackwell Scientific Publications Ltd (Oxford).]

6. Paleoenvironmental Implications

Clearly the importance of fossil silicification is that it is essentially an early process which renders fossils relatively resistant to subsequent diagenesis. However, because of this, paleoenvironmental inferences based on assemblages of silicified biotas have to be treated with caution. The exceptional preservation of Proterozoic cyanobacteria, Cretaceous skeletal debris, or Eocene wood may distract us from investigating associated less well-preserved assemblages.

Hofmann and Schopf (1983) suggested that the resilient sheaths and envelopes of Proterozoic cyanobacteria possess a greater fossilization potential than the cell contents and cell walls of the organisms. This may lead to differing interpretations of features observed in thin section as cells delineated by walls, or as remnants of decay-resistant mucilagenous envelopes lacking internal contents. Even if the structures are shown to be cells, they may represent only the more decay-resistant elements of the original microbial community and thus provide a biased sample (Horodyski and Donaldson, 1983). Knoll and Golubic (1979) reached such a conclusion in an analysis of a Precambrian stromatolite from the Bitter Springs Formation (Australia). They described five distinct lithological fabrics within the stromatolite, each one associated with differential preservation of microfossils. Post-mortem degradation caused an apparent increase in the diversity of chroococcoid algae, while virtually all traces of entophysalidacean and filamentous cyanophytes were removed. Francis *et al.* (1978) pointed out that it is unwise to rely on the external morphological characters of cells for evolutionary systematics. Caution should be exercised in identifying Proterozoic microfossils as eukaryotes on the basis of size alone, since their experimental work showed that degradation of cells prior to and during silicification can lead to the appearance of a "cell wall" distinctly larger than that of the original living organism. Degradation of cellular material may form "nuclei" when the cell contents collapse and pull away from the cell wall (Francis *et al.*, 1978). In the fossil record, such structures may be interpreted incorrectly as true membrane-bound nuclei, and thus give a false age for the earliest eukaryote. Hofmann and Schopf (1983) noted that most of the well-preserved microbiotas are from shallow-water stromatolitic carbonate facies that have undergone early diagenetic silicification. Such chert assemblages are from a restricted environment, and may comprise an ecologically biased and "unrepresentative sample" of early Proterozoic life. Horodyski and Donaldson (1983) also noted that microbial communities originally encased in amorphous silica and those present in silicified sulfates should be atypical, having grown under harsh conditions. Conversely, microbiotas found in silicified carbonates should be more representative of a "normal" Proterozoic benthic population.

Although the extent of environmental bias within Proterozoic silicified biotas may not be known (Knoll, 1985), there is considerable bias in Phanerozoic silicified peats. This is because plants that colonize swamp habitats were adapted to wet, acidic, and oxygen- and nutrient-deficient environments. Land plants on the contemporaneous floodplain evolved adaptations to cope with different factors such as variations (limited or seasonal) in water supply and competition for sunlight. Thus, well-established swamp biotas were relatively resistant to invasion from adjacent environments, and diversified much more slowly than floodplain floras, taking on a relictual character. Consequently, the preservation of swamp floras within silicified peats may lead to an unrepresentative view of general diversity (although this would not apply to allochthonous elements within the peat).

Despite the need for caution when investigating phylogenetic relationships within ancient silicified biotas, there is no doubt that they may provide exceptionally detailed preservation of anatomical structure (Knoll, 1985), as well as information on the contemporaneous macro- and microenvironments and timing of diagenetic events. Barghoorn and Tyler (1965) identified eight genera of Middle Precambrian microorganisms from the "algal" Gunflint Chert, Ontario. Several of these were comparable to present-day cyanobacteria, while the remainder showed no recent analogue. They suggested that the organisms were photosynthetic on the basis of the gross similarity of the stromatolitic domes to extant biostromes and of the low $\delta^{13}C$ values of the organic matter (indicative of photosynthetic activity). Since the mineralogical associations of the organic remains indicate a largely reducing environment, Barghoorn and Tyler (1965) concluded that locally oxidizing conditions only developed adjacent to the algal colonies, in response to photosynthesis. Thus, the biotas of that time may indicate a transitional period in the evolution of a well-oxygenated atmosphere.

The coexistence of *in situ* silicified wood and evaporites from the Upper Jurassic of southern England is evidence for a seasonal climate of warm, wet winters and hot, arid summers (Francis, 1984). Francis (1984) suggested that, following submergence of the forests by hypersaline lagoon water, the pH would have been raised by photosynthesizing algae and seasonal evaporation. Values of 9–10 may have been attained, promoting the dissolution of detrital silica. Subsequent deposition of silica would have taken place preferentially within the decaying organic matter of tree stumps, where the pH would have been considerably reduced (Francis, 1984). Silicification is restricted to the centers of roots and branches greater than 6–10 cm in diameter. This indicates that below this size there was insufficient restriction of fluids, preventing the decrease of pH and the deposition of silica (J. E. Francis, 1982).

The value of silicification in interpreting diagenetic events is also illustrated in the formation of flints. The formation of chert (flints) within Upper Cretaceous chalks (Section 3.3) has been related to the presence of an ichnofauna, especially *Thalassinoides*. Although the size of the crustacean forming *Thalassinoides* is unlikely to have varied to any great extent, there is considerable variation in the size and morphology of flints. This may be ascribed largely to the activity of burrowing organisms during differing periods of sediment nondeposition (omission). During a substantial sedimentary break, prolonged compaction of the host sediment occurred while burrows were kept open. This led to an enhanced permeability contrast, causing the silicification to be constrained by the burrow walls, producing a digitate flint. Where the period of omission was shorter, less permeability contrast led to the development of irregular nodules, frequently overgrowing the original burrow. Where bioturbation was absent, negligible permeability contrast caused H_2S–O_2 mixing to occur along a smooth boundary, forming continuous tabular flint (Clayton, 1984).

As well as elucidating the conditions of sediment deposition, flint may be used to infer subsequent events. Paramoudras form as a direct consequence of the presence of *Bathichnus* burrows (Section 3.3), and these in turn are believed to have been used as escape structures, following extremely rapid sedimentation. The presence of paramoudras is thus indicative of allochthonous sediment deposition within a chalk sequence (Bromley and Ekdale, 1987).

7. Conclusions

Silicification fabrics are found in a diverse range of paleoenvironments, and the factors involved in silicification have varied with facies and time. In reality, these factors are numerous and complex. However, an understanding of silicification may be gained by considering the interplay of the source of silica and the concentration mechanism (Fig. 16). Possible sources of the silica are volcanic/tectonic, hydrothermal, biogenic, and detrital. In many cases, the *ultimate* source of the silica may be equivocal (for example, was siliceous plankton the source for numerous Mesozoic silicification fabrics, or did a high marine silica concentration initiate the plankton blooms?).

7.1. Concentration Mechanisms

Concentration mechanisms may be divided broadly into evaporation, meteoric interaction, redox/pH reactions, and depositional saturation.

7.1.1. Evaporation

Evaporation encourages silicification by increasing pH and thus mobilizing silica. For example, an increase in pH promoted by evaporation (along with photosynthesizing algae) dissolved *detrital* quartz, thus enabling the silicification of wood in the Portlandian of southern England (Francis, 1984).

During the Pleistocene in alkaline lakes in east Africa, the initial source of the silica was a high concentration in the lake related to the *tectonic*

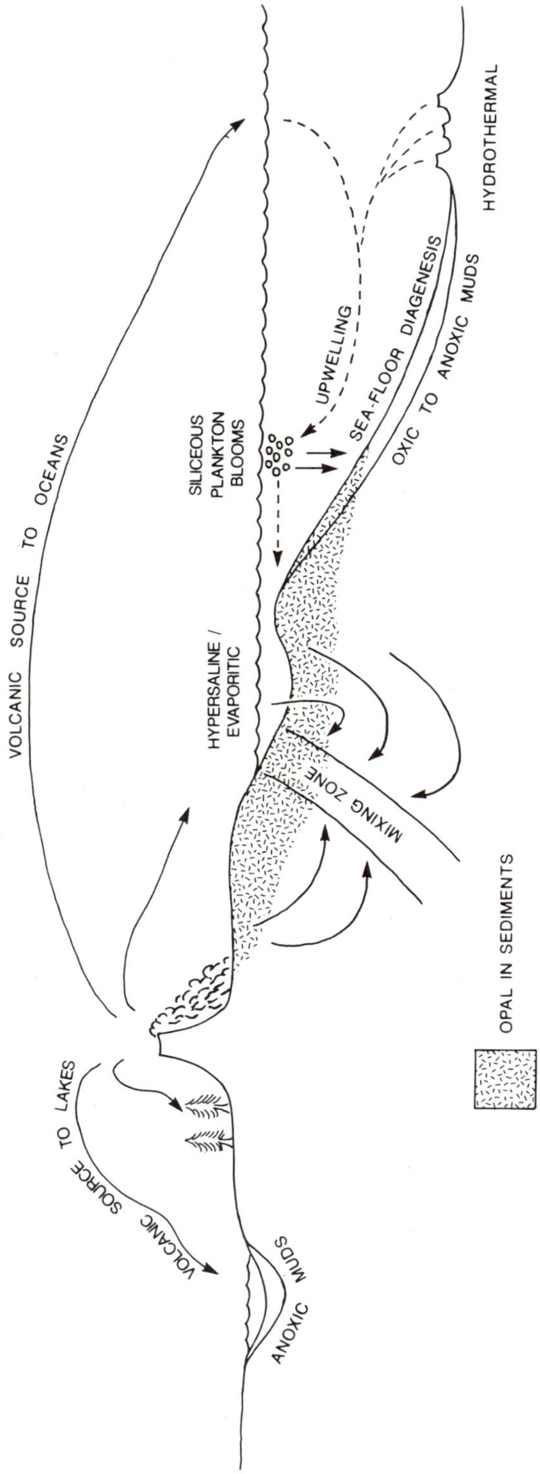

FIGURE 16. The principal sources of silica and environments of silicification throughout geologic time.

setting. Chert was formed by subsequent evaporation/dehydration of sodium silicate precursors (Eugster, 1967; Hay, 1968).

7.1.2. Meteoric Interaction

Mixing of two solutions saturated with calcite but of differing CO_2 and/or temperature results in an undersaturated solution. Where waters are also saturated with silica (as in zones of marine–meteoric mixing), replacement of carbonate by silica may take place.

All cases involving a meteoric mechanism for silica concentration are related to a *biogenic* source of silica. Examples range from the formation of Ordovician and Carboniferous cherts (Meyers, 1977; Meyers and James, 1978; Geeslin and Chafetz, 1982) to the silicification of biotas from the Portlandian of southern England (Astin, 1987).

7.1.3. Redox/pH Reactions

Acidic conditions decrease silica solubility and thus favor silicification. A low pH may exist due to stagnation (aquatic environments) or be due to bacterially-mediated oxidation of organic matter (within sediments).

Lowering of pH within local swamp environments of the Triassic of Arizona caused mineralization of wood, the source of the silica being *volcanic* ash (Sigleo, 1978).

A *biogenic* source of silica, coupled with early diagenetic bacterial redox reactions within carbonate host rocks, accounts for many examples of ichnofaunal (Clayton, 1986; Zijlstra, 1987) and bioclast (Holdaway and Clayton, 1982; Carson, 1987) silicification during the Mesozoic.

7.1.4. Depositional Saturation

As well as the "diagenetic" processes listed above, silica supersaturation may be achieved by a local increase in the availability of silica.

An increase in silica concentration due to the weathering of *volcanic* ash has been the most favored model for the silicification of wood throughout the Phanerozoic (Murata, 1940; Karowe and Jefferson, 1987).

The formation of cherts in Precambrian iron formations may have been due to a much greater concentration of silica in the oceans than prevails today. This was due to *hydrothermal* waters discharging at the sea floor (Simonson, 1985a,b).

The dissolution of *biogenic* opal (Radiolaria) within Mesozoic sediments of the Mediterranean raised the concentration of dissolved silica, promoting the formation of Radiolarian cherts (McBride and Folk, 1979).

Silica saturation levels were also raised by the dissolution of *detrital* silica enabling the formation of Oligocene quartzites in the Paris basin (Thiry *et al.*, 1988) and the silicification of a Devonian stromatoporoid (Henderson, 1984).

These processes are by no means mutually exclusive. For instance, silica which formed replacement chert during the Proterozoic may have been concentrated initially in broad tidal flats and lagoons, and subsequently dissolved and further concentrated in meteoric ground-waters (Knoll, 1985). Geochemical evidence (Kolodony *et al.*, 1980) points to initial silicification in a marine/hypersaline environment, followed by involvment of meteoric water.

Much of our knowledge of silicification is based on the detailed work on chert formation rather than the direct study of silicified fossils. By drawing together examples from Proterozoic microbiotas and Phanerozoic plant material and skeletal debris, certain common features are apparent.

7.2. Features of Silicification

7.2.1. Organic Matter

Organic matter is important in the silicification of both Proterozoic microorganisms and Phanerozoic plants. The relatively limited quantity of organic matter associated with skeletal debris and burrows is vital to create a microenvironment conducive to silicification. In the form of associated microorganisms, the presence of organic matter may also explain the "random" location of many silicification fabrics, such as botryoidal chalcedony on detrital quartz and within shell fragments (Fig. 5) (Carson, 1988).

7.2.2. Timing

In nearly all cases, silicification occurred at a relatively early stage in the burial history of the enclosing sediment, at depths from 0 to 10 m (e.g., Green *et al.*, 1989). Silicification of fossils *initiated* during deeper burial is rarely documented, although silicification may of course *continue* to greater depths. Bioclast silicification within a mixing-zone environment (Knauth, 1979) is an exception, since the seawater/freshwater interface may be encountered at considerable depth within the sediment [e.g., 70 m on the Yucatan Peninsula of Mexico (Back and Hanshaw, 1970)].

7.2.3. Other Elements

Other elements, notably iron, are frequently associated with silicified skeletal debris, and are probably necessary in the preservation of microorganisms (Ferris *et al.*, 1988). Microorganisms themselves may well play a much greater role in the silicification of macrofossils than is currently appreciated.

Atmospheric and oceanic chemistries evolved continuously through the Proterozoic. Major changes in macro- and microenvironments at the end of the Proterozoic played an important part in determining sources of silica and the types and rates of processes involved in silicification throughout the Phanerozoic.

ACKNOWLEDGMENTS. I am indebted to Julian Andrews and Ian Fairchild for their comments and useful suggestions for improving this chapter. Any inadequacies that remain are, of course, my own responsibility. I am grateful to Steve Crowley for use of his photographs, along with Greg ''Skanking'' Samways and a benefactor who needs to remain anonymous, for providing some drafting material. Jenny, Charlotte, and Amy maintained the chaos at an acceptable level during the completion of this work.

References

Altschuler, Z. S., Dwornik, E. J., and Kramer, H., 1963, Transformation of montmorillonite to kaolinite during weathering, *Science* **141**:148–152.

Amos, G. L., 1952, *Silica in timbers,* CSIRO Aust. Bull. No. 267.

Astin, T. R., 1987, Petrology (including fluorescence microscopy) of cherts from the Portlandian of Wiltshire, U.K.—Evidence of an episode of meteoric water circulation, in: Diagenesis of Sedimentary Sequences (J. D. Marshall, ed.), *Geol. Soc. Lond. Spec. Publ.* **36**:73–85.

Back, W., and Hanshaw, B. B., 1970, Comparison of chemical hydrogeology of the carbonate peninsulas of Florida and Yucatan, *J. Hydrol.* **10**:330–368.

Badiozamani, K., 1973, The Dorag dolomitization model—Application to the middle Ordovician of Wisconsin, *J. Sediment. Petrol.* **43**:965–984.

Barghoorn, E. S., and Tyler, S. A., 1965, Microorganisms from the Gunflint Chert, *Science* **147**:563–577.

Becker, R. H., and Clayton, R. N., 1976, Oxygen isotope study of a Precambrian banded iron-formation, Hamersley Range, Western Australia, *Geochim. Cosmochim. Acta* **40**:1153–1165.

Berner, R. A., 1971, *Principles of Chemical Sedimentology,* McGraw-Hill, New York.

Berry, L. G., 1974, *Selected Powder Diffraction Data for Minerals,* Joint Committee on Powder Diffraction Standards, Swarthmore, Pennsylvania.

Brasier, M. D., 1980, *Microfossils,* George Allen & Unwin, London.

Bromley, R. G., 1967, Some observations on burrows of thalassinidean Crustacea in Chalk hardgrounds, *Proc. Geol. Soc. Lond.* No. **1643**:235–236.

Bromley, R. G., and Ekdale, A. A., 1986, Flint and fabric in the European chalk, in: *The Scientific Study of Flint and Chert* (G. de G. Sieveking and M. B. Hart, eds.), Cambridge University Press, Cambridge, pp.71–82.

Bromley, R. G., and Ekdale, A. A., 1987, Mass transport in European Cretaceous chalk; fabric criteria for its recognition, *Sedimentology* **34**:1079–1092.

Bromley, R. G., and Gale, A. S., 1982, The lithostratigraphy of the English Chalk Rock, *Cretaceous Res.* **3**:273–306.

Bromley, R. G., Schulz, M-G., and Peake, N. B., 1975, Paramoudras: Giant flints, long burrows and the early diagenesis of chalks, *K. Dan. Vidensk. Selsk. Biol. Skr.* **20**(10):1–31.

Brown, G., Catt, J. A., Hollyer, S. E., and Ollier, C. D., 1969, Partial silicification of chalk fossils from the Chilterns, *Geol. Mag.* **106**:583–586.

Brunton, C. H. C., 1966, Silicified productoids from the Visean of County Fermanagh, *Bull. Br. Mus. Nat. Hist. (Geol.)* **12**:173–243.

Brunton, C. H. C., 1968, Silicified brachiopods from the Visean of County Fermanagh (II), *Bull. Br. Mus. Nat. Hist. (Geol.)* **16**:1–70.

Brunton, C. H. C., 1984, Silicified brachiopods from the Visean of County Fermanagh, Ireland (III). Rhynchonellids, spiriferids and terebratulids, *Bull. Br. Mus. Nat. Hist. (Geol.)* **38**:27–130.

Buurman, P., and Van der Plas, L., 1971, The genesis of Belgian and Dutch flints and cherts, *Geol. Mijn.* **50**:9–28.

Buurman, P., van Breemen, N., and Henstra, S., 1973, Recent silicification of plant remains in acid sulphate soils, *Neves Jahrb. Mineral. Monatsh.* **3**:117–124.

Calvert, S. E., 1966, Accumulation of diatomaceous silica in the sediments of the Gulf of California, *Bull. Geol. Soc. Am.* **77**:569–596.

Calvert, S. E., 1974, Deposition and diagenesis of silica in marine sediments, in: Pelagic Sediments on Land and under the Sea (K. J. Hsu and H. C. Jenkyns, eds.), *Int. Assoc. Sedimentol. Spec. Publ.* **1**:273–299.

Carson, G. A., 1987, Silicification fabrics from the Cenomanian and basal Turonian of Devon, England: Isotopic results, in: Diagenesis of Sedimentary Sequences (J. D. Marshall, ed.), *Geol. Soc. Lond. Spec. Publ.* **36**:87–102.

Carson, G. A., 1988, A diagenetic and palaeooceanographic study of the mid-Cretaceous of southern England, Ph. D. thesis, University of Liverpool, Liverpool, England.

Chester, R., 1990, *Marine Geochemistry*, Unwin Hyman, London.

Clayton, C. J., 1982, Growth history and microstructure of flint, in: *International Association of Sedimentologists, 3rd European Meeting, Copenhagen*, pp. 105–107.

Clayton, C. J., 1984, The geochemistry of chert formation in Upper Cretaceous Chalks, Ph. D. thesis, University of London, London, England.

Clayton, C. J., 1986, The chemical environment of flint formation in Upper Cretaceous chalks, in: *The Scientific Study of Flint and Chert* (G. de G. Sieveking and M. B. Hart, eds.), Cambridge University Press, Cambridge, pp. 43–54.

Cloud, P., Awramik, S. M., Morrison, K., and Hadley, D. G., 1979, Earliest Phanerozoic or latest Proterozoic fossils from the Arabian Shield, *Precambrian Res.* **10**:73–93.

Coleman, M. L., 1985, Geochemistry of diagenetic non-silicate minerals: kinetic considerations, *Phil. Trans. R. Soc. Lond. A* **315**:39–56.

Coleman, M. L., and Raiswell, R., 1981, Carbon, oxygen and sulphur isotope variations in concretions from the Upper Lias of N.E. England, *Geochim. Cosmochim. Acta* **45**:329–340.

Crowley, S. F., 1988, Aspects of the diagenesis of the Great Limestone Cyclothem, Ph. D. thesis, University of Liverpool, Liverpool, England.

Curry, G. B., 1986, Fossils and tectonics along the Highland Boundary Fault in Scotland, *J. Geol. Soc. Lond.* **143**:193–198.

Deer, W. A., Howie, R. A., and Zussman, J., 1966, *An Introduction to the Rock-Forming Minerals*, Longman, London.

Dewers, T., and Ortoleva, P., 1990, Force of crystallization during the growth of siliceous concretions, *Geology* **18**:204–207.

Dietrich, R. V., Hobbs, C. R. B., and Lowry, W. D., 1963, Dolomitization interrupted by silicification, *J. Sediment. Petrol.* **33**:646–663.

Dorf, E., 1964, The petrified forests of Yellowstone Park, *Sci. Am.* **210**:107–114.

Eley, B. E., and Jull, R. K., 1982, Chert in the middle Silurian Fossil Hill Formation of Mantitoulin Island, Ontario, *Bull. Can. Petrol. Geol.* **30**:208–215.

Elorza, J. J., 1987, Early silicification in carbonate turbidites of the middle Eocene Flysch (Punta Gale, NE Bilbao, Spain), in: *International Association of Sedimentologists, 8th Regional Meeting Tunis*, p. 219.

Elorza, J., and Orue-Etxebarria, X., 1985, An example of silicification in *Gryphaea* sp. shells from Lano (south of Vitoria, Spain), in: *International Association of Sedimentologists, 6th Regional Meeting Lleida*, pp. 556–559.

Eugster, H. P., 1967, Hydrous sodium silicates from Lake Magadi, Kenya: Precursors of bedded chert, *Science* **157**:1177–1180.

Eugster, H. P., 1969, Inorganic bedded cherts from the Magadi area, Kenya, *Contrib. Mineral. Petrol.* **22**:1–31.

Ferris, F. G., Beveridge, T. J., and Fyfe, W. S., 1986, Iron-silica crystallite nucleation by bacteria in a geothermal sediment, *Nature* **320**:609–611.

Ferris, F. G., Fyfe, W. S., and Beveridge, T. J., 1988, Metallic ion binding by *Bacillus substilis:* Implications for the fossilization of microorganisms, *Geology* **16**:149–152.

Folk, R. L., and Pittman, J. S., 1971, Length slow chalcedony: A new testament for vanished evaporites, *J. Sediment. Petrol.* **41**:1045–1058.

Folk, R. L., and Weaver, C. E., 1952, A study on the texture and composition of chert, *Am. J. Sci.* **250:**498–510.

Francis, J. E., 1982, The fossil forests of the basal Purbeck Formation (Upper Jurassic) of Dorset, southern England, unpublished Ph. D. thesis, University of Southampton, Southampton, England.

Francis, J. E., 1983, The dominant conifer of the Jurassic Purbeck Formation, England, *Palaeontology* **26:**277–294.

Francis, J. E., 1984, The seasonal environment of the Purbeck (Upper Jurassic) fossil forests, *Palaeogeogr. Palaeoclimatol. Palaeoecol.* **48:**285–307.

Francis, S., Margulis, L., and Barghoorn, E. S., 1978, On the experimental silicification of microorganisms. II. On the time of appearance of eukaryotic organisms in the fossil record, *Precambrian Res.* **6:**65–100.

Froelich, P. N., Klinkhammer, G. P., Bender, M. L., Luedtke, N. A., Heath, G. R., Cullen, D., Dauphin, P., Hammond, D., Hartman, B., and Maynard, V., 1979, Early oxidation of organic matter in pelagic sediments of the eastern equatorial Atlantic: Suboxic diagenesis, *Geochim. Cosmochim. Acta* **43:**1075–1090.

Garrison, R. E., 1974, Radiolarian cherts, pelagic limestones, and igneous rocks in eugeosynclinal assemblages, in: Pelagic Sediments on Land and under the Sea (K. J. Hsu and H. C. Jenkyns, eds.), *Int. Assoc. Sedimentol. Spec. Publ.* **1:**367–399.

Geeslin, J. H., and Chafetz, H. S., 1982, Ordovician Aleman ribbon cherts: An example of silicification prior to carbonate lithification, *J. Sediment. Petrol.* **52:**1283–1293.

Golubic, S., 1976, Organisms that build stromatolites, in: *Stromatolites* (M. R. Walter, ed.), Elsevier, Amsterdam, pp. 113–126.

Green, J. W., Knoll, A. H., and Swett, K., 1989, Microfossils from silicified stromatolitic carbonates of the Upper Proterozoic Limestone-Dolomite "Series", central East Greenland, *Geol. Mag.* **126:**567–585.

Håkansson, E., Bromley, R., and Perch-Nielsen, K., 1974, Maastrichtian chalk of north-west Europe—A pelagic shelf sediment, in: Pelagic Sediments: on Land and under the Sea (K. Hsu and H. Jenkyns, eds.), *Int. Assoc. Sedimentol. Spec. Publ.* **1:**211–233.

Hardie, L. A., 1987, Dolomitization: A critical view of some current views, *J. Sediment. Petrol.* **57:**166–183.

Hay, R. L., 1968, Chert and its sodium-silicate precursors in sodium carbonate lakes of East Africa, *Contrib. Mineral. Petrol.* **17:**255–274.

Hedges, E. S., 1932, *Liesegang Rings and Other Periodic Structures*, Champman & Hall, London.

Henderson, R. A., 1984, Diagenetic growth of euhedral megaquartz in the skeleton of a stromatoporoid, *J. Sediment. Petrol.* **54:**1138–1146.

Herbold, D. R., and Glaser, L., 1975, *Bacillus subtilis* N-acetylmuramic acid L-alanine amidase, *J. Biol. Chem.* **250:**1676–1682.

Hofmann, H., and Schopf, J. W., 1983, Early Proterozoic microfossils, in: *Earth's Earliest Biosphere* (J. W. Schopf, ed.), Princeton University Press, Princeton, New Jersey, pp. 321–360.

Holdaway, H. K., and Clayton, C. J., 1982, Preservation of shell microstructure in silicified brachiopods from the Upper Cretaceous Wilmington Sands of Devon, *Geol. Mag.* **119:**371–382.

Horodyski, R. J., and Donaldson, J. A., 1983, Distribution and significance of microfossils in cherts of the middle Proterozoic Dismal Lakes Group, district of Mackenzie, Northwest Territories, Canada, *J. Paleontol.* **57:**271–288.

Iijima, A., and Tada, R., 1981, Silica diagenesis of Neogene diatomaceous and volcaniclastic sediments in northern Japan, *Sedimentology* **28:**185–200.

Iler, R. K., 1979, *The Chemistry of Silica; Solubility, Polymerisation, Colloid and Surface Properties and Biochemistry*, Wiley, Chichester, England.

Ireland, B. J., Curtis, C. D., and Whiteman, J. A., 1983, Compositional variation within some glauconites and illites and implications for their stability and origins, *Sedimentology* **30:**769–786.

Irwin, H., Curtis, C. D., and Coleman, M. L., 1977, Isotopic evidence for source of diagenetic

carbonates formed during burial of organic-rich sediments, *Nature* **269**:209–213.
Jacka, A. D., 1974, Replacement of fossils by length-slow chalcedony and associated dolomitization, *J. Sediment. Petrol.* **44**:421–427.
James, N. P., and Choquette, P. W., 1984, Limestones—The meteoric diagenetic environment, *Geosci. Can.* **11**:161–194.
Jarvis, I., Gale, A. S., and Clayton, C., 1982, Litho and biostratigraphical observations on the type sections of the Craie de Villedieu Formation (Upper Cretaceous, western France), *Newsl. Stratigr.* **11**:64–82.
Jones, J. B., and Segnit, E. R., 1971, The nature of opal: 1. Nomenclature and constituent phases, *J. Geol. Soc. Aust.* **18**:57–68.
Jones, J. B., and Segnit, E. R., 1972, Genesis of cristobalite and tridymite at low temperatures, *J. Geol. Soc. Aust.* **18**:419–422.
Karowe, A. L., and Jefferson, T. H., 1987, Burial of trees by eruptions of Mount St Helens, Washington: Implications for the interpretation of fossil forests, *Geol. Mag.* **124**:191–204.
Kastner, M., 1980, Length-slow chalcedony: The end of the new testament, *Amer. Geophys. Union (EOS)* **61**:399.
Kastner, M., 1981, Authigenic silicates in deep-sea sediments: Formation and diagenesis, in: *The Sea*, Volume 7, *The Oceanic Lithosphere* (C. Emiliani, ed.), Wiley, New York, pp. 915–980.
Kempe, S., and Degens, E. T., 1985, An early soda ocean?, *Chem. Geol.* **53**:95–108.
Kennedy, W. J., and Garrison, R. E., 1975, Morphology and genesis of nodular chalks and hardgrounds in the Upper Cretaceous of southern England, *Sedimentology* **22**:311–386.
Kenrick, P., and Edwards, D., 1988, The anatomy of Lower Devonian *Gosslingia breconensis* Heard based upon pyritized axes, with some comments on the permineralization process, *Bot. J. Linn. Soc.* **97**:95–123.
Khoury, H. N., 1987, Tripolization of chert in Jordan, *Sediment. Geol.* **53**:305–310.
Knauth, L. P., 1979, A model for the origin of chert in limestone, *Geology* **7**:274–277.
Knauth, L. P., and Epstein, S., 1975, Hydrogen and oxygen isotope ratios in silica from the JOIDES DSDP, *Earth Planet. Sci. Lett.* **25**:1–10.
Knauth, L. P., and Epstein, S., 1976, Hydrogen and oxygen isotope ratios in nodular and bedded cherts, *Geochim. Cosmochim. Acta* **40**:1095–1108.
Knauth, L. P., and Epstein, S., 1982, The nature of water in hydrous silica, *Am. Mineral.* **67**:510–520.
Knauth, L. P., and Lowe, D. R., 1978, Oxygen isotope geochemistry of cherts from the Onverwacht Group (3.4 billion years), Transvaal, South Africa, with implications for secular variations in the isotopic composition of cherts, *Earth Planet. Sci. Lett.* **41**:209–222.
Knoll, A. H., 1985, Exceptional preservation of photosynthetic organisms in silicified carbonates and silicified peats, *Phil. Trans. R. Soc. Lond. B* **311**:111–122.
Knoll, A. H., and Golubic, S., 1979, Anatomy and taphonomy of a Precambrian algal stromatolite, *Precambrian Res.* **10**:115–151.
Knoll, A. H., and Simonson, B., 1981, Early Proterozoic microfossils and penecontemporaneous quartz cementation in the Sokoman Iron Formation, Canada, *Science* **211**:478–480.
Kolodony, Y., Taraboulos, A., and Frieslander, U., 1980, Participation of fresh water in chert diagenesis: Evidence from oxygen isotopes and boron α-track mapping, *Sedimentology* **27**:305–316.
Labeyrie, L. D., and Juillet, A., 1982, Oxygen isotope exchangeability of diatom valve silica; interpretation and consequences for paleoclimatic studies, *Geochim. Cosmochim. Acta* **46**:967–975.
Land, L. S., 1979, Chert-chalk diagenesis: The Miocene island slope of north Jamaica, *J. Sediment. Petrol.* **49**:223–232.
Leduc, M., Kasra, R., and Van Heijenoort, J., 1982, Introduction and control of the autolytic system of *Escherichia coli*, *J. Bacteriol.* **152**:26–34.
Leo, R. F., and Barghoorn, E. S., 1976, Silicification of wood, *Bot. Mus. Leaflts Harv. Univ.* **25**:1–47.
Loope, D. B., and Watkins, D. K., 1989, Pennsylvanian fossils replaced by red chert: Early

oxidation of pyritic precursors, *J. Sediment. Petrol.* **59**:375–386.
Maliva, R. G., and Siever, R., 1988, Mechanism and controls of silicification of fossils in limestones, *J. Geol.* **96**:387–398.
Matheney, R. K., and Knauth, L. P., 1989, Oxygen-isotope fractionation between marine biogenic silica and seawater, *Geochim. Cosmochim. Acta* **53**:3207–3214.
McBride, E. F., and Folk, R. L., 1979, Features and origin of Italian Jurassic radiolarites deposited on continental crust, *J. Sediment. Petrol.* **49**:837–868.
Meyers, W. J., 1977, Chertification in the Mississippian Lake Valley Formation, Sacremento Mountains, New Mexico, *Sedimentology* **24**:75–105.
Meyers, W. J., and James, A. T., 1978, Stable isotopes of cherts and carbonate cements in the Lake Valley Formation (Mississippian), Sacremento Mts, New Mexico, *Sedimentology* **25**:105–124.
Mitchell, R. S., and Tufts, S., 1973, Wood opal—A tridymite-like mineral, *Am. Mineral.* **58**:717–720.
Murata, K. J., 1940, Volcanic ash as a source of silica for the silicification of wood, *Am. J. Sci.* **238**:586–596.
Murata, K. J., and Nakata, J. K., 1974, Cristobalitic stage in the diagenesis of diatomaceous shale, *Science* **184**:567–568.
Murata, K. J., and Norman, M. B., 1976, An index of crystallinity for quartz, *Am. J. Sci.* **276**:1120–1130.
Murata, K. J., Friedman, I., and Gleason, J. D., 1977, Oxygen isotope relations between diagenetic silica minerals in Monterey Shale, Temblor Range, California, *Am. J. Sci.* **277**:259–272.
Newell, N. D., and Boyd, D. W., 1970, Oyster-like Permian Bivalvia, *Bull. Am. Mus. Nat. Hist.* **143**:217–282.
Oehler, J. H., 1976, Experimental studies in Precambrian paleontology: Structural and chemical changes in blue-green algae during simulated fossilization in synthetic chert, *Bull. Geol. Soc. Am.* **68**:117–129.
Oehler, J. H., and Schopf, J. W., 1971, Artificial microfossils: Experimental studies of permineralization of blue-green algae in silica, *Science* **174**:1229–1231.
Onishi, H., 1972, Antimony, in: *Handbook of Geochemistry* (K. H. Wedepohl, ed.), Springer, New York.
Orme, G. R., 1974, Silica in the Viséan limestones of Derbyshire, England, *Proc. Yorks. Geol. Soc.* **40**:63–104.
Oskvarek, J. D., and Perry, E. C., Jr., 1976, Temperature limits on the early Archean ocean from oxygen isotope variations in the Isua supercrustal sequence, West Greenland, *Nature* **259**:192–194.
Perry, E. C., Jr., 1967, The oxygen isotope geochemistry of ancient cherts, *Earth Planet. Sci. Lett.* **3**:62–66.
Perry, E. C., Jr., and Tan, F. C., 1972, Significance of oxygen and carbon isotope variations in early Precambrian cherts and carbonate rocks of Southern Africa, *Bull. Geol. Soc. Am.* **83**:647–664.
Perry, E. C., Jr., Ahmad, S. N., and Swulius, T. M., 1978, The oxygen isotope composition of 3,800 my old metamorphosed chert and iron formation from Isukasia, west Greenland, *J. Geol.* **86**:223–239.
Pingitore, N. E., Jr., 1982, The role of diffusion during carbonate diagenesis, *J. Sediment. Petrol.* **52**:27–39.
Pittman, J. S., 1959, Silica in Edwards Limestone, Travis County, Texas, in: Silica in Sediments (H. A. Ireland, ed.), *Soc. Econ. Paleontol. Mineral. Spec. Publ.* **7**:121–134.
Read, C. B., and Brown, R. W., 1937, American Cretaceous ferns of the genus *Tempskya*, *U. S. Geol. Surv. Prof. Pap.* **186-F**:105–132.
Rogers, A. F., 1917, A review of the amorphous minerals, *J. Geol.* **25**:515–535.
Sahni, B., 1941, Indian silicified plants, *Proc. Ind. Acad. Sci.* B **14**:489–501.
Sandberg, P. A., and Hudson, J. D., 1983, Aragonite relic preservation in Jurassic calcite-replaced bivalves, *Sedimentology* **30**:879–892.
Schmitt, J. G., and Boyd, D. W., 1981, Patterns of silicification in Permian pelecypods and

brachiopods from Wyoming, *J. Sediment. Petrol.* **51**:1297–1308.
Scurfield, G., and Segnit, E. R., 1984, Petrifaction of wood by silica minerals, *Sediment. Geol.* **39**:149–167.
Scurfield, G., Anderson, C. A., and Segnit, E. R., 1974, Silica in woody stems, *Aust. J. Bot.* **22**:211–229.
Siedlecka, A., 1972, Length-slow chalcedony and relicts of sulphates—Evidences of evaporitic environments in the Upper Carboniferous and Permian beds of Bear Island, Svalbard, *J. Sediment. Petrol.* **42**:812–816.
Sigleo, A. C., 1978, Organic geochemistry of silicified wood, Petrified Forest National Park, Arizona, *Geochim. Cosmochim. Acta* **42**:1397–1405.
Sigleo, A. C., 1979, Geochemistry of silicified wood and associated sediments, Petrified Forest National Park, Arizona, *Chem. Geol.* **26**:151–163.
Simonson, B. M., 1985a, Sedimentological constraints on the origins of Precambrian iron-formations, *Bull. Geol. Soc. Am.* **96**:244–252.
Simonson, B. M., 1985b, Sedimentology of cherts in the Early Proterozoic Wishart Formation, Quebec–Newfoundland, Canada, *Sedimentology* **32**:23–40.
Simonson, B. M., 1987, Early silica cementation and subsequent diagenesis in arenites from four early Proterozoic iron formations of North America, *J. Sediment. Petrol.* **57**:494–511.
Stein, C. L., 1982, Silica recrystallization in petrified wood, *J. Sediment. Petrol.* **52**:1277–1282.
Tada, R., and Iijima, A., 1983, Identification of mixtures of opaline silica phases and its implication for silica diagenesis, in: *Siliceous Deposits in the Pacific Region* (A. Iijima, J. R. Hein, and R. Siever, eds.), Elsevier, Amsterdam, pp. 229–246.
Thiry, M., Bertrand Ayrault, M., and Grisoni, J.-C., 1988, Ground-water silicification and leaching in sands: Example of the Fontainbleau Sand (Oligocene) in the Paris Basin, *Bull. Geol. Soc. Am.* **100**:1283–1290.
Tresise, G. R., 1961, The nature and origin of chert in the Upper Greensand of Wessex, *Proc. Geol. Assoc.* **72**:333–356.
Voigt, E., and Häntzschel, W., 1956, Die grauen Bänder in der Schreibkreide Nordwest-Deutschlands und ihre Deutung als Lebensspuren, *Mitt. Geol. Staatsinst. Hamburg* **25**:104–22.
Walter, L. M., 1985, Relative reactivity of skeletal carbonates during dissolution: Implications for diagenesis, in: Carbonate Cements (N. Schneidermann, and P. M. Harris, eds.), *Soc. Econ. Paleontol. Mineral. Spec. Publ.* **36**:3–16.
Williams, L. A., Parks, G. A., and Crerar, D. A., 1985, Silica diagenesis, I. Solubility controls, *J. Sediment. Petrol.* **55**:301–311.
Wilson, R. C. L., 1966, Silica diagenesis in Upper Jurassic limestones of southern England, *J. Sediment. Petrol.* **36**:1036–1049.
Wise, S. W., and Kelts, K. R., 1972, Inferred diagenetic history of a weakly silicified deep-sea chalk, *Trans. Gulf Coast Assoc. Geol. Soc.* **22**:177–203.
Wise, S. W., Weaver, F. M., and Guven, N., 1973, Early silica diagenesis in volcanic and sedimentary rocks: Devitrification and replacement phenomena, in: *31st Annual Proceedings Electron Microscopists America*, pp. 206–207.
Zijlstra, H. J. P., 1987, Early diagenetic silica precipitation in relation to redox boundaries and bacterial metabolism, in late Cretaceous of the Maastrichtian type locality, *Geol. Mijn.* **66**:343–355.

Chapter 11

Taphofacies Controls

Background and Episodic Processes in Fossil Assemblage Preservation

STEPHEN E. SPEYER and CARLTON E. BRETT

STEPHEN E. SPEYER • Department of Geosciences, University of Arizona, Tucson, Arizona 85721; *present address:* Department of Geology, Arizona State University, Tempe, Arizona 85287. CARLTON E. BRETT • Department of Geological Sciences, University of Rochester, Rochester, New York 14627.

Taphonomy: Releasing the Data Locked in the Fossil Record, Volume 9 of *Topics in Geobiology*, edited by Peter A. Allison and Derek E. G. Briggs, Plenum Press, New York, 1991.

1. Introduction

Taphonomic comparisons between skeletal deposits have become a recent focus of paleoecological and stratigraphic research. The taphonomic attributes which characterize fossil and modern shell accumulations may be quantified, providing indices for convenient stratigraphic comparisons (e.g., Norris, 1986; Speyer and Brett, 1986; Meldahl, 1987; Meldahl and Flessa, 1990). Many studies have demonstrated a correlation between taphonomic attributes and environmental conditions (Schäfer, 1972; Brett and Baird, 1986; Kidwell and Behrensmeyer, 1988; Kidwell, 1988; Speyer and Brett, 1988). This correlation has encouraged an emphasis upon paleoenvironmental reconstruction and related facies research; a majority of studies incorporating taphonomic analyses have been designed to evaluate the character and distribution of environmental settings among strata (e.g., Seilacher *et al.*, 1985; Parsons *et al.*, 1988).

The relationship between taphonomic process and environmental setting has been corroborated by actualistic investigations of modern process–product associations (Staff *et al.*, 1986; Fürsich and Flessa, 1987; Davies *et al.*, 1989). These data define sedimentary conditions and aid in paleoenvironmental reconstruction, but actualistic investigations fail to address the relationship between taphonomic process and time. The way in which taphonomic processes interact at varied temporal scales and intensities remains obscure and has not been investigated beyond the level of preliminary taphofacies models (Johnson, 1960; Speyer and Brett, 1986, 1988). Models of this sort often overlook the important interaction between cooperating and contrasting factors. There exists a clear duality to the stratigraphic record; sedimentary events leave conspicuous signatures that are superimposed upon background patterns of deposition and stratification, and vice versa.

Episodic sedimentary phenomena, although pervasive in many strata and critical to formatting the stratigraphic record of carbonate and siliciclastic-starved depositional settings, operate in conjunction with background processes that occur on a day-to-day basis. The stratigraphic expression of neither process system is truly independent of the other; this is the essence of "dynamic stratigraphy." It is therefore necessary to incorporate both these perspectives into our stratigraphic reconstructions. Taphonomic processes are tightly correlated to environmental conditions and therefore provide important insight concerning the stratigraphic interaction between environmental processes operating at these two temporal scales.

The objectives of this chapter are: (1) to evaluate the taphonomic imprint of episodic and background environmental conditions, (2) to consider the stratigraphic interaction between these taphonomic processes as indicated in the rock record, and (3) to discuss the impact of these interactions with respect to taphofacies models, using examples from the Middle Devonian of New York State.

2. Taphonomy, Taphofacies, and Taphofacies Models

Comparative taphonomy has emerged during the past several years as an important tool in facies analysis (Behrensmeyer and Kidwell, 1985; Brett and Baird, 1986). Taphonomic properties are sensitive to changes in gross paleoenvironmental setting, but also incorporate biological data (i.e., organism distributions) necessary to understand paleoecological phenomena that correspond to inferred paleoenvironments. Taphonomic gradients [e.g., rare to abundant skeletal fragmentation (Norris, 1986)] reflect environmental gradients. Thus, vertical facies associations may be related to lateral environmental transitions (Walther's law).

Taphofacies (Brett and Baird, 1986; Speyer and Brett, 1986) provide a conceptual basis for evaluating the characteristics of a particular environmental setting and comparing facies attributes among such settings within an environmental system. Taphofacies can be discriminated on the basis of quantified taphonomic attributes (Speyer and Brett, 1986), or less rigorously defined taphofacies models can be derived deductively and allow for more comprehensive comparisons (Speyer and Brett, 1988).

3. Background and Episodic Taphonomic Processes

Seven general taphofacies are recognized within the Hamilton Group exposed across western New York State (Fig. 1A) and are based on

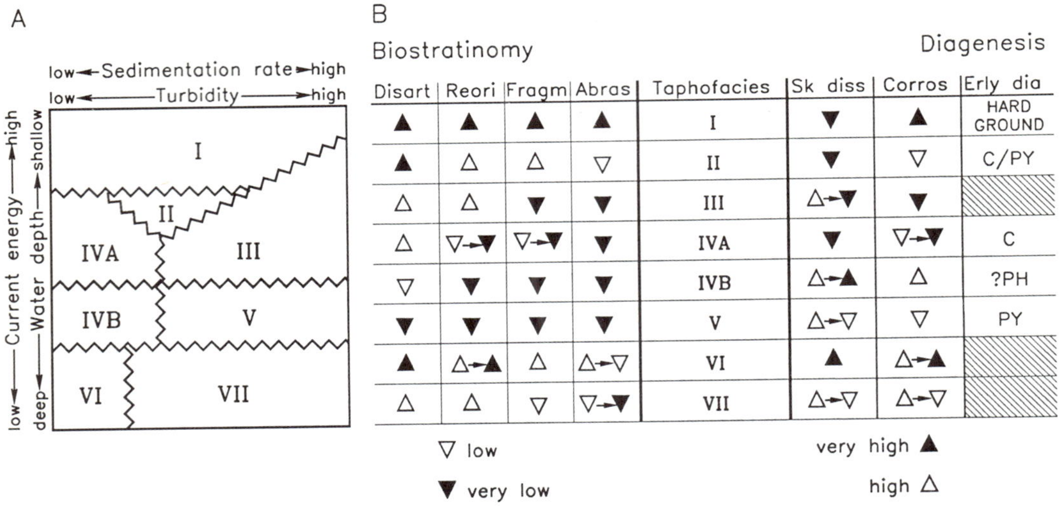

Disart	Reori	Fragm	Abras	Taphofacies	Sk diss	Corros	Erly dia
▲	▲	▲	▲	I	▼	▲	HARD GROUND
▲	△	△	▽	II	▼	▽	C/PY
△	△	▼	▼	III	△→▼	▼	
△	▽→▼	▽→▼	▼	IVA	▼	▽→▼	C
▽	▼	▼	▼	IVB	△→▲	△	?PH
▼	▼	▼	▼	V	△→▽	▽	PY
▲	△→▲	△	△→▽	VI	▲	△→▲	
△	△	▽	▽→▼	VII	△→▽	△→▽	

FIGURE 1. Generalized taphofacies mosaic and corresponding attributes. (A) Seven taphofacies are distributed according to lateral gradients in background sedimentation rate and turbidity, and vertical gradients in current energy and water depth. (B) Each taphofacies illustrated corresponds to a specific set of taphonomic attributes. Biostratinomic traits (Disarticulation, Reorientation, Fragmentation, Abrasion) are distributed according to current energy, and duration of surface exposure (a function of sedimentation rate) (see Fig. 2). Diagenetic attributes (Skeletal dissolution, Corrosion, Early Diagenesis) are largely the product of seafloor and/or interstitial oxygenation, and the distribution of chemical species (see Fig. 6). Early diagenetic mineralization may include hardground, carbonate (c), pyrite (py) or phosphate (ph). Disart, disarticulation; Reori, reorientation; Fragm, fragmentation; Abras, abrasion; Sk diss, skeletal dissolution; Corros, corrosion; Erly dia, early diagenesis. C, carbonate; PY, pyrite; PH, phosphate.

representative fossil assemblages that recur throughout the Ludlowville and Moscow Formations. These fossil assemblages, categorized according to diagnostic suites of taphonomic attributes (Fig. 1B), portray the superimposed effects of processes operating at two distinct temporal levels: background processes span relatively long periods of time and are interrupted by episodic phenomena which are essentially instantaneous.

3.1. Background Taphonomic Processes

Background taphonomic phenomena comprise all preservational processes that operate over long periods of time (see Fig. 2). Background processes are intrinsically destructive and account for the majority of biases that characterize the fossil record (Davies *et al.*, 1989; Powell *et al.*, in press). Three interrelated processes mediate the intensity of background processes: rate of burial, environmental energy (currents), and bioturbation (Fig. 3). Of these, the first is clearly the most influential and governs the

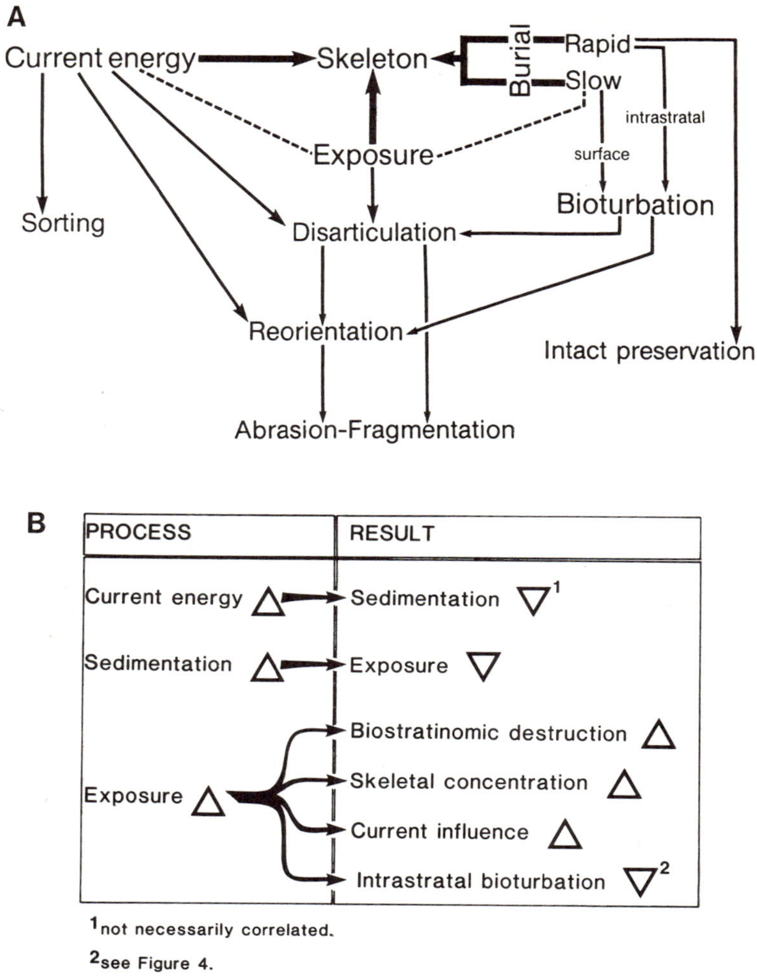

FIGURE 2. Background taphonomic histories. (A) Flow chart of interrelated processes that interact to produce specific background taphonomic signatures. (B) Chart comparing the relationships between processes and consequences for the major components in Part A. The level of current energy, duration of surface exposure, rate of background burial, and manner of bioturbation are primarily responsible for the intensity of taphonomic processes (e.g., disarticulation, reorientation, sorting, abrasion, and fragmentation). Surface exposure is directly related to the rate of background burial, which in turn affects the mode of bioturbation (i.e., surface or intrastratal). Particularly high rates of sedimentation may result in intact preservation.

taphonomic impact of mechanical processes, including disarticulation, sorting, reorientation, fragmentation, and abrasion. Background burial refers simply to the persistent accumulation of sediment in a particular setting (Schindel, 1982); background burial may be rapid (in excess of 100 cm/1000 years), intermediate (10–100 cm/1000 years), slow (up to 10 cm/1000 years), or nonexistent. These processes are primarily the result of current energy and surface exposure, but may reflect secondarily the influence of biogenic scavenging and/or deposit feeding.

Kidwell and Jablonski (1983) clearly demonstrated the relationship between sedimentation, shell bed accumulation, and bioturbation. Decreased net deposition results in increased shell concentration and thus increases the number of available colonization substrates and inhibits intrastratal bioturbation (taphonomic feedback). Depressed sediment input also prolongs the duration of surface exposure and thereby enhances the intensity of mechanical destruction. Simply stated, the longer a skeleton remains at the seafloor surface, the more likely it is that it will become disarticulated, reoriented, sorted, abraded, and/or corroded (see Fig. 2).

Conversely, persistently high sediment input may favor intrastratal bioturbation (note that, to a point, bioturbation rates will apparently increase as sediment input is depressed), because shelly accumulation is diluted, and "rapid burial" inhibits prolonged surface processing (Fig. 3). Although a high sedimentation rate diminishes the effects of mechanical taphonomic process (e.g., fragmentation, abrasion), greater deep bioturbation increases the rate and scope of fluid flow through the sediment. Depending upon gross environmental regime, the rate of fluid flow within the sediment influences the intensity of skeletal corrosion and dissolution (Flessa and Brown, 1982). Intrastratal deposit feeding by macroinfauna results in disarticulation and characteristic patterns of reorientation (Salizar-Jiminéz *et al.*, 1982). Hence, although the impact of mechanical taphonomic

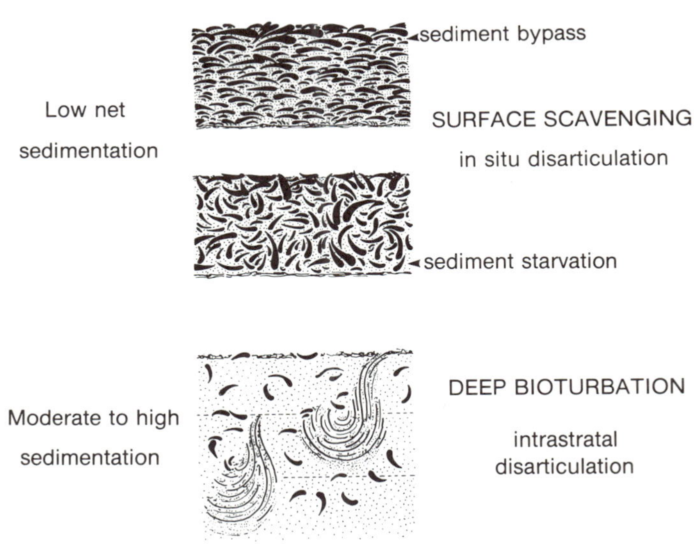

FIGURE 3. The relationships among sedimentation rate, current energy, and bioturbation directly influence shell bed fabric. Low net sedimentation may result in an organized (moderate to high current energy; sediment bypass) or a chaotic (low energy; sediment starvation) shell bed. Moderate to high rates of sedimentation, regardless of ambient energy levels, dilute the concentration of shelly remains and favor intrastratal bioturbation [see Kidwell and Jablonski (1983) for discussion of "taphonomic feedback")].

processes is minimized by high rates of sedimentation, biogenic and chemical processes may be greatly enhanced.

Sedimentation rate and current energy are interdependent (Fig. 2); low sediment input may be the result of sediment bypass due to winnowing currents or, conversely, may reflect sediment starvation in settings characterized by low turbidity, depressed current energy, and nondeposition (Fig. 3). The result in terms of biogenic interaction is much the same, although the stratigraphic pattern may be diagnostically different. Current-winnowed sediment will generate a hydrodynamically stable packing of shelly debris [see stacked and nested clusters in a ''sedimentologic skeletal accumulation'' (Kidwell *et al.*, 1986)], whereas shell distribution and orientation will be more affected by biogenic influences as environmental energy decreases [see oblique, loosely packed assemblage in an ''extrinsic biogenic skeletal accumulation'' (Kidwell *et al.*, 1986)].

A gradient exists that corresponds to a relationship between sedimentation and burial rates. Certain modes of bioturbation characterize the end members of this continuum as well as the midrange. Deep bioturbation pervades settings characterized by moderately high depositional rates, whereas surface scavenging predominates across hiatal surfaces (in regions of nondeposition). Bioturbation may be absent or nonindicative in areas characterized by very rapid background deposition.

Many studies, actualistic (e.g., Fürsich and Flessa, 1987; Cutler, 1989; Davies *et al.*, 1989; Meldahl and Flessa, 1990) as well as stratigraphic (e.g., Miller *et al.*, 1988), have demonstrated the importance of exposure time in determining the intensity of taphonomic effects. The distribution and relative intensity of taphonomic bias are systematically distributed throughout the taphofacies mosaic, indicating that paleoecological resolution and quality of fossil preservation vary with environmental setting. This variation corresponds primarily to background taphonomic phenomena, but is overprinted by event-related phenomena (Brett and Speyer, 1989). Clearly, taphonomic bias is primarily the result of background taphonomic conditions.

3.2. Episodic Taphonomic Processes

Taphonomic processes that operate during short, but catastrophic time intervals generate an array of features that may superficially resemble background taphonomic effects (Fig. 4). Sedimentary events occur at a temporal level that is extremely brief in the context of geological time (on the order of hours to days). Nevertheless, the results of such phenomena may be all important in the accumulation of strata, particularly in settings otherwise typified by low background rates of sedimentation or nondeposition. Sedimentary events interrupt background processes and generate a varied spectrum of deposits across a shore-to-basin transect of environments (Aigner, 1985; Brett *et al.*, 1986b; Norris, 1986). Nearshore settings, above normal wave base, are represented by amalgamated shell deposits that grade

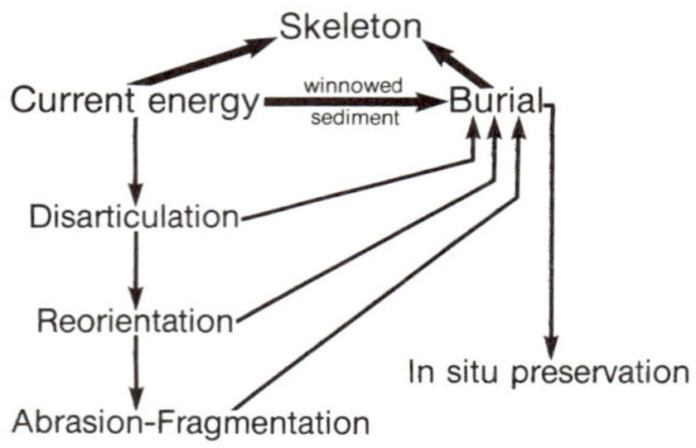

FIGURE 4. Episodic taphonomic histories. Current energy and timing of final burial are the two most critical factors that influence the episodic taphonomy of skeletal remains. Current winnowing in proximal environmental settings results in severe taphonomic processing, but may provide the sediment load that will enhance *in situ* preservation in a more distal setting.

downslope into thin, graded shelly sands (tempestites) between normal and storm wave bases, and smothered bottom assemblages below storm wave base.

Event-related taphonomic processes are the interactive result of abruptly increased current energy and/or skeletal burial (Fig. 4). As discussed above, currents winnow sediment, thereby concentrating autochthonous shelly remains in lag deposits. Winnowed sediment is carried in temporary suspension or as traction load until deposited during current wane. Deposition may result in *in situ* preservation in particular environmental settings and/or may preserve the accumulated effects of event processing of skeletal remains (e.g., disarticulation, reorientation, abrasion–fragmentation). The resulting assemblage is generally a complex mixture of allochthonous and autochthonous components, the classic thanatocoenosis concisely described by Brouwer (1966). Rapid episodic or catastrophic burial is distinguished from rapid background sedimentation on the basis of amount of sediment deposited over unit time. The most important difference is the time unit involved. Whereas background rates of sedimentation are measured on a scale of thousands of years, episodic burial occurs over a span of hours to days.

Shell-rich deposits (Seilacher *et al.*, 1985) are of two types: *Konservat-Lagerstätten* and *Konzentrat-Lagerstätten*. *Konzentrat-Lagerstätten* (''concentration'' deposits) are biogenically and mechanically concentrated assemblages which have been buried as a result of event-related phenomena, thereby terminating lag accumulation. In contrast, exceptionally well preserved biotas, with or without soft-part anatomy intact, are referred to as *Konservat-Lagerstätten* (''conserved'' deposits) and reflect particular, often unique, sedimentary circumstances (Allison, 1988). Seilacher *et al.* (1985) discussed the sedimentological and ecological implications of the varied Lagerstätten in terms of environment of deposition and burial.

Obrution deposits (a type of conservation deposit) represent rapidly buried or ''smothered'' assemblages. Varying types of obrution deposits are distributed in predictable patterns among environmental facies (Brett and Speyer, 1989; Speyer and Brett, 1990); e.g., lenticular obrution deposits occur in high-energy, shallow-water facies, whereas ''smothered bottom'' horizons (mud tempestites) occur below storm wave base. Similarly, this pattern of occurrence indicates a systematically variable level of paleoecological resolution among obrution *Lagerstätten*. Thus, there is great interest in the recognition and interpretation of episodic sedimentary deposits (Seilacher, 1982). Numerous studies detail the character and distribution of these event deposits (e.g., Kumar and Sanders, 1976; Kreisa, 1981; Aigner, 1982; Brett, 1983; Brett *et al.*, 1986b; Johnson, 1989) and usually relate observed features to a proximality spectrum of some sort (*sensu* Aigner, 1982; Brett *et al.*, 1986b). Taphonomic properties have figured importantly in the synthesis of such facies models (Kreisa and Bambach, 1982; Norris, 1986; Speyer and Brett, 1986). Although critical to understanding stratigraphic dynamics, sedimentary events are nevertheless superimposed on background processes that were clearly more influential in governing organism distribution and ecology. Paleoecological stratigraphic research seeks, in large part, a better understanding of the interaction between processes representing these two categories. Both suites of processes respond to ambient environmental parameters in specific and predictable manners and each exerts effects on the other.

3.3. Superposition of Taphonomic Processes

Taphofacies represent the composite record of both background and episodic environmental conditions. This interaction between background and episodic taphonomic processes provides additional data useful in discriminating taphofacies. Superimposed background and episodic taphonomic signatures vary in the relative importance of each component and may be categorized according to the product of their interaction: discrete, conflicting, constructive, or cooperating (Fig. 5).

In certain stratigraphic settings the taphonomic signature of both short-term and long-term processes may be clearly recognized. Barren, laminated, graded, or structureless deposits, with or without evidence of bioturbation, may be situated between densely fossiliferous interbeds. Regular alternation of these microfacies indicates punctuated changes in depositional setting which are not associated with lateral (environmental) shifts, but represent differences between episodic and background environmental conditions in the same setting. Many apparently rhythmic packages of strata may represent deposition at different temporal levels without significant change in gross environmental setting (Speyer and Jones, 1988). Long-term skeletal concentration in regions characterized by low background sedimentation may be unpredictably perturbed by episodes of dramatically increased sedimentation. The counterintuitive result is that long intervals of time are represented by little stratal thickness, but short, catastrophic intervals may be represented by a comparatively thick section (Parsons *et al.*, 1988).

Discrete signatures range between the extremes represented by superimposed episodic and background taphonomic signatures. Because discrete signatures constitute a continuum, the relative stratigraphic influence of

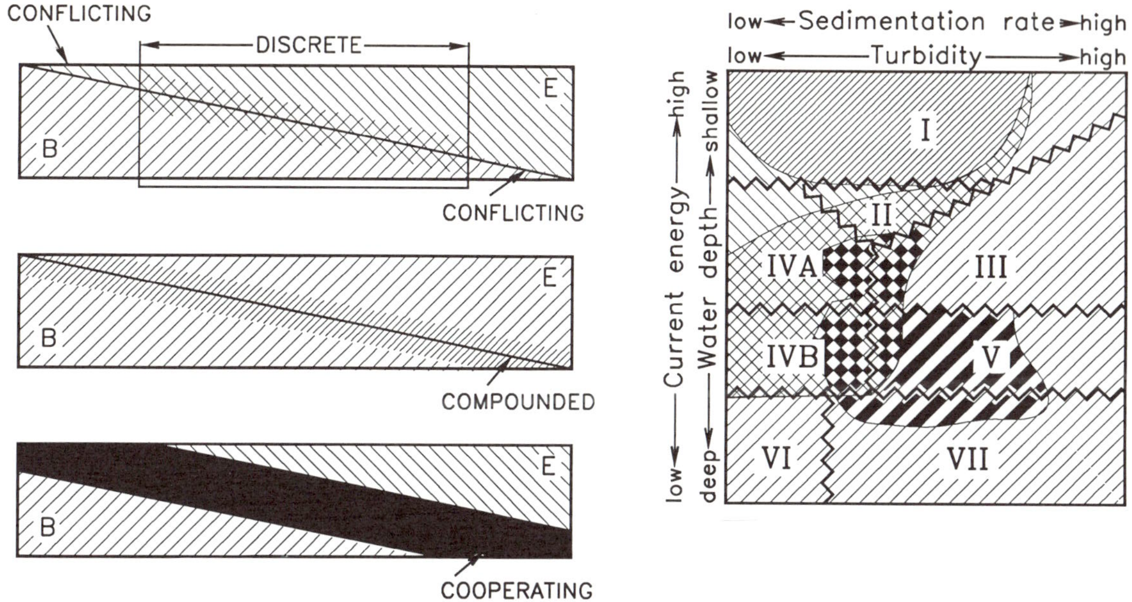

FIGURE 5. The stratigraphic interaction of background and episodic taphonomic processes results in four distinct signatures that grade according to the relative influence of each component and the circumstances of interaction. *Conflicting* signatures display the predominant influence of either background (B) or episodic (E) processes in formatting the stratigraphic sequence. *Discrete* signatures represent the gradient between these two extremes; background and episodic processes are both indicated by corresponding attributes in a shell bed. *Compounded* signatures display the cumulative result of similar background and episodic phenomena. Neither signature is discretely represented. *Cooperating* signatures indicate the unique interaction of specific background processes and episodic processes (typically rapid burial). Syngenetic mineralization is characteristically the result of rapid burial and the decay of biogenic remains within the context of a background-chemical setting (see Fig. 6). The environmental distribution of these superimposed signatures is indicated as a modified version of Figure 1A. See text for accompanying discussion.

each component should vary according to environment, background processes dominating at one end of the gradient, episodic processes most influential at the other. In general, a particular taphofacies can be characterized as *event-dominated* or *background-dominated*, depending upon the relative influence of event versus background processes. In reality, both temporal scales of processes have shaped the appearance of fossil accumulations, although one may mask the effects of the other. This gives rise to the concept of conflicting signatures.

Conflicting signatures represent an overprinting of background processes by episodic processes or vice versa. If the signature of taphonomic processes operating at one temporal level is modified or obscured by taphonomic processes characterizing the other, a conflicting signature is generated. The orientation of shells deposited and buried during a sedimentary event, for example, may be obscured by subsequent intrastratal bioturbation. Conversely, a sedimentary event may result in the preservation of articulated remains in a setting where none normally occurs. Although difficult to recognize, conflicting signatures may indicate the relative influence of background versus episodic factors in the stratification process and thereby provide additional information useful in environmental reconstructions.

Compounded signatures result when the superimposed effects of background and episodic processes enhance the taphonomic signature generated independently by either phase. In essence, the observed taphonomy is an additive result of superimposing virtually identical signatures; e.g., in nearshore regions, the proportion of fragmented remains generated in a particular setting by ''background'' processes is added to during a sedimentary event. Processes operating on a day-to-day basis are enhanced by event-related processes and vice versa. Typically, the resolution of individual events is exceedingly poor and discrete substrate levels cannot be recognized.

Cooperating signatures reflect the constructive interaction of background phenomena and episodic processes. The result is a specific ''taphonomic'' effect that supersedes the diagnostic importance of signatures generated independently by either taphonomic phase. The constructive interaction of background and episodic processes is well illustrated by early diagenetic phenomena (Fig. 6). Background sediment chemistry is regulated, in part, by sedimentation rate and bottom-water chemistry, which, in turn, govern the intensity of bioturbation and density of skeletal accumulations. As previously detailed [and more completely discussed in Kidwell and Jablonski (1983)], skeletal concentrations limit the depth of intrastratal bioturbation and provide a chemical buffer to skeletal dissolution (Speyer and Brett, 1988).

Background processes set the stage for early diagenesis by providing the requisite chemical conditions. Mineralization would not occur if episodic burial did not isolate chemical microenvironments. In short, episodes of rapid burial act as diagenetic catalysts to initiate diagenesis in those settings

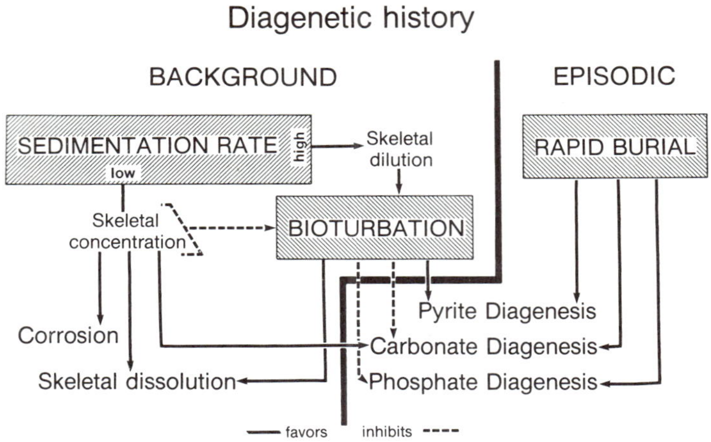

FIGURE 6. Early diagenesis is typically the result of episodic processes that have been superimposed on background processes. The result is a unique, cooperative taphonomic signature. Sedimentation rate and intensity of bioturbation set the stage for specific diagenetic pathways. High rates of sedimentation favor intrastratal bioturbation, which in turn increases the fluid flow through the sediment. This results in increased skeletal dissolution and reduces the amount of calcium that can accumulate interstitially. Bioturbation, however, increases the flux of oxidized sulfate into the sediment where it can be reduced to sulfide during pyrite diagenesis. In contrast, depressed rates of sediment input favor skeletal buildup, inhibit deep bioturbation [cf. ''taphonomic feedback'' (Kidwell and Jablonski, 1983)], and provide a calcium sink necessary for carbonate diagenesis. In all cases, rapid burial isolates decomposing biogenic remains from external processes and diffusion, and thereby provides a ''chemical microenvironment'' within which early diagenesis is initiated.

where background conditions have concentrated chemical species or in some other fashion set the stage necessary for authigenic mineralization.

In most cases, early diagenesis coincides with organism death and decay. Decay and necrolysis substantially change the chemical conditions (in particular pH) within restricted ''microenvironments'' and thus favor the precipitation of specific diagenetic products (e.g., phosphate, carbonate, pyrite, silica). Different background conditions favor different diagenetic pathways (Fig. 6; see Sections 4.4 and 4.6).

4. Integrated Taphofacies Models for Shell Beds

Taphofacies discussed in the following sections illustrate the duality of taphonomic phenomena and provide a strong argument that strictly uniformitarian- or event-biased approaches to fossil assemblage analysis are in themselves inadequate. An understanding of the important interactions between superimposed background and episodic phenomena is critical to interpreting the stratigraphic distribution of environmental facies.

The Middle Devonian Hamilton Group provides an appropriate study area for investigating taphonomic dynamics because it yields a diversity of marine faunas that occur in facies representing a broad range of paleoenvironmental settings (Speyer and Brett, 1986; Miller *et al.*, 1988; Parsons *et al.*, 1988). These strata are mixed carbonate and fine-grained siliciclastic lithologies and represent deposition across the western flank of the Middle Paleozoic Appalachian Basin (Brett *et al.*, 1986b; Fig. 1).

Detailed descriptions of specific localities and stratigraphic horizons are beyond the scope of this chapter; the interested reader is referred to discussions in the literature cited under the headings of ''Reference localities,'' ''Typical lithofacies,'' and ''Typical biofacies'' for each taphofacies. All reference sections are from the Middle Devonian (Givetian) Hamilton Group.

4.1. Taphofacies Model I: Nearshore Amalgamated Shell Beds

Reference examples: Centerfield Limestone, Schaeffer Creek Submember, Genesee and Livingston Counties, western New York State (Gray, 1985; Savarese *et al.*, 1986); Jaycox Member, Hills Gulch Bed, Erie County, New York (Mayer, 1989); Tichenor Limestone Member, western New York (Baird, 1979; Baird and Brett, 1981).

Typical lithofacies: Crinoid grainstone; amalgamated tempestites (Brett *et al.*, 1986b).

Typical biofacies: Favosites hamiltoniae (Brett *et al.*, 1990).

4.1.1. General Features

This taphofacies is characterized by thick, often tabular shell beds. Discrete bedding has been obscured largely by recurring scour and fill. Locally, bedding surfaces are recognized on the basis of shaley partings with more abundant and sometimes better-articulated fossils.

In general, shells are highly abraded, fragmented to comminuted (unidentifiable), and oriented in a hydrodynamically stable posture. Massive skeletal components are locally abundant and may be toppled; e.g., large tabulate coral heads, massive trepostome bryozoans. Taphofacies Model I beds display a variable proportion of complete, nonfragmented remains (1–5%). Bivalved components typically include robust brachiopod valves which are disarticulated, fragmented, and abraded. Fragile arborescent remains are rare or altogether absent.

Crinoid remains are abundant, but are represented by disarticulated plates and columnal sections. Trilobites are uncommon, but when present are typically disarticulated with individual sclerites fragmented. Rarely, multielement and bivalved components may be articulated. Robust, heavily calcified brachiopods make up a significant proportion of the skeletal deposit. In general, shelly remains that are taphonomically resistant dominate: in particular, massive and robust brachiopod valves with varying amounts of disarticulated pelmatozoan material. Arthropods, more fragile brachiopods and mollusks, and arborescent corals and bryozoans are typically less abundant.

4.1.2. Background Taphonomy

Large, upright favositid coral heads typically show abundant evidence of prolonged exposure and surface corrosion (Brett and Baird, 1986). Shell beds are variable, but comprise broken and abraded valves, typically oriented convex upward. Fragmentation is coupled with abrasion, accounting for an abundance of rounded and subrounded edges on shell fragments.

In many cases shells and shell fragments are oriented in stable concave-down postures suggesting the influence of persistent laminar currents. Strata are generally grain-supported with little or no mud matrix, further corroborating the idea that background processes included persistent reorientation and reworking (Fig. 7A). Background taphonomic processes in this setting are clearly destructive, accounting for a great deal of stratigraphic condensation and taxonomic bias. Furthermore, the persistent influence of current winnowing and skeletal destruction makes a substantial contribution to the amount of sediment available for transportation into deeper-water, less turbulent settings.

In rare instances, decreased substrate mobility and prolonged hiatus resulted in early interstitial cementation and hardground genesis. Early carbonate diagenesis stabilized the sediment substrate permitting colonization by varied encrusting organisms (e.g., pelmatozoan echinoderms; Fig. 7A). These organisms, in particular the echinoderms, contributed abundant skeletal remains to accumulating deposits. In many cases echinoderm debris comprises a major component of the skeletal lag (encrinite); these remains are typically more resistant to mechanical and chemical destruction than are other calcareous remains.

4.1.3. Episodic Taphonomy

The influence of event-related processes in this taphonomic setting is indicated by the rare occurrence of completely articulated echinoderm and trilobite remains. Overturned and otherwise reoriented favositid coral heads

FIGURE 7. *Taphofacies Model I* taphonomic history. (A) Shallow-water (near fair weather wave base), marine settings; shelly debris accumulates as a result of persistent winnowing. Relatively high energy and long periods of surface exposure conspire to degrade shells. Consequently, disarticulation is exceptionally high, fragmentation and abrasion are abundant, and shells are typically comminuted beyond taxonomic recognition. (B) Intermittent, but relatively common storm-wave events (Speyer and Brett, 1988, Fig. 12) may overturn massive colonial skeletons (e.g., favositid coral head) and reorganize the "background" fabric of the shell bed. (C) The result is an amalgamated shell accumulation that shows a compounded taphonomic signature. In certain instances, shell lags that are conserved from background processing may display a chaotic fabric indicative of resedimentation of debris during episodic storms. Symbols: Cr, Camerate crinoid; F, Favositid coral; t, trilobite *Phacops*.

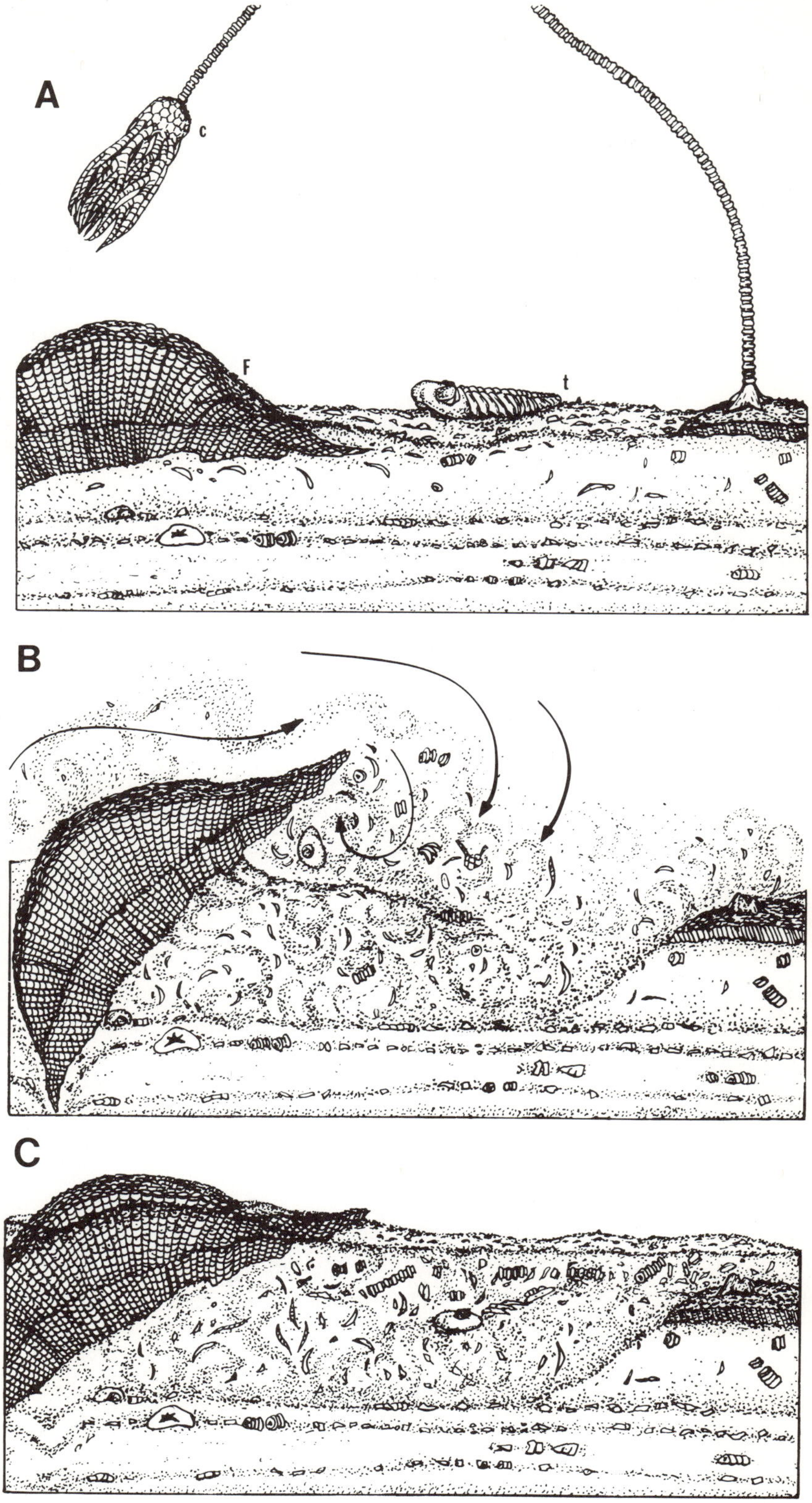
A
c
F
t
B
C

(Figs. 7B and 7C) indicate sedimentary events during which current energy greatly exceeded levels that characterized background processes. These data are substantiated by the occurrence of intraformational clasts, derived from exhumed and broken hardground surfaces.

Chaotically oriented skeletal remains are present in most large burrow traces and scour-enhanced channels (gutters), suggesting rapid accumulation as tubular tempestites (Wanless, 1988) and very little hydrodynamic organization (Brett *et al.*, 1986b; Fig. 5). This fabric characterizes many event-related processes in nearshore settings, but is likely reorganized by more persistent current influence during background processing (see Section 4.1.4). Scoured depressions display contrasting sedimentary fabrics and therefore record the imprint of episodic debris deposition.

4.1.4. Discussion

Figure 1A illustrates Taphofacies Model I in a shallow-water, high-current-energy setting above normal wave base [cf. Taphofacies 1A, Speyer and Brett (1986); Taphofacies Model 1, Speyer and Brett (1988)]. Clearly, the stratigraphic signature of processes in this facies indicates the destructive nature of this setting. The effects of background and episodic processes are *compounded;* therefore, it is not always possible to distinguish between episodic and background processes. However, the abundance of abraded skeletal material attests to the preeminent influence of background processes on these assemblages. The undisturbed signatures of single sedimentary events in this setting are reflected by the rare intact preservation of articulated remains. Day-to-day homogenization of deposits, however, has largely destroyed this signature. The result is a thoroughly amalgamated deposit which reflects the compounded effects of background and episodic processes (Seilacher, 1982).

4.2. Taphofacies Model II: Current-Winnowed Shell Beds

Reference examples: Wanakah Shale Member, *Pleurodictyum* beds, Erie County, New York (Miller, 1986); Kashong Shale Member, Erie and Genesee Counties, New York (Lukasik, 1985); Windom Shale Member, Tauntan Beds (Parsons *et al.*, 1988).

Typical lithofacies: Argillaceous packestones and grainstones; proximal tempestites (Brett *et al.*, 1986b).

Typical biofacies: Diverse brachiopods (*Pseudoatrypa, Mediospirifer*); prominent *Pleurodictyum* coral beds; fistuliporoid bryozoans (Brett *et al.*, 1983, 1986a).

4.2.1. General Features

Strata assigned to Taphofacies Model II comprise thin-bedded accumulations of shelly remains that alternate with sparsely fossiliferous mudstone beds of variable thickness (1–20 cm). Shell lags, although thin, are densely packed and may include a high-diversity fauna. Intercalated mudstone beds lack abundant skeletal remains. Fossils in thin shell deposits display varied modes of preservation ranging from complete articulation of bivalved (brachiopod) and multielement (trilobite and crinoid) skeletons to disarticulation and reorientation of strongly concavoconvex elements.

In most instances, shell beds display a diagnostic microstratigraphy; the shell bed matrix comprises disarticulated, fragmented, and oriented shell remains and contrasts with the bed surface, which displays a greater abundance of articulated skeletons and *in situ* assemblages.

4.2.2. Background Taphonomy

Arborescent bryozoans display variable intensities of fragmentation and preferred alignment; massive skeletal components (e.g., tabulate corals and trepostomate bryozoans) may be upended or in life orientation. In contrast to Taphofacies Model I, shelly remains are less severely abraded and typically more intact (Fig. 1B). Shells and skeletal remains within the main body of the accumulation typically show convex-up orientations. As suggested for Taphofacies Model I, this fabric indicates that the shell bed accumulated as a result of frequent winnowing by currents which served to orient the shells in hydrodynamically stable postures (see Fig. 3). Shells exposed at the lag surface are characteristically encrusted, but show far less abrasion than Taphofacies Model I. Concavoconvex brachiopods (e.g., the orthid *Tropidoleptus*) commonly have been reoriented and then served as a substrate for encrusting epibionts [e.g., bryozoans, *Cornulites,* rarely pelmatozoan holdfasts; see Speyer and Brett (1988, Fig. 5a)]. Clearly, these examples indicate current influence prior to final burial.

Faunas within the shell beds are identical to faunas at the bed surface. This indicates little or no transport and corroborates autochthonous accumulation. The dense packing of shelly remains therefore is the result of winnowing and bypass (Kidwell and Jablonski, 1983) which produced "sedimentary shell lags" (Fig. 8A). Surface assemblages include pavements of tabulate corals (Parsons *et al.*, 1988; Fig. 7B), intermittent bryozoan (trepostome) colonies and associated pelmatozoans, and *in situ* brachiopod clusters.

4.2.3. Episodic Taphonomy

These shell deposits display abundant evidence for rapid and permanent burial of the shell lag upper surface. Although internal structures may indicate persistent background currents and winnowing of fine-grained sediment (e.g., convex-upward orientations), and perhaps recurrent storm currents, well-preserved and articulated remains at the substrate surface argue strongly for authochthonous (*in situ*) or parautochthonous (transport of living or recently dead individuals) preservation. Complete crinoids and

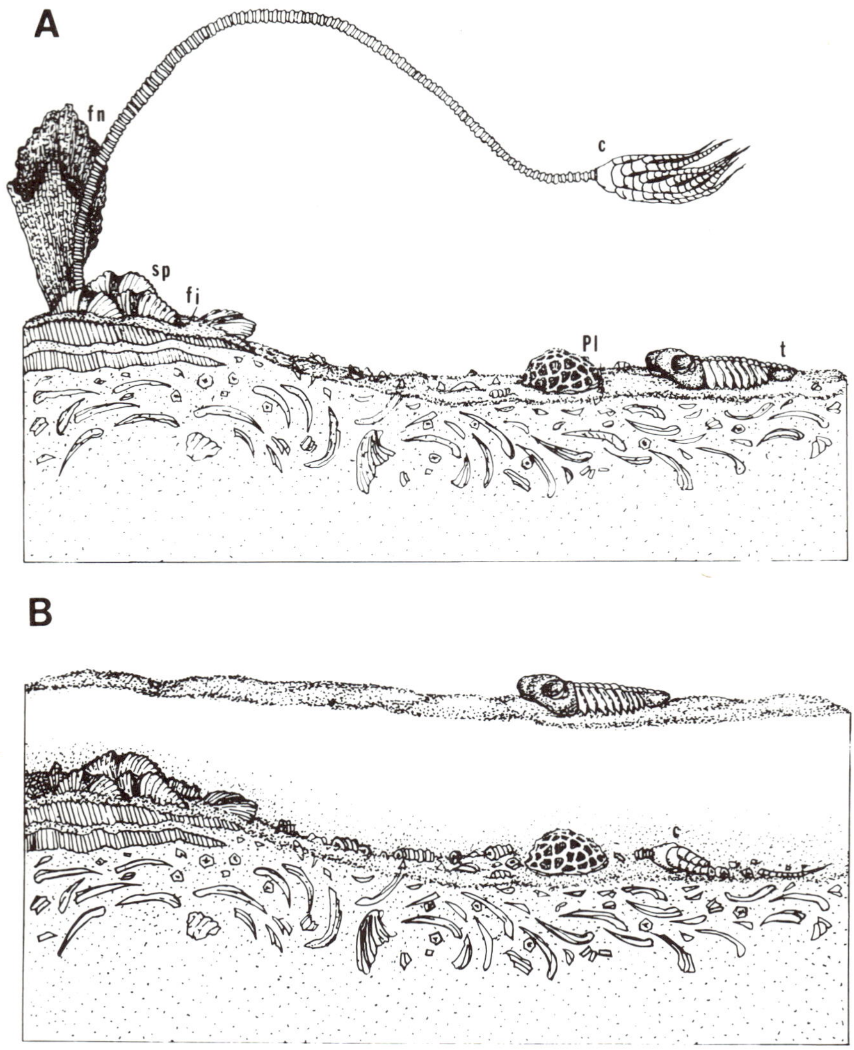

FIGURE 8. *Taphofacies Model II* taphonomic history. (A) Shallow sea floors between fairweather and storm wave base; frequent winnowing by bottom (gradient?) currents concentrates a sedimentary shell lag. Shells are commonly convex-up to imbricated, and display varying degrees of fragmentation and abrasion. Shell bed surfaces are commonly colonized by brachiopods, corals, crinoids, and bryozoans, which also occur within the shell bed proper. (B) Rapid burial beneath a sediment blanket results in two discrete taphonomic signatures. The shell bed proper is characterized by densely packed shell material. The shell bed surface, however, displays very well preserved remains, often in life position; note crinoid crown (right of center); in many instances, vagrant benthos (e.g., trilobites) are not preserved, due to postburial escape. The bed surface is covered by a layer of barren mudstone which represents the burial layer itself. These differing modes of preservation constitute discrete taphonomic signatures; episodic and background phenomena are both represented in the stratigraphic example. Symbols: C, inadunate crinoid; fi, fistuliporoid bryozoan; fn, fenestrate bryozoan; Pl, tabulate coral *Pleurodictyum*; Sp, spiriferid brachiopods; t, trilobite *Phacops*.

trilobite molt remains are associated with in-place and articulated brachiopod spats and minor bryozoan "microbioherms" (Archer and Feldman, 1986) providing a well-resolved view of faunal associations and paleocommunity organization.

Taphofacies Model II deposits show little evidence for transport of shells, but may lie on a continuum between classic "tempestite" deposits with well-defined grading of allochems (*sensu* Aigner, 1982) and more obviously autochthonous accumulations in distal settings. Aigner (1982) discussed the range of occurrences between amalgamated shell concentrations (cf. Taphofacies Model I, Section 4.1) and "distal tempestites" in which the proportion of mud matrix systematically increases as allochem proportion decreases. The distribution and comparative taphonomy of Hamilton Group shell beds along this continuum are discussed elsewhere (Parsons *et al.*, 1988; Miller *et al.*, 1988).

Although rapid burial is indicated by relatively abundant articulated skeletons, sessile organisms were clearly predisposed to preservation. Vagrant benthos (e.g., trilobites) likely escaped through sediment layers that were insufficiently thick to prohibit such biasing of the assemblage (Fig. 8B). Pelmatozoan echinoderms and varied colonial organisms (corals and bryozoans) were smothered, killed, and preserved. In certain cases, a record of burial is preserved within the trepostome bryozoans that comprise microbioherms across the substrate/bedding plane surface; colony rejuvenation following sediment burial indicates that burial layers were often thin and did not completely engulf the resident fauna (Parsons *et al.*, 1988; Fig. 8). Fauna that perished during burial remained undisturbed because intrastratal bioturbation and disturbance was prohibited by dense, time-accrued shell lags.

4.2.4. Discussion

Stark differences in the taphonomy of "within-bed" and "bed-surface" remains correspond to different phases of shell bed history: processes operating during long periods of time versus those operating over relatively short (geologically instantaneous) intervals. Two *discrete* signatures are evident. Thin but laterally persistent shell beds reflect frequent current winnowing and background skeletal accumulation (see Fig. 5). In contrast, surface assemblages commonly display articulated crinoids and *in situ* brachiopods, the result of rapid burial. These signatures roughly correspond to the two *Lagerstätten* types, *Konzentrat-Lagerstätten* and *Konservat-Lagerstätten* (Seilacher, 1970). Concentration deposits record long-term accumulation of skeletal remains (in this case, due to winnowing) and constitute the bulk of each shell bed.

Sediment bypass implies a degree of turbidity which is corroborated by the ecological composition of the fauna: large corallite tabulate corals (*Pleurodictyum*) are common, but *Favosites* (with smaller corallite diameter) and solitary rugose corals are rare or absent. Bivalves and many brachiopods (e.g., *Tropidoleptus, Chonetes*) typically display morphologies indicative of soft-substrate life. Similarly, discoidal corals (e.g., *Pleurodictyum*) possess "snowshoe-like" basal coenostea that provide soft-substrate stabilization. Nevertheless, the beds themselves are commonly tightly packed shell aggregations. This seemingly contradictory evidence suggests event condensation. Intervals of mud deposition and colonization by soft-substrate-adapted organisms alternated with periods of winnowing which removed fine-grained sediment and amalgamated various generations of parautochthonous skeletal material. To this degree a conflicting signature may exist within the shell beds themselves.

Conservation deposits (articulated pelmatozoan remains, *in situ* brachiopod spats, complete molt ensembles) are typically preserved on top of the shell bed proper, the result of rapid burial (Fig. 8B). In rare instances, faunal remains have been pyritized, indicating a *cooperative* signature; most early pyrite and carbonate diagenesis, however, is limited to more distal settings of this taphofacies model.

Barren, relatively featureless mudstones occur between shell beds and represent episodes of mud deposition that smothered and terminated shell lag accumulation. The distinct alternation of lithologies and taphonomic properties produces a misleading stratigraphic "rhythm" (Speyer and Jones, 1988). This "rhythm" appears to display a specific tempo that corresponds to gross environmental setting, i.e., the relative thickness of intervening burial layers and internal anatomy of the shell bed proper vary in a predictable manner. Relative thicknesses remain consistent for considerable stratigraphic intervals suggesting a constancy in environmental setting (Speyer and Brett, 1986; Brett *et al.*, 1986b; Miller *et al.*, 1988; Parsons *et al.*, 1988). Gradual changes in the relative thickness proportions (between shelly beds and barren intervals) may indicate shifts in environmental setting or event intensity or regularity; e.g., less frequent event burial results in thicker shell accumulations, whereas greater storm intensity (or closer proximity to sediment source) generates a thicker burial layer.

Kidwell and Jablonski (1983) discussed a variety of shell beds and detailed four modes of generation with correspondingly diagnostic features. Sedimentary shell beds, as discussed here, correspond to "sediment bypass" situations and reflect frequent surface winnowing and skeletal concentration. A gradient corresponding to sedimentary shell beds, starvation shell accumulations, and biogenic shell beds (Kidwell *et al.*, 1986) exists across Taphofacies Models II, IVA, and IVB, respectively.

4.3. Taphofacies Model III: Bioturbated Deposits with Shelly Patches

Reference examples: Centerfield Member, calcareous mudstones, Cayuga County, New York (Gray, 1985); Deep Run Member, lower and middle mudstones, Ontario County, New York (Baird and Brett, 1981); Kashong

Shale Member, lower and middle mudstones; western New York (Lukasik, 1985).

Typical lithofacies: Calcareous, silty, massive, bioturbated mudstone with argillaceous skeletal wackestones and packstones.

Typical biofacies: Zoophycos; Tropidoleptus brachiopods (Brett *et al.*, 1990).

4.3.1. General Features

Taphofacies Model III represents a complex environmental setting with a distinct two-part anatomy: burrow-homogenized and sparsely fossiliferous mudstone deposits which comprise 80–90% of the facies, and small patches (1–2 m) or "microbioherms" (*sensu* Archer and Feldman, 1986) including corals, bryozoans, and brachiopods with varied encrusting epifauna. Taphonomic properties vary greatly between these two divisions at any given time, but interact to produce a diagnostic stratigraphic fabric. Minor shell layers of disarticulated pelmatozoan remains, broken shells, and isolated, articulated trilobite remains [characteristically enrolled; see Taphofacies 2A of Speyer and Brett (1986)] as well as rare, *in situ* endobyssate bivalves occur sporadically throughout the bioturbated mudstone interval.

Encrusters include solitary rugose corals, varied bryozoan colonies, and pelmatozoans with cemented holdfast structures; crinoid and blastoid remains display varied levels of disarticulation. Articulated trilobite remains occur along the flanks of shelly patches and within interstitial mudstones of fistuliporoid mounds.

4.3.2. Background Taphonomy

The intensity and thoroughness of *Zoophycos* bioturbation indicate that the substrate was thoroughly pelleted and thus relatively unstable (Kotake, 1989). Minor shell accumulations were typically disrupted to some degree by deposit-feeding organisms.

Intermittent shelly stringers, dominated by pelmatozoan debris, divide an otherwise monotonous mudstone interval. These shell deposits, in part, are correlative with biohermal accumulations, suggesting a genetic relationship. Episodic accumulations of skeletal debris provided shelly pavements too thick to be reworked by infauna [negative taphonomic feedback (Kidwell and Jablonski, 1983)]. These layers offered a stabilized substrate and a potential foundation for subsequent colonization (Alberstadt and Walker, 1976; Miller *et al.*, 1988).

Brachiopod and bivalve shells, diffusely distributed throughout the mudstone deposits, are typically disarticulated and display random, commonly oblique orientations (Salizar-Jiménez *et al.*, 1982). Small brachiopods and other shelly debris are concentrated as peripheral lags along the distal margins of *Zoophycos* spreiten (Fig. 9A). Abundant evidence of disarticulation without significant fragmentation, and random patterns of shell

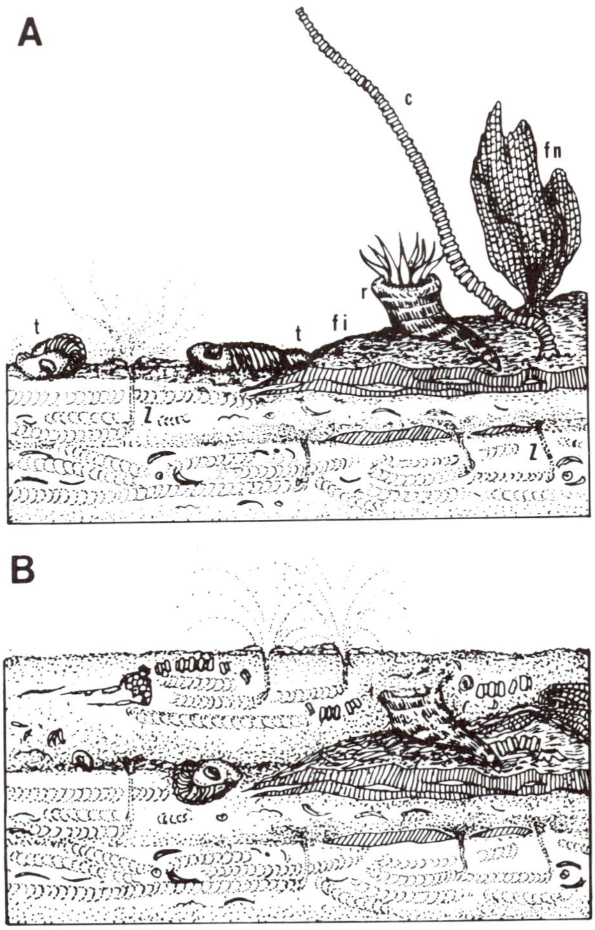

FIGURE 9. *Taphofacies Model III* taphonomic history. (A) Shallow muddy sea floors near the lower limit of storm wave base; moderate to high rates of background sedimentation dilute the accumulation of shelly remains and favor deep, intrastratal bioturbation. Shell lags that are not homogenized provide a foundation for colonization and stabilization. Early-colonizing biota include fistuliporoid bryozoans. Later-appearing fauna (e.g., varied pelmatozoan echinoderms, fenestrate bryozoans) attach to the "microbioherm" and are later overgrown. *Zoophycos* bioturbation has homogenized the intermound regions and resulted in mixed shell debris. (B) Rapid burial may or may not terminate mound growth. Internal bedding indicates that certain bryozoan structures, although perturbed, were able to grow around and over some burial layers. Buried fauna (e.g., trilobites and crinoids) may be disarticulated and scavenged by postevent bioturbation or may be preserved intact; note crinoid column and calyx partially disrupted and partially intact. Enrolled trilobite near skeletal pavement also escaped disruption due to the absence of burrowing in this area. Symbols: c, crinoid; fi, matlike colonies of fistuliporoid bryozoans (note internal mud layers); fn, fenestrate bryozoan; r, rugose coral cemented to bryozoan colony; t, trilobite *Phacops*; Z, pervasive chevronlike spreiten structures of *Zoophycos*.

orientation (Fig. 1B), indicate that intrastratal bioturbation was a dominant factor controlling the taphonomy of preserved remains.

Skeletal debris may have been derived from immediate sources (i.e., accumulated at the periphery of already established microbioherms) or brought in from other regions (i.e., storm-deposited; see below). Once established, faunal diversity increased in proportion to patch size; larger biostromes and microbioherms signify longer periods of uninterrupted growth. Although very little corrosion is evident on shells composing patches, macroepibiont bioerosion is pervasive (Vogel *et al.*, 1987).

4.3.3. Episodic Taphonomy

The initiation of local patch colonization was apparently contingent on the stabilization of an otherwise thixotropic and unstable substrate. Thin deposits of shelly material often occur as discontinuous lenses and layers within the mudstone units characterizing this taphofacies; thicker shell deposits may be graded and display a ''tempestite'' fabric (Aigner, 1982, 1985; Aigner and Reineck, 1982). The influence of event-related deposition is indicated by the presence of winnowed shell layers and corroborated by the occurrence of completely articulated trilobite, crinoid, and bivalve remains within these beds. Although more severe events deposited lags that prohibited intrastratal mixing, many minor events likely remain unnoticed due to postevent bioturbation (Fig. 9B).

Miller (1986) and Parsons *et al.* (1988) noted that fistuliporoid bryozoan colonies often record a history of burial events within the core of the colony. Layers represent instances of rapid, but nonlethal burial due to event phenomena. ''Laminites'' present in microbioherms in the Waldron Shale (Silurian) probably represent similar phenomena (Archer and Feldman, 1986). Miller *et al.* (1988) discussed the ecological impact of these thin mud pulses in disrupting growth in colonial organisms.

Articulated trilobite remains occur in two manners: as grouped molt ensembles at the periphery of skeletal patches, or as isolated enrolled carcasses in undifferentiated mudstone strata. The preservation of intact fenestrate bryozoans, trilobites, blastoids, and, less frequently, complete crinoids within interstitial muds in bryozoan thickets provides strong evidence for rapid and final burial (see Fig. 9). Strong evidence for the influence of episodic sedimentary phenomena, although scattered, clearly exists within strata assigned to this taphofacies.

4.3.4. Discussion

A moderate rate of background sedimentation diluted skeletal concentrations and favored deposit-feeding bioturbation. Stratigraphic details suggest that this sediment was derived from and/or bypassed deposition in environments represented by Taphofacies Models I and II (Fig. 1). These taphofacies intergrade, decreased sedimentation resulting in more pronounced skeletal accumulations, but increased sediment input resulting in shelly layers forming laterally discontinuous lenses and mounds (Fig. 9A). At an extreme, where sedimentation rate was greatest [not illustrated on Fig. 1, but roughly equivalent to Taphofacies 2B of Speyer and Brett (1986); see Fig. 2] and skeletal concentrations minimal, lenticular buildups did not occur and the strata were completely homogenized by infauna. Background conditions of moderately high sedimentation rate and pervasive bioturbation result in a taphofacies with a strong background overprint. There is a close relationship connecting the rate of sediment input, skeletal concentration, and intensity of intrastratal bioturbation (Section 3.1; see Figs. 2 and 3). Evidence that might have existed to indicate episodic processes was obscured by deep burrowing (see Fig. 9). The scattered occurrences of articulated bivalve, trilobite, or echinoderm remains, however, indicate that this environmental setting did experience episodic burial. The concentration of these occurrences around shelly buildups provides a strong argument that event stratification was elsewhere destroyed by postevent bioturbation; the buildups prohibited burrowing by virtue of negative feedback.

The superimposed signature, then is *conflicting;* postevent, background processes obscure the imprint of episodic burial, yet details of biostratinomy prove the existence of such phenomena. In more proximal settings, or where background sedimentation rate is diminished, event deposits [cf. tempestites, *sensu* Aigner (1982)] are more conspicuous because graded shell lags inhibit intrastratal mixing by infauna. This gradient exists across the range of environments represented by Taphofacies Models II and III. The extent to which the stratigraphic signature is discrete or conflicting varies across this spectrum according to event dynamics (i.e., traction or suspension deposit) and rate of background deposition.

4.4. Taphofacies Model IVA: Sediment-Starved Shell Beds (Proximal)

Reference examples: Centerfield Limestone Member, Brown's Creek Bed, Genessee and Ontario Counties, New York (Brett *et al.*, 1986a); Wanakah Shale Member, Murder Creek Bed (Speyer and Brett, 1985, 1986); Windom Shale Member, Smoke Creek and Spezzano Beds (Speyer and Brett, 1985, 1986; Parsons *et al.*, 1988).

Typical lithofacies: Calcareous, fossiliferous mudstone and argillaceous, concretionary skeletal wackestone (Brett *et al.*, 1986b).

Typical biofacies: Stereolasma (coral), diverse brachiopods (in part), abundant *Phacops* (Brett *et al.*, 1990).

4.4.1. General Features

Many of the classic and fossil-rich beds in the Hamilton Group are included within the spectrum of Taphofacies Model IV [e.g., "Wanakah

Shale Trilobite Beds'' (Grabau, 1898–1899)]. Figure 1A shows that Taphofacies IV, which is characterized by low sedimentation rates, occurs across a wide bathymetric spectrum.

4.4.2. Background Taphonomy

Strata that are included within Taphofacies Model IV characteristically display discrete shell concentrations that resemble those discussed under Taphofacies Model II. Important differences are evident upon comparisons of specific taphonomic details, in particular, levels of fragmentation and reorientation (Fig. 1B). Bivalved and multielement skeletons are typically disarticulated. Individual skeletal elements are generally not fragmented or preferentially oriented (shells are approximately equally distributed convex up and down).

Encruster and endobiont (boring) faunas are relatively diverse [10–15 species (Brett and Bordeaux, 1988; Bordeaux and Brett, 1990)] and bioerosion is considerably more pervasive among shells in these environmental settings due to diminished turbidity and prolonged stability (no reorientation). Epibionts that encrust both surfaces of the skeleton corroborate low or insignificant current energies (cf. Taphofacies Model II, in which encrusters are preferentially positioned on convex-up shell surfaces). Vogel *et al.* (1987) described two of their three most diverse endolith associations from Hamilton Group strata that correspond to Taphofacies Model IV. Although mechanical background processes have little effect upon shell preservation in Taphofacies Model IV, biogenic effects, particularly bioerosion, are clearly a destructive factor.

Each skeletal bed displays a great deal of taphonomic variety that systematically changes from poorly preserved, corroded material at the base to well-preserved and even completely articulated remains at the bed upper surface. Less frequently, carbonate-rich lenses display fossil remains that appear identical (taxonomically and taphonomically) to remains in other, less calcareous beds, but are more diffusely distributed (see Fig. 10C and 10D). These occurrences provide important insight regarding the original shell bed and its stratigraphic character; early interstitial carbonate diagenesis (see below) preserved a truer depositional and shell accumulation thickness.

Diagenetically enhanced shell accumulations (Brett *et al.*, 1986b) also display pervasive, fine threadlike burrows that otherwise have been obliterated by compaction (Fig. 10A), but larger-scale bioturbation structures (e.g., *Zoophycos*) are rare or absent. Invariably, shell accumulation is terminated (and truncated) by a barren homogeneous mudstone.

4.4.3. Episodic Taphonomy

Mudstone layers that terminate shell accumulations reflect episodic burial and are usually associated with well-preserved skeletal remains at the

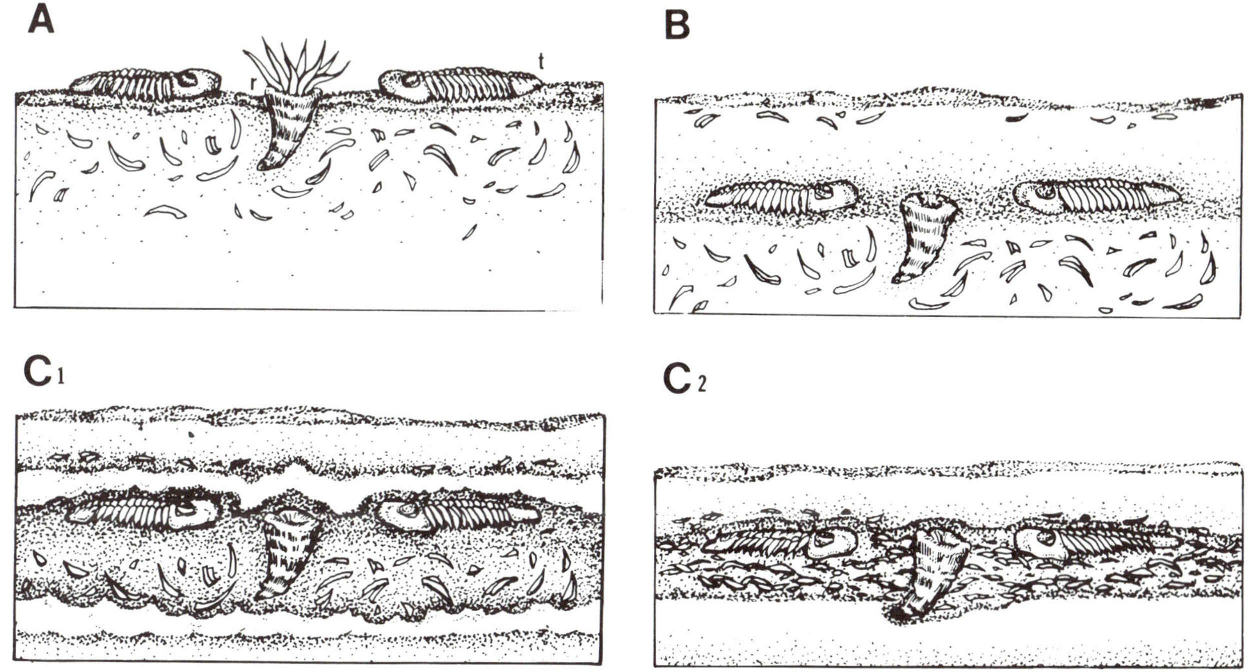

FIGURE 10. *Taphofacies Model IVA* taphonomic history. (A) Deeper-water, muddy sea floors, below storm wave base; background conditions are characterized by slow to moderate sedimentation and long periods of exposure of skeletal debris at the substrate surface. Sediment starvation results in a skeletal lag within which shells display a variety of taphonomic conditions; corroded shells may cooccur with well-preserved, but disarticulated materials. Fragmentation and abrasion are generally low due to quiet water conditions. (B) Rapid burial of the sea floor results in mass mortality. Vagrant benthos (e.g., trilobites) along with sessile faunas are smothered, provided that the burial layer thickness prohibits escape (Kranz, 1974). (C_1, C_2) Two modes of fossil assemblage preservation: (C_1) Provided that appropriate geochemical conditions are present, rapid burial of the substrate may initiate carbonate diagenesis, resulting in a carbonate-enhanced shell bed (stippled area around trilobites; see Fig. 7). (C_2) Without early precipitation of carbonate to preserve interstitial pore space, the surrounding sediment is subject to considerable compaction by overburden dewatering. The result is a compressed fauna and a shell bed that appears denser than it was at the time of burial. Symbols: r, small rugose coral *Stereolasma*; t, trilobite *Phacops*.

top of shell lags (Fig. 10B). Articulated trilobite and echinoderm (crinoid and blastoid) remains signify rapid and final burial, the effects of a sedimentary event. In addition, these intact remains and assemblages include small rugose corals in life position (i.e., prototheca down) and articulated brachiopods with commissure positioned upward. Spectacular clusters of trilobites and trilobite exuviae argue strongly that burial was episodic (Speyer and Brett, 1985).

In many instances individual shell beds with these attributes can be traced across tens of kilometers, indicating widespread effects of burial and the lateral distribution of condensed conditions (see also Waage, 1964). Burial is likely the result of suspension cloud fallout and may be associated with high-density, water-saturated, pelletal sediment flows that engulf, but do not disturb living faunas (Fig. 10B). Articulated remains indicate that burial was rapid, complete, and widespread.

4.4.4. Discussion

Low rates of background sedimentation result in skeletal lags which accumulate as organisms die. These shell concentrations may provide a chemical buffer against carbonate dissolution, which consequently is relatively minor [Kriesa and Bambach (1982); see Sjoberg and Rickard (1985); Speyer and Brett (1988) for discussion of caveats]. This deposit of calcium carbonate also provides a source for early, postburial carbonate diagenesis (concretion and steinkern). Depressed sedimentation rates result in long-term accumulation of loosely packed shells that show varied, apparently random postures. *In situ* dismemberment and disintegration appear to be the main processes. Surface scavenging may have reoriented some elements, but shells remained more or less in living sites (Fig. 10A). Individual skeletal components served as substrates for a variety of epibionts and endoliths.

The obvious dichotomy between shell beds and intervening burial layers provides a clear indication of both background and episodic signatures; i.e., *discrete* taphonomic signatures. The taphonomic effects of background processes are provided in "time-lapse" as well as "snapshot" fashion; the main body of the skeletal lag records the progressive deterioration of shelly remains, whereas the tops of many shell beds display an intact and largely undisturbed faunal assemblage. The interactive relationship between long-term hiatus and shell concentration and episodic burial sometimes resulted in a diagnostic *cooperating* taphonomic signature: early carbonate diagenesis (Fig. 10C). Concentrations of shelly remains and enhanced pore-water concentrations due to aragonitic shell dissolution provide a sink of dissolved ions (Aller, 1982) which are mobilized following rapid burial and isolated necrolysis. The release of ammonia (during decay) raises the pH in an isolated chemical microenvironment and favors the precipitation of calcium carbonate in the immediate vicinity (Waage, 1964; Berner, 1968). Laterally continuous carbonate lenses may result where appropriate chemical and sedimentological conditions exist at the time of burial (Brett *et al.*, 1986b; Speyer and Brett, 1985).

More often, the shell beds were not diagenetically overprinted. The result is a compressed horizon of shelly material that displays taphonomic details similar to diagenetic carbonate beds, but in addition displays overburden compaction (Fig. 10D). Carbonate diagenesis requires two obvious conditions (Fig. 6). First, background conditions must favor calcium carbonate concentration within the sediment. This requires a depressed rate of background sedimentation (to concentrate calcareous skeletal remains) and limited or nonexistent fluid flow through the sediment [so that dissolved ions are not flushed from the immediate environment (Sjoberg and Rickard, 1985)]. Second, diagenesis requires that the chemical environment be more or less isolated from ambient diffusion and that a chemical gradient or series of interacting chemical gradients be present. This situation is initiated by rapid, event-related burial of the substrate and living fauna. Decomposition of organic matter sets up the necessary chemical gradients which then result in early interstitial precipitation. Both components of this system must be present in order for diagenesis to occur; other factors undoubtedly influence the extent of diagenesis and quality of preservation.

4.5. Taphofacies Model IVB: Sediment-Starved Shell Beds (Distal)

Reference examples: Levanna Shale Member (in part), Genesee and Livingston Counties, New York (C. E. Brett and G. C. Baird, unpublished results); Upper Ledyard Shale Member, Erie and Cayuga Counties, New York (S. E. Speyer, 1985, unpublished results); Upper Windom Shale Member (in part), Livingston County, New York.

Typical lithofacies: Gray, slightly calcareous mudstone and shale with thin shell horizons.

Typical biofacies: Auloporid corals, *Ambocoelia*-chonetid brachiopods (generally small), abundant *Phacops* remains.

4.5.1. General Features

Shelly remains in laterally discontinuous patches represent a slightly deeper water variant of the condensed taphofacies spectrum (Taphofacies Models I, IVA, and IVB). Patchiness is largely a function of biogenic productivity and duration of accumulation. There is a complete gradient among Taphofacies Model IV shell beds that indicates the combined influence of these two components.

Skeletal remains within these concentrations are typically disarticulated, but may display a relict position that directly relates to the original organism posture (Speyer, 1985). In Hamilton strata, Taphofacies Model IVB is represented by small to moderately large accumulations of diminutive brachiopods and trilobite exuviae with associated faunas of varying, but typically low diversity. The associated fauna includes auloporid tabulate corals and, infrequently, dense bundles of articulated inadunate crinoid columns. Diversity is positively correlated with the overall size of the patch and displays a systematic gradient that may reflect an ecological autogenic type of succession.

Patches occur along horizons that are traceable within a single outcrop; seven to eight such horizons may occur within ½ m or so thickness of this facies.

4.5.2. Background Taphonomy

The overall biostratinomic overprint in this facies is least severe of all the taphofacies models discussed, but skeletal dissolution may be an important destructive taphonomic agent (see Fig. 1B). Although abundant trilobite remains occur throughout the interval, a wide range of preservation is observed; completely disarticulated remains, partially disarticulated but reasonably undisturbed material, and completely articulated molts cooccur across the patch surface. Brachiopods are only rarely disarticulated and auloporid coralla are characteristically intact. Crinoid columns occur in bundles and show little or no disarticulation.

Taphofacies Model IVB is characterized by patches of trilobite molt remains that reflect site-specific ecdysis through an unknown period of time. Speyer (1985) described "molt accumulations" and drew a contrast with "molt clusters" (Speyer and Brett, 1985). Molt accumulations represent the time-accrued lag of successive molts in a particular site and contrast with synchronized ecdysis indicated by molt cluster biostratinomy. Consequently, trilobite exuviae display a varied quality of preservation and instar size. As exuviae accumulated, skeletal patches broadened and provided a suitable substrate for other fauna, notably corals and brachiopods (Fig. 11).

4.5.3. Episodic Taphonomy

The high proportion of articulated, nonfragmented, and *in situ* remains indicates that patches were episodically buried beneath mud blankets (Fig. 11). Speyer and Brett (1986) noted that these accumulations (discussed as Taphofacies Model IVA) contain a high proportion of articulated molt remains, but very few carcass remains; this is attributed to thin burial and trilobite escape potential (Kranz, 1974). Conversely, virtually all brachiopods and auloporid corals are intact and reasonably well preserved. Particularly important is the articulated preservation of crinoid remains associated with the larger patches. These data indicate burial in sediment thick enough to smother sessile fauna, but insufficient to trap and kill vagrant benthos.

Following episodic burial, background conditions and biogenic accumulation resumed, reestablishing molt patches across the homogeneous sea floor.

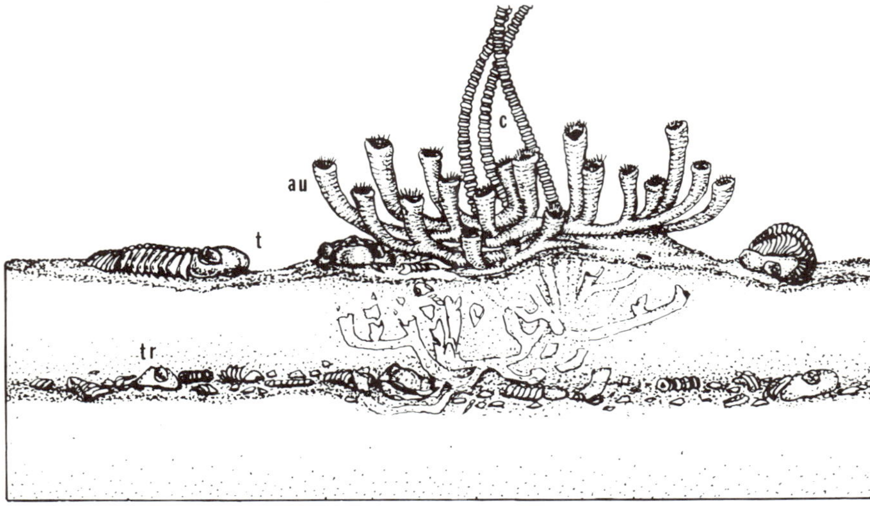

FIGURE 11. *Taphofacies Model IVB* taphonomic history. Deeper marine settings below storm wave base, lower aerobic zone; sedimentation rate is exceptionally low, sediment starvation results in shell lags that show the effects of long-term exposure. In many instances, these environments support a low-diversity fauna making shell accumulation unusual. However, thin skeletal accumulations (e.g., exuviae of trilobites) may provide substrates for colonization by later-appearing fauna, such as auloporid corals. Trilobite remains typically display a wide range of preservational grades (thoroughly disarticulated to essentially intact), corroborating the idea that these are long-term accumulations. Episodic burial of these patchy horizons provides a unique taphonomic contrast. Sessile faunal elements are preserved intact (auloporid corals are not fragmented). Vagrant organisms, however, are poorly represented except as nonliving remnants (e.g., trilobite molt remains). This dichotomy provides an important clue regarding the thickness of the burial layer; living trilobites were able to escape, whereas burial was thick enough to kill corals. Symbols: au, auloporid corals; c, long-stemmed small crinoids, primarily the flexible *Eutaxocrinus* and the inadunate *Anamesocrinus*; tr, trilobite remains, mainly exuviae; t, trilobite *Phacops*.

4.5.4. Discussion

Taphofacies Model IVB displays attributes that indicate long periods of nondeposition punctuated by short-duration pulses of fine-grained sediment. The nonrandom accumulation of trilobite exuviae and associated epifauna suggests that biogenic processes significantly influenced the taphonomy of these assemblages. Biogenic shell lags accumulate as a result of specific biological or behavioral phenomena. Kidwell *et al.* (1986) discussed biogenic accumulations as the result of extrinsic (i.e., bioturbation) or intrinsic (i.e., behavioral or preferential colonization) factors.

The biology and ecology of these exuvial accumulations are discussed elsewhere (Speyer, 1985), but it is important to note that molt patches are derived from behavioral concentrations of exuviae and that these concentrations stabilize an otherwise thixotropic and oxygen-deficient mud substratum (Fig. 11). Patch diameter, faunal diversity, and inferred duration of accumulation are all positively correlated. The absence of fragmented, abraded, and reoriented remains circumstantially corroborates an interpretation of exceedingly low current energy. The low diversity and composition of faunas in these strata reflect a particularly severe environmental setting probably associated with depressed oxygenation, as would be consistent with a stratified basin (Rhoads and Morse, 1971) (e.g., epeiric sea) and long-term condensation.

Obrution deposits in regions typified by low background sedimentation and situated in a setting below normal and storm-wave influence are best suited for preservation and event-related resolution (Speyer and Brett, 1988; Brett and Speyer, 1989). Burial is rapid and nondisruptive; original faunal associations are preserved. The number of events per unit stratigraphic thickness is relatively low because of limited burial sedimentation (Speyer and Brett, 1988). Presumably, skeletal remains accumulate continuously in deeper-water settings where rates of production are high enough to exceed ambient dissolution.

Oxygen levels were marginally aerobic in surface waters and likely anaerobic interstitially, thereby prohibiting extensive macrobioturbation. Although surface scavenging may have been a factor in disarticulation, abundant evidence suggests that *in situ* dismemberment and disintegration occurred without significant biogenic influence. Taphofacies Model IVB displays a clear distinction between background and episodic taphonomic processes and therefore is characterized by a *discrete* superimposed signature (Fig. 11).

In instances of extreme stratigraphic condensation (e.g., minimal sediment input, episodic or otherwise) trilobite molt remains are exposed for longer periods of time and are therefore more severely affected by surface biostratinomic and corroding conditions. In these settings, trilobites are thoroughly disarticulated and show reorientation patterns that indicate some bottom current processing. Patches are less well developed and overall diversity is drastically diminished; in most cases only the enigmatic fossil *Styliolina* cooccurs with the *Phacops* remains. In these cases lag concentrations of the univalved *Styliolina* are current-aligned and accumulate around concave-down cephala which provide mud shadows for underlying sediment.

4.6. Taphofacies Model V: Dysaerobic, Mud-Supported Shell Beds

Reference examples: Ledyard Shale Member, Alden Pyrite Beds, Erie and Genesee Counties, New York (Dick and Brett, 1986; Babcock and Speyer, 1987); middle Wanakah Shale Member, Erie County, New York (Dick and Brett, 1986); middle Windom Shale Member, Penn Dixie and Big Tree Pyrite Beds, Erie County, New York (Dick and Brett, 1986; Brett *et al.*, 1990).

Typical lithofacies: Medium to dark gray, bioturbated mudstone with thin, sparse shell pavements, pyrite nodules, and carbonate concretions (Brett *et al.*, 1986a, 1990, and in press).

Typical biofacies: Ambocoelia-chonetid, chonetid-nuculid bivalve, *Athyris* (in part), diminutive brachiopod (rarely) (Brett *et al.*, 1990).

4.6.1. General Features

Many of the medium and dark gray mudstones in the Hamilton Group sequence are diffusely fossiliferous, contain a low-diversity fauna of opportunistic taxa, and display relatively abundant evidence for early pyritization (Dick and Brett, 1986). Articulated skeletons are usually intact or display minimum post-mortem disturbance; bivalves may be splayed and intact molt ensembles are not uncommon. Benthic faunas are associated with the remains of nektonic organisms (e.g., nautiloids and ammonoids; Fig. 12A).

Deposits show evidence for discrete burrow construction where burrows have been diagenetically enhanced (pyritized). Pyritized horizons delineate a crude bedding which otherwise is obscure and poorly defined; shell layers are inconspicuous, laterally discontinuous, and relatively thin.

4.6.2. Background Taphonomy

Fossils in these strata are scattered and diffusely distributed, but display a rather high level of articulation (see Fig. 1B). Splayed bivalves (particularly nuculid clams) indicate that death was followed shortly by burial and conservation; decay of the adductor muscles and contraction of the hingeline ligaments resulted in splay, followed by burial [see Allmon (1985) and Linsley (1988) for slightly differing views]. The relative abundance of splayed bivalves increases in strata that display evidence of higher rates of background sedimentation [cf. Taphofacies 4C of Speyer and Brett (1986)].

FIGURE 12. *Taphofacies Model V* taphonomic history. (A) Deeper mud bottom, dysaerobic environmental settings typified by moderate to high rates of sedimentation; these support a low-diversity vagrant epifauna and deposit-feeding infauna. Shell accumulations are thin or absent in these strata; shells that accumulate with time are diluted by sedimentation. (B) Episodic burial of the sea floor is often obscured due to a paucity of shelly remains that might demark the original substrate, and the high rates of sedimentation that characterize background conditions. A contrast may be drawn, however, between mortal remains that were preserved as a result of catastrophic burial (e.g., enrolled trilobites) and those that present "background" death (e.g., splayed bivalve shells). Crosses indicate development of geochemical concentration halos (e.g., H_2S) around entombed organisms due to anaerobic bacterial decay of organic matter. (C) In organic-poor (nonsulfidic) environments early diagenetic pyritization may occur around buried organisms. Diagenetically enhanced, pyritized faunas provide important clues as to the distribution of episodic depositional conditions; pyritization requires concurrent death and burial (see Fig. 5). If the buried fauna has not been enhanced by early diagenesis, the signature of episodic conditions is obscured, if not altogether absent, due to compaction, dewatering, and skeletal dissolution. Symbols: l, lingulid brachiopod; n, nuculid bivalves; t, trilobite *Phacops*; w, tube-dwelling polychaete worm (hypothetical producers of pyritic burrows). [Figure modified from drawing by G. C. Baird.]

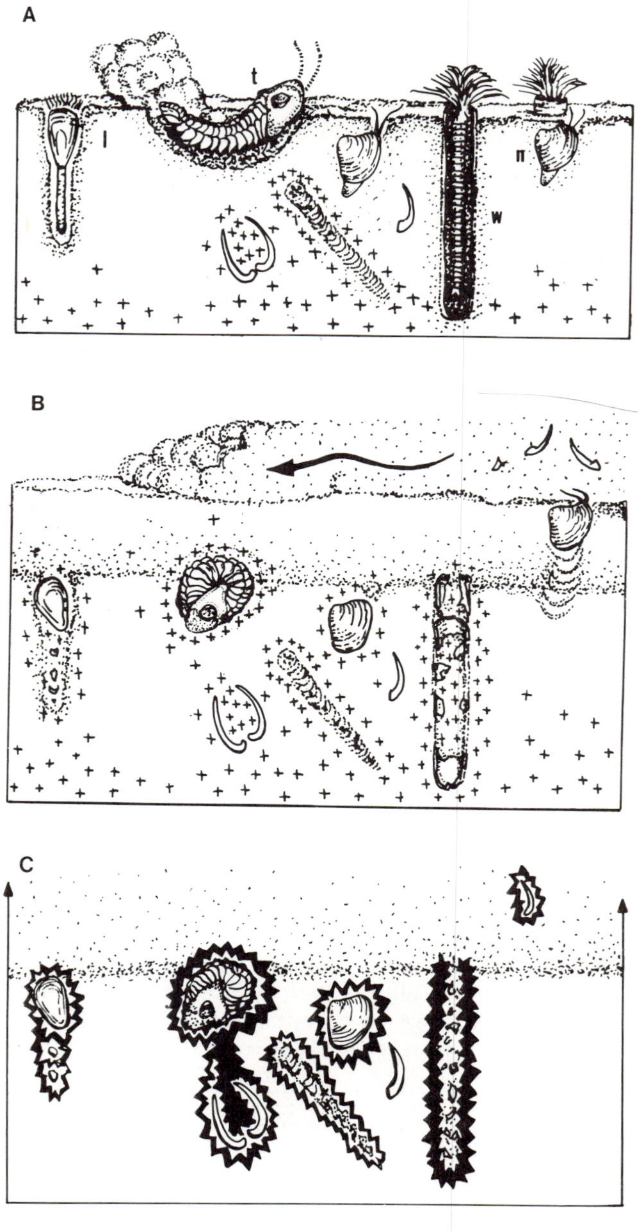
A
t
l
w
B
C

Fossils display little fragmentation or abrasion; valve ratios (pedicle/brachial) and proportions of skeletal elements (pygidium/cephalon) are very nearly 1 (Speyer and Brett, 1986; Brett *et al.*, 1991). Approximately equal numbers of valves are convex up and down (Dick and Brett, 1986); minor preferred orientations are noted among elongate fossils (e.g., orthocone nautiloids).

Burrows suggest some degree of bioturbation, but preservation indicates that most traces were vertical domiciles with an organic sheath to prevent burrow collapse.

4.6.3. Episodic Taphonomy

Poorly defined layers which display a relative abundance of articulated trilobite, blastoid, and brachiopod remains likely represent discrete event beds (Fig. 12B). These horizons are occasionally accentuated by event-related diagenesis, but otherwise are cryptic because carbonate skeletons are distorted, smeared, and dissolved during compaction and dewatering (Figs. 12C) (Speyer and Brett, 1988, Fig. 4C).

In most instances, articulated remains are associated with various nektonic organisms, most notably cephalopods. This association suggests a catastrophic circumstance that affected deep-swimming as well as benthic fauna. The event signature, however, is largely burial and lacks indication of significant current influence; i.e., low fragmentation and abrasion indices (Fig. 12B). Articulated remains, including particularly vulnerable trilobite exuviae (Speyer, 1987), mark these horizons and contrast with poorly defined levels which preserve splayed bivalves, *Styliolina* shells, or no fossil remains at all.

4.6.4. Discussion

The paucity and diffuse distribution of fossil remains in this taphofacies may be the result of excess sedimentation compared to rate of shell contribution and/or environmental constraints on diversity and productivity. The general absence of fragmentation, abrasion, sorting, and reorientation provides a strong argument for gradual, but undisturbed accumulation.

Taphofacies Model VI is characterized by a background-dominated, but *cooperating* taphonomic signature; background and episodic processes interact to produce a taphonomic character particular to that setting. Pyrite diagenesis is discussed in detail elsewhere (Raiswell, 1982; Raiswell and Canfield, this volume, Chapter 7; Berner, 1984); the point emphasized here is the necessary interaction of background and event-related processes to effect diagenesis. Background bioturbation and burrowing provide sulfate to zones of decomposition, rich with sulfate-reducing bacteria. Mortality of benthic and other faunal elements, as a consequence of rapid burial, generates a chemical microenvironment in which mobilized iron, sulfate, and sulfate-reducing bacteria lead to pyrite diagenesis which may preserve the buried skeletal components and mucous-lined burrows (Fig. 12C). In the absence of any of these components, the event-buried assemblage is not pyritized and is more poorly preserved. In this case, the stratigraphic pattern would be obscured by *conflicting* signatures; the moderate rate of background sedimentation would mask any increased input due to depositional events.

4.7. Taphofacies Model VI: Exaerobic, Sediment-Starved Shell Beds

Reference examples: Upper Levanna Shale Member, western New York (Savarese *et al.*, 1986); Ledyard Shale Member, Seneca and Cayuga Counties, New York; Middle and upper Windom Shale Member, Livingston and Cayuga Counties, New York (Brett *et al.*, 1990).

Typical lithofacies: Dark gray to nearly black, laminated to vaguely bioturbated fissile shale (Brett *et al.*, 1990).

Typical biofacies: Diminutive brachiopods; "*Leiorhynchus*" (Brett *et al.*, 1990, 1991).

4.7.1. General Features

Dark gray and black shales in the Hamilton Group are characterized by a very distinctive taphofacies comprising poorly preserved, highly compressed fossil assemblages. Shelly remains are normally restricted to discrete bedding planes in an otherwise barren shale. Fossils consist of mixed pelagic forms (e.g., cephalopods, styliolinids, rare shark teeth and placoderm fish armor, tasmanitid spores, and wood) and a low-diversity benthic fauna of bivalves, rare gastropods, and small, thin-shelled brachiopods including chonetids, leiorhynchids, and lingulids.

4.7.2. Background Taphonomy

Fossils in this taphofacies are typically poorly preserved. Most bivalved shells are disarticulated; articulated splayed bivalves are uncommon, but do occur on certain bedding planes. Valves of brachiopods may be intact and oriented convex upward on the bedding surface. Some surfaces display an abundance of fragmented remains, always parallel to the bedding plane. Most shelly remains show evidence of intense corrosion due to prolonged surface exposure (Fig. 13).

Calcitic shells are commonly thin and partly to completely decalcified; aragonitic shells are uniformly represented by molds that rarely show traces of the carbonized periostracal film. Wood is carbonized or represented by coalified compressions. Fish bones and conodonts are well preserved as brown to black phosphate-impregnated hard parts. Discontinuity surfaces in these black shales are typified by accumulated lags of conodont, fish teeth,

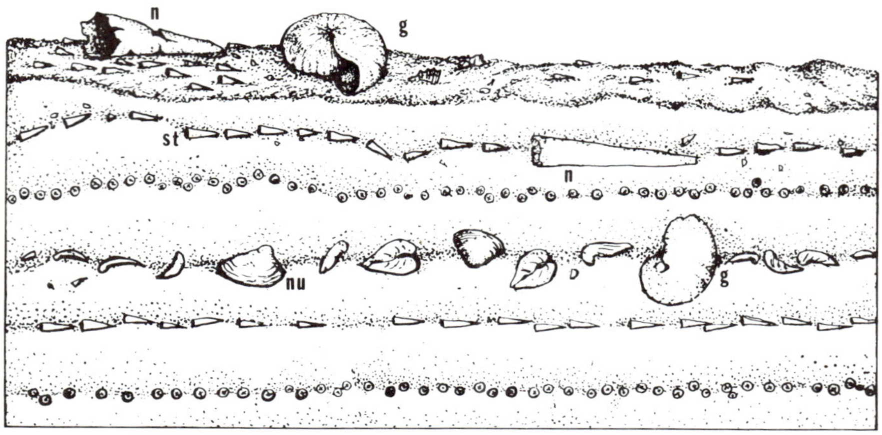

FIGURE 13. *Taphofacies Model VI* taphonomic history. Relatively deep, lower dysaerobic to anaerobic water. Low oxygen prevents colonization of the sea floor during much of the time; skeletal accumulations consist primarily of shells of pelagic organisms, including nautiloid and ammonoid cephalopods, and styliolinids. Slow burial permits corrosion and collapse of shells exposed in acidic seafloor environments. General absence of burrowing infauna facilitates the preservation of sediment lamination and minor current alignment of shells on some bedding planes. Rarely, however, minor oxygenation events permit ephemeral colonization of the sea floor by oxygen-tolerant nuculoid bivalves (middle layer). Symbols: g, goniatite ammonoid; n, nautiloid; nu, nuculoid bivalves; st, styliolinids.

and phosphatized brachiopod, trilobite, and rare conulariid remains (phosphatization occurred prior to or during lag concentration); less frequently, reworked pyrite nodules and pyritized remains (see Section 4.6) occur in lags.

Multielement skeletons are uncommon, represented by rare crinoids, trilobites, vertebrates, and conodont apparatuses. When present, these remains are typically entirely disarticulated and scattered. Cephalopods include phragmocone and body chamber fragments; more complete orthocones as well as styliolinids may show a weak preferred alignment (Fig. 13). In general, fossils are highly compressed and often show a wrinkled surface indicating plastic deformation. There is little or no early diagenetic overprint except for rare pyrite patinas on shells.

Bioturbation is minor or absent; hence primary lamination is partly or completely preserved (Fig. 13). Ichnofossils are restricted to rare, minute threadlike traces and *Chondrites*. Epibionts are exceedingly rare. Endoliths of one or two types occur sparingly on shells.

4.7.3. Episodic Taphonomy

Evidence for sedimentation events is rare in this facies. Beds of complete, well-preserved cephalopods (typically in calcareous nodules) occur sporadically. In addition, rare bedding planes display abundant leiorhynchid brachiopods mainly oriented hingeline down, as in presumed life position.

Finally, articulated fish and conodont apparatuses have been reported from a few bedding planes. These examples indicate occasional mass mortalities associated with pulses of burying sediment. Minor disturbance events are suggested by thin windrows of convex-upward brachiopod valves, shell debris, and aligned styliolinids (Fig. 13).

4.7.4. Discussion

The absence of oxygen in the sediment precludes the development of any benthic macroinfauna; consequently fossil assemblages are highly restricted. Bioturbation, however, is similarly absent or restricted, allowing for preservation of minute, thin laminae, the most distal effects of storm or turbidity current events. The presence of persistent and/or episodic bottom currents and the absence of reorienting infauna allow the preservation of delicate, thin-shelled taxa (e.g., styliolinids, cephalopods) in aligned lags.

In siliciclastic deposits of this facies, fossil preservation is typically poor due to a high degree of dissolution. Dissolution is likely the result of low pH (presumably the buildup of carbonic or sulfuric acids during organic decay at or near the sediment–water interface) which results in carbonate undersaturation. The organic periostracum of cephalopod shells may be preserved in this chemical setting, but it appears as films on molds. Calcitic remains, when buried, suffer dissolution and become highly compressed during deposit dewatering; the result is shells that display a wrinkled and nonfractured skeletal impression (Speyer and Brett, 1988).

The uniform distribution of organic matter within the sediment prevents the establishment of pronounced chemical gradients and inhibits localized mineralization of organism remains. Nevertheless, discontinuity surfaces may be marked by accumulations of chemically resistant fossil particles (e.g., conodont elements, fish remains) and phosphatic nodule lags. Although scavenging and bioturbation were absent, the low rates of sedimentation yielded assemblages comprised entirely of disarticulated remains.

A *conflicting* strongly background-dominated taphonomic signature characterizes this taphofacies. Long periods of insignificant deposition were interrupted by pulses of fine-grained sedimentation that represent the most distal influence of storm-generated currents. There is little evidence of burial events, presumably because benthic fauna were sparse and burial layers were very thin.

The disarticulated and sometimes abraded nature of shells on some bedding planes simply indicate the sediment-starved nature of this environmental setting and the prolonged periods of surface exposure. Following leaching, shells were very fragile and vulnerable to breakage during the slightest disturbance. It is important to note that shell preservation in these most distal, sediment-starved settings is similarly poor and shows some of the characteristic taphonomic attributes of the most proximal taphofacies. For example, both Taphofacies Models I and VI show high disarticulation and shell fragmentation and tend to contain only the most chemically resistant skeletal remains. Both display background-dominated taphonomic signatures with an obscure record of sedimentary events.

The taphonomic similarities between Taphofacies Models I and VI are superficial and the processes responsible for each are fundamentally different. The key point of similarity is the emphasis upon exposure time and duration of skeletal residence on the sea floor. Taphofacies Model VI represents a low-energy setting with only minor current disturbance, whereas Model I records high-energy phenomena and is typified by persistent reworking. Abrasion and corrosion are typical biasing agents in Taphofacies Model I, whereas geochemical dissolution without significant abrasion typifies Model VI. The proximal Taphofacies Model I was agitated and presumably well-oxygenated, but Taphofacies Model VI strata accumulated in low-energy, oxygen-depleted water (dysaerobic to exaerobic).

4.8. Taphofacies Model VII: Dysaerobic, Sediment-Dominated Beds (''Barren Shales'')

Reference examples: Mt. Marion Shale Member (in part), Hudson Valley, New York (Ver Straeten, 1989); Cardiff Shale Member (in part), central New York (Grasso, 1986); Levanna Shale Member (in part), Cayuga County, New York (G. C. Baird and C. E. Brett, unpublished results); middle Windom Shale Member, Erie County, New York (Brett and Baird, 1982).

Typical lithofacies: Medium to dark gray, silty shales and shaley mudstones with intercalated thin siltstones (distal tempestites or turbidites).

Typical biofacies: Nuculid bivalves, chonetid brachiopods, *Leiorhynchus* (rarely) (Brett *et al.*, 1990, 1991).

4.8.1. General Features

Thick sequences of dark gray, nearly barren silty shales typify much of the lower Hamilton Group exposed in central and, more rarely, western New York State. These beds may display well-preserved lamination, as in Taphofacies Model VI, with which this facies is typically interbedded. Taphofacies Model VII strata are distinctly depleted in organic matter and show an exceedingly depauperate fauna. Scattered fossils include many of the taxa present in Taphofacies Model VI strata, particularly pelagic forms.

4.8.2. Background Taphonomy

The scarcity and scattered distribution of fossils are the primary characteristics of this facies. Preservation is generally very poor, although the frequency of articulated (particularly splayed) bivalves is notably higher than in Taphofacies Model VI. The aragonitic shells of mollusks may be relatively complete or fragmented, but typically display a "ghosting" that results from early dissolution and plastic deformation of mud molds. Shells are randomly positioned, with proportions of convex up and down approximately equal.

There is little evidence of early diagenetic mineralization; pyritic coatings or patinas occur very rarely. Large carbonate concretions are uncommon, but horizons of small concretions occur in some sections. Although most nodules are barren, some contain well-preserved, uncompacted mollusk shells.

4.8.3. Episodic Taphonomy

There is very little taphonomic evidence for episodic processes in this facies. Nevertheless, there may be some indication from purely sedimentological features (e.g., thin, sharp-based laminated siltstone layers, carbonate nodule layers). In general, fossils are too sparse to use diagnostically in evaluating tempo of sedimentation. In addition, background rates of sedimentation obscure increases that may be attributed to episodic phenomena. However, scattered shell-rich bedding planes may represent minor winnowing events. Infrequently, small complete articulated crinoids (or long column lengths) or *in situ* articulated bivalves may provide indications of burial events.

4.8.4. Discussion

As in more condensed black shale facies (see Taphofacies Model VI), this taphofacies displays a background-dominated *conflicting signature*. In this case, however, taphonomic and sedimentological evidence indicate a higher rate of deposition. The absence or paucity of bioturbation probably reflects a combination of rapid deposition and marginally dysaerobic bottom conditions (cf. Taphofacies Models III and V). This combination of stresses made this environmental setting inhospitable to benthic faunas. Indeed, these strata are very nearly barren in some intervals.

The fact that splayed bivalves may predominate over closed, articulated shells or even single valves suggests relatively rapid *background* as opposed to episodic rates of sedimentation. However, as noted, there are indications of minor episodes of event-related deposition. Even these event beds (e.g., millimeter-thick siltstone laminae) are preserved because of the lack of bioturbation.

The absence of ion concentration gradients and the short residence time of buried organisms in a near-surface zone may have precluded widespread early diagenesis. A general absence of carbonate buffers in siliciclastic sediments and perhaps low pH conditions caused very early dissolution of aragonite and calcite, leaving poorly preserved molds that were subsequently deformed during dewatering and compaction.

5. Conclusions and Prospectus

The majority of shelly deposits present in the stratigraphic record are attributed to episodic sedimentary conditions. Sedimentary events leave conspicuous signatures superimposed upon "normal" background conditions. Consequently, there is an emphasis on the recognition and interpretation of episodic sedimentary deposits in stratigraphic research (Aigner, 1979, 1982; Seilacher, 1982; Brett *et al.*, 1986b). Background processes (e.g., gradual sedimentation, winnowing, bioturbation, and current amalgamation), however, affect the preservation of these event-stratified deposits. Hence, the stratigraphic record is a composite of background and event effects; both correspond to ambient environmental gradients and each is influenced by the other.

Davies *et al.* (1989) argued that shell dissolution greatly exceeds shell production in marine settings, making background-accumulated shell beds unlikely. Mollusk shells in Texas Gulf Coast shelf environments appear to have a surprisingly short residence time on the sea floor (shell half-lives of less than 1 year) (Powell *et al.*, in press; Cummins *et al.*, 1986); durations are based upon amino acid racemization, a technique that remains controversial (Ernst, 1989; Brooks *et al.*, 1990). Although it is unlikely that the rates of dissolution documented by Powell *et al.* (in press), Cummins *et al.* (1986), and Davies *et al.* (1989) apply to all marine settings, much less throughout Phanerozoic time, they underscore the importance of burial in shell preservation. However, the interrelated roles of "stratigraphic disorder" [rearrangement of temporally organized shelly remains (Flessa *et al.*, 1989)] and "biogenic stratification" [reverse stratigraphic organization of shell remains as a result of bioturbation (Meldahl, 1987)] must also be considered. Flessa *et al.* (1989) and Meldahl (1987) recorded a wide range of shell ages in shell beds (based upon C-14 radiometric dating), some in excess of 1000 years. These results provide important perspectives on the Kidwell and Jablonski (1983) "taphonomic feedback" model.

A critical interplay exists between episodic burial and background processes, and taphonomic histories are inherently complex. Taphofacies mosaics change according to temporal circumstances, here summarized as background and episodic. Thus, although the accumulation of substantial shell layers requires extended periods of time, such layers are only preserved in the long term by episodic burial. Indeed, many seemingly simple shell layers have likely undergone multiple episodes of burial and exhumation.

Although the Hamilton Group is not a template for all future taphonomic studies, there are certain generalizations that may apply to other strata. Taphonomic signatures clearly vary across stratigraphic facies and predictably. They represent changes in surface exposure, frequency of episodic burial, and intensity of current reworking (see Fig. 3) (Kidwell, 1988; Parsons *et al.*, 1988; Speyer and Brett, 1988). Clearly, strict uniformitarian or episodic approaches to facies analysis are in themselves inadequate unless modified to incorporate the imprint of reciprocal phenomena.

Realizing the dual nature of the stratigraphic record in our investigations of taphonomic process and taphofacies pattern will enable a glimpse into the "black box" of stratigraphy and diminish our vulnerability to biased presumptions. Documenting the stratigraphic interaction and facies distribution of background versus episodic signatures also provides a new source of data for time-resolution analysis (e.g., Schindel, 1982; Behrensmeyer and Schindel, 1983) and enhances our understanding of the caliber of paleoecological information preserved in specific paleoenvironmental settings.

ACKNOWLEDGMENTS. We are very grateful to Susan Reilein Hill (University of Arizona) for carefully and patiently enduring the tedious revision, reorganization, and last minute changes involved with the manuscript. Heidi Jacob (University of Rochester) also assisted in the typing and revision of earlier versions. The artistic taphofacies diagrams (Figs. 7–13) were drawn in ink by Kathryn White.

The present chapter incorporates many excellent comments provided by Gordon Baird, David Lehmann, and Ronald West, and reviews by the editors of this volume. The project has benefitted greatly from discussions with workers too numerous to list; our sincere thanks for their input and continued interest in the taphofacies idea.

Financial support for this project is gratefully acknowledged from NSF (EAR-8816856 awarded to C. E. B.), and the University of Arizona.

References

Aigner, T., 1979, Schill-Tempestite im oberen Muschelkalk (Trias, SW-Deutschland), *Neues Jahrb. Geol. Paläontol. Abh.* **157**:193–217.

Aigner, T., 1982, Calcareous tempestites: Storm-dominated stratification in Upper Muschelkalk limestones (Middle Triassic, SW-Germany), in: *Cyclic and Event Stratification* (G. Einsele and A. Seilacher, eds.), Springer-Verlag, New York, pp. 180–198.

Aigner, T., 1985, *Storm Depositional Systems: Dynamic Stratigraphy in Modern and Ancient Shallow Marine Sequences*, Springer-Verlag, New York.

Aigner, T., and Reineck, H. E., 1982, Proximality trends in modern storm sands from the Heligoland Bight (North Sea) and their implications for basin analysis, *Senkenb. Marit.* **14**:183–215.

Alberstadt, L. P., and Walker, K. W., 1976, A receptaculitid-echinoderm pioneer community in a Middle Ordovician reef, *Lethaia* **9**:261–272.

Aller, R. C., 1982, Carbonate dissolution in nearshore terrigenous muds: The role of physical and biological reworking, *J. Geol.* **90**:79–95.

Allison, P. A., 1988, Konservat-Lagerstätten: Cause and classification, *Paleobiology* **14**:331–345.

Allmon, R. A., 1985, "Butterflied" bivalves as paleoenvironmental indicators, *Geol. Soc. Am. Abstr. Progr.* **17**(7):512.

Archer, A. W., and Feldman, H. R., 1986, Microbioherms of the Waldron Shale (Silurian, Indiana): Implications for organic framework in Silurian reefs of the Great Lakes area, *Palaios* **1**:133–140.

Babcock, L. E., and Speyer, S. E., 1987, Enrolled trilobites from the Alden Pyrite Bed, Ledyard Shale (Middle Devonian) of western New York, *J. Paleontol.* **61**:539–548.

Baird, G. C., 1979, Sedimentary relationships of Portland Point and associated Middle Devonian rocks in central and western New York, *N. Y. State Mus. Bull.* **433:**1–24.

Baird, G. C., and Brett, C. E., 1981, Submarine discontinuities and sedimentary condensation in the upper Hamilton Group: Examination of paleoslope deposits in the Cayuga Valley, in: *N. Y. State Geological Association Guidebook—53rd Annual Meeting, Binghamton, N. Y.*, pp. A1–A33.

Behrensmeyer, A. K., and Kidwell, S. M., 1985, Taphonomy's contributions to paleobiology, *Paleobiology* **11:**105–119.

Behrensmeyer, A. K., and Schindel, D., 1983, Resolving time in paleobiology, *Paleobiology* **9:**1–8.

Berner, R. A., 1968, Calcium carbonate concretions formed by the decomposition of organic matter, *Science* **159:**195–197.

Berner, R. A., 1984, Sedimentary pyrite, an update, *Geochim. Cosmochim. Acta* **48:**605–615.

Bordeaux, Y. L., and Brett, C. E., 1990, Substrate specificity and associations of epibionts on Middle Devonian brachiopods, *Hist. Biol.* **4:**203–220.

Brett, C. E., 1983, Sedimentology, facies and depositional environments of the Rochester Shale (Silurian; Wenlockian) in western New York and Ontario, *J. Sediment. Petrol.* **53:**947–971.

Brett, C. E., and Baird, G. C., 1982, Upper Moscow-Genesee stratigraphic relationships in western New York: Evidence for regional erosive beveling in the Late Middle Devonian, in: *N. Y. State Geological Association Guidebook—54th Annual Meeting Buffalo, N. Y.*, pp. 217–245.

Brett, C. E., and Baird, G. C., 1986, Comparative taphonomy: A key to paleoenvironmental interpretation based on fossil preservation, *Palaios* **1:**207–227.

Brett, C. E., and Bordeaux, Y., 1988, Epibionts on Devonian brachiopods: Paleoecologic implications, *Geol. Soc. Am. Abstr. Progr.* **20:**10.

Brett, C. E., and Speyer, S. E., 1989, Facies distribution of obrution deposits: Towards a predictive model for fossil burial events, in: *28th International Geological Congress (Washington, D.C.), Abstracts*, Volume 1, p. 197.

Brett, C. E., Baird, G. C., Gray, L. M., and Savarese, M. L., 1983, Middle Devonian (Givetian) Coral associations of western and central New York State, in: *Silurian and Devonian Corals and Stromatoporoids of New York* (J. E. Sorauf and W. A. Oliver, Jr., eds.), International Association for the Study of Fossil Cnidaria, 4th International Symposium, Washington, D.C.

Brett, C. E., Miller, K. B., and Baird, G. C., 1986a, Sedimentary cycles and lateral facies gradients across a Middle Devonian shelf-to-basin ramp, Ludlowville Formation, Cayuga Valley, N.Y., in: *N. Y. State Geological Association Guidebook—58th Annual Meeting, Ithaca, N. Y.*, pp. 70–128.

Brett, C. E., Speyer, S. E., and Baird, G. C., 1986b, Storm-generated sedimentary units: Tempestite proximality and event stratification in the Middle Devonian Hamilton Group of New York State, in: Dynamic Stratigraphy and Depositional Environments of the Hamilton Group (Middle Devonian) in New York State, Part 1 (C. E. Brett, ed.), *N. Y. State Mus. Bull.* **457:**129–156.

Brett, C. E., Miller K. B., and Baird, G. C., 1990, A temporal hierarchy of paleoecological processes in a Middle Devonian epeiric sea, in: Paleo Community Temporal Dynamics: The Long-Term Development of Multispecies Assemblages (W. Miller III, ed.), *Paleontol. Soc. Spec. Publ.* **5:**178–209.

Brett, C. E., Dick, V. B., and Baird, G. C., 1991, Comparative taphonomy and paleoecology of middle Devonian dark grey and black shales from western New York, in: Dynamic Stratigraphy and Depositional Environments of the Hamilton Group (Middle Devonian) in New York State, Part 2 (E. Landing and C. E. Brett, eds.), *N. Y. State Mus. Bull.* **469:**5–36.

Brooks, A. S., Hare, P. E., Kokis, J. E., Miller, G. H., Ernst, R. D., and Wendorf, F., 1990, Dating Pleistocene archeological sites by protein diagenesis in ostrich eggshell, *Science* **248:**60–64.

Brouwer, A., 1966, *General Paleontology*, University of Chicago Press, Chicago.

Cummins, H., Powell, E. N., Stanton, Jr., R., and Staff, G., 1986, The rate of taphonomic loss in modern benthic habitats: How much of the potentially preservable community is preserved?, *Palaeogeog. Palaeoclimatol. Palaeoecol.* **52**:291–320.

Cutler, A. H., 1989, Shells survive—Loss, persistence and accumulation of hard parts in shallow marine sediments, *Geol. Soc. Am. Abstr. Progr.* **21**(6):71.

Davies, D. J., Powell, E. N., and Stanton, R. J., Jr., 1989, Relative rates of shell dissolution and net sediment accumulation—A commentary: Can shell beds form by the gradual accumulation of biogenic debris on the sea floor?, *Lethaia* **22**:207–212.

Dick, V. B., and Brett, C. E., 1986, Petrology, taphonomy and sedimentary environments of pyritic fossil beds from the Hamilton Group (Middle Devonian) of western New York, in Dynamic Stratigraphy and Depositional Environments of the Hamilton Group (Middle Devonian) in New York State, Part 1 (C. E. Brett, ed.), *N. Y. State Mus. Bull.* **457**:102–128.

Ernst, R. D., 1989, Reaction kinetics of protein hydrolysis, amino acid decomposition, and isoleucine epimerization in eggshell of the African ostrich, *Struthio camelus,* Masters thesis, Department of Geosciences, University of Arizona, Tucson, Arizona.

Flessa, K., and Brown, T. J., 1982, Selective solution of macroinvertebrate calcareous hard parts: A laboratory study, *Lethaia* **16**:193–205.

Flessa, K. W., Cutler, A. H., Meldahl, K. H., and Fürsich, F. T., 1989, Taphonomic processes and stratigraphic disorder, in: *28th International Geological Congress,* Volume 1, pp. 493–494.

Fürsich, F. T., and Flessa, K., 1987, Taphonomy of tidal flat molluscs in the Northern Gulf of California: Paleoenvironmental analysis despite the perils of preservation, *Palaios* **2**:543–559.

Grabau, A. W., 1898–1899, Geology and paleontology of Eighteen-mile Creek and the lake shore sections of Erie County, New York, *Buffalo Soc. Nat. Sci. Bull.* **6**:1–403.

Grasso, T. X., 1986, Redefinition, stratigraphy and depositional environments of the Mottville Member (Hamilton Group) in central and eastern New York, in: Dynamic Stratigraphy and Depositional Environments of the Hamilton Group (Middle Devonian) in New York State, Part 1 (C. E. Brett, ed.), *N. Y. State Mus. Bull.* **457**:5–31.

Gray, L. M., 1985, Stratigraphy and depositional environments of the Centerfield and Chenango members (Middle Devonian) in western and central New York State, Ph. D. thesis, University of Rochester, Rochester, New York.

Johnson, R. G., 1960, Models and methods for analysis of the mode of formation of fossil assemblages, *Bull. Geol. Soc. Am.* **71**:1075–1086.

Johnson, M. E., 1989, Tempestites recorded as variable *Pentamerus* layers in the Lower Silurian of southern Norway, *J. Paleontol.* **63**:195–205.

Kidwell, S. M., 1988, Taphonomic comparison of passive and active continental margins: Neogene shellbeds of the Atlantic Coastal Plain and Northern Gulf of California, *Palaeogeogr. Palaeoclimatol. Palaeoecol.* **63**:201–223.

Kidwell, S. M., and Behrensmeyer, A. K., 1988, Overview: Ecological and evolutionary implications of taphonomic processes, *Palaeogeogr. Palaeoclimatol. Palaeoecol.* **63**:1–13.

Kidwell, S. M., and Jablonski, D., 1983, Taphonomic feedback: Ecologic consequences of shell accumulation, in: *Biotic Interactions in Recent and Fossil Benthic Communities* (M. J. S. Teresz and P. L. McCall, eds.), Plenum Press, New York, pp. 195–248.

Kidwell, S. M., Fürsich, F. T., and Aigner, T., 1986, Conceptual framework for the analysis and classification of fossil concentrations, *Palaios* **1**:228–238.

Kotake, N., 1989, Paleoecology of the *Zoophycos* producers, *Lethaia* **22**:327–341.

Kranz, P., 1974, Anastrophic burial of bivalves and its paleoecologic significance, *J. Geol.* **82**:237–265.

Kreisa, R. D., 1981, Storm-generated sedimentary structures in subtidal marine facies with examples from the Middle and Upper Ordovician of southwestern Virginia, *J. Sediment. Petrol.* **51**:823–848.

Kreisa, R. D., and Bambach, R. K., 1982, The role of storm processes in generating shell beds in Paleozoic shelf environments, in: *Cyclic and Event Statification* (G. Einsele and A. Seilacher, eds.), Springer-Verlag, New York, pp. 200–207.

Kumar, N., and Sanders, J. E., 1976, Characteristics of shoreface storm deposits, *J. Sed. Petrol.* **46**:145–162.

Linsley, D. M., 1988, Autecology, based on mode of preservation, of some Middle Devonian bivalves, *N. Y. State Mus. Bull.* **462**:11–12.

Lukasik, D. M., 1985, Stratigraphy, sedimentology, and paleoecology of the Kashong Shale (Middle Devonian) of New York, M.Sc. Thesis, University of Cincinnati, Cincinnati, Ohio.

Mayer, S., 1989, Stratigraphy and paleontology of the Jaycox Shale Member, Hamilton Group of the Finger Lakes region New York State, M.A. thesis, SUNY College Fredonia, New York.

Meldahl, K. H., 1987, Sedimentologic and taphonomic implications of biogenic stratification, *Palaios* **2**:350–358.

Meldahl, K. H., and Flessa, K., 1990, Taphonomic pathways and comparative biofacies and taphofacies in a Recent intertidal/shallow shelf environment, *Lethaia* **23**:43–60.

Miller, K. B., 1986, Depositional environments and sequences, "*Pleurodictyum* Zone", Ludlowville Formation of western New York, in: Dynamic Stratigraphy and Depositional Environments of the Hamilton Group (Middle Devonian) of New York State, Part 1 (C. E. Brett, ed.), *N. Y. State Mus. Bull.* **457**:57–77.

Miller, K. B., Brett, C. E., and Parsons, K., 1988, The paleoecologic significance of storm-generated disturbance within a Middle Devonian muddy epeiric sea, *Palaios* **3**:35–52.

Norris, R. D., 1986, Taphonomic gradients in shelf fossil assemblages, Pliocene Pursima Formation, California, *Palaios* **1**:256–270.

Parsons, K., Brett, C. E., and Miller, K. B., 1988, Taphonomy and depositional dynamics of Devonian shell-rich mudstones, *Palaeogeogr. Palaeoclimatol. Palaeoecol.* **63**:109–139.

Powell, E. N., Staff, G. M., Davies, D. J., and Callender, W. R., in press, Macrobenthic death assemblages in modern marine environments: Formation, interpretation and application, *Rev. Aquat. Sci.*

Raiswell, R., 1982, Pyrite texture, isotopic composition and the availability of iron, *Am. J. Sci.* **282**:1244–1263.

Rhoads, D. C., and Morse, J. W., 1971, Evolutionary and ecologic significance of oxygen-stratified marine basins, *Lethaia* **4**:413–428.

Salazar-Jiménez, A., Frey, R. W., and Howard, J. D., 1982, Concavity orientations of bivalve shells in estuarine and nearshore shelf sediments, *J. Sediment. Petrol.* **52**:565–586.

Savarese, M. L., Gray L. M., and Brett, C. E., 1986. Faunal and lithologic cyclicity in the Centerfield Member (Middle Devonian: Hamilton Group of western New York), in: Dynamic Stratigraphy and Depositional Environments of the Hamilton Group (Middle Devonian) in New York State, Part 1 (C. E. Brett, ed.), *N. Y. State Mus. Bull.* **457**:32–56.

Schäfer, W., 1972, *Ecology and Paleoecology of Marine Environments*, University of Chicago Press, Chicago.

Schindel, D., 1982, Resolution analysis: A new approach to the gaps in the fossil record, *Paleobiology* **8**:340–353.

Seilacher, A., 1970, Begriff und Bedeutung der Fossil-Lagerstätten, *Neues Jahrb. Geol. Paläontol. Abh.* **1970**:34–39.

Seilacher, A., 1982, General remarks about event deposits, in: *Cyclic and Event Stratification*, (G. Einsele and A. Seilacher, eds.), Springer-Verlag, New York, pp. 161–174.

Seilacher, A., Reif, W. E., and Westphal, F., 1985, Sedimentological, ecological and temporal patterns of fossil *Lagerstätten*, *Phil. Trans. R. Soc. Lond. B* **311**:5–23.

Sjoberg, E. L., and Rickard, D. T., 1985, Calcite dissolution kinetics: Surface speciation and the origin of variable pH dependence, *Chem. Geol.* **42**:119–136.

Speyer, S. E., 1985, Moulting in phacopid trilobites, *Trans. R. Soc. Ed. Earth Sci.* **76**:239–254.

Speyer, S. E., 1987, Comparative taphonomy and paleoecology of trilobite *Lagerstätten*, *Alcheringa* **11**:205–232.

Speyer, S. E., and Brett, C. E., 1985, Clustered trilobite assemblages in the Middle Devonian Hamilton Group, *Lethaia* **18**:85–103.

Speyer, S. E., and Brett, C. E., 1986, Trilobite taphonomy and Middle Devonian taphofacies, *Palaios* **1**:312–327.

Speyer, S. E., and Brett, C. E., 1988, Taphofacies models for epeiric sea environments: Middle Paleozoic examples, *Palaeogeogr. Palaeoclimatol. Palaeoecol.* **63**:225–262.

Speyer, S. E., and Brett, C. E., 1990, Environmental distribution of marine obrution deposits: facies models and paleoecological resolution, *Geol. Soc. Amer. Abst. Prog.* **22**(7):A81.

Speyer, S. E., and Jones, B., 1988, "Rhythmically-bedded" carbonates; a superposition of cyclic and event sedimentary processes, in: *SEPM Mid-Year Meeting (Columbus), Abstracts with Programs,* Volume 5, p. 51.

Staff, G. M., Stanton, R. J., Jr., Powell, E., and Cummins, H., 1986, Time-averaging, taphonomy, and their impact on paleocommunity reconstruction: Death assemblages in Texas bays, *Geol. Soc. Am. Bull.* **97**:428–443.

Ver Straeten, C., 1989, Microstratigraphy and depositional environments of an Acadian synorogenic basin: The Mount Marion Formation in eastern New York, *Geol. Soc. Am. Abstr. Progr.* **21**(2):73.

Vogel, K., Golubic, S., and Brett, C. E., 1987, Endolith associations and their relation to facies distribution in the Middle Devonian of New York State, U.S.A., *Lethaia* **20**:263–290.

Waage, K. M., 1964, Origin of repeated fossiliferous concretion layers in the Fox Hills Formation (Cretaceous, SD), *Kansas Geol. Surv. Bull.* **169**:541–563.

Wanless, H. R., 1988, Production of subtidal tubular and surficial tempestites by Hurricane Kate, Caicos Platform, British West Indies, *J. Sediment. Petrol.* **58**:739–750.

Index